Beecher MyLab Math Digital Update

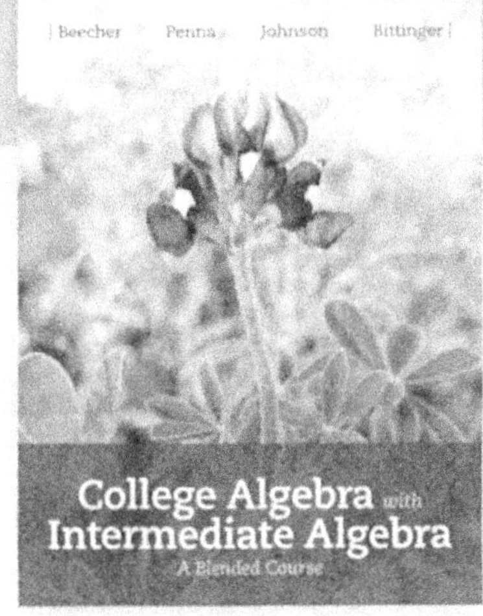

In math courses today the needs of instructors and students are changing. Since *College Algebra with Intermediate Algebra: A Blended Course* first published, corequisite courses have continued to increase across the nation. We have added new features in the MyLab™ Math course and within the Video Notebook to further support students and instructors.

Corequisite Courses Require Flexibility

▼ **Flexible MyLab Platform** Your course is unique. Whether you'd like to build your own assignments, structure students' work with a **learning path**, or set prerequisites, you have the flexibility to easily create *your* course to fit *your* needs. Need two MyLab Math courses for your corequisite course? You can do that, too—while students still pay only once.

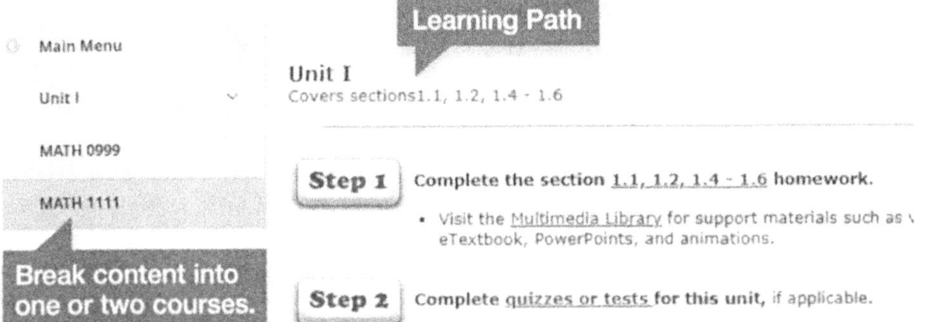

Personalized Learning Each student learns at a different pace. With **Personalized Homework**, students take a quiz or test and receive a subsequent personalized homework assignment based on their performance. This way, students can focus on just the topics they have not yet mastered—no more, no less.

Instructor Spotlight

> **"** I had my best results when I reviewed the Video Notebook for completion, and it was part of the students' grade. **"**

Mari Menard, Lone Star College – Kingwood

Course ▼	Math 0314 – NCBO for College Algebra; Math 1314 – College Algebra
Course Format ▼	M/W Corequisite Support and T/TH College Algebra (12 week)
Homework ▼	MyLab Math Homework as bonus to assist students in learning the content; all learning aids available except for "View an Example." I disable the Study Plan so that students focus their learning and practice using the Video Notebook and bonus homework assignments.
Quizzes ▼	Each quiz covers multiple sections with usually no more than 10–12 questions that generate from a question pool; the quiz has a time limit with 3 attempts and the student must review the quiz immediately after submission.
Exams ▼	Tests are part of the grade for College Algebra. The corequisite course has a comprehensive diagnostic assessment at semester end.
Other ▼	I use Mindset and study skills in discussion boards, projects, and the Video Notebook.

Beecher MyLab Math Digital Update

Course Calendar Just-in-Time Sample

WEEK	CONTENT
1	Introduction to Functions (Sections 2.1–2.4)
2	Linear Functions (Sections 2.5–2.7)
3	Polynomial Expressions and Equations (Chapter 4)
4	Functions: Increasing, Decreasing, Piecewise, Symmetry, transformations (Sections 6.8, 7.1, 7.2)
5	Quadratic Functions (Sections 7.3–7.5)
6	Polynomial Functions (Sections 8.1–8.4)
7	Rational Equations and Expressions (Sections 6.1–6.5)
8	Radical Equations and Functions (Sections 6.1–6.7)

Instructor Support

NEW ▶ The **Implementation Guide**, found within the MyLab Math course, includes helpful hints about setting up and teaching a corequisite course, including sample syllabi, correlation guides, and suggestions for how to provide student support and equity in your classroom.

Premade assignments in the MyLab make it easier to build a corequisite course; simply assign the sample assignments for objectives you'd like to include.

Other Instructor Resources:

- **NEW** **Prebuilt MyLab Math Courses** are available for your specific corequisite course approach—for instance, a just-in-time support approach. Video assignments are included and are perfect for online, hybrid, or flipped classroom. Course IDs to copy are found within the Implementation Guide.

- **NEW** **Personal Inventory Assessment Manual** provides a brief summary of each online exercise designed to promote self-reflection. The manual includes suggested learning activities, assignments, and discussions for you to improve your students' metacognition.

- **NEW** **Mindset Instructor Guide** provides suggestions on how to implement the Growth Mindset Videos and assessment questions.

- **Instructor's Solution Manual** and **PowerPoint Lecture Slides**

Instructor Spotlight

❝ I integrate the remedial topics into the College Algebra topics, as needed, to create a seamless course. ❞

Tina Ragsdale, West Kentucky Community and Technical College

Course	MAT 150 College Algebra with MAT 100 Workshop for College Algebra
Course Format	4 days a week (40-minutes) for 3-credit class hours and 2-credit (25-mintues) workshop.
Homework	All homework has a due date, but without penalties. Students may repeat homework and improve their grade. Video Notebook must be completed as part of homework grade.
Quizzes	Some in-class quizzes may be given—small point values—usually formative in nature.
Exams	One exam for each of the six modules. Students may retake 1 exam during the last week of the course for a higher grade.
Other	By allowing students to retest their lowest test the last week of the semester, it was an unexpected bonus that these students were studying the material they were weakest in directly before the final exam.

Beecher MyLab Math Digital Update Features

Videos provide an engaging learning environment where students can learn at their own pace. Pair the videos with the Video Notebook and students have an excellent study guide for review and test preparation.

Videos

NEW ▶ **Preview Videos** walk students through introductory examples of what is coming up in the chapter. Preview worksheets are available for students to work through the video on their own.

$$\text{Check:}\quad \begin{array}{ll} \text{Write the equation.} & x = \sqrt{x+4}+2 \\ \text{Substitute 5 for } x. & 5 \mid \sqrt{5+4}+2 \\ \text{Simplify.} & \sqrt{9}+2 \\ & 3+2 \\ & 5 \mid 5 \qquad \text{True} \end{array}$$

We get a true equation, $5 = 5$, so 5 is a solution.

NEW ◀ **College Algebra Essential Topic Videos** cover the crucial topics your students need in College Algebra and teach the concepts just like you would in the classroom. Corresponding worksheets are available to keep your students actively participating.

Section Videos offer lectures for each section of the text.

Chapter Test Prep Videos are step-by-step solutions to the questions in the Chapter Test.

NEW **Example Solution Videos** provide a detailed walk-through of the solution process for key examples. Videos that feature graphing have been updated with GeoGebra Graphing Utility. Prefer Graphing Calculators? We retained those videos, too.

Video Notebook

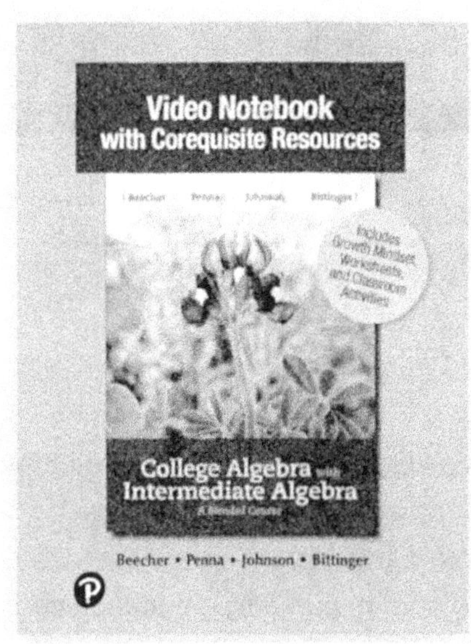

REVISED ◀ **Video Notebook** now includes more resources to support your students' understanding.

- **Your Turn exercises** now appear throughout the entire Video Notebook, allowing students an opportunity to practice on their own.

- **Mathematical Growth Mindset Activities** expand and enrich the way students look at mathematics.

- **Preview Worksheets** provide hands-on practice with introductory examples and exercises that are perfect to introduce before starting a new topic.

- **Activities** are application-based to enhance classroom time and promote peer-to-peer learning.

Beecher MyLab Math Digital Update Features

Active Learning Figures and **Guided Visualizations** in MyLab Math allow students to examine visual representations of concepts through guided and open-ended explorations.

NEW The **correlation guides** in the Video Notebook pair the Preview Worksheet and text section with the corresponding Active Learning Figures, either in class or as a MyLab Math assignment.

NEW ▶ **GeoGebra Exercises** are gradable computational and graphing exercises. Students interact with the graphing exercises as they would on paper. It's a quick and easy way for them to show you what they know.

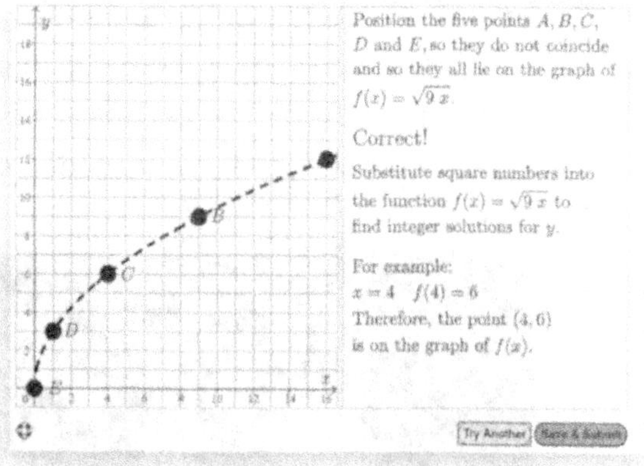

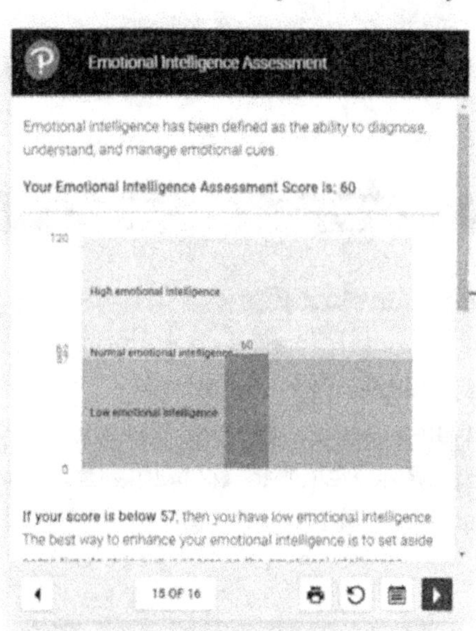

◀ NEW **Personal Inventory Assessments** are a collection of online exercises designed to promote student self-reflection and metacognition. These 33 assessments include topics such as Stress Management, Diagnosing Poor Performance and Enhancing Motivation, and Time Management. An accompanying Instructor Manual, curated by author Barbara Johnson, includes classroom activities, assignments, and discussion topics.

NEW ▶ **Growth Mindset Videos** are assignable with corresponding open-ended exercises to foster your students' growth mindset. This material encourages students to maintain a positive attitude about learning, value their own ability to grow, and view mistakes as learning opportunities—so often a hurdle for math students.

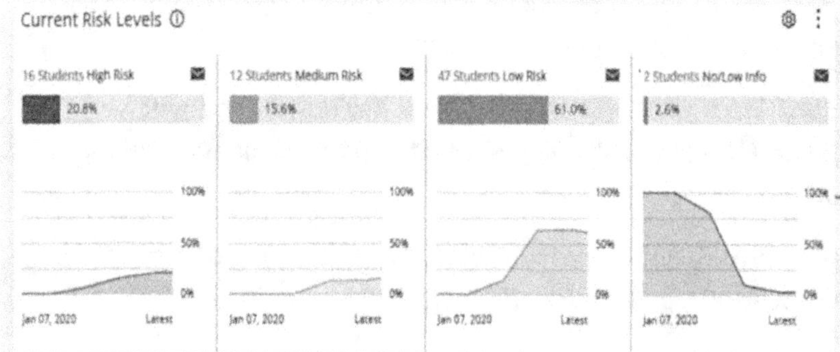

▲ NEW **Early Alerts** use predictive analytics based on students' work in MyLab Math such as assignment scores and correct answers on first try. This information helps identify struggling students as early as possible—even if their assignment scores are not a cause for concern. With this insight, you can provide informed feedback and support at the very moment students need it, so students can stay—and succeed—in your course.

Video Notebook with Corequisite Resources

College Algebra with Intermediate Algebra: A Blended Course

Judith A. Beecher

Judith A. Penna

Barbara L. Johnson
Indiana University Purdue University Indianapolis

Marvin L. Bittinger
Indiana University Purdue University Indianapolis

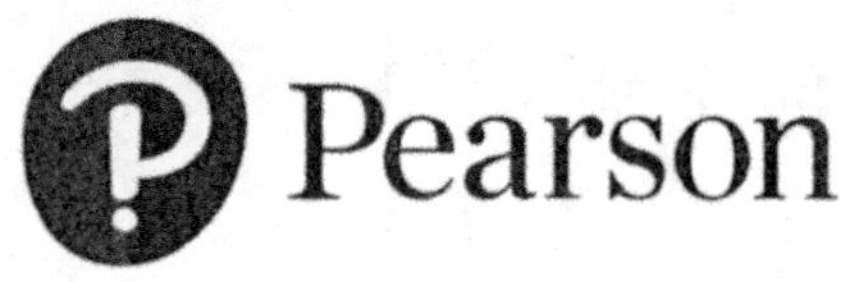

ISBN-13: 978-0-13-693828-6
ISBN-10: 0-13-693828-0

Contents

A Note to the Student

Realizing that there are many different circumstances facing those who take this course, we want to provide the best resources to make productive and efficient use of your time. Drawing on our many years of teaching experience, research in mathematics education, and our awareness of the challenges facing today's students, we developed this *Video Notebook with Corequisite Resources* to accompany our textbook *College Algebra with Intermediate Algebra: A Blended Course*, by Beecher, Penna, Johnson, and Bittinger. In this Notebook, you will find a variety of elements that will contribute to your understanding of the course material.

First you will see the **Mathematical Growth Mindset Activities** that will serve to expand and enrich the way you look at mathematics and to give you confidence in your ability to master the concepts in the course. Next you will find **Interactive Preview Worksheets** that will prepare you to master the topics in your course. Your filled-out Preview Worksheets will be a valuable resource for study and review.

There are also **Video Worksheets** that provide an interactive experience as you work through the section level videos that accompany the course. The interaction will serve to heighten your understanding of the concepts presented. And, finally you will find **Student Activities**, many of which are group activities. Research and our experience indicate that, when you work with other students, the opportunities to discuss concepts and to explain your reasoning to others in the group serve to increase your understanding.

We want you to be successful in this course and hope that we have given you the best chance to do that. If you invest an adequate amount of time in the learning process, you will have a positive experience. We wish you the best.

Judy Beecher
Judy Penna
Barbara Johnson

Correlation Guides

The **Mindset Activities** in this Notebook accompany *College Algebra with Intermediate Algebra*, by Beecher/Penna/Johnson/Bittinger. The following table correlates the Mindset Activities with sections in the text and with the Interactive Preview Worksheets.

Mindset Activity	Section in the Text	Preview Worksheet
1 Reasoning	R.2 Operations with Real Numbers	1 Operations on the Real Numbers
2 Creativity and Innovation	3.4 Solving Applied Problems: Two Equations 3.6 Solving Applied Problems: Three Equations	
3 Multiple Representations	5.6 Applications and Proportions	24 Solving Proportions
4 Exploration	6.8 Increasing, Decreasing, and Piecewise Functions: Applications	29 Graphing Piecewise Functions
5 Curiosity and Inquiry	7.5 Analyzing Graphs of Quadratic Functions	32 Introduction to Quadratic Functions
6 Locus of Control	8.3 Polynomial Division: The Remainder Theorem and the Factor Theorem 8.4 Theorems about Zeros of Polynomial Functions	
7 "Yet"	9.6 Solving Exponential Equations and Logarithmic Equations	37 Solving Exponential Equations

The **Interactive Preview Worksheets** in this Notebook accompany *College Algebra with Intermediate Algebra* by /Beecher/Penna/Johnson/Bittinger. The following table correlates the Preview Worksheets with the sections in the text.

Preview Worksheet		Section in the Text	
1	Operations on the Real Numbers	R.2	Operations with Real Numbers
2	Adding, Subtracting, Multiplying, and Dividing Fractions	R.2	Operations with Real Numbers
3	Order of Operations	R.3	Exponential Notation and Order of Operations
4	Integers as Exponents	R.7	Properties of Exponents and Scientific Notation
5	Solving Linear Equations	1.1	Solving Equations
6	Interval Notation	1.4	Sets, Inequalities, and Interval Notation
7	Solving Linear Inequalities	1.4	Sets, Inequalities, and Interval Notation
8	Solving Equations and Inequalities with Absolute Value	1.6	Absolute-Value Equations and Inequalities
9	Graphing Linear Equations	2.1	Graphs of Equations
10	Function Values; Domain and Range	2.3	Finding Domain and Range
11	Determine the Domain and the Range of a Function	2.3	Finding Domain and Range
12	Introduction to Polynomials	4.1	Introduction to Polynomials and Polynomial Functions
13	Adding and Subtracting Polynomials	4.1	Introduction to Polynomials and Polynomial Functions
14	Multiplying Binomials	4.2	Multiplication of Polynomials
15	Factoring by Grouping	4.3	Introduction to Factoring
16	Factoring Trinomials: $ax^2 + bx + c$	4.5	Factoring Trinomials: $ax^2 + bx + c,\ a \neq 1$
17	Factoring Trinomial Squares and Differences of Squares	4.6	Special Factoring
18	The Principle of Zero Products	4.8	Applications of Polynomial Equations and Functions
19	The Pythagorean Theorem	4.8	Applications of Polynomial Equations and Functions
		6.7	Applications Involving Powers and Roots
20	Simplifying Rational Expressions	5.1	Rational Expressions and Functions: Multiplying, Dividing, and Simplifying
21	Finding the LCM of Algebraic Expressions	5.2	LCMs, LCDs, Addition and Subtraction
22	Simplifying Complex Rational Expressions	5.4	Complex Rational Expressions

(continued)

Preview Worksheet		**Section in the Text**	
23	Solving Rational Equations	5.5	Solving Rational Equations
24	Solving Proportions	5.6	Applications and Proportions
25	Simplifying Radical Expressions	6.3	Simplifying Radical Expressions
26	Multiplying Radical Expressions	6.3	Simplifying Radical Expressions
27	Rationalizing Denominators	6.5	More on Division of Radical Expressions
28	Checking Solutions of Radical Equations	6.6	Solving Radical Equations
29	Graphing Piecewise Functions	6.8	Increasing, Decreasing, and Piecewise Functions: Applications
30	Transformations	7.2	Transformations
31	Completing the Square	7.4	Quadratic Equations, Functions, Zeros, and Models
32	Introduction to Quadratic Functions	7.5	Analyzing Graphs of Quadratic Functions
33	Zeros of Polynomial Functions	8.1	Polynomial Functions and Models
34	Asymptotes of Rational Functions	8.5	Rational Functions
35	Graphing Inverse Functions	9.2	Inverse Functions
36	Introduction to Logarithms	9.4	Logarithmic Functions and Graphs
37	Solving Exponential Equations	9.6	Solving Exponential Equations and Logarithmic Equations
38	Solving Logarithmic Equations	9.6	Solving Exponential Equations and Logarithmic Equations
39	Using an Inverse Matrix to Solve a System of Equations	10.3	Inverses of Matrices
40	The Ellipse	11.2	The Circle and the Ellipse
41	The Hyperbola	11.3	The Hyperbola
42	Classifying Equations of Conic Sections	11.3	The Hyperbola

The **Student Activities** in this Notebook accompany *College Algebra with Intermediate Algebra* by Beecher/Penna/Johnson/Bittinger. They are located in Tools for Success within MyLab Math. The following table correlates the Student Activities with sections in the text.

Student Activity		Section in the Text	
1	American Football	R.2	Operations with Real Numbers
2	Finding the Magic Number	R.4	Introduction to Algebraic Expressions
3	Pets in the United States	1.3	Applications and Problem Solving
4	Mobile Data	2.4	The Algebra of Functions
5	Linear Regression on a Graphing Utility: TV Viewership and Frozen Entrée Sales	2.7	Finding Equations of Lines; Applications
6	Going Beyond High School	2.7	Finding Equations of Lines; Applications
7	Fruit Juice Consumption	3.4	Systems of Equations in Two Variables
8	Waterfalls	3.7	Systems of Inequalities and Linear Programming
9	Construction	3.7	Systems of Inequalities and Linear Programming
10	Visualizing Factoring	4.4	Factoring Trinomials: $x^2 + bx + c$
11	Matching Factorizations	4.5	Factoring Trinomials: $ax^2 + bx + c,\ a \neq 1$
12	Data and Downloading	5.8	Variation and Applications
13	Firefighting Formulas	6.3	Simplifying Radical Expressions
14	Music Downloads	7.5	Analyzing Graphs of Quadratic Functions
15	Let's Go to the Movies	7.5	Analyzing Graphs of Quadratic Functions
16	Teaching about Zeros: Creating a Video	8.4	Theorems about Zeros of Polynomial Functions
17	Earthquake Magnitude	9.4	Logarithmic Functions and Graphs
18	A Cosmic Path	11.2	The Circle and the Ellipse
19	Bargaining for a Used Car	12.3	Geometric Sequences and Series
20	Interest Rates on Credit Cards	Left to the instructor	
21	Home Loans	Left to the instructor	

The **Interactive Figures** listed below accompany *College Algebra with Intermediate Algebra* by Beecher/Penna/Johnson/Bittinger. They are located in the Video and Resource Library within MyLab Math. The following table correlates these figures with the sections in the text and the Interactive Preview Worksheets.

Interactive Figure	Section in the Text	Interactive Preview Worksheet
Graphing Inequalities	R.1	
Order on the Number Line	R.1	
Negative Exponents	R.3	
Absolute-Value Equations and Inequalities	1.6	Preview 8
Graphing Linear Equations	2.1	Preview 9
Graphing Functions	2.2	Preview 9
Finding Function Values	2.2	Preview 10
Domain and Range of Functions	2.3	Preview 10 Preview 11
Sum and Difference of Two Functions	2.4	
Product and Quotient of Two Functions	2.4	
Slope	2.5	
Slope of a Line	2.5	
Slope-Intercept Form	2.5	
Equations of Lines: Slope-Intercept Form	2.6	
Equations of Lines: Point-Slope Form	2.7	
Equation of Lines: General Form	2.7	
Equations of Lines: Parallel and Perpendicular	2.7	
Modeling Data Regression	2.7	
Mixture Problems	3.4	
Linear Inequalities in Two Variables	3.7	
Special Products	4.2	Preview 14
Difference Quotients	5.4	
Motion Problems	5.6	
Graphs of Radical Functions	6.1	
Increasing and Decreasing Functions	6.8	
Graphs of Piecewise Functions	6.8	Preview 29
Application: Volume of a Box	6.8	
Symmetry of Functions: Even and Odd	7.1	
Transformation of Functions	7.2	Preview 30
Intercepts and Solutions	7.4	Preview 33
Quadratic Functions and Their Graphs	7.5	Preview 32
Graphs of Quadratic Functions	7.5	Preview 32
Application: Height of a Baseball	7.5	Preview 32

(continued)

Interactive Figure	Section in the Text	Interactive Preview Worksheet
Graphs of Power Functions	8.1	
Leading-Term Test and Polynomials	8.1	
Zeros of Polynomial Functions	8.1	Preview 33
The Intermediate-Value Theorem	8.2	
Graphs of Rational Functions: Part I	8.5	Preview 34
Graphs of Rational Functions: Part II	8.5	Preview 34
Graphs of Rational Functions: Oblique Asymptotes	8.5	
Polynomial and Rational Inequalities	8.6	
Composite Functions	9.1	
Graphing Functions and Their Inverses	9.2	Preview 35
Graphs of Inverse Functions	9.2	Preview 35
Graphs of Exponential Functions	9.3	
Graphs of Logarithmic Functions	9.4	
Exponential Growth Models	9.7	
Graphs of Logistic Functions	9.7	
The Distance Formula	11.1	
Equations of Circles	11.2	Preview 42
Graphing Hyperbolas	11.3	Preview 41
Conic Sections: Rectangular Form	11.3	Preview 42
Geometric Sequences and Series	12.3	

Mathematical Growth Mindset Activities

Correlation Guide

The **Mindset Activities** in this Notebook accompany ***College Algebra with Intermediate Algebra***, by Beecher/Penna/Johnson/Bittinger. The following table correlates the Mindset Activities with sections in the text and with the Interactive Preview Worksheets.

Mindset Activity	Section in the Text	Preview Worksheet
1 Reasoning	R.2 Operations with Real Numbers	1 Operations on the Real Numbers
2 Creativity and Innovation	3.4 Solving Applied Problems: Two Equations 3.6 Solving Applied Problems: Three Equations	
3 Multiple Representations	5.6 Applications and Proportions	24 Solving Proportions
4 Exploration	6.8 Increasing, Decreasing, and Piecewise Functions; Applications	29 Graphing Piecewise Functions
5 Curiosity and Inquiry	7.5 Analyzing Graphs of Quadratic Functions	32 Introduction to Quadratic Functions
6 Locus of Control	8.3 Polynomial Division; The Remainder Theorem and the Factor Theorem 8.4 Theorems about Zeros of Polynomial Functions	
7 "Yet"	9.6 Solving Exponential Equations and Logarithmic Equations	37 Solving Exponential Equations

Developing a Mathematical Growth Mindset: Reasoning
This task can be completed as an introduction to Interactive Preview Worksheet 1:
OPERATIONS ON THE REAL NUMBERS.

For each of the following, determine the value of the expression. Then write an argument you could use to convince a student who does not know the rules for operations on the real numbers that your answer is correct. Use a model or a real-life example.

1.   Expression: $8 + (-12)$

    a) $8 + (-12) = $ _______

    b) Convincing argument:

2.   Expression: $-3 - (-2)$

    a) $-3 - (-2) = $ _______

    b) Convincing argument:

3.   Expression: $(-5) \cdot (-2)$

    a) $(-5) \cdot (-2) = $ _______

    b) Convincing argument:

4. Form groups of three. Each member should choose one of the three expressions above and use their argument to try to convince the others that their answer is correct. The remaining group members should evaluate the argument to see if it is convincing. If possible, offer suggestions to make the argument more sound.

5. Reflect on the task you have just completed.

   - Which of the expressions was the most difficult to model and explain? Why do you think that is true?

   - Why do you think reasoning is an important part of mathematics?

   - How can you develop your ability to justify and explain your conclusions?

REASONING AND CONVINCING ARGUMENTS

A big part of mathematics development is justifying the conclusions reached. In order for new mathematical discoveries to be published, other mathematicians must agree that the author's work is valid.

Your ability to explain why your work is correct will grow as your mathematical vocabulary, conceptual understanding, and experience expands. The logic and reasoning developed through growth in mathematics can be used to find solutions to problems in many other areas of work and study.

Developing a Mathematical Growth Mindset: Creativity and Innovation
This task can be completed in conjunction with Section 3.4 or Section 3.6:
SOLVING APPLIED PROBLEMS WITH SYSTEMS.

As a group of two or more, choose a problem from Exercise Set 3.4 or Exercise Set 3.6, or use the problem chosen for you by your instructor.

1. Split your group in half. One half of the group should approach the problem analytically and one creatively.

    *   For an analytical approach, look at the examples given in your notes or in the textbook section. Look for similarities between your problem and the examples, and model your solution after an existing solution of a similar problem.

    *   For a creative approach, try understanding and solving the problem in a way not yet presented in class. Document the approaches that you use.

2. As a group, compare your solutions. Did both approaches lead to a solution? Were the answers the same? Which approach do you prefer?

3. Both analytical and creative approaches to problem solving are authentic and effective ways to find solutions. Some individuals prefer or excel in one approach. Take the Personal Inventory Assessment in MyLab Math titled "Problem Solving, Creativity, and Innovation." How did you score in analytical problem solving? In creative problem solving? In overall appropriate problem solving?

4. With your group, develop a scenario in which analytical problem solving might be more appropriate and a second scenario in which creative problem solving might be more appropriate.

5. As a group or in the entire class, discuss how creativity in problem solving could be fostered in a classroom or work environment.

<table>
<tr><td>

CREATIVITY AND INNOVATION

Individuals with a growth mindset understand that their ability to learn can develop. This means that individuals do not always have to follow existing procedures, but they can experiment and develop their own innovative ways to understand and approach solutions to problems.

It is important to understand and be able to apply existing approaches to problem solving, but don't be afraid to try new ways to solve problems that fit your abilities and interests. As you combine increased mathematical knowledge with creativity and innovation, you will be able to solve problems in many areas that before were challenging or impossible.

</td></tr>
</table>

Developing a Mathematical Growth Mindset: Multiple Representations
This task can be completed as an introduction to Interactive Preview Worksheet 24:
SOLVING PROPORTIONS.

Material: Colored pens or pencils; graph paper if desired

PART I: Adrian and Bailey are comparing the number of photos they have on their phones. Adrian has 1000 photos on his phone, and 100 of them are selfies. Bailey has 500 photos on her phone, and 50 of them are selfies. Adrian claims that, although he has more selfies, the fractional part of his photos that are selfies is the same as the fractional part of Bailey's photos that are selfies.

1. Is Adrian correct? Instead of using fraction or decimal notation, justify your answer with a drawing. That is, make a visual representation of the fraction of photos that are selfies for both Adrian and Bailey.

2. Discuss your drawing with at least one other class member. Do they agree that your representation justifies your response to Adrian's claim?

3. Compare your drawing with others' drawings. How are your visualizations similar? How are they different?

Because the fractions $\dfrac{100}{1000}$ and $\dfrac{50}{500}$, are equal, we say that the pairs 100, 1000 and 50, 500 are **proportional**. We can see whether pairs of numbers are proportional by comparing using areas of models of the pairs.

PART II: Adrian has 100 selfies on his phone and 65 screenshots. Bailey has 50 selfies on her phone and 35 screenshots.

4. Is the ratio of selfies to screenshots the same on both phones? That is, are the pairs 100, 65 and 50, 35 proportional? Justify your answer with a drawing.

5. Discuss your drawing with at least one other class member. Do they agree that your representation justifies your response?

6. Compare your drawing with others' drawings. How are your visualizations similar? How are they different?

PART III: Form groups of two or three, and find information on your phone about the number of different types of media that are stored on the phone.

    7.   Develop your own proportion question, similar to those in Parts I and II, that compares ratios of media on your phone with the other phones in your group.

    8.   Compare the ratios visually. Use rounded numbers to make the drawings.

    9.   Justify your conclusion to another group.

PART IV: Reflect on the task you have just completed.

    10.   How many different types of visualizations did you make note of in your class? Which visualization seemed to make the most sense to you?

    11.   Did you use the same type of model in all three parts of this activity? Why or why not?

    12.   Did you note any visualizations which did not seem to model a ratio accurately? How did they differ from more appropriate models?

MULTIPLE REPRESENTATIONS

There are often many ways to represent, or model, a situation mathematically. While, in algebra, we often use expressions and equations to model and solve problems, graphs and other visual representations are often just as important. As you may have noted in this activity, there can be more than one way to represent a situation visually as well.

Multiple representations in mathematics allow us to view a problem from different perspectives. These representations often result from differences in the backgrounds and interests of those modeling a situation. The diverse experiences, abilities, and creativity that you bring to problem-solving contribute to your growth, and to others' growth, in mathematical thinking.

Developing a Mathematical Growth Mindset: Exploration
This task can be completed as an introduction to Interactive Preview Worksheet 29:
GRAPHING PIECEWISE FUNCTIONS.

Material: Graph paper and scissors

Preparation: Choose a classmate to work with. Each of you should use graph paper with identical axes and units. For best results, make sure that both the horizontal and vertical axes show units at least from -10 to 10.

1. Each participant should write a linear function of the form $y = mx + b$ (like $f(x) = 2x$ or $f(x) = \dfrac{1}{2}x + 3$). Your functions should be different. Choose numbers between -3 and 3 for m and for b.

2. Each participant should graph their chosen function on their own piece of graph paper.

3. Discuss:
   - What is the domain of each function?
   - Do both functions have the same domain? Why or why not?

4. Between yourselves, choose a number between -5 and 5 at which to split the domain of your function. Carefully draw a vertical line from the top of the graph paper to the bottom of the graph paper through that value of the domain.

5. Cut each graph apart on the vertical line that you drew. The graph of your linear function will now be in two *pieces*.

6. Swap the left sides of your graph with your classmate.

7. Now place the left side of your classmate's graph next to the right side of your own graph and discuss the following questions:
   - Is your graph *continuous*? (That is, you can draw it from left to right without picking up your pencil.)
   - Is there an x-value at which there is no clearly-defined y-value? How is this pictured by the graph?
   - Is your graph the graph of a function? Why or why not?

8.   If your graph has a break in it, the break will occur at the point at which you split the domain. Choose the y-value of the endpoint on the left side of the new graph as the function value at that point. Draw a small solid circle at that point. Then draw a small open circle at the endpoint on the right side of the new graph.

9.   What you have just created is a *piecewise function*, where $a =$ the value at which you split your domain. This function can be described as follows

$$f(x) = \begin{cases} \underline{\hspace{4cm}}, & \text{if } x \le \underline{\hspace{1cm}} \\ \text{partner function definition} & a \\[1em] \underline{\hspace{4cm}}, & \text{if } x > \underline{\hspace{1cm}} \\ \text{your function definition} & a \end{cases}$$

10.   Reflect on the task you have just completed.

- Compare your graphs with those of other groups. What differences do you notice?

- If you did not know at what point you split the domain, how could you tell from the graph?

- If another group handed you a graph without a description of the equations used, could you describe the piecewise function from that graph? What would you need to know?

EXPLORE MATHEMATICS

When you're in the middle of a mathematics course, it's easy to look at mathematics as a series of definitions to learn and skills to master. In reality, mathematics is a science of exploration and discovery. In a mathematics course, you are studying the discoveries of earlier mathematicians, but in the process you can also explore questions, connections, and relationships that you encounter.

If you are learning a new topic, it is always helpful to ask "what if" and "why" and look for patterns and other relationships. When you see patterns and make connections yourself, you are growing mathematically.

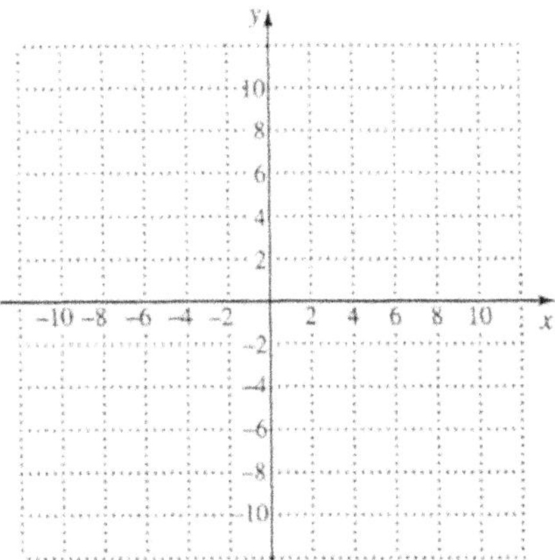

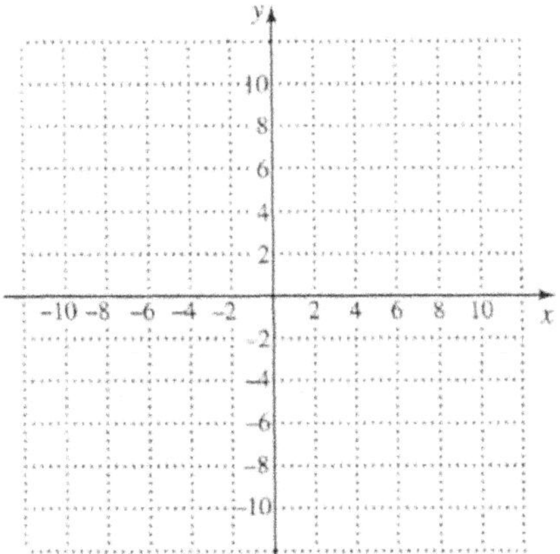

Developing a Mathematical Growth Mindset: Curiosity and Inquiry
This task can be completed before Interactive Preview Worksheet 32:
INTRODUCTION TO QUADRATIC FUNCTIONS.

Alex and Ryann plan to install a fence in their backyard. During their research, they came across the following tables.

Material and Labor	Cost
5-ft fencing and posts	$6 per linear foot
6-ft fencing and posts	$7 per linear foot
5-ft driveway gate	$400
6-ft driveway gate	$450
Equipment rental	$25 per day (4 days)
Labor costs	$40 per hour (30 hours)

City	Cost of 400-ft fence, 5 ft high, installed
New York City, NY	$8800
Philadelphia, PA	$7600
Washington, DC	$6800
Atlanta, GA	$6000
Miami, FL	$8400
Minneapolis, MN	$6800
Chicago, IL	$8000
Houston, TX	$8400
Denver, CO	$5200
San Jose, CA	$7600
Seattle, WA	$5600

1. If you were Alex or Ryann, what questions would you have? Write down 2 – 4 questions that reading this information has raised in your mind.

2. What additional information do you need in order to answer your questions?

3. Alex and Ryann have a budget of $5000. What kinds of decisions will they have to make in order to stay within this budget while building the fence?

4. Design a fence that might be within a budget of $5000. As you work, consider the following questions as well as those that you raised earlier:

   - What shape should the fenced-in area take? Does the shape of the area affect the cost of the fence?
   - Where should the fence be located in relation to the house? Does the location of the fence affect its cost?

5. Using graph paper on the next page, sketch the fence you have designed. Be sure to decide on the scale you are using before beginning the sketch. Explain to a classmate why you designed the fence the way you did.

6. Estimate the cost of installing your designed fence.

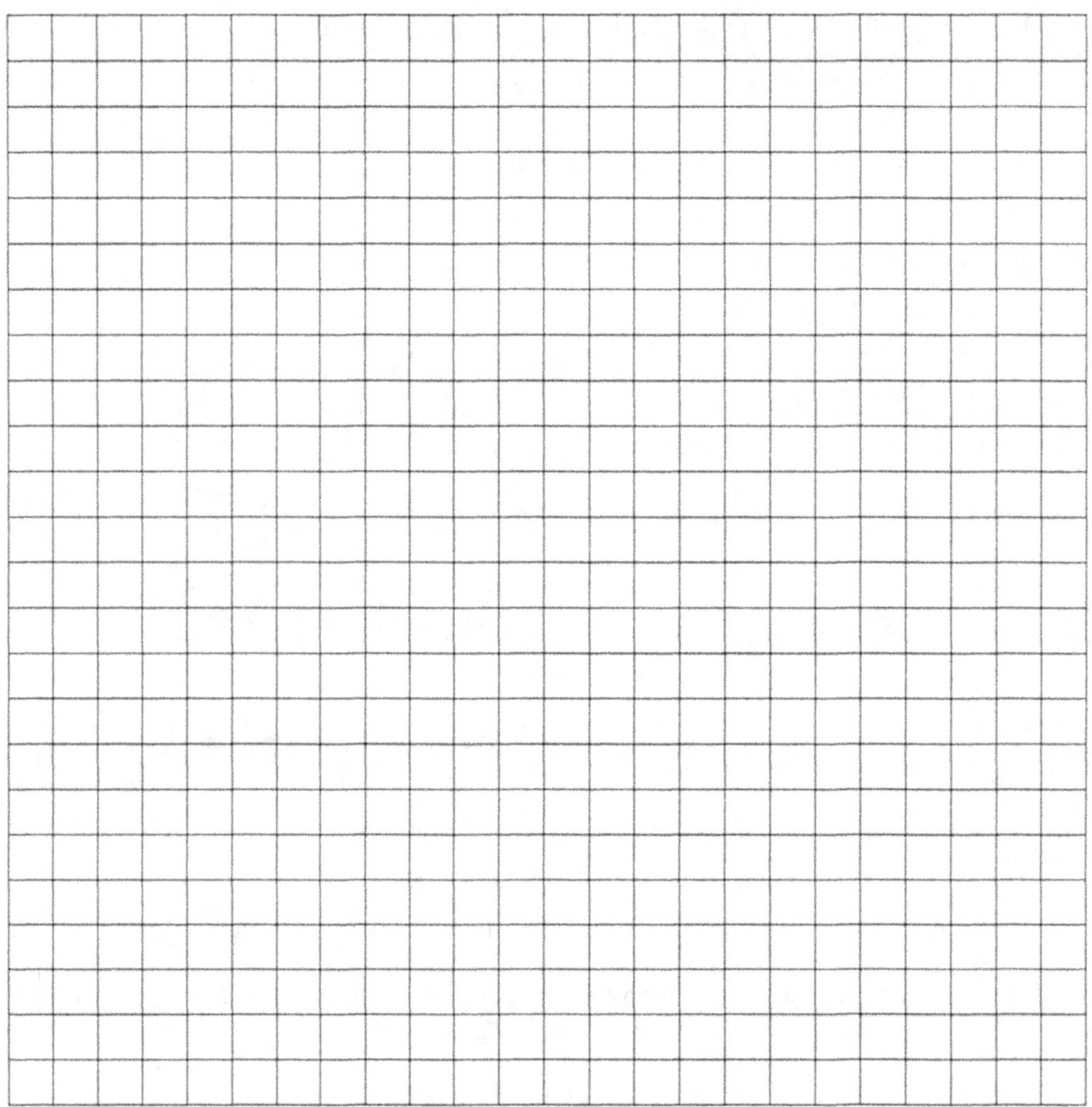

Curiosity and Inquiry

Many problems in mathematics classes are clearly defined with one correct answer. In reality, problems and data encountered in mathematics and elsewhere often raise more questions than answers. Like the fencing design problem above, mathematics is often open-ended. Mathematics itself grows through an inquiry process of asking questions such as "Why?" "How?" or "What if?" The process of learning mathematics is also open-ended. Students with a growth mindset understand that their ability to learn is not fixed but can grow. One essential part of having a growth mindset is curiosity – and interest in and a willingness to ask questions. When you engage in inquiry while studying mathematics, not only will you learn more about mathematics, but you will also expand your ability to learn mathematics.

Developing a Mathematical Growth Mindset: Locus of Control
This task can be completed in conjunction with Sections 8.3 and 8.4:
THEOREMS ABOUT POLYNOMIAL FUNCTIONS.

There are seven theorems about polynomials in Sections 8.3 and 8.4. They are set apart from
the rest of the text in boxes with colored backgrounds. Your task is to develop a plan for
understanding and remembering these theorems and to begin work on that plan.

1. Copy each theorem at the top of a piece of paper. You can write them out or cut them
   out from the next page of this workbook and affix each at the top of one piece of paper.

2. Find out from your instructor the level at which you are expected to know these
   theorems. For example, should you be able to tell what the theorem states when given
   the name of the theorem? Will you be given the theorems and asked to apply them?

3. Think about how you have been successful understanding and learning concepts in the
   past. Have you rewritten statements in your own words? Have you developed examples
   that illustrate an idea? Do you learn best when you begin by memorizing a theorem
   word for word?

4. Using past successful methods, develop a plan for understanding, remembering, and
   being able to apply the theorems from these sections.

5. Choose one theorem and follow your plan for that theorem. After you have worked on
   that theorem for 10 – 15 minutes, reflect on your approach. Do you think it will be
   successful? Are there any changes you would like to make?

6. When you believe that you are responsible for and capable of understanding and remembering mathematical concepts such as these theorems about polynomials, you are demonstrating an *internal locus of control*. Take the Personal Inventory Assessment in MyLab Math titled "Locus of Control Scale." Does your score indicate that you have a high external locus of control or a high internal locus of control?

7. If you would like to increase your internal locus of control, reflect on a recent experience of academic challenge or other life event. Identify what you think caused that struggle or event and then outline how you could have had more control.

LOCUS OF CONTROL

Because individuals with a growth mindset believe that they are capable of learning anything, they tend to have an internal locus of control. If they succeed, they credit their own hard work, and if they fail, they blame themselves for not having done more.

Believing that you are in control of your own learning is a powerful step toward success in mathematics. However, in every life there are circumstances that are outside of one's control. Understanding what you can control and what you cannot helps provide a healthy balance between an internal and an external locus of control.

THE REMAINDER THEOREM

If a number c is substituted for x in the polynomial $f(x)$ then the result $f(c)$ is the remainder that would be obtained by dividing $f(x)$ by $x-c$. That is, if $f(x)=(x-c)\cdot Q(x)+R$, then $f(c)=R$.

THE FACTOR THEOREM

For a polynomial $f(x)$, if $f(c)=0$, then $x-c$ is a factor of $f(x)$.

THE FUNDAMENTAL THEOREM OF ALGEBRA

Every polynomial function of degree n, with $n\geq 1$, has at least one zero in the set of complex numbers.

NONREAL ZEROS: $a+bi$ and $a-bi$, $b\neq 0$

If a complex number $a+bi$, $b\neq 0$, is a zero of a polynomial function $f(x)$ with *real* coefficients, then its conjugate, $a-bi$, is also a zero. For example, if $2+7i$ is a zero of a polynomial function $f(x)$ with real coefficients, then its conjugate, $2-7i$, is also a zero. (Nonreal zeros occur in conjugate pairs.)

IRRATIONAL ZEROS: $a+c\sqrt{b}$ and $a-c\sqrt{b}$, b IS NOT A PERFECT SQUARE

If $a+c\sqrt{b}$, where a, b, and c are rational and b is not a perfect square, is a zero of a polynomial function $f(x)$ with *rational* coefficients, then its conjugate, $a-c\sqrt{b}$, is also a zero. For example, if $-3+5\sqrt{2}$ is a zero of a polynomial function $f(x)$ with rational coefficients, then its conjugate, $-3-5\sqrt{2}$, is also a zero. (Irrational zeros occur in conjugate pairs.)

THE RATIONAL ZEROS THEOREM

Let $P(x)=a_{n}x^{n}+a_{n-1}x^{n-1}+\cdots+a_{1}x+a_{0}$, where all the coefficients are integers. Consider a rational number denoted by p/q, where p and q are relatively prime (having no common factor besides -1 and 1). If p/q is a zero of $P(x)$, then p is a factor of a_{0} and q is a factor of a_{n}.

DESCARTES' RULE OF SIGNS

Let $P(x)$, written in descending order or ascending order, be a polynomial function with real coefficients and a nonzero constant term. The number of positive real zeros of $P(x)$ is either:

1. The same as the number of variations of sign in $P(x)$, or
2. Less than the number of variations of sign in $P(x)$ by a positive even integer.

The number of negative real zeros of $P(x)$ is either:

3. The same as the number of variations of sign in $P(-x)$, or
4. Less than the number of variations of sign in $P(-x)$ by a positive even integer.

A zero of multiplicity m must be counted m times.

Developing a Mathematical Growth Mindset: "Yet"
This task can be completed as an introduction to Interactive Preview Worksheet 37:
SOLVING EXPONENTIAL EQUATIONS.

Complete Steps 1 – 3 for each of the equations in the table. Discuss the solution of each equation before completing the solution of the next equation. If you cannot solve an equation, simply move to the next one. When you have solved as many equations in the table as you can, complete the reflection in Step 4.

For each equation, do the following.
1. On your own or with another student, solve the equation and check your answer. Show each step of the process and be ready to explain your thinking.
2. Write the number of solutions that you found as well as the solution(s) in the chart.
3. As a group, compare your solution(s). Did you all find the same solution(s)? Did you all follow the same process? How was the process you used to solve this equation similar to the process you used to solve the previous equation? How was it different?

	Number of solutions	**Solution(s)**
$3x - 5 = 7$		
$(x - 7)(3x - 5) = 0$		
$\sqrt{x + 1} = 9$		
$|x + 1| = 9$		
$(x + 1)^2 = 9$		
$x^2 = -64$		
$x^3 = 64$		
$x^4 = 16$		
$2^x = 16$		
$2^x = 5$		

4. Reflect on the task you have just completed.

- Which of the equations you were successfully able to solve do you think you would not have been able to solve before beginning this course?

- Did other students in your group solve any of these equations using a different approach than you did? If so, what is one thing you learned from their approach?

- Did you make a mistake in solving any of the equations? If so, what is one thing you learned from a mistake?

- Which equation was the most challenging to solve for you or for your group? Were there any you were unable to solve?

YET

One of the most important words in developing a growth mindset is the word "Yet." As you solved equations in this task, you probably realized that your equation-solving ability has grown during this course. And, if there were some equations you could not solve, you may have become frustrated. However, you will learn how to solve all of these equations in this course.

Whenever you realize that you cannot do something in math, add the word "yet" to the end of any statement describing your ability: "I cannot solve this equation *yet*." As you work diligently and grow mathematically, you will be able to do or understand what seems now difficult or confusing.

Interactive Preview Worksheets

(continued)

Correlation Guide

The **Interactive Preview Worksheets** in this Notebook accompany *College Algebra with Intermediate Algebra* by /Beecher/Penna/Johnson/Bittinger. The following table correlates the Preview Worksheets with the sections in the text.

Preview Worksheet		Section in the Text	
1	Operations on the Real Numbers	R.2	Operations with Real Numbers
2	Adding, Subtracting, Multiplying, and Dividing Fractions	R.2	Operations with Real Numbers
3	Order of Operations	R.3	Exponential Notation and Order of Operations
4	Integers as Exponents	R.7	Properties of Exponents and Scientific Notation
5	Solving Linear Equations	1.1	Solving Equations
6	Interval Notation	1.4	Sets, Inequalities, and Interval Notation
7	Solving Linear Inequalities	1.4	Sets, Inequalities, and Interval Notation
8	Solving Equations and Inequalities with Absolute Value	1.6	Absolute-Value Equations and Inequalities
9	Graphing Linear Equations	2.1	Graphs of Equations
10	Function Values; Domain and Range	2.3	Finding Domain and Range
11	Determine the Domain and the Range of a Function	2.3	Finding Domain and Range
12	Introduction to Polynomials	4.1	Introduction to Polynomials and Polynomial Functions
13	Adding and Subtracting Polynomials	4.1	Introduction to Polynomials and Polynomial Functions
14	Multiplying Binomials	4.2	Multiplication of Polynomials
15	Factoring by Grouping	4.3	Introduction to Factoring
16	Factoring Trinomials: $ax^2 + bx + c$	4.5	Factoring Trinomials: $ax^2 + bx + c, \, a \neq 1$
17	Factoring Trinomial Squares and Differences of Squares	4.6	Special Factoring
18	The Principle of Zero Products	4.8	Applications of Polynomial Equations and Functions
19	The Pythagorean Theorem	4.8	Applications of Polynomial Equations and Functions
		6.7	Applications Involving Powers and Roots
20	Simplifying Rational Expressions	5.1	Rational Expressions and Functions: Multiplying, Dividing, and Simplifying

(continued)

Preview Worksheet		**Section in the Text**	
21	Finding the LCM of Algebraic Expressions	5.2	LCMs, LCDs, Addition and Subtraction
22	Simplifying Complex Rational Expressions	5.4	Complex Rational Expressions
23	Solving Rational Equations	5.5	Solving Rational Equations
24	Solving Proportions	5.6	Applications and Proportions
25	Simplifying Radical Expressions	6.3	Simplifying Radical Expressions
26	Multiplying Radical Expressions	6.3	Simplifying Radical Expressions
27	Rationalizing Denominators	6.5	More on Division of Radical Expressions
28	Checking Solutions of Radical Equations	6.6	Solving Radical Equations
29	Graphing Piecewise Functions	6.8	Increasing, Decreasing, and Piecewise Functions: Applications
30	Transformations	7.2	Transformations
31	Completing the Square	7.4	Quadratic Equations, Functions, Zeros, and Models
32	Introduction to Quadratic Functions	7.5	Analyzing Graphs of Quadratic Functions
33	Zeros of Polynomial Functions	8.1	Polynomial Functions and Models
34	Asymptotes of Rational Functions	8.5	Rational Functions
35	Graphing Inverse Functions	9.2	Inverse Functions
36	Introduction to Logarithms	9.4	Logarithmic Functions and Graphs
37	Solving Exponential Equations	9.6	Solving Exponential Equations and Logarithmic Equations
38	Solving Logarithmic Equations	9.6	Solving Exponential Equations and Logarithmic Equations
39	Using an Inverse Matrix to Solve a System of Equations	10.3	Inverses of Matrices
40	The Ellipse	11.2	The Circle and the Ellipse
41	The Hyperbola	11.3	The Hyperbola
42	Classifying Equations of Conic Sections	11.3	The Hyperbola

Interactive Preview Worksheet 1: OPERATIONS ON THE REAL NUMBERS

Adding Real Numbers

1. *Positive numbers*: Add the numbers. The result is positive.
2. *Negative numbers*: Add absolute values. Make the answer negative.
3. *A positive number and a negative number*:
   - If the numbers have the same absolute value, the answer is 0.
   - If the numbers have different absolute values, subtract the smaller absolute value from the larger. Then:
     a) If the positive number has the greater absolute value, make the answer positive.
     b) If the negative number has the greater absolute value, make the answer negative.
4. *One number is zero*: The sum is the other number.

Examples Add.

1. $5 + 8 = 13$

2. $-5 + 8 = 3$

3. $5 + (-8) = -3$

4. $-5 + (-8) = -13$

5. $-8 + 0 = -8$

6. $21 + (-7) = 14$

7. $-21 + (-7) = -28$

8. $-23 + 23 = 0$

9. $6.2 + (-17.1) = -10.9$

10. $-\dfrac{5}{6} + \left(-\dfrac{5}{8}\right) = -\dfrac{5}{6} \cdot \dfrac{4}{4} + \left(-\dfrac{5}{8}\right) \cdot \dfrac{3}{3}$

$$= -\dfrac{20}{24} + \left(-\dfrac{15}{24}\right) = -\dfrac{35}{24}$$

Subtracting Real Numbers

For any real numbers a and b, $a - b = a + (-b)$.

We can subtract by adding the **opposite (additive inverse)** of the number being subtracted. (Two numbers whose sum is 0 are called opposites, or additive inverses. For example, the opposite of -2 is 2 and the opposite of $\dfrac{4}{7}$ is $-\dfrac{4}{7}$.)

Examples Subtract. Rewrite each subtraction as an addition and then use the rules for adding real numbers.

11. $24 - 9 = 24 + (-9) = 15$

12. $24 - (-9) = 24 + 9 = 33$

13. $-24 - 9 = -24 + (-9) = -33$

14. $-24 - (-9) = -24 + 9 = -15$

15. $0 - 9 = 0 + (-9) = -9$

16. $0.13 - 0.7 = 0.13 + (-0.7) = -0.57$

17. $\dfrac{2}{3} - \left(-\dfrac{5}{3}\right) = \dfrac{2}{3} + \dfrac{5}{3} = \dfrac{7}{3}$

18. $-4.3 - 10 = -4.3 + (-10) = -14.3$

19. $-\dfrac{7}{8} - \left(-\dfrac{2}{5}\right) = -\dfrac{7}{8} + \dfrac{2}{5}$

$$= -\dfrac{7}{8} \cdot \dfrac{5}{5} + \dfrac{2}{5} \cdot \dfrac{8}{8}$$

$$= -\dfrac{35}{40} + \dfrac{16}{40} = -\dfrac{19}{40}$$

20. $\dfrac{11}{60} - \dfrac{5}{24} = \dfrac{11}{60} + \left(-\dfrac{5}{24}\right) = \dfrac{11}{60} \cdot \dfrac{2}{2} + \left(-\dfrac{5}{24}\right) \cdot \dfrac{5}{5}$

$$= \dfrac{22}{120} + \left(-\dfrac{25}{120}\right) = -\dfrac{3}{120}$$

$$= -\dfrac{1 \cdot 3}{40 \cdot 3} = -\dfrac{1}{40} \cdot \dfrac{3}{3} = -\dfrac{1}{40} \cdot 1 = -\dfrac{1}{40}$$

Multiplying or Dividing Real Numbers

1. Multiply or divide the absolute values.
2. If the signs are the same, then the answer is positive.
3. If the signs are different, then the answer is negative.

Examples Multiply.

21. $-3 \times 10 = -30$

22. $-3 \times (-10) = 30$

23. $\dfrac{7}{11} \cdot \dfrac{9}{2} = \dfrac{63}{22}$

24. $0.5 \times (-1.2) = -0.6$

25. $400(-300) = -120,000$

26. $-\dfrac{5}{8} \cdot \dfrac{3}{20} = -\dfrac{15}{160} = -\dfrac{5 \cdot 3}{5 \cdot 32}$

$$= \dfrac{5}{5} \cdot \left(-\dfrac{3}{32}\right) = 1 \cdot \left(-\dfrac{3}{32}\right) = -\dfrac{3}{32}$$

Examples Divide.

27. $\dfrac{-42}{-6} = 7$

28. $500 \div (-50) = -10$

29. $\dfrac{-33}{11} = -3$

30. $-1.44 \div (-1.2) = 1.2$

31. $-\dfrac{1}{3} \div 3 = -\dfrac{1}{3} \cdot \dfrac{1}{3} = -\dfrac{1}{9}$

32. $-2 \div \left(-\dfrac{3}{4}\right) = -\dfrac{2}{1} \left(-\dfrac{4}{3}\right) = \dfrac{8}{3}$

33. $\dfrac{4}{15} \div \left(-\dfrac{3}{10}\right) = \dfrac{4}{15} \cdot \left(-\dfrac{10}{3}\right)$

$$= -\dfrac{40}{45} = -\dfrac{5 \cdot 8}{5 \cdot 9}$$

$$= \dfrac{5}{5} \cdot \left(-\dfrac{8}{9}\right) = 1 \cdot \left(-\dfrac{8}{9}\right) = -\dfrac{8}{9}$$

Check Your Understanding

Match each expression with an equivalent expression from choices A – H.

1. $-10 - (-31)$

2. $-\dfrac{1}{4} \div \left(-\dfrac{2}{5}\right)$

A. $-10 + (-31)$

E. $-\dfrac{1}{4} \cdot \left(-\dfrac{5}{2}\right)$

3. $-\dfrac{1}{4} \div 5$

4. $10 - 31$

B. $10 + (-31)$

F. $4 \cdot \dfrac{5}{2}$

5. $\dfrac{2}{5} \div (-4)$

6. $10 - (-31)$

C. $-10 + 31$

G. $\dfrac{2}{5} \cdot \left(-\dfrac{1}{4}\right)$

7. $-10 - 31$

8. $4 \div \dfrac{2}{5}$

D. $10 + 31$

H. $-\dfrac{1}{4} \cdot \dfrac{1}{5}$

Exercises Perform the indicated operation.

1. $-3+17$

2. $-11-(-6)$

3. $-80 \cdot (-3)$

4. $\dfrac{42}{-21}$

5. $\dfrac{1}{4}-\dfrac{5}{2}$

6. $-\dfrac{2}{5} \div 5$

7. $-5.14-(-0.3)$

8. $0-8$

9. $-24 \div (-3)$

10. $-9+(-25)$

11. $-21 \div (0.3)$

12. $-\dfrac{4}{9} \cdot \dfrac{5}{2}$

13. $-3 \div \left(-\dfrac{11}{12}\right)$

14. $4.5(-1.2)$

15. $\dfrac{15}{8}-\left(-\dfrac{5}{3}\right)$

16. $\dfrac{1}{4} \div (-4)$

17. $-15-12$

18. $-\dfrac{14}{9} \div \left(-\dfrac{7}{3}\right)$

19. $3.16+(-5.2)$

20. $8-(-20)$

Interactive Preview Worksheet 2: ADDING, SUBTRACTING, MULTIPLYING, AND DIVIDING FRACTIONS

Multiplying Fractions

To multiply fractions, we multiply the numerators to get the new numerator, and we multiply the denominators to get the new denominator. Then we simplify, if possible.

Example 1 Multiply and simplify: $\dfrac{5}{6} \cdot \dfrac{9}{25}$.

Multiply numerators and denominators.	$= \dfrac{5 \cdot 9}{6 \cdot 25}$
Factor the numerator and the denominator.	$= \dfrac{5 \cdot 3 \cdot 3}{2 \cdot 3 \cdot 5 \cdot 5}$
Remove a factor of 1: $\dfrac{5 \cdot 3}{5 \cdot 3} = 1$.	$= \dfrac{\cancel{5} \cdot \cancel{3} \cdot 3}{2 \cdot \cancel{3} \cdot \cancel{5} \cdot 5} = \dfrac{3}{2 \cdot 5}$
Simplify.	$= \dfrac{3}{10}$

Adding Fractions

When denominators are the same, we add fractions by adding the numerators and keeping the same denominator.

Example 2 Add and simplify: $\dfrac{7}{8} + \dfrac{5}{8}$.

The common denominator is 8. We add the numerators and keep the denominator.	$= \dfrac{7+5}{8} = \dfrac{12}{8}$
Factor the numerator and the denominator.	$= \dfrac{2 \cdot 2 \cdot 3}{2 \cdot 2 \cdot 2}$
Remove a factor of 1: $\dfrac{2 \cdot 2}{2 \cdot 2} = 1$.	$= \dfrac{\cancel{2} \cdot \cancel{2} \cdot 3}{\cancel{2} \cdot \cancel{2} \cdot 2}$
Simplify.	$= \dfrac{3}{2}$

When denominators are not the same, we first find the least common multiple (LCM) of the denominators. Then we multiply by 1 to express fractions in terms of the least common denominator (LCD).

Example 3 Add and simplify: $\dfrac{3}{8} + \dfrac{5}{12}$.

The least common denominator is 24. We multiply $\dfrac{3}{8}$ by $\dfrac{3}{3}$ and $\dfrac{5}{12}$ by $\dfrac{2}{2}$.	$= \dfrac{3}{8} \cdot \dfrac{3}{3} + \dfrac{5}{12} \cdot \dfrac{2}{2}$
Multiply.	$= \dfrac{9}{24} + \dfrac{10}{24}$
Add.	$= \dfrac{19}{24}$

Subtracting Fractions

Example 4 Subtract and simplify: $\dfrac{9}{26} - \dfrac{5}{26}$.

The common denominator is 26. We subtract the numerators and keep the denominator.
$$= \frac{9-5}{26} = \frac{4}{26}$$

Factor the numerator and the denominator.
$$= \frac{2\cdot 2}{2\cdot 13}$$

Remove a factor of 1: $\dfrac{2}{2} = 1$.
$$= \frac{\cancel{2}\cdot 2}{\cancel{2}\cdot 13}$$

Simplify.
$$= \frac{2}{13}$$

Example 5 Subtract and simplify: $\dfrac{18}{35} - \dfrac{3}{10}$.

The LCD is 70. We multiply $\dfrac{18}{35}$ by $\dfrac{2}{2}$ and $\dfrac{3}{10}$ by $\dfrac{7}{7}$.
$$= \frac{18}{35}\cdot\frac{2}{2} - \frac{3}{10}\cdot\frac{7}{7}$$

Multiply.
$$= \frac{36}{70} - \frac{21}{70}$$

Subtract.
$$= \frac{36-21}{70} = \frac{15}{70}$$

Factor the numerator and the denominator.
$$= \frac{3\cdot 5}{2\cdot 5\cdot 7}$$

Remove a factor of 1: $\dfrac{5}{5} = 1$.
$$= \frac{3\cdot\cancel{5}}{2\cdot\cancel{5}\cdot 7}$$

Simplify.
$$= \frac{3}{14}$$

Dividing Fractions

Two numbers whose product is 1 are called **reciprocals**, or **multiplicative inverses**. All real numbers, except 0, have reciprocals.

Examples Find the reciprocal.

6. The reciprocal of $\dfrac{2}{11}$ is $\dfrac{11}{2}$ because $\dfrac{2}{11}\cdot\dfrac{11}{2} = \dfrac{22}{22} = 1$.

7. The reciprocal of 9 is $\dfrac{1}{9}$ because $\dfrac{9}{1}\cdot\dfrac{1}{9} = \dfrac{9}{9} = 1$.

8. The reciprocal of $\dfrac{1}{40}$ is 40 because $\dfrac{1}{40}\cdot\dfrac{40}{1} = \dfrac{40}{40} = 1$.

To divide fractions, we multiply by the reciprocal of the divisor.

Example 9   Divide and simplify: $\dfrac{2}{3} \div \dfrac{4}{9}$.

The reciprocal of $\dfrac{4}{9}$ is $\dfrac{9}{4}$. Multiply by $\dfrac{9}{4}$.
$$= \dfrac{2}{3} \cdot \dfrac{9}{4} = \dfrac{2 \cdot 9}{3 \cdot 4}$$

Factor the numerator and the denominator.
$$= \dfrac{2 \cdot 3 \cdot 3}{3 \cdot 2 \cdot 2}$$

Remove a factor of 1: $\dfrac{2 \cdot 3}{2 \cdot 3} = 1$.
$$= \dfrac{\cancel{2} \cdot \cancel{3} \cdot 3}{\cancel{3} \cdot \cancel{2} \cdot 2}$$

Simplify.
$$= \dfrac{3}{2}$$

Example 10   Divide and simplify: $\dfrac{5}{6} \div 30$.

The reciprocal of 30 is $\dfrac{1}{30}$. Multiply by $\dfrac{1}{30}$.
$$= \dfrac{5}{6} \cdot \dfrac{1}{30} = \dfrac{5 \cdot 1}{6 \cdot 30}$$

Factor.
$$= \dfrac{5 \cdot 1}{2 \cdot 3 \cdot 2 \cdot 3 \cdot 5}$$

Remove a factor of 1: $\dfrac{5}{5} = 1$.
$$= \dfrac{\cancel{5} \cdot 1}{2 \cdot 3 \cdot 2 \cdot 3 \cdot \cancel{5}}$$

Simplify.
$$= \dfrac{1}{36}$$

Example 11   Divide and simplify: $16 \div \dfrac{8}{3}$.

The reciprocal of $\dfrac{8}{3}$ is $\dfrac{3}{8}$. Multiply by $\dfrac{3}{8}$.
$$= \dfrac{16}{1} \cdot \dfrac{3}{8} = \dfrac{16 \cdot 3}{1 \cdot 8}$$

Factor.
$$= \dfrac{2 \cdot 2 \cdot 2 \cdot 2 \cdot 3}{1 \cdot 2 \cdot 2 \cdot 2}$$

Remove a factor of 1: $\dfrac{2 \cdot 2 \cdot 2}{2 \cdot 2 \cdot 2} = 1$.
$$= \dfrac{\cancel{2} \cdot \cancel{2} \cdot \cancel{2} \cdot 2 \cdot 3}{1 \cdot \cancel{2} \cdot \cancel{2} \cdot \cancel{2}}$$

Simplify.
$$= \dfrac{6}{1} = 6$$

Exercises. Compute and simplify.

1. $78 \div \dfrac{1}{6}$

2. $\dfrac{15}{4} \cdot \dfrac{3}{4}$

3. $\dfrac{11}{12} - \dfrac{2}{5}$

4. $\dfrac{4}{9} + \dfrac{13}{18}$

5. $\dfrac{8}{9} \div \dfrac{4}{15}$

6. $\dfrac{1}{9} + \dfrac{5}{9}$

7. $\dfrac{15}{16} - \dfrac{5}{12}$

8. $\dfrac{3}{4} \div 10$

9. $\dfrac{12}{5} - \dfrac{2}{5}$

10. $\dfrac{1}{3} \cdot \dfrac{1}{4}$

11. $\dfrac{13}{12} \div \dfrac{39}{5}$

12. $\dfrac{11}{12} - \dfrac{3}{8}$

13. $\dfrac{15}{16} \cdot \dfrac{8}{5}$

14. $\dfrac{1}{20} \div \dfrac{1}{5}$

15. $\dfrac{147}{50} - 2$

16. $\dfrac{3}{10} + \dfrac{8}{15}$

17. $\dfrac{1}{4} + \dfrac{1}{3}$

18. $\dfrac{10}{11} \cdot \dfrac{11}{10}$

19. $\dfrac{3}{4} \div \dfrac{3}{7}$

20. $\dfrac{9}{8} + \dfrac{7}{12}$

21. $\dfrac{1}{30} \div 30$

22. $\dfrac{3}{16} - \dfrac{1}{18}$

23. $5 \cdot \dfrac{2}{3}$

24. $\dfrac{7}{30} + \dfrac{5}{12}$

Interactive Preview Worksheet 3: ORDER OF OPERATIONS

Rules for Order of Operations

1. Do all the calculations within grouping symbols, like parentheses, before operations outside.
2. Evaluate all exponential expressions.
3. Do all multiplications and divisions in order from left to right.
4. Do all additions and subtractions in order from left to right.

Example 1 Simplify: $9 + 4(7 - 1)$.

Subtract $7 - 1$.	$= 9 + 4(6)$
Multiply $4(6)$.	$= 9 + 24$
Add $9 + 24$.	$= 33$

Example 2 Simplify: $-48 \div (-6) \div (-2)$.

Divide $-48 \div (-6)$.	$= 8 \div (-2)$
Divide $8 \div (-2)$.	$= -4$

Example 3 Simplify: $-3 \cdot 15 + 25 \div 5$.

Multiply $-3 \cdot 15$.	$= -45 + 25 \div 5$
Divide $25 \div 5$.	$= -45 + 5$
Add $-45 + 5$.	$= -40$

Example 4 Simplify: $24 - 16 \div 4$.

Divide $16 \div 4$.	$= 24 - 4$
Subtract $24 - 4$.	$= 20$

Example 5 Simplify: $8 \cdot 11 - 3^2 - 2^3$.

Evaluate 3^2.	$= 8 \cdot 11 - 9 - 2^3$
Evaluate 2^3.	$= 8 \cdot 11 - 9 - 8$
Multiply $8 \cdot 11$.	$= 88 - 9 - 8$
Subtract $88 - 9$.	$= 79 - 8$
Subtract $79 - 8$.	$= 71$

Example 6 Simplify: $27+(9-12)^2 \div 3$.

Subtract $9-12$.	$=27+(-3)^2 \div 3$
Evaluate $(-3)^2$.	$=27+9 \div 3$
Divide $9 \div 3$.	$=27+3$
Add $27+3$.	$=30$

Example 7 Simplify: $20-5(10 \div 2)+5 \cdot 50$.

Divide $10 \div 2$.	$=20-5(5)+5 \cdot 50$
Multiply $5(5)$.	$=20-25+5 \cdot 50$
Multiply $5 \cdot 50$.	$=20-25+250$
Subtract $20-25$.	$=-5+250$
Add $-5+250$.	$=245$

Example 8 Simplify: $9 \cdot 10^2 - 4^3 + 40 \div 4$.

Evaluate 10^2.	$=9 \cdot 100 - 4^3 + 40 \div 4$
Evaluate 4^3.	$=9 \cdot 100 - 64 + 40 \div 4$
Multiply $9 \cdot 100$.	$=900 - 64 + 40 \div 4$
Divide $40 \div 4$.	$=900 - 64 + 10$
Subtract $900-64$.	$=836+10$
Add $836+10$.	$=846$

Check Your Understanding

Select the correct step-by-step solution. Explain the error in the incorrect solution.

1. Simplify: $18 \div 6 + 3 - 5$.

   A. $18 \div 6 + 3 - 5$

        $= 3 + 3 - 5$ (1)

        $= 6 - 5$ (2)

        $= 1$ (3)

   B. $18 \div 6 + 3 - 5$

        $= 18 \div 9 - 5$ (1)

        $= 2 - 5$ (2)

        $= -3$ (3)

2. Simplify: $80 - 4^2 \cdot 5 \div 5(10-2)$.

   A. $80 - 4^2 \cdot 5 \div 5(10-2)$

        $= 80 - 4^2 \cdot 5 \div 5(8)$ (1)

        $= 80 - 16 \cdot 5 \div 5(8)$ (2)

        $= 80 - 80 \div 5(8)$ (3)

        $= 80 - 80 \div 40$ (4)

        $= 80 - 2$ (5)

        $= 78$ (6)

   B. $80 - 4^2 \cdot 5 \div 5(10-2)$

        $= 80 - 4^2 \cdot 5 \div 5(8)$ (1)

        $= 80 - 16 \cdot 5 \div 5(8)$ (2)

        $= 80 - 80 \div 5(8)$ (3)

        $= 80 - 16(8)$ (4)

        $= 80 - 128$ (5)

        $= -48$ (6)

Exercises In Exercises 1 and 2, describe the calculation in each step of the solution.

1. Simplify: $6-3\cdot5+8\div2^2$.

$$\begin{array}{l|l}
\underline{\hspace{3cm}} & =6-3\cdot5+8\div4 \\
\underline{\hspace{3cm}} & =6-15+8\div4 \\
\underline{\hspace{3cm}} & =6-15+2 \\
\underline{\hspace{3cm}} & =-9+2 \\
\underline{\hspace{3cm}} & =-7
\end{array}$$

2. Simplify: $32\div4\cdot2^2-2(5-3)$.

$$\begin{array}{l|l}
\underline{\hspace{3cm}} & =32\div4\cdot2^2-2\cdot2 \\
\underline{\hspace{3cm}} & =32\div4\cdot4-2\cdot2 \\
\underline{\hspace{3cm}} & =8\cdot4-2\cdot2 \\
\underline{\hspace{3cm}} & =32-2\cdot2 \\
\underline{\hspace{3cm}} & =32-4 \\
\underline{\hspace{3cm}} & =28
\end{array}$$

Simplify.

3. $9-2(7+4)$

4. $9\cdot8-2\cdot6$

5. $-2+2\cdot3$

6. $300\div15\div5$

7. $8\div2(5-3)^2$

8. $3\cdot8^2-5^2\div5$

9. $-12-6\div2-(-4)$

10. $35-5(14\div2)+7\cdot10$

11. $24-(2-4)^2 \div 2$

12. $4 \cdot \left(\dfrac{1}{2}\right)^2 -(3-8)$

13. $\dfrac{1}{3}(12-3)^2$

14. $6 \cdot 7^2 + 2^3 - 50 \div 5$

15. $6^3 + 25 \cdot 6 \div \left(20 + \dfrac{1}{2} \cdot 10\right)$

16. $8(7-1) \div 2 - 3(-2)^2$

17. $-81 + 9 \cdot 3 \div (-3)^2 - 27$

18. $5 \cdot 5^2 - 5 \div 5 + 5(5-2)$

19. $\left[2(11-8)\right]^2$

20. $\left[18 \div (2+1)\right]^3$

Interactive Preview Worksheet 4: INTEGERS AS EXPONENTS

<table>
<tr><td colspan="2">Definitions and Rules for Exponents</td></tr>
<tr><td>Exponent of 1:</td><td>$a^1 = a$</td></tr>
<tr><td>Exponent of 0:</td><td>$a^0 = 1,\ a \neq 0$</td></tr>
<tr><td>Negative exponents:</td><td>$a^{-n} = \dfrac{1}{a^n},\ \dfrac{1}{a^{-n}} = a^n,\ a \neq 0$</td></tr>
<tr><td>Product Rule:</td><td>$a^m \cdot a^n = a^{m+n}$</td></tr>
<tr><td>Quotient Rule:</td><td>$\dfrac{a^m}{a^n} = a^{m-n},\ a \neq 0$</td></tr>
<tr><td>Power Rule:</td><td>$\left(a^m\right)^n = a^{mn}$</td></tr>
<tr><td>Raising a product to a power:</td><td>$(ab)^n = a^n b^n$</td></tr>
<tr><td>Raising a quotient to a power:</td><td>$\left(\dfrac{a}{b}\right)^n = \dfrac{a^n}{b^n},\ b \neq 0$</td></tr>
</table>

Examples Simplify.

1. $5^0 = 1$

2. $18^1 = 18$

3. $y^1 = y$

4. $t^0 = 1,\ t \neq 0$

Examples Express using positive exponents. Then simplify.

5. $a^{-6} = \dfrac{1}{a^6}$

6. $\dfrac{1}{4^{-2}} = 4^2 = 16$

7. $\dfrac{1}{x^{-3}} = x^3$

8. $3^{-4} = \dfrac{1}{3^4} = \dfrac{1}{81}$

Examples Simplify using the product rule.

9. $8^5 \cdot 8^2 = 8^{5+2} = 8^7$

10. $b^{-4} \cdot b^7 = b^{-4+7} = b^3$

11. $x \cdot x^{12} = x^{1+12} = x^{13}$

12. $2^{-10} \cdot 2^3 = 2^{-10+3} = 2^{-7} = \dfrac{1}{2^7}$

Examples Simplify using the quotient rule.

13. $\dfrac{a^8}{a^3} = a^{8-3} = a^5$

14. $\dfrac{7^2}{7^5} = 7^{2-5} = 7^{-3} = \dfrac{1}{7^3}$

15. $\dfrac{y^6}{y^{10}} = y^{6-10} = y^{-4} = \dfrac{1}{y^4}$

16. $\dfrac{13^9}{13^4} = 13^{9-4} = 13^5$

Examples Simplify using the power rule.

17. $\left(9^2\right)^6 = 9^{2 \cdot 6} = 9^{12}$

18. $\left(x^{-3}\right)^5 = x^{-3 \cdot 5} = x^{-15} = \dfrac{1}{x^{15}}$

19. $\left(3^2\right)^{-3} = 3^{2(-3)} = 3^{-6} = \dfrac{1}{3^6}$

20. $\left(y^4\right)^8 = y^{4 \cdot 8} = y^{32}$

Examples Simplify using the rule for raising a product to a power.

21. $\left(3^5 y^2\right)^4 = \left(3^5\right)^4 \cdot \left(y^2\right)^4 = 3^{5 \cdot 4} \cdot y^{2 \cdot 4} = 3^{20} y^8$

22. $\left(2z^6\right)^3 = 2^3 \cdot \left(z^6\right)^3 = 8 \cdot z^{6 \cdot 3} = 8z^{18}$

Examples Simplify using the rule for raising a quotient to a power.

23. $\left(\dfrac{2}{5}\right)^3 = \dfrac{2^3}{5^3} = \dfrac{8}{125}$

24. $\left(\dfrac{x^2}{4^5}\right)^6 = \dfrac{\left(x^2\right)^6}{\left(4^5\right)^6} = \dfrac{x^{2 \cdot 6}}{4^{5 \cdot 6}} = \dfrac{x^{12}}{4^{30}}$

Exercises Simplify. Express answer using positive exponents.

1. 3^{-5}

2. $a^7 a^{11}$

3. $\dfrac{x^{16}}{x^4}$

4. $\left(2^7\right)^3$

5. $\left(\dfrac{5}{3}\right)^3$

6. 100^1

7. 3^0

8. $x^3 \cdot x^{-5}$

9. $\dfrac{1}{y^{-10}}$

10. $\left(7^2 x^4\right)^3$

11. $4^2 \cdot 4^9$

12. $\dfrac{1}{3^{-4}}$

13. $\left(t^3\right)^{-4}$

14. z^{-2}

15. $\dfrac{2^4}{2^9}$

16. $\left(s^2\right)^3$

17. p^1

18. $y^0,\ y \neq 0$

19. $\left(\dfrac{y^3}{2^5}\right)^4$

20. $\dfrac{10^{15}}{10^5}$

21. $w \cdot w^8$

22. $\dfrac{c^{-4}}{c^{-11}}$

23. $\left(3x^2\right)^2$

24. $y^{-2} \cdot y^{-4}$

Equation Solving Principles

For any real numbers a, b, and c:

The Addition Principle: If $a = b$ is true, then $a + c = b + c$ is true.

The Multiplication Principle: If $a = b$ is true, then $ac = bc$ is true.

Using the Addition Principle

Example 1   Solve: $y - 3 = 20$.

Add 3 on both sides.	$y - 3 + 3 = 20 + 3$
Simplify.	$y + 0 = 23$
Identity property of 0: $a + 0 = a$	$y = 23$

Example 2   Solve: $8 + x = 5$.

Add -8, or subtract 8, on both sides.	$8 + x - 8 = 5 - 8$
Simplify.	$x + 0 = -3$
	$x = -3$

Using the Multiplication Principle

Example 3   Solve: $3z = -12$.

Multiply by $\dfrac{1}{3}$, or divide by 3, on both sides.	$\dfrac{3z}{3} = \dfrac{-12}{3}$
Simplify.	$1 \cdot z = -4$
Identity property of 1: $1 \cdot a = a$	$z = -4$

Example 4   Solve: $\dfrac{3}{2}x = 9$.

Multiply by $\dfrac{2}{3}$, the reciprocal of $\dfrac{3}{2}$.	$\dfrac{2}{3} \cdot \dfrac{3}{2}x = \dfrac{2}{3} \cdot 9$
Simplify.	$1 \cdot x = \dfrac{18}{3}$
	$x = 6$

Example 5 Solve: $-a = 1.4$.

$-a = -1 \cdot a$.	$-1 \cdot a = 1.4$
Divide by -1 on both sides.	$\dfrac{-1 \cdot a}{-1} = \dfrac{1.4}{-1}$
Simplify.	$a = -1.4$

Exercises Solve.

1. $w + 8 = 11$

2. $\dfrac{3}{5} - t = -\dfrac{2}{5}$

3. $-9z = 288$

4. $\dfrac{2}{7}y = 42$

5. $-x = 2.5$

6. $x - 4 = -10$

Using the Principles Together

Example 6   Solve: $5x - 6 = 29$.

Add 6 on both sides.	$5x - 6 + 6 = 29 + 6$
Simplify.	$5x = 35$
Divide by 5 on both sides.	$\dfrac{5x}{5} = \dfrac{35}{5}$
Simplify.	$x = 7$

Example 7   Solve: $3 + \dfrac{1}{2}y = -13$.

Subtract 3 on both sides.	$3 + \dfrac{1}{2}y - 3 = -13 - 3$
Simplify.	$\dfrac{1}{2}y = -16$
Multiply by 2 on both sides.	$2 \cdot \dfrac{1}{2}y = 2(-16)$
Simplify.	$y = -32$

Exercises Solve.

7. $9t + 4 = -104$

8. $4x - 7 = 81$

9. $11 = 2a - 15$

10. $5 - \dfrac{2}{3}y = -4$

Collecting Like Terms

Example 8   Solve: $6x + 5 - 7x = 10 - 4x + 3$.

Collect like terms on each side.	$-x + 5 = 13 - 4x$
Add $4x$.	$3x + 5 = 13$
Subtract 5.	$3x = 8$
Divide by 3.	$x = \dfrac{8}{3}$

Removing Parentheses

Example 9   Solve: $7(3x + 6) = 11 - (x + 2)$.

Remove parentheses using the distributive law.	$21x + 42 = 11 - x - 2$
Collect like terms on the right.	$21x + 42 = 9 - x$
Add x.	$22x + 42 = 9$
Subtract 42.	$22x = -33$
Divide by 22.	$x = \dfrac{-33}{22}$
Simplify.	$x = -\dfrac{3}{2}$

Exercises Solve.

11. $6y + 20 = 10 + 3y + y$

12. $5y - (2y - 10) = 25$

13. $80 = 10(3t + 2)$

14. $0.9x - 0.7x = 4.2$

15. $5 - 4a = a - 13$

16. $4(2x - 7) = 5 - (x + 3)$

Clearing Fractions

Example 10   Solve: $\dfrac{5}{6}y = \dfrac{1}{3}y - 7$

Multiply by 6, the least common multiple of 6 and 3.

$$6\left(\dfrac{5}{6}y\right) = 6\left(\dfrac{1}{3}y - 7\right)$$

Remove parentheses. $5y = 2y - 42$

Subtract $2y$. $3y = -42$

Divide by 3. $y = -14$

Clearing Decimals

Example 11   Solve: $1.4 - 1.05x = 0.7x$

Greatest number of decimal places is 2. Multiply by 100.

$$100(1.4 - 1.05x) = 100(0.7x)$$

Remove parentheses. $140 - 105x = 70x$

Add $105x$. $140 = 175x$

Divide by 175. $\dfrac{140}{175} = x$

Simplify. $\dfrac{4}{5} = x$

Exercises Solve.

17. $5.4x + 2.04 = 1.5$

18. $\dfrac{5}{12}a - \dfrac{3}{8}a = \dfrac{1}{4}$

19. $\dfrac{3}{5}t - \dfrac{1}{10} = \dfrac{4}{15}t$

20. $6.7 + 0.001y = 9.82$

21. $2.1x + 45.2 = 3.2 - 8.4x$

22. $\dfrac{2}{7}x - \dfrac{1}{2}x = \dfrac{3}{4}x + 1$

Interactive Preview Worksheet 6: INTERVAL NOTATION

The graph of an inequality is a drawing that represents its solutions. The following graph illustrates an inequality whose solution set is all real numbers less than 5.

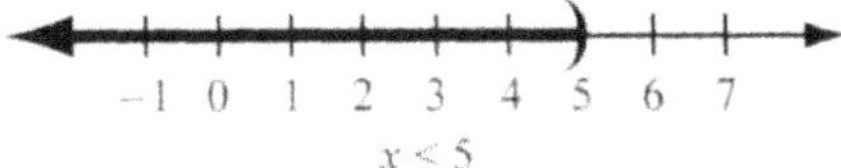

The solution set can be written in set-builder notation, $\{x \mid x < 5\}$. This is read "The set of all x such that x is less than 5." The solution set can be also written in interval notation, $(-\infty, 5)$.

Interval notation uses parentheses ( ) and brackets [ ]. In Examples 1-9, intervals of the types (a, b), $[a, b]$, $[a, b)$, $(a, b]$, (a, ∞), $[a, \infty)$, $(-\infty, b)$, $(-\infty, b]$, and $(-\infty, \infty)$ are illustrated. The points a and b are endpoints of the interval. A parenthesis indicates that the endpoint is not included in the graph. A bracket indicates that the endpoint is included in the graph. Some intervals extend without bound in one or both directions. We use the symbol ∞, read "infinity," and $-\infty$, read "negative infinity," to name these intervals.

Examples

Graph	Set Notation	Interval Notation
1.	$\{x \mid -5 < x < -2\}$	$(-5, -2)$
2.	$\{x \mid -1 \le x \le 3\}$	$[-1, 3]$
3.	$\{x \mid 0 \le x < 4\}$	$[0, 4)$
4.	$\{x \mid 2 < x \le 5\}$	$(2, 5]$
5.	$\{x \mid x > 1\}$	$(1, \infty)$
6.	$\{x \mid x \ge -3\}$	$[-3, \infty)$
7.	$\{x \mid x < -4\}$	$(-\infty, -4)$

8.   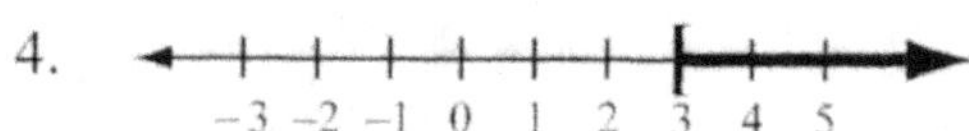 $\{x \mid x \le 2\}$ $(-\infty, 2]$

9.   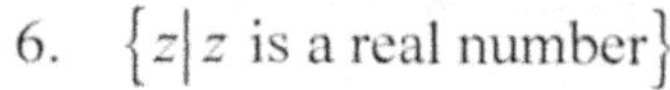 $\{x \mid x \text{ is a real number}\}$ $(-\infty, \infty)$

Check Your Understanding

For each set, consider the endpoints of the interval and choose from choices A – D, the format for interval notation.

A. $(\ , \ )$ B. $[\ , \ ]$ C. $[\ , \ )$ D. $(\ , \ ]$

1. $\{q \mid 0 < q \le 9\}$ 2. $\{x \mid -4 \le x < 1\}$

3. $\left\{y \mid y \ge \dfrac{3}{4}\right\}$ 4. $\{t \mid 0.7 < t < 2.5\}$

5. $\{x \mid -10 \le x \le 13\}$ 6. $\{p \mid p < -2\}$

Exercises

Write interval notation for the given set or graph.

1. 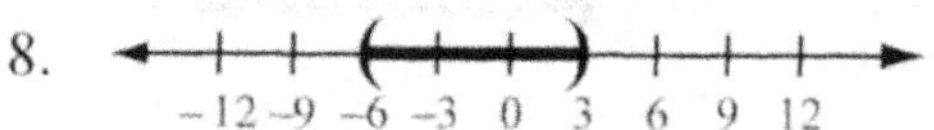   2. $\{c \mid -4.1 \le c < 5.6\}$

3. $\{x \mid -8 < x \le 21\}$ 4.

5. $\{w \mid w > 0\}$ 6. $\{z \mid z \text{ is a real number}\}$

7.   8.

9. $\left\{y \mid y < \dfrac{2}{5}\right\}$ 10. $\{d \mid d \ge 17\}$

11.   12.

Given a solution set expressed in interval notation, graph it on the number line.

13. $[3, 6)$

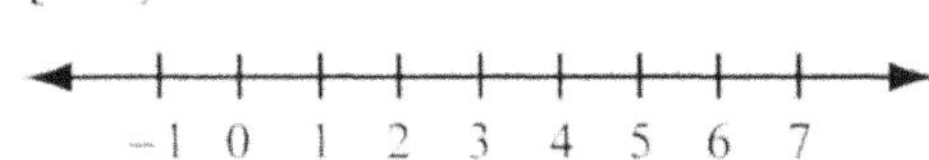

14. $[4, \infty)$

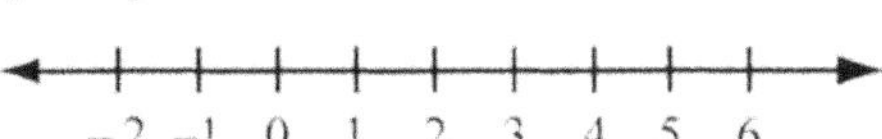

15. $(-1, \infty)$

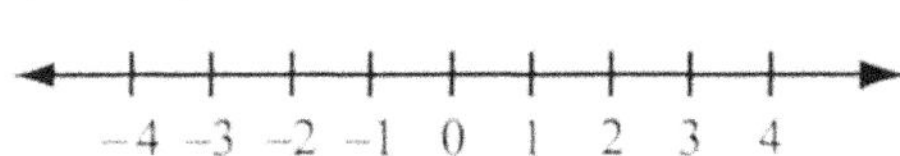

16. $(-1, 4]$

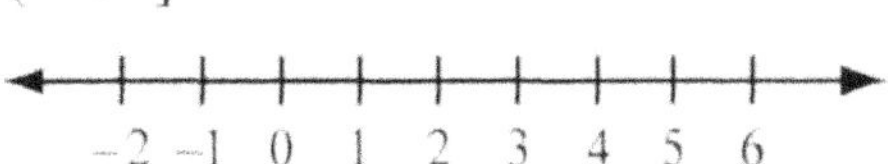

17. $[-5, -1]$

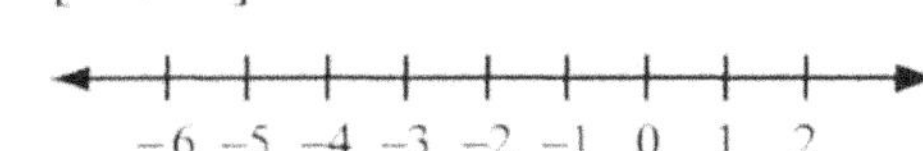

18. $(-\infty, \infty)$

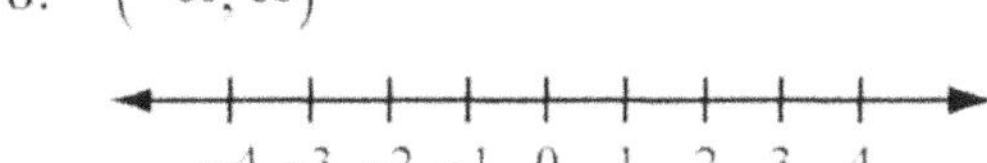

19. $(-\infty, -1.5)$

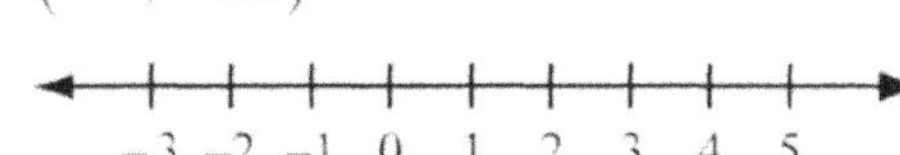

20. $(-\infty, 2]$

21. $[-12, 1)$

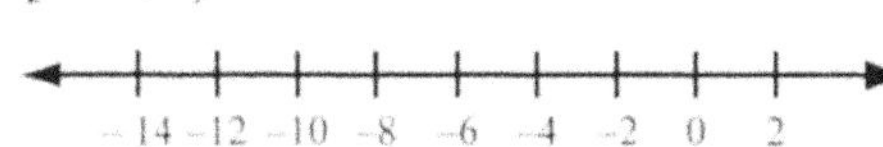

22. $(0, 20)$

23. $(-15, \infty)$

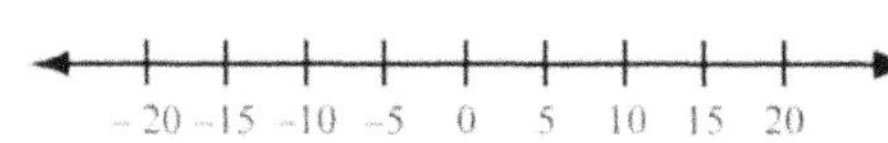

24. $(-3, 5]$

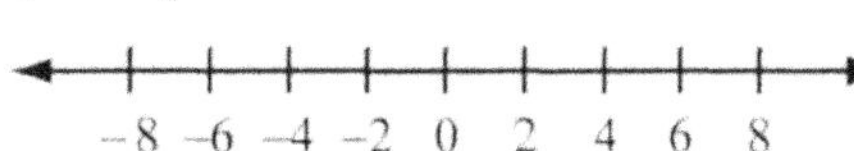

Notes:

Interactive Preview Worksheet 7: SOLVING LINEAR INEQUALITIES

Principles for Solving Inequalities

For any real numbers a, b, and c:

The Addition Principle for Inequalities:

If $a < b$ is true, then $a + c < b + c$ is true.

The Multiplication Principle for Inequalities:

a) If $a < b$ and $c > 0$ are true, then $ac < bc$ is true.

b) If $a < b$ and $c < 0$ are true, then $ac > bc$ is true.

   (When both sides of an inequality are multiplied by a negative number, the inequality sign must be reversed.)

Similar statements hold for $a \le b$.

Example 1 Solve: $x - 2 \ge -3$.

$$\begin{array}{ll} \text{Add 2.} & x - 2 + 2 \ge -3 + 2 \\ \text{Simplify.} & x \ge -1 \end{array}$$

The solution set is $\{x \mid x \ge -1\}$, or $[-1, \infty)$.

Graph of solution:

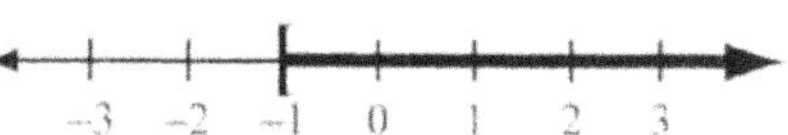

Example 2 Solve: $y + \dfrac{1}{3} < \dfrac{5}{4}$.

$$\begin{array}{ll} \text{Subtract } \dfrac{1}{3}. & y + \dfrac{1}{3} - \dfrac{1}{3} < \dfrac{5}{4} - \dfrac{1}{3} \\[2ex] \begin{array}{l}\text{Multiply by 1 to}\\ \text{obtain a common}\\ \text{denominator.}\end{array} & y < \dfrac{5}{4} \cdot \dfrac{3}{3} - \dfrac{1}{3} \cdot \dfrac{4}{4} \\[2ex] & y < \dfrac{15}{12} - \dfrac{4}{12} \\[2ex] & y < \dfrac{11}{12} \end{array}$$

The solution set is $\left\{ y \mid y < \dfrac{11}{12} \right\}$, or $\left(-\infty, \dfrac{11}{12} \right)$.

Graph of solution:

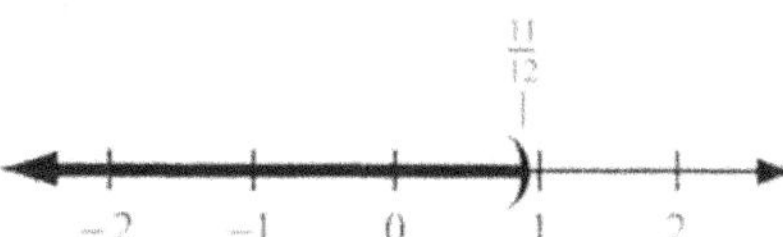

Example 3 Solve: $3x > -15$.

$$\begin{array}{ll} \text{Divide by 3.} & \dfrac{3x}{3} > \dfrac{-15}{3} \\[2ex] \text{Simplify.} & x > -5 \end{array}$$

The solution set is $\{x \mid x > -5\}$, or $(-5, \infty)$.

Graph of solution:

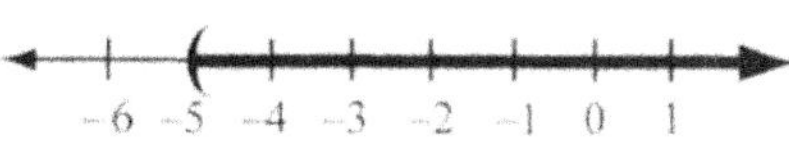

Example 4 Solve: $-4x \le 8$.

Divide by -4. $\left|\; \dfrac{-4x}{-4} \ge \dfrac{8}{-4} \right.$ ← The symbol must be reversed.

Simplify. $\left|\; x \ge -2 \right.$

The solution set is $\{x \mid x \ge -2\}$, or $[-2, \infty)$.

Graph of solution:

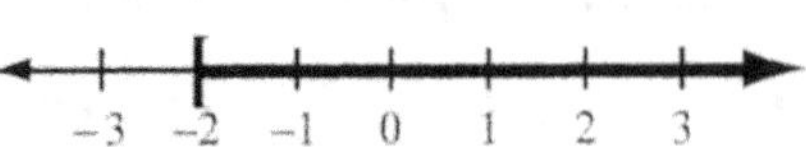

Example 5 Solve: $-\dfrac{2}{3}a > -\dfrac{1}{10}$.

Multiply by $-\dfrac{3}{2}$. $\left|\; -\dfrac{3}{2}\left(-\dfrac{2}{3}a\right) < -\dfrac{3}{2}\left(-\dfrac{1}{10}\right) \right.$ ← The symbol must be reversed.

Simplify. $\left|\; a < \dfrac{3}{20} \right.$

The solution set is $\left\{a \mid a < \dfrac{3}{20}\right\}$, or $\left(-\infty, \dfrac{3}{20}\right)$.

Example 6 Solve: $10 - 6x > 3x - 8$.

Add 8. $\left|\; 18 - 6x > 3x \right.$

Add $6x$. $\left|\; 18 > 9x \right.$

Divide by 9. $\left|\; 2 > x \right.$

The solution set is $\{x \mid 2 > x\}$, or $\{x \mid x < 2\}$, or $(-\infty, 2)$.

Example 7 Solve: $2(t - 5) - 8 \ge 3t - 7$.

Remove parentheses. $\left|\; 2t - 10 - 8 \ge 3t - 7 \right.$

Collect like terms. $\left|\; 2t - 18 \ge 3t - 7 \right.$

Subtract $3t$. $\left|\; -1t - 18 \ge -7 \right.$

Add 18. $\left|\; -t \ge 11 \right.$

Multiply by -1 and reverse the symbol. $\left|\; t \le -11 \right.$

The solution set is $\{t \mid t \le -11\}$, or $(-\infty, -11]$.

Exercises Solve.

1. $y + 8 \ge 5$

2. $x - \dfrac{4}{5} > \dfrac{3}{4}$

3. $2c < -12$

4. $-7q \le -42$

5. $-\dfrac{5}{8}x > \dfrac{5}{12}$

6. $4z - 10 < z + 26$

7. $x - 1 \ge 1 - x$

8. $23 - 5x \ge 2x - 12$

9. $8t - 6 \le 2(t + 9)$

10. $-3(x + 1) - 2 < 3 - x$

Interactive Preview Worksheet 8: SOLVING EQUATIONS AND INEQUALITIES WITH ABSOLUTE VALUE

Let's look at graphs of solution sets of equations and inequalities with absolute value.

Examples

		Equivalent Statement or Inequality	Graph of Solution Set

1. $|x| = 3$ $x = -3$ *or* $x = 3$ 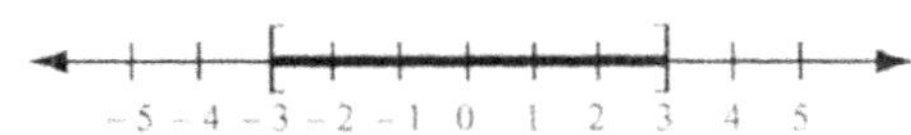

The solutions are -3 and 3.

2. $|x| < 3$ $-3 < x < 3$ 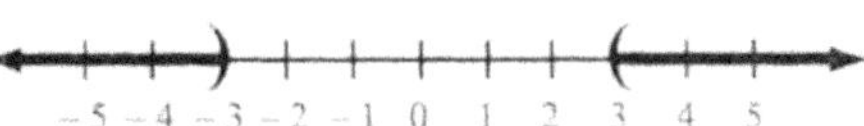

The solutions are all numbers between -3 and 3. To indicate that -3 and 3 are not solutions, we use a parenthesis at -3 and at 3. In interval notation, the solution set is $(-3, 3)$.

3. $|x| \le 3$ $-3 \le x \le 3$ 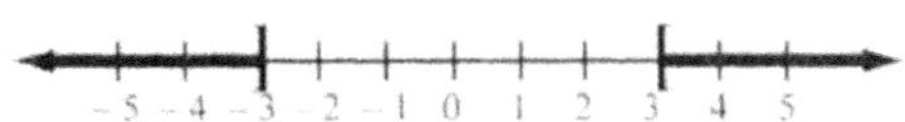

The solutions are all numbers from -3 to 3, including -3 and 3. To indicate that -3 and 3 are solutions, we use a bracket at -3 and at 3. In interval notation, the solution set is $[-3, 3]$.

4. $|x| > 3$ $x < -3$ *or* $x > 3$

The solutions are all numbers less than -3 and all numbers greater than 3. Parentheses indicate that -3 and 3 are not solutions. In interval notation, the solution set is $(-\infty, -3) \cup (3, \infty)$.

5. $|x| \ge 3$ $x \le -3$ *or* $x \ge 3$

The solutions are all numbers less than or equal to -3 and all numbers greater than or equal to 3. Brackets indicate that -3 and 3 are solutions. In interval notation, the solution set is $(-\infty, -3] \cup [3, \infty)$.

Exercises

Graph the solution set, then state the solution. For inequalities, express the solution set in interval notation. Remember: Always use a parenthesis with $-\infty$ and ∞.

		Graph of Solution Set	Solution Set

1. $|x| = 1$ 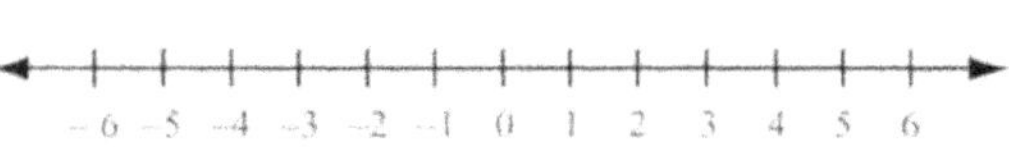

____

2. $|z| < 2$ 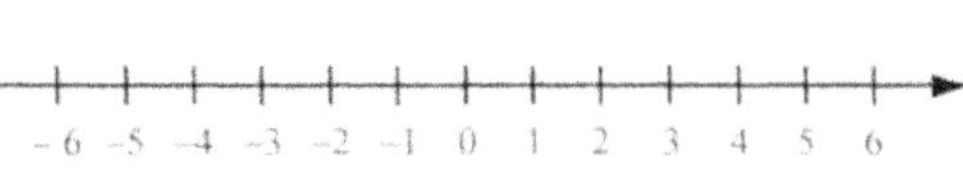

____

3. $|t| \le 5$ 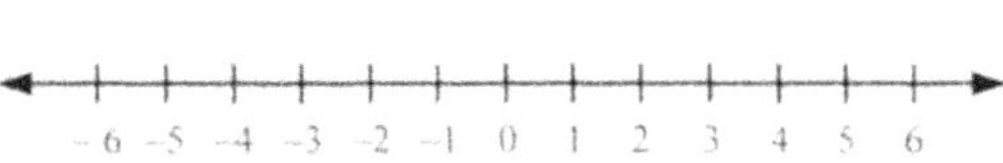

____

4. $|y| > 4$

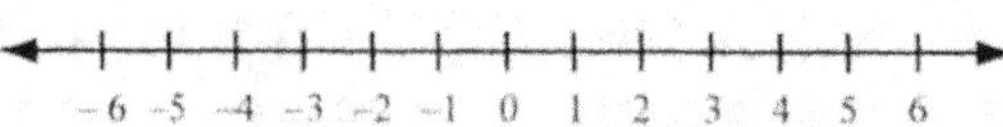

5. $|y| \geq 3.5$

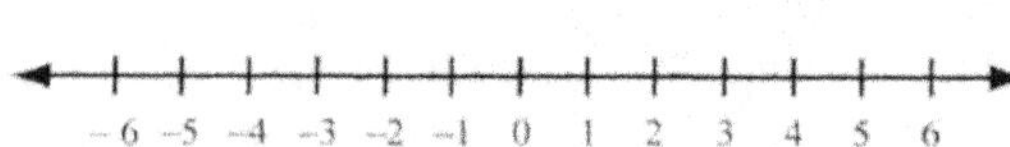

The first step in solving equations and inequalities with absolute value is to write an equivalent statement or inequality that does not contain absolute value notation.

Examples Write an equivalent statement or inequality.

		Equivalent Statement or Inequality
6.	$|w| = 2$	$w = -2 \quad or \quad w = 2$
7.	$|y + 5| = 14$	$y + 5 = -14 \quad or \quad y + 5 = 14$
8.	$|t| < 9$	$-9 < t < 9$
9.	$|2s - 3| \leq 10$	$-10 \leq 2s - 3 \leq 10$
10.	$|z| \geq 29$	$z \leq -29 \quad or \quad z \geq 29$
11.	$|4 - x| > 8$	$4 - x < -8 \quad or \quad 4 - x > 8$

Exercises Write an equivalent statement or inequality.

Equivalent Statement or Inequality

6. $|q| = 8$

7. $|p + 5| \leq 2$

8. $|0.5 - x| > 6.5$

9. $|7y - 1| = 12$

10. $\left| \dfrac{3}{4}t + 6 \right| < 21$

11. $|18q| \geq 9$

Examples Solve and graph the solution set.

12. $|x-2|=1$

Write an equivalent statement.	$x-2=-1 \quad or \quad x-2=1$
Add 2.	$x=1 \quad or \quad x=3$
State the solution set.	The solutions are 1 and 3.

Graph the solution set.

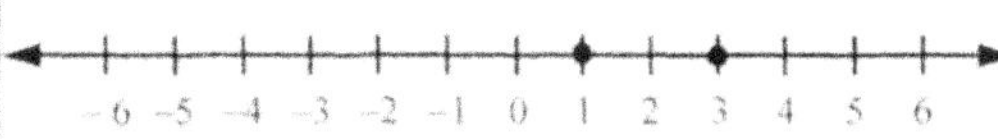

13. $|x+1|\geq 3$

Write an equivalent inequality.	$x+1\leq -3 \quad or \quad x+1\geq 3$
Subtract 1.	$x\leq -4 \quad or \quad x\geq 2$
State the solution set.	$(-\infty, -4]\cup[2, \infty)$

Graph the solution set.

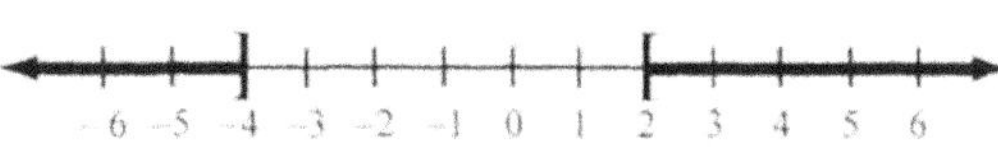

14. $|x-3|<1$

Write an equivalent inequality.	$-1<x-3<1$
Add 3.	$2<x<4$
State the solution set.	$(2, 4)$

Graph the solution set.

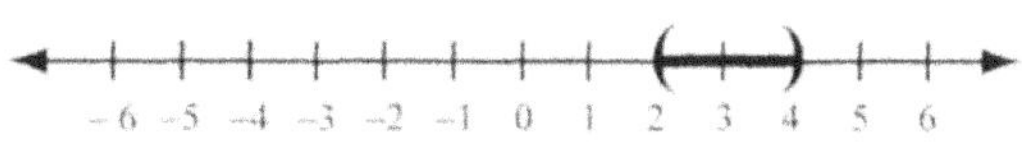

Exercises Solve and graph the solution set. In Exercises 12 and 13, fill in the blanks in key steps.

12. $|w+10|\leq 30$

$$\boxed{}\leq w+10\leq 30$$

$$-40\leq w\leq \boxed{}$$

Solution set: $\left[\boxed{}, 20\right]$

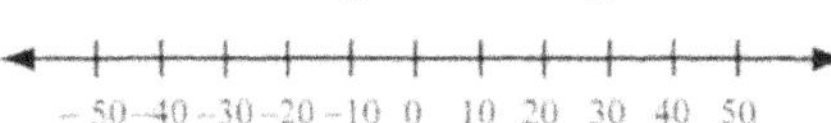

13. $|y-4|>6$

$$y-4<\boxed{} \quad or \quad y-4>6$$

$$y<-2 \quad or \quad y>\boxed{}$$

Solution set: $\left(\boxed{}, -2\right)\cup\left(\boxed{}, \infty\right)$

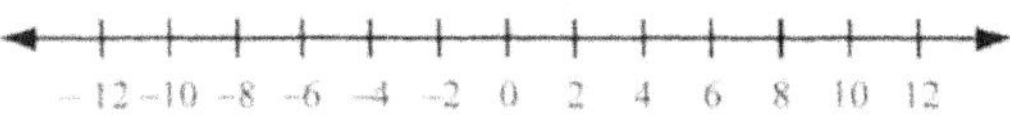

14. $|t-2|=3$

15. $|x-4|<1$

Solutions: _________________

Solution set: _________________

16. $|b+5|\geq 5$

17. $\left|y+\dfrac{4}{5}\right|=\dfrac{1}{5}$

Solution set: _________________

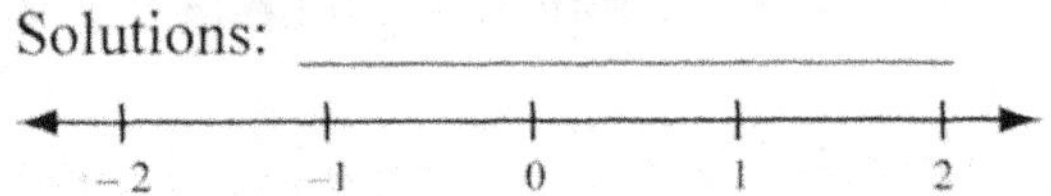

Solutions: _________________

18. $|z+1|>1$

19. $\left|x-\dfrac{1}{2}\right|\leq\dfrac{5}{2}$

Solution set: _________________

Solution set: _________________

Examples Solve and graph the solution set.

15. $|2x+4|>6$

Write an equivalent inequality.	$2x+4<-6 \quad or \quad 2x+4>6$
Subtract 4.	$2x<-10 \quad or \quad 2x>2$
Divide by 2.	$x<-5 \quad or \quad x>1$
State the solution set.	$(-\infty,-5)\cup(1,\infty)$
Graph the solution set.	

16. $|3-t|\le 2$

Write an equivalent inequality.	$-2\le 3-t\le 2$
Subtract 3.	$-5\le -t\le -1$
Multiply by -1; reverse the inequality symbols.	$5\ge t\ge 1$
Rearrange.	$1\le t\le 5$
State the solution set.	$[1,5]$
Graph the solution set.	

Exercises Solve and graph the solution set.

20. $|3x-6|\le 9$

21. $|2-t|>4$

Solution set: _______________________

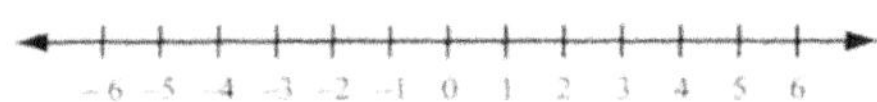

Solution set: _______________________

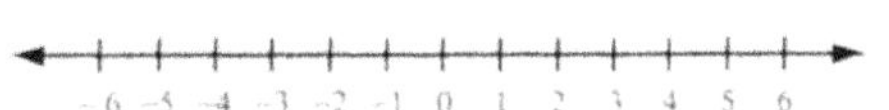

22. $\left|5-t\right|=15$

23. $\left|4x+2\right|\geq 6$

Solutions: ________________

$$\longleftarrow\ |\ \ |\ \ |\ \ |\ \ |\ \ |\ \ |\ \ |\ \ |\ \ |\ \ |\ \longrightarrow$$
$$-50\ -40\ -30\ -20\ -10\ \ 0\ \ 10\ \ 20\ \ 30\ \ 40\ \ 50$$

Solution set: ________________

$$\longleftarrow\ |\ \ |\ \ |\ \ |\ \ |\ \ |\ \ |\ \ |\ \ |\ \ |\ \ |\ \ |\ \ |\ \longrightarrow$$
$$-6\ -5\ -4\ -3\ -2\ -1\ \ 0\ \ 1\ \ 2\ \ 3\ \ 4\ \ 5\ \ 6$$

24. $\left|\dfrac{1}{4}x+\dfrac{1}{2}\right|<\dfrac{3}{2}$

25. $\left|30-10x\right|\leq 40$

Solution set: ________________

$$\longleftarrow\ |\ \ |\ \ |\ \ |\ \ |\ \ |\ \ |\ \ |\ \ |\ \ |\ \ |\ \ |\ \longrightarrow$$
$$-10\ -8\ -6\ -4\ -2\ \ 0\ \ 2\ \ 4\ \ 6\ \ 8\ \ 10$$

Solution set: ________________

$$\longleftarrow\ |\ \ |\ \ |\ \ |\ \ |\ \ |\ \ |\ \ |\ \ |\ \ |\ \ |\ \ |\ \longrightarrow$$
$$-2\ -1\ \ 0\ \ 1\ \ 2\ \ 3\ \ 4\ \ 5\ \ 6\ \ 7\ \ 8$$

Notes:

Interactive Preview Worksheet 9: GRAPHING LINEAR EQUATIONS

Equations of the form $y = mx + b$, $Ax + By = C$, $x = a$ and $y = b$ are said to be *linear* because the graph of each equation is a straight line. The following equations are examples of linear equations.

$$y = 7x - \frac{3}{4}, \qquad\qquad 3x + 9y = 20, \qquad\qquad x = -5, \qquad\qquad y = 8$$

The graph of an equation is a drawing that represents all of its solutions. The solutions are ordered pairs with coordinates listed in alphabetical order of the variables.

Linear Equations of the Form $y = mx + b$

Example 1 Graph $y = 2x + 3$.

If $x = -2$, $y = 2(-2) + 3 = -4 + 3 = -1$.

If $x = 0$, $y = 2(0) + 3 = 0 + 3 = 3$.

If $x = 1$, $y = 2(1) + 3 = 2 + 3 = 5$.

x	y
-2	-1
0	3
1	5

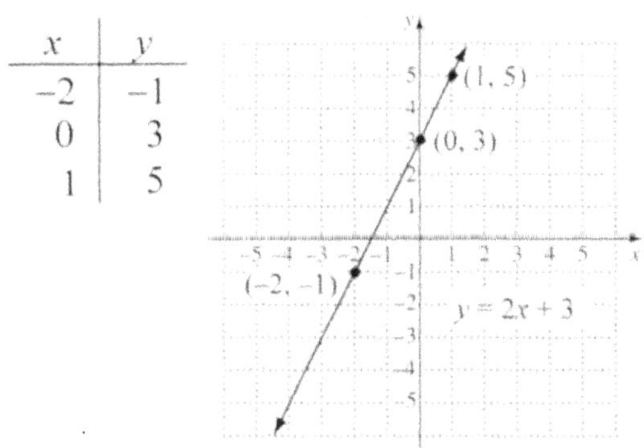

Example 2 Graph $y = -\frac{2}{5}x$.

If $x = -5$, $y = -\frac{2}{5}(-5) = 2$.

If $x = 0$, $y = -\frac{2}{5}(0) = 0$.

If $x = 5$, $y = -\frac{2}{5}(5) = -2$.

x	y
-5	2
0	0
5	-2

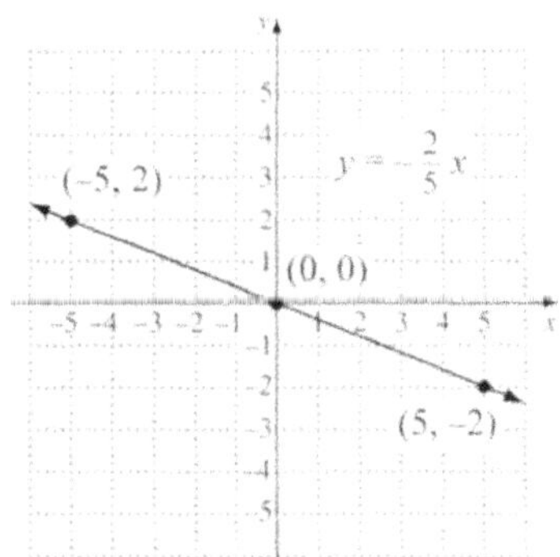

Linear Equations of the Form $Ax + By = C$

Example 3 Graph $4y + 3x = -8$.

We first solve for y to find an equivalent form $y = mx + b$.

	$4y + 3x = -8$
Subtract $3x$ on both sides.	$4y = -3x - 8$
Multiply by $\frac{1}{4}$ on both sides.	$\frac{1}{4}(4y) = \frac{1}{4}(-3x - 8)$
Simplify.	$y = -\frac{3}{4}x - 2$

If $x = 4$, $y = -\frac{3}{4}(4) - 2 = -3 - 2 = -5$.

If $x = 0$, $y = -\frac{3}{4}(0) - 2 = 0 - 2 = -2$.

If $x = -4$, $y = -\frac{3}{4}(-4) - 2 = 3 - 2 = 1$.

x	y
4	-5
0	-2
-4	1

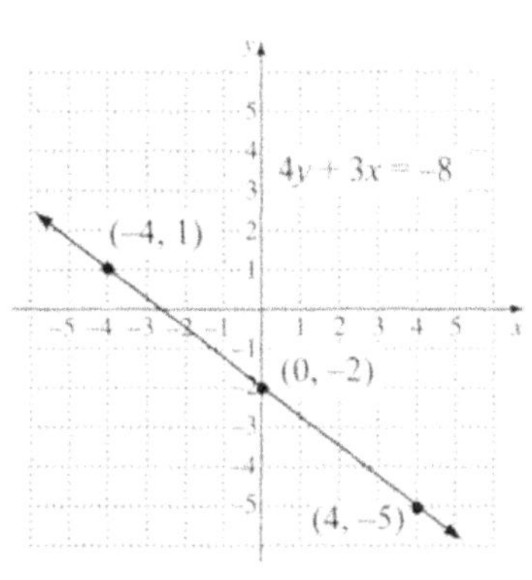

When the coefficients of the x and y-terms in $Ax + By = C$ are factors of the constant C, it is easier to graph using intercepts.

Intercepts

The **y-intercept** is $(0, b)$. To find b, let $x = 0$ and solve the equation for y.

The **x-intercept** is $(a, 0)$. To find a, let $y = 0$ and solve the equation for x.

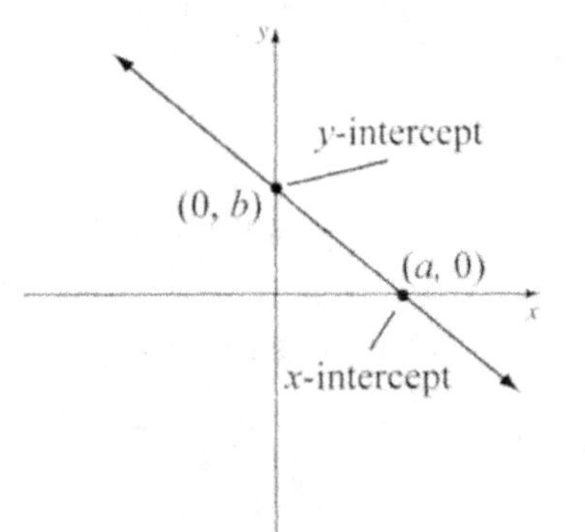

Example 4 Graph $-2x + 3y = 6$.

To find the y-intercept, we let $x = 0$ and solve for y.

$$-2x + 3y = 6$$
$$-2(0) + 3y = 6$$
$$3y = 6$$
$$y = 2$$

$(0, 2)$ is the y-intercept.

To find the x-intercept, we let $y = 0$ and solve for x.

$$-2x + 3y = 6$$
$$-2x + 3(0) = 6$$
$$-2x = 6$$
$$x = -3$$

$(-3, 0)$ is the x-intercept.

Find a third point as a check. Here we let $x = 3$.

$$-2x + 3y = 6$$
$$-2(3) + 3y = 6$$
$$-6 + 3y = 6$$
$$3y = 12$$
$$y = 4$$

$(3, 4)$ is also on the graph.

x	y
0	2
-3	0
3	4

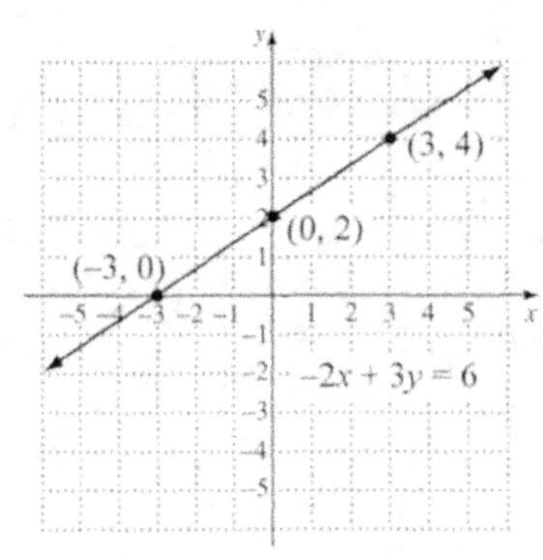

  Copyright © 2022 Pearson Education, Inc.

Linear Equations of the Form $x = a$ and $y = b$

Example 5 Graph $y = -4$.

Think of $y = -4$ as $0 \cdot x + y = -4$.

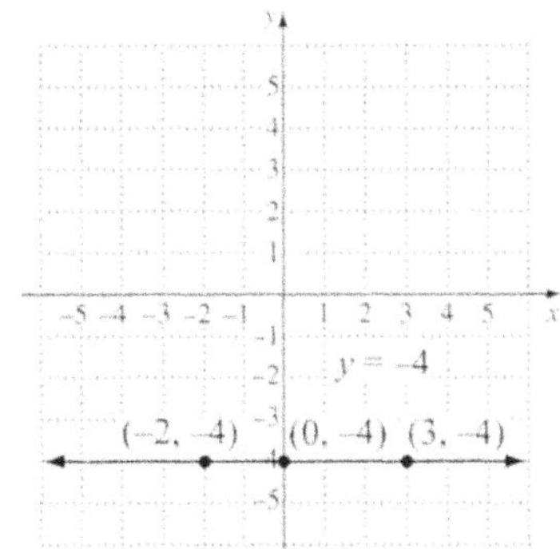

x	y
-2	-4
0	-4
3	-4

x can be any number

y must be -4.

Example 6 Graph $x = 3$.

Think of $x = 3$ as $x + 0 \cdot y = 3$.

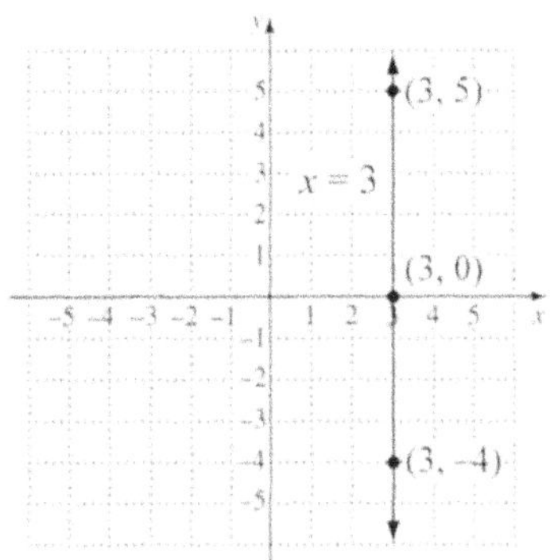

x	y
3	-4
3	0
3	5

x must be 3.

y can be any number.

Horizontal Lines and Vertical Lines

The graph of $y = b$ is a horizontal line. The y-intercept is $(0, b)$.
The graph of $x = a$ is a vertical line. The x-intercept is $(a, 0)$.

Check Your Understanding

Fill-in the coordinates of each intercept.

1. 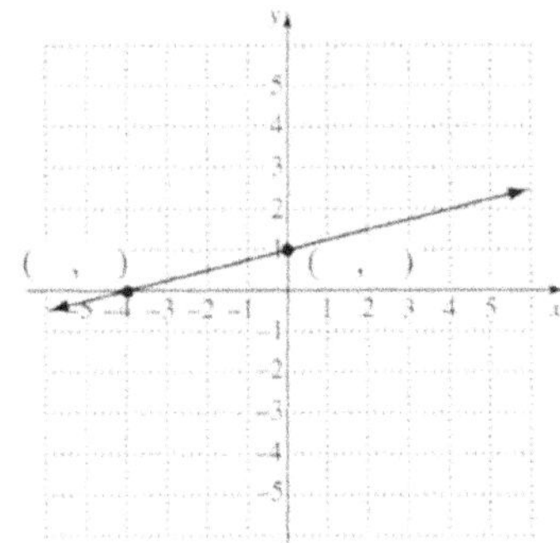

2. 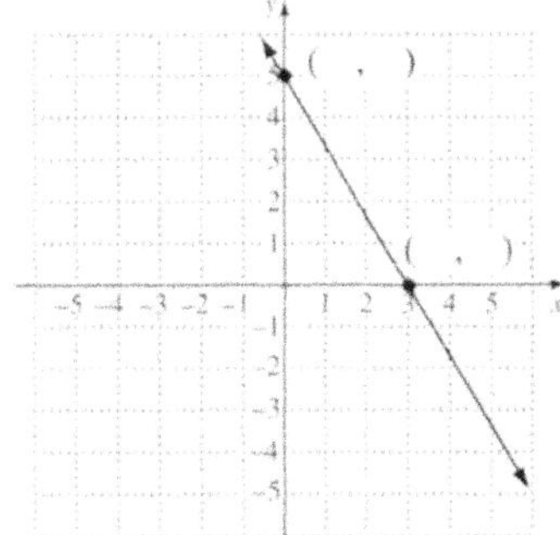

3. 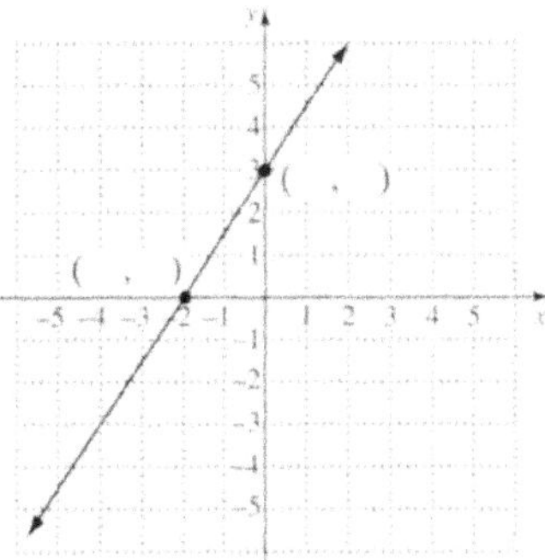

Determine whether the statement is true or false.

4. To find the x-intercept of the graph of $2x - 7y = -14$, let $x = 0$.

5. The second coordinate of each point of the graph of $y = -1$ is -1.

6. The graph of a linear equation is always a straight line.

Match each equation with its graph from choices A – D.

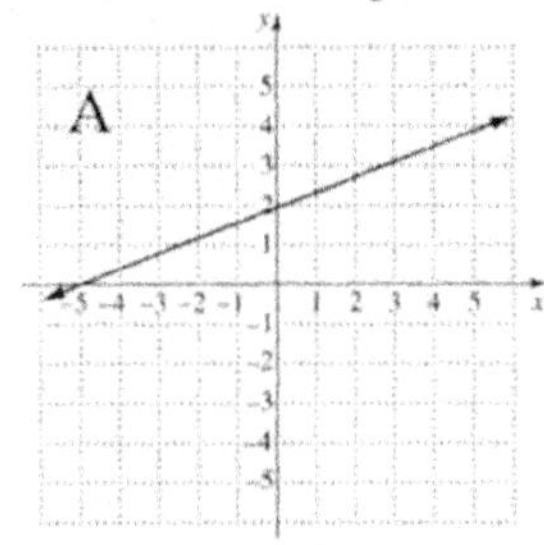

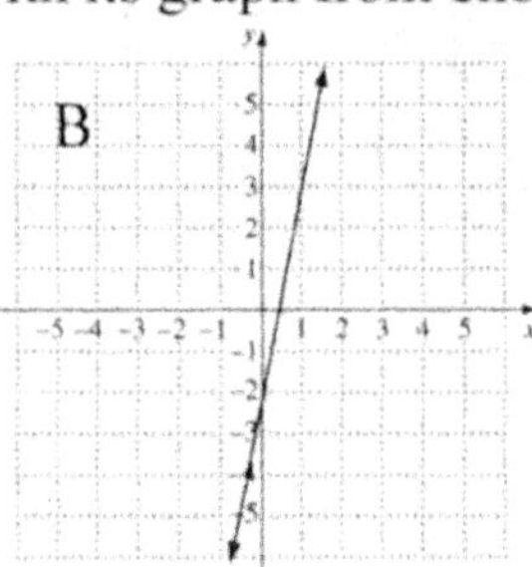

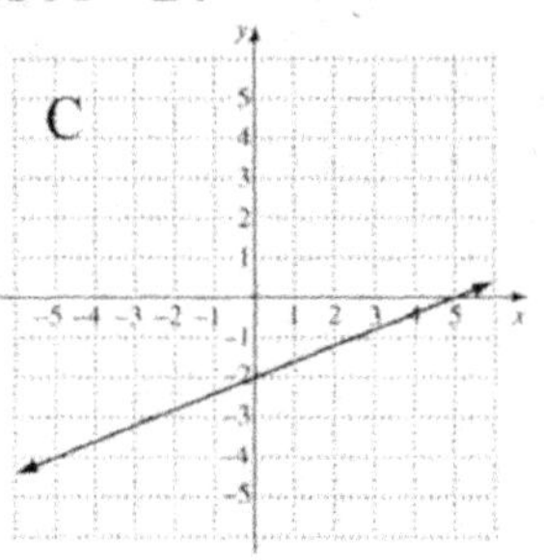

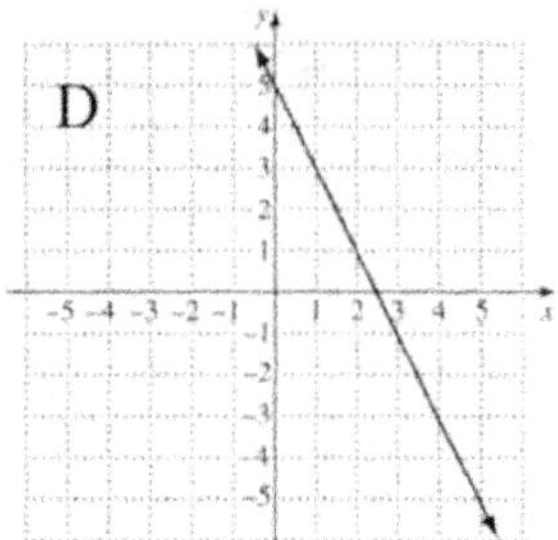

7. $2x - 5y = 10$

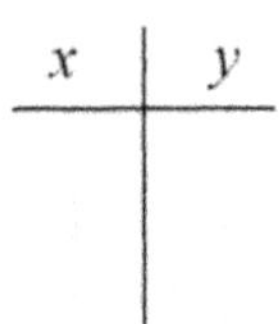

x	y

8. $y = 5x - 2$

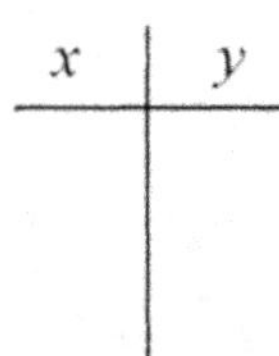

x	y

9. $y = -2x + 5$

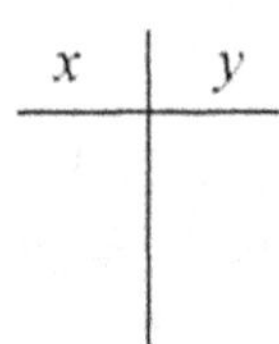

x	y

10. $5y - 2x = 10$

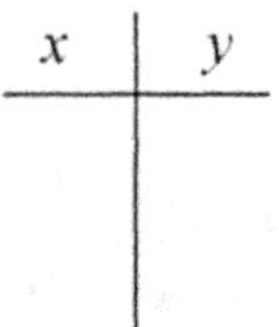

x	y

Exercises
Graph.

1. $y = -x$

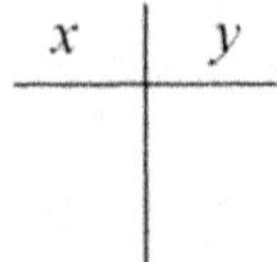

x	y

2. $y = -4x + 5$

x	y

3. $2x - 3y = 9$

x	y

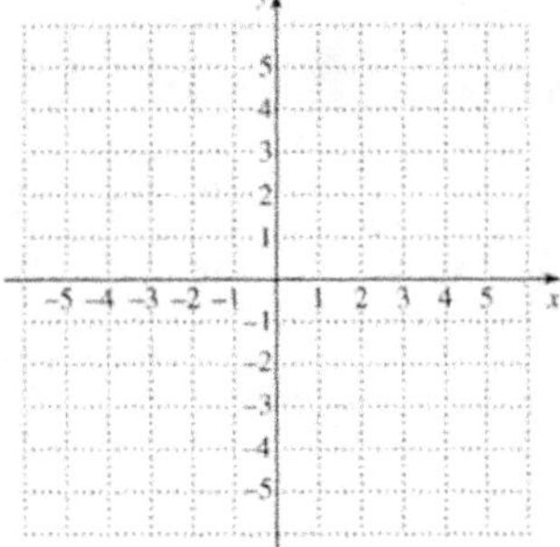

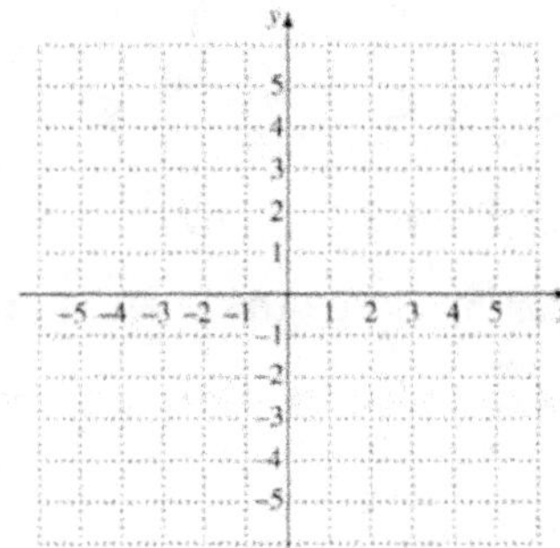

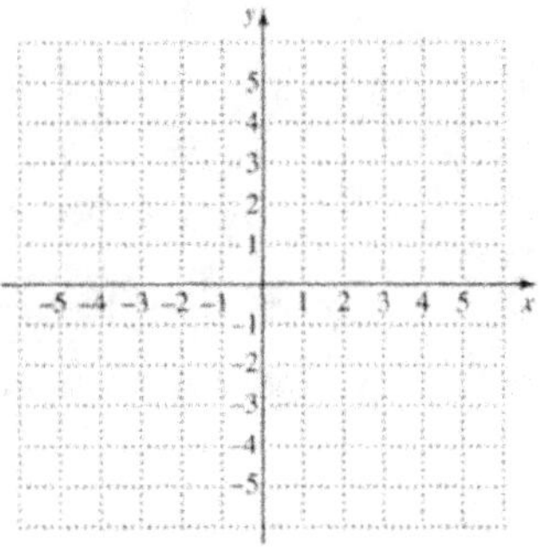

4. $5x - 4y = 20$

5. $x = -1$

6. $3x - 5y = 10$

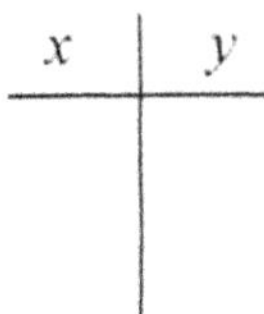

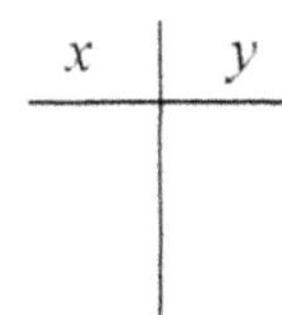

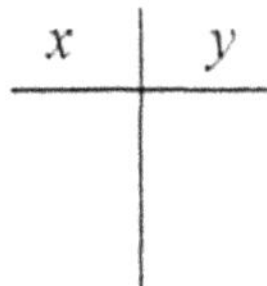

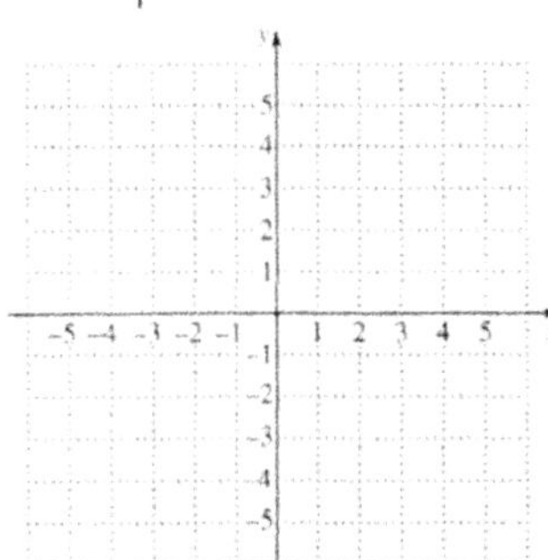

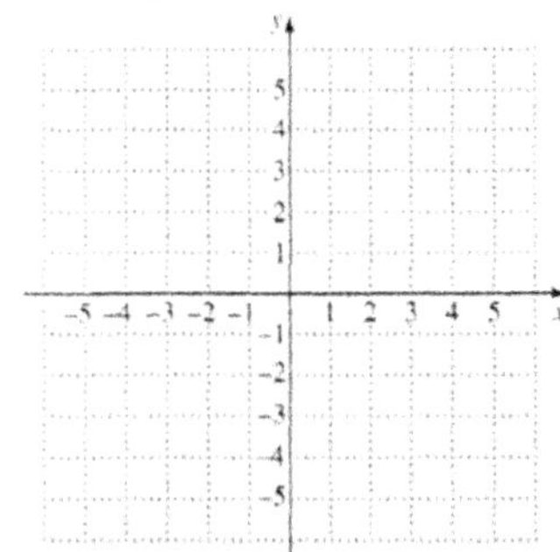

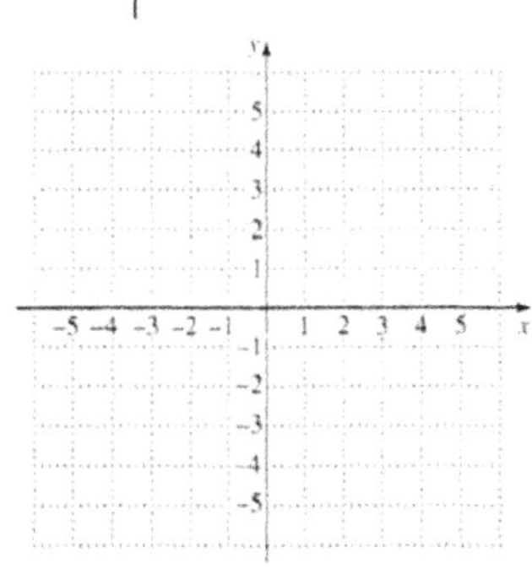

7. $y = \dfrac{2}{3}x - 4$

8. $x + y = 5$

9. $3x - 2y = 6$

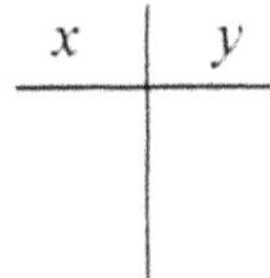

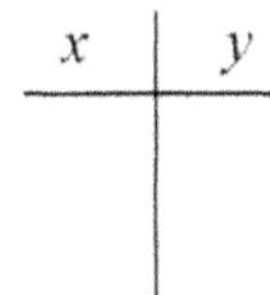

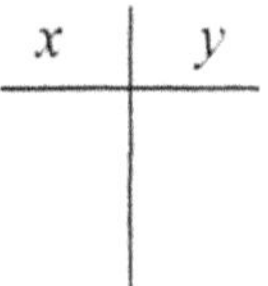

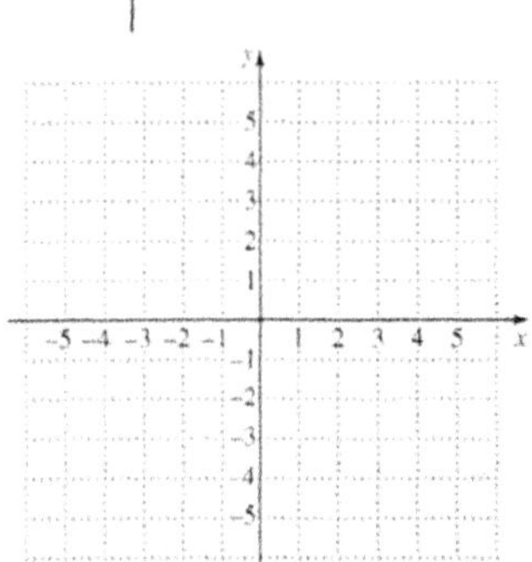

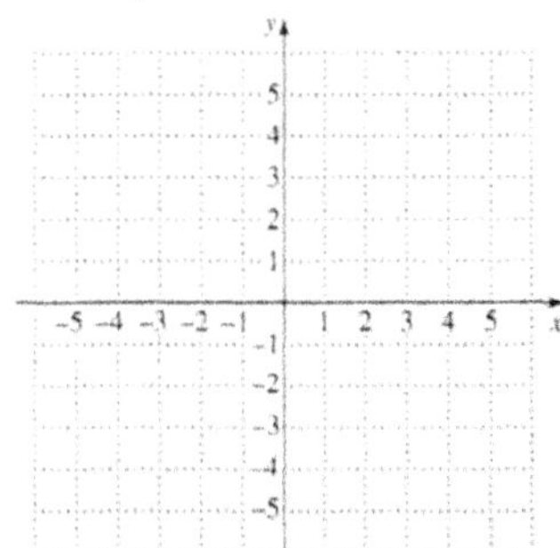

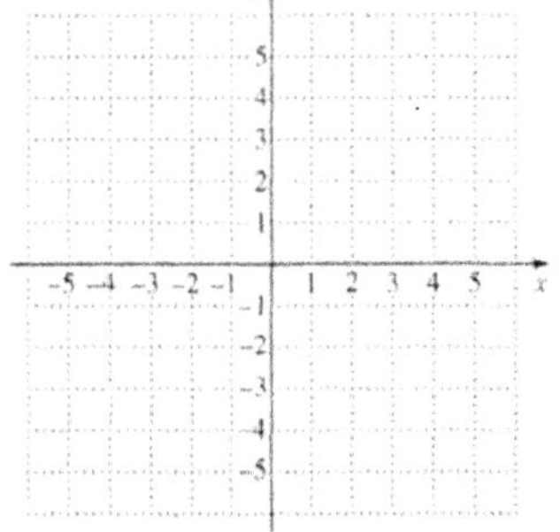

10. $y = 2$

11. $y + \dfrac{1}{2}x = 1$

12. $3x + y = 7$

x	y

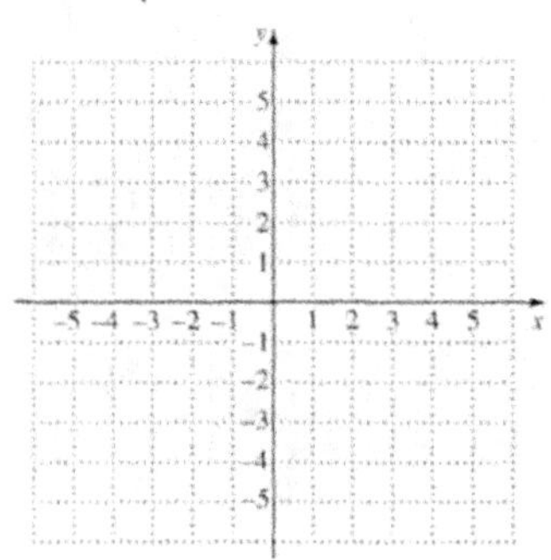

x	y

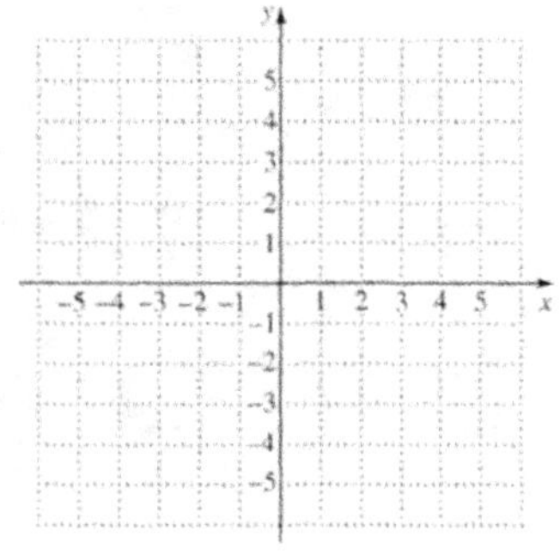

x	y

Notes:

Interactive Preview Worksheet 10: FUNCTION VALUES; DOMAIN AND RANGE

Example 1 Consider the graph of $f(x) = x^2 + 3x - 4$. Find $f(2)$ and $f(-3)$.

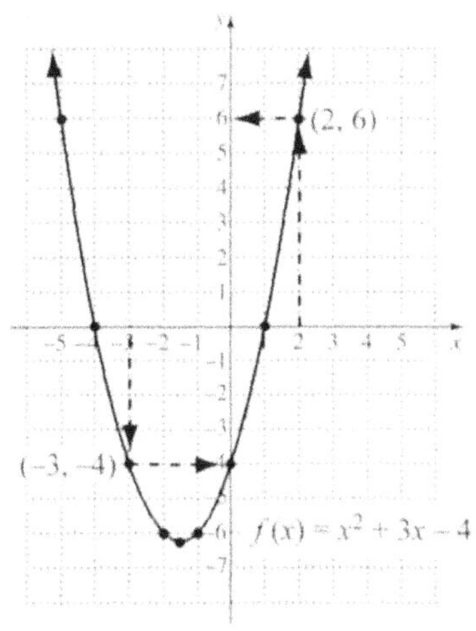

To find the function value $f(2)$ from the graph, we locate the input 2 on the x-axis, move vertically to the graph of the function, and then move horizontally to find the output on the y-axis. We see that $f(2) = 6$. The ordered pair label for the point on the graph is $(2, 6)$.

To find the function value $f(-3)$ from the graph, we locate the input -3 on the x-axis, move vertically to the graph of the function, and then move horizontally to find the output on the y-axis. We see that $f(-3) = -4$. The ordered-pair label for the point on the graph is $(-3, -4)$. We also see that $f(0) = -4$, $f(1) = 0$, and $f(-5) = 6$.

Exercises

Use the graph of the function to determine function values and fill in the coordinates of the ordered-pair labels.

1.

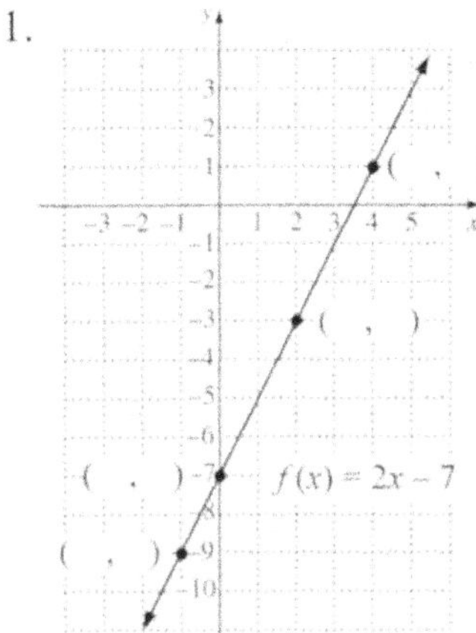

$f(2) = $ _______

$f(-1) = $ _______

$f(0) = $ _______

$f(4) = $ _______

2.

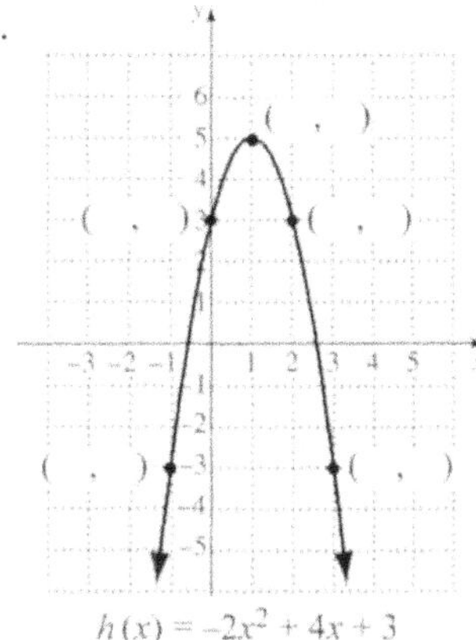

$h(0) = $ _______

$h(1) = $ _______

$h(-1) = $ _______

$h(3) = $ _______

$h(2) = $ _______

3.

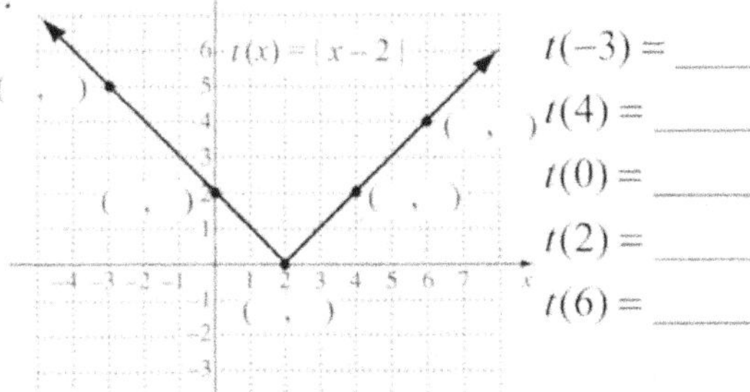

$t(-3) = $ _______

$t(4) = $ _______

$t(0) = $ _______

$t(2) = $ _______

$t(6) = $ _______

4.

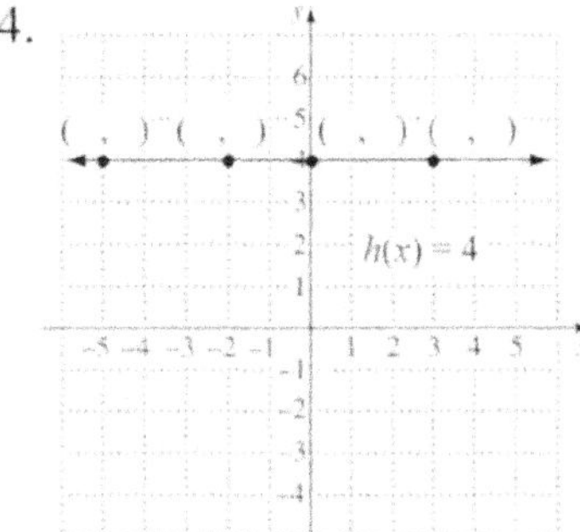

$h(3) = $ _______

$h(-2) = $ _______

$h(-5) = $ _______

$h(0) = $ _______

5.

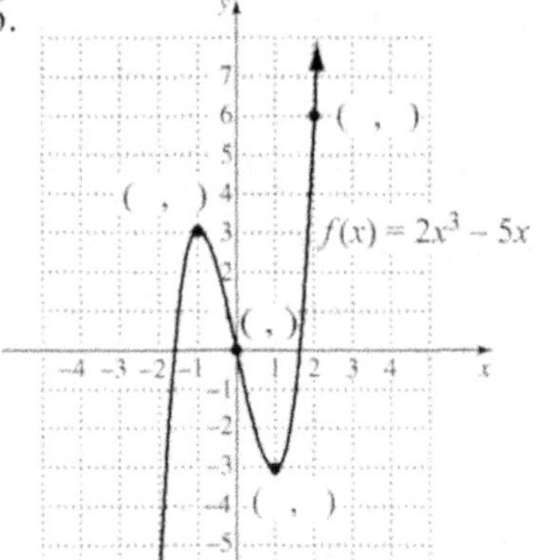

$f(-3) =$ ______

$f(-2) =$ ______

$f(1) =$ ______

$f(6) =$ ______

6.

$f(-1) =$ ______

$f(2) =$ ______

$f(-2) =$ ______

$f(0) =$ ______

$f(1) =$ ______

Example 2 Using the graph, find the domain and the range of the function. Shade the domain (inputs) on the x-axis and the range (outputs) on the y-axis.

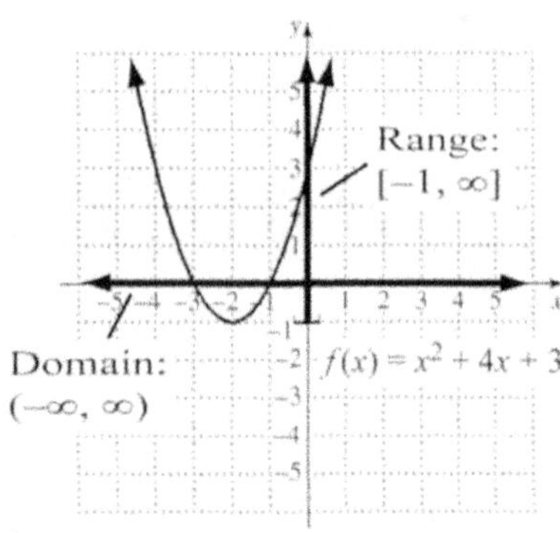

To determine the domain of a function, we look for inputs on the x-axis that correspond to a point on the graph. We see that they include the entire set of real numbers. The domain is $(-\infty, \infty)$. To find the range, we look for outputs on the y-axis that correspond to a point on the graph. We see that they include -1 and all real numbers greater than -1. The bracket at -1 indicates that -1 is included in the interval. The range is $\{y \mid y \geq -1\}$, or $[-1, \infty)$.

Exercises

Using the graphs of the functions shown in Exercises 1-6, find the domain and the range of each function (Hint: Shade the domain and the range of the function on the x- and y- axes).

7. $f(x) = 2x - 7$

(See Exercise 1.)

Domain: ____________
Range: ____________

8. $h(x) = -2x^2 + 4x + 3$

(See Exercise 2.)

Domain: ____________
Range: ____________

9. $t(x) = |x - 2|$

(See Exercise 3.)

Domain: ____________
Range: ____________

10. $h(x) = 4$

(See Exercise 4.)

Domain: ____________
Range: ____________

11. $f(x) = \sqrt{x + 3}$

(See Exercise 5.)

Domain: ____________
Range: ____________

12. $f(x) = 2x^3 - 5x$

(See Exercise 6.)

Domain: ____________
Range: ____________

Interactive Preview Worksheet 11: DETERMINE THE DOMAIN AND THE RANGE OF A FUNCTION

Recall the following regarding the graph of a function.

Domain = the set of a function's *inputs*,
found on the horizontal axis (*x*-axis).

Range = the set of a function's *outputs*,
found on the vertical axis (*y*-axis).

Example 1 Using the graph of the function, determine the domain and the range of the function.

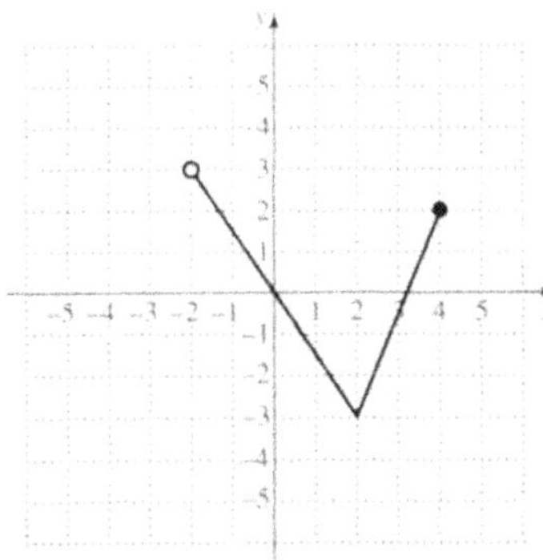

We see that the inputs of the function include all real numbers between -2 and 4. The open circle at -2 indicates that -2 is not in the domain. The closed circle at 4 indicates that 4 is in the domain. The outputs of the function include -3 and all numbers between -3 and 3. Thus,

the **domain** is $(-2, 4]$, and

the **range** is $[-3, 3)$.

Example 2 Using the graph of the function, determine the domain and the range of the function.

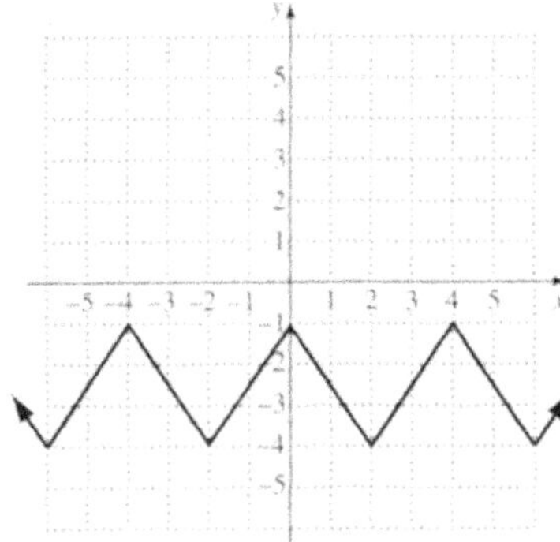

The inputs of the function include the entire set of real numbers. The outputs of the function include all real numbers between -4 and -1 and the numbers -4 and -1. Thus,

the **domain** is $(-\infty, \infty)$, and

the **range** is $[-4, -1]$.

Example 3 Using the graph of the function, determine the domain and the range of the function.

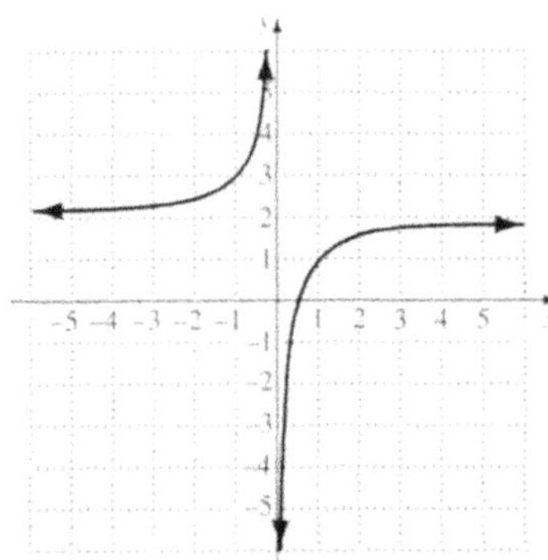

The inputs of the function include all real numbers except 0. The outputs of the function include all real numbers except 2. Thus,

the **domain** is $(-\infty, 0) \cup (0, \infty)$, and

the **range** is $(-\infty, 2) \cup (2, \infty)$.

Exercises Using the graph of the function, determine the domain and the range of the function.

1.

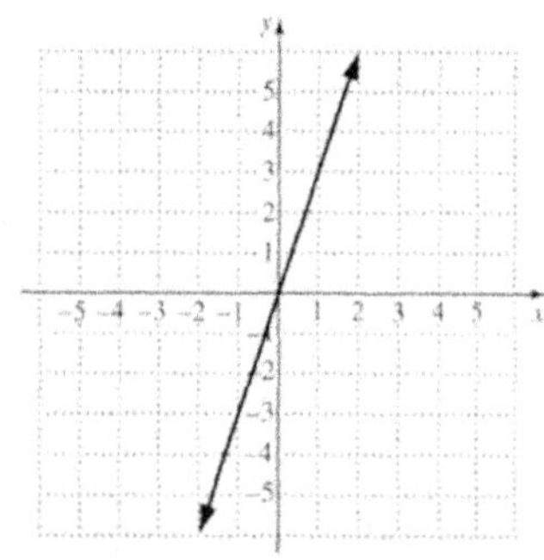

Domain = _______
Range = _______

2.

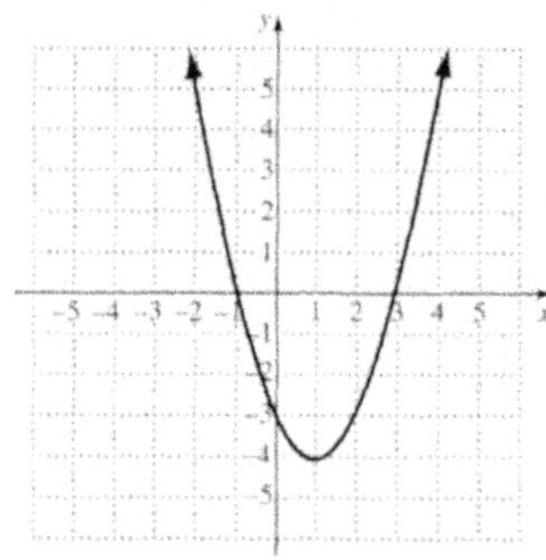

Domain = _______
Range = _______

3.

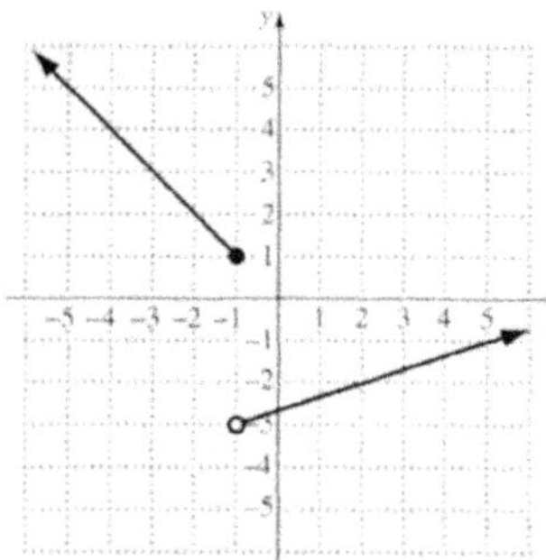

Domain = _______
Range = _______

4.

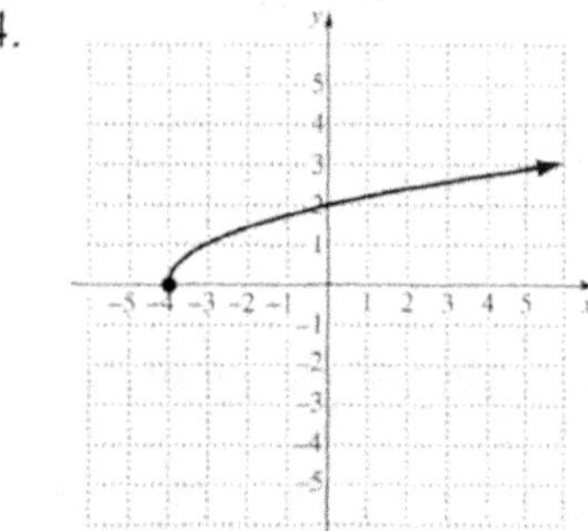

Domain = _______
Range = _______

5.

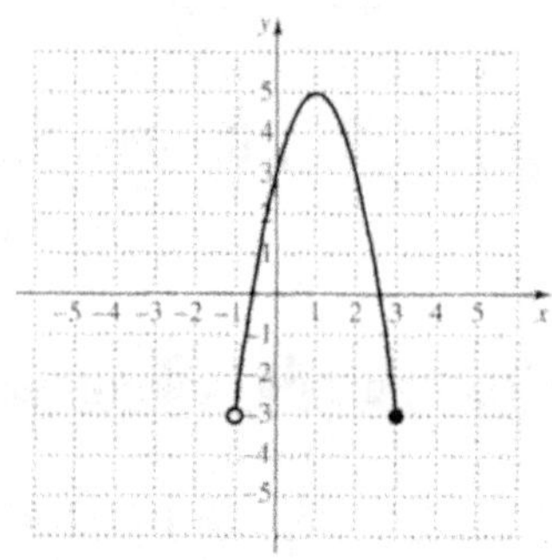

Domain = _______
Range = _______

6.

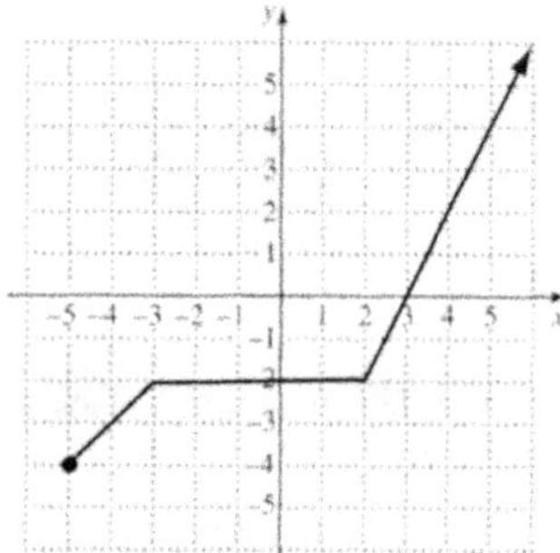

Domain = _______
Range = _______

7.

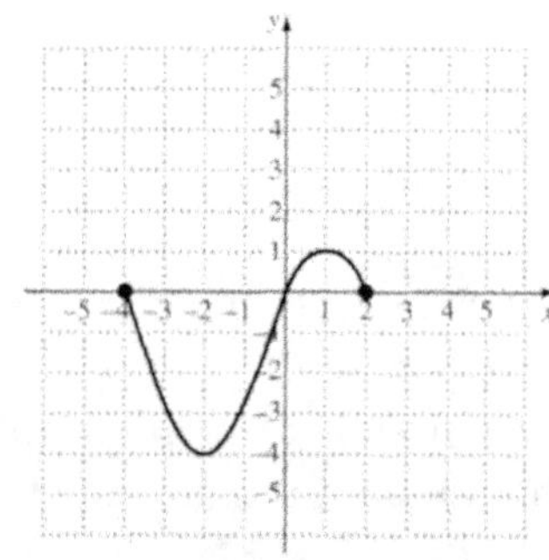

Domain = _______
Range = _______

8.

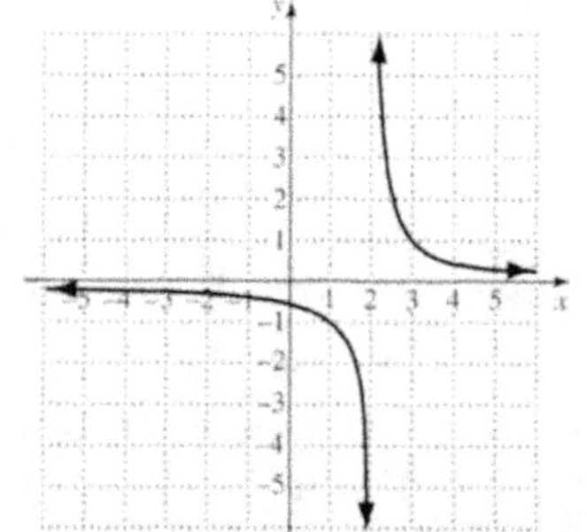

Domain = _______
Range = _______

9.

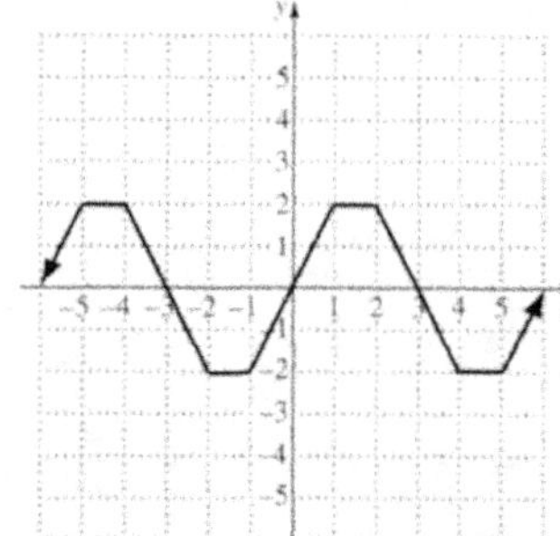

Domain = _______
Range = _______

Interactive Preview Worksheet 12: INTRODUCTION TO POLYNOMIALS

A **monomial** is a constant or a constant times some variable or variables raised to powers that are nonnegative integers. A **polynomial** is a monomial or a combination of sums and/or differences of monomials.

The following are examples of *monomials*:

$$7, \qquad z, \qquad \frac{1}{2}x, \qquad 4a^2, \qquad -2.8y^3, \qquad 5ab^4.$$

Expressions like these are *polynomials*:

$$-6, \qquad t-9, \qquad x^2+5x+6, \qquad 5x^3+\frac{3}{4}x^2-x, \qquad c^2-d^2.$$

The following are algebraic expressions that are *not* polynomials:

$$(1) \ \frac{y-2}{y+3}, \qquad (2) \ x^5-x^3+\frac{1}{x}, \qquad (3) \ w^{1/2}.$$

Expression (1) is not a polynomial because it represents a quotient. In expression (2), although $1/x$ can be written as x^{-1}, this is not a monomial because the exponent is negative. The exponent in expression (3) is not a nonnegative integer, so $w^{1/2}$ is not a monomial.

In this lesson, we will consider only polynomials in one variable.

Exercises Determine if the expression is a polynomial. Answer yes or no.

1. $4y-\dfrac{1}{y}$
2. 9
3. $2x^2+4x-5$

4. $8a^3$
5. $z^{1/2}+10$
6. $\dfrac{t}{t+6}$

The **terms** of a polynomial are separated by + signs. The polynomial

$$5x^4-2x^3-x+8, \text{ or } 5x^4+\left(-2x^3\right)+\left(-x\right)+8,$$

has four terms:

$$5x^4, \qquad -2x^3, \qquad -x, \qquad \text{and} \quad 8.$$

The **coefficients** of the terms are 5, -2, -1, and 8. The term 8 is called a **constant term**.

The **degree of a term** is the exponent of the variable or the sum of the exponents of the variables, if there are variables For example,

the degree of the term $4y^7$ is 7,

the degree of the term $3x^2 y^3$ is $2+3$, or 5, and

the degree of the term 9, or $9x^0$, is 0.

Because we can express 0 as $0 = 0x^5 = 0x^8$, and so on, using any exponent we wish, the term 0 has *no* degree.

The **degree of a polynomial** is the same as the degree of its term of highest degree. For example,

the degree of the polynomial $2 - 9x^2 + x^6 + 4x$ is 6.

The **leading term** of a polynomial is the term of highest degree. Its coefficient is called the **leading coefficient**. For example,

the leading term of $9x^2 - 5x^3 + x - 10$ is $-5x^3$ and

the leading coefficient is -5.

Example 1 Identify the terms, the degree of each term, and the degree of the polynomial $2x^3 + 8x^2 - 17x - 3$. Then identify the leading term, the leading coefficient, and the constant term.

Term	$2x^3$	$8x^2$	$-17x$	-3
Degree of Term	3	2	1	0
Degree of Polynomial	3			
Leading Term	$2x^3$			
Leading Coefficient	2			
Constant Term	-3			

Exercises Identify the terms, the degree of each term, and the degree of the polynomial. Then identify the leading term, the leading coefficient, and the constant term.

7. $3y^4 - 6y^3 + 8y^2 - \dfrac{2}{3}y + 5$

Term				
Degree of Term				
Degree of Polynomial				
Leading Term				
Leading Coefficient				
Constant Term				

8. $x - 3x^4 + 4x^3 - 13 + 7x^2 - 6x^5$

Term					
Degree of Term					
Degree of Polynomial					
Leading Term					
Leading Coefficient					
Constant Term					

The following are some names for certain types of polynomials.

Type	Definition	Examples
Monomial	One term	$17,\ -4x^5$
Binomial	Two terms	$a^2 + 10,\ w - 1$
Trinomial	Three terms	$y^2 - 6y + 8,\ b^6 + 4b^3 - 9b$

Exercises Classify the polynomial as a monomial, a binomial, or a trinomial.

9. $x^2 - 36$

10. $w^2 - 7w - 44$

11. -18

12. $5 - t^2 + t^8$

13. y

14. $q^4 + 16$

We generally arrange polynomials in one variable in **descending order** so that the exponents *decrease* from left to right. Sometimes they may be written so that the exponents *increase* from left to right, which is **ascending order**. In general, if an exercise is written in a particular order, we write the answer in that same order.

Example 2 Consider $12 + x^2 - 7x$. Arrange in descending order and then in ascending order.

Descending order: $x^2 - 7x + 12$ Ascending order: $12 - 7x + x^2$

Example 3 Consider $y^4 + 5y - 20 - 2y^5 - 3y^2$. Arrange in descending order and then in ascending order.

$\qquad$ Descending order: $-2y^5 + y^4 - 3y^2 + 5y - 20$

$\qquad$ Ascending order: $-20 + 5y - 3y^2 + y^4 - 2y^5$

Exercises Arrange the polynomial in descending order and then in ascending order.

15. $\quad y + 2y^3 - 8y^2$

16. $\quad 3 - a^4 + a^3$

17. $\quad -5t + 3t^2 - t^3 + 3$

18. $\quad 12x + 5 + 8x^5 - 4x^3$

19. $\quad 23 - 9b^3 + 4b - 5b^2$

20. $\quad \dfrac{3}{5}q + 5q^7 - \dfrac{1}{3}$

Notes:

Adding Polynomials

When two terms have the same variable(s) raised to the same power(s), they are called **like terms**, or **similar terms**, and they can be "collected," or "combined," using the distributive laws. The sum of two polynomials can be found by writing a plus sign between them and then collecting like terms to simplify the expression.

Example 1   Add: $\left(3x^3 + 4x^2 - 7x - 2\right) + \left(-7x^3 - 2x^2 + 3x + 4\right)$.

Collect like terms. $\quad = (3-7)x^3 + (4-2)x^2 + (-7+3)x + (-2+4)$

Simplify. $\quad = -4x^3 + 2x^2 - 4x + 2$

Example 2   Add: $\left(8t^4 - t + 3t^3 - 8\right) + \left(5t - 3t^2 - t^3 + 2t^4\right)$.

Collect like terms. $\quad = (8+2)t^4 + (3-1)t^3 - 3t^2 + (-1+5)t - 8$

Simplify. $\quad = 6t^4 + 2t^3 - 3t^2 + 4t - 8$

Subtracting Polynomials

To subtract a real number, we add its *opposite*.

$$7 - (-3) = 7 + 3 = 10 \qquad\qquad -9 - 4 = -9 + (-4) = -13$$

To subtract a polynomial, we also add its *opposite*. We can find the opposite of a polynomial by changing the sign of *each* term. The opposite of $5x^2 - 4x - 3$ is $-5x^2 + 4x + 3$.

$$-\left(5x^2 - 4x - 3\right) \text{ is equivalent to } -5x^2 + 4x + 3.$$

Example 3 Subtract: $\left(2x^2 - 5x + 3\right) - \left(x^2 - 9x + 10\right)$.

Add the opposite. $\quad = \left(2x^2 - 5x + 3\right) + \left(-x^2 + 9x - 10\right)$

Collect like terms. $\quad = (2-1)x^2 + (-5+9)x + (3-10)$

Simplify. $\quad = x^2 + 4x - 7$

Example 4 Subtract: $\left(y^3 - 1\right) - \left(y^3 + 2y + 6\right)$.

Add the opposite. $\quad = \left(y^3 - 1\right) + \left(-y^3 - 2y - 6\right)$

Collect like terms. $\quad = \left[1 + (-1)\right]y^3 - 2y + \left[-1 + (-6)\right]$

Simplify. $\quad = -2y - 7$

Example 5 Subtract: $\left(-9x^3 + x^2 - 5x\right) - \left(-3x^4 + 2x^3 - 7x - 15\right)$.

Add the opposite. $\quad = \left(-9x^3 + x^2 - 5x\right) + \left(3x^4 - 2x^3 + 7x + 15\right)$

Collect like terms. $\quad = 3x^4 + \left[-9 + (-2)\right]x^3 + x^2 + (-5+7)x + 15$

Simplify. $\quad = 3x^4 - 11x^3 + x^2 + 2x + 15$

Check Your Understanding

For each subtraction, choose an equivalent expression from choices a) – l).

a) $\left(4y^3 + 2y^2\right) + \left(y^3 + 8y - 4\right)$

b) $(w - 3) + (-2w + 11)$

c) $\left(p^5 - p^3 - p\right) + (3p + 5)$

d) $\left(7x^2 + x + 4\right) + \left(-5x^2 + 2x - 1\right)$

e) $\left(7x^2 - x - 4\right) + \left(5x^2 - 2x - 1\right)$

f) $(w - 3) + (-2w - 11)$

g) $(w + 3) + (2w - 11)$

h) $\left(4y^3 - 2y^2\right) - \left(-y^3 + 8y - 4\right)$

i) $\left(-4y^3 + 2y^2\right) + \left(-y^3 + 8y - 4\right)$

j) $\left(p^5 - p^3 - p\right) + (-3p + 5)$

k) $\left(p^5 - p^3 - p\right) + (3p - 5)$

l) $\left(7x^2 - x - 4\right) + \left(5x^2 - 2x + 1\right)$

1. $(w - 3) - (2w + 11)$

2. $\left(7x^2 - x - 4\right) - \left(-5x^2 + 2x - 1\right)$

3. $\left(-4y^3 + 2y^2\right) - \left(y^3 - 8y + 4\right)$

4. $\left(p^5 - p^3 - p\right) - (-3p - 5)$

Exercises Perform the indicated operation and simplify.

1. $\left(t^2 - 6t + 5\right) + \left(-3t^2 - 8\right)$

2. $(5x - 2) - (x - 7)$

3. $\left(2x^2 - 3x - 4\right) - \left(-7x^2 + 12\right)$

4. $\left(11y^2 + 6y - 3\right) + \left(9y^2 - 2y + 9\right)$

5. $\left(8b^3 - 11b\right) - \left(-2b^3 + b^2 + 3\right)$

6. $\left(3x^4 - 4x^2 + x - 3\right) - \left(5x^3 + 7x^2 + 1\right)$

7. $\left(15x^2 + 12x + 1\right) - \left(-x^2 + 3x + 2\right)$

8. $\left(7a^3 - 5a + 10\right) + \left(2a^4 - 9a^2 + a - 3\right)$

9. $\left(-10x^4 + x^3 - 2\right) + \left(-5x^3 + 2x - 6\right)$

10. $\left(w^5 - w^4 + w^2 - 1\right) - \left(-w^4 + 5w^3 - w^2 - 5\right)$

Interactive Preview Worksheet 14: MULTIPLYING BINOMIALS

The Product of Two Binomials Using the FOIL Method

Consider $(x+7)(x+4)$, the product of two binomials. We multiply each term of $(x+7)$ by each term of $(x+4)$:

$$(x+7)(x+4) = x \cdot x + x \cdot 4 + 7 \cdot x + 7 \cdot 4.$$

The multiplication illustrates a pattern that occurs whenever two binomials are multiplied:

$$\begin{array}{cccc} \text{First} & \text{Outside} & \text{Inside} & \text{Last} \\ \text{terms} & \text{terms} & \text{terms} & \text{terms} \end{array}$$

$$(x+7)(x+4) = x \cdot x + x \cdot 4 + 7 \cdot x + 7 \cdot 4 = x^2 + 11x + 28.$$

This special method of multiplying is called the **FOIL method**. Keep in mind that this method is based on the distributive law.

The Foil Method

To multiply two binomials, $A+B$ and $C+D$, multiply the **F**irst terms AC, the **O**utside terms AD, the **I**nside terms BC, and then the **L**ast terms BD. Then collect like terms, if possible.

$$(A+B)(C+D) = AC + AD + BC + BD$$

1. Multiply **F**irst terms: AC.
2. Multiply **O**utside terms: AD.
3. Multiply **I**nside terms: BC.
4. Multiply **L**ast terms: BD.

$$\downarrow$$

FOIL

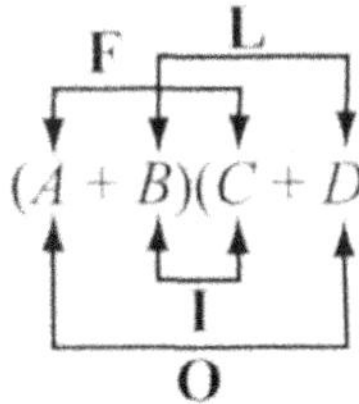

Example 1   Multiply: $(x+8)(x-5)$.

	F O I L
Use the FOIL method.	$= x \cdot x + x(-5) + 8 \cdot x + 8(-5)$
Simplify.	$= x^2 - 5x + 8x - 40$
Collect like terms.	$= x^2 + 3x - 40$

Example 2   Multiply: $(3x-2)(6x-7)$.

	F O I L
Use the FOIL method.	$= 3x \cdot 6x + 3x(-7) + (-2)(6x) + (-2)(-7)$
Simplify.	$= 18x^2 - 21x - 12x + 14$
Collect like terms.	$= 18x^2 - 33x + 14$

Exercises Multiply.

1. $(y-2)(y-3)$

2. $(2b-5)(8b+1)$

3. $(w+5)(w-2)$

4. $(z-9)(z+1)$

5. $(4x+3)(5x-6)$

6. $(3q+5)(q+20)$

7. $(x-4)(x+13)$

8. $(10b-7)(3b+4)$

9. $(5w-1)(5w+8)$

10. $(a+6)(a+11)$

Squares of Binomials

We use the FOIL method to develop special products for the square of a binomial:

$$(x+5)^2 = (x+5)(x+5) = x^2 +5x+5x+5^2 = x^2 +10x+25,$$
$$(y-5)^2 = (y-5)(y-5) = y^2 -5y-5y+(-5)^2 = y^2 -10y+25.$$

Square of a Binomial

The **square of a binomial** is the square of the first term, plus twice the product of the two terms, plus the square of the last term.

$$(A+B)^2 = A^2 +2AB+B^2;$$
$$(A-B)^2 = A^2 -2AB+B^2$$

Example 3   Multiply: $(y-7)^2$.

Use $(A-B)^2 = A^2 -2AB+B^2$. $= y^2 -2\cdot y\cdot 7+7^2$

Simplify. $= y^2 -14y+49$

Example 4   Multiply: $(3x+4)^2$.

$$\text{Use } (A+B)^2 = A^2 + 2AB + B^2. \quad = (3x)^2 + 2\cdot 3x \cdot 4 + 4^2$$
$$\text{Simplify.} \quad = 9x^2 + 24x + 16$$

Exercises Multiply.

11. $(x+6)^2$

12. $(2x-1)^2$

13. $(3a+2)^2$

14. $(t+10)^2$

15. $(z-12)^2$

16. $(4w+5)^2$

17. $(b-3)^2$

18. $(6y-3)^2$

Products of Sums and Differences

Another special case of a product of two binomials is the product of a sum and a difference. Note the following:

$$(x+9)(x-9) = x^2 - 9x + 9x + 9(-9) = x^2 - 81.$$

Product of a Sum and a Difference

The product of the sum and the difference of the same two terms is the square of the first term minus the square of the second term (the difference of their squares).

$$(A+B)(A-B) = A^2 - B^2 \quad \text{This is called a \textbf{difference of squares}.}$$

Example 5   Multiply: $(x+4)(x-4)$.

Use $(A+B)(A-B)=A^2-B^2$. $=x^2-4^2$

                Simplify. $=x^2-16$

Example 6   Multiply: $(5z+8)(5z-8)$.

Use $(A+B)(A-B)=A^2-B^2$. $=(5z)^2-8^2$

                Simplify. $=25z^2-64$

Exercises Multiply.

19. $(y+10)(y-10)$ 20. $(3c+4)(3c-4)$ 21. $(t-6)(t+6)$

22. $(w+1)(w-1)$ 23. $(2z+7)(2z-7)$ 24. $(10y-5)(10y+5)$

     Try to multiply polynomials mentally. When several types are mixed, first check to see what types of polynomials are to be multiplied. Then use the quickest method. Sometimes we might use more than one method to find a product. Remember that FOIL *always* works for multiplying binomials!

Exercises Multiply.

25. $(x-3)(x+10)$ 26. $(4x+9)(4x-9)$

27. $(3x-1)(7x-4)$ 28. $(y+20)(y+5)$

29. $(s-3)(s+3)$ 30. $(a-8)^2$

31. $(5y+4)^2$ 32. $(6a-2)(a+11)$

Interactive Preview Worksheet 15: FACTORING BY GROUPING

In some polynomials, a *binomial* is a common factor.

Example 1 Factor: $y^2(y+2)+3(y+2)$.

The common factor is the binomial $y+2$. We factor out the common factor, $y+2$.

$$y^2(y+2)+3(y+2)=(y+2)(y^2+3)$$

In Example 2, we factor a polynomial with four terms. We begin by grouping pairs of terms.

Example 2   Factor: $x^3+5x^2+7x+35$.

Group before factoring.	$=(x^3+5x^2)+(7x+35)$
Factor each binomial.	$=x^2(x+5)+7(x+5)$
Factor out the common factor, $x+5$.	$=(x+5)(x^2+7)$

This is called **factoring by grouping**.

Example 3   Factor: $5t^3-15t^2-t+3$.

Consider the first two terms.	$5t^3-15t^2$
Factor out the greatest common factor, or GCF, $5t^2$.	$=5t^2(t-3)$
Consider the last two terms.	$-t+3$
Factor out -1 so one of the factors is $t-3$. Note: $-t+3=-1(t-3)$.	$=-1(t-3)$

Now we can factor out $t-3$ in the polynomial, as shown below.

	$5t^3-15t^2-t+3$
Group before factoring.	$=(5t^3-15t^2)+(-t+3)$
Factor each binomial. Note: $-t+3=-1(t-3)$.	$=5t^2(t-3)-1(t-3)$
Factor out the common factor, $t-3$.	$=(t-3)(5t^2-1)$

Example 4   Factor: $2a^3 + 12a^2 - 5a - 30$.

Group before factoring. $= \left(2a^3 + 12a^2\right) + \left(-5a - 30\right)$

Factor each binomial.
Note: $-5a - 30 = -5(a+6)$. $= 2a^2(a+6) - 5(a+6)$

Factor out the common factor, $a+6$. $= (a+6)\left(2a^2 - 5\right)$

Example 5   Factor: $4z^3 - 9 + z - 36z^2$.

Arrange in descending order. $= 4z^3 - 36z^2 + z - 9$

Group before factoring. $= \left(4z^3 - 36z^2\right) + (z-9)$

Factor each binomial. $= 4z^2(z-9) + 1(z-9)$

Factor out the common factor, $z-9$. $= (z-9)\left(4z^2 + 1\right)$

Example 6   Factor: $3x^7 - 15x^6 + 6x - 30$.

Factor out the common factor, 3. $= 3\left(x^7 - 5x^6 + 2x - 10\right)$

Group before factoring. $= 3\left[\left(x^7 - 5x^6\right) + (2x - 10)\right]$

Factor each binomial. $= 3\left[x^6(x-5) + 2(x-5)\right]$

Factor out the common factor, $x-5$. $= 3(x-5)\left(x^6 + 2\right)$

Check Your Understanding

Not all polynomials with four terms can be factored by grouping. For example, $x^3 + x^2 + 3x - 3 = x^2(x+1) + 3(x-1)$ and cannot be factored by grouping.

Determine if the polynomial can be factored by grouping. Answer with "yes" or "no".

1. $y^3 + 8y^2 - 5y - 40$

2. $a^3 - 2a^2 + 5a + 10$

3. $2x^3 + 3x^2 - 8x + 12$

4. $t^3 - 6t + 9t^2 - 54$

Exercises Factor by grouping, if possible. In Exercises 1 and 2, fill-in the blanks in key steps.

1. $y^3 + 4y^2 + 9y + 36$

$= \left(y^3 + 4y^2\right) + \left(9y + 36\right)$

$= \boxed{}(y + 4) + 9\left(\boxed{}\right)$

$= \left(\boxed{}\right)\left(y^2 + 9\right)$

2. $3s^3 - 15s^2 - 7s + 35$

$= \left(3s^3 - 15s^2\right) + \left(\boxed{}\right)$

$= 3s^2\left(\boxed{}\right) + \boxed{}(s - 5)$

$= (s - 5)\left(\boxed{}\right)$

3. $c^3 + 6c^2 - 5c - 30$

4. $x^3 - 7x^2 + 4x - 28$

5. $3x^3 - 30x^2 - x - 10$

6. $18y - 2y^2 - 18 + 2y^3$

7. $x^3 - 5x^2 + x - 5$

8. $2y^3 + 3y^2 - 6y - 9$

9. $xy - xz + wy - wz$

10. $2x^4 + 6x^2 - 5x^2 - 15$

Notes:

Interactive Preview Worksheet 16: FACTORING TRINOMIALS: $ax^2 + bx + c$

Let's begin by reviewing the FOIL method of multiplying two binomials.

$$\overset{\text{F} \quad \text{O} \quad \text{I} \quad \text{L}}{\text{Multiply:} \ (x+p)(x+q) = x^2 + qx + px + pq}$$

$$= x^2 + (p+q)x + pq$$

Example: $(x+4)(x+7) = x^2 + 7x + 4x + 4 \cdot 7$

$$= x^2 + (4+7)x + 28 = x^2 + 11x + 28$$

To factor a trinomial, we think of FOIL in reverse.

Factor: $x^2 + (p+q)x + pq = (x+p)(x+q)$

Example: $x^2 + 11x + 28 = x^2 + (4+7)x + 4 \cdot 7$

$$= (x+4)(x+7)$$

We look for two numbers whose product is 28 and whose sum is 11. Those numbers are 4 and 7.

Example 1 Factor: $x^2 + 8x + 15$.

The first term is x^2, so the first term in each binomial is x.

$$x^2 + 8x + 15 = (x \qquad)(x \qquad)$$

Constant, 15, is *positive*; coefficient of middle term, 8, is *positive*.

Find two numbers whose product is 15 and whose sum is 8.

Pairs of factors	Sums of factors
1, 15	16
−1, −15	−16
3, 5	8 ←
−3, −5	−8

The numbers we need are 3 and 5. $\quad x^2 + 8x + 15 = (x+3)(x+5)$

Example 2 Factor: $x^2 - 8x + 15$.

The first term is x^2, so the first term in each binomial is x.

$$x^2 - 8x + 15 = (x \qquad)(x \qquad)$$

Constant, 15, is *positive*; coefficient of middle term, −8, is *negative*.

Find two numbers whose product is 15 and whose sum is −8. (See the table in Example 1.)

The numbers we need are −3 and −5. $\quad x^2 - 8x + 15 = (x-3)(x-5)$

When the constant term of a trinomial is positive, we look for two factors with the same sign (both positive or both negative). The sign is that of the middle term.

Example 3 Factor: $y^2 + 9y - 36$.

The first term in each binomial is y.	$y^2 + 9y - 36 = (y \quad)(y \quad)$

Constant, -36, is *negative*; coefficient of middle term, 9, is *positive*.

Find two numbers, whose product is -36 and whose sum is 9.

Pairs of factors	Sums of factors
1, -36	-35
-1, 36	35
2, -18	-16
-2, 18	16
3, -12	-9
-3, 12	9 ←
4, -9	-5
-4, 9	5
6, -6	0

The numbers we need are -3 and 12.

$$y^2 + 9y - 36 = (y - 3)(y + 12)$$

Example 4 Factor: $y^2 - 9y - 36$.

The first term in each binomial is y.

$$y^2 - 9y - 36 = (y \quad)(y \quad)$$

Constant, -36, is *negative*; coefficient of middle term, -9, is *negative*.

Find two numbers whose product is -36 and whose sum is -9. (See the table in Example 3.) The numbers we need are 3 and -12.

$$y^2 - 9y - 36 = (y + 3)(y - 12)$$

> When the constant term of a trinomial is negative, we look for two factors whose product is negative. One of them must be positive and the other negative. Their sum must be the coefficient of the middle term.

Exercises Factor.

1. $x^2 + 5x - 36$

2. $y^2 - 10y + 9$

3. $y^2 - 3y - 10$

4. $s^2 - 9s - 90$ 5. $x^2 + 7x - 8$ 6. $b^2 + 8b + 12$

7. $x^2 - x - 56$ 8. $a^2 - 3a - 54$ 9. $y^2 - 5y - 14$

10. $w^2 - 10w - 11$ 11. $q^2 + 19q + 90$ 12. $x^2 - 14x + 40$

13. $x^2 + 120x + 2000$ 14. $z^2 - 10z + 25$

In Examples 1-4, we factored trinomials of the type $x^2 + bx + c$. Now let's review factoring trinomials of the type $ax^2 + bx + c$, $a \neq 1$.

Example 5 Factor: $3x^2 + 10x - 8$.

Factor the first term. The only possibility is $3x \cdot x$.	$3x^2 + 10x - 8 = (3x\quad)(x\quad)$
Factor the last term, -8, which is negative.	The possibilities are $1(-8)$, $(-1)8$, $(2)(-4)$, and $(-2)(4)$. Each can be written in either order.
Look for combinations of factors such that the sum of the outside and the inside products is the middle term $10x$.	There are 8 combinations of factors. $(3x+1)(x-8) = 3x^2 - 23x - 8$ $(3x-8)(x+1) = 3x^2 - 5x - 8$ $(3x-1)(x+8) = 3x^2 + 23x - 8$ $(3x+8)(x-1) = 3x^2 + 5x - 8$ $(3x+2)(x-4) = 3x^2 - 10x - 8$ $(3x-4)(x+2) = 3x^2 + 2x - 8$ $(3x-2)(x+4) = 3x^2 + 10x - 8 \quad\leftarrow$ $(3x+4)(x-2) = 3x^2 - 2x - 8$
Correct factorization.	$3x^2 + 10x - 8 = (3x-2)(x+4)$.

Example 6 Factor $21t^2 - 32t + 12$.

Factor the first term, $21t^2$. There are two possibilities: $21t \cdot t$ or $7t \cdot 3t$.	$21t^2 - 32t + 12 = (21t \quad)(t \quad)$ or $21t^2 - 32t + 12 = (7t \quad)(3t \quad)$
Factor the last term, 12, which is positive.	The possibilities are $12 \cdot 1, \ -12(-1), \ 6 \cdot 2, \ -6(-2), \ 4 \cdot 3, \ \text{and} \ -4(-3)$.
Since the middle term, $-32t$, is negative, consider only three of the six possibilities.	$-12(-1), \ -6(-2), \ \text{and} \ -4(-3)$. Each can be written in either order.
Begin by using $(7t \quad)(3t \quad)$. If a correct factorization is not found, then consider $(21t \quad)(t \quad)$. Look for combinations of factors such that the sum of the outside and the inside products is the middle term, $-32t$.	$(7t-12)(3t-1) = 21t^2 - 43t + 12$ $(7t-1)(3t-12) = 21t^2 - 87t + 12$ $(7t-6)(3t-2) = 21t^2 - 32t + 12 \ \leftarrow$ $(7t-2)(3t-6) = 21t^2 - 48t + 12$ $(7t-4)(3t-3) = 21t^2 - 33t + 12$ $(7t-3)(3t-4) = 21t^2 - 37t + 12$
Correct factorization.	$21t^2 - 32t + 12 = (7t-6)(3t-2)$.

Exercises Factor.

15. $6x^2 + 17x - 14$

16. $4w^2 - 4w - 15$

17. $30x^2 + 7x - 2$

18. $24y^2 - 29y - 4$

19. $20x^2 + 44x - 15$ 20. $40c^2 + 31c + 6$

When the three terms of a trinomial have a common factor, we begin by factoring it out.

Example 7 Factor: $z^3 - 14z^2 + 33z$.

Factor out the common factor z.	$z^3 - 14z^2 + 33z = z\left(z^2 - 14z + 33\right)$
Now factor the trinomial $z^2 - 14z + 33$. The first term is z^2, so the first term of each binomial is z.	$z^2 - 14z + 33 = (z \quad)(z \quad)$

Find two numbers whose product is 33 and whose sum is -14.

Pairs of factors	Sum of factors
1, 33	34
$-1, -33$	-34
3, 11	14
$-3, -11$	-14 ←

The numbers we need are -3 and -11.	$z^2 - 14z + 33 = (z - 3)(z - 11)$
Don't forget the common factor, z.	$z^3 - 14z^2 + 33z = z(z - 3)(z - 11)$

Example 8 Factor: $3x^4 + 18x^3 - 21x^2$.

Factor out the common factor $3x^2$.	$3x^4 + 18x^3 - 21x^2 = 3x^2\left(x^2 + 6x - 7\right)$
Now factor the trinomial $x^2 + 6x - 7$. The first term is x^2, so the first term in each binomial is x.	$x^2 + 6x - 7 = (x \qquad)(x \qquad)$
Find two numbers whose product is -7 and whose sum is 6. Those numbers are -1 and 7.	$x^2 + 6x - 7 = (x - 1)(x + 7)$
Don't forget the common factor, $3x^2$.	$3x^4 + 18x^3 - 21x^2 = 3x^2(x - 1)(x + 7)$

Exercises Factor.

21. $x^3 - 5x^2 + 6x$

22. $b^3 - 4b^2 - 32b$

23. $2y^4 + 26y^3 - 60y^2$

24. $5q^3 + 65q^2 + 180q$

**Interactive Preview Worksheet 17: FACTORING TRINOMIAL SQUARES
AND DIFFERENCES OF SQUARES**

Trinomial Squares

When the factorization of a trinomial is the square of a binomial, the trinomial is called a **trinomial square**, or perfect-square trinomial.

Examples:

$$x^2 + 14x + 49 = (x+7)(x+7) = (x+7)^2, \text{ and}$$

$$z^2 - 20z + 100 = (z-10)(z-10) = (z-10)^2$$

> How to recognize a **trinomial square**, $A^2 + 2AB + B^2$ or $A^2 - 2AB + B^2$:
> **a)** The two expressions A^2 and B^2 must be squares.
> **b)** There must be no minus sign before either A^2 or B^2.
> **c)** Multiplying A and B (expressions whose squares are A^2 and B^2) and doubling the result gives either the *remaining term* or its *opposite*.

> **Factoring Trinomial Squares**
>
> $$A^2 + 2AB + B^2 = (A+B)^2$$
> $$A^2 - 2AB + B^2 = (A-B)^2$$

Example 1 Factor: $y^2 + 8y + 16$.

y^2 and 16 are squares: y^2 and 4^2. $(A = y, \; B = 4)$

No minus sign before either y^2 or 16.

The remaining term $8y = 2 \cdot y \cdot 4$.
$ 2 \cdot A \cdot B$

$y^2 + 8y + 16$ is a trinomial square.

$$y^2 + 8y + 16 = (y+4)^2$$

Example 2 Factor: $25x^2 - 30x + 9$.

$25x^2$ and 9 are squares: $(5x)^2$ and 3^2. $(A = 5x, \; B = 3)$

No minus sign before either $25x^2$ or 9.

The remaining term $-30x = -(2 \cdot 5x \cdot 3)$
$ = -(2 \cdot A \cdot B)$ Think: The opposite of $2 \cdot A \cdot B$.

$25x^2 - 30x + 9$ is a trinomial square.

$$25x^2 - 30x + 9 = (5x - 3)^2.$$

Differences of Squares

The following are differences of squares:

$$x^2 - 81, \ 16 - 9y^2, \text{ and } 100w^2 - 1.$$

Before factoring a difference of squares, let's review finding the product of the sum and the difference of the same two terms.

$$(x+9)(x-9) = x^2 - 9x + 9x - 81 = x^2 - 81$$

We see that two terms $-9x$ and $9x$ are opposites. They add to 0 and "drop out." Looking at this product in reverse, we see that $x^2 - 81 = (x+9)(x-9)$.

Factoring A Difference or Squares

$$A^2 - B^2 = (A+B)(A-B)$$

Example 3 Factor: $x^2 - 36$.

$$x^2 - 36 = x^2 - 6^2$$
$$= (x+6)(x-6)$$

Example 4 Factor: $121c^2 - 4$.

$$121c^2 - 4 = (11c)^2 - 2^2$$
$$= (11c+2)(11c-2)$$

Check Your Understanding

Classify each of the following as a trinomial square, a difference of squares, or neither of these.

1. $100 - 9w^2$
2. $x^2 - 2x + 4$
3. $y^2 - 6y + 9$
4. $36x^2 - 1$
5. $4x^2 + 49$
6. $16x^2 + 40x + 25$

Exercises Factor.

1. $64x^2 - 1$
2. $x^2 - 12x + 36$
3. $81y^2 - 16$

4. $w^2 - 4w + 4$
5. $49x^2 + 42x + 9$
6. $25a^2 - 36$

7. $49 - z^2$
8. $x^2 - 225$
9. $y^2 + 16y + 64$

10. $4x^2 - 20x + 25$
11. $5x^2 - 45$
12. $3x^2 + 24x + 48$

Interactive Preview Worksheet 18: THE PRINCIPLE OF ZERO PRODUCTS

The principle of zero products gives us a method for solving polynomial equations.

The Principle of Zero Products

For any real numbers a and b:

If $ab = 0$, then $a = 0$ or $b = 0$ (or both).

If $a = 0$ or $b = 0$, then $ab = 0$.

Example 1 Solve: $x^2 - x - 6 = 0$.

Factor. $\qquad (x-3)(x+2) = 0$

Set each factor equal to 0. $\qquad x - 3 = 0 \quad or \quad x + 2 = 0$

Solve separately. $\qquad x = 3 \quad or \qquad x = -2$

Check.

$$x^2 - x - 6 = 0$$
$$\begin{array}{c|c} 3^2 - 3 - 6 & 0 \\ 9 - 3 - 6 & \\ 0 & 0 \quad \text{True} \end{array}$$

$$x^2 - x - 6 = 0$$
$$\begin{array}{c|c} (-2)^2 - (-2) - 6 & 0 \\ 4 + 2 - 6 & \\ 0 & 0 \quad \text{True} \end{array}$$

The numbers 3 and -2 are both solutions.

We must have 0 on one side of the equation in order to use the principle of zero products.

Example 2 Solve: $7y + 3y^2 = -2$.

Get 0 on one side and write in descending order.
$$3y^2 + 7y + 2 = 0$$

Factor.
$$(3y + 1)(y + 2) = 0$$

Use the principle of zero products.
$$3y + 1 = 0 \quad or \quad y + 2 = 0$$

Solve separately.
$$3y = -1 \quad or \qquad y = -2$$

$$y = -\frac{1}{3} \quad or \qquad y = -2$$

Check.

$$7y + 3y^2 = -2$$
$$\begin{array}{c|c} 7\left(-\dfrac{1}{3}\right) + 3\left(-\dfrac{1}{3}\right)^2 & -2 \\[2mm] 7\left(-\dfrac{1}{3}\right) + 3\left(\dfrac{1}{9}\right) & \\[2mm] -\dfrac{7}{3} + \dfrac{1}{3} & \\[2mm] -\dfrac{6}{3} & \\[2mm] -2 & -2 \quad \text{True} \end{array}$$

$$7y + 3y^2 = -2$$
$$\begin{array}{c|c} 7(-2) + 3(-2)^2 & -2 \\[2mm] 7(-2) + 3(4) & \\[2mm] -14 + 12 & \\[2mm] -2 & -2 \quad \text{True} \end{array}$$

The solutions are $-\dfrac{1}{3}$ and -2.

Example 3 Solve: $5b^2 = 10b$.

$$
\begin{array}{r|l}
\text{Get 0 on one side.} & 5b^2 - 10b = 0 \\
\text{Factor.} & 5b(b-2) = 0 \\
\text{Use the principle of zero products.} & 5b = 0 \quad or \quad b - 2 = 0 \\
\text{Solve separately.} & b = 0 \quad or \quad b = 2 \\
& \text{The solutions are 0 and 2.}
\end{array}
$$

Example 4 Solve: $6x - x^2 = 9$.

$$
\begin{array}{r|l}
\text{Get 0 on one side and the leading coefficient on the other side positive.} & 0 = x^2 - 6x + 9 \\
\text{Factor.} & 0 = (x-3)(x-3) \\
\text{Use the principle of zero products.} & x - 3 = 0 \quad or \quad x - 3 = 0 \\
\text{Solve separately.} & x = 3 \quad or \quad x = 3 \\
& \text{There is only one solution, 3.}
\end{array}
$$

Example 5 Solve: $3x^3 - 9x^2 = 30x$.

$$
\begin{array}{r|l}
\text{Get 0 on one side.} & 3x^3 - 9x^2 - 30x = 0 \\
\text{Factor out a common factor.} & 3x(x^2 - 3x - 10) = 0 \\
\text{Factor the trinomial.} & 3x(x+2)(x-5) = 0 \\
\text{Use the principle of zero products.} & 3x = 0 \quad or \quad x + 2 = 0 \quad or \quad x - 5 = 0 \\
\text{Solve separately.} & x = 0 \quad or \quad x = -2 \quad or \quad x = 5 \\
& \text{The solutions are 0, } -2 \text{, and 5.}
\end{array}
$$

Check Your Understanding

Determine whether each statement is true or false.

1. If $(s-6)(s+8) = 0$, then both $s-6$ and $s+8$ must equal 0.

2. If $(x+12)(x+9) = 0$, then $x + 12 = 0$ or $x + 9 = 0$.

3. If $(y+7)(y-3) = 21$, then $y + 7 = 21$ or $y - 3 = 21$.

Exercises

Solve using the principle of zero products. In Exercises 1 and 2, fill-in the blanks in key steps.

1.
$$x^2 - 7x = -10$$
$$x^2 - 7x + \boxed{} = 0$$
$$\left(x - \boxed{}\right)(x - 2) = 0$$
$$x - 5 = \boxed{} \quad or \quad x - 2 = \boxed{}$$
$$x = 5 \quad or \quad x = \boxed{}$$
The solutions are $\boxed{}$ and $\boxed{}$.

2.
$$15z^2 = -3z$$
$$15z^2 + 3z = 0$$
$$\boxed{}(5z + 1) = 0$$
$$3z = 0 \quad or \quad 5z + 1 = 0$$
$$z = \boxed{} \quad or \quad z = -\frac{1}{5}$$
The solutions are $\boxed{}$ and $\boxed{}$.

3. $y^2 + 2y = 63$

4. $18x^2 = 9x$

5. $9x + x^2 + 20 = 0$

6. $x^2 + 20x + 100 = 0$

7. $32 + 4x - x^2 = 0$

8. $11x + 4x^2 = -6$

9. $10 - r - 21r^2 = 0$

10. $2x^3 - 2x^2 = 12x$

Notes:

Interactive Preview Worksheet 19: THE PYTHAGOREAN THEOREM

A **right triangle** is a triangle with a $90°$ angle, as shown here.
In a right triangle, the longest side is called the **hypotenuse**.
It is the side opposite the right angle. The other two sides are
called **legs**. We generally use the letters a and b for the lengths
of the legs and c for the length of the hypotenuse. They are
related as follows.

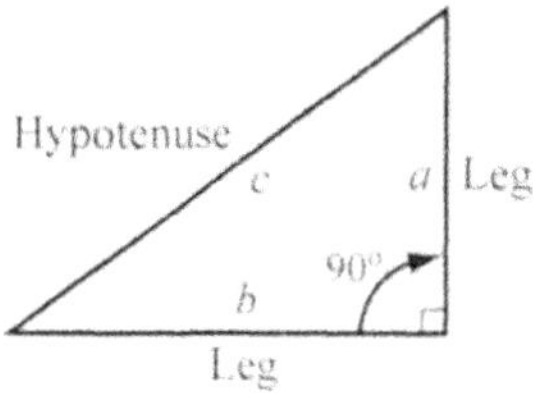

The Pythagorean Theorem

In any right triangle, if a and b are the lengths of the legs and c is the length of the
hypotenuse, then

$$a^2 + b^2 = c^2, \text{ or}$$

$$(\text{Leg})^2 + (\text{Other leg})^2 = (\text{Hypotenuse})^2.$$

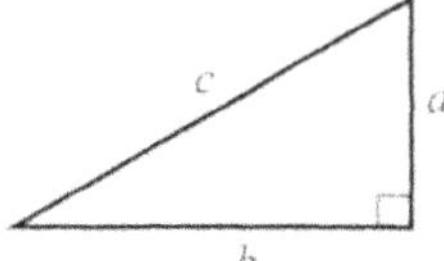

If we know the lengths of any two sides of a right triangle, we can use the Pythagorean
theorem to determine the length of the third side.

Example 1 Find the length of the hypotenuse of this right triangle.

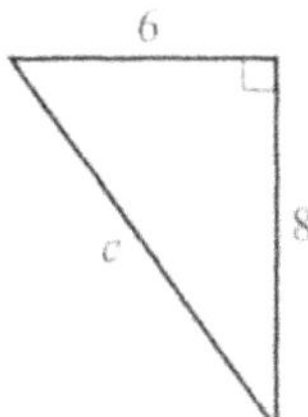

$$a^2 + b^2 = c^2$$
$$6^2 + 8^2 = c^2 \qquad \text{Substituting}$$
$$36 + 64 = c^2$$
$$100 = c^2$$

The solution of this equation is the square root of 100, which is 10.

$$c = \sqrt{100} = 10.$$

Example 2 Find the length b for the right triangle shown. Give an exact answer and an
approximation to three decimal places.

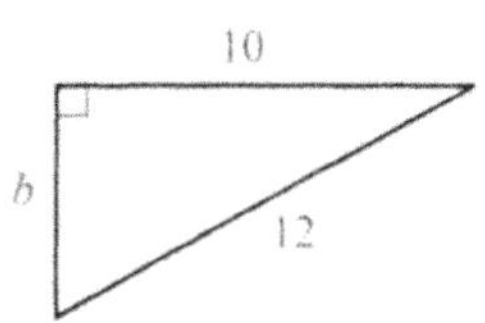

$$a^2 + b^2 = c^2$$
$$10^2 + b^2 = 12^2 \qquad \text{Substituting}$$
$$100 + b^2 = 144.$$
$$100 + b^2 - 100 = 144 - 100 \qquad \text{Subtracting 100 on both sides}$$
$$b^2 = 44$$

Exact answer: $b = \sqrt{44}$

Approximation: $b \approx 6.633.$ Using a calculator

Exercises Find the length of the third side of each right triangle. Give an exact answer and, when appropriate, an approximation to three decimal places.

1. 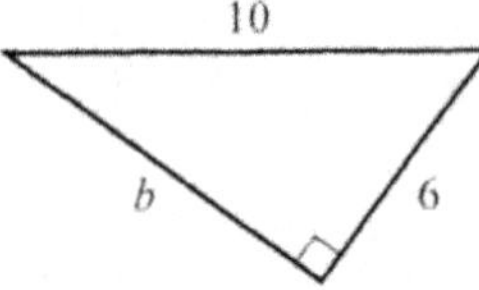

2. 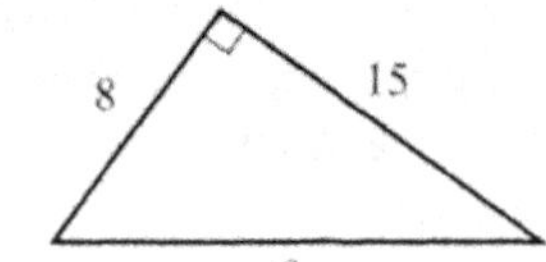

3. 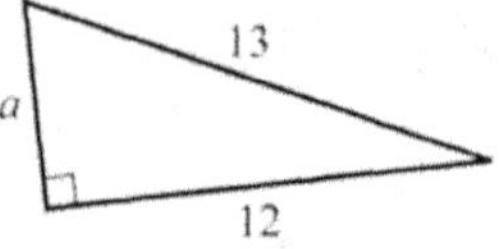

4. 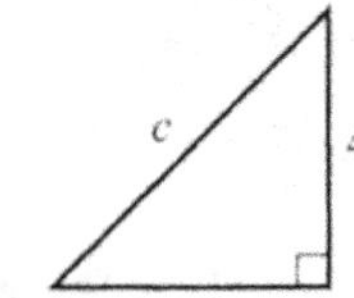

5. 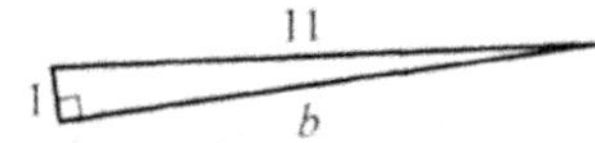

6. 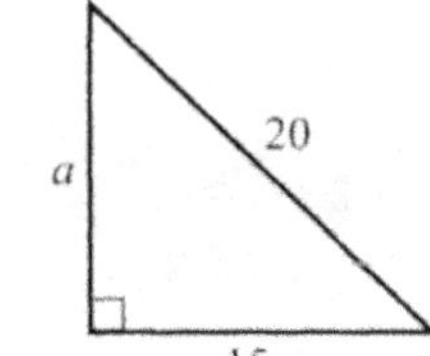

7. 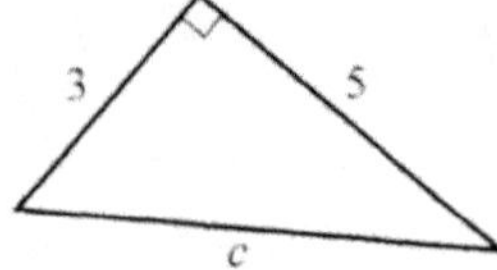

8. 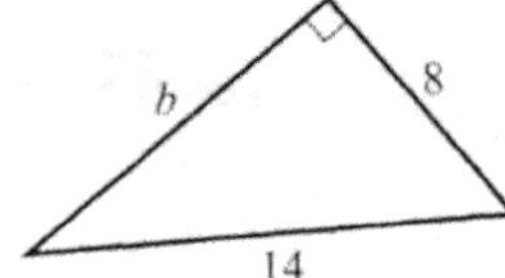

9. 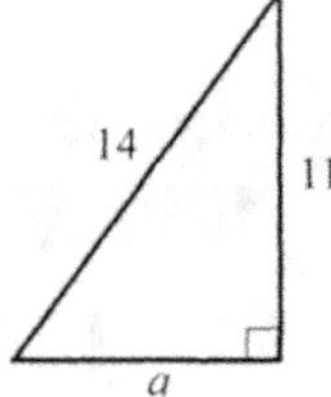

10. 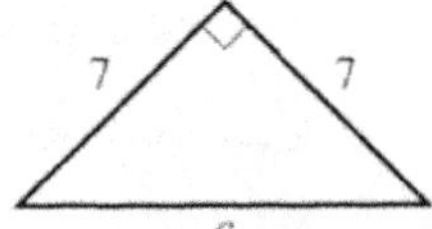

Interactive Preview Worksheet 20: SIMPLIFYING RATIONAL EXPRESSIONS

We simplify rational expressions using the identity property of 1 in reverse. That is, we "remove" factors that are equal to 1. We first factor the numerator and the denominator and then factor the rational expression, so that a factor is equal to 1.

Example 1   Simplify: $\dfrac{120}{320}$.

Factor the numerator and the denominator.	$= \dfrac{40 \cdot 3}{40 \cdot 8}$
Factor the rational expression.	$= \dfrac{40}{40} \cdot \dfrac{3}{8}$
Remove a factor of 1, $\dfrac{40}{40} = 1$.	$= 1 \cdot \dfrac{3}{8} = \dfrac{3}{8}$

Example 2   Simplify: $\dfrac{5x^2}{x}$.

Factor the numerator and the denominator.	$= \dfrac{5x \cdot x}{1 \cdot x}$
Factor the rational expression.	$= \dfrac{5x}{1} \cdot \dfrac{x}{x}$
Remove a factor of 1, $\dfrac{x}{x} = 1$.	$= 5x \cdot 1 = 5x$

Example 3   Simplify: $\dfrac{2x^2 + 4x}{6x^2 + 2x}$.

Factor the numerator and the denominator.	$= \dfrac{2x(x+2)}{2x(3x+1)}$
Factor the rational expression.	$= \dfrac{2x}{2x} \cdot \dfrac{x+2}{3x+1}$
Remove a factor of 1, $\dfrac{2x}{2x} = 1$.	$= 1 \cdot \dfrac{x+2}{3x+1} = \dfrac{x+2}{3x+1}$

Exercises Simplify.

1. $\dfrac{60}{75}$

2. $\dfrac{6w^3}{2w^2}$

3. $\dfrac{4x^2 + 16x}{36x^2 + 4x}$

4. $\dfrac{x^3 + 10x^2}{x^3 - 9x^2}$

5. $\dfrac{y}{9y^7}$

6. $\dfrac{12z^3 - 15z^2}{3z^3 + 45z^2}$

Example 4 Simplify: $\dfrac{x^2+2x}{3x+6}$.

Factor the numerator and the denominator.	$=\dfrac{x(x+2)}{3(x+2)}$
Factor the rational expression.	$=\dfrac{x}{3}\cdot\dfrac{x+2}{x+2}$
Remove a factor of 1, $\dfrac{x+2}{x+2}=1$.	$=\dfrac{x}{3}\cdot 1=\dfrac{x}{3}$

Example 5 Simplify: $\dfrac{q^2-5q-66}{q^2+15q+54}$.

Factor the numerator and the denominator.	$=\dfrac{(q+6)(q-11)}{(q+6)(q+9)}$
Factor the rational expression.	$=\dfrac{q+6}{q+6}\cdot\dfrac{q-11}{q+9}$
Remove a factor of 1, $\dfrac{q+6}{q+6}=1$.	$=\dfrac{q-11}{q+9}$

Example 6 Simplify: $\dfrac{x-3}{3-x}$.

Rewrite $3-x$ as $-(x-3)$.	$=\dfrac{x-3}{-(x-3)}$
Factor the numerator and the denominator.	$=\dfrac{1(x-3)}{-1(x-3)}$
Simplify.	$=-1$

Exercises Simplify.

7. $\dfrac{y^2-49}{y^2+11y+28}$

8. $\dfrac{6x^2-6x}{3x^3-3x^2}$

9. $\dfrac{a^2-a-6}{a^2-11a+24}$

10. $\dfrac{8y+24}{y^2+3y}$

11. $\dfrac{2-t}{t-2}$

12. $\dfrac{3x^2+11x-4}{2x^2+9x+4}$

Interactive Preview Worksheet 21: FINDING THE LCM OF ALGEBRAIC EXPRESSIONS

Equations containing rational expressions are called rational equations. To solve rational equations such as

$$\frac{5}{42} + \frac{7}{6} = \frac{x}{12}, \quad \text{and} \quad \frac{3}{x^2-4} - \frac{1}{5x+10} = \frac{4}{x-2},$$

the first step is to clear the equation of fractions. To do this, multiply all terms on both sides of the equation by the least common multiple, or LCM, of all the denominators. To find the LCM of two or more algebraic expressions, we factor them. Then we use each factor the greatest number of times that it occurs in any one factorization.

Example 1 Find the LCM of 42, 6 and 12.

$$\left.\begin{array}{l} 42 = 2\cdot 3\cdot 7 \\ 6 = 2\cdot 3 \\ 12 = 2\cdot 2\cdot 3 \end{array}\right\} \text{The LCM is } 2\cdot 2\cdot 3\cdot 7, \text{ or } 84.$$

Example 2 Find the LCM of y^2-7y+6 and y^2-5y-6.

$$\left.\begin{array}{l} y^2-7y+6 = (y-6)(y-1) \\ y^2-5y-6 = (y-6)(y+1) \end{array}\right\} \text{The LCM is } (y-6)(y-1)(y+1).$$

Example 3 Find the LCM of $3x$, $6x^2$, and $15x+18$.

$$\left.\begin{array}{l} 3x = 3\cdot x \\ 6x^2 = 2\cdot 3\cdot x\cdot x \\ 15x+18 = 3(5x+6) \end{array}\right\} \text{The LCM is } 2\cdot 3\cdot x\cdot x\cdot(5x+6), \text{ or } 6x^2(5x+6).$$

Example 4 Find the LCM of x^2-4, $5x+10$, and $x-2$.

$$\left.\begin{array}{l} x^2-4 = (x+2)(x-2) \\ 5x+10 = 5(x+2) \\ x-2 = x-2 \end{array}\right\} \text{The LCM is } 5(x+2)(x-2).$$

Example 5 Find the LCM of a^2-9 and a^2-6a+9.

$$\left.\begin{array}{l} a^2-9 = (a+3)(a-3) \\ a^2-6a+9 = (a-3)(a-3) \end{array}\right\} \text{The LCM is } (a+3)(a-3)(a-3), \text{ or } (a+3)(a-3)^2.$$

Example 6 Find the LCM of $4x^5 + 4x^4$ and $8x^4 - 8x^3 - 16x^2$.

$4x^5 + 4x^4 = 4x^4(x+1) = 2 \cdot 2 \cdot x \cdot x \cdot x \cdot x(x+1)$

$8x^4 - 8x^3 - 16x^2 = 8x^2(x^2 - x - 2) = 2 \cdot 2 \cdot 2 \cdot x \cdot x(x-2)(x+1)$

The LCM is $2 \cdot 2 \cdot 2 \cdot x \cdot x \cdot x \cdot x \cdot (x+1)(x-2)$, or $8x^4(x+1)(x-2)$.

Exercises Find the LCM by factoring.

1. $45, 54$

2. $12, 18, 48$

3. $3x^3,\ x,\ 9x^2$

4. $4x - 24,\ x + 6,\ x^2 - 36$

5. $a,\ a - 8,\ a + 8$

6. $y^2 - 7y - 30,\ y^2 + 13y + 30$

7. $x^2 - 4,\ x^2 + 5x + 6$

8. $5t,\ t^2,\ 10t - 40,\ 10t^3$

9. $z^7,\ z^2 - 1,\ z^3 - 2z^2 + z$

10. $a^2 - 4a - 21,\ a^2 - 6a - 7$

11. $30x^2 - 30,\ 105x - 105$

12. $9y^2,\ 9y^3,\ 24y - 27$

13. $9q^7 + 9q^6,\ 6q^6 - 30q^5 - 36q^4$

14. $35c^2,\ c + 7,\ 70c,\ c^2 + 2c - 35$

Interactive Preview Worksheet 22: SIMPLIFYING COMPLEX RATIONAL EXPRESSIONS

A **complex rational expression** is a rational expression that contains rational expressions within its numerator and/or its denominator. Here are some examples:

$$\frac{\dfrac{3}{10}-\dfrac{1}{6}}{\dfrac{7}{8}}, \qquad \frac{\dfrac{1}{t}+6}{\dfrac{1}{t-5}}, \qquad \frac{y-\dfrac{6}{y}}{y+\dfrac{6}{y}}, \qquad \frac{1+\dfrac{1}{x}}{1-\dfrac{1}{x^2}}, \qquad \frac{\dfrac{1}{x+h}-\dfrac{1}{x}}{h}$$

Example 1 Simplify: $\dfrac{\dfrac{3}{10}-\dfrac{1}{6}}{\dfrac{7}{8}}$.

The LCD in the numerator is 30. We multiply by 1. Then subtract in the numerator.

$$=\frac{\dfrac{3}{10}\cdot\dfrac{3}{3}-\dfrac{1}{6}\cdot\dfrac{5}{5}}{\dfrac{7}{8}}=\frac{\dfrac{9}{30}-\dfrac{5}{30}}{\dfrac{7}{8}}=\frac{\dfrac{4}{30}}{\dfrac{7}{8}}$$

Multiply by the reciprocal of the denominator.

$$=\frac{4}{30}\cdot\frac{8}{7}=\frac{4\cdot8}{30\cdot7}=\frac{32}{210}$$

Factor.

$$=\frac{2\cdot16}{2\cdot105}=\frac{2}{2}\cdot\frac{16}{105}$$

Remove a factor of 1.

$$=\frac{16}{105}$$

Example 2 Simplify: $\dfrac{\dfrac{1}{t}+6}{\dfrac{1}{t}-5}$.

The LCD in the numerator is t. The LCD in the denominator is t. We multiply by 1.

$$=\frac{\dfrac{1}{t}+6\cdot\dfrac{t}{t}}{\dfrac{1}{t}-5\cdot\dfrac{t}{t}}=\frac{\dfrac{1}{t}+\dfrac{6t}{t}}{\dfrac{1}{t}-\dfrac{5t}{t}}$$

Add in the numerator and subtract in the denominator.

$$=\frac{\dfrac{1+6t}{t}}{\dfrac{1-5t}{t}}$$

Multiply by the reciprocal of the denominator.

$$=\frac{1+6t}{t}\cdot\frac{t}{1-5t}=\frac{t(1+6t)}{t(1-5t)}$$

Factor.

$$=\frac{t}{t}\cdot\frac{1+6t}{1-5t}$$

Remove a factor of 1.

$$=\frac{1+6t}{1-5t}$$

Exercises Simplify.

1. $\dfrac{\frac{y}{2}}{7} =$

2. $\dfrac{\frac{3}{x}}{\frac{2}{x}} =$

3. $\dfrac{\frac{3}{8} - \frac{1}{12}}{\frac{5}{6}} =$

4. $\dfrac{\frac{2}{3} + \frac{4}{5}}{\frac{3}{4} - \frac{1}{2}} =$

5. $\dfrac{\frac{1}{x} - 9}{\frac{1}{x} + 2} =$

6. $\dfrac{3 - \frac{5}{z}}{4 + \frac{3}{z}} =$

Example 3 Simplify: $\dfrac{y - \frac{6}{y}}{y + \frac{6}{y}}$.

The LCD in the numerator is y. The LCD in the denominator is y. We multiply by 1.	$= \dfrac{\frac{y}{1} \cdot \frac{y}{y} - \frac{6}{y}}{\frac{y}{1} \cdot \frac{y}{y} + \frac{6}{y}} = \dfrac{\frac{y^2}{y} - \frac{6}{y}}{\frac{y^2}{y} + \frac{6}{y}}$
Subtract in the numerator and add in the denominator.	$= \dfrac{\frac{y^2 - 6}{y}}{\frac{y^2 + 6}{y}}$
Multiply by the reciprocal of the denominator.	$= \dfrac{y^2 - 6}{y} \cdot \dfrac{y}{y^2 + 6} = \dfrac{y(y^2 - 6)}{y(y^2 + 6)}$
Factor and remove a factor of 1.	$= \dfrac{y}{y} \cdot \dfrac{y^2 - 6}{y^2 + 6} = \dfrac{y^2 - 6}{y^2 + 6}$

Example 4   Simplify: $\dfrac{1+\dfrac{1}{x}}{1-\dfrac{1}{x^2}}$.

The LCD in the numerator is x. The LCD in the denominator is x^2. We multiply by 1.

$$= \frac{1\cdot\dfrac{x}{x}+\dfrac{1}{x}}{1\cdot\dfrac{x^2}{x^2}-\dfrac{1}{x^2}} = \frac{\dfrac{x}{x}+\dfrac{1}{x}}{\dfrac{x^2}{x^2}-\dfrac{1}{x^2}}$$

Add in the numerator and subtract in the denominator.

$$= \frac{\dfrac{x+1}{x}}{\dfrac{x^2-1}{x^2}}$$

Multiply by the reciprocal of the denominator.

$$= \frac{x+1}{x}\cdot\frac{x^2}{x^2-1} = \frac{x^2(x+1)}{x(x^2-1)}$$

Factor and remove a factor of 1.

$$= \frac{x\cdot x(x+1)}{x(x+1)(x-1)} = \frac{x(x+1)}{x(x+1)}\cdot\frac{x}{x-1} = \frac{x}{x-1}$$

Exercises

7. $\dfrac{p-\dfrac{5}{p}}{p+\dfrac{5}{p}} =$

8. $\dfrac{\dfrac{1}{a}+7a}{\dfrac{1}{a}+a} =$

9. $\dfrac{4-\dfrac{1}{x^2}}{2+\dfrac{1}{x}} =$

10. $\dfrac{5-\dfrac{1}{y}}{25-\dfrac{1}{y^2}} =$

11. $\dfrac{\dfrac{1}{a}+\dfrac{1}{b}}{\dfrac{1}{a^2}-\dfrac{1}{b^2}} =$

Example 5 Simplify: $\dfrac{\dfrac{1}{x+h}-\dfrac{1}{x}}{h}$.

The LCD in the numerator is $x(x+h)$. We multiply by 1.

$$= \frac{\dfrac{1}{x+h}\cdot\dfrac{x}{x}-\dfrac{1}{x}\cdot\dfrac{x+h}{x+h}}{h} = \frac{\dfrac{x}{x(x+h)}-\dfrac{x+h}{x(x+h)}}{h}$$

Subtract in the numerator; $-(x+h)=-x-h$. It helps to rewrite the h in the denominator as $\dfrac{h}{1}$.

$$= \frac{\dfrac{x-(x+h)}{x(x+h)}}{\dfrac{h}{1}} = \frac{\dfrac{x-x-h}{x(x+h)}}{\dfrac{h}{1}} = \frac{\dfrac{-h}{x(x+h)}}{\dfrac{h}{1}}$$

Multiply by the reciprocal of the denominator.

$$= \frac{-h}{x(x+h)}\cdot\frac{1}{h} = \frac{-h}{x(x+h)h}$$

Factor and remove a factor of 1.

$$= \frac{-1}{x(x+h)}\cdot\frac{h}{h} = \frac{-1}{x(x+h)}, \text{ or } -\frac{1}{x(x+h)}$$

Exercises Simplify.

12. $\dfrac{\dfrac{1}{z+h}-\dfrac{1}{z}}{h} =$

13. $\dfrac{\dfrac{2}{x+h}-\dfrac{2}{x}}{h} =$

Interactive Preview Worksheet 23: SOLVING RATIONAL EQUATIONS

Equations containing rational expressions are called **rational equations**. To solve a rational equation, the first step is to *clear the equation of fractions*. To do this, multiply all terms on both sides of the equation by the least common denominator (LCD) of all the rational expressions. The LCD is the least common multiple (LCM) of the denominators.

Example 1   Solve: $\dfrac{1}{4} - \dfrac{5}{6} = \dfrac{1}{a}$.

The LCM of 4, 6, and a is $12a$. Multiply on both sides by the LCD, $12a$.

$$12a\left(\frac{1}{4} - \frac{5}{6}\right) = 12a \cdot \frac{1}{a}$$

Remove parentheses.

$$12a \cdot \frac{1}{4} - 12a \cdot \frac{5}{6} = 12a \cdot \frac{1}{a}$$

Simplify.

$$3a - 10a = 12$$

Collect like terms.

$$-7a = 12$$

Divide by -7.

$$a = -\frac{12}{7}$$

The solution is $-\dfrac{12}{7}$.

Check:

$$\frac{1}{4} - \frac{5}{6} = \frac{1}{a}$$

$$\frac{\dfrac{1}{4} \cdot \dfrac{3}{3} - \dfrac{5}{6} \cdot \dfrac{2}{2}}{} \,\Bigg|\, \dfrac{1}{-\dfrac{12}{7}}$$

$$\frac{3}{12} - \frac{10}{12} \,\Bigg|\, 1 \cdot \left(-\frac{7}{12}\right)$$

$$-\frac{7}{12} \,\Bigg|\, -\frac{7}{12} \quad \text{True}$$

Example 2   Solve: $\dfrac{x-1}{15} - \dfrac{x+2}{10} = 0$.

Multiply on both sides by the LCD, 30.

$$30\left(\frac{x-1}{15} - \frac{x+2}{10}\right) = 30 \cdot 0$$

Remove parentheses.

$$30 \cdot \frac{x-1}{15} - 30 \cdot \frac{x+2}{10} = 30 \cdot 0$$

Simplify.

$$2(x-1) - 3(x+2) = 0$$

$$2x - 2 - 3x - 6 = 0$$

Collect like terms.

$$-x - 8 = 0$$

Add x.

$$-8 = x$$

The solution is -8.

Check:

$$\frac{x-1}{15} - \frac{x+2}{10} = 0$$

$$\frac{-8-1}{15} - \frac{-8+2}{10} \,\Bigg|\, 0$$

$$-\frac{9}{15} - \frac{-6}{10} \,\Bigg|$$

$$-\frac{3}{5} + \frac{3}{5} \,\Bigg|$$

$$0 \,\Big|\, 0 \quad \text{True}$$

Exercises Solve.

1. $\dfrac{5}{8} - \dfrac{2}{5} = \dfrac{1}{x}$

2. $\dfrac{t+3}{4} - \dfrac{t-3}{5} = 2$

Example 3 Solve: $\dfrac{7}{5y-2} = \dfrac{5}{y-1}.$

Multiply on both sides by the LCD, $(5y-2)(y-1)$.	$(5y-2)(y-1)\cdot\dfrac{7}{5y-2} = (5y-2)(y-1)\cdot\dfrac{5}{y-1}$
Simplify.	$7(y-1) = 5(5y-2)$
Remove parentheses.	$7y-7 = 25y-10$
Subtract $7y$ and add 10.	$3 = 18y$
Divide by 18.	$\dfrac{3}{18} = \dfrac{18y}{18}$
Simplify.	$\dfrac{1}{6} = y$

The solution is $\dfrac{1}{6}.$

Exercises Solve.

3. $\dfrac{4}{x-1} = \dfrac{3}{x+2}$

4. $\dfrac{-2}{c+6} = \dfrac{8}{3c-1}$

Example 4   Solve: $\dfrac{2t}{t-1}=\dfrac{7}{t-4}$.

Multiply on both sides by the LCD. $(t-1)(t-4)$.	$(t-1)(t-4)\cdot\dfrac{2t}{t-1}=(t-1)(t-4)\cdot\dfrac{7}{t-4}$
Simplify.	$2t(t-4)=7(t-1)$
Remove parentheses.	$2t^{2}-8t=7t-7$
Subtract $7t$ and add 7 to get 0 on one side.	$2t^{2}-15t+7=0$
Factor.	$(2t-1)(t-7)=0$
Use the principle of zero products.	$2t-1=0 \ \ or \ \ t-7=0$
Solve the two equations separately.	$2t=1 \ \ or \quad t=7$
	$t=\dfrac{1}{2} \ \ or \quad t=7$

Both $\dfrac{1}{2}$ and 7 check and are the solutions.

Example 5   Solve: $x+\dfrac{20}{x}=9$.

Multiply on both sides by the LCD, x.	$x\left(x+\dfrac{20}{x}\right)=x\cdot 9$
Remove parentheses.	$x\cdot x+x\cdot\dfrac{20}{x}=x\cdot 9$
Simplify.	$x^{2}+20=9x$
Subtract $9x$ to get 0 on one side.	$x^{2}-9x+20=0$
Factor.	$(x-5)(x-4)=0$
Use the principle of zero products.	$x-5=0 \ \ or \ \ x-4=0$
Solve the two equations separately.	$x=5 \ \ or \quad x=4$

Checks.

$$x+\dfrac{20}{x}=9 \qquad\qquad x+\dfrac{20}{x}=9$$

$$\begin{array}{c|c} 5+\dfrac{20}{5} & 9 \\ 5+4 & \\ \hline 9 & 9 \end{array}\ \text{True} \qquad \begin{array}{c|c} 4+\dfrac{20}{4} & 9 \\ 4+5 & \\ \hline 9 & 9 \end{array}\ \text{True}$$

The solutions are 5 and 4.

5. $\dfrac{3y}{5y+8} = \dfrac{2}{y-4}$

6. $z + \dfrac{22}{z} = -13$

When we multiply all terms on both sides of an equation by the LCD, the resulting equation might yield numbers that are *not* solutions of the original equation. Thus, we must always check possible solutions in the original equation.

Example 6   Solve: $\dfrac{y^2}{y-6} = \dfrac{36}{y-6}$.

Multiply on both sides by the LCD, $y-6$.
$$(y-6) \cdot \dfrac{y^2}{y-6} = (y-6) \cdot \dfrac{36}{y-6}$$

Simplify.
$$y^2 = 36$$

Use the principle of square roots.
$$y = -6 \;\; or \;\; y = 6$$

Checks.

$$\dfrac{y^2}{y-6} = \dfrac{36}{y-6}$$

$(-6)^2$	36
$-6-6$	$-6-6$
$\dfrac{36}{-12}$	$\dfrac{36}{-12}$

True

$$\dfrac{y^2}{y-6} = \dfrac{36}{y-6}$$

$(6)^2$	36
$6-6$	$6-6$
$\dfrac{36}{0}$	$\dfrac{36}{0}$

Not defined

Since division by 0 is not defined, 6 is *not* a solution. The number -6 checks, so it is the solution.

7. $\dfrac{y^2}{y-9} = \dfrac{81}{y-9}$

8. $\dfrac{8x^2}{x+5} = \dfrac{200}{x+5}$

Interactive Preview Worksheet 24: SOLVING PROPORTIONS

When two pairs of numbers, such as 3, 2 and 6, 4, have the same ratio, we say that they are **proportional**. The equation

$$\frac{3}{2} = \frac{6}{4} \qquad \text{(Note: } 3 \cdot 4 = 2 \cdot 6.\text{)}$$

states that the pairs 3, 2 and 6, 4 are proportional. Such an equation is called a **proportion**.

Two fractions are equal if their **cross products** are equal. One way to solve a proportion, such as $\frac{x}{7} = \frac{15}{35}$, is to use cross products. Then we can divide on both sides to get the variable alone.

Example 1 Solve the proportion $\frac{x}{7} = \frac{15}{35}$.

Equate the cross products.	$x \cdot 35 = 7 \cdot 15$
Divide by 35 on both sides.	$\dfrac{x \cdot 35}{35} = \dfrac{7 \cdot 15}{35}$
Simplify.	$x = \dfrac{7 \cdot 3 \cdot 5}{5 \cdot 7} = \dfrac{7 \cdot 5}{7 \cdot 5} \cdot 3$
	$x = 3$
Check by replacing x with 3.	$\dfrac{3}{7} = \dfrac{15}{35}$
Find cross products.	$3 \cdot 35 = 7 \cdot 15$
True	$105 = 105$
	The solution is 3.

Example 2 Solve the proportion $\frac{8}{y} = \frac{5}{3}$.

Equate the cross products.	$8 \cdot 3 = y \cdot 5$
Divide by 5.	$\dfrac{8 \cdot 3}{5} = \dfrac{y \cdot 5}{5}$
Simplify.	$\dfrac{24}{5} = y$, or
	$4.8 = y$
Check by replacing y with 4.8.	$\dfrac{8}{4.8} = \dfrac{5}{3}$
Find cross products.	$8 \cdot 3 = 4.8(5)$
True	$24 = 24$
	The solution is $\dfrac{24}{5}$, or 4.8.

Example 3 Solve the proportion $\dfrac{\frac{3}{4}}{\frac{9}{20}} = \dfrac{\frac{1}{2}}{h}$.

Equate the cross products.	$\dfrac{3}{4} \cdot h = \dfrac{9}{20} \cdot \dfrac{1}{2}$
Multiply on the right.	$\dfrac{3}{4} \cdot h = \dfrac{9}{40}$
Divide by $\dfrac{3}{4}$.	$\dfrac{\frac{3}{4}h}{\frac{3}{4}} = \dfrac{\frac{9}{40}}{\frac{3}{4}}$
Multiply on the right side by the reciprocal of the divisor.	$h = \dfrac{9}{40} \cdot \dfrac{4}{3} = \dfrac{9 \cdot 4}{40 \cdot 3}$
Factor and simplify.	$h = \dfrac{\cancel{3} \cdot 3 \cdot \cancel{2} \cdot \cancel{2}}{\cancel{2} \cdot \cancel{2} \cdot 2 \cdot 5 \cdot \cancel{3}} = \dfrac{3}{10}$

The solution is $\dfrac{3}{10}$.

Check by replacing h with $\dfrac{3}{10}$:

$$\dfrac{\frac{3}{4}}{\frac{9}{20}} = \dfrac{\frac{1}{2}}{\frac{3}{10}}$$

$$\dfrac{3}{4} \cdot \dfrac{3}{10} = \dfrac{9}{20} \cdot \dfrac{1}{2} \quad \text{Equating cross products}$$

$$\dfrac{9}{40} = \dfrac{9}{40} \quad \text{True}$$

Exercises. Solve using cross products.

1. $\dfrac{x}{8} = \dfrac{9}{6}$

2. $\dfrac{8}{12} = \dfrac{20}{a}$

3. $\dfrac{80}{y} = \dfrac{20}{7}$

4. $\dfrac{1}{8} = \dfrac{7}{t}$

5. $\dfrac{8}{10} = \dfrac{n}{5}$

6. $\dfrac{0.2}{x} = \dfrac{6}{15}$

7. $\dfrac{5}{6} = \dfrac{t}{12}$

8. $\dfrac{6}{x} = \dfrac{18}{15}$

9. $\dfrac{1.2}{4} = \dfrac{x}{9}$

10. $\dfrac{b}{\frac{1}{2}} = \dfrac{\frac{3}{4}}{\frac{2}{9}}$

11. $\dfrac{0.9}{6.3} = \dfrac{w}{0.7}$

12. $\dfrac{5}{\frac{1}{3}} = \dfrac{3}{c}$

Interactive Preview Worksheet 25: SIMPLIFYING RADICAL EXPRESSIONS

For any nonnegative radicands A and B,

$$\sqrt{A} \cdot \sqrt{B} = \sqrt{A \cdot B}$$

The product of square roots is the square root of the product of the radicands. To factor radical expressions, we can use the product rule for radicals in reverse.

$$\sqrt{AB} = \sqrt{A}\sqrt{B}$$

When simplifying a square-root radical expression, if the radicand is not a perfect square, we determine whether it has perfect-square factors. If so, the radicand is then factored and the radical expression simplified using the preceding rule. A square-root radical expression is simplified when its radicand has no factors that are perfect squares.

Example 1   Simplify: $\sqrt{18}$.

Identify a perfect-square factor and factor the radicand.	$= \sqrt{9 \cdot 2}$
Factor into a product of radicals.	$= \sqrt{9} \cdot \sqrt{2}$
Simplify $\sqrt{9}$.	$= 3\sqrt{2}$

Exercises Simplify by factoring.

1. $\sqrt{12}$ 2. $\sqrt{75}$ 3. $\sqrt{80}$ 4. $\sqrt{72}$

5. $\sqrt{363}$ 6. $\sqrt{450}$ 7. $\sqrt{320}$ 8. $\sqrt{600}$

Example 2   Simplify: $\sqrt{48x}$.

Identify a perfect-square factor and factor the radicand.	$= \sqrt{16 \cdot 3 \cdot x}$
Factor into a product of radicals.	$= \sqrt{16} \cdot \sqrt{3x}$
Simplify.	$= 4\sqrt{3x}$

For our work here, we assume that expressions under radicals do no represent the square of a negative number. Thus, absolute-value signs are not necessary.

Example 3   Simplify: $\sqrt{20t^2}$.

Factor the radicand.	$= \sqrt{4 \cdot 5 \cdot t^2}$
Factor into a product of radicals.	$= \sqrt{4} \cdot \sqrt{t^2} \cdot \sqrt{5}$
Simplify.	$= 2t\sqrt{5}$

Example 4 Simplify: $\sqrt{x^2 - 8x + 16}$.

$$\text{Factor the radicand.} \quad = \sqrt{(x-4)^2}$$
$$\text{Simplify.} \quad = x - 4$$

Example 5 Simplify: $\sqrt{a^{10}}$.

$$\text{Factor the radicand; note that } a^{10} = a^5 \cdot a^5. \quad = \sqrt{(a^5)^2}$$
$$\text{Simplify.} \quad = a^5$$

Example 6 Simplify: $\sqrt{24x^{17}}$.

$$\text{Factor the radicand; note that } x^{17} = x^{16} \cdot x. \quad = \sqrt{4 \cdot 6 \cdot x^{16} \cdot x}$$
$$\text{Factor into a product of radicals. Note that } x^{16} = x^8 \cdot x^8. \quad = \sqrt{4} \cdot \sqrt{(x^8)^2} \cdot \sqrt{6x}$$
$$\text{Simplify.} \quad = 2x^8 \sqrt{6x}$$

Exercises Simplify by factoring.

9. $\sqrt{25x}$

10. $\sqrt{y^{22}}$

11. $\sqrt{36x^2}$

12. $\sqrt{a^2 + 2a + 1}$

13. $\sqrt{84t^4}$

14. $\sqrt{x^2 - 10x + 25}$

15. $\sqrt{w^9}$

16. $\sqrt{32x^2}$

17. $\sqrt{x^2 - 24x + 144}$

18. $\sqrt{3y^2 + 6y + 3}$

19. $\sqrt{1400b^6}$

20. $\sqrt{81x^7}$

Product Rule for Radicals

For any nonnegative radicands A and B,

$$\sqrt{A}\cdot\sqrt{B}=\sqrt{A\cdot B}.$$

(The product of square roots is the square root of the product of the radicands.)

Example 1 Multiply and simplify: $\sqrt{2}\sqrt{14}$.

Multiply radicands.	$=\sqrt{2\cdot14}$
Factor.	$=\sqrt{2\cdot2\cdot7}$
Look for pairs of factors.	$=\sqrt{2\cdot2}\sqrt{7}$
Simplify.	$=2\sqrt{7}$

Example 2 Multiply and simplify: $\sqrt{45}\sqrt{15}$.

Multiply radicands.	$=\sqrt{45\cdot15}$
Factor.	$=\sqrt{5\cdot3\cdot3\cdot3\cdot5}$
Look for pairs of factors.	$=\sqrt{5\cdot5}\sqrt{3\cdot3}\sqrt{3}$
Simplify	$=5\cdot3\cdot\sqrt{3}$
	$=15\sqrt{3}$

In this course, we will assume that no radicands are formed by raising negative quantities to even powers. Thus, absolute value signs are not necessary.

Example 3 Multiply and simplify: $\sqrt{3}\sqrt{2y+5}$.

Multiply radicands. We cannot factor further.	$=\sqrt{3(2y+5)}$
Simplify.	$=\sqrt{6y+15}$

Example 4 Multiply and simplify: $\sqrt{6t}\sqrt{30t}$

Multiply radicands.	$=\sqrt{6t\cdot30t}$
Factor.	$=\sqrt{2\cdot3\cdot t\cdot2\cdot3\cdot5\cdot t}$
Look for pairs of factors.	$=\sqrt{2\cdot2}\sqrt{3\cdot3}\sqrt{t\cdot t}\sqrt{5}$
	$=2\cdot3\cdot t\sqrt{5}$
Simplify.	$=6t\sqrt{5}$

Example 5 Multiply and simplify: $\sqrt{x-7}\sqrt{x-7}$.

$$\text{Multiply radicands.} \quad = \sqrt{(x-7)^2}$$
$$\text{Simplify.} \quad = x-7$$

Exercises Multiply and simplify.

1. $\sqrt{3}\sqrt{18}$

2. $\sqrt{5}\sqrt{60}$

3. $\sqrt{15}\sqrt{90}$

4. $\sqrt{18}\sqrt{14x}$

5. $\sqrt{12}\sqrt{18a}$

6. $\sqrt{13}\sqrt{13}$

7. $\sqrt{23}\sqrt{23y}$

8. $\sqrt{24w}\sqrt{40w}$

9. $\sqrt{2t}\sqrt{2t}$

10. $\sqrt{z+9}\sqrt{z+9}$

11. $\sqrt{7}\sqrt{4x+1}$

12. $\sqrt{125}\sqrt{200}$

13. $\sqrt{27x}\sqrt{6x}$

14. $\sqrt{x^5}\sqrt{x^8}$

15. $\sqrt{y^{12}}\sqrt{y^4}$

Interactive Preview Worksheet 27: RATIONALIZING DENOMINATORS

The expressions $\sqrt{\dfrac{2}{7}}$ and $\dfrac{\sqrt{14}}{7}$ are equivalent.

The expressions $\dfrac{1}{\sqrt{5}}$ and $\dfrac{\sqrt{5}}{5}$ are equivalent.

In each example, the second expression does not have a radical expression in the denominator. The procedure for finding such an expression is called **rationalizing the denominator**. We carry this out by multiplying by 1.

Example 1 Rationalize the denominator: $\sqrt{\dfrac{10}{3}}$.

We multiply by 1, using $\dfrac{\sqrt{3}}{\sqrt{3}}$. We do this so the denominator of the radicand is a perfect square.

$$\sqrt{\frac{10}{3}} = \frac{\sqrt{10}}{\sqrt{3}} \cdot \frac{\sqrt{3}}{\sqrt{3}} = \frac{\sqrt{10} \cdot \sqrt{3}}{\sqrt{3} \cdot \sqrt{3}} = \frac{\sqrt{30}}{\sqrt{3^2}} \quad \leftarrow \text{The radicand is a perfect square.}$$

$$= \frac{\sqrt{30}}{3}$$

Example 2 Rationalize the denominator: $\dfrac{1}{\sqrt{7}}$.

We multiply by 1, using $\dfrac{\sqrt{7}}{\sqrt{7}}$. We do this so the radicand of the denominator is a perfect square.

$$\frac{1}{\sqrt{7}} = \frac{1}{\sqrt{7}} \cdot \frac{\sqrt{7}}{\sqrt{7}} = \frac{1 \cdot \sqrt{7}}{\sqrt{7} \cdot \sqrt{7}} = \frac{\sqrt{7}}{\sqrt{7^2}} \quad \leftarrow \text{The radicand is a perfect square.}$$

$$= \frac{\sqrt{7}}{7}$$

Example 3 Rationalize the denominator: $\dfrac{5}{\sqrt{21}}$.

$$\frac{5}{\sqrt{21}} = \frac{5}{\sqrt{21}} \cdot \frac{\sqrt{21}}{\sqrt{21}} = \frac{5\sqrt{21}}{\sqrt{21^2}} = \frac{5\sqrt{21}}{21}$$

Example 4 Rationalize the denominator: $\dfrac{4\sqrt{5}}{9\sqrt{2}}$.

$$\frac{4\sqrt{5}}{9\sqrt{2}} = \frac{4\sqrt{5}}{9\sqrt{2}} \cdot \frac{\sqrt{2}}{\sqrt{2}} = \frac{4\sqrt{10}}{9\sqrt{2^2}} = \frac{4\sqrt{10}}{9 \cdot 2} = \frac{4\sqrt{10}}{18} = \frac{2\sqrt{10}}{9}$$

Example 5 Rationalize the denominator: $\sqrt[3]{\dfrac{7}{25}}$.

To get a perfect cube in the denominator, we consider the index 3 and the factors of the denominator.

$$\sqrt[3]{\frac{7}{25}} = \frac{\sqrt[3]{7}}{\sqrt[3]{25}} = \frac{\sqrt[3]{7}}{\sqrt[3]{5 \cdot 5}}$$

We have 2 factors of 5, and we need 3 factors of 5. We achieve this by multiplying by 1, using $\dfrac{\sqrt[3]{5}}{\sqrt[3]{5}}$.

$$\frac{\sqrt[3]{7}}{\sqrt[3]{5 \cdot 5}} = \frac{\sqrt[3]{7}}{\sqrt[3]{5 \cdot 5}} \cdot \frac{\sqrt[3]{5}}{\sqrt[3]{5}} = \frac{\sqrt[3]{35}}{\sqrt[3]{5^3}} \quad \leftarrow \text{The radicand is a perfect cube.}$$

$$= \frac{\sqrt[3]{35}}{5}$$

Exercises Rationalize the denominator.

1. $\sqrt{\dfrac{13}{5}}$

2. $\dfrac{1}{\sqrt{5}}$

3. $\dfrac{2}{\sqrt{3}}$

4. $\dfrac{2\sqrt{13}}{9\sqrt{10}}$

5. $\sqrt[3]{\dfrac{2}{9}}$

6. $\dfrac{1}{\sqrt{13}}$

7. $\dfrac{3\sqrt{2}}{8\sqrt{11}}$

8. $\dfrac{7}{\sqrt{2}}$

9. $\dfrac{1}{\sqrt{21}}$

10. $\sqrt[4]{\dfrac{5}{8}}$

Interactive Preview Worksheet 28: CHECKING SOLUTIONS OF RADICAL EQUATIONS

A **radical equation** has a variable in one or more radicands. For example,

$$\sqrt{2x+3}=4 \quad \text{and} \quad \sqrt{x+2}-\sqrt{x+3}=8$$

are radical equations. We use the principle of powers to solve radical equations.

The Principle of Powers

For any natural number n, if an equation $a=b$ is true, then $a^n=b^n$ is true.

For an even integer n, if an equation $a^n=b^n$ is true, it *might not* be true that $a=b$. For example, $5^2=(-5)^2$, but $5\neq-5$. Thus, we must check the possible solutions of radical equations containing even roots.

Example 1 Using the principle of powers, we find that the *possible* solution of $\sqrt{x}+6=3$ is 9. Determine if this is the solution.

Check: Write the equation.
Substitute 9 for x.
Simplify.

$$
\begin{array}{c|c}
\multicolumn{2}{c}{\sqrt{x}+6=3} \\
\hline
\sqrt{9}+6 & 3 \\
3+6 & \\
9 & 3 \quad \text{False}
\end{array}
$$

We get a false equation, $9=3$, so 9 is not the solution. This equation has no solution.

Example 2 Using the principle of powers, we find that the *possible* solutions of $x=\sqrt{x+7}+5$ are 2 and 9. Determine if these are the solutions.

For 2:

Check: Write the equation.
Substitute 2 for x.
Simplify.

$$
\begin{array}{c|c}
\multicolumn{2}{c}{x=\sqrt{x+7}+5} \\
\hline
2 & \sqrt{2+7}+5 \\
& \sqrt{9}+5 \\
& 3+5 \\
2 & 8 \qquad \text{False}
\end{array}
$$

We get a false equation, $2=8$, so 2 is not a solution.

Next we will check 9.

For 9:

Check: Write the equation.

Substitute 9 for x.

Simplify.

$$\begin{array}{c|l}
 & x = \sqrt{x+7}+5 \\ \hline
9 & \sqrt{9+7}+5 \\
 & \sqrt{16}+5 \\
 & 4+5 \\
9 & 9 \qquad \text{True}
\end{array}$$

We get a true equation, $9 = 9$, so 9 is a solution.

The solution of $x = \sqrt{x+7}+5$ is 9.

Example 3 Using the principle of powers, we find that the *possible* solutions of $\sqrt{2m-3}+2 = \sqrt{m+7}$ are 2 and 42. Determine if these are the solutions.

For 2:

Check: Write the equation.

Substitute 2 for m.

Simplify.

$$\begin{array}{l|l}
\multicolumn{2}{c}{\sqrt{2m-3}+2 = \sqrt{m+7}} \\ \hline
\sqrt{2\cdot 2-3}+2 & \sqrt{2+7} \\
\sqrt{4-3}+2 & \sqrt{9} \\
\sqrt{1}+2 & 3 \\
1+2 & \\
3 & 3 \qquad \text{True}
\end{array}$$

We get a true equation, $3 = 3$, so 2 is a solution.

For 42:

Check: Write the equation.

Substitute 42 for m.

Simplify.

$$\begin{array}{l|l}
\multicolumn{2}{c}{\sqrt{2m-3}+2 = \sqrt{m+7}} \\ \hline
\sqrt{2\cdot 42-3}+2 & \sqrt{42+7} \\
\sqrt{84-3}+2 & \sqrt{49} \\
\sqrt{81}+2 & 7 \\
9+2 & \\
11 & 7 \qquad \text{False}
\end{array}$$

We get a false equation, $11 = 7$, so 42 is not a solution.

The number 2 checks, but 42 does not, so the solution is 2.

Exercises For each equation, the possible solutions that are found using the principle of powers are given. Determine if each number is a solution.

1. $\sqrt{y} = 5$; 25

2. $\sqrt{x} + 4 = 2$; 4

3. $\sqrt{2x - 1} = 5$; 13

4. $\sqrt{6z + 3} = \sqrt{4z + 9}$; 3

5. $\sqrt{y + 2} + \sqrt{3y + 4} = 2$; -1, 7

6. $x - 7 = 2\sqrt{x+1}$; 3, 15

7. $\sqrt{2y+7} = y+2$; 1, −3

8. $\sqrt{x+2} - \sqrt{2x+2} + 1 = 0$; 7, −1

9. $\sqrt{2z-5} = 1 + \sqrt{z-3}$; 3, 7

Interactive Preview Worksheet 29: GRAPHING PIECEWISE FUNCTIONS

When a function is defined by different equations for various parts of its domain, it is said to be defined **piecewise**.

Example 1 Graph the function

$$f(x) = \begin{cases} \dfrac{1}{2}x, & \text{if } x \le -1 \qquad \text{Equation 1} \\ x+3, & \text{if } x > -1 \quad\ \text{Equation 2} \end{cases}$$

This function is defined using two different equations: $f(x) = \dfrac{1}{2}x$ and $f(x) = x+3$. We graph each equation separately. Let's visualize the domain of each.

Domain

Equation 1: $(-\infty, -1]$

Equation 2: $(-1, \infty)$

Equation 1: Choose x-values less than or equal to -1.

If $x = -1$, $y = \dfrac{1}{2}(-1) = -\dfrac{1}{2}$.

If $x = -2$, $y = \dfrac{1}{2}(-2) = -1$.

If $x = -4$, $y = \dfrac{1}{2}(-4) = -2$.

$x \le -1$	$f(x) = \dfrac{1}{2}x$
-1	$-\dfrac{1}{2}$
-2	-1
-4	-2

Equation 2: Choose x-values greater than -1.

If $x = 0$, $y = 0 + 3 = 3$.

If $x = 1$, $y = 1 + 3 = 4$.

If $x = 2$, $y = 2 + 3 = 5$.

$x > -1$	$f(x) = x+3$
0	3
1	4
2	5

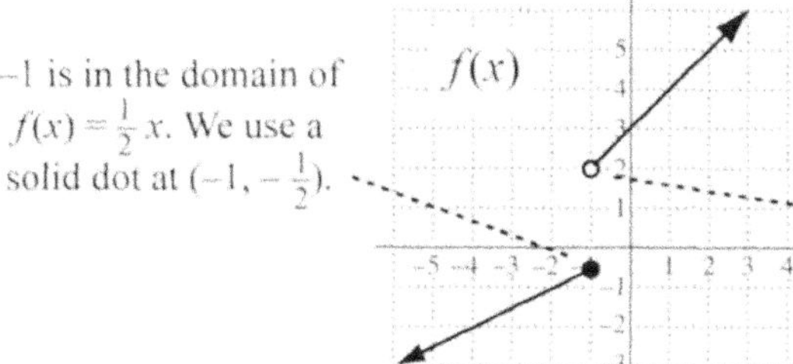

-1 is in the domain of $f(x) = \dfrac{1}{2}x$. We use a solid dot at $\left(-1, -\dfrac{1}{2}\right)$.

For $f(x) = x + 3$, if $x = -1$, $f(-1) = -1 + 3 = 2$. But -1 is **not** in the domain of $f(x) = x + 3$. Thus we use an open circle at $(-1, 2)$.

Example 2 Graph the function

$$g(x) = \begin{cases} x^2, & \text{if } x < 0 & \quad \text{Equation 1} \\ -2, & \text{if } 0 \le x < 3 & \quad \text{Equation 2} \\ 4 - x, & \text{if } x \ge 3 & \quad \text{Equation 3} \end{cases}$$

Domain

Equation 1: $(-\infty, 0)$

Equation 2: $[0, 3)$

Equation 2: $[3, \infty)$

Equation 1

$x < 0$	$g(x) = x^2$
$-\dfrac{1}{2}$	$\dfrac{1}{4}$
-1	1
-2	4

Equation 2

$0 \le x < 3$	$g(x) = -2$
0	-2
1	-2
2	-2

- For $g(x) = x^2$, if $x = 0$, $g(0) = 0^2 = 0$. But 0 **is not** in the domain of $g(x) = x^2$. Thus, we use an open circle at $(0, 0)$.

- 0 **is** in the domain of $g(x) = -2$. We use a solid dot at $(0, -2)$.

- For $g(x) = -2$, if $x = 3$, $g(3) = -2$. But 3 **is not** in the domain of $g(x) = -2$. Thus, we use an open circle at $(3, -2)$.

Equation 3

$x \ge 3$	$g(x) = 4 - x$
3	1
4	0
5	-1

- 3 **is** in the domain of $g(x) = 4 - x$. We use a solid dot at $(3, 1)$.

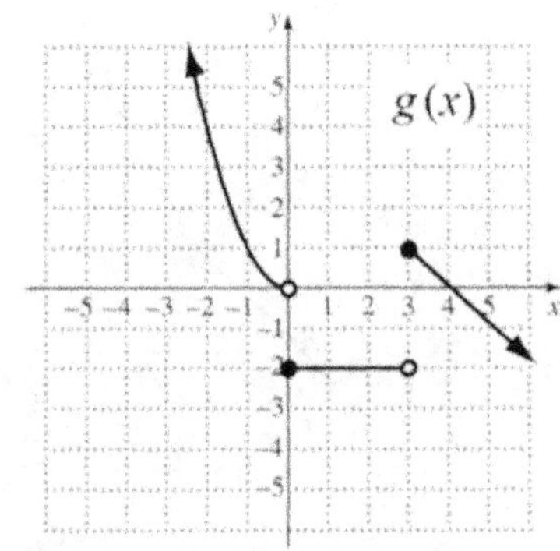

Exercises Graph the piecewise function. In Exercises 1 and 2, fill in the blanks at key steps in the graphing process. In Exercises 3-6, you need to show only the graph.

1. $f(x) = \begin{cases} 3x, & \text{if } x < 2 \quad (1) \\ x-1, & \text{if } x \geq 2 \quad (2) \end{cases}$

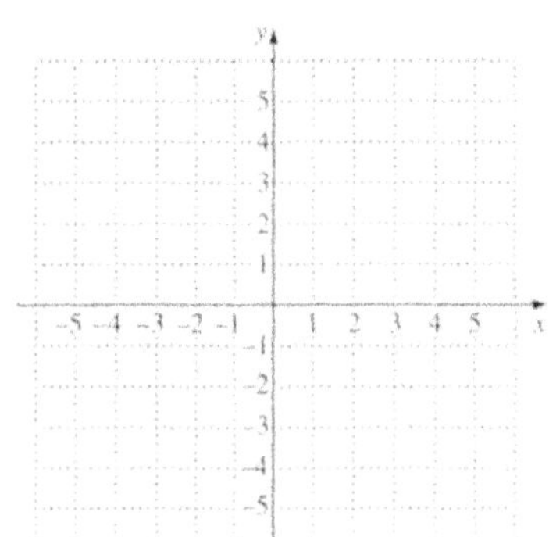

Domain of Equation 1: $\left(\boxed{}, 2 \right)$

Domain of Equation 2: $\left[\boxed{}, \infty \right)$

Equation 1

$x < 2$	$f(x) = 3x$
1	$\boxed{}$
0	$\boxed{}$
-1	$\boxed{}$

Equation 2

$x \geq 2$	$f(x) = x - 1$
2	$\boxed{}$
4	$\boxed{}$
5	$\boxed{}$

- If $f(x) = 3x$ and $x = 2$, $f(2) = 3 \cdot \boxed{} = 6$. Since 2 ________ in the domain of
  $\underset{\text{is / is not}}{}$
  $f(x) = 3x$, we use a(an) ________ at $\left(2, \boxed{}\right)$.
  $\underset{\text{solid dot / open circle}}{}$

- If $f(x) = x - 1$ and $x = 2$, $f(2) = \boxed{} - 1 = 1$. Since 2 ________ in the domain
  $\underset{\text{is / is not}}{}$
  of $f(x) = x - 1$ we use a(an) ________ at $\left(2, \boxed{}\right)$.
  $\underset{\text{solid dot / open circle}}{}$

2. $g(x) = \begin{cases} -x+1, & \text{if } x < -1 \quad (1) \\ -3, & \text{if } -1 \leq x < 4 \quad (2) \\ 2x-4, & \text{if } x \geq 4 \quad (3) \end{cases}$

Domain of Equation 1: $\left(-\infty, \boxed{} \right)$

Domain of Equation 2: $\left[\boxed{}, \boxed{} \right)$

Domain of Equation 3: $\left[4, \boxed{} \right)$

Equation 1

$x < -1$	$g(x) = -x + 1$
-2	$\boxed{}$
-3	$\boxed{}$
-4	$\boxed{}$

Equation 2

$-1 \leq x < 4$	$g(x) = -3$
-1	$\boxed{}$
1	$\boxed{}$
3	$\boxed{}$

Equation 3

$x \geq 4$	$g(x) = 2x - 4$
4	$\boxed{}$
5	$\boxed{}$
6	$\boxed{}$

- If $g(x) = -x + 1$ and $x = -1$, $g(-1) = -\left(\boxed{}\right) + 1 = 2$.

  Since -1 ________ in the domain of $g(x) = -x + 1$,

  <u>is / is not</u>

  we use a(an) ________________ at $\left(-1, \boxed{}\right)$.

  <u>solid dot / open circle</u>

- If $g(x) = -3$ and $x = -1$, $g(-1) = -3$. Since -1

  ________ in the domain of $g(x) = -3$ we use a(an)

  <u>is / is not</u>

  ________________ at $\left(-1, \boxed{}\right)$.

  <u>solid dot / open circle</u>

- If $g(x) = -3$ and $x = 4$, $g(4) = -3$. Since 4

  ________ in the domain of $g(x) = -3$ we use a(an)

  <u>is / is not</u>

  ________________ at $\left(4, \boxed{}\right)$.

  <u>solid dot / open circle</u>

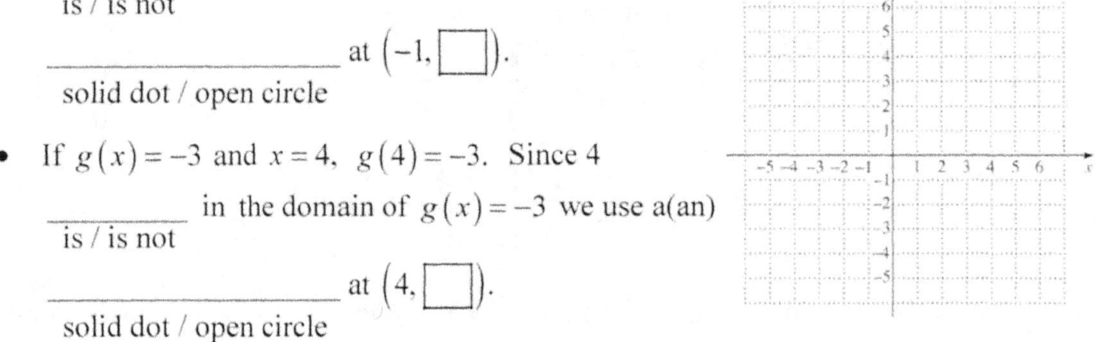

- If $g(x) = 2x - 4$ and $x = 4$, $g(4) = 2 \cdot \boxed{} - 4 = 4$.

  Since 4 ________ in the domain of $g(x) = 2x - 4$ we

  <u>is / is not</u>

  use a(an) ________________ at $\left(4, \boxed{}\right)$.

  <u>solid dot / open circle</u>

3. $h(x) = \begin{cases} 4, & \text{if } x \le 0 \quad (1) \\ 2 - x, & \text{if } x > 0 \quad (2) \end{cases}$

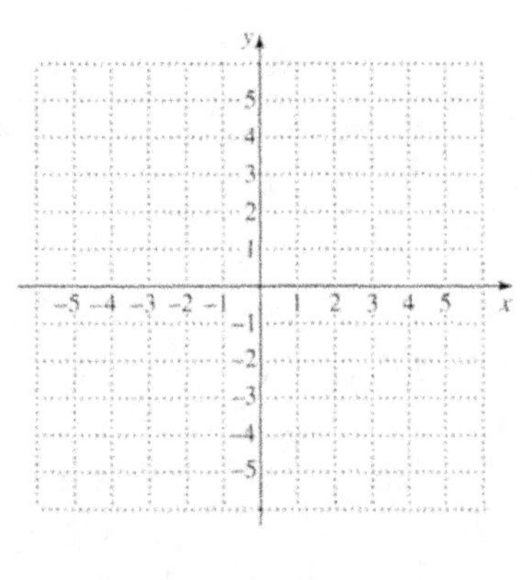

Domain of Equation 1: ________

Domain of Equation 2: ________

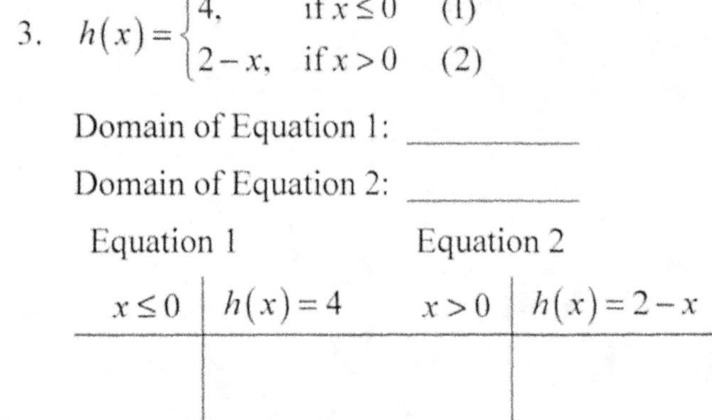

Equation 1		Equation 2	
$x \le 0$	$h(x) = 4$	$x > 0$	$h(x) = 2 - x$

4. $f(x) = \begin{cases} x^2, & \text{if } x \le -1 \quad (1) \\ 2x - 3, & \text{if } x > -1 \quad (2) \end{cases}$

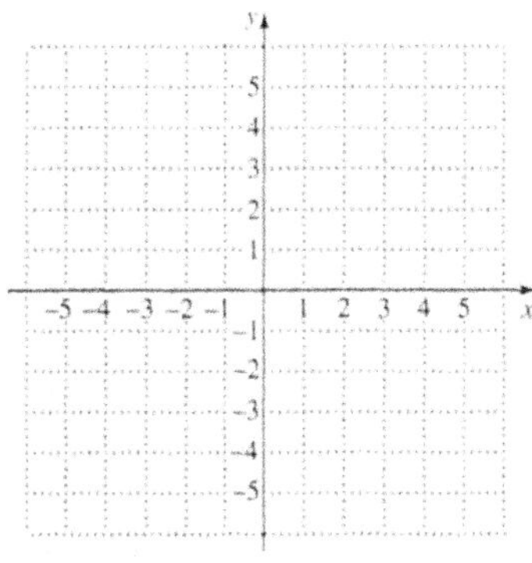

Domain of Equation 1: _____________

Domain of Equation 2: _____________

Equation 1		Equation 2	
$x \le -1$	$f(x) = x^2$	$x > -1$	$f(x) = 2x - 3$

5. $t(x) = \begin{cases} |x|, & \text{if } x < -2 \quad (1) \\ \dfrac{1}{2}x - 1, & \text{if } x \ge -2 \quad (2) \end{cases}$

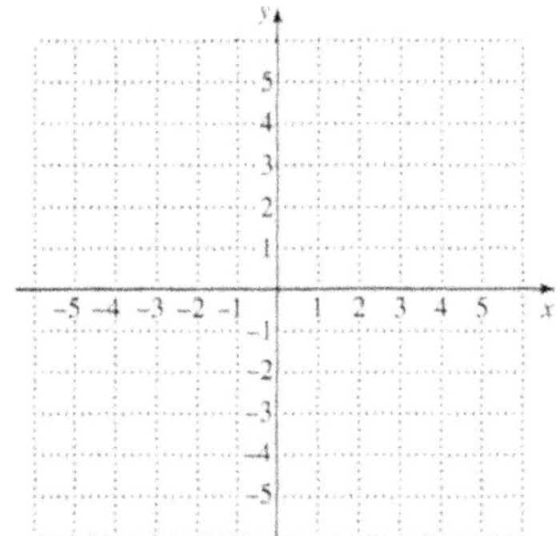

Domain of Equation 1: _____________

Domain of Equation 2: _____________

Equation 1		Equation 2	
x	$t(x)$	x	$t(x)$

6. $g(x) = \begin{cases} -x, & \text{if } x < -5 \quad (1) \\ x+3, & \text{if } -5 \leq x < 2 \quad (2) \\ \frac{1}{4}x^2 - 6, & \text{if } x \geq 2 \quad (3) \end{cases}$

Domain of Equation 1: ___________

Domain of Equation 2: ___________

Domain of Equation 3: ___________

Equation 1		Equation 2		Equation 3	
x	$g(x)$	x	$g(x)$	x	$g(x)$

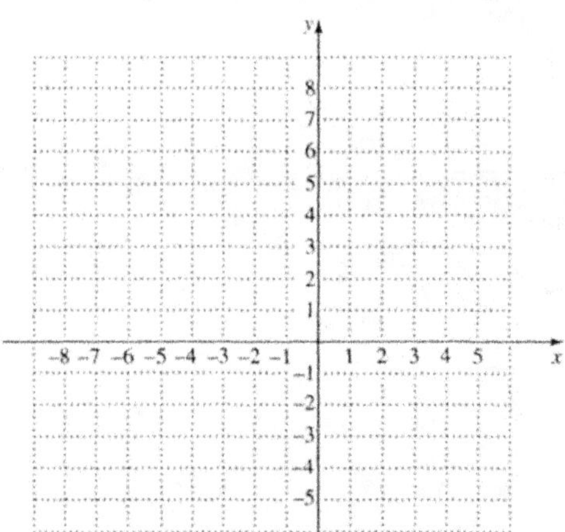

Notes:

Interactive Preview Worksheet 30: TRANSFORMATIONS

The squaring function $f(x) = x^2$ is shown below. We can create graphs of additional functions by

- shifting the graph of $f(x) = x^2$ horizontally or vertically,
- reflecting the graph of $f(x) = x^2$ across an axis, and
- stretching or shrinking the graph of $f(x) = x^2$.

The new graphs are **transformations** of $f(x) = x^2$.

$f(x) = x^2$

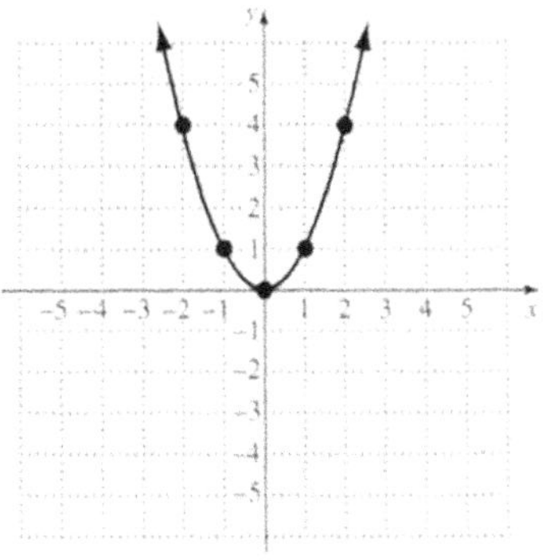

Vertical Translations

$g(x) = x^2 - 3$

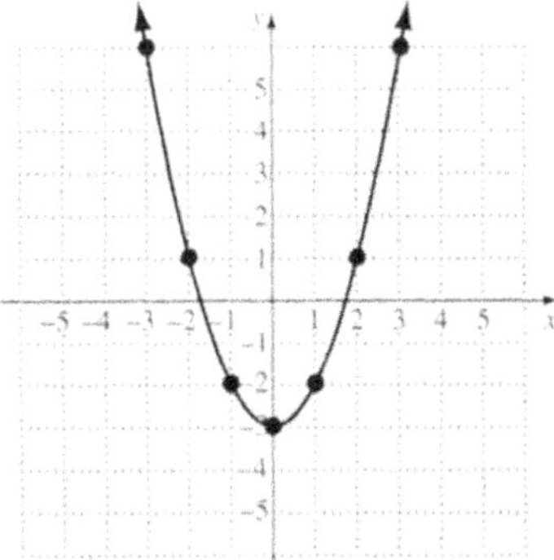

$g(x) = x^2 + 2$

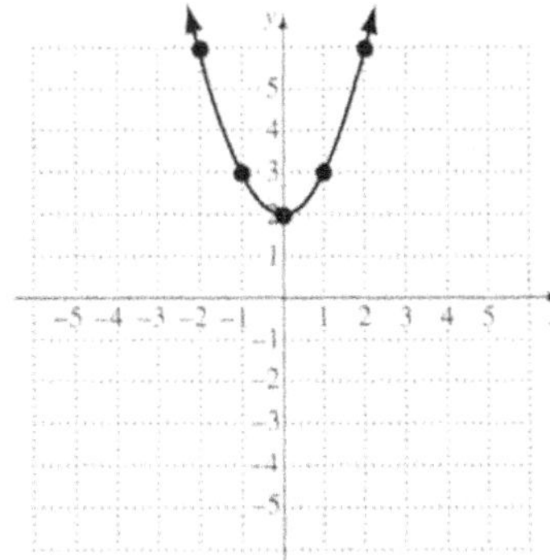

For the graph of $g(x) = x^2 - 3$, the graph of $f(x) = x^2$ is *shifted down* 3 units. We say that $g(x) = f(x) - 3$.

For the graph of $g(x) = x^2 + 2$, the graph of $f(x) = x^2$ is *shifted up* 2 units. We say that $g(x) = f(x) + 2$.

Horizontal Translations

$$g(x) = (x-2)^2$$

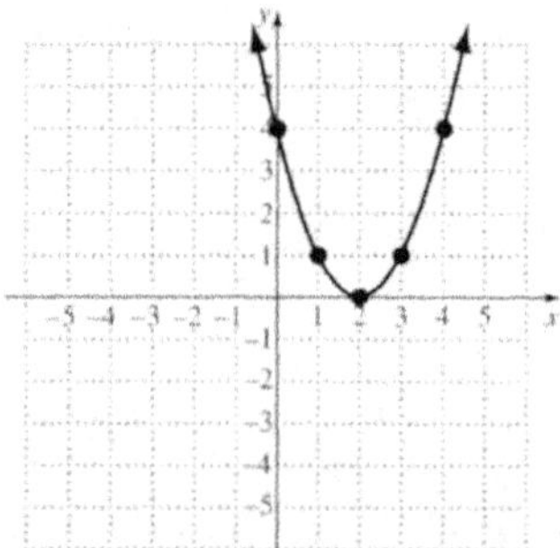

$$g(x) = (x+3)^2$$

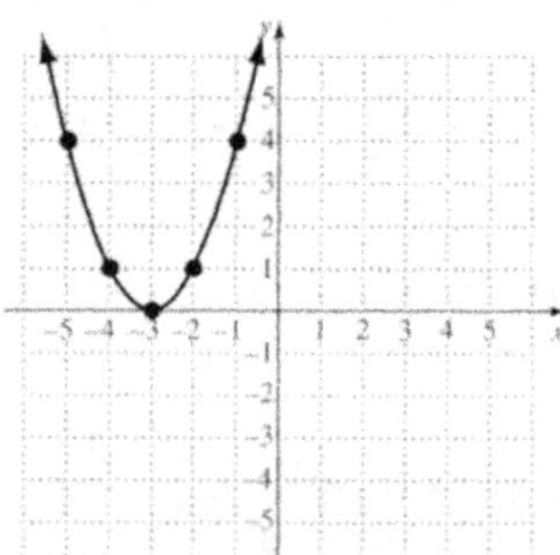

For the graph of $g(x) = (x-2)^2$, the graph of $f(x) = x^2$ is *shifted right* 2 units. We say that $g(x) = f(x-2)$.

For the graph of $g(x) = (x+3)^2$, the graph of $f(x) = x^2$ is *shifted left* 3 units. We say that $g(x) = f(x+3)$.

Vertical Stretching

$$g(x) = 2x^2$$

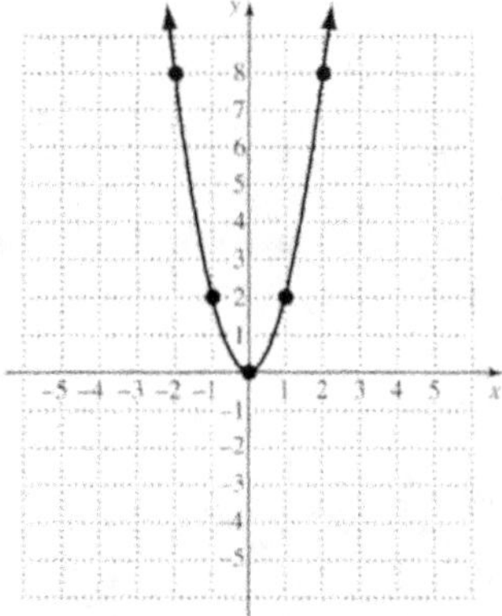

Vertical Shrinking

$$g(x) = \frac{1}{2}x^2$$

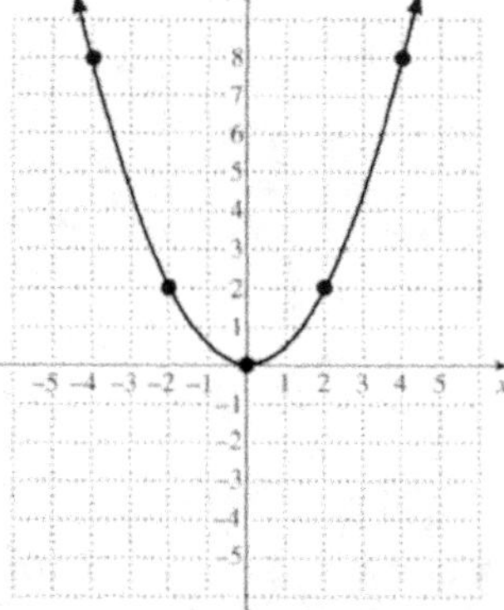

The graph of $g(x) = 2x^2$ is a *vertical stretching* of $f(x) = x^2$ by a factor of 2.

We say that $g(x) = 2f(x)$.

The graph of $g(x) = \frac{1}{2}x^2$ is a *vertical shrinking* of $f(x) = x^2$ by a factor of $\frac{1}{2}$.

We say that $g(x) = \frac{1}{2}f(x)$.

Horizontal Stretching

$$g(x) = \left(\frac{1}{2}x\right)^2$$

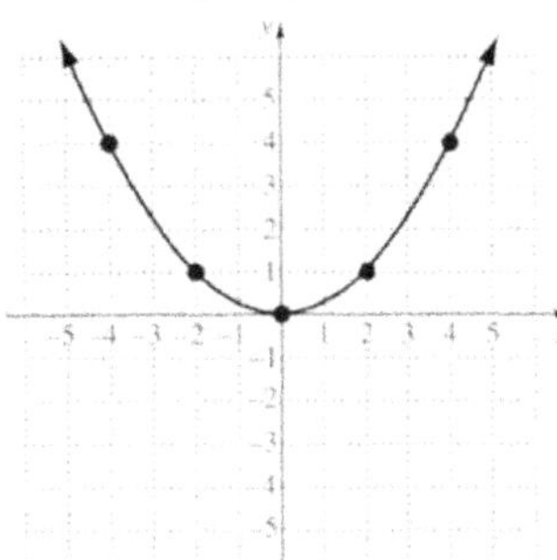

The graph of $g(x) = \left(\frac{1}{2}x\right)^2$ is
a *horizontal stretching* of $f(x) = x^2$.
The points of $g(x)$ can be found by
dividing the x-coordinates of the
points of $f(x)$ by $\frac{1}{2}$ (which is the
same as multiplying by 2). We
have $g(x) = f\left(\frac{1}{2}x\right)$.

Horizontal Shrinking

$$g(x) = (2x)^2$$

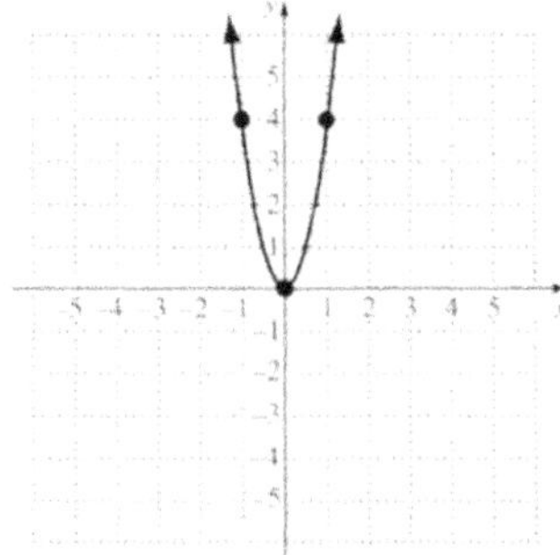

The graph of $g(x) = (2x)^2$ is
a *horizontal shrinking* of $f(x) = x^2$.
The points of $g(x)$ can be found by
dividing the x-coordinates of the
points of $f(x)$ by 2. We have
$$g(x) = f(2x).$$

Reflection Across the x-Axis

$$g(x) = -x^2$$

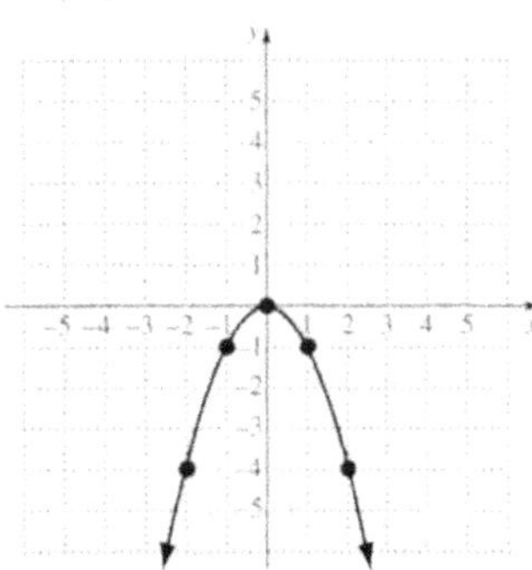

The graph of $g(x) = -x^2$ is a
reflection of the graph of $f(x)$
across the x-axis. We have
$$g(x) = -f(x).$$

Reflection Across the y-Axis

$$g(x) = (-x)^2$$

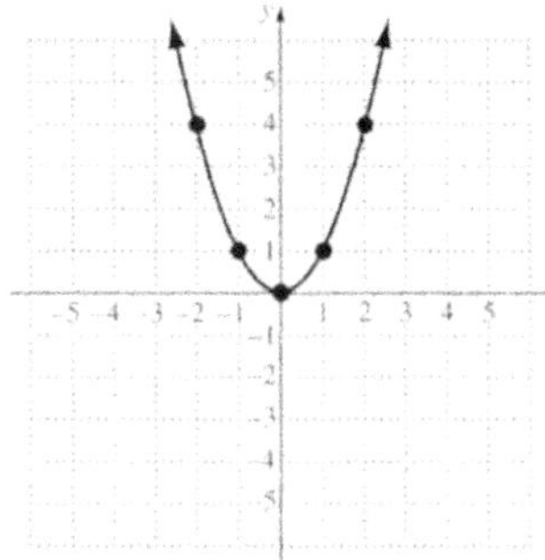

The graph of $g(x) = (-x)^2$ is a
reflection of the graph of $f(x)$
across the y-axis. Note: $(-x)^2 = x^2$,
so $g(x) = f(-x) = f(x)$. The graph
of $f(x) = x^2$ is symmetric with respec
to the y-axis; thus $f(x) = f(-x)$.

Exercises

The graph of $f(x) = |x|$ (the absolute-value function) is shown in figure (a) below. In Exercises 1-10, match the function g with one of the graphs (a) – (h) that follow. Some graphs will be used more than once.

(a)

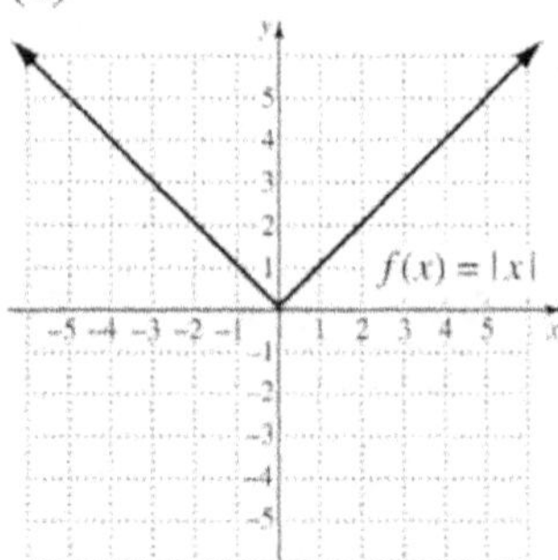

(b)

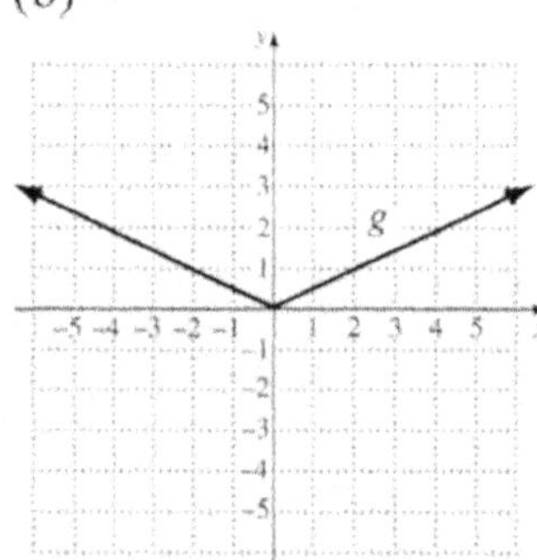

(c)

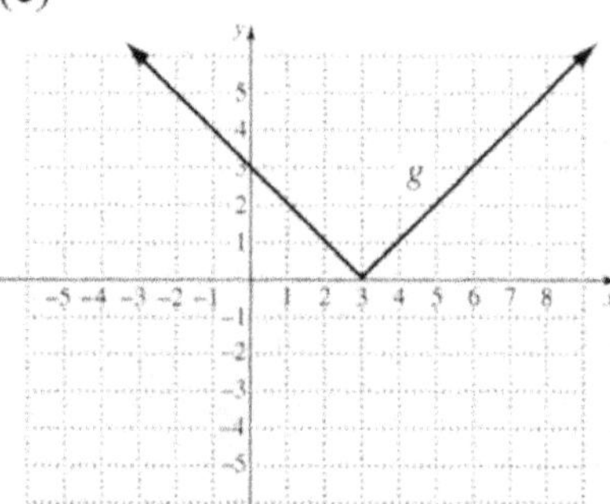

(d)

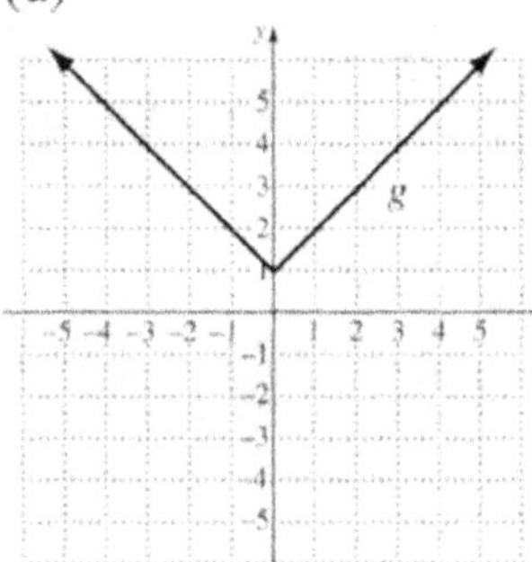

(e)

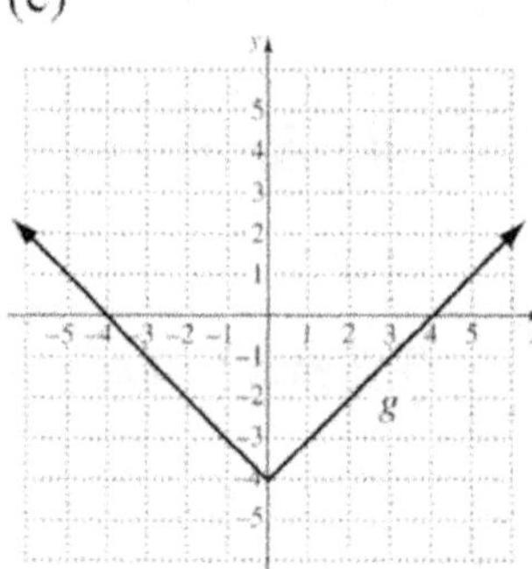

(f)

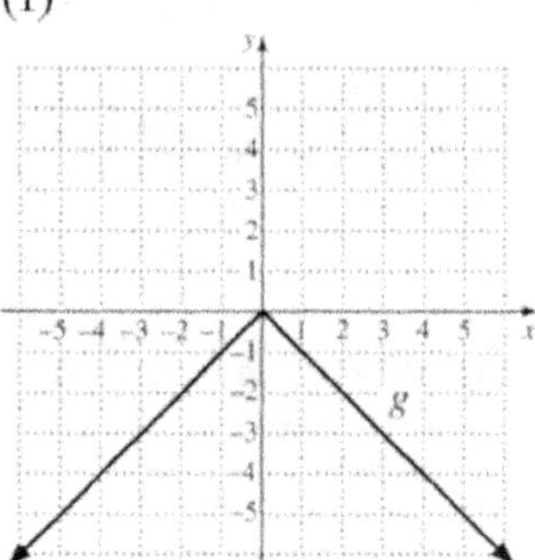

(g)

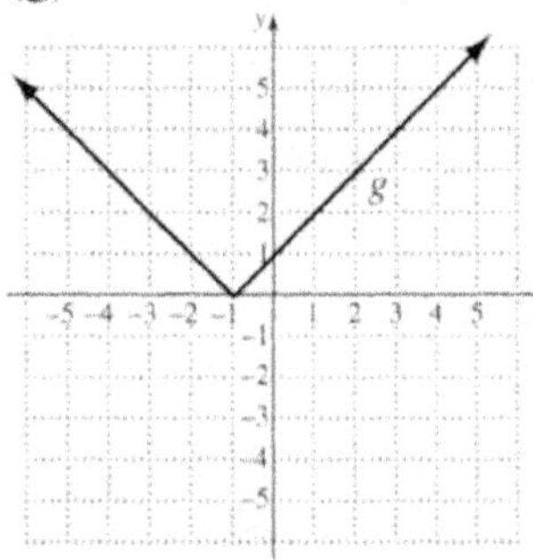

(h)

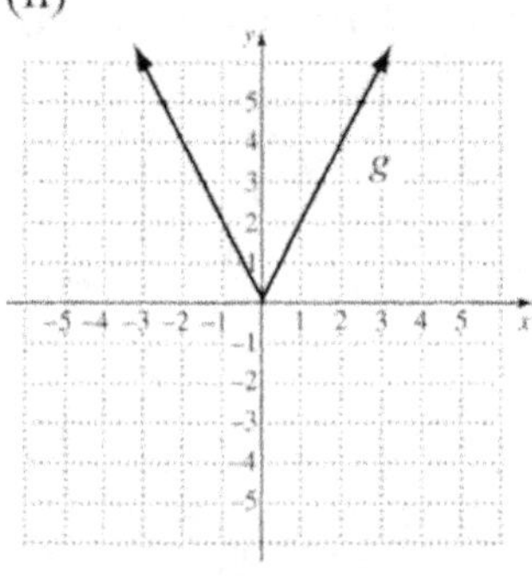

1. $g(x) = |x+1|$

2. $g(x) = 2|x|$

3. $g(x) = |x| + 1$

4. $g(x) = |-x|$

5. $g(x) = \frac{1}{2}|x|$

6. $g(x) = |x-3|$

7. $g(x) = \left|\frac{1}{2}x\right|$

8. $g(x) = |x| - 4$

9. $g(x) = |2x|$

10. $g(x) = -|x|$

The graph of the function f is shown in figure (a) below. In Exercises 11-20, match the function g with one of the graphs (b) – (k) that follow.

(a)

(b)

(c)

(d)

(e)

(f)

(g)

(h)

(i)

(j)

(k)

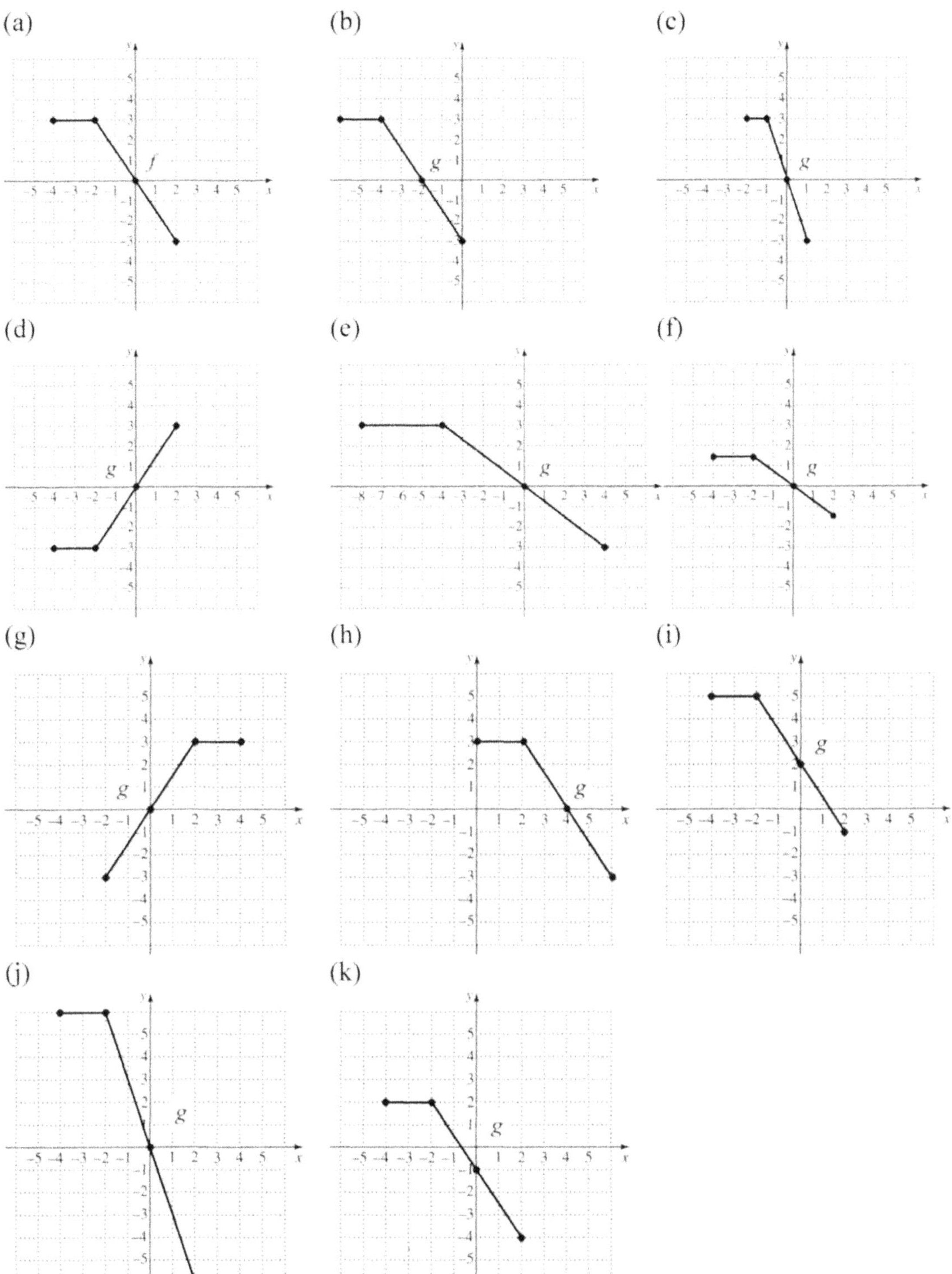

11. $g(x) = f\left(\dfrac{1}{2}x\right)$

12. $g(x) = f(-x)$

13. $g(x) = f(x-4)$

14. $g(x) = f(x)-1$

15. $g(x) = f(2x)$

16. $g(x) = f(x)+2$

17. $g(x) = f(x+2)$

18. $g(x) = 2f(x)$

19. $g(x) = -f(x)$

20. $g(x) = \dfrac{1}{2}f(x)$

Notes:

Interactive Preview Worksheet 31: COMPLETING THE SQUARE

A quadratic equation of the form $(x+c)^2 = d$ can be solved using the *principle of square roots*. Let's work through two examples.

$$x^2 = 81 \qquad\qquad\qquad (x-8)^2 = 7$$

$$x = \sqrt{81} \quad or \quad x = -\sqrt{81} \qquad\qquad x - 8 = \sqrt{7} \qquad or \quad x - 8 = -\sqrt{7}$$

$$x = 9 \quad or \quad x = -9 \qquad\qquad\qquad x = 8 + \sqrt{7} \quad or \qquad x = 8 - \sqrt{7}$$

The solutions are ± 9. $\qquad\qquad\qquad$ The solutions are $8 \pm \sqrt{7}$.

Any quadratic equation, such as $x^2 - 10x + 3 = 0$, can be put in the form $(x+c)^2 = d$ by *completing the square*. We can add a number on both sides of an equation in order to make one side of the equation the square of a binomial. To determine the number to add, we take half the x-coefficient and square it.

Example 1 $\qquad$ Solve: $x^2 - 10x + 3 = 0$ by completing the square.

Subtract 3 on both sides. $\quad$ $x^2 - 10x \quad\quad = -3$

Add 25 on both sides. $\quad$ $x^2 - 10x + 25 = -3 + 25 \quad$ Adding 25: $\dfrac{1}{2}(-10) = -5$, and $(-5)^2 = 25$

Express the left side as the square of a binomial and add on the right.

$$(x-5)^2 = 22$$

Use the principle of square roots.

$$x - 5 = \sqrt{22} \qquad or \quad x - 5 = -\sqrt{22}$$

Simplify.

$$x = 5 + \sqrt{22} \quad or \qquad x = 5 - \sqrt{22}$$

The solutions are $5 \pm \sqrt{22}$.

Example 2 $\qquad$ Solve: $x^2 + 3x - 2 = 0$ by completing the square.

Add 2 on both sides. $\quad$ $x^2 + 3x \quad\quad = 2$

Add $\dfrac{9}{4}$ on both sides. $\quad$ $x^2 + 3x + \dfrac{9}{4} = 2 + \dfrac{9}{4} \quad$ Adding $\dfrac{9}{4}$: $\dfrac{1}{2} \cdot 3 = \dfrac{3}{2}$, and $\left(\dfrac{3}{2}\right)^2 = \dfrac{9}{4}$

Express the left side as the square of a binomial and add on the right.

$$\left(x + \dfrac{3}{2}\right)^2 = \dfrac{8}{4} + \dfrac{9}{4}$$

$$\left(x + \dfrac{3}{2}\right)^2 = \dfrac{17}{4}$$

Use the principle of square roots.

$$x + \dfrac{3}{2} = \sqrt{\dfrac{17}{4}} \qquad or \quad x + \dfrac{3}{2} = -\sqrt{\dfrac{17}{4}}$$

Simplify.

$$x = -\dfrac{3}{2} + \dfrac{\sqrt{17}}{2} \quad or \qquad x = -\dfrac{3}{2} - \dfrac{\sqrt{17}}{2}$$

The solutions are $-\dfrac{3}{2} \pm \dfrac{\sqrt{17}}{2}$.

Check Your Understanding

Fill in the blanks with the number that completes the square on the left side.

1. $x^2 + 8x + \boxed{} = 3 + \boxed{}$

2. $y^2 - 6y + \boxed{} = -1 + \boxed{}$

3. $t^2 - 5t + \boxed{} = -4 + \boxed{}$

4. $x^2 + x + \boxed{} = 2 + \boxed{}$

Exercises Solve by completing the square.

1. $t^2 - 2t - 1 = 0$

2. $x^2 + 14x + 26 = 0$

3. $x^2 + 5x - 3 = 0$

4. $y^2 - y - 8 = 0$

5. $a^2 - 8a + 3 = 0$

6. $x^2 + 6x + 4 = 0$

7. $y^2 + 11y - 2 = 0$

8. $x^2 + \dfrac{3}{2}x - 5 = 0$

Interactive Preview Worksheet 32: INTRODUCTION TO QUADRATIC FUNCTIONS

A **quadratic function** f is a function that can be written in the form

$$f(x) = ax^2 + bx + c, \quad a \neq 0,$$

where a, b, and c are real numbers.

Graphs of Quadratic Functions

- The graph of a quadratic function is called a **parabola**.

- The point (h, k) at which the graph turns is called the **vertex**.

- If $a > 0$ (the graph opens up), the second coordinate k of the vertex (h, k) is a **minimum value** of the function.

- If $a < 0$ (the graph opens down), the second coordinate k of the vertex (h, k) is a **maximum value** of the function.

- The **axis of symmetry**, $x = h$, is a vertical line that passes through the vertex (h, k).

- The **zeros of the function** are the first coordinates of the x-intercepts of the graph.

Example:

$$f(x) = x^2 - 2x - 8$$
$$= (x + 2)(x - 4)$$

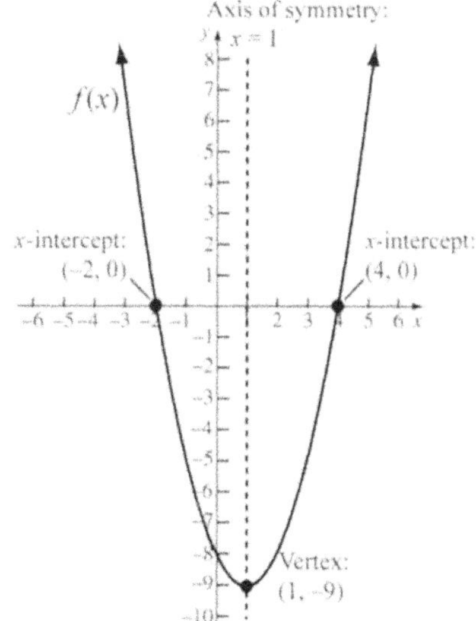

$a = 1$;

$a > 0$, thus the graph opens up.

The zeros of $f(x)$ are -2 and 4.

The minimum value of $f(x)$ is -9.

Exercises Label the vertex, the axis of symmetry, and the x-intercepts of the graph of the quadratic function. Then fill in the blanks in the statements below the graph.

1. $h(x) = x^2 - 8x + 15$

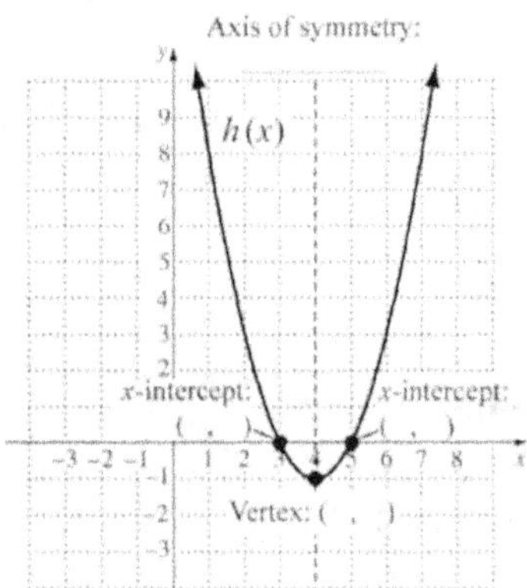

- $a = 1$; $a \,\square\, 0$, thus the graph opens $\underline{\hspace{2cm}}$.
  up / down

- The zeros of $h(x)$ are $\square$ and $\square$.

- The $\underline{\hspace{3cm}}$
  maximum / minimum
  value of $h(x)$ is $\square$.

2. $f(x) = -x^2 - 4x - 3$

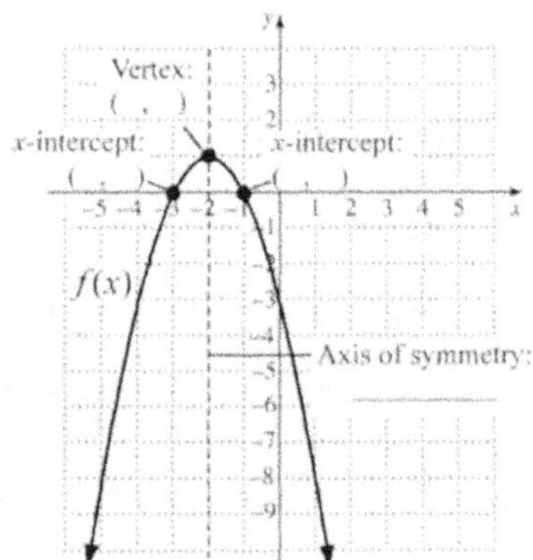

- $a = -1$; $a \,\square\, 0$, thus the graph opens $\underline{\hspace{2cm}}$.
  up / down

- The zeros of $f(x)$ are $\square$ and $\square$.

- The $\underline{\hspace{3cm}}$
  maximum / minimum
  value of $f(x)$ is $\square$.

3. $g(x) = -x^2 - 2x + 3$

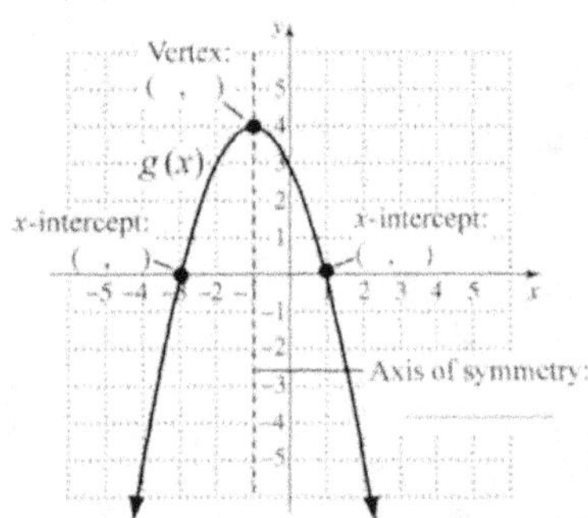

- $a = -1$; $a \,\square\, 0$, thus the graph opens $\underline{\hspace{2cm}}$.
  up / down

- The zeros of $g(x)$ are $\square$ and $\square$.

- The $\underline{\hspace{3cm}}$
  maximum / minimum
  value of $g(x)$ is $\square$.

4. $f(x) = x^2 - 4x$

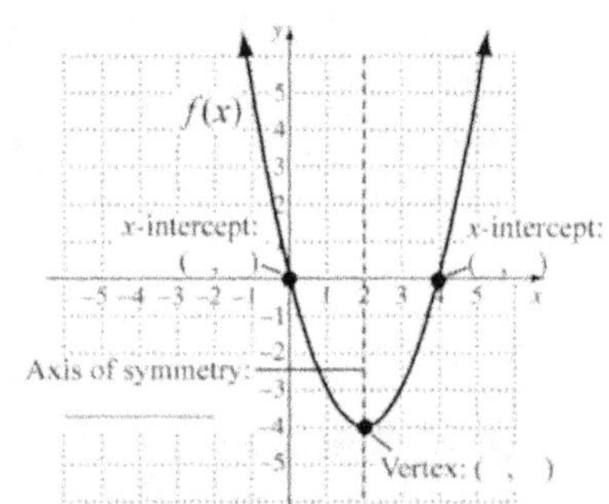

- $a = 1$; $a \,\square\, 0$, thus the graph opens $\underline{\hspace{2cm}}$.
  up / down

- The zeros of $f(x)$ are $\square$ and $\square$.

- The $\underline{\hspace{3cm}}$
  maximum / minimum
  value of $f(x)$ is $\square$.

Interactive Preview Worksheet 33: ZEROS OF POLYNOMIAL FUNCTIONS

The **zeros of a function** $y = f(x)$ are also the **solutions of the equation** $f(x) = 0$, and the *real-number* zeros are the **first coordinates of the x-intercepts** of the graph of the function.

Examples

1. Linear Function

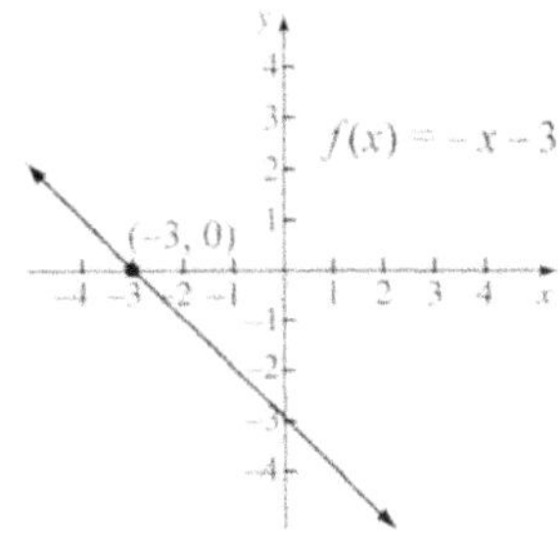

The *x-intercept* of the graph of
$f(x) = -x - 3$ is
$$(-3, 0).$$
The *solution* of the equation
$-x - 3 = 0$ is
$$-3.$$
The *zero* of the function
$f(x) = -x - 3$ is
$$-3.$$

2. Quadratic Function

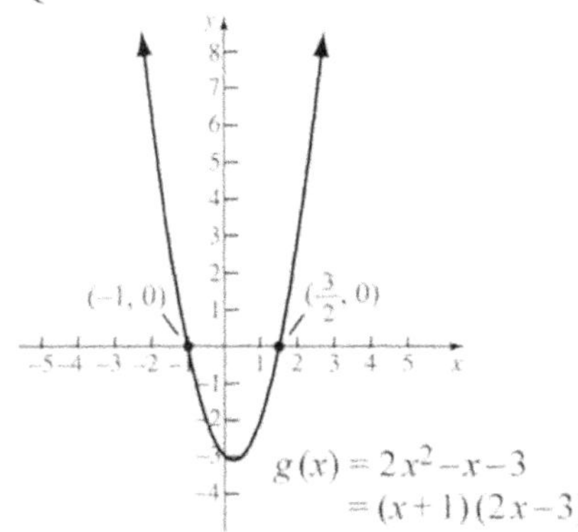

The *x-intercepts* of the graph of
$g(x) = 2x^2 - x - 3$ are
$$(-1, 0) \text{ and } \left(\frac{3}{2}, 0\right).$$
The *solutions* of the equation
$2x^2 - x - 3 = 0$, or $(x+1)(2x-3) = 0$, are
$$-1 \text{ and } \frac{3}{2}.$$
The *zeros* of the function
$g(x) = 2x^2 - x - 3$, or $g(x) = (x+1)(2x-3)$, are
$$-1 \text{ and } \frac{3}{2}.$$

3. Cubic Function

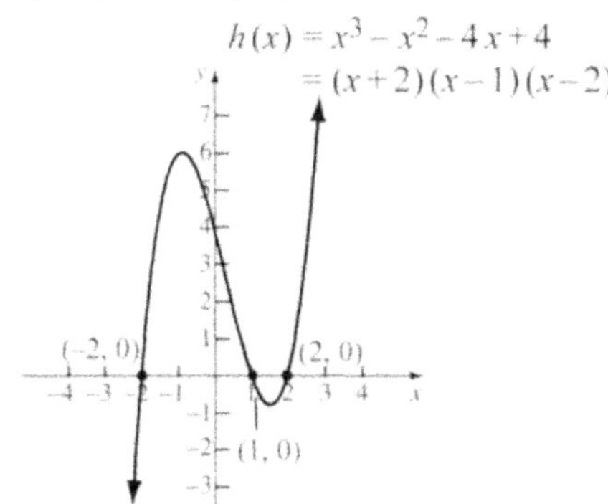

The *x-intercepts* of the graph of
$h(x) = x^3 - x^2 - 4x + 4$ are
$$(-2, 0), \ (1, 0), \text{ and } (2, 0).$$
The *solutions* of the equation
$x^3 - x^2 - 4x + 4 = 0$, or $(x+2)(x-1)(x-2) = 0$, are
$$-2, \ 1, \text{ and } 2.$$
The *zeros* of the function
$h(x) = x^3 - x^2 - 4x + 4$, or
$h(x) = (x+2)(x-1)(x-2)$ are
$$-2, \ 1, \text{ and } 2.$$

Exercises Label the x-intercepts of each graph. Then complete the statements with the x-intercepts of the graph of the function $f(x)$, the solutions of the equation $f(x) = 0$, and the zeros of the function $f(x)$.

1. 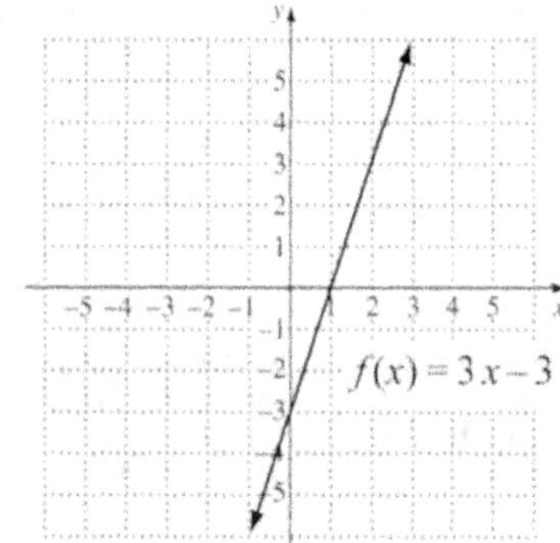

The x-intercept of the graph of
$$f(x) = 3x - 3 \text{ is } \underline{\hspace{2cm}}.$$

The solution of the equation
$$3x - 3 = 0 \text{ is } \underline{\hspace{2cm}}.$$

The zero of the function
$$f(x) = 3x - 3 \text{ is } \underline{\hspace{2cm}}.$$

2. 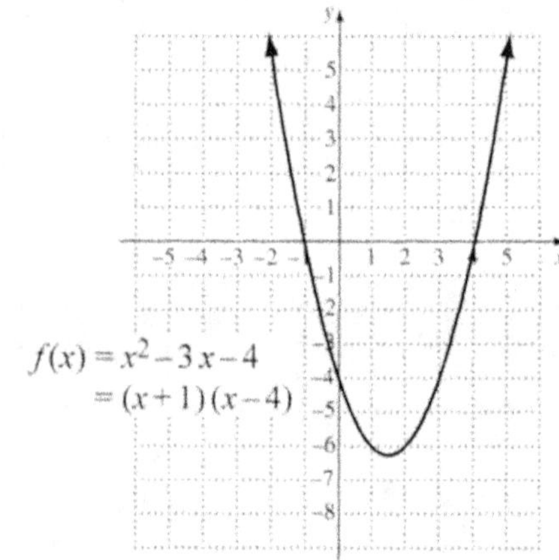

The x-intercepts of the graph of
$$f(x) = x^2 - 3x - 4 \text{ are } \underline{\hspace{2cm}} \text{ and } \underline{\hspace{2cm}}.$$

The solutions of the equation
$x^2 - 3x - 4 = 0$, or $(x+1)(x-4) = 0$, are
$$\underline{\hspace{2cm}} \text{ and } \underline{\hspace{2cm}}.$$

The zeros of the function
$f(x) = x^2 - 3x - 4$, or $f(x) = (x+1)(x-4)$ are
$$\underline{\hspace{2cm}} \text{ and } \underline{\hspace{2cm}}.$$

3. 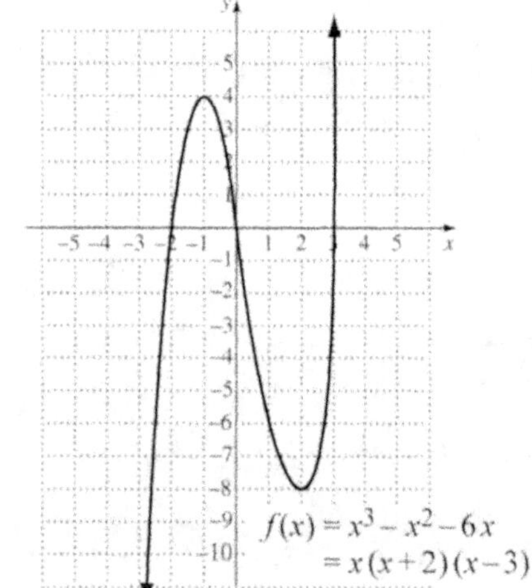

The x-intercepts of the graph of
$f(x) = x^3 - x^2 - 6x$ are $\underline{\hspace{1.5cm}}$, $\underline{\hspace{1.5cm}}$, and $\underline{\hspace{1.5cm}}$.

The solutions of the equation
$x^3 - x^2 - 6x = 0$, or $x(x+2)(x-3) = 0$, are
$\underline{\hspace{1.5cm}}$, $\underline{\hspace{1.5cm}}$, and $\underline{\hspace{1.5cm}}$.

The zeros of the function
$f(x) = x^3 - x^2 - 6x$, or $f(x) = x(x+2)(x-3)$,
are $\underline{\hspace{1.5cm}}$, $\underline{\hspace{1.5cm}}$, and $\underline{\hspace{1.5cm}}$.

Examples Using the *principle of zero products,* find the zeros of the functions in Examples 1-3.

4. $f(x) = -x - 3$ (See Example 1.)

   We solve $f(x) = 0$.

$f(x) = 0$	$-x - 3 = 0$
Add x on both sides.	$-3 = x$
	The zero of $f(x)$ is -3.

5. $g(x) = 2x^2 - x - 3$ (See Example 2.)

   $\qquad = (x+1)(2x-3)$

   We solve $g(x) = 0$.

$g(x) = 0$	$2x^2 - x - 3 = 0$
	$(x+1)(2x-3) = 0$
Use the principle of zero products.	$x + 1 = 0$ *or* $2x - 3 = 0$
Solve the equations separately.	$x = -1$ *or* $2x = 3$
	$x = -1$ *or* $x = \dfrac{3}{2}$
	The zeros of $g(x)$ are -1 and $\dfrac{3}{2}$.

6. $h(x) = x^3 - x^2 - 4x + 4$ (See Example 3.)

   $\qquad = (x+2)(x-1)(x-2)$

   We solve $h(x) = 0$.

$h(x) = 0$	$x^3 - x^2 - 4x + 4 = 0$
	$(x+2)(x-1)(x-2) = 0$
Use the principle of zero products.	$x + 2 = 0$ *or* $x - 1 = 0$ *or* $x - 2 = 0$
Solve the equations separately.	$x = -2$ *or* $x = 1$ *or* $x = 2$
	The zeros of $h(x)$ are -2, 1, and 2.

Exercises Using the principle of zero products, find the zeros of the functions in Exercises 1-3. Show your work.

4. $f(x) = 3x - 3$ (See Exercise 1.)

5. $f(x) = x^2 - 3x - 4$ (See Exercise 2.)
$$= (x+1)(x-4)$$

6. $f(x) = x^3 - x^2 - 6x$ (See Exercise 3.)
$$= x(x+2)(x-3)$$

Rational Function

A rational function f is a function that is a quotient of two polynomials

$$f(x) = \frac{p(x)}{q(x)},$$

where $p(x)$ and $q(x)$ are polynomials and $q(x)$ is not the zero polynomial.

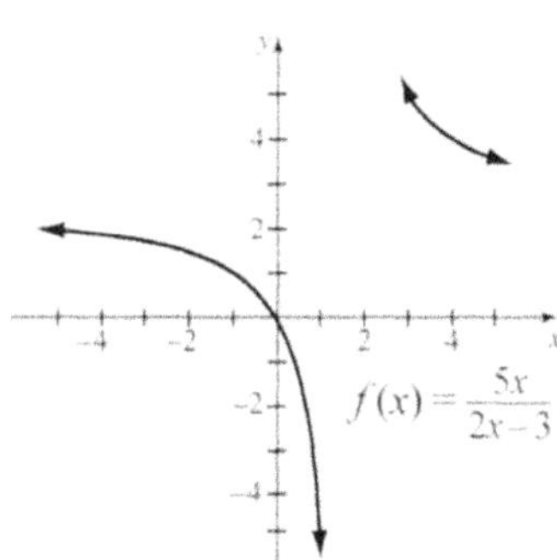

When graphing rational functions, it is helpful to first sketch the asymptotes. Asymptotes can be vertical, horizontal, or oblique (slanted). In this worksheet, we will consider only rational functions with horizontal and vertical asymptotes.

Let's begin by reviewing graphs of horizontal and vertical lines. The graph of an equation of the form $x = a$ is a vertical line with x-intercept $(a, 0)$. The graph of an equation of the form $y = b$ is a horizontal line with y-intercept $(0, b)$. The graphs of $x = -3$, $y = 2$, $x = 0$, and $y = 0$ are shown below.

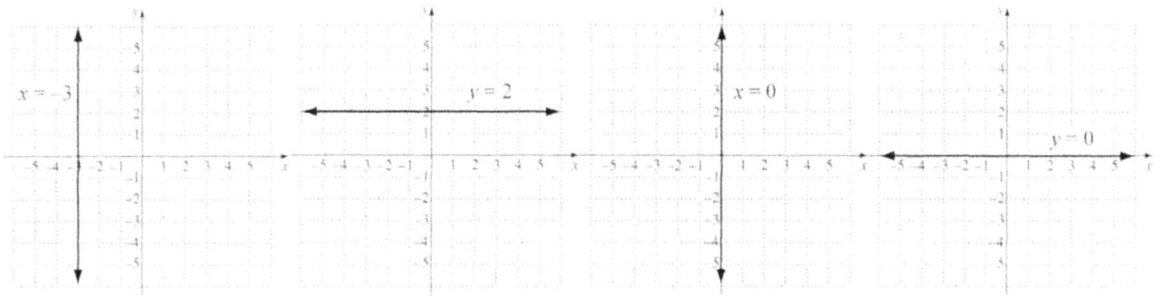

Example 1 Consider the graph of the rational function $f(x) = \dfrac{-6x-1}{3x-3}$ and observe the equations of the asymptotes.

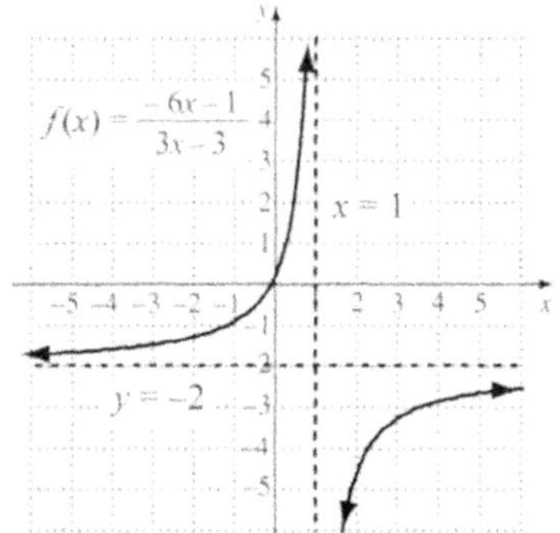

As x-values get closer to 1 from the left, the function values (y-values) approach positive infinity. Also, as the x-values get closer to 1 from the right, the function values (y-values) approach negative infinity. The vertical line $x = 1$ is the **vertical asymptote** for this curve.

As x-values approach positive infinity, function values approach -2. Likewise, as x-values approach negative infinity, function values approach -2. The horizontal line $y = -2$ is the **horizontal asymptote** for this curve.

Exercises

Label with equations all vertical and horizontal asymptotes of the graph.

1. $f(x) = \dfrac{5x-1}{5x-10}$

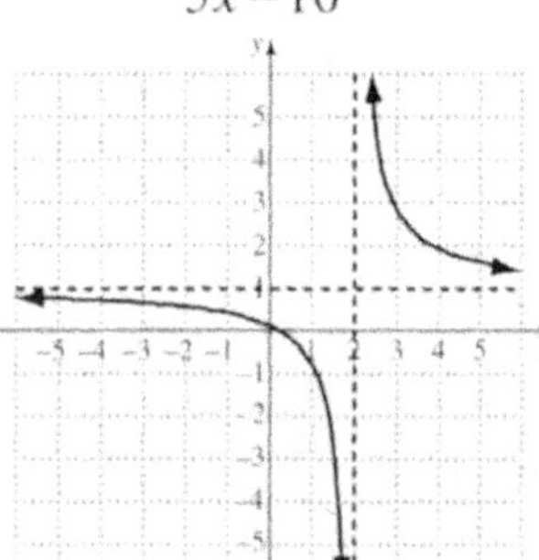

2. $f(x) = \dfrac{1}{x^2}$

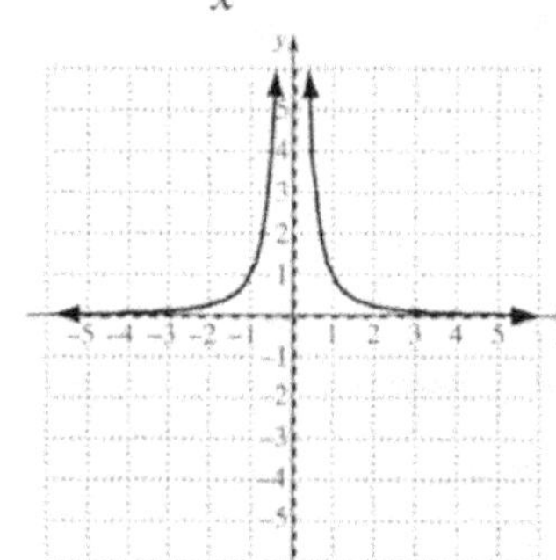

3. $f(x) = \dfrac{4}{x+2}$

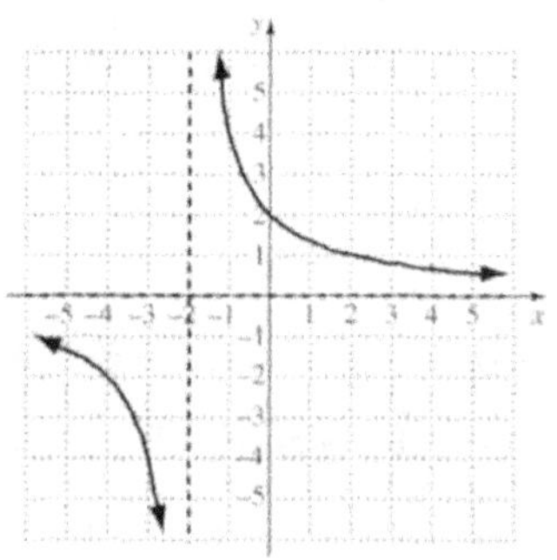

4. $f(x) = \dfrac{1+6x}{6+3x}$

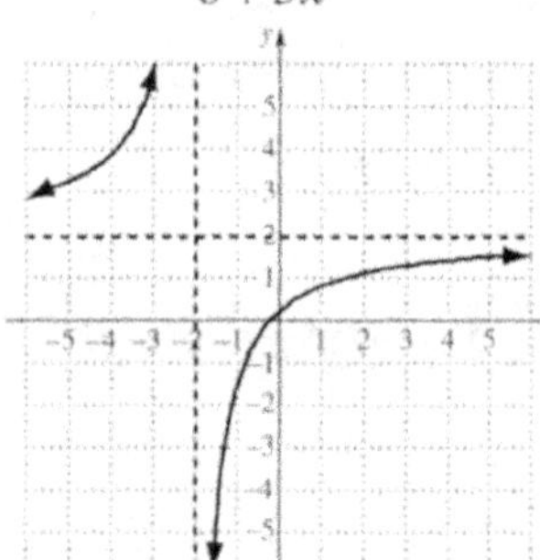

5. $f(x) = \dfrac{5}{x^2+3x}$

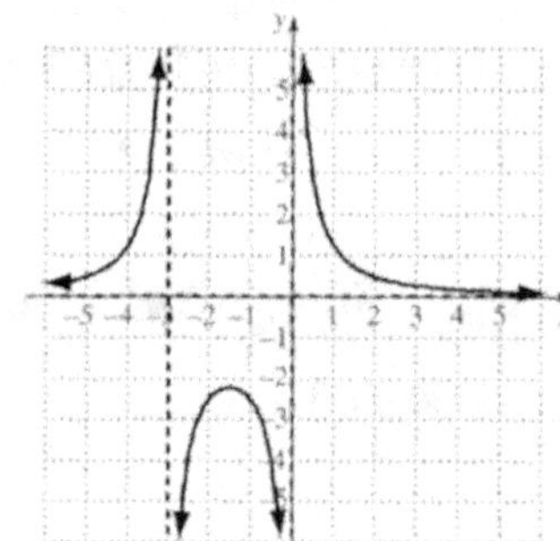

6. $f(x) = \dfrac{-2x^2+4x+3}{x^2-x-2}$

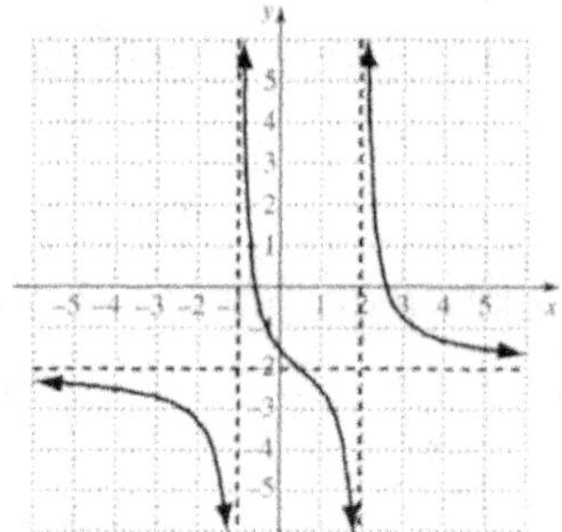

Determining Vertical Asymptotes

For a rational function $f(x) = p(x)/q(x)$, where $p(x)$ and $q(x)$ are polynomials with no common factors other than constants, if a is a zero of the denominator, then the line $x = a$ is a **vertical asymptote** for the graph of the function.

Example 2 Determine algebraically the vertical asymptotes for the graph of

$$f(x) = \frac{-6x-1}{3x-3}.$$

The numerator and the denominator do not have a common factor, so we need to determine only the zeros of the denominator $3x - 3$.

$$3x - 3 = 0$$
$$3x = 3$$
$$x = 1$$

Solving $3x - 3 = 0$, we see that 1 is a zero of the denominator. Thus, $x = 1$ is a **vertical asymptote**, as shown in the graph in Example 1.

     Copyright © 2022 Pearson Education, Inc.

Exercises

Determine algebraically the vertical asymptote(s) for the graph of the function. The numerator and the denominator of each function in Exercises 7-12 do not have a common factor so we need to find only the zeros of the denominator.

7. $f(x) = \dfrac{5x-1}{5x-10}$

   Solve: $5x - 10 = 0$.

   Zero(s) of the denominator: ________
   Vertical asymptote(s): __________
   (View the graph in Exercise 1.)

8. $f(x) = \dfrac{1}{x^2}$

   Solve: $x^2 = 0$.

   Zero(s) of the denominator: ________
   Vertical asymptote(s): __________
   (View the graph in Exercise 2.)

9. $f(x) = \dfrac{4}{x+2}$

   Solve: $x + 2 = 0$.

   Zero(s) of the denominator: ________
   Vertical asymptote(s): __________
   (View the graph in Exercise 3.)

10. $f(x) = \dfrac{1+6x}{6+3x}$

    Solve: $6 + 3x = 0$.

    Zero(s) of the denominator: ________
    Vertical asymptote(s): __________
    (View the graph in Exercise 4.)

11. $f(x) = \dfrac{5}{x^2+3x}$

    Solve: $x^2 + 3x = 0$.

    Zero(s) of the denominator: ________
    Vertical asymptote(s): __________
    (View the graph in Exercise 5.)

12. $f(x) = \dfrac{-2x^2+4x+3}{x^2-x-2}$

    Solve: $x^2 - x - 2 = 0$.

    Zero(s) of the denominator: ________
    Vertical asymptote(s): __________
    (View the graph in Exercise 6.)

Determining Horizontal Asymptotes

- **Degree of Numerator = Degree of Denominator**
  When the numerator and the denominator of a rational function have the same degree, the line $y = a/b$ is the **horizontal asymptote**, where a and b are the leading coefficients of the numerator and the denominator, respectively. (For example, see the graphs in Exercises 1, 4, and 6.)

- **Degree of Numerator < Degree of Denominator**
  When the degree of the numerator of a rational function is less than the degree of the denominator, the x-axis, or $y = 0$, is the **horizontal asymptote**. (For example, see the graphs in Exercises 2, 3, and 5.)

- **Degree of Numerator > Degree of Denominator**
  When the degree of the numerator of a rational function is greater than the degree of the denominator, there is no horizontal asymptote.

Example 3 Determine algebraically the horizontal asymptote for the graph of

$$f(x) = \frac{-6x-1}{3x-3}.$$ Degree of numerator: 1
Degree of denominator: 1

The numerator and the denominator have the *same* degree. The leading coefficient of the numerator is −6. The leading coefficient of the denominator is 3. The ratio of the coefficients is −6/3, or −2. The *horizontal asymptote* is $y = -2$, as shown in the graph in Example 1.

Exercises

Determine algebraically the horizontal asymptote for the graph of the function. In Exercises 13-18, we consider only functions in which the numerator and the denominator have the same degree, or the degree of the numerator is less than the degree of the denominator.

13. $f(x) = \dfrac{5x-1}{5x-10}$

Degree of numerator: _________
Degree of denominator: _________
Degree of numerator is _____________________ degree of denominator.
     same as / less than
Leading coefficient of numerator: _________
Leading coefficient of denominator: _________
Ratio of leading coefficients: _________
Horizontal asymptote: _________
(View the graph in Exercise 1.)

14. $f(x) = \dfrac{1}{x^2}$

Degree of numerator: _________
Degree of denominator: _________
Degree of numerator is _____________________ degree of denominator.
     same as / less than
Horizontal asymptote: _________
(View the graph in Exercise 2.)

15. $f(x) = \dfrac{4}{x+2}$

Horizontal asymptote: _________
(View the graph in Exercise 3.)

16. $f(x) = \dfrac{1+6x}{6+3x}$

Horizontal asymptote: _________
(View the graph in Exercise 4.)

17. $f(x) = \dfrac{5}{x^2+3x}$

Horizontal asymptote: _________
(View the graph in Exercise 5.)

18. $f(x) = \dfrac{-2x^2+4x+3}{x^2-x-2}$

Horizontal asymptote: _________
(View the graph in Exercise 6.)

Interactive Preview Worksheet 35: GRAPHING INVERSE FUNCTIONS

A function is one-to-one if each output has exactly one input. The function $f(x) = 3x - 4$ is a one-to-one function. Its inverse is $f^{-1}(x) = \dfrac{x+4}{3}$. The graphs of f and f^{-1} are shown below.

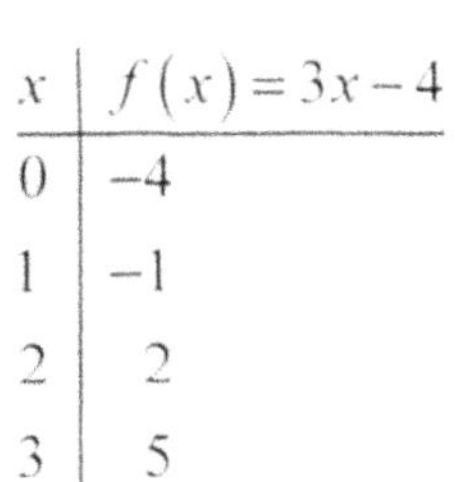

x	$f(x) = 3x - 4$
0	-4
1	-1
2	2
3	5

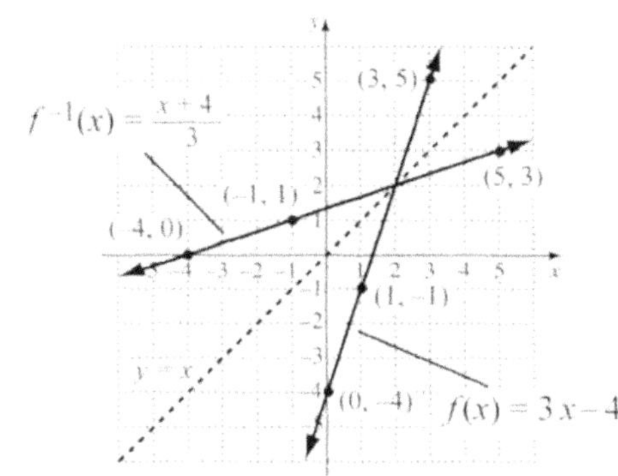

x	$f^{-1}(x) = \dfrac{x+4}{3}$
-4	0
-1	1
2	2
5	3

We can make some observations:

- The ordered-pair solutions of the inverse function, $f^{-1}(x) = \dfrac{x+4}{3}$, can be found by interchanging the first and second coordinates of each ordered-pair solution of the original function, $f(x) = 3x - 4$. For example, $(3, 5)$ is a solution of f, thus $(5, 3)$ is a solution of f^{-1}.

- The graph of $f^{-1}(x) = \dfrac{x+4}{3}$ is the reflection of the graph of $f(x) = 3x - 4$ across the line $y = x$.

One-To-One Functions and Inverses

- If a function f is one-to-one, then its inverse f^{-1} is a function.
- The domain of f is the range of f^{-1}.
- The range of f is the domain of f^{-1}.
- The solutions of f^{-1} can be found from the solutions of f by interchanging the first and second coordinates of each ordered pair.
- The graph of f^{-1} is a reflection of the graph of f across the line $y = x$.

Example 1 Graph the one-to-one function $g(x) = -2x + 3$ and its inverse $g^{-1}(x) = \dfrac{3-x}{2}$ on the same set of axes.

Begin by listing a few solutions of $g(x)$ in a table. Then plot the points and draw the graph.

x	$g(x) = -2x + 3$
-1	5
0	3
2	-1
3	-3

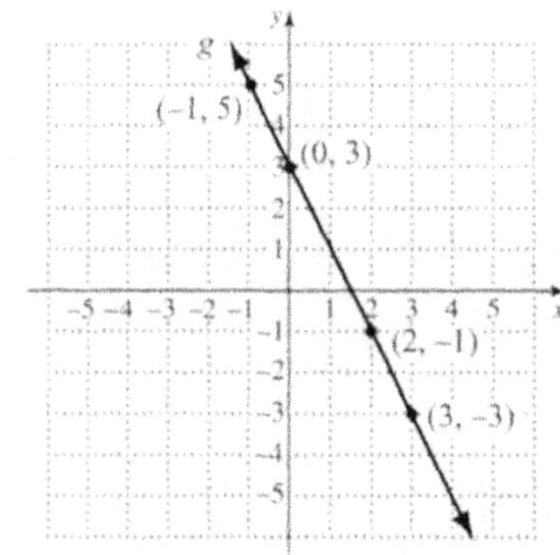

Next, create a table of solutions of $g^{-1}(x)$ by *interchanging* the x and y-values in the table for $g(x)$. Then plot the points and draw the graph of $g^{-1}(x)$ on the same set of axes as the graph of $g(x)$.

x	$g^{-1}(x) = \dfrac{3-x}{2}$
5	-1
3	0
-1	2
-3	3

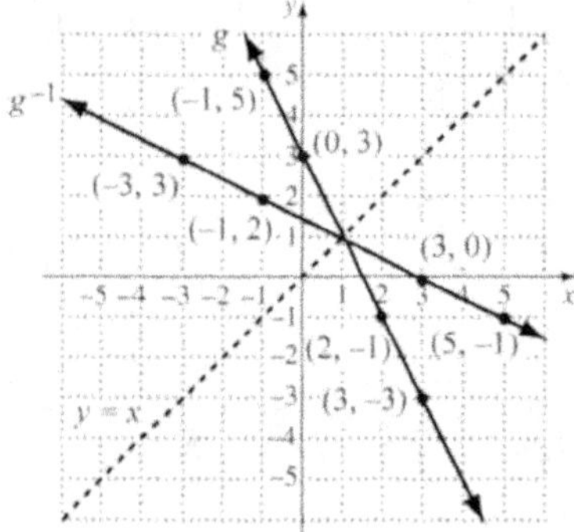

The graph of $g^{-1}(x)$ is a reflection of the graph of $g(x)$ across the line $y = x$.

When the inverse of a function is not a function, the domain of the function can be restricted to allow the inverse to be a function. Consider the graphs of $h(x) = x^2 + 1$ and its inverse $y = \pm\sqrt{x-1}$. The function $h(x) = x^2 + 1$ is not one-to-one because some outputs

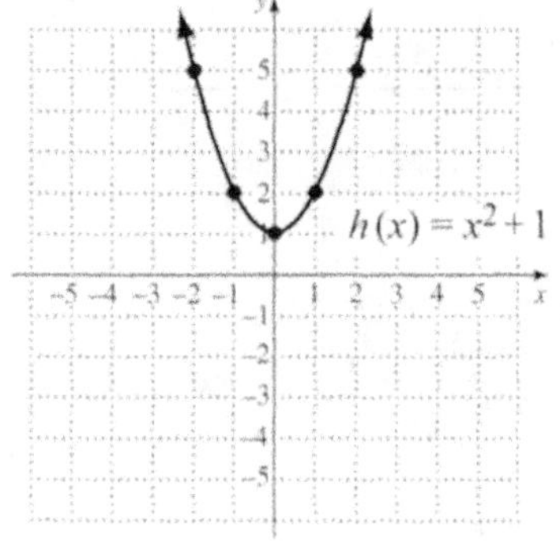

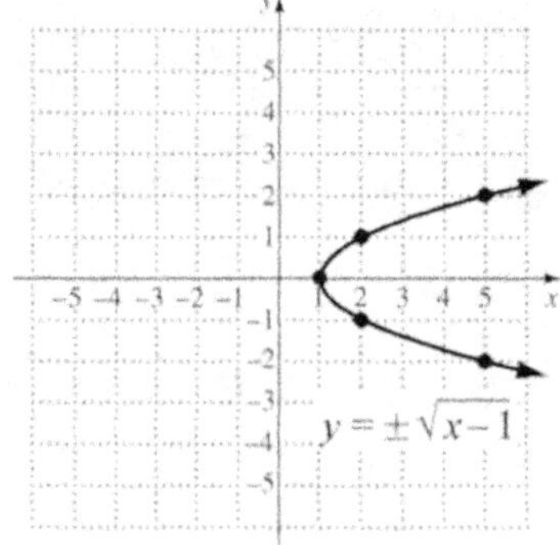

correspond to more than one input. Thus its inverse $y = \pm\sqrt{x-1}$ is *not* a function. However, if we restrict the domain of $h(x) = x^2 + 1$ to nonnegative numbers, $x \ge 0$, then its inverse is a function. (See Example 2.)

Example 2 Graph the one-to-one function $h(x) = x^2 + 1$, $x \geq 0$ and its inverse $h^{-1}(x) = \sqrt{x-1}$ on the same set of axes.

We first list solutions of $h(x) = x^2 + 1$ with the restricted domain $x \geq 0$. Then we list solutions for $h^{-1}(x)$ by interchanging the x and y-values in the table for $h(x)$ and we graph $h(x)$ and $h^{-1}(x)$ on the same set of axes.

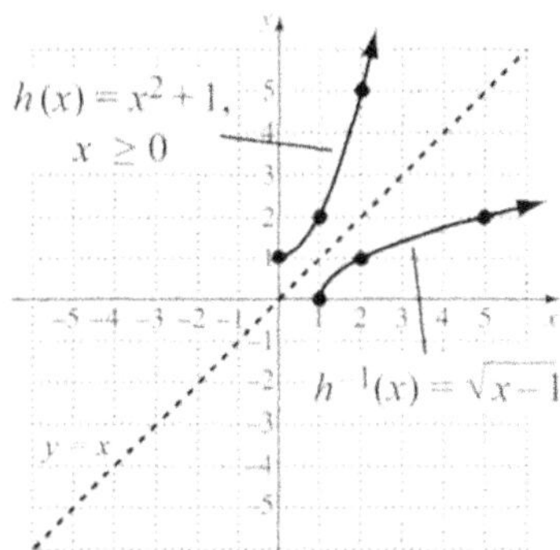

x	$h(x) = x^2 + 1$, $x \geq 0$
0	1
1	2
2	5

x	$h^{-1}(x) = \sqrt{x-1}$
1	0
2	1
5	2

The graph of $h^{-1}(x)$ is a reflection of the graph of $h(x)$, $x \geq 0$, across the line $y = x$.

Exercises Complete tables of solutions for the one-to-one function and its inverse. Then graph both functions on the same set of axes and observe the reflection across the line $y = x$.

1. $f(x) = 2x + 4$; $f^{-1}(x) = \dfrac{x-4}{2}$

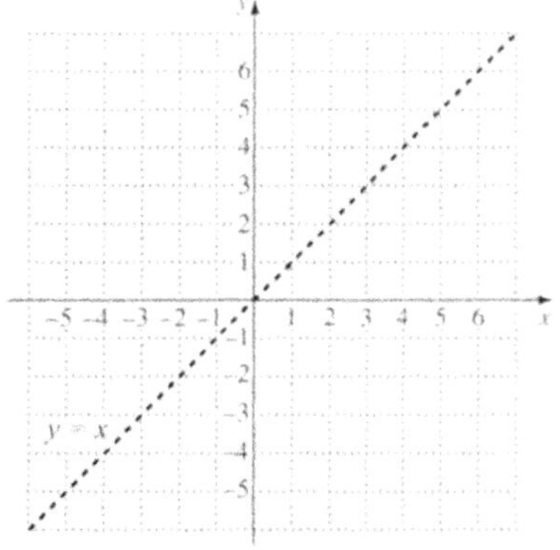

x	$f(x) = 2x + 4$
-4	
-3	
-1	
0	

x	$f^{-1}(x) = \dfrac{x-4}{2}$

2. $h(x) = 2 - 3x$; $h^{-1}(x) = \dfrac{2-x}{3}$

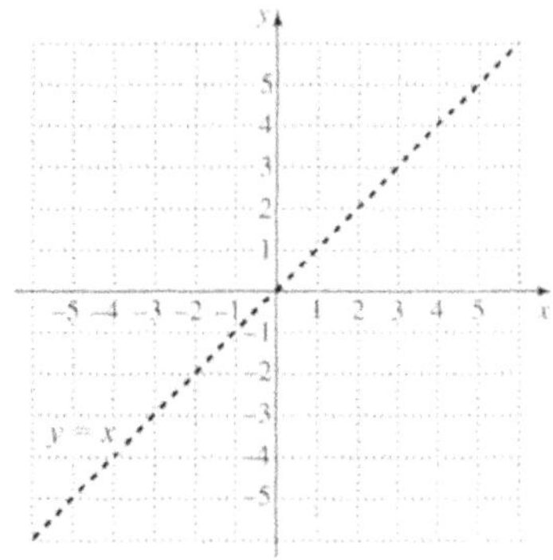

x	$h(x) = 2 - 3x$
-1	
0	
1	
2	

x	$h^{-1}(x) = \dfrac{2-x}{3}$

3. $g(x) = x^3 - 3$; $g^{-1}(x) = \sqrt[3]{x+3}$

x	$g(x) = x^3 - 3$
-2	
-1	
0	
1	
2	

x	$g^{-1}(x) = \sqrt[3]{x+3}$

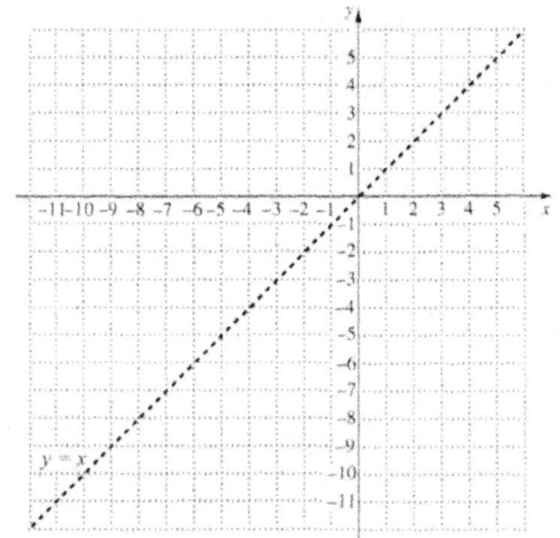

4. $f(x) = x^2 + 4,\ x \ge 0$; $f^{-1}(x) = \sqrt{x-4}$

x	$f(x) = x^2 + 4,\ x \ge 0$
0	
1	
2	
3	

x	$f^{-1}(x) = \sqrt{x-4}$

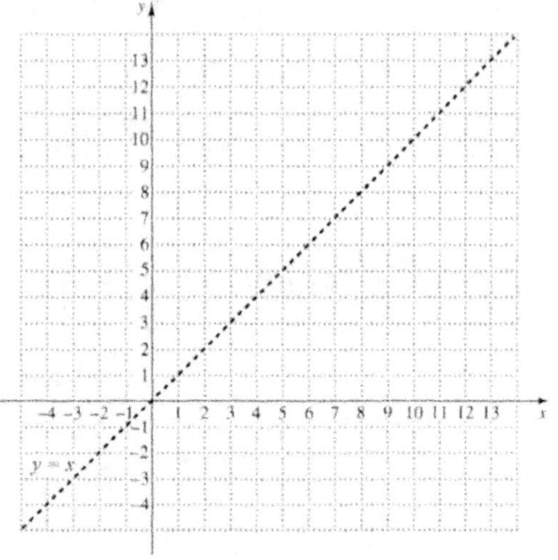

5. $h(x) = x^2 - 3,\ x \ge 0$; $h^{-1}(x) = \sqrt{x+3}$

x	$h(x) = x^2 - 3,\ x \ge 0$
0	
1	
2	
3	

x	$h^{-1}(x) = \sqrt{x+3}$

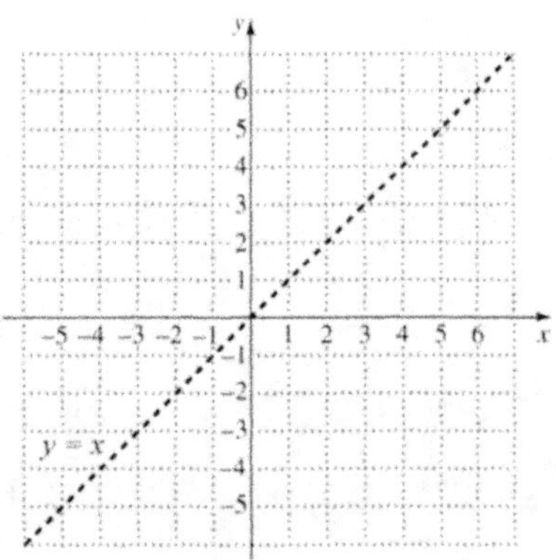

  Copyright © 2022 Pearson Education, Inc.

Interactive Preview Worksheet 36: INTRODUCTION TO LOGARITHMS

Logarithmic functions are inverses of exponential functions and have applications in fields such as business, science, psychology, and sociology.

Logarithmic Function, Base a

- We define $y = \log_a x$ as that number y such that $x = a^y$, where $x > 0$ and a is a positive constant other than 1.

$$\log_a x = y \text{ is equivalent to } x = a^y.$$

- A logarithm is an EXPONENT!
- We read $\log_a x$ as "the logarithm, base a, of x."

Example 1 Find $\log_2 8$.

We read $\log_2 8$ as "the logarithm, base 2, of 8."

Let $y = \log_2 8$. This equation is equivalent to $2^y = 8$. The power to which we raise 2 to get 8 is 3; thus $\log_2 8 = 3$. Check: $2^3 = 8$.

Example 2 Find $\log_{10} 100{,}000$.

Let $y = \log_{10} 100{,}000$, or $10^y = 100{,}000$. The power to which we raise 10 to get $100{,}000$ is 5; thus $\log_{10} 100{,}000 = 5$. Check: $10^5 = 100{,}000$.

Base-10 logarithms are called **common logarithms**. The abbreviation **log**, with no *base* written, is used for the common logarithm. Thus, $\log a = \log_{10} a$. For example, $\log 100{,}000 = \log_{10} 100{,}000$.

It is helpful to review some definitions and rules of exponents.

- For any real number a that is nonzero and any integer n,

$$a^{-n} = \frac{1}{a^n}. \qquad \left(\text{Example: } 9^{-2} = \frac{1}{9^2} \right)$$

- $a^1 = a$, for any number a. $\qquad \left(\text{Example: } 4^1 = 4 \right)$

- $a^0 = 1$, for any nonzero number a. $\qquad \left(\text{Example: } 8^0 = 1 \right)$

Example 3 Find $\log 0.0001$. $\left(\text{Remember: } \log 0.0001 = \log_{10} 0.0001 \right)$

$$0.0001 = \frac{1}{10{,}000} = \frac{1}{10^4} = 10^{-4}$$

$$\log 0.0001 = \log 10^{-4}$$

We let $y = \log_{10} 10^{-4}$, or $10^y = 10^{-4}$. The power to which we raise 10 to get 0.0001 is -4; thus $\log 0.0001 = -4$.

Example 4 Find $\log_5 \dfrac{1}{25}$.

$$\frac{1}{25} = \frac{1}{5^2} = 5^{-2}$$

$$\log_5 \frac{1}{25} = \log_5 5^{-2}$$

We let $y = \log_5 5^{-2}$, or $5^y = 5^{-2}$. The power to which we raise 5 to get $\dfrac{1}{25}$ is -2; thus

$$\log_5 \frac{1}{25} = -2.$$

Example 5 Find $\log_3 1$.

Let $y = \log_3 1$, or $3^y = 1$. We know that $3^0 = 1$. Substituting 3^0 for 1 in $3^y = 1$, we have $3^y = 3^0$. The power to which we raise 3 to get 1 is 0; thus $\log_3 1 = 0$.

Example 6 Find $\log_7 7$.

Let $y = \log_7 7$, or $7^y = 7$. We know that $7^1 = 7$. Substituting 7^1 for 7 in $7^y = 7$, we have $7^y = 7^1$. The power to which we raise 7 to get 7 is 1; thus $\log_7 7 = 1$.

•	$\log_a 1 = 0$	The logarithm, base a, of 1 is always 0.
•	$\log_a a = 1$	The logarithm, base a, of a is always 1.

Exercises Find each of the following. Remember: a logarithm is an exponent.

1. $\log_6 36$

2. $\log 1000$

3. $\log_{19} 19$

4. $\log_8 1$

5. $\log \dfrac{1}{10}$

6. $\log 1,000,000$

7. $\log_4 4$

8. $\log_5 625$

9. $\log 0.01$

10. $\log_7 \dfrac{1}{49}$

11. $\log 0.00001$

12. $\log 1$

Interactive Preview Worksheet 37: SOLVING EXPONENTIAL EQUATIONS

Equations with variables in the exponents, such as

$$7^x = 49, \quad 3^{4x-15} = 243, \quad 6^y = 13, \text{ and } 10e^{0.1t} = 50,$$

are called **exponential equations**.

We use the following property to solve exponential equations.

Base-Exponent Property

For any $a > 0$, $a \neq 1$,

if $a^x = a^y$, then $x = y$, and

if $x = y$, then $a^x = a^y$.

Example 1 Solve: $\quad 7^x = 49.$

		Check:
Note that $49 = 7^2$. We can write each side as a power of 7.	$7^x = 7^2$	$7^x = 49$
Use the base-exponent property.	$x = 2$	$7^2 \mid 49$
	The solution is 2.	$49 \mid 49 \quad$ True

Example 2 Solve: $\quad 3^{4x-15} = 243.$

		Check:
Note that $243 = 3^5$. We can write each side as a power of 3.	$3^{4x-15} = 3^5$	$3^{4x-15} = 243$
Use the base-exponent property.	$4x - 15 = 5$	$3^{4 \cdot 5 - 15} \mid 243$
Add 15.	$4x = 20$	3^{20-15}
Divide by 4.	$x = 5$	3^5
	The solution is 5.	$243 \mid 243 \quad$ True

Another property that is used when solving exponential equations is the property of logarithmic equality.

Property of Logarithmic Equality

For any $M > 0$, $N > 0$, $a > 0$, and $a \neq 1$,

if $\log_a M = \log_a N$, then $M = N$, and

if $\quad M = N$, the $\log_a M = \log_a N$.

Example 3 Solve: $\qquad 6^x = 13.$

Take the common logarithm on both sides.	$\log 6^x = \log 13$
Use the power rule: $\log_a M^p = p \log_a M.$	$x \log 6 = \log 13$
Divide by $\log 6.$	$x = \dfrac{\log 13}{\log 6}$
Approximate using a calculator.	$x \approx 1.4315$

Check:

$$\dfrac{6^x = 13}{6^{1.4315} \,\big|\, 13}$$
$$\approx 13 \,\big|\, 13 \quad \text{True}$$

The solution is about 1.4315.

Example 4 Solve: $\qquad e^x = 8.$

Take the natural logarithm on both sides.	$\ln e^x = \ln 8$
Use the power rule.	$x \cdot \ln e = \ln 8$
Substitute 1 for $\ln e$: $\ln e = 1.$	$x \cdot 1 = \ln 8$
	$x = \ln 8$
Approximate using a calculator.	$x \approx 2.0794$

Check:

$$\dfrac{e^x = 8}{e^{2.0794} \,\big|\, 8}$$
$$\approx 8 \,\big|\, 8 \quad \text{True}$$

The solution is about 2.0794.

Example 5 Solve: $\qquad 4e^{3x} = 720.$

Divide by 4.	$e^{3x} = 180$
Take the natural logarithm on both sides.	$\ln e^{3x} = \ln 180$
Use the power rule.	$3x \ln e = \ln 180$
$\ln e = 1.$	$3x = \ln 180$
Divide by 3.	$x = \dfrac{\ln 180}{3}$
Approximate using a calculator.	$x \approx 1.7310$

Check:

$$\dfrac{4e^{3x} = 720}{4e^{3(1.7310)} \,\big|\, 720}$$
$$\approx 720 \,\big|\, 720 \quad \text{True}$$

The solution is about 1.7310.

Exercises

In Exercises 1-4, describe the calculation in each step of the solution. (See Examples 1-4.)

1. Solve: $\qquad$ $5^x = 625$. (See Example 1.)

$$5^x = 5^4$$
$$x = 4$$

The solution is 4.

2. Solve: $\qquad$ $2^{3x+1} = 128$. (See Example 2.)

$$2^{3x+1} = 2^7$$
$$3x + 1 = 7$$
$$3x = 6$$
$$x = 2$$

The solution is 2.

3. Solve: $\qquad$ $8^x = 25$. (See Example 3.)

$$\log 8^x = \log 25$$
$$x \log 8 = \log 25$$
$$x = \frac{\log 25}{\log 8}$$
$$x \approx 1.5480$$

The solution is about 1.5480.

4. Solve: $\qquad$ $e^x = 350$. (See Example 4.)

$$\ln e^x = \ln 350$$
$$x \ln e = \ln 350$$
$$x = \ln 350$$
$$x \approx 5.8579$$

The solution is about 5.8579.

Solve. Approximate the answer to 4 decimal places.

5. $10^x = 1,000,000$ 6. $3^{5x} = 243$ 7. $e^x = 12$

8. $5^{3x} = 625$ 9. $9^x = 2$ 10. $e^{5x} = 120$

11. $2^x = 256$ 12. $4^{5x+2} = 64$ 13. $6^{4x-3} = 216$

14. $10^x = 43$ 15. $6e^{3x} = 78$ 16. $4^x = 17$

17. $4^{0.5x} = 16$ 18. $e^x = 10,000$ 19. $3^{x-10} = 729$

20. $15,000e^{0.02x} = 60,000$ 21. $e^{-x} = 0.8$ 22. $3^x = \dfrac{1}{27}$

Interactive Preview Worksheet 38: SOLVING LOGARITHMIC EQUATIONS

Equations containing variables in logarithmic expressions, such as

$$\log_2 x = -5, \quad \log_4 (7 - 3x) = 2,$$

$$\log_3 x + \log_3 (x + 8) = 2, \text{ and } \log(3x + 2) - \log(3x - 1) = 1,$$

are called **logarithmic equations**.

To solve logarithmic equations algebraically, we first try to obtain a single logarithmic expression on one side and then write an equivalent exponential equation.

Example 1 Solve: $\log_2 x = -5$.

		Check:
Convert to an exponential equation.	$2^{-5} = x$	$\log_2 x = -5$
Write with a positive exponent.	$\dfrac{1}{2^5} = x$	$\log_2 \left(\dfrac{1}{32} \right) \mid -5$
	$\dfrac{1}{32} = x$	$\log_2 2^{-5}$
		$-5 \mid -5$ True

The solution is $\dfrac{1}{32}$.

Example 2 Solve: $\log_4 (7 - 3x) = 2$.

		Check:
Convert to an exponential equation.	$4^2 = 7 - 3x$	$\log_4 (7 - 3x) = 2$
	$16 = 7 - 3x$	$\log_4 \left[7 - 3(-3) \right] \mid 2$
Subtract 7.	$9 = -3x$	$\log_4 16$
Divide by -3.	$-3 = x$	$2 \mid 2$ True

The solution is -3.

Example 3 Solve: $\log x = 0$.

A common logarithm, the base is 10.	$\log_{10} x = 0$
Convert to an exponential equation.	$10^0 = x$
$a^0 = 1, \ a \neq 0$	$1 = x$

The solution is 1.

Example 4 Solve: $\log_6 x = 1$.

Convert to an exponential equation.	$6^1 = x$
$a^1 = a$	$6 = x$

The solution is 6.

$\log_a 1 = 0$, for any logarithmic base a.

$\log_a a = 1$, for any logarithmic base a.

Exercises Solve.

1. $\log_3 x = 5$ 2. $\log_8 x = 0$ 3. $\log_5(18-x)=3$ 4. $\log_2(3x-7)=4$

5. $\log_{100} x = 1$ 6. $\log(3x+28)=2$ 7. $\log x = -3$ 8. $\log_7\left(\dfrac{1}{2}x+3\right)=2$

Example 5 Solve: $\log_4(5x+12) - \log_4(2x-6) = 2$.

Use the quotient rule.	$\log_4 \dfrac{5x+12}{2x-6} = 2$
Convert to an exponential equation.	$4^2 = \dfrac{5x+12}{2x-6}$
	$16 = \dfrac{5x+12}{2x-6}$
Multiply on both sides by the LCD, $2x-6$.	$(2x-6)\cdot 16 = (2x-6)\cdot \dfrac{5x+12}{2x-6}$
Simplify.	$32x - 96 = 5x + 12$
Subtract $5x$ and add 96.	$27x = 108$
Divide by 27.	$x = 4$
Check.	

$$\log_4(5x+12) - \log_4(2x-6) = 2$$

$$\begin{array}{c|c} \log_4(5\cdot 4+12) - \log_4(2\cdot 4 - 6) & 2 \\ \log_4 32 - \log_4 2 & \\ \log_4 \dfrac{32}{2} & \\ \log_4 16 & \\ 2 & 2 \quad \text{True} \end{array}$$

The solution is 4.

Exercises Solve.

9. $\log_6 x - \log_6(x+10) = -1$

10. $\log(x+7) - \log(x-2) = 1$

11. $\log_3(x-1) - \log_3(3x+4) = -2$

12. $\log_5(3x+5) - \log_5 x = 2$

Example 6 Solve: $\log_3 x + \log_3(x+8) = 2$.

Use the product rule.	$\log_3\left[x(x+8)\right] = 2$
Convert to an exponential equation.	$x(x+8) = 3^2$
	$x^2 + 8x = 9$
	$x^2 + 8x - 9 = 0$
Factor.	$(x+9)(x-1) = 0$
Use the principle of zero products.	$x+9 = 0 \quad or \quad x-1 = 0$
	$x = -9 \quad or \quad x = 1$
	(Checks for -9 and 1 are on the following page.)

Example 6 (continued)

Check for -9.

$$\log_3 x + \log_3(x+8) = 2$$

$$\log_3(-9) + \log_3(-9+8) \mid 2$$

The number -9 is **not** a solution because negative numbers do not have real-number logarithms.

Check for 1.

$$\log_3 x + \log_3(x+8) = 2$$

$$
\begin{array}{c|c}
\log_3 1 + \log_3(1+8) & 2 \\
\log_3 1 + \log_3 9 & \\
0 + 2 & \\
2 & 2 \quad \text{True}
\end{array}
$$

The solution is 1.

Exercises Solve.

13. $\log_{12}(x-5) + \log_{12}(x+5) = 2$

14. $\log_8 x + \log_8(x-7) = 1$

15. $\log(x-21) + \log x = 2$

16. $\log_3(x+1) + \log_3(x-1) = 3$

Interactive Preview Worksheet 39: USING AN INVERSE MATRIX TO SOLVE A SYSTEM OF EQUATIONS

We begin by recalling the multiplicative inverse property and the definitions of an identity matrix and the inverse of a matrix.

Multiplicative Inverse Property

For every nonzero real number a, there is a **multiplicative inverse**, $\dfrac{1}{a}$, or a^{-1}, such that

$$a \cdot \frac{1}{a} = \frac{1}{a} \cdot a = 1, \text{ or } a \cdot a^{-1} = a^{-1} \cdot a = 1.$$

Example: 6 and $\dfrac{1}{6}$ are multiplicative inverses.

$$6 \cdot \frac{1}{6} = \frac{1}{6} \cdot 6 = 1, \text{ or } 6 \cdot 6^{-1} = 6^{-1} \cdot 6 = 1.$$

Identity Matrix

For any positive integer n, the $n \times n$ **identity matrix**, $\mathbf{I}$, is an $n \times n$ matrix with 1's on the main diagonal and 0's elsewhere.

Examples:
$$\mathbf{I} = \begin{bmatrix} 1 & 0 \\ 0 & 1 \end{bmatrix} \qquad \text{and} \qquad \mathbf{I} = \begin{bmatrix} 1 & 0 & 0 \\ 0 & 1 & 0 \\ 0 & 0 & 1 \end{bmatrix}$$

2×2 identity matrix $\qquad\qquad$ 3×3 identity matrix

$\mathbf{A} \times \mathbf{I} = \mathbf{I} \times \mathbf{A} = \mathbf{A}$ for any $n \times n$ matrix $\mathbf{A}$.

Example:
$$\begin{bmatrix} 2 & -9 \\ -4 & 8 \end{bmatrix} \cdot \begin{bmatrix} 1 & 0 \\ 0 & 1 \end{bmatrix} = \begin{bmatrix} 1 & 0 \\ 0 & 1 \end{bmatrix} \cdot \begin{bmatrix} 2 & -9 \\ -4 & 8 \end{bmatrix} = \begin{bmatrix} 2 & -9 \\ -4 & 8 \end{bmatrix}$$
$$\mathbf{A} \quad \cdot \quad \mathbf{I} \quad = \quad \mathbf{I} \quad \cdot \quad \mathbf{A} \quad = \quad \mathbf{A}$$

Inverse of a Matrix

For an $n \times n$ matrix $\mathbf{A}$, if there is a matrix $\mathbf{A}^{-1}$ for which $\mathbf{A}^{-1} \cdot \mathbf{A} = \mathbf{A} \cdot \mathbf{A}^{-1} = \mathbf{I}$, then $\mathbf{A}^{-1}$ is the inverse of $\mathbf{A}$.

Example: $\begin{bmatrix} 8 & -5 \\ -3 & 2 \end{bmatrix}$ is the inverse of $\begin{bmatrix} 2 & 5 \\ 3 & 8 \end{bmatrix}$ because

$$\begin{bmatrix} 8 & -5 \\ -3 & 2 \end{bmatrix}\begin{bmatrix} 2 & 5 \\ 3 & 8 \end{bmatrix} = \begin{bmatrix} 2 & 5 \\ 3 & 8 \end{bmatrix}\begin{bmatrix} 8 & -5 \\ -3 & 2 \end{bmatrix} = \begin{bmatrix} 1 & 0 \\ 0 & 1 \end{bmatrix}$$
$$\mathbf{A}^{-1} \quad \cdot \quad \mathbf{A} \qquad \mathbf{A} \quad \cdot \quad \mathbf{A}^{-1} \quad = \quad \mathbf{I}$$

Let's compare solving an equation using the multiplicative inverse property with solving a system of equations using an inverse matrix and an identity matrix.

Example 1   Solve: $\dfrac{2}{5}x = -8$.

Multiply on both sides by $\dfrac{5}{2}$, the multiplicative inverse of $\dfrac{2}{5}$.	$\dfrac{5}{2} \cdot \dfrac{2}{5}x = \dfrac{5}{2}(-8)$
Use the multiplicative inverse property: $\dfrac{5}{2} \cdot \dfrac{2}{5} = 1$.	$1 \cdot x = -\dfrac{40}{2}$
Simplify.	$x = -20$

The solution is -20.

We can write a system of n linear equations in n variables as a matrix equation $\mathbf{AX} = \mathbf{B}$. If $\mathbf{A}$ has an inverse, then the system of equations has a unique solution that can be found by solving for $\mathbf{X}$.

Example 2   Solve: $2x + 5y = -1,$ The inverse of $\begin{bmatrix} 2 & 5 \\ 3 & 8 \end{bmatrix}$ is $\begin{bmatrix} 8 & -5 \\ -3 & 2 \end{bmatrix}$.

$\phantom{\text{Example 2}\quad\text{Solve: }}3x + 8y = 2.$

Write an equivalent matrix equation.	$\begin{bmatrix} 2 & 5 \\ 3 & 8 \end{bmatrix} \cdot \begin{bmatrix} x \\ y \end{bmatrix} = \begin{bmatrix} -1 \\ 2 \end{bmatrix}$
Multiply on the left on both sides by the inverse of $\begin{bmatrix} 2 & 5 \\ 3 & 8 \end{bmatrix}$.	$\begin{bmatrix} 8 & -5 \\ -3 & 2 \end{bmatrix} \cdot \begin{bmatrix} 2 & 5 \\ 3 & 8 \end{bmatrix} \cdot \begin{bmatrix} x \\ y \end{bmatrix} = \begin{bmatrix} 8 & -5 \\ -3 & 2 \end{bmatrix} \cdot \begin{bmatrix} -1 \\ 2 \end{bmatrix}$
Use $\mathbf{A}^{-1} \cdot \mathbf{A} = \mathbf{I}$ on the left side and multiply on the right side.	$\begin{bmatrix} 1 & 0 \\ 0 & 1 \end{bmatrix} \cdot \begin{bmatrix} x \\ y \end{bmatrix} = \begin{bmatrix} -18 \\ 7 \end{bmatrix}$
Use $\mathbf{I} \cdot \mathbf{A} = \mathbf{A}$.	$\begin{bmatrix} x \\ y \end{bmatrix} = \begin{bmatrix} -18 \\ 7 \end{bmatrix}$

Check.

$$\begin{array}{c|c} 2x + 5y = -1 & \\ \hline 2(-18) + 5 \cdot 7 & -1 \\ -36 + 35 & \\ -1 & -1 \quad \text{True} \end{array}$$

Substitute -18 for x and 7 for y.

$$\begin{array}{c|c} 3x + 8y = 2 & \\ \hline 3(-18) + 8 \cdot 7 & 2 \\ -54 + 56 & \\ 2 & 2 \quad \text{True} \end{array}$$

The solution is $(-18, 7)$.

Exercises In Exercises 1-4, fill in the blanks at key steps in the problem-solving process.

1. Solve: $\dfrac{4}{9}y = -16$.

$$\boxed{} \cdot \dfrac{4}{9}y = \boxed{} \cdot (-16)$$

$$1 \cdot y = -\dfrac{\boxed{}}{4}$$

$$y = \boxed{}$$

2. Solve: $-\dfrac{5}{2}t = -\dfrac{7}{10}$.

$$\boxed{} \cdot \left(-\dfrac{5}{2}t\right) = \boxed{} \cdot \left(-\dfrac{7}{10}\right)$$

$$\boxed{} \cdot t = \dfrac{14}{\boxed{}}$$

$$t = \dfrac{\boxed{}}{25}$$

3. Solve: $-3x + 4y = -9$,
$5x - 7y = 16$.

The inverse of $\begin{bmatrix} -3 & 4 \\ 5 & -7 \end{bmatrix}$ is $\begin{bmatrix} -7 & -4 \\ -5 & -3 \end{bmatrix}$.

Write an equivalent matrix equation, $\mathbf{AX} = \mathbf{B}$.

$$\begin{bmatrix} \\ \end{bmatrix} \cdot \begin{bmatrix} x \\ y \end{bmatrix} = \begin{bmatrix} \\ \end{bmatrix}$$

Multiply on both sides by the inverse of $\mathbf{A}$.

$$\begin{bmatrix} \\ \end{bmatrix} \cdot \begin{bmatrix} -3 & 4 \\ 5 & -7 \end{bmatrix} \cdot \begin{bmatrix} x \\ y \end{bmatrix} = \begin{bmatrix} \\ \end{bmatrix} \cdot \begin{bmatrix} -9 \\ 16 \end{bmatrix}$$

Simplify.

$$\begin{bmatrix} \\ \end{bmatrix} \cdot \begin{bmatrix} x \\ y \end{bmatrix} = \begin{bmatrix} \\ \end{bmatrix}$$

Use $\mathbf{I} \cdot \mathbf{A} = \mathbf{A}$.

$$\begin{bmatrix} \\ \end{bmatrix} = \begin{bmatrix} \\ \end{bmatrix}$$

Check:

$-3x + 4y = -9$		$5x - 7y = 16$	
$-3 \cdot \boxed{} + 4 \cdot \boxed{}$	-9	$5 \cdot \boxed{} - 7 \cdot \boxed{}$	16
$\boxed{} + \boxed{}$		$\boxed{} - \boxed{}$	
$\boxed{}$	-9	$\boxed{}$	16

The solution is $\left(\boxed{}, \boxed{}\right)$.

4. Solve: $5a - 4b = -6$,
$\qquad 7a - 6b = -10$.

The inverse of $\begin{bmatrix} 5 & -4 \\ 7 & -6 \end{bmatrix}$ is $\begin{bmatrix} 3 & -2 \\ \dfrac{7}{2} & -\dfrac{5}{2} \end{bmatrix}$.

Write an equivalent matrix equation, $\mathbf{AX} = \mathbf{B}$.	$\begin{bmatrix} 5 & -4 \\ 7 & -6 \end{bmatrix} \cdot \begin{bmatrix} \ \ \\ \ \ \end{bmatrix} = \begin{bmatrix} \ \ \\ \ \ \end{bmatrix}$
Multiply by the inverse of $\mathbf{A}$.	$\begin{bmatrix} \ \ \\ \ \ \end{bmatrix} \cdot \begin{bmatrix} 5 & -4 \\ 7 & -6 \end{bmatrix} \cdot \begin{bmatrix} \ \ \\ \ \ \end{bmatrix} = \begin{bmatrix} \ \ \\ \ \ \end{bmatrix} \cdot \begin{bmatrix} \ \ \\ \ \ \end{bmatrix}$
Simplify.	$\begin{bmatrix} \ \ \\ \ \ \end{bmatrix} \begin{bmatrix} a \\ b \end{bmatrix} = \begin{bmatrix} \ \ \\ \ \ \end{bmatrix}$
Use $\mathbf{I} \cdot \mathbf{A} = \mathbf{A}$.	$\begin{bmatrix} a \\ b \end{bmatrix} = \begin{bmatrix} \ \ \\ \ \ \end{bmatrix}$

The solution is $\left(\boxed{}, \boxed{} \right)$.

5. Solve: $-3x = \dfrac{6}{11}$.

6. Solve: $\dfrac{3}{4}y = -450$.

7. Solve: $2s - 3t = 16$,
$\qquad s - t = 7$.

The inverse of $\begin{bmatrix} 2 & -3 \\ 1 & -1 \end{bmatrix}$ is $\begin{bmatrix} -1 & 3 \\ -1 & 2 \end{bmatrix}$.

8. Solve: $\quad 2x + 3y = 1$,
$\qquad 16x + 25y = 5$.

The inverse of $\begin{bmatrix} 2 & 3 \\ 16 & 25 \end{bmatrix}$ is $\begin{bmatrix} \dfrac{25}{2} & -\dfrac{3}{2} \\ -8 & 1 \end{bmatrix}$.

Interactive Preview Worksheet 40: THE ELLIPSE

In this Preview, we will consider only ellipses with the center at the origin.

Example 1

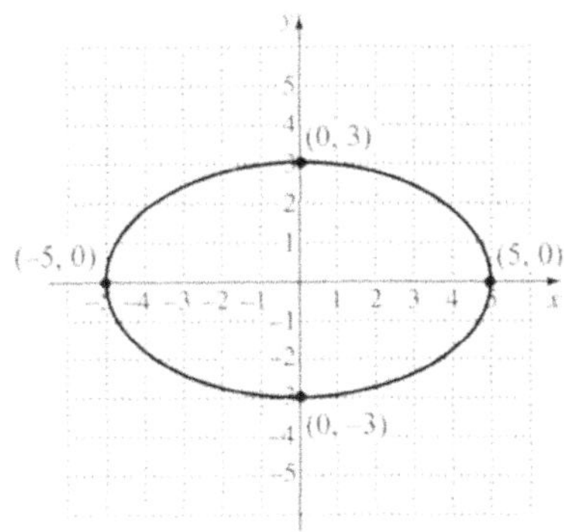

Equation: $\dfrac{x^2}{5^2} + \dfrac{y^2}{3^2} = 1$, or $\dfrac{x^2}{25} + \dfrac{y^2}{9} = 1$

$\lfloor 5 > 3 \rfloor$ and 5^2 is the denominator of $\dfrac{x^2}{5^2}$;

thus, the *major axis* is on the x-axis.

Center: $(0, 0)$

Endpoints of the **major axis**: $(-5, 0)$ and $(5, 0)$

These points are the **vertices** of the ellipse. They are also the *x-intercepts* of the ellipse. The major axis is *horizontal*.

Endpoints of the **minor axis**: $(0, -3)$ and $(0, 3)$

These points are the *y-intercepts* of the ellipse. The minor axis is *vertical*.

Example 2

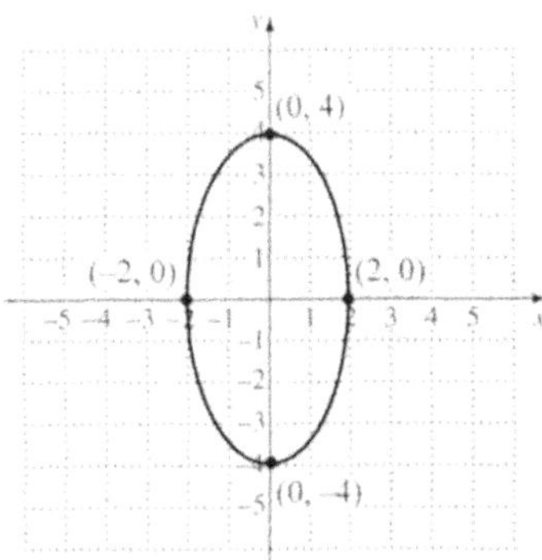

Equation: $\dfrac{x^2}{2^2} + \dfrac{y^2}{4^2} = 1$, or $\dfrac{x^2}{4} + \dfrac{y^2}{16} = 1$

$\lfloor 4 > 2 \rfloor$ and 4^2 is the denominator of $\dfrac{y^2}{4^2}$;

thus, the *major axis* is on the y-axis.

Center: $(0, 0)$

Endpoints of the **major axis**: $(0, -4)$ and $(0, 4)$

These points are the **vertices** of the ellipse. They are also the *y-intercepts* of the ellipse. The major axis is *vertical*.

Endpoints of the **minor axis**: $(-2, 0)$ and $(2, 0)$

These points are the *x-intercepts* of the ellipse. The minor axis is *horizontal*.

Graphs of Ellipses with Center at the Origin

- The longer axis is the **major axis**. The shorter axis is the **minor axis**.
- The endpoints of the major axis are the **vertices** of the ellipse.
- If the major axis is *horizontal*, the endpoints of the axis are the x-intercepts. Then the endpoints of the minor axis are the y-intercepts.
- If the major axis is *vertical*, the endpoints of the axis are the y-intercepts. Then the endpoints of the minor axis are the x-intercepts.

Exercises Fill in the coordinates of the *x*- and *y*-intercepts of the graph of the ellipse. Then fill in the blanks to complete the description of the graph.

1.

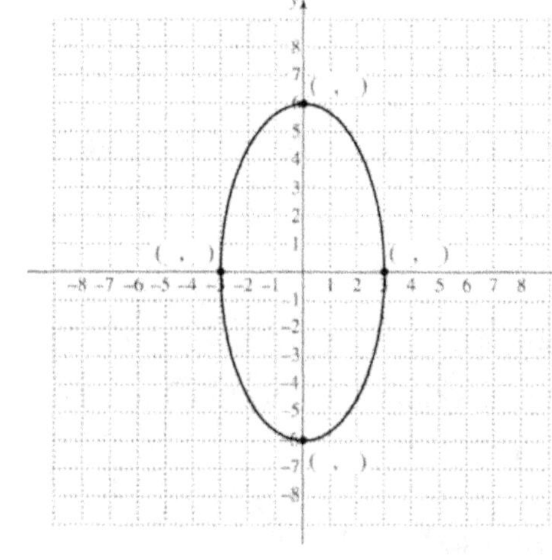

Center: (,   )

The major axis is _____________.
horizontal / vertical

The minor axis is _____________.
horizontal / vertical

Vertices: (,   ) and (,   )

Endpoints of major axis: (,   ) and (,   )

Endpoints of minor axis: (,   ) and (,   )

x-intercepts: (,   ) and (,   )

y-intercepts: (,   ) and (,   )

Equation: $\dfrac{x^2}{\Box^2}+\dfrac{y^2}{\Box^2}=1$, or $\dfrac{x^2}{\Box}+\dfrac{y^2}{\Box}=1$

2.

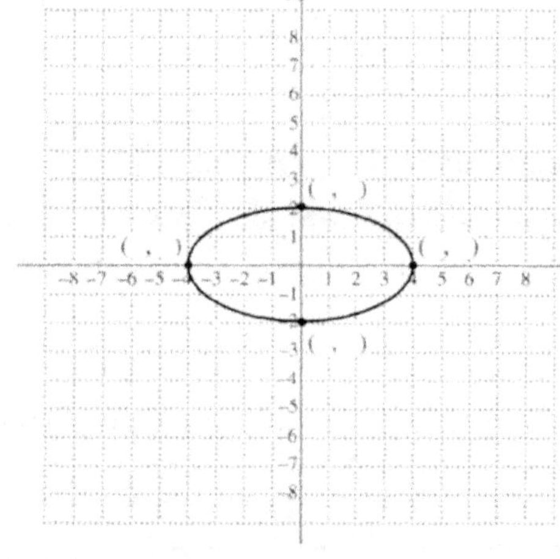

Center: (,   )

The major axis is _____________.
horizontal / vertical

The minor axis is _____________.
horizontal / vertical

Vertices: (,   ) and (,   )

Endpoints of major axis: (,   ) and (,   )

Endpoints of minor axis: (,   ) and (,   )

x-intercepts: (,   ) and (,   )

y-intercepts: (,   ) and (,   )

Equation: $\dfrac{x^2}{\Box^2}+\dfrac{y^2}{\Box^2}=1$, or $\dfrac{x^2}{\Box}+\dfrac{y^2}{\Box}=1$

Interactive Preview Worksheet 41: THE HYPERBOLA

In this Preview, we will consider only hyperbolas with the center at the origin.

Example 1

Equation: $\dfrac{x^2}{4} - \dfrac{y^2}{25} = 1$, or $\dfrac{x^2}{2^2} - \dfrac{y^2}{5^2} = 1$

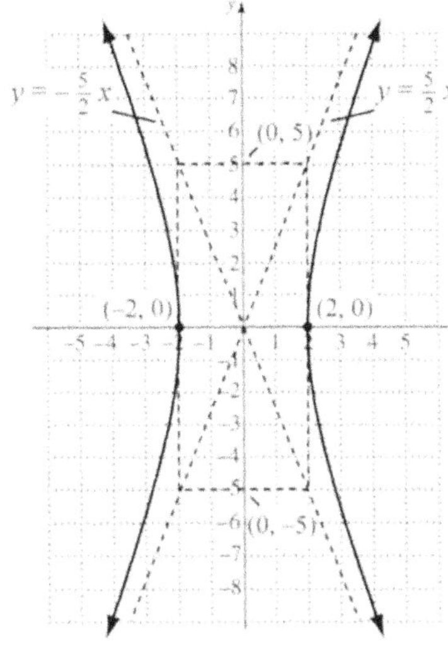

The x^2-term is the first term, thus

- the **transverse axis** is horizontal and is on the x-axis,
- the endpoints of the transverse axis are the **x-intercepts** $(-2, 0)$ and $(2, 0)$,
- the **vertices** of the hyperbola are the endpoints of the transverse axis, $(-2, 0)$ and $(2, 0)$, and
- the graph opens left and right.

The y^2-term is the second term, thus

- the **conjugate axis** is vertical and is on the y-axis, and
- the endpoints of the conjugate axis are $(0, -5)$ and $(0, 5)$.

The lines $y = \dfrac{5}{2}x$ and $y = -\dfrac{5}{2}x$ are the asymptotes of the hyperbola. As $|x|$ gets larger, the graph of the hyperbola gets closer and closer to the asymptotes.

Example 2

Equation: $\dfrac{y^2}{9} - \dfrac{x^2}{16} = 1$, or $\dfrac{y^2}{3^2} - \dfrac{x^2}{4^2} = 1$

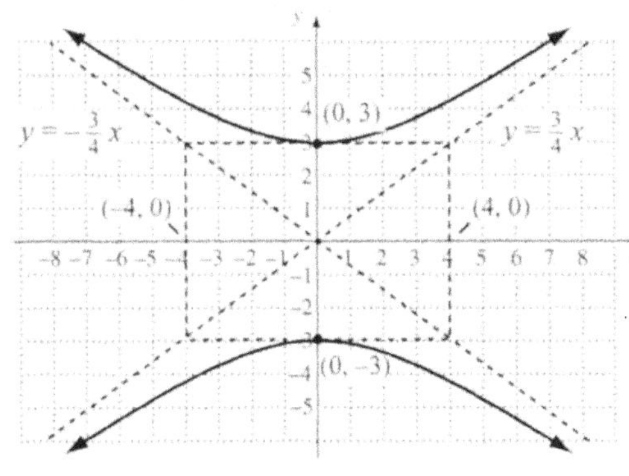

The y^2-term is the first term, thus

- the **transverse axis** is vertical and is on the y-axis,
- the endpoints of the transverse axis are the **y-intercepts** $(0, -3)$ and $(0, 3)$,
- the **vertices** of the hyperbola are the endpoints of the transverse axis, $(0, -3)$ and $(0, 3)$, and
- the graph opens up and down.

The x^2-term is the second term, thus

- the **conjugate axis** is horizontal and is on the x-axis, and
- the endpoints of the conjugate axis are $(-4, 0)$ and $(4, 0)$.

The lines $y = \dfrac{3}{4}x$ and $y = -\dfrac{3}{4}x$ are the asymptotes of the hyperbola. As $|y|$ gets larger, the graph of the hyperbola gets closer and closer to the asymptotes.

To draw a graph of a hyperbola, it helps to first sketch the asymptotes. The asymptotes of the hyperbola are the extended diagonals of the rectangle formed by sides passing through the endpoints of the transverse axis and the conjugate axis.

Example 3 Graph: $\dfrac{y^2}{16} - \dfrac{x^2}{49} = 1$, or $\dfrac{y^2}{4^2} - \dfrac{x^2}{7^2} = 1$.

The y^2-term is the first term. The transverse axis is vertical. The endpoints of the transverse axis are $(0, -4)$ and $(0, 4)$. The endpoints of the conjugate axis are $(-7, 0)$ and $(7, 0)$. We sketch a rectangle whose sides pass through these four points. (See Figure 1 below.)

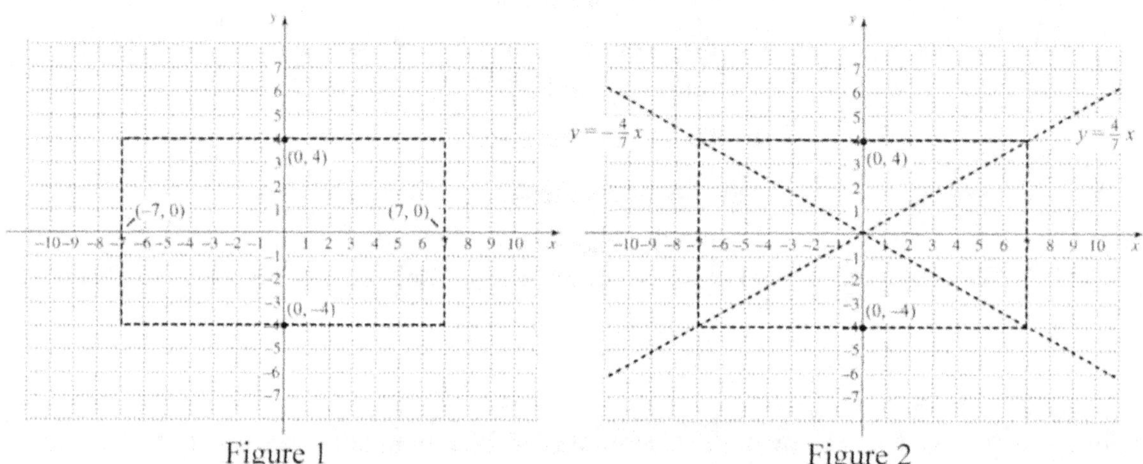

Figure 1
Figure 2

Next, we sketch the asymptotes which are the extended diagonals of the rectangle. (See Figure 2 above.) The equations of the asymptotes are $y = \dfrac{4}{7}x$ and $y = -\dfrac{4}{7}x$. The vertices $(0, 4)$ and $(0, -4)$ are on the y-axis. The graph opens up and down. Finally, we draw the branches of the hyperbola outward from the vertices toward the asymptotes. It can be helpful to plot four more points.

If $y = 5$ or $y - 5$, then

$$\frac{25}{16} - \frac{x^2}{49} = 1.$$

We solve for x:

$$-\frac{x^2}{49} = -\frac{9}{16}$$

$$x^2 = \frac{9 \cdot 49}{16}$$

$$x = \pm 5.25$$

Four points on the graph are $(5.25, 5)$, $(5.25, -5)$, $(-5.25, 5)$, and $(-5.25, -5)$.

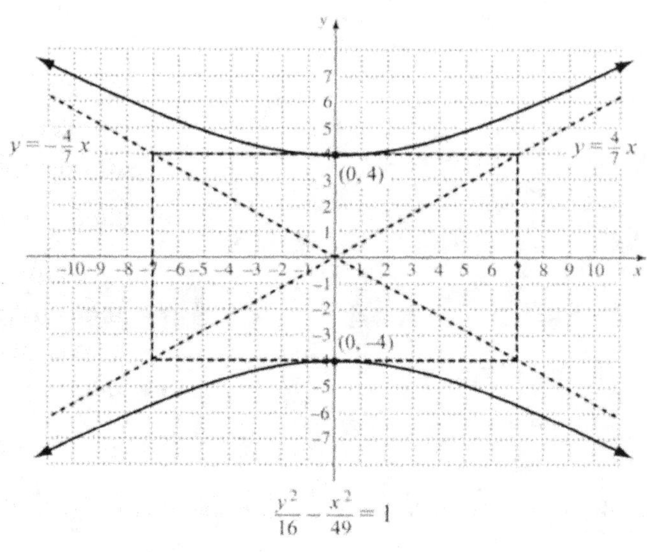

Exercises

1. Graph: $\dfrac{x^2}{36} - \dfrac{y^2}{9} = 1$, or $\dfrac{x^2}{6^2} - \dfrac{y^2}{3^2} = 1$. Begin by completing the statements below.

    - The ___-term is the first term, thus the transverse axis is on the ___-axis.

    - The endpoints of the transverse axis are _________ and _________.

    - The endpoints of the conjugate axis are _________ and _________.

    - The vertices are _________ and _________.

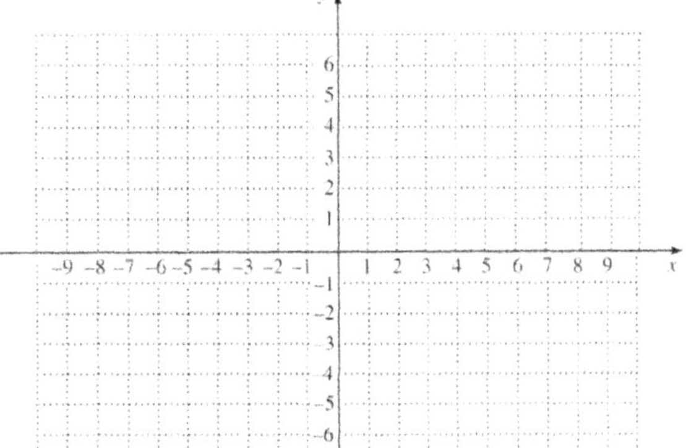

2. Graph: $\dfrac{y^2}{4} - \dfrac{x^2}{25} = 1$, or $\dfrac{y^2}{2^2} - \dfrac{x^2}{5^2} = 1$. Begin by completing the statements below.

    - The ___-term is the first term, thus the transverse axis is on the ___-axis.

    - The endpoints of the transverse axis are _________ and _________.

    - The endpoints of the conjugate axis are _________ and _________.

    - The vertices are _________ and _________.

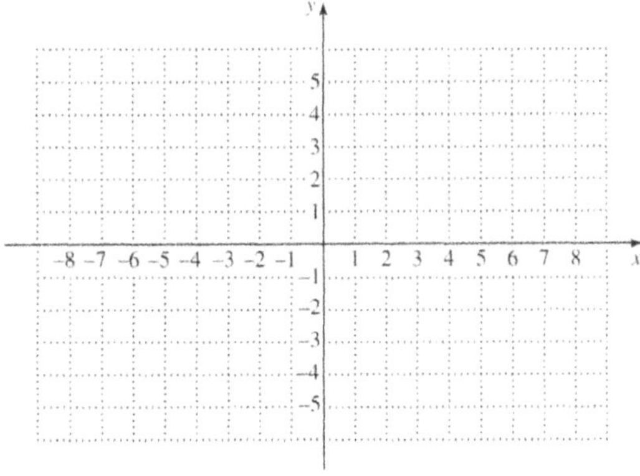

In Exercises 3-8, match each equation with its graph from choices A-F.

A.
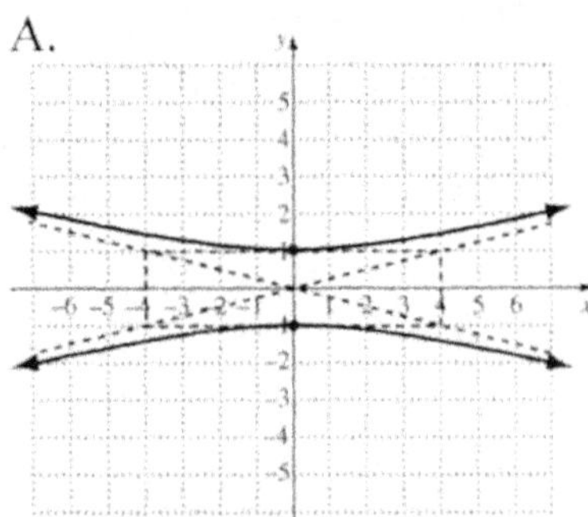

B.
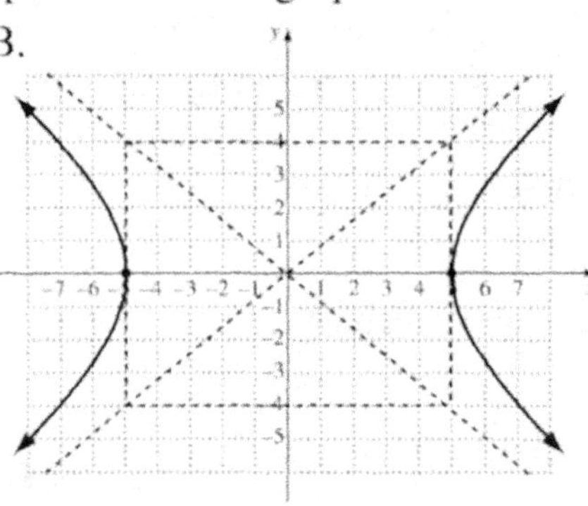

C.
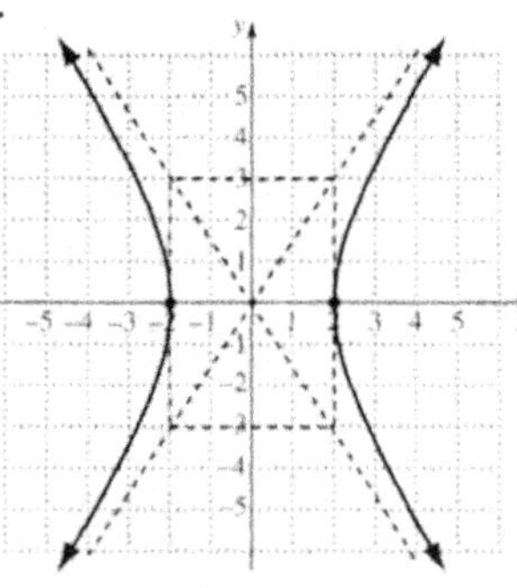

D.
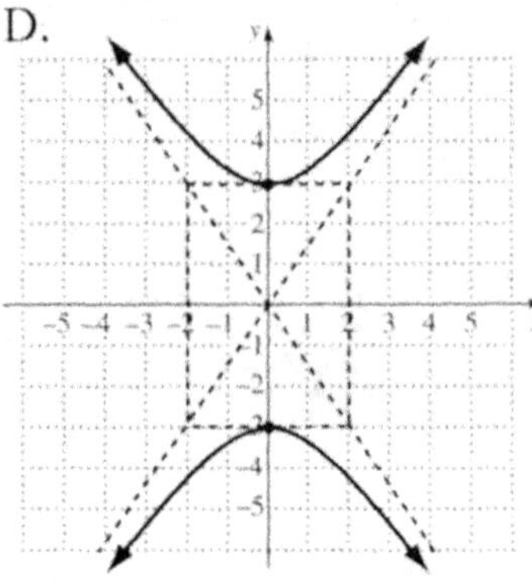

E.
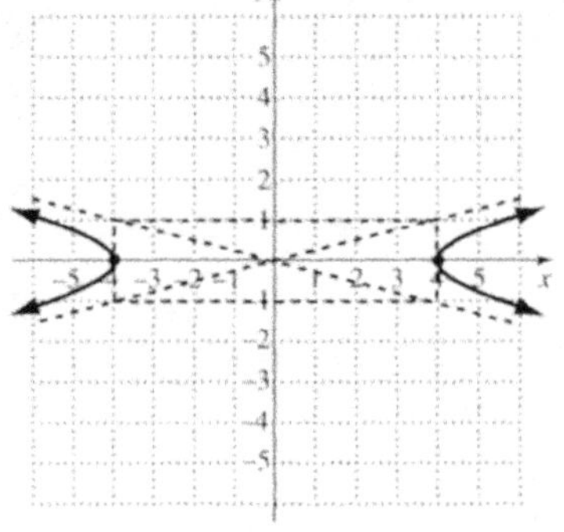

F.
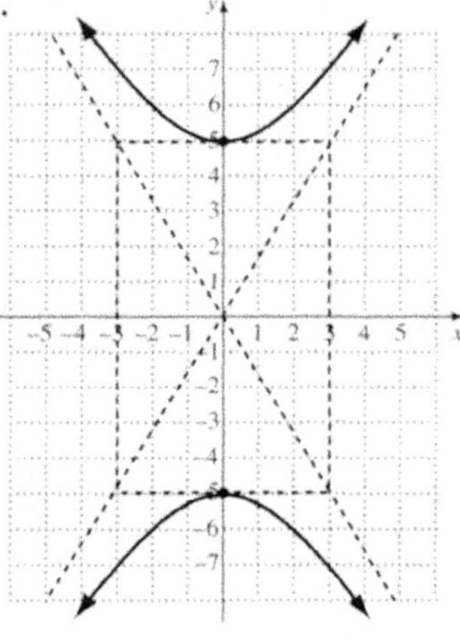

3. $\dfrac{x^2}{16} - \dfrac{y^2}{1} = 1$

4. $\dfrac{y^2}{9} - \dfrac{x^2}{4} = 1$

5. $\dfrac{y^2}{1} - \dfrac{x^2}{16} = 1$

6. $\dfrac{x^2}{25} - \dfrac{y^2}{16} = 1$

7. $\dfrac{y^2}{25} - \dfrac{x^2}{9} = 1$

8. $\dfrac{x^2}{4} - \dfrac{y^2}{9} = 1$

Interactive Preview Worksheet 42: CLASSIFYING EQUATIONS OF CONIC SECTIONS

Conic sections can be described algebraically using second-degree equations of the form $Ax^2 + Bxy + Cy^2 + Dx + Ey + F = 0$. Circles, ellipses, hyperboles, and parabolas are conic sections. In this Preview, the circles, ellipses, and hyperbolas we consider are centered at the origin.

Circles:

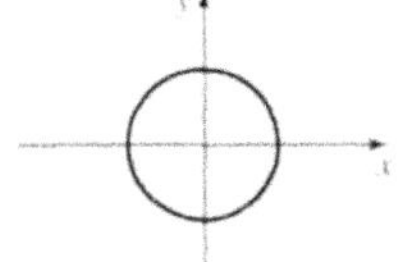

- Can be written in the form $x^2 + y^2 = r^2$, $r > 0$.

- Both variables, x and y, are squared.

- The squared terms, x^2 and y^2, are added.

- The coefficients of x^2 and y^2 are the same.

Examples of equations of circles:

Equivalent equation

$x^2 + y^2 = 30$	$x^2 + y^2 = \left(\sqrt{30}\right)^2$
$4x^2 + 4y^2 = 49$	$x^2 + y^2 = \left(\dfrac{7}{2}\right)^2$
$y^2 - 23 = 2 - x^2$	$x^2 + y^2 = 5^2$

Ellipses:

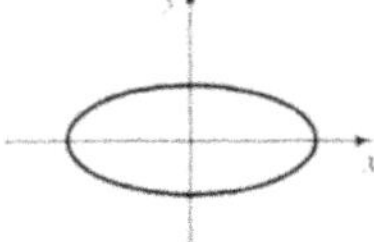

- Can be written in the form $\dfrac{x^2}{t^2} + \dfrac{y^2}{s^2} = 1$, t and $s > 0$.

- Both variables, x and y, are squared.

- The squared terms, x^2 and y^2, are added.

- The coefficients of x^2 and y^2 are not the same.

Examples of equations of ellipses:

Equivalent equation

$\dfrac{x^2}{9} + \dfrac{y^2}{16} = 1$	$\dfrac{x^2}{3^2} + \dfrac{y^2}{4^2} = 1$
$25y^2 + 4x^2 = 100$	$\dfrac{x^2}{5^2} + \dfrac{y^2}{2^2} = 1$
$4y^2 = 16 - x^2$	$\dfrac{x^2}{4^2} + \dfrac{y^2}{2^2} = 1$

Hyperbolas:

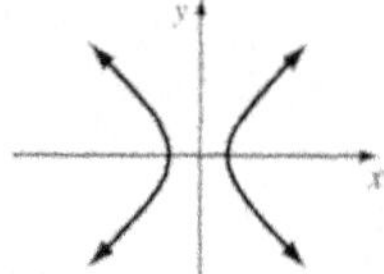

- Can be written in the form $\dfrac{x^2}{r^2} - \dfrac{y^2}{w^2} = 1$ or $\dfrac{y^2}{w^2} - \dfrac{x^2}{r^2} = 1$, r and $w > 0$.

- Both variables, x and y, are squared.

- The squared terms, x^2 and y^2, are not added.

Examples of equations of hyperbolas:

Equivalent equation

$\dfrac{x^2}{4} - \dfrac{y^2}{25} = 1$	$\dfrac{x^2}{2^2} - \dfrac{y^2}{5^2} = 1$
$4y^2 - 9x^2 = 36$	$\dfrac{y^2}{3^2} - \dfrac{x^2}{2^2} = 1$
$9x^2 - 225 = 25y^2$	$\dfrac{x^2}{5^2} - \dfrac{y^2}{3^2} = 1$

Parabolas:

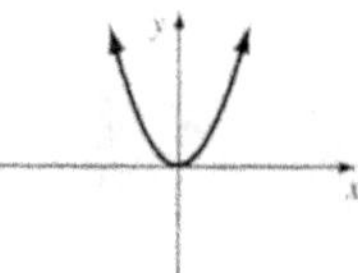

- Can be written in the form $y = ax^2 + bx + c$ or $x = ay^2 + by + c$.

- Only one of the variables, x and y, is squared.

Examples of equations of parabolas:

$y = 3x^2 + 13x - 10$

$x = y^2 - 5y + 4$

$3y = x^2 \qquad \left(\text{Equivalent equation: } y = \dfrac{1}{3}x^2\right)$

$y - 10x = x^2 + 24 \qquad \left(\text{Equivalent equation: } y = x^2 + 10x + 24\right)$

Examples Classify the equation as an equation of a circle, an ellipse, a hyperbola, or a parabola.

1. $6x^2 + 6y^2 = 216$

   - An equivalent equation is $x^2 + y^2 = 36$, or $x^2 + y^2 = 6^2$.
   - Both variables are squared, so this cannot be a parabola.
   - The squared terms are added, so this cannot be a hyperbola.
   - The coefficients of x^2 and y^2 are the same.
   - This is an equation of a circle.

2. $16x^2 + 25y^2 = 400$

- An equivalent equation is $\dfrac{x^2}{25} + \dfrac{y^2}{16} = 1$, or $\dfrac{x^2}{5^2} + \dfrac{y^2}{4^2} = 1$.
- Both variables are squared, so this cannot be a parabola.
- The squared terms are added, so this cannot be a hyperbola.
- The coefficients of x^2 and y^2 are not the same.
- This is an equation of an ellipse.

3. $49y^2 - 196 = 4x^2$

- An equivalent equation is $\dfrac{y^2}{4} - \dfrac{x^2}{49} = 1$, or $\dfrac{y^2}{2^2} - \dfrac{x^2}{7^2} = 1$.
- Both variables are squared, so this cannot be a parabola.
- The squared terms are not added, so this cannot be a circle or an ellipse.
- This is an equation of a hyperbola.

4. $y - 7x = x^2 - 18$

- An equivalent equation is $y = x^2 + 7x - 18$.
- Only one variable is squared, so this cannot be a circle, an ellipse, or a hyperbola.
- This is an equation of a parabola.

Exercises Classify the equation as an equation of a circle, an ellipse, a hyperbola, or a parabola.

1. $y = 3x^2$

2. $\dfrac{x^2}{100} + \dfrac{y^2}{25} = 1$

3. $\dfrac{x^2}{81} - \dfrac{y^2}{4} = 1$

4. $36y^2 - 9x^2 = 324$

5. $y^2 = 16 - x^2$

6. $x = y^2 + 6y - 27$

7. $y = x^2 - 2x$

8. $9x^2 + 4x^2 = 36$

9. $\dfrac{x^2}{16} - \dfrac{y^2}{25} = 1$

10. $5x^2 + 5y^2 = 20$

11. $y - 16x = x^2 + 55$

12. $4x^2 - 100 = 25y^2$

13. $25y^2 = 10 - x^2$

14. $\dfrac{1}{2}x^2 + \dfrac{1}{2}y^2 = 1$

15. $\dfrac{x^2}{3} + \dfrac{y^2}{5} = 1$

16. $x = y^2 + 4$

Video Worksheets

Set Notation and the Set of Real Numbers

ESSENTIALS

A **set** is a collection of objects. A **subset** is a set contained within another set. **Roster notation** is a way of designating a set by enclosing a list of its elements in braces.

Natural numbers (or **counting numbers**) are those numbers used for counting:
$$\{1, 2, 3, \ldots\}.$$

Whole numbers are the set of natural numbers with 0 included: $\{0, 1, 2, 3, \ldots\}.$

Integers are the set of all whole numbers and their opposites:
$$\{\ldots, -4, -3, -2, -1, 0, 1, 2, 3, 4, \ldots\}.$$

The **opposite** of a number is found by reflecting it across the number 0 on the number line. The natural numbers are called **positive integers**. The opposites of the natural numbers are called **negative integers**.

Set-builder notation is a way of designating a set that specifies conditions under which a number is in a set.

A **rational number** can be expressed as an integer divided by a nonzero integer. The set of rational numbers is

$$\left\{ \frac{p}{q} \,\middle|\, p \text{ is an integer, } q \text{ is an integer, and } q \neq 0 \right\}.$$

Rational numbers are numbers whose decimal representation either terminates or has a repeating block of digits. **Irrational numbers** are numbers whose decimal representation neither terminates nor has a repeating block of digits. They cannot be represented as the quotient of two integers. The set of all rational numbers, combined with the set of all irrational numbers, gives us the set of **real numbers**. The set of real numbers is
$$\{x \mid x \text{ is a rational number } or \text{ } x \text{ is an irrational number}\}.$$

Examples

- Represent the set consisting of the first 6 odd natural numbers.

  Roster notation: $\{1, 3, 5, 7, 9, 11\}$

  Set-builder notation: $\{n \mid n \text{ is an odd number between 0 and 13}\}$

- List the set(s) of numbers to which 3.5 belongs.

  Since 3.5 is a terminating decimal, it is a rational number, and it is also a real number.

- π and $\sqrt{13}$ are examples of irrational numbers.

GUIDED LEARNING   🔲 **Textbook**   👤 **Instructor**   ▶ **Video**

EXAMPLE 1	YOUR TURN 1
Find the opposite of 4.	Find the opposite of -6.
When 4 is reflected across 0 on the number line, we find the opposite. The opposite is the number ☐ .	

EXAMPLE 2	YOUR TURN 2
Use roster notation to write the set of natural numbers greater than 20.	Use roster notation to write the set of whole numbers between 6 and 15.
$\{21, 22, \boxed{}, 24, \ldots\}$	

EXAMPLE 3	YOUR TURN 3
Use set-builder notation to write the set $\{-5, -4, -3, -2, -1, 0, 1\}$.	Use set-builder notation to write the set $\{7, 8, 9, 10\}$.
$\{x \mid x$ is an integer greater than $\boxed{}$ and less than $\boxed{}\}$	

EXAMPLE 4	YOUR TURN 4
Which numbers in the list are (a) natural numbers? (b) whole numbers? (c) integers? (d) rational numbers? (e) irrational numbers? (f) real numbers?	Which numbers in the list are (a) natural numbers? (b) whole numbers? (c) integers? (d) rational numbers? (e) irrational numbers? (f) real numbers?
$5, -\dfrac{1}{3}, -4, \pi, 3.6, 0, \sqrt{2}$	$0.2, 0, -16, -\dfrac{1}{9}, \sqrt{7}, 125, 1$
a) Natural number: 5	
b) Whole numbers: 5, ☐	
c) Integers: 5, ☐, ☐	
d) Rational numbers: 5, ☐, -4, ☐, 0	
e) Irrational numbers: ☐, $\sqrt{2}$	
f) Real numbers: $5, -\dfrac{1}{3}, -4, \boxed{}, \boxed{}, 0, \sqrt{2}$	

YOUR NOTES   Write your questions and additional notes.

Order for the Real Numbers

ESSENTIALS

Inequality Symbols

$<$ means "is less than."

$>$ means "is greater than."

$\leq$ means "is less than or equal to."

$\geq$ means "is greater than or equal to."

If x is a positive real number, then $x > 0$.

If x is a negative real number, then $x < 0$.

$a < b$ also has the meaning $b > a$.

Sentences containing $<$, $>$, $\leq$, or $\geq$ are called **inequalities**.

Examples

- $-6 > -9$ because -6 is to the right of -9 on the number line.

- The inequalities $x > 8$ and $8 < x$ have the same meaning.

- Write the meaning of the inequality and determine whether it is a true statement:
$-4 \geq -2\frac{1}{4}$.

  "-4 is greater than or equal to $-2\frac{1}{4}$" is false because -4 is to the left of $-2\frac{1}{4}$ on the number line.

GUIDED LEARNING   📖 **Textbook**   👤 **Instructor**   ▶ **Video**

EXAMPLE 1	YOUR TURN 1
Use either $<$ or $>$ for ☐ to write a true sentence. 7 ☐ 2 Since 7 is to the right of 2 on the number line, 7 is greater than 2, so $7 > 2$.	Use either $<$ or $>$ for ☐ to write a true sentence. -11 ☐ -6
EXAMPLE 2	YOUR TURN 2
Use either $<$ or $>$ for ☐ to write a true sentence. -8 ☐ 3 Since -8 is to left of 3 on the number line, -8 is less than 3, so $-8 < 3$.	Use either $<$ or $>$ for ☐ to write a true sentence. 1 ☐ -15

EXAMPLE 3	YOUR TURN 3
Use either $<$ or $>$ for $\boxed{}$ to write a true sentence. $6.24\ \boxed{}\ 6.56$ Since 6.24 is to the left of 6.56 on the number line, 6.24 is less than 6.56, so $6.24 < 6.56$.	Use either $<$ or $>$ for $\boxed{}$ to write a true sentence. $\dfrac{2}{5}\ \boxed{}\ \dfrac{4}{7}$
EXAMPLE 4	YOUR TURN 4
Write a different inequality with the same meaning as $3 \ge x$. $\quad x\ \boxed{}\ 3$	Write a different inequality with the same meaning as $y < 10$.
EXAMPLE 5	YOUR TURN 5
Determine whether the following is true or false. $\quad -6 \ge -1.8$ The statement is _______________ since neither $-6 > -1.8$ nor $-6 = -1.8$ is true.	Determine whether the following is true or false. $3 \ge -7\dfrac{3}{4}$

YOUR NOTES   Write your questions and additional notes.

Graphing Inequalities on the Number Line

ESSENTIALS

A replacement that makes an inequality true is called a **solution**. The set of all solutions is called the **solution set**. A **graph** of an inequality is a drawing that represents its solution set.

Examples

- Graph $x \leq -2$.

  -6 -5 -4 -3 -2 -1 0 1 2 3 4 5 6

- Graph $x > 3$.

  -6 -5 -4 -3 -2 -1 0 1 2 3 4 5 6

	Textbook	Instructor	Video

GUIDED LEARNING

EXAMPLE 1	YOUR TURN 1
Graph: $x > 1$.	Graph: $x < 2$.

EXAMPLE 1

Graph: $x > 1$.

The solutions consist of all real numbers ___________ 1.
less than / greater than

They are shown on the number line by shading all numbers to the ___________ of 1.
left / right

The parenthesis at 1 indicates that 1 ___________ a
is / is not
solution.

-6 -5 -4 -3 -2 -1 0 1 2 3 4 5 6

YOUR TURN 1

Graph: $x < 2$.

-6 -5 -4 -3 -2 -1 0 1 2 3 4 5 6

EXAMPLE 2

Graph: $x \leq -3$.

The graph consists of 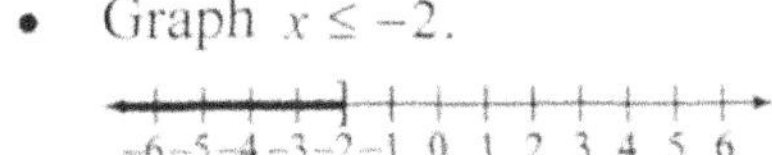 as well as the numbers less than -3.

They are shown on the number line by shading all the numbers to the ___________ of -3.
left / right

The bracket at -3 indicates that -3 ___________ a
is / is not
solution.

-6 -5 -4 -3 -2 -1 0 1 2 3 4 5 6

YOUR TURN 2

Graph: $x \leq 4$.

-6 -5 -4 -3 -2 -1 0 1 2 3 4 5 6

YOUR NOTES   Write your questions and additional notes.

Absolute Value

ESSENTIALS

The **absolute value** of a number is its distance from 0 on the number line. We use the symbol $|x|$ to represent the absolute value of a number x.

Examples

- $|-9| = 9$ The distance of -9 from 0 is 9.

- $|20| = 20$ The distance of 20 from 0 is 20.

- $|0| = 0$ The distance of 0 from 0 is 0.

GUIDED LEARNING   🔖 **Textbook**   👤 **Instructor**   ▶ **Video**

EXAMPLE 1	YOUR TURN 1
Find the absolute value: $|14|$. The distance of 14 from 0 is ☐, so $|14| = $ ☐.	Find the absolute value: $|-39|$.
EXAMPLE 2	YOUR TURN 2
Find the absolute value: $|-8.1|$. The distance of -8.1 from 0 is ☐, so $|-8.1| = $ ☐.	Find the absolute value: $\left|\dfrac{7}{8}\right|$.

YOUR NOTES   Write your questions and additional notes.

Practice Exercises

Readiness Check

Determine if each statement is true or false.

1. _________________ The number $2.61324189\ldots$ is a rational number.

2. _________________ The number 0 is a natural number.

3. _________________ All rational numbers and irrational numbers are real numbers.

4. _________________ Absolute value is never negative.

Set Notation and the Set of Real Numbers

Which numbers in the list provided are (a) natural numbers? (b) whole numbers? (c) integers? (d) rational numbers? (e) irrational numbers? (f) real numbers?

5. $-3.1,\ -1,\ 0,\ \sqrt{10},\ \dfrac{75}{2},\ 4$

6. $-\dfrac{3}{8},\ \dfrac{1}{2},\ \sqrt{12},\ \dfrac{43}{2},\ 51$

Use roster notation to write each set.

7. The set of letters in the word "history"

8. The set of all natural numbers that are multiples of 5

Use set-builder notation to write each set.

9. The set of all multiples of 10 between 20 and 100.

10. $\{72, 74, 76, 78, 80\}$

Order for the Real Numbers

Use either < or > for ☐ *to write a true sentence.*

11. $1 \,\boxed{}\, -15$ **12.** $-42 \,\boxed{}\, -21$ **13.** $-13.2 \,\boxed{}\, -15.1$

14. Write a second inequality with the same meaning as $x \ge -7$.

Write true or false.

15. $0 \le 2$ **16.** $6 \ge 6$ **17.** $-2 < -10\dfrac{1}{3}$

Graphing Inequalities on the Number Line

Graph on the number line.

18. $x < -4$ **19.** $x \ge 1$

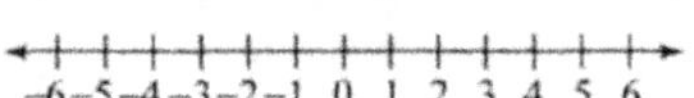

Absolute Value

Find the absolute value.

20. $|5|$ **21.** $|-3.8|$ **22.** $\left|\dfrac{0}{5}\right|$

Addition

ESSENTIALS

To find $a + b$ using the number line, we start at 0, move to a, and then move according to b.

- If b is positive, move to the right.
- If b is negative, move to the left.
- If b is 0, stay at a.

Rules for Addition of Real Numbers

1. *Positive numbers*: Add the numbers. The result is positive.
2. *Negative numbers*: Add absolute values. Make the answer negative.
3. *A positive number and a negative number*:
   - If the numbers have the same absolute value, the answer is 0.
   - If the numbers have different absolute values, subtract the smaller absolute value from the larger. Then:
     a) If the positive number has the greater absolute value, make the answer positive.
     b) If the negative number has the greater absolute value, make the answer negative.
4. *One number is zero*: The sum is the other number. This rule is known as the **identity property of 0**. It says that for any real number a, $a + 0 = a$.

Examples

- To add $5 + (-8)$ on the number line, we start at 0 and move 5 units right. Since -8 is negative, we then move 8 units left. $\qquad 5 + (-8) = -3$

- Add: $-7 + (-5)$.

  Both numbers are negative. Add absolute values, 7 and 5, and make the answer negative: $-7 + (-5) = -12$.

- Add: $12 + (-12)$.

  One number is positive and one is negative. The numbers have the same absolute value. The answer is 0: $12 + (-12) = 0$.

	Textbook		Instructor		Video

GUIDED LEARNING

EXAMPLE 1	YOUR TURN 1
Add: $-3 + 7$.	Add: $-2 + 5$.
Start at 0 and move ☐ units left. Then move ☐ units __________.	
$-3 + 7 = $ ☐	

EXAMPLE 2	YOUR TURN 2
Add: $-10+(-28)$.	Add: $-6+(-11)$.
Both numbers are _____________. _____________ absolute values, 10 and 28, and make the answer negative. $-10+(-28)=\boxed{}$	

EXAMPLE 3	YOUR TURN 3
Add: $-9+0$.	Add: $-13+13$.
One number is zero. The sum is the other number. $-9+0=\boxed{}$	

EXAMPLE 4	YOUR TURN 4
Add: $-2.5+10.3$.	Add: $-0.5+8.5$.
One number is positive and one is _________. _____________ the smaller absolute value from the greater absolute value: $10.3-2.5=7.8$. Since $|10.3|>|-2.5|$, the answer is _____________. $-2.5+10.3=\boxed{}$	

EXAMPLE 5	YOUR TURN 5
Add: $\dfrac{1}{4}+\left(-\dfrac{2}{3}\right)$.	Add: $\dfrac{3}{10}+\left(-\dfrac{1}{5}\right)$.
One number is _________ and one is negative. _____________ the smaller absolute value from the greater absolute value: $\dfrac{2}{3}-\dfrac{1}{4}=\dfrac{8}{12}-\dfrac{3}{12}=\dfrac{5}{12}$. Since $\left|-\dfrac{2}{3}\right|>\left|\dfrac{1}{4}\right|$, the answer is _____________. $\dfrac{1}{4}+\left(-\dfrac{2}{3}\right)=\boxed{}$	

YOUR NOTES   Write your questions and additional notes.

Opposites, or Additive Inverses

ESSENTIALS

Two numbers whose sum is 0 are called **opposites**, or **additive inverses**, of each other. For any real number a, the opposite, or additive inverse, of a, which is denoted $-a$, is such that

$$a + (-a) = 0.$$

A negative number is sometimes said to have a "negative sign." A positive number is said to have a "positive sign." When we replace a number with its opposite, or additive inverse, we can say that we have *"changed the sign."*

For any real number a, the absolute value of a, denoted $|a|$, is given by

$$|a| = \begin{cases} a, & \text{if } a \geq 0. \\ -a, & \text{if } a < 0. \end{cases}$$

For example, $|6| = 6$ and $|0| = 0$.
For example, $|-9| = -(-9) = 9$.

(The absolute value of a is a if a is nonnegative. The absolute value of a is the opposite of a if a is negative.)

Examples

- The opposite of 27 is -27 because $27 + (-27) = 0$.

- The opposite of $-\dfrac{3}{5}$ is $\dfrac{3}{5}$ because $-\dfrac{3}{5} + \dfrac{3}{5} = 0$.

- The opposite of 0 is 0 because $0 + 0 = 0$.

	🔖 **Textbook**	👤 **Instructor**	▶ **Video**

GUIDED LEARNING

EXAMPLE 1	YOUR TURN 1
Find the opposite, or additive inverse, of 11. $-(11) = \boxed{}$ The opposite of 11 is $\boxed{}$.	Find the opposite, or additive inverse, of -15.
EXAMPLE 2	**YOUR TURN 2**
Evaluate $-x$ and $-(-x)$ when $x = -12$. If $x = -12$, then $-x = -(-12) = \boxed{}$. If $x = -12$, then $-(-x) = -(-(-12)) = \boxed{}$.	Evaluate $-x$ and $-(-x)$ when $x = 7$.

EXAMPLE 3	YOUR TURN 3
Change the sign of $-\dfrac{3}{14}$. (Find the opposite, or additive inverse.) The opposite of $-\dfrac{3}{14}$ is $\boxed{}$.	Change the sign of $\dfrac{2}{25}$. (Find the opposite, or additive inverse.)

YOUR NOTES   Write your questions and additional notes.

Subtraction

ESSENTIALS

The difference $a - b$ is the number c for which $a = b + c$.

For any real numbers a and b, $a - b = a + (-b)$. (We can subtract by adding the opposite, or additive inverse, of the number being subtracted.)

Examples

- $8 - 9 = 8 + (-9) = -1$ Changing the sign of 9 and adding

- $-3 - (-7) = -3 + 7 = 4$ Changing the sign of -7 and adding

GUIDED LEARNING	📖 **Textbook**	👤 **Instructor**	▶ **Video**

EXAMPLE 1	YOUR TURN 1
Subtract: $5 - 9$.	Subtract: $1 - 6$.
Change the sign of 9 and add.	
$5 - 9 = 5 + \left(\boxed{}\right) = \boxed{}$	
EXAMPLE 2	YOUR TURN 2
Subtract: $-2.6 - (-9.1)$.	Subtract: $\dfrac{1}{4} - \left(-\dfrac{1}{2}\right)$.
Change the sign of -9.1 and add.	
$-2.6 - (-9.1) = -2.6 + \boxed{} = \boxed{}$	

YOUR NOTES   Write your questions and additional notes.

Multiplication

ESSENTIALS

To multiply a positive number and a negative number, multiply their absolute values. Then make the answer negative.

To multiply two negative numbers, multiply their absolute values. The answer is positive.

Examples

- $2(-5) = -10$

- $(-4)(-5) = 20$

- $-2(-1.8) = 3.6$

GUIDED LEARNING   📘 **Textbook**   👤 **Instructor**   ▶ **Video**

EXAMPLE 1	YOUR TURN 1
Multiply: $(-3)(20)$.	Multiply: $-8(11)$.

Since one number is negative and one is positive, the answer is ____________.

positive / negative

$(-3)(20) = \boxed{}$

EXAMPLE 2	YOUR TURN 2
Multiply: $\left(-\dfrac{2}{7}\right)\left(-\dfrac{3}{7}\right)$.	Multiply: $\left(-\dfrac{4}{5}\right)\left(-\dfrac{1}{3}\right)$.

Since both numbers are negative, the answer is ____________.

positive / negative

$\left(-\dfrac{2}{7}\right)\left(-\dfrac{3}{7}\right) = \boxed{}$

YOUR NOTES   Write your questions and additional notes.

Division

The quotient $a \div b$, or $\dfrac{a}{b}$, where $b \neq 0$, is that unique real number c for which $a = b \cdot c$.

To multiply or divide two real numbers:

1. Multiply or divide the absolute values.
2. If the signs are the same, then the answer is positive.
3. If the signs are different, then the answer is negative.

Division by 0 is not defined and is not possible.

Two numbers whose product is 1 are **reciprocals** (or **multiplicative inverses**). Every nonzero real number a has a **reciprocal** (or **multiplicative inverse**) $1/a$. The reciprocal of a positive number is positive. The reciprocal of a negative number is negative.

For any real numbers a and b, $b \neq 0$,

$$a \div b = \frac{a}{b} = a \cdot \frac{1}{b}.$$

(To divide, we can multiply by the reciprocal of the divisor.)

For any numbers a and b, $b \neq 0$,

$$\frac{-a}{b} = \frac{a}{-b} = -\frac{a}{b} \quad \text{and} \quad \frac{-a}{-b} = \frac{a}{b}.$$

Examples

- $\dfrac{8}{-2} = -4$ $\qquad\qquad$ $\dfrac{-8}{2} = -4$ $\qquad\qquad$ $-8 \div (-2) = 4$

- $\dfrac{2}{0}$ is not defined. $\qquad$ $0 \div 2 = 0$ $\qquad\qquad$ $\dfrac{2}{x - x}$ is not defined.

- The reciprocal of $-\dfrac{3}{4}$ is $-\dfrac{4}{3}$ because $\left(-\dfrac{3}{4}\right)\left(-\dfrac{4}{3}\right) = 1$.

- The reciprocal of 6 is $\dfrac{1}{6}$ because $6\left(\dfrac{1}{6}\right) = 1$.

- $\dfrac{2}{5} \div \left(-\dfrac{3}{10}\right) = \dfrac{2}{5} \cdot \left(-\dfrac{10}{3}\right) = -\dfrac{20}{15}$, or $-\dfrac{4}{3}$

GUIDED LEARNING　　　📖 **Textbook**　　👤 **Instructor**　　▶ **Video**	
EXAMPLE 1	YOUR TURN 1
Divide: $4.2 \div (-3)$. Do the long division $3\overline{)4.2}$ with quotient 1.4. The answer is negative. $4.2 \div (-3) = \boxed{}$	Divide: $-14.8 \div (-2)$.
EXAMPLE 2	YOUR TURN 2
Divide, if possible: $\dfrac{-8}{0}$. $\dfrac{a}{0}$ is not defined for any real number a, so $\dfrac{-8}{0}$ is ___________ .	Divide, if possible: $\dfrac{0}{4}$.
EXAMPLE 3	YOUR TURN 3
Find the reciprocal of $\dfrac{-1}{6}$. The reciprocal of $\dfrac{-1}{6}$ is $\dfrac{\boxed{}}{\boxed{}}$, or $\boxed{}$.	Find the reciprocal of $-\dfrac{4}{7}$.
EXAMPLE 4	YOUR TURN 4
Divide by multiplying by the reciprocal of the divisor: $\left(-\dfrac{3}{8}\right) \div \left(-\dfrac{1}{2}\right)$. $\left(-\dfrac{3}{8}\right) \div \left(-\dfrac{1}{2}\right) = \left(-\dfrac{3}{8}\right) \cdot \left(-\dfrac{\boxed{}}{\boxed{}}\right) = \dfrac{6}{8}$ $= \dfrac{3 \cdot \boxed{}}{4 \cdot \boxed{}} = \dfrac{3}{4} \cdot \dfrac{\boxed{}}{\boxed{}} = \dfrac{\boxed{}}{\boxed{}}$	Divide by multiplying by the reciprocal of the divisor: $\dfrac{1}{3} \div \left(-\dfrac{5}{9}\right)$.

YOUR NOTES — Write your questions and additional notes.

Practice Exercises

Readiness Check

Choose the word or phrase below the blank that will make the statement true.

1. To add two negative numbers, we _____________ their absolute values and make the answer negative.
   add / subtract

2. To multiply two negative numbers, multiply their absolute values. The answer is _____________.
   positive / negative

3. Division by zero is _____________.
   defined to be 0 / not defined

4. To divide by a fraction, _____________ by its reciprocal.
   divide / multiply

Addition

Add.

5. $-11+(-12)$

6. $11.6+(-31.8)$

7. $-\dfrac{3}{7}+\dfrac{9}{14}$

Opposites, or Additive Inverses

Evaluate $-a$ for each of the following.

8. $a=13.6$

9. $a=-\dfrac{7}{2}$

10. $a=0$

11. Evaluate $-(-x)$ when $x=14$.

12. Find the opposite of -2.5. (Change the sign.)

Subtraction

Subtract.

13. $0-(-32)$

14. $-90.6-(-43.2)$

15. $\dfrac{2}{3}-\dfrac{4}{5}$

Multiplication

Multiply.

16. $-5\cdot3$

17. $-8\cdot(-6.1)$

18. $\left(\dfrac{3}{4}\right)\cdot\left(-\dfrac{1}{2}\right)$

Division

Divide, if possible.

19. $-12\div(-6)$

20. $\dfrac{-77}{11}$

21. $5\div0$

Find the reciprocal of each number, if it exists.

22. $-\dfrac{7}{5}$

23. -8

Divide.

24. $-\dfrac{4}{3}\div\left(-\dfrac{5}{6}\right)$

25. $\dfrac{3}{4}\div\left(-\dfrac{3}{16}\right)$

Exponential Notation

ESSENTIALS

Using **exponential notation**, we can write the product $2 \cdot 2 \cdot 2 \cdot 2 \cdot 2$ as 2^5.

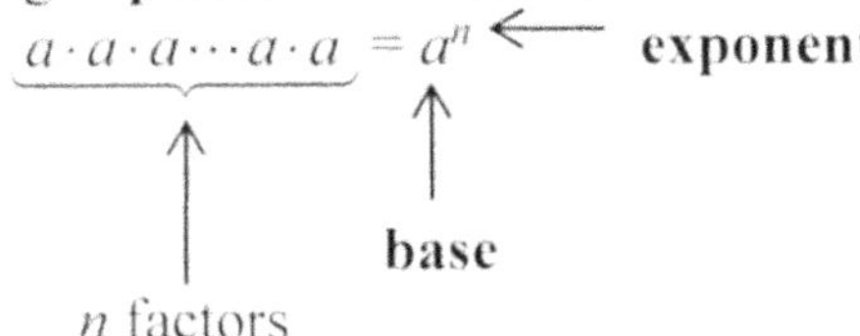

We read a^n as "a to the nth power," or simply "a to the nth." We can read a^2 as "a-squared" and a^3 as "a-cubed."

For any number a, we agree that a^1 means a. For any nonzero number a, we agree that a^0 means 1.

Examples

- $4^3 = 4 \cdot 4 \cdot 4 = 64$

- $4^1 = 4$

- $3^0 = 1$

GUIDED LEARNING   📖 **Textbook**   👤 **Instructor**   ▶ **Video**

EXAMPLE 1	YOUR TURN 1
Write exponential notation: $x \cdot x \cdot x \cdot x$. There are 4 factors of x. $x \cdot x \cdot x \cdot x = x^{\square}$	Write exponential notation: $m \cdot m \cdot m \cdot m \cdot m \cdot m$.
EXAMPLE 2	YOUR TURN 2
Evaluate: $(-6)^2$. $(-6)^2 = \boxed{} \cdot \boxed{} = \boxed{}$	Evaluate: $(-7)^2$.

EXAMPLE 3	YOUR TURN 3
Evaluate: $\left(\dfrac{3}{5}\right)^3$.	Evaluate: $\left(\dfrac{5}{8}\right)^2$.
$\left(\dfrac{3}{5}\right)^3 = \dfrac{3}{5} \cdot \boxed{} \cdot \boxed{} = \dfrac{27}{125}$	

EXAMPLE 4	YOUR TURN 4
Evaluate: $(4.6)^3$.	Evaluate: $(10.1)^2$.
$(4.6)^3 = \boxed{} \cdot \boxed{} \cdot \boxed{} = \boxed{}$	

EXAMPLE 5	YOUR TURN 5
Evaluate: $-(2)^4$.	Evaluate: $(-2)^4$.
$-(2)^4 = -\boxed{} \cdot \boxed{} \cdot \boxed{} \cdot \boxed{} = -16$	

YOUR NOTES   Write your questions and additional notes.

Negative Integers as Exponents

ESSENTIALS

For any real number a that is nonzero and any integer n,

$$a^{-n} = \frac{1}{a^n}.$$

A negative exponent does not necessarily indicate that an answer is negative!

Examples

- $n^{-3} = \dfrac{1}{n^3}$

- $\dfrac{1}{x^{-6}} = x^6$

GUIDED LEARNING	📖 Textbook	👤 Instructor	▶ Video

EXAMPLE 1	YOUR TURN 1
Rewrite using a positive exponent: b^{-4}. $b^{-4} = \dfrac{1}{\boxed{}}$.	Rewrite using a positive exponent: t^{-9}.
EXAMPLE 2	YOUR TURN 2
Rewrite using a positive exponent: $(-6)^{-3}$. Evaluate, if possible. $(-6)^{-3} = \dfrac{1}{(-6)^{\boxed{}}} = \dfrac{1}{(-6)(-6)(-6)} = -\dfrac{1}{\boxed{}}$	Rewrite using a positive exponent: $(5)^{-2}$. Evaluate, if possible.
EXAMPLE 3	YOUR TURN 3
Rewrite using a positive exponent: $\left(\dfrac{2}{3}\right)^{-2}$. Evaluate, if possible. $\left(\dfrac{2}{3}\right)^{-2} = \dfrac{1}{\left(\dfrac{2}{3}\right)^2} = \dfrac{1}{\boxed{}} = 1 \cdot \dfrac{9}{4} = \dfrac{\boxed{}}{\boxed{}}$	Rewrite using a positive exponent: $\left(\dfrac{3}{5}\right)^{-3}$. Evaluate, if possible.

EXAMPLE 4	YOUR TURN 4
Rewrite using a negative exponent: $\dfrac{1}{y^7}$. $\dfrac{1}{y^7} = y^{\square}$	Rewrite using a negative exponent: $\dfrac{1}{(-9)^2}$.

YOUR NOTES   Write your questions and additional notes.

Order of Operations

ESSENTIALS

Rules for Order of Operations

1. Do all the calculations within grouping symbols, like parentheses, before operations outside.
2. Evaluate all exponential expressions.
3. Do all multiplications and divisions in order from left to right.
4. Do all additions and subtractions in order from left to right.

When parentheses occur within parentheses, **computations in the innermost ones are to be done first**.

In addition to parentheses, brackets, and braces, a fraction bar and absolute-value signs can act as grouping symbols.

Example

- $$7 - 6^2 \div 2(-3) - (1-5) = 7 - 6^2 \div 2(-3) - (-4)$$
$$= 7 - 36 \div 2(-3) - (-4)$$
$$= 7 - 18(-3) - (-4)$$
$$= 7 + 54 - (-4)$$
$$= 61 - (-4)$$
$$= 65$$

	Textbook	Instructor	Video

GUIDED LEARNING

EXAMPLE 1	YOUR TURN 1
Simplify: $4 - 15 \div 3 - 7$.	Simplify: $3 - 4 \cdot 5 + 9$.
There are no grouping symbols or exponential expressions, so we begin with the division.	

$$4 - 15 \div 3 - 7 = 4 - \boxed{} - 7$$
$$= \boxed{} - 7$$
$$= \boxed{}$$

EXAMPLE 2	YOUR TURN 2
Simplify: $18 \div 2 - \left\{ 7 - \left[(1-4)^2 + 3 \right] \right\}$.	Simplify: $-5 - \left\{ \left[(9+7) \div 2^3 \right] + 6 \right\} + 4$.

We do the calculations in the innermost grouping symbols first.

$$18 \div 2 - \left\{ 7 - \left[(1-4)^2 + 3 \right] \right\}$$
$$= 18 \div 2 - \left\{ 7 - \left[(\boxed{})^2 + 3 \right] \right\}$$
$$= 18 \div 2 - \left\{ 7 - \left[\boxed{} + 3 \right] \right\}$$
$$= 18 \div 2 - \left\{ 7 - \boxed{} \right\}$$
$$= 18 \div 2 - \left\{ \boxed{} \right\}$$
$$= 18 \div 2 + \boxed{}$$
$$= \boxed{} + 5 = \boxed{}$$

EXAMPLE 3	YOUR TURN 3
Simplify: $\dfrac{2 - 4\left[1 - (5-7)^3 \right]}{6^2 - 5^2}$.	Simplify: $\dfrac{12 - \left[5 - (2^2 - 3^2) \right]}{7 - (3-4)^5}$.

We simplify the numerator and the denominator separately.

$$\frac{2 - 4\left[1 - (5-7)^3 \right]}{6^2 - 5^2} = \frac{2 - 4\left[1 - (\boxed{})^3 \right]}{\boxed{} - \boxed{}}$$
$$= \frac{2 - 4\left[1 - (\boxed{}) \right]}{\boxed{}}$$
$$= \frac{2 - 4\left[\boxed{} \right]}{11}$$
$$= \frac{2 - \boxed{}}{11} = \frac{\boxed{}}{11} = \boxed{}$$

YOUR NOTES   Write your questions and additional notes.

Practice Exercises

Readiness Check

In each of Exercises 1-6, name the operation that should be performed first. Do not calculate.

1. $2(12 \div 3 + 7 - 9)$

2. $14 - 5 \cdot 3 + 8$

3. $3[4(24 \div 12) - 15]$

4. $(9 - 5) \cdot 4 + 16$

5. $4 \cdot 3 - 9 \cdot 5 \div 3$

6. $12 - 3(7 + 4)$

Exponential Notation

Write exponential notation.

7. $t \cdot t \cdot t$

8. $\dfrac{5}{11} \cdot \dfrac{5}{11} \cdot \dfrac{5}{11} \cdot \dfrac{5}{11} \cdot \dfrac{5}{11}$

9. $(-54.9)(-54.9)(-54.9)(-54.9)$

Evaluate.

10. 2^4

11. $(-5)^3$

12. 1.2^1

13. $(-6)^0$

Negative Integers as Exponents

Rewrite using a positive exponent. Evaluate, if possible.

14. $\left(\dfrac{3}{2}\right)^{-3}$

15. $\dfrac{1}{y^{-5}}$

16. -3^{-3}

Rewrite using a negative exponent.

17. $\dfrac{1}{5^3}$

18. $\dfrac{1}{x^5}$

19. $\dfrac{1}{(-6)^3}$

Order of Operations

Simplify.

20. $14 - 7 \cdot 3 + 8$

21. $6 - 2(4-9)^2 + 1$

22. $72 \div 3(1-5) - 2^2 - 8$

23. $\left[4 - (-11 - 3)\right] \div 3^2 - 9$

24. $\dfrac{(-2)^3 + 3^2}{4 \cdot 5 - 6^2 + 2 \cdot 7}$

25. $\dfrac{-22 + 4^3}{6 - (3-4)^5}$

Translating to Algebraic Expressions

ESSENTIALS

When a letter is used to represent various numbers, it is called a **variable**. If a letter represents one particular number, it is called a **constant**.

An **algebraic expression** consists of variables, numbers, and operation signs, such as $+$, $-$, $\cdot$, $\div$. When an equals sign, $=$, is placed between two expressions, an **equation** is formed.

Key Words

Addition	Subtraction	Multiplication	Division
add	subtract	multiply	divide
sum	difference	product	quotient
plus	minus	times	divided by
total	decreased by	twice	ratio
increased by	less than	of	per
more than			

Examples

Translate to an algebraic expression.

- Nine less than some number: $w - 9$

- Ten more than three times a number: $3y + 10$, or $10 + 3y$

- Four less than twenty-three percent of some number: $23\%x - 4$, or $0.23x - 4$

- The sum of a number and 83: $b + 83$, or $83 + b$

- Some number divided by 0.7: $\dfrac{x}{0.7}$, or $x \div 0.7$

	🔲 **Textbook**	👤 **Instructor**	▶ **Video**

GUIDED LEARNING

EXAMPLE 1	YOUR TURN 1
Translate the phrase to an algebraic expression.	Translate the phrase to an algebraic expression.
The sum of four and a number	Five less than some number
Let $n =$ the number.	Let $s =$ the number.
Expression: _______________	Expression: _______________

EXAMPLE 2	YOUR TURN 2
Translate the phrase to an algebraic expression.	Translate the phrase to an algebraic expression.
Eight more than three times some number	One-third of a number minus four
Let x = the number.	Let n = the number.
Expression: _____________	(Think of this as four less than one third of a number.)
	Expression: _____________

EXAMPLE 3	YOUR TURN 3
Translate the phrase to an algebraic expression.	Translate the phrase to an algebraic expression.
Fifteen percent of some number	Forty-two percent of a number
Let y = the number.	Let w = the number.
Expression: _____________	Expression: _____________

EXAMPLE 4	YOUR TURN 4
Translate the phrase to an algebraic expression.	Translate the phrase to an algebraic expression.
Five less than the product of two numbers	Two more than the quotient of two numbers
Let x and y = the numbers.	Let a and b = the numbers.
Expression: _____________	Expression: _____________

YOUR NOTES   Write your questions and additional notes.

Evaluating Algebraic Expressions

ESSENTIALS

When we replace a variable with a number, we say that we are **substituting** for the variable. Carrying out the resulting calculation is called **evaluating the expression**. The result is called the **value** of the expression.

Example

- Evaluate $7x + y^2$ for $x = 5$ and $y = 4$.

$$7x + y^2 = 7(5) + 4^2 = 7(5) + 16 = 35 + 16 = 51$$

GUIDED LEARNING	🅣 **Textbook** 🅐 **Instructor**	🅿 **Video**

EXAMPLE 1	YOUR TURN 1
Evaluate $2x - 5y$ for $x = 12$ and $y = -3$. $$2x - 5y = 2\left(\boxed{}\right) - 5(-3)$$ $$= 24 - \left(\boxed{}\right)$$ $$= 24 + \boxed{}$$ $$= \boxed{}$$	Evaluate $11rs - 2t$ for $r = 4$, $s = -2$, and $t = 21$.
EXAMPLE 2	YOUR TURN 2
The area of a triangle with base of length b and height of length h is given by the formula $A = \dfrac{1}{2}bh$. Find the area when $b = 2.1$ m and $h = 5$ m. $$A = \frac{1}{2}bh$$ $$= \frac{1}{2}(2.1)(5)$$ $$= 1.05(5)$$ $$= \boxed{} \text{ sq m, or } \boxed{} \text{ m}^2$$	The area of a triangle with base of length b and height of length h is given by the formula $A = \dfrac{1}{2}bh$. Find the area when $b = 3.6$ ft and $h = 7$ ft.

EXAMPLE 3	YOUR TURN 3
Evaluate $12 - x^5 + 14 \div 2y^3$ for $x = 1$ and $y = 2$. $12 - x^5 + 14 \div 2y^3 = 12 - (1)^5 + 14 \div 2(2)^3$ $\qquad = 12 - 1 + 14 \div 2(\boxed{})$ $\qquad = 12 - 1 + \boxed{}(8)$ $\qquad = 12 - 1 + 56$ $\qquad = 11 + \boxed{}$ $\qquad = \boxed{}$	Evaluate $18 \div 3y^2 + 15 - xy$ for $x = 4$ and $y = 3$.

YOUR NOTES　　Write your questions and additional notes.

　　Copyright © 2022 Pearson Education, Inc.

Practice Exercises

Readiness Check

Match each key word(s) to the corresponding symbol.

1. product
2. increased by
3. less than
4. quotient

a) $+$

b) $-$

c) $\times$

d) $\div$

Translating to Algebraic Expressions

Translate each phrase to an algebraic expression.

5. Three times d

6. Five more than twice x

7. Six more than the difference of a and b

8. Seven less than twenty percent of y

9. 10 increased by m

10. The product of $\dfrac{1}{5}$ and three times q

11. 1 less than 25 times y

12. 34 subtracted from r

13. What is the price of a refrigerator after a 15% reduction if the price before the reduction was P?

14. Jen's part-time job pays $8.65 per hour. How much does she earn for working h hours?

Evaluating Algebraic Expressions

Evaluate.

15. $5(x-8)+1$, for $x=9$

16. $(n-12)^2 -10$, for $n=20$

17. $12a-(b+a^2)$, for $a=2$ and $b=-6$

18. $x^3 +8-y$, for $x=-3$ and $y=14$

19. $3a^3b-4b$, for $a=2$ and $b=-1$

20. $7x \div (15-y+2)$, for $x=6$ and $y=14$

21. Find the area of a triangular fireplace with a base of 2.1 ft and a height of 9 ft. Use $A=\dfrac{1}{2}bh.$

Equivalent Expressions

ESSENTIALS

Two expressions that have the same value for all *allowable* replacements are called **equivalent expressions**.

Example

- Complete the following table by evaluating each of the expressions for the given values. Then look for expressions that appear to be equivalent.

VALUE	$4x - 2x$	$2x$	$3x + x$
$x = -3$			
$x = 2$			
$x = 0$			

We substitute and find the value of each expression. For example, for $x = -3$,

$$4x - 2x = 4(-3) - 2(-3) = -12 + 6 = -6,$$

$$2x = 2(-3) = -6, \quad \text{and}$$

$$3x + x = 3(-3) + (-3) = -9 - 3 = -12.$$

VALUE	$4x - 2x$	$2x$	$3x + x$
$x = -3$	-6	-6	-12
$x = 2$	4	4	8
$x = 0$	0	0	0

Because the values of $4x - 2x$ and $2x$ are the same for all given values of x, and are in fact the same for any allowable real-number replacement of x, the expressions $4x - 2x$ and $2x$ are **equivalent**. The expressions $4x - 2x$ and $3x + x$ are not equivalent, and the expressions $2x$ and $3x + x$ are not equivalent, since values are not the same for *all* x.

 **Textbook**    **Instructor**    **Video**

GUIDED LEARNING

EXAMPLE 1	YOUR TURN 1

EXAMPLE 1

Complete the table by evaluating each expression for the given values. Then look for expressions that may be equivalent.

VALUE	x^3	$3x$	$2x+x$
$x=-5$			
$x=2$			
$x=2.1$			

For $x=-5$,

$$x^3 = (-5)^3 = -125,$$

$$3x = 3(-5) = \boxed{}, \quad \text{and}$$

$$2x+x = 2(-5)+(-5) = \boxed{} - 5 = -15.$$

Also evaluate the expressions for $x=2$ and for $x=2.1$.

VALUE	x^3	$3x$	$2x+x$
$x=-5$	-125	-15	-15
$x=2$	8	$\boxed{}$	6
$x=2.1$	$\boxed{}$	$\boxed{}$	6.3

The expressions $\boxed{}$ and $2x+x$ are equivalent.

YOUR TURN 1

Complete the table by evaluating each expression for the given values. Then tell if the expressions are equivalent.

VALUE	x^2+3	$5x$
$x=-4$		
$x=1$		
$x=3.4$		

YOUR NOTES   Write your questions and additional notes.

Equivalent Fraction Expressions

ESSENTIALS

The Identity Property of 1

For any real number a,

$$a \cdot 1 = 1 \cdot a = a.$$

(The number 1 is the **multiplicative identity**.)

Examples

- Use multiplying by 1 to find an expression equivalent to $\dfrac{5}{8}$ with a denominator of $16a$.

  Because $16a = 8 \cdot 2a$, we multiply by 1, using $\dfrac{2a}{2a}$ as a name for 1:

  $$\frac{5}{8} = \frac{5}{8} \cdot 1 = \frac{5}{8} \cdot \frac{2a}{2a} = \frac{10a}{16a}.$$

- Simplify: $-\dfrac{27z}{18z}$.

  $$-\frac{27z}{18z} = -\frac{3 \cdot 9z}{2 \cdot 9z} = -\frac{3}{2} \cdot \frac{9z}{9z} = -\frac{3}{2} \cdot 1 = -\frac{3}{2}$$

GUIDED LEARNING   📖 **Textbook**   👤 **Instructor**   ▶ **Video**

EXAMPLE 1	YOUR TURN 1
Write a fraction expression equivalent to $\dfrac{2}{3}$ with a denominator of $15y$. $$\frac{2}{3} = \frac{2}{3} \cdot 1 = \frac{2}{3} \cdot \frac{\Box}{\Box} = \frac{\Box}{15y}$$	Write a fraction expression equivalent to $\dfrac{4}{7t}$ with a denominator of $21t$.

EXAMPLE 2	YOUR TURN 2
Simplify: $\dfrac{13s}{52s}$. $$\frac{13s}{52s} = \frac{\Box \cdot 1}{\Box \cdot 4} = \frac{\Box}{\Box} \cdot \frac{1}{4}$$ $$= 1 \cdot \frac{1}{4} = \frac{\Box}{\Box}$$	Simplify: $\dfrac{11ab}{55a}$.

EXAMPLE 3	YOUR TURN 3
Simplify: $-\dfrac{36b}{8b}$.	Simplify: $-\dfrac{63x}{18x}$.

$$-\frac{36b}{8b} = -\frac{9 \cdot \boxed{}}{2 \cdot \boxed{}} = -\frac{9}{2} \cdot \frac{\boxed{}}{\boxed{}}$$

$$= -\frac{9}{2} \cdot 1 = -\frac{\boxed{}}{\boxed{}}$$

YOUR NOTES   Write your questions and additional notes.

The Commutative Laws and the Associative Laws

ESSENTIALS

The Commutative Laws

Addition. For any numbers a and b,

$$a + b = b + a.$$

(We can change the order when adding without affecting the answer.)

Multiplication. For any numbers a and b,

$$ab = ba.$$

(We can change the order when multiplying without affecting the answer.)

The Associative Laws

Addition. For any numbers a, b, and c,

$$a + (b + c) = (a + b) + c.$$

(Numbers can be grouped in any manner for addition.)

Multiplication. For any numbers a, b, and c,

$$a \cdot (b \cdot c) = (a \cdot b) \cdot c.$$

(Numbers can be grouped in any manner for multiplication.)

Examples

- Evaluate $m + n$ and $n + m$ when $m = 12$ and $n = 4$.

$$m + n = 12 + 4 = 16; \qquad n + m = 4 + 12 = 16$$

- Evaluate mn and nm when $m = 8$ and $n = 3$.

$$mn = 8 \cdot 3 = 24; \qquad nm = 3 \cdot 8 = 24$$

- Evaluate $r + (s + t)$ and $(r + s) + t$ when $r = 5$, $s = 7$, and $t = 2$.

$$
\begin{aligned}
r + (s + t) &= 5 + (7 + 2) & (r + s) + t &= (5 + 7) + 2 \\
&= 5 + 9 & &= 12 + 2 \\
&= 14; & &= 14
\end{aligned}
$$

- Evaluate $r \cdot (s \cdot t)$ and $(r \cdot s) \cdot t$ when $r = 3$, $s = 5$, and $t = 4$.

$$
\begin{aligned}
r \cdot (s \cdot t) &= 3 \cdot (5 \cdot 4) & (r \cdot s) \cdot t &= (3 \cdot 5) \cdot 4 \\
&= 3 \cdot 20 & &= 15 \cdot 4 \\
&= 60; & &= 60
\end{aligned}
$$

GUIDED LEARNING   🛈 **Textbook**   🛈 **Instructor**   ▶ **Video**

EXAMPLE 1	YOUR TURN 1
Evaluate $x+y$ and $y+x$ when $x=-2$ and $y=6$. $$x+y=-2+\boxed{}=\boxed{};$$ $$y+x=6+(-2)=\boxed{}$$	Evaluate xy and yx when $x=10$ and $y=-7$.

EXAMPLE 2	YOUR TURN 2
Evaluate $a\cdot(b\cdot c)$ and $(a\cdot b)\cdot c$ when $a=9$, $b=3$, and $c=2$. $$a\cdot(b\cdot c)=\boxed{}\cdot(3\cdot 2)$$ $$=9\cdot\boxed{}$$ $$=54;$$ $$(a\cdot b)\cdot c=(9\cdot 3)\cdot 2$$ $$=\boxed{}\cdot 2$$ $$=\boxed{}$$	Evaluate $a+(b+c)$ and $(a+b)+c$ when $a=2$, $b=6$, and $c=7$.

EXAMPLE 3	YOUR TURN 3
Use the commutative laws and the associative laws to write at least three expressions equivalent to $(c+1)+d$. Answers may vary. $$(c+1)+d=c+\left(\boxed{}+d\right)$$ $$(c+1)+d=\boxed{}+(c+1)$$ $$(c+1)+d=\left(1+\boxed{}\right)+d$$	Use the commutative laws and the associative laws to write at least three expressions equivalent to $x\cdot(3\cdot y)$. Answers may vary.

YOUR NOTES   Write your questions and additional notes.

The Distributive Laws

ESSENTIALS

The Distributive Law of Multiplication over Addition

For any numbers a, b, and c,

$$a(b+c)=ab+ac, \quad \text{or} \quad (b+c)a=ba+ca.$$

(We can add and then multiply, or we can multiply and then add.)

The Distributive Law of Multiplication over Subtraction

For any real numbers a, b, and c,

$$a(b-c)=ab-ac, \quad \text{or} \quad (b-c)a=ba-ca.$$

(We can subtract and then multiply, or we can multiply and then subtract.)

The reverse of multiplying is called **factoring**. Factoring an expression involves factoring its terms. **Terms** of algebraic expressions are the parts separated by addition signs.

To **factor** an expression is to find an equivalent expression that is a product. If $N = a \cdot b$, then a and b are **factors** of N.

Examples

- Evaluate $6(x-y)$ and $6x-6y$ when $x=8$ and $y=1$.

$$
\begin{aligned}
6(x-y) &= 6(8-1) & 6x-6y &= 6\cdot 8-6\cdot 1 \\
&= 6(7) & &= 48-6 \\
&= 42; & &= 42
\end{aligned}
$$

- $4(x-2)=4\cdot x-4\cdot 2=4x-8$

- The terms of $-5x-2y+3z=-5x+(-2y)+3z$ are $-5x$, $-2y$, and $3z$.

- The factorization of $3a+3y$ is $3(a+y)$.

GUIDED LEARNING	🛈 Textbook	🙎 Instructor	▶ Video

EXAMPLE 1	YOUR TURN 1
Multiply: $4(x-3)$.	Multiply: $5(x+2)$.
$4(x-3)$ $=4\cdot \boxed{}-4\cdot \boxed{}$ Using the distributive law of multiplication over subtraction $=4x-\boxed{}$	

EXAMPLE 2	YOUR TURN 2
Multiply: $-8(3x+4y-z)$. $-8(3x+4y-z)$ $=-8\cdot(\boxed{})+(-8)\cdot(\boxed{})-(-8)\cdot\boxed{}$ $=-24x-\boxed{}+\boxed{}$	Multiply: $-2(5m-7n+9p)$.

EXAMPLE 3	YOUR TURN 3
Factor: $12x-15y+6$. $12x-15y+6$ $=\boxed{}\cdot 4x-\boxed{}\cdot 5y+\boxed{}\cdot 2$ $=\boxed{}(4x-5y+2)$	Factor: $8a-12b+20$.

EXAMPLE 4	YOUR TURN 4
Factor: $ax+bx-cx$. $ax+bx-cx$ $=x\cdot\boxed{}+x\cdot b-x\cdot\boxed{}$ $=x\left(a+\boxed{}-\boxed{}\right)$	Factor: $mx+my+mz$.

YOUR NOTES   Write your questions and additional notes.

Practice Exercises

Readiness Check

Choose from the column on the right an equation that illustrates the law.

1. Commutative law of multiplication
2. Commutative law of addition
3. Associative law of multiplication
4. Associative law of addition
5. Distributive law of multiplication over subtraction

a) $5(6-2) = 5 \cdot 6 - 5 \cdot 2$

b) $9 \cdot \left(\dfrac{1}{3} \cdot 7 \right) = \left(9 \cdot \dfrac{1}{3} \right) \cdot 7$

c) $4 \cdot 7 = 7 \cdot 4$

d) $2 + 10 = 10 + 2$

e) $2(7+4) = 2 \cdot 7 + 2 \cdot 4$

f) $3 + \left(1 + \dfrac{1}{4} \right) = (3+1) + \dfrac{1}{4}$

Equivalent Expressions

Complete the table by evaluating each expression for the given values. Then look for expressions that are equivalent.

6.

VALUE	$4(x+5)$	$4x+5$	$4x+20$
$x=-2$			
$x=1.8$			
$x=0$			

Equivalent Fraction Expressions

Use multiplying by 1 to find an equivalent expression with the given denominator.

7. $\dfrac{3}{7}$; $7y$

8. $\dfrac{2}{5}$; $15a$

Simplify.

9. $-\dfrac{64xy}{24xy}$

10. $\dfrac{45ab}{20b}$

The Commutative Laws and the Associative Laws

Use a commutative law to find an equivalent expression.

11. $6+n$

12. xy

Use an associative law to find an equivalent expression.

13. $(u+v)+3$

14. $(5 \cdot c) \cdot p$

Use the commutative laws and the associative laws to write three equivalent expressions.

15. $5+(y+z)$

16. $(x \cdot y) \cdot 4$

The Distributive Laws

Multiply.

17. $3(x+10)$

18. $-3(7x-2y-z)$

List the terms of each expression.

19. $5x-3y$

20. $-3a-b+18c$

Factoring

Factor.

21. $7a+7b$

22. $4x-2$

23. $6m+12n-15$

24. $2 \cdot \pi \cdot r \cdot h + 2 \cdot \pi \cdot r \cdot r$

Collecting Like Terms

ESSENTIALS

If two terms have the same letter or letters, we say that they are **like terms**, or **similar terms**. (If powers, or exponents, are involved, then like terms must have the same letters raised to the same powers.) If two terms are simply numbers with no letters, they are also like terms.

We can simplify by **collecting**, or **combining**, **like terms**.

Examples

Collect like terms.

- $6x - 2x = (6 - 2)x = 4x$

- $5x + 4y + x - 8y$

$$= 5 \cdot x + 4 \cdot y + 1 \cdot x - 8 \cdot y$$
$$= 5 \cdot x + 1 \cdot x + 4 \cdot y + (-8 \cdot y)$$
$$= (5 + 1)x + (4 - 8)y$$
$$= 6x - 4y$$

GUIDED LEARNING   📘 **Textbook**   👤 **Instructor**   ▶ **Video**

EXAMPLE 1	YOUR TURN 1
Collect like terms: $9x + 13x$. $9x + 13x = \left(\boxed{} + \boxed{}\right)x$ $= \boxed{}\, x$	Collect like terms: $14x - x$.
EXAMPLE 2	YOUR TURN 2
Collect like terms: $11.2x - 6.3y - 9.8x + 2.7y$. $11.2x - 6.3y - 9.8x + 2.7y$ $= 11.2x + (-6.3y) + (-9.8x) + 2.7y$ $= \left(\boxed{} - 9.8\right)x + \left(-6.3 + \boxed{}\right)y$ $= \boxed{}\, x - 3.6y$	Collect like terms: $7x + 1.3 - 2x + 5.6$.

EXAMPLE 3	YOUR TURN 3
Collect like terms: $\dfrac{3}{4}x + \dfrac{5}{6}y + \dfrac{1}{3}x - \dfrac{2}{3}y$.	Collect like terms: $\dfrac{4}{7}x - \dfrac{7}{12}y - \dfrac{1}{2}x + \dfrac{3}{8}y$.

$$\dfrac{3}{4}x + \dfrac{5}{6}y + \dfrac{1}{3}x - \dfrac{2}{3}y$$

$$= \left(\dfrac{3}{4} + \dfrac{\boxed{}}{\boxed{}}\right)x + \left(\dfrac{5}{6} - \dfrac{2}{3}\right)y$$

$$= \left(\dfrac{9}{12} + \dfrac{\boxed{}}{12}\right)x + \left(\dfrac{\boxed{}}{\boxed{}} - \dfrac{4}{6}\right)y$$

$$= \dfrac{\boxed{}}{\boxed{}}x + \dfrac{1}{6}y$$

YOUR NOTES   Write your questions and additional notes.

Multiplying by –1 and Removing Parentheses

ESSENTIALS

The Property of –1

For any number a,

$$-1 \cdot a = -a.$$

(Negative 1 times a is the opposite of a. In other words, changing the sign is the same as multiplying by -1.)

The Opposite of a Difference

For any real numbers a and b,

$$-(a-b) = b-a.$$

(The opposite of $a-b$ is $b-a$.)

Examples

- $-(6x) = -1(6x)$

$$= (-1 \cdot 6)x$$

$$= -6x$$

- $-(-14y) = -1(-14y)$

$$= \left[-1(-14)\right]y$$

$$= 14y$$

- $-(2x-4y+9) = -1(2x-4y+9)$

$$= -2x+4y-9$$

- $3y+(4x+5)-(10y-2) = 3y+4x+5-10y+2$

$$= -7y+4x+7$$

GUIDED LEARNING   🔖 **Textbook**   👤 **Instructor**   ▶ **Video**

EXAMPLE 1	YOUR TURN 1
Find an equivalent expression without parentheses: $-(21a-16b+28)$.	Find an equivalent expression without parentheses: $-(-19a-8b-12)$.
$-(21a-16b+28)$ $= -21a \boxed{} - 28$	

EXAMPLE 2	YOUR TURN 2
Remove parentheses and simplify: $2x - (4x + 6)$. $\quad 2x - (4x + 6) = 2x - 4x \; \boxed{}$ $\qquad\qquad\qquad = \boxed{} - 6$	Remove parentheses and simplify: $7x - (5y - 8) + (3x - 14)$.

EXAMPLE 3	YOUR TURN 3
Simplify: $8c - \left\{ 3\left[6(2c - 1) - 4(2c + 4) + 5 \right] - 7 \right\}$. $\quad 8c - \left\{ 3\left[6(2c - 1) - 4(2c + 4) + 5 \right] - 7 \right\}$ $\quad = 8c - \left\{ 3\left[12c - 6 - 8c - \boxed{} + 5 \right] - 7 \right\}$ $\quad = 8c - \left\{ 3[4c - 17] - 7 \right\}$ $\quad = 8c - \left\{ 12c - \boxed{} - 7 \right\}$ $\quad = 8c - \left\{ 12c - 58 \right\}$ $\quad = 8c - 12c + \boxed{}$ $\quad = \boxed{}\, c + 58$	Simplify: $4d - \left\{ 9d - \left[(d + 8) - (d - 2) \right] \right\}$.

YOUR NOTES   Write your questions and additional notes.

Practice Exercises

Readiness Check

Determine whether each set of expressions are equal or opposite.

1. $-(-4x), \quad 4x$

                     ____

2. $x - 5, \quad 5 - x$

                     ____

3. $2x - 4y + 15, \quad -2x + 4y - 15$

                     ____

4. $7y - (-2y), \quad 9y$

                     ____

Collecting Like Terms

Collect like terms.

5. $2m + 7m$

6. $8n - 11n$

7. $14p - p$

8. $10a + 13b - 4a$

9. $4.3s - 1.1t - 6.5t + s$

10. $\dfrac{1}{3}x - \dfrac{4}{5}x - \dfrac{2}{5}y + \dfrac{1}{2}y$

Multiplying by -1 and Removing Parentheses

Find an equivalent expression without parentheses.

11. $-(-5a)$

12. $-(d + 14)$

13. $-(3x + 2y - 8)$

14. $-(-3f+4g+2h-j)$

15. $-\left(-x-\dfrac{1}{5}w-4.7y+15.4z\right)$

Simplify by removing parentheses and collecting like terms.

16. $b+(8b-7)$

17. $9x-(x+23)$

18. $6y-(3y+9)+4(2y-12)$

19. $\dfrac{1}{3}(9a-6)-\dfrac{1}{4}(16a+20)-5$

Simplify.

20. $3\{-2-6[4x+1-3(5x+7)]\}$

21. $[16-4(8y+3)]-[2(y+1)-4]$

Multiplication and Division

ESSENTIALS

The Product Rule

For any number a and any integers m and n,

$$a^m \cdot a^n = a^{m+n}.$$

(When multiplying with exponential notation, add the exponents if the bases are the same.)

The Quotient Rule

For any nonzero number a and any integers m and n,

$$\frac{a^m}{a^n} = a^{m-n}.$$

(When dividing with exponential notation, subtract the exponent of the denominator from the exponent of the numerator if the bases are the same.)

Examples

- Multiply and simplify: $\left(2a^2b^{-3}\right)(3ab)$.

$$\left(2a^2b^{-3}\right)(3ab) = 2 \cdot 3 \cdot a^2 \cdot a \cdot b^{-3} \cdot b$$

$$= 6a^{2+1}b^{-3+1} = 6a^3b^{-2} = \frac{6a^3}{b^2}$$

- Divide and simplify: $\dfrac{-8x^{-7}y^5}{2x^2y^3}$.

$$\frac{-8x^{-7}y^5}{2x^2y^3} = -\frac{8}{2} \cdot x^{-7-2} \cdot y^{5-3} = -4x^{-9}y^2 = -\frac{4y^2}{x^9}$$

GUIDED LEARNING   📍 **Textbook**   👤 **Instructor**   ▶ **Video**

EXAMPLE 1	YOUR TURN 1
Multiply and simplify: $\left(7y^{2n}\right)\left(3y^n\right)$.	Multiply and simplify: $\left(-3x^{4n}\right)\left(8x^{2n}\right)$.
$\left(7y^{2n}\right)\left(3y^n\right) = 7 \cdot \boxed{} \cdot y^{2n+\boxed{}}$	
$\qquad = 21y^{\boxed{}}$	

EXAMPLE 2	YOUR TURN 2
Multiply and simplify: $\left(-5x^{-5}y^{10}\right)\left(4x^3y^7\right)$. $$\left(-5x^{-5}y^{10}\right)\left(4x^3y^7\right)=-5\cdot4\cdot x^{-5+3}y^{\square+\square}$$ $$=\boxed{}$$	Multiply and simplify: $\left(-7a^{10}b^{-2}\right)\left(2ab^3\right)$.
EXAMPLE 3	YOUR TURN 3
Divide and simplify: $\dfrac{24x^{4n}}{6x^{5n}}$. $$\frac{24x^{4n}}{6x^{5n}}=\frac{\boxed{}}{6}\cdot\frac{x^{4n}}{x^{5n}}$$ $$=4x^{\square-5n}=4x^{\square}=\frac{4}{x^{n}}$$	Divide and simplify: $\dfrac{-27y^{8n}}{9y^{12n}}$.
EXAMPLE 4	YOUR TURN 4
Divide and simplify: $\dfrac{-24m^6n^8}{-8m^5n^2}$. $$\frac{-24m^6n^8}{-8m^5n^2}=\frac{-24}{-8}m^{\square-\square}n^{8-2}$$ $$=\boxed{}$$	Divide and simplify: $\dfrac{-18m^{10}n^2}{2m^7n}$.

YOUR NOTES   Write your questions and additional notes.

Raising Powers to Powers and Products and Quotients to Powers

ESSENTIALS

The Power Rule

For any real number a and any integers m and n,

$$\left(a^m\right)^n = a^{mn}.$$

(To raise a power to a power, multiply the exponents.)

Raising a Product to a Power

For any real numbers a and b and any integer n,

$$\left(ab\right)^n = a^n b^n.$$

(To raise a product to the nth power, raise each factor to the nth power.)

Raising a Quotient to a Power

For any real numbers a and b and any integer n,

$$\left(\frac{a}{b}\right)^n = \frac{a^n}{b^n}, b \neq 0; \quad \text{and} \quad \left(\frac{a}{b}\right)^{-n} = \left(\frac{b}{a}\right)^n = \frac{b^n}{a^n}, a \neq 0, b \neq 0.$$

(To raise a quotient to the nth power, raise the numerator to the nth power and divide by the denominator to the nth power.)

Examples

- $\left(y^2\right)^6 = y^{2 \cdot 6} = y^{12}$

- $\left(-3x^4 y^{-2}\right)^4 = (-3)^4 \left(x^4\right)^4 \left(y^{-2}\right)^4 = 81x^{16} y^{-8} = \frac{81x^{16}}{y^8}$

- $\left(\frac{3x^5}{y^2}\right)^{-2} = \left(\frac{y^2}{3x^5}\right)^2 = \frac{\left(y^2\right)^2}{\left(3x^5\right)^2} = \frac{y^4}{9x^{10}}$

GUIDED LEARNING

| | Textbook | Instructor | Video |

EXAMPLE 1	YOUR TURN 1
Simplify: $\left(x^3\right)^{-7s}$.	Simplify: $\left(x^{-6}\right)^{4p}$.
$\left(x^3\right)^{-7s} = x^{3(-7s)} = x^{\boxed{}}$	

EXAMPLE 2	YOUR TURN 2
Simplify: $\left(-2x^6 y^{-2} z\right)^{-3}$.	Simplify: $\left(-3x^{-7} y^{-2} z^5\right)^{-4}$.

$$\left(-2x^6 y^{-2} z\right)^{-3} = \left(-2\right)^{-3} \cdot \left(x^6\right)^{\square} \cdot \left(y^{-2}\right)^{\square} \cdot z^{-3}$$

$$= \frac{1}{\left(-2\right)^3} \cdot x^{-18} \cdot y^6 \cdot z^{-3}$$

$$= \boxed{} \cdot \frac{1}{x^{18}} \cdot y^6 \cdot \frac{1}{z^3}$$

$$= -\frac{y^6}{8x^{18} z^{\square}}$$

EXAMPLE 3	YOUR TURN 3
Simplify: $\left(\dfrac{3x^5}{y^2}\right)^{-2}$.	Simplify: $\left(\dfrac{x}{2y^{-3}}\right)^{-4}$.

$$\left(\frac{3x^5}{y^2}\right)^{-2} = \left(\frac{y^2}{3x^5}\right)^{2}$$

$$= \frac{\left(y^2\right)^{\square}}{\left(3x^5\right)^{\square}}$$

$$= \frac{y^4}{\square\, x^{\square}}$$

YOUR NOTES   Write your questions and additional notes.

Scientific Notation

ESSENTIALS

Scientific notation for a number is an expression of the type

$$M \times 10^n,$$

where n is an integer, M is greater than or equal to 1 and less than 10 ($1 \leq M < 10$), and M is expressed in decimal notation. 10^n is also considered to be scientific notation when $M = 1$.

A positive exponent in scientific notation indicates a large number (greater than or equal to 10) and a negative exponent indicates a small number (between 0 and 1).

Examples

- Convert 129,040,000 to scientific notation.

  $$129{,}040{,}000. \qquad\qquad 129{,}040{,}000 = 1.2904 \times 10^8$$

  We must move the decimal point 8 places. The number is large so the exponent is positive.

- Convert 1.903×10^{-4} to decimal notation.

  $$0.0001.903 \qquad\qquad 1.903 \times 10^{-4} = 0.0001903$$

  The exponent is negative, so the answer is a small number. Move the decimal point 4 places.

- Multiply: $\left(2.1 \times 10^9\right)\left(5.4 \times 10^7\right)$

  $$\left(2.1 \times 10^9\right)\left(5.4 \times 10^7\right) = (2.1 \times 5.4)\left(10^9 \times 10^7\right)$$
  $$= 11.34 \times 10^{16}$$
  $$= \left(1.134 \times 10^1\right) \times 10^{16}$$
  $$= 1.134 \times 10^{17}$$

- Divide: $\dfrac{1.8 \times 10^{-11}}{1.5 \times 10^{-17}}$.

  $$\frac{1.8 \times 10^{-11}}{1.5 \times 10^{-17}} = \frac{1.8}{1.5} \times \frac{10^{-11}}{10^{-17}}$$
  $$= 1.2 \times 10^6$$

GUIDED LEARNING   🔵 **Textbook**   🔵 **Instructor**   🔵 **Video**

EXAMPLE 1	YOUR TURN 1
Convert 0.000046 to scientific notation. 0.000046 (continued)	Convert 70,100,000,000,000 to scientific notation.

We move the decimal point ☐ places.

The number is small so the exponent is

___________.

$0.000046 = \boxed{} \times 10^{\boxed{}}$

EXAMPLE 2	YOUR TURN 2
Convert 5.47×10^6 to decimal notation. The exponent is ___________, so the answer is a large number. Move the decimal point ☐ places. $5.\underset{\curvearrowright}{470000}.$ $5.47 \times 10^6 = \boxed{}$	Convert 4.078×10^{-3} to decimal notation.
EXAMPLE 3	YOUR TURN 3
Multiply: $\left(3.3 \times 10^8\right)\left(2.5 \times 10^{-13}\right)$. Write scientific notation for the result. $\left(3.3 \times 10^8\right)\left(2.5 \times 10^{-13}\right) = 3.3 \times 2.5 \times 10^8 \times 10^{-13}$ $= \boxed{} \times 10^{\boxed{}}$	Multiply: $\left(3.6 \times 10^{-12}\right)\left(2.2 \times 10^{15}\right)$. Write scientific notation for the result.
EXAMPLE 4	YOUR TURN 4
Divide: $\dfrac{3.6 \times 10^{30}}{4.8 \times 10^{50}}$. Write scientific notation for the result. $\dfrac{3.6 \times 10^{30}}{4.8 \times 10^{50}} = \dfrac{3.6}{4.8} \times \dfrac{10^{30}}{10^{50}}$ $= \boxed{} \times 10^{-20}$ $= \left(\boxed{} \times 10^{\boxed{}}\right) \times 10^{-20}$ $= 7.5 \times 10^{\boxed{}}$	Divide: $\dfrac{1.2 \times 10^{-7}}{1.5 \times 10^{-18}}$. Write scientific notation for the result.

YOUR NOTES   Write your questions and additional notes.

Practice Exercises

Readiness Check

Choose the word from the list below that will make each statement true. Not all words will be used.

negative   positive   multiply   add   scientific

1. When multiplying with exponential notation, _______________ the exponents if the bases are the same.

2. To raise a power to a power, _______________ the exponents and leave the base unchanged.

3. The number 3.1×10^{-4} is written in _______________ notation.

4. In scientific notation, _______________ exponents are used to represent small numbers between 0 and 1.

Multiplication and Division

Multiply and simplify.

5. $3^9 \cdot 3^2$

6. $8x^0 \cdot 4x^4$

7. $\left(-2a^{-5}\right)\left(9a^3\right)$

Divide and simplify.

8. $\dfrac{x^{10}}{x^3}$

9. $\dfrac{20x^{-12}}{4x^8}$

10. $\dfrac{-45x^9 y^5}{5x^6 y}$

Raising Powers to Powers and Products and Quotients to Powers

Simplify.

11. $\left(-6xy^2\right)^2$

12. $\left(2x^{-2}y^3\right)^{-2}$

13. $\left(\dfrac{3x^4 y^7}{4xy^8} \right)^{3}$

14. $\left(\dfrac{2x^4 y^{-3}}{3x^{-7} y^{-9}} \right)^{-4}$

Scientific Notation

Convert each number to scientific notation.

15. 0.047

16. 501,790,000,000

17. Before the Dalles Dam was built, Celilo Falls was the largest waterfall on the Columbia River. Its average discharge was 190,000 cubic feet of water per second. Express this number in scientific notation.

Convert each number to decimal notation.

18. 8.13×10^{7}

19. 2.04×10^{-3}

20. There are about 2.4×10^{4} grains of rice in a pound. Express this number in decimal notation.

Multiply or divide and write scientific notation for the result.

21. $\left(2 \times 10^{5} \right)\left(3 \times 10^{4} \right)$

22. $\left(1.5 \times 10^{3} \right)\left(8.7 \times 10^{-5} \right)$

23. $\dfrac{9.2 \times 10^{-3}}{2.3 \times 10^{17}}$

24. $\dfrac{2.4 \times 10^{6}}{9.6 \times 10^{-10}}$

Equations and Solutions

ESSENTIALS

An **equation** is a number sentence that says that the expressions on either side of the equals sign, =, represent the same number.

The replacements for the variable that make an equation true are called the **solutions** of the equation. The set of all solutions is called the **solution set** of the equation. When we find all the solutions, we say that we have **solved** the equation.

Equations with the same solutions are called **equivalent equations**.

Examples

- $2 + 5 = 7$ The equation is *true*.

- $9 - 3 = 3$ The equation is *false*.

- $x - 8 = 11$ The equation is *neither* true nor false, because we do not know what number x represents.

- Determine whether 3 is a solution of $2x + 2 = 9$.

$$\begin{array}{ll} 2x + 2 = 9 & \text{Writing the equation} \\ 2 \cdot 3 + 2 \ ? \ 9 & \text{Substituting 3 for } x \\ 6 + 2 & \\ 8 & \text{FALSE} \end{array}$$

The number 3 is not a solution of the equation.

GUIDED LEARNING   📖 **Textbook**   👤 **Instructor**   ▶ **Video**

EXAMPLE 1	YOUR TURN 1
Determine whether the equation is true, false, or neither.	Determine whether the equation is true, false, or neither.
$4 - 6 = 2$	$5 - 9 = -4$
The equation is ______________. true / false / neither	
EXAMPLE 2	YOUR TURN 2
Determine whether the equation is true, false, or neither.	Determine whether the equation is true, false, or neither.
$13 + 7 = 5 + 15$	$12 + 4 = 7 + 7$
The equation is ______________. true / false / neither	

EXAMPLE 3	YOUR TURN 3
Determine whether the equation is true, false, or neither. $x + 5 = 14$ The equation is ____________. true / false / neither	Determine whether the equation is true, false, or neither. $7 + 3 = x$

EXAMPLE 4	YOUR TURN 4

Determine whether -6 is a solution of $10 - y = 16$.
 	
$10 - y = 16$ Writing the equation
 	
$10 - \left(\boxed{} \right) \overset{?}{=} 16$ Substituting -6 for y
 	
$\boxed{}$ TRUE
 	
The statement $16 = 16$ is __________.
      true / false
 	
-6 __________ a solution of $10 - y = 16$.
  is / is not
	Determine whether -22 is a solution of $x + 2 = 20$.

EXAMPLE 5	YOUR TURN 5
Determine whether $a = 13$ and $2a + 11 = 39$ are equivalent. $2a + 11 = 39$ $2(13) + 11 = 39$ Substituting 13 for a $26 + 11 = 39$ $37 = 39$ The statement $37 = 39$ is __________. true / false The equations $a = 13$ and $2a + 11 = 39$ __________ equivalent. are / are not	Determine whether $x = -4$ and $-3x - 9 = 3$ are equivalent.

YOUR NOTES   Write your questions and additional notes.

The Addition Principle

ESSENTIALS

The Addition Principle

For any real numbers a, b, and c,

$a = b$ is equivalent to $a + c = b + c$.

Examples

- Solve: $w - 9 = 2$.

$$w - 9 = 2$$
$$w - 9 + 9 = 2 + 9$$
$$w + 0 = 11$$
$$w = 11$$

Check:
$$\frac{w - 9 = 2}{11 - 9 \,?\, 2}$$
$$2 \mid \text{TRUE}$$

- Solve: $x + 3 = -7$.

$$x + 3 = -7$$
$$x + 3 + (-3) = -7 + (-3)$$
$$x + 0 = -10$$
$$x = -10$$

Check:
$$\frac{x + 3 = -7}{-10 + 3 \,?\, -7}$$
$$-7 \mid \text{TRUE}$$

GUIDED LEARNING | 📖 **Textbook**   👤 **Instructor**   ▶ **Video**

EXAMPLE 1	YOUR TURN 1
Solve: $y + 13 = -2$.	Solve: $x + 9 = -15$.

EXAMPLE 1

Solve: $y + 13 = -2$.

$$y + 13 = -2$$
$$y + 13 + \left(\boxed{}\right) = -2 + \left(\boxed{}\right) \quad \text{Adding } -13 \text{ on both sides}$$
$$y + 0 = -15$$
$$y = \boxed{} \quad \text{Identity property of 0}$$

Check:
$$\frac{y + 13 = -2}{-15 + 13 \,?\, -2} \quad \text{Substituting } -15 \text{ for } y$$
$$-2 \mid \text{TRUE}$$

The solution is -15.

YOUR TURN 1

Solve: $x + 9 = -15$.

EXAMPLE 2	YOUR TURN 2
Solve: $-5.3 = b - 2.4$.	Solve: $-7.6 = x - 4.2$.

$$-5.3 = b - 2.4$$

$$-5.3 + \boxed{} = b - 2.4 + \boxed{}$$

$$\boxed{} = b$$

Check:

$$-5.3 = b - 2.4$$

$$-5.3 \;?\; -2.9 - 2.4$$

$$-5.3 \qquad \text{TRUE}$$

The solution is $\boxed{}$.

EXAMPLE 3	YOUR TURN 3
Solve: $\dfrac{1}{2} + x = -\dfrac{2}{5}$.	Solve: $\dfrac{2}{3} + w = -\dfrac{3}{4}$.

$$\frac{1}{2} + x = -\frac{2}{5}$$

$$\frac{1}{2} - \boxed{} + x = -\frac{2}{5} - \boxed{}$$

$$x = -\frac{2}{5} \cdot \frac{\boxed{}}{2} - \frac{1}{2} \cdot \frac{5}{\boxed{}}$$

$$x = -\frac{4}{10} - \frac{5}{10}$$

$$x = -\frac{\boxed{}}{\boxed{}}$$

The number $-\dfrac{9}{10}$ checks.

The solution is $\boxed{}$.

YOUR NOTES   Write your questions and additional notes.

The Multiplication Principle

The Multiplication Principle

For any real numbers a, b, and c, $c \neq 0$,

$$a = b \text{ is equivalent to } a \cdot c = b \cdot c.$$

In a product like $4x$, the number in front of the variable is called the **coefficient**.

Examples

- Solve: $\dfrac{2}{5}x = 84$.

$$\dfrac{2}{5}x = 84$$
$$\dfrac{5}{2} \cdot \dfrac{2}{5}x = \dfrac{5}{2} \cdot 84$$
$$1 \cdot x = 210$$
$$x = 210$$

Check:
$$\dfrac{2}{5}x = 84$$
$$\dfrac{2}{5} \cdot 210 \,?\, 84$$
$$84 \,\bigg|\, \text{TRUE}$$

- Solve: $-7x = 42$.

$$-7x = 42$$
$$\dfrac{-7x}{-7} = \dfrac{42}{-7}$$
$$1 \cdot x = -6$$
$$x = -6$$

Check:
$$-7x = 42$$
$$-7(-6) \,?\, 42$$
$$42 \,\bigg|\, \text{TRUE}$$

 **Textbook**    **Instructor**    **Video**

GUIDED LEARNING

EXAMPLE 1	YOUR TURN 1
Solve: $\dfrac{5}{9} = -\dfrac{4}{3}x$.	Solve: $\dfrac{3}{8}y = -\dfrac{9}{4}$.

$$\dfrac{5}{9} = -\dfrac{4}{3}x$$

$$-\dfrac{\square}{\square} \cdot \dfrac{5}{9} = -\dfrac{\square}{\square} \cdot \left(-\dfrac{4}{3}x\right)$$

$$-\dfrac{5}{12} = 1 \cdot x$$

$$-\dfrac{5}{12} = x$$

The number $-\dfrac{5}{12}$ checks and is the solution.

EXAMPLE 2	YOUR TURN 2
Solve: $6x = 84$.	Solve: $7y = 91$.

$$6x = 84$$

$$\frac{6x}{\boxed{}} = \frac{84}{\boxed{}}$$

$$1 \cdot x = 14$$

$$x = \boxed{}$$

The number 14 checks and is the solution.

EXAMPLE 3	YOUR TURN 3
Solve: $2.43y = 9963$.	Solve: $1.97x = 4334$.

$$2.43y = 9963$$

$$\frac{2.43y}{\boxed{}} = \frac{9963}{\boxed{}}$$

$$1 \cdot y = 4100$$

$$y = \boxed{}$$

The number 4100 checks and is the solution.

EXAMPLE 4	YOUR TURN 4
Solve: $-x = 15$.	Solve: $-a = -32$.

$$-x = 15$$

$$\frac{-x}{\boxed{}} = \frac{15}{\boxed{}}$$

$$1 \cdot x = -15$$

$$x = \boxed{}$$

The solution is $\boxed{}$.

YOUR NOTES   Write your questions and additional notes.

Using the Principles Together

ESSENTIALS

Example

- Solve: $-2x + 4 = 58$.

$$-2x + 4 = 58$$

$$-2x + 4 - 4 = 58 - 4 \qquad \text{Using the addition principle: subtracting 4}$$

$$-2x = 54 \qquad \text{Simplifying}$$

$$\frac{-2x}{-2} = \frac{54}{-2} \qquad \text{Using the multiplication principle: dividing by } -2$$

$$x = -27 \qquad \text{Simplifying}$$

The solution is -27.

Check:

$$-2x + 4 = 58$$

$$-2(-27) + 4 \;?\; 58$$

$$54 + 4$$

$$58 \;\Big|\; \text{TRUE}$$

GUIDED LEARNING

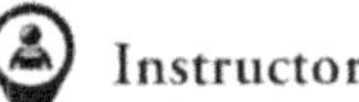 **Textbook**    **Instructor**   ▶ **Video**

EXAMPLE 1	YOUR TURN 1

EXAMPLE 1

Solve: $\dfrac{3}{4}x - \dfrac{1}{3} + \dfrac{1}{2}x = 2x - \dfrac{4}{3}$.

The number $\boxed{}$ is the LCM of all the denominators.

$$\boxed{}\left(\frac{3}{4}x - \frac{1}{3} + \frac{1}{2}x\right) = \boxed{}\left(2x - \frac{4}{3}\right) \qquad \text{Multiplying by 12 on both sides}$$

$$12\cdot\frac{3}{4}x - 12\cdot\frac{1}{3} + 12\cdot\frac{1}{2}x = 12\cdot 2x - 12\cdot\frac{4}{3}$$

$$9x - 4 + 6x = 24x - 16$$

$$\boxed{} - 4 = 24x - 16 \qquad \text{Collecting like terms}$$

$$15x - 4 - \boxed{} = 24x - 16 - \boxed{} \qquad \text{Subtracting } 15x$$

$$-4 = 9x - 16$$

$$-4 + \boxed{} = 9x - 16 + \boxed{} \qquad \text{Adding 16}$$

$$12 = 9x$$

$$\frac{12}{\boxed{}} = \frac{9x}{\boxed{}} \qquad \text{Dividing by 9}$$

$$\boxed{} = x$$

The number $\dfrac{4}{3}$ checks. The solution is $\boxed{}$.

YOUR TURN 1

Solve:

$$3x - \frac{4}{5}x = \frac{2}{5} + 2x + \frac{1}{2}.$$

EXAMPLE 2	YOUR TURN 2
Solve: $-15.3 + 4.2b = -3.96$. We multiply on both sides by 10^2, or 100, because the greatest number of decimal places is $\boxed{}$. $$-15.3 + 4.2b = -3.96$$ $$\boxed{}(-15.3 + 4.2b) = \boxed{}(-3.96)$$ $$100(-15.3) + 100(4.2b) = -396 \quad \text{Using the distributive law}$$ $$-1530 + 420b = -396$$ $$-1530 + 420b + 1530 = -396 + 1530 \quad \text{Adding 1530}$$ $$420b = \boxed{}$$ $$\frac{420b}{\boxed{}} = \frac{1134}{\boxed{}} \quad \text{Dividing by 420}$$ $$b = \frac{1134}{420}, \text{ or } \frac{27}{10}$$ The solution is $\frac{27}{10}$.	Solve: $17.2 - 5.4x = -4.94$.
EXAMPLE 3	YOUR TURN 3
Solve: $7 - 3x = 6 - 3(x + 1)$. $$7 - 3x = 6 - 3(x + 1)$$ $$7 - 3x = 6 - \boxed{} - \boxed{} \quad \text{Using the distributive law}$$ $$7 - 3x = 3 - 3x$$ $$7 - 3x + \boxed{} = 3 - 3x + \boxed{} \quad \text{Adding } 3x$$ $$7 = 3 \quad \text{False}$$ There is *no* solution.	Solve: $14 + 5x = 7 + 5(x - 9)$.
EXAMPLE 4	YOUR TURN 4
Solve: $6x + 5 = 17 + 6(x - 2)$. $$6x + 5 = 17 + 6(x - 2)$$ $$6x + 5 = 17 + \boxed{} - 12 \quad \text{Using the distributive law}$$ $$6x + 5 = 6x + 5$$ $$\boxed{} + 6x + 5 = \boxed{} + 6x + 5 \quad \text{Adding } -6x$$ $$5 = 5 \quad \text{True for all real numbers}$$ The equation has *infinitely* many solutions. All real numbers are solutions.	Solve: $-8x + 7 = 15 - 8(x + 1)$.

YOUR NOTES Write your questions and additional notes.

Practice Exercises

Readiness Check

Choose from the column on the right the most appropriate first step in solving each equation.

1. $-4x = 20$

2. $\dfrac{1}{4}x = 20$

3. $20x = -4$

4. $4 = -\dfrac{1}{20}x$

a) Multiply by 4 on both sides.

b) Multiply by 20 on both sides.

c) Divide by 4 on both sides.

d) Divide by 20 on both sides.

e) Multiply by -20 on both sides.

f) Divide by -4 on both sides.

Equations and Solutions

Determine whether the given number is a solution of the given equation.

5. $14; \ x + 19 = 43$

6. $24; \ \dfrac{x}{6} = 4$

7. $-4; \ 6(y - 3) = 42$

The Addition Principle

Solve using the addition principle. Don't forget to check.

8. $x + 7 = 9$

9. $-10 + y = 3$

10. $6.2 = -4.7 + y$

11. $-\dfrac{3}{4} + x = -\dfrac{5}{6}$

The Multiplication Principle

Solve using the multiplication principle. Don't forget to check.

12. $9x = 72$

13. $-x = 17$

14. $-1.6x = -1.44$

15. $-\dfrac{3}{4}c = \dfrac{9}{8}$

Using the Principles Together

Solve using the principles together. Don't forget to check.

16. $3x - 7 = 11$

17. $\dfrac{3}{5}x - \dfrac{3}{2} = -\dfrac{4}{5} + 2x$

18. $8.2x - 5.9x = 17.48$

19. $4x - 9 = 3 + 4(x - 3)$

20. $3x - 11 = 7x - 4(x - 11)$

21. $7(x + 5) - 3 = 2\left[9 + 2(x - 1)\right]$

Evaluating and Solving Formulas

ESSENTIALS

A **formula** is an equation that represents or models a relationship between two or more quantities.

Example

- Solve the formula $C = \dfrac{5}{9}(F - 32)$ for F.

$$C = \frac{5}{9}(F - 32)$$

$$\frac{9}{5} \cdot C = \frac{9}{5} \cdot \frac{5}{9}(F - 32) \quad \text{Multiplying on both sides by } \frac{9}{5}$$

$$\frac{9}{5}C = F - 32$$

$$\frac{9}{5}C + 32 = F - 32 + 32 \quad \text{Adding 32 on both sides}$$

$$\frac{9}{5}C + 32 = F$$

	📱 **Textbook**	👤 **Instructor**	▶ **Video**

GUIDED LEARNING

EXAMPLE 1	YOUR TURN 1
The area of a triangle with base b and height h is given by $A = \dfrac{1}{2}bh$. Find the area of a triangle with base 4 in. and height 9 in. $A = \dfrac{1}{2} \cdot \boxed{} \cdot \boxed{}$ Substituting $A = \boxed{}$ Multiplying The area of the triangle is $\boxed{}$.	The distance d that a car will travel at a rate, or speed, r in time t is given by $d = rt$. A car travels at 65 miles per hour (mph) for 3.5 hr. How far does it travel?

EXAMPLE 2	YOUR TURN 2

EXAMPLE 2

Solve for s: $W = \dfrac{s-t}{3}$.

$$W = \frac{s-t}{3}$$

$$W \cdot \boxed{} = \frac{s-t}{3} \cdot \boxed{} \qquad \text{Multiplying by 3 to clear the fraction}$$

$$3W = s - t$$

$$3W + \boxed{} = s - t + \boxed{} \qquad \text{Adding } t$$

$$\boxed{} = s$$

YOUR TURN 2

Solve for y: $A = \dfrac{x+y}{2}$.

This is a formula for the average of two numbers.

EXAMPLE 3

Solve $B = 3m + 4n$ for m.

$$B = 3m + 4n$$

$$B - \boxed{} = 3m + 4n - 4n \qquad \text{Subtracting } 4n$$

$$B - 4n = 3m$$

$$\frac{B - 4n}{3} = \frac{3m}{3} \qquad \text{Dividing by 3}$$

$$\frac{B - 4n}{3} = \boxed{}$$

YOUR TURN 3

Solve $B = 3m + 4n$ for n.

EXAMPLE 4

Solve $pr + pq = m$ for p.

$$pr + pq = m$$

$$\boxed{}(r + q) = m \qquad \text{Factoring out } p \text{ (collecting like terms)}$$

$$\frac{p(r + q)}{r + q} = \frac{m}{r + q} \qquad \text{Dividing both sides by } r + q$$

$$\boxed{} = \frac{m}{r + q}$$

YOUR TURN 4

Solve $abc + b = 2a$ for b.

YOUR NOTES   Write your questions and additional notes.

Practice Exercises

Readiness Check

Determine whether each statement is true or false.

1. To solve the formula $V = lwh$ for w you would subtract l and h on both sides of the equation.

2. When two or more terms on the same side of a formula contain the letter for which we are solving, we can factor so that the letter is only written once.

3. A formula is an equation that uses letters to represent a relationship between two or more quantities.

4. The first step in solving the equation $T = 0.3(I - 12{,}000)$ for I is to multiply on both sides by 0.3.

Evaluating and Solving Formulas

5. The wavelength w, in meters per cycle, of a musical note is given by $w = \dfrac{r}{f}$, where r is the speed of the sound, in meters per second, and f is the frequency, in cycles per second. The speed of sound in air is 344 m/sec. What is the wavelength of a note whose frequency in air is 32 cycles per second?

Solve for the given letter.

6. $d = 60t$, for t

7. $y = 26 - x$, for x

8. $T = \dfrac{a}{b}$, for a

9. $3y = 2x$, for x

10. $y = mx + b$, for x

11. $A = \dfrac{p + q + r}{3}$, for q

12. *Perimeter of a rectangle:*

$P = 2l + 2w$, for w

13. *Surface area of a sphere:*

$S = 4\pi r^2$, for r^2

14. $\dfrac{d}{r} = t$, for r

15. $x = ya^2 + yz$, for y

16. The number of calories, K, needed each day by a moderately active woman who weighs w pounds, is h inches tall, and is a years old can be estimated by the formula

$$K = 917 + 6(w + h - a).$$

Allyson is moderately active, weighs 110 lb, and is 28 years old. If Allyson needs 1781 calories per day to maintain her weight, how tall is she?

17. A garden is constructed in the shape of a triangle. Using the formula for the area of a triangle, $A = \dfrac{1}{2}bh$, find the height of the garden if the base of the triangle is 9 ft and the garden will have an area of 54 ft^2.

Five Steps for Problem Solving

Five Steps for Problem Solving
1. *Familiarize* yourself with the problem situation.
2. *Translate* the problem to an equation.
3. *Solve* the equation.
4. *Check* the answer in the original problem.
5. *State* the answer to the problem clearly.

Consecutive integers like 11, 12, 13, 14 or -25, -24, -23, -22 can be represented in the form x, $x+1$, $x+2$, $x+3$, and so on. **Consecutive even integers** like 12, 14, 16, 18 or -24, -22, -20, -18 can be represented in the form x, $x+2$, $x+4$, $x+6$, and so on, as can **consecutive odd integers** like 11, 13, 15, 17 or -25, -23, -21, -19.

GUIDED LEARNING    **Textbook**    **Instructor**    **Video**

EXAMPLE 1	YOUR TURN 1
Of the top ten rice-producing countries, Brazil is the only country not in Asia. In 2012, the Philippines produced 11,000 thousand metric tons of rice. This was 4640 thousand metric tons less than two times as much as Brazil. Find the amount of rice produced in Brazil.	In 2012, India produced 99,000 thousand metric tons of rice. This was 17,000 thousand metric tons more than four times as much as Thailand. Find the amount of rice produced in Thailand.

1. **Familiarize.** We let $b =$ the amount of rice produced in Brazil.

2. **Translate.** We translate as follows.

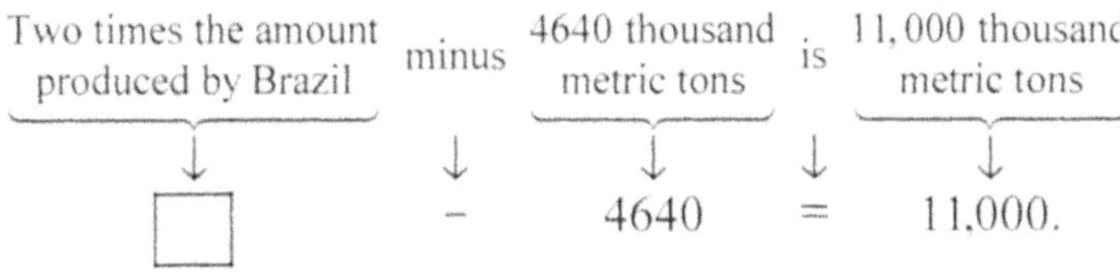

3. **Solve.** We solve the equation as follows.

$$2b - 4640 = 11,000$$
$$2b - 4640 + 4640 = 11,000 + 4640$$
$$2b = \boxed{}$$
$$\frac{2b}{\boxed{}} = \frac{15,640}{\boxed{}}$$
$$b = 7820$$

(continued)

4. **Check.** If Brazil produced 7820 thousand metric tons of rice, then the Philippines produced $2(7820)-4640$, or 11,000, thousand metric tons of rice. The amount checks.

5. **State.** In 2012, Brazil produced 7820 thousand metric tons of rice.

EXAMPLE 2	YOUR TURN 2

The sum of three consecutive integers is 90. Find the integers.

The sum of three consecutive integers is -36. Find the integers.

1. **Familiarize.** Let x, $x+1$, and $x+2$ represent the consecutive integers.

2. **Translate.**

$$\underbrace{\text{The sum of three consecutive integers}}_{\downarrow} \quad \underset{\downarrow}{\text{is}} \quad \underset{\downarrow}{90}$$

$$x+(x+1)+(x+2) \quad = \quad 90.$$

3. **Solve.** Solve the equation.

$$x+(x+1)+(x+2)=90$$

$$\boxed{}+3=90$$

$$3x=\boxed{}$$

$$x=\boxed{}$$

The other two numbers are $x+1=29+1=30$ and $x+2=29+2=\boxed{}$.

4. **Check.** The numbers 29, 30, and 31 are consecutive integers. Their sum is $29+30+31=90$. The answer checks.

5. **State.** The numbers are $\boxed{}$ $\boxed{}$, and $\boxed{}$.

YOUR NOTES   Write your questions and additional notes.

Basic Motion Problems

ESSENTIALS

The Motion Formula

Distance = Rate (or speed) · Time

$$d = rt$$

Example

- An airplane traveling 415 km/h in still air encounters a 25 km/h headwind. How long will it take the plane to travel 650 km into the wind?

  1. **Familiarize.** Because the airplane is traveling into the headwind, the speed of the wind can be subtracted from the speed of the airplane to determine the airplane's speed with the headwind. Let t represent the time, in hours, required for the plane to travel 650 km into the wind.

Distance to be traveled	650 km
Airplane's speed in still air	415 km/h
Speed of the headwind	25 km/h
Airplane's speed with the headwind	390 km/h
Time required	t

  2. **Translate.** Using the motion formula, $d = rt$, we have

$$d = rt$$
$$650 = 390 \cdot t.$$

  3. **Solve.** We solve the equation:

$$650 = 390 \cdot t$$
$$\frac{650}{390} = \frac{390 \cdot t}{390}$$
$$\frac{5}{3} = t.$$

  4. **Check.** At a speed of 390 km/h and in a time of 5/3, or $1\frac{2}{3}$, hours, the airplane would travel $d = 390 \cdot \frac{5}{3} = 650$ km. This answer checks.

  5. **State.** The airplane will travel the distance of 650 km in 5/3, or $1\frac{2}{3}$, hours.

GUIDED LEARNING    **Textbook**    **Instructor**   ▶ **Video**

EXAMPLE 1	YOUR TURN 1

EXAMPLE 1

Jill rides her bicycle 15 miles to the shore at a speed of 22 mph on a still day. If riding with a 14 mph tailwind, how long will it take to bike the 15 mi?

1. **Familiarize.** We let $t =$ the time, in hours, it will take to bike to the shore. Because Jill is biking in the same direction as the tailwind, the two speeds can be added for total speed.

Distance to be traveled	15 mi
Jill's speed on a still day	22 mph
Speed of the tailwind	14 mph
Jill's speed with the tailwind	36 mph
Time required	t

2. **Translate.** Using the motion formula,

$$d = rt$$

$$\boxed{} = 36 \cdot t.$$

3. **Solve.** We solve the equation.

$$15 = 36 \cdot t$$

$$\frac{15}{\boxed{}} = \frac{36 \cdot t}{\boxed{}}$$

$$\frac{15}{36}, \text{ or } \frac{5}{\boxed{}} = t$$

4. **Check.** At a speed of 36 mph and a time of 5/12 hours, Jill would travel $d = 36 \cdot \dfrac{5}{12} = 15$ mi. This answer checks.

5. **State.** Jill will travel the 15 mi in 5/12 hr, or 25 min.

YOUR TURN 1

Jinger rides her bicycle 22 mi to the shore at a speed of 25 mph. If she is biking with an 8 mph tailwind, how long will the ride take?

YOUR NOTES   Write your questions and additional notes.

Practice Exercises

Readiness Check

Fill in the steps in the five steps for problem solving in the correct order. All steps are listed below.

Check   Translate   Familiarize   State   Solve

1. __________

2. __________

3. __________

4. __________

5. __________

Five Steps for Problem Solving

Solve.

6. A gas station offers a 5% discount to anyone who pays cash for gas. Sue paid $65.25 in cash for gas. What would she have paid if she did not pay with cash?

7. The sum of two integers is fourteen. The larger integer is four less than the twice the smaller integer. Find the integers.

8. A worker on a production line is paid a base salary of $240 per week plus $0.81 for each unit produced. One week the worker earned $414.15. How many units were produced?

9. The sum of the page numbers on the facing pages of a book is 85. What are the page numbers?

10. In a triangle, the measure of the second angle is 5 times the measure of the first angle, and the measure of the third angle is 12° less than 10 times the measure of the first angle. Find the measure of each angle.

Basic Motion Problems

Solve.

11. A moving walkway in an airport is 880 ft long and it moves at a rate of 5 ft/sec. If Mike walks at a rate of 3 ft/sec, how long will it take him to walk its entire length?

Inequalities

ESSENTIALS

An **inequality** is a sentence containing $<$, $>$, $\leq$, $\geq$, or $\neq$.

The replacement or value for the variable that makes an inequality true is called a **solution** of the inequality. The set of all solutions is called the **solution set**. When all the solutions of an inequality have been found, we say that we have **solved** the inequality.

Examples

- 1 is a solution of the inequality $x + 5 < 7$ because $1 + 5 < 7$, or $6 < 7$, is a *true* sentence.

- -4 is a solution of the inequality $x + 9 \leq 5$ because $-4 + 9 \leq 5$, or $5 \leq 5$, is a *true* sentence.

- 0 is not a solution of the inequality $2x > x + 8$ because $2(0) > 0 + 8$, or $0 > 8$, is a *false* sentence.

GUIDED LEARNING	📖 **Textbook** 👤 **Instructor** ▶ **Video**

EXAMPLE 1	YOUR TURN 1
Determine whether the given number is a solution of the inequality: $x - 7 < 9;\ 12$. $\boxed{} - 7 < 9$ $\boxed{} < 9$ 5 is less than 9 is a *true* sentence so 12 is a solution of the inequality.	Determine whether the given number is a solution of the inequality: $x + 3 \geq 4;\ -1$.
EXAMPLE 2	YOUR TURN 2
Determine whether the given number is a solution of the inequality: $2x + 3 > 6;\ -3$. $2\left(\boxed{}\right) + 3 > 6$ $\boxed{} > 6$ -3 is greater than 6 is a *false* sentence, so -3 is not a solution of the inequality.	Determine whether the given number is a solution of the inequality: $3x + 1 > 4;\ -5$.

EXAMPLE 3	YOUR TURN 3
Determine whether the given number is a solution of the inequality:	Determine whether the given number is a solution of the inequality:
$2x - 2 \geq x + 5; \; 8.$	$3x - 9 \leq x + 1; \; 5.$
$2\left(\boxed{}\right) - 2 \geq \boxed{} + 5$	
$\boxed{} \geq \boxed{}$	
14 is greater than or equal to 13 is a *true* sentence, so 8 is a solution of the inequality.	

YOUR NOTES   Write your questions and additional notes.

Inequalities and Interval Notation

ESSENTIALS

There are three ways to represent the solutions of an inequality such as $a<x<b$:

- A **graph**, or drawing, represents the solutions.

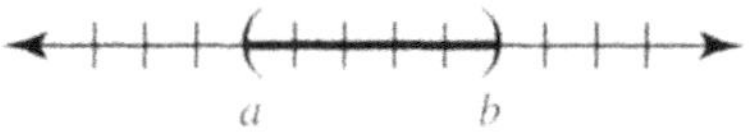

- **Set-builder notation** describes the solution set of an inequality.

  $\{x \mid a<x<b\}$

  This is read as "the set of all x such that x is both greater than a and less than b."

- **Interval notation** uses parentheses ( ) and brackets [ ].

  (a,b)

  The **endpoints** of this interval are a and b. The parentheses indicate a and b are *not* included in the interval. Brackets are used when endpoints *are* included. When an interval extends without bound in one or both directions, we use the symbols ∞ and $-\infty$, read "infinity" and "negative infinity." Parentheses are always used with ∞ and $-\infty$.

Examples

Interval Notation	Set Notation	Graph
$[3,\infty)$	$\{x \mid x \geq 3\}$	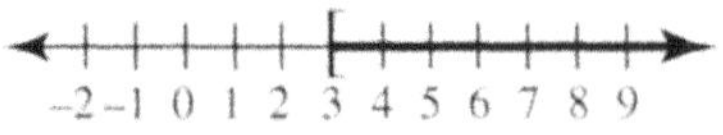
$(-\infty,-5)$	$\{x \mid x<-5\}$	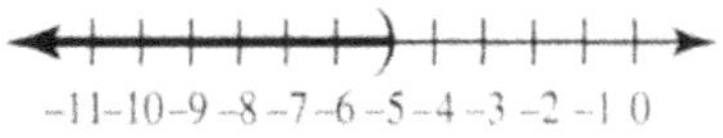
$(-1,4]$	$\{x \mid -1<x \leq 4\}$	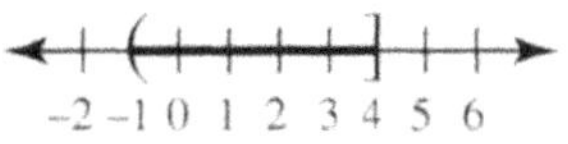
$(-\infty,\infty)$	$\{x \mid x \text{ is a real number}\}$	

GUIDED LEARNING **Textbook** **Instructor** **Video**

EXAMPLE 1	YOUR TURN 1
Write interval notation for the set:	Write interval notation for the set:
$\{x \mid 0 \le x < 7\}$.	$\{x \mid -4 < x \le 3\}$.
Interval notation: $[\,\square\,,\square\,)$	

EXAMPLE 2	YOUR TURN 2
Write interval notation for the set:	Write interval notation for the set:
$\{x \mid x < 4\}$.	$\{x \mid x < -7\}$.
Interval notation: $\square\,-\infty, 4\,\square$	

EXAMPLE 3	YOUR TURN 3
Write interval notation for the graph.	Write interval notation for the graph.

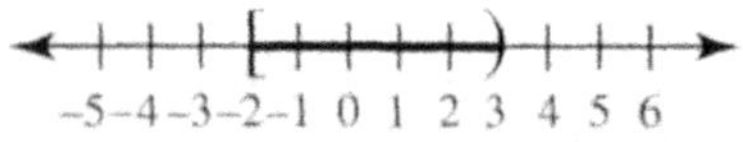

EXAMPLE 4	YOUR TURN 4
Write interval notation for the graph.	Write interval notation for the graph.

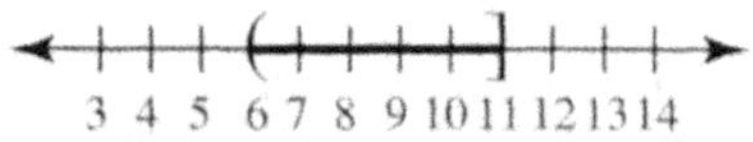

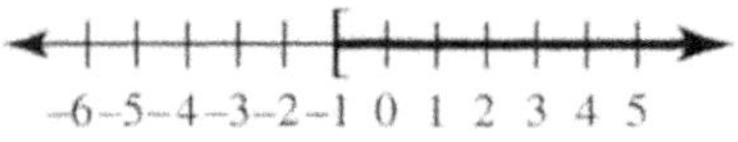

Interval notation (Example 3): $[\,\square\,,\square\,)$

Interval notation (Example 4): $\square\,-1, \infty\,\square$

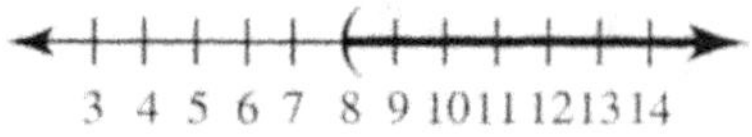

YOUR NOTES   Write your questions and additional notes.

Solving Inequalities

ESSENTIALS

Two inequalities are **equivalent** if they have the same solution set.

The Addition Principle for Inequalities
For any real numbers a, b, and c:

$a < b$ is equivalent to $a + c < b + c$;

$a > b$ is equivalent to $a + c > b + c$.

Similar statements hold for $\leq$ and $\geq$.

The Multiplication Principle for Inequalities
For any real numbers a and b, and any *positive* number c:

$a < b$ is equivalent to $ac < bc$;

$a > b$ is equivalent to $ac > bc$.

For any real numbers a and b, and any *negative* number c:

$a < b$ is equivalent to $ac > bc$;

$a > b$ is equivalent to $ac < bc$.

Similar statements hold for $\leq$ and $\geq$.

Examples

- Solve $x - 3 \geq 5$.

$$x - 3 \geq 5$$
$$x - 3 + 3 \geq 5 + 3$$
$$x \geq 8$$

The solution set is $\{x \mid x \geq 8\}$, or $[8, \infty)$.

- Solve $\dfrac{1}{2}y \leq 24$.

$$\frac{1}{2}y \leq 24$$
$$2 \cdot \frac{1}{2}y \leq 2 \cdot 24$$
$$y \leq 48$$

The solution is $\{y \mid y \leq 48\}$, or $(-\infty, 48]$.

- Solve $-2x > 24$.

$$-2x > 24$$
$$\frac{-2x}{-2} < \frac{24}{-2}$$
$$x < -12$$

The solution is $\{x \mid x < -12\}$, or $(-\infty, -12)$.

GUIDED LEARNING **Textbook** **Instructor** **Video**

EXAMPLE 1	YOUR TURN 1
Solve and graph: $x+7\le 9$.	Solve and graph: $x+8>2$.

$$x+7\le 9$$
$$x+7-\square\le 9-\square$$
$$x\le\square$$

The solution set is $\{x\,|\,x\le\square\}$, or $(-\infty,\square]$.

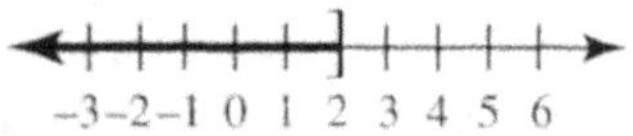

EXAMPLE 2	YOUR TURN 2
Solve and graph: $2x-3<3x-9$.	Solve and graph: $-4x+6<-3x-1$.

$$2x-3<3x-9$$
$$2x-3+\square<3x-9+\square$$
$$2x+\square<3x$$
$$2x+6-\square<3x-\square$$
$$6<\square,\text{ or } x>6$$

The solution set is $\{x\,|\,x>\square\}$, or $(\square,\infty)$.

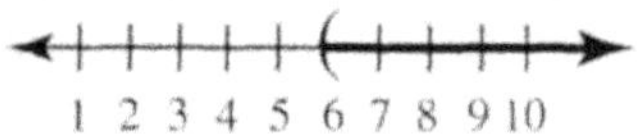

EXAMPLE 3	YOUR TURN 3
Solve and graph: $12x>36$.	Solve and graph: $5x\ge 25$.

$$12x>36$$
$$\frac{\square}{\square}\cdot 12x>\frac{\square}{\square}\cdot 36$$ Multiplying by $\frac{1}{12}$; the symbol stays the same.
$$\square>\square$$

The solution set is $\{x\,|\,\square>\square\}$, or $(\square,\infty)$.

EXAMPLE 4	YOUR TURN 4
Solve and graph: $-2x > 8$.	Solve and graph: $-3x \geq -27$.

EXAMPLE 4

Solve and graph: $-2x > 8$.

$$-2x > 8$$

$$\frac{-2x}{\boxed{}} < \frac{8}{\boxed{}} \qquad \text{Dividing by } -2 \text{ and reversing the inequality symbol}$$

$$x < \boxed{}$$

The solution set is $\left\{ x \mid x < \boxed{} \right\}$, or $\left(\boxed{}, \boxed{} \right)$.

$$-9\;-8\;-7\;-6\;-5\;-4\;-3\;-2\;-1\;\;0$$

YOUR TURN 4

Solve and graph: $-3x \geq -27$.

EXAMPLE 5

Solve and graph: $4x - 12 \leq 18 + 10x$.

$$4x - 12 \leq 18 + 10x$$

$$4x - 12 + \boxed{} \leq 18 + 10x + \boxed{}$$

$$4x \leq \boxed{} + 10x$$

$$4x - \boxed{} \leq 30 + 10x - \boxed{}$$

$$\boxed{} \leq 30$$

$$\frac{-6x}{\boxed{}} \geq \frac{30}{\boxed{}} \qquad \text{Dividing by } -6; \text{ the symbol must be reversed.}$$

$$x \boxed{} -5$$

The solution set is $\left\{ x \mid x \boxed{}\,\boxed{} \right\}$, or $\left[-5, \boxed{} \right)$.

$$-10\;-9\;-8\;-7\;-6\;-5\;-4\;-3\;-2\;-1$$

YOUR TURN 5

Solve and graph:
$20 - 3x \geq 4x - 8$.

YOUR NOTES   Write your questions and additional notes.

Applications and Problem Solving

Translating "At Least and "At Most"

A quantity x is **at least** some amount q: $x \geq q$.

(If x is at least q, it cannot be less than q.)

A quantity x is **at most** some amount q: $x \leq q$.

(If x is at most q, it cannot be more than q.)

Examples

A few common phrases that are used in application problems involving inequalities are underlined in the examples below.

Sample Sentence	Translation
The temperature of a fully cooked chicken <u>is at least</u> 165°F.	$t \geq 165$
Today's high temperature <u>is at most</u> 75°F.	$t \leq 75$
Your speed on the highway <u>cannot exceed</u> 65 mph.	$s \leq 65$
Your grade <u>must exceed</u> 69.5.	$g > 69.5$
The recommended tire pressure <u>is between</u> 28 psi and 32 psi.	$28 < p < 32$
He can withdraw <u>no more than</u> $1500 from his account.	$w \leq 1500$
The bank requires that you deposit <u>more than</u> $50.	$d > 50$

GUIDED LEARNING

 **Textbook**   **Instructor**   **Video**

EXAMPLE 1	YOUR TURN 1
Greg is playing in a golf tournament. After the first three days of the tournament his scores are 68, 67, and 71. The record low score for this particular golf course is a four-day total of 276. What scores could Greg make on the fourth day in order for his four-day total to be less than the course record?	Sharon manages a small retail clothing store whose 3-day weekend (Friday, Saturday, Sunday) sales record is $5072.53. She has made $1235.67 in sales on Friday and $2790.84 on Saturday. How much money must Sharon make in sales on Sunday so that her total weekend sales will be more than the current weekend record?

1., 2. Familiarize and Translate.

Let f = Greg's score on day # ☐. We want Greg's total score to be ___________
less than / greater than
the record. Thus, total score < course record.

This translates to $68 + 67 + 71 + f$ ☐ 276.

3. **Solve.** We solve the inequality:

$$68 + 67 + 71 + f < 276$$

$$\boxed{} + f < 276$$

$$f < \boxed{}.$$

4. **Check.** As a partial check, substitute a number less than 70 for f. We choose 69. Replacing f with 69, we get $68 + 67 + 71 + 69 = 275$ which is less than the record low of 276.

5. **State.** Greg must score less than 70 on the fourth day of the tournament to have his four day total score less than the course record of 276. The solution set is $\{f \mid f < 70\}$.

EXAMPLE 2	YOUR TURN 2
Samantha is considering two different catering companies to provide the food for her upcoming wedding. Company A charges a flat fee of \$500 plus \$10 per guest attending the wedding. Company B charges a flat fee of \$300 plus \$14 per guest. For what number of guests will Company A cost less?	John and Katie are planning a birthday party for their son. Location A charges \$8 per guest, whereas location B charges \$50 plus \$6 per guest. For what number of guests will location B cost less?

1. **Familiarize.** Total cost includes the flat fee plus the cost for guests.

   Company A total:

   $$\$500 \ + \ (\$10)(\text{Number of guests})$$

   Company B total:

   $$\boxed{} \ + \ (\$14)(\text{Number of guests})$$

2. **Translate.** Let $G =$ the number of guests attending. Thus we want to find all values of G such that

$$\underbrace{\begin{array}{c}\text{Cost of}\\\text{Company A}\end{array}}_{\downarrow} \quad \underbrace{\begin{array}{c}\text{is less}\\\text{than}\end{array}}_{\downarrow} \quad \underbrace{\begin{array}{c}\text{Cost of}\\\text{Company B}\end{array}}_{\downarrow}$$

$$500+10G \quad < \quad 300+14G.$$

3. **Solve.** We solve the inequality:

$$500+10G < 300+14G$$
$$\boxed{}+10G < 14G$$
$$200 < \boxed{}$$
$$\boxed{} < G, \text{ or } G > \boxed{}$$

4. **Check.** For $G = 50$, the cost of Company A is $500+10\cdot 50$, or \$1000 and the cost of Company B is $300+14\cdot 50$, or \$1000. Thus, for 50 guests, the cost of both plans is the same. We can substitute numbers greater than 50 to make a partial check of our result.

5. **State.** The solution set is $\{G \mid G > 50\}$.

   Company A will cost less if more than 50 guests are invited.

YOUR NOTES   Write your questions and additional notes.

Practice Exercises

Readiness Check

For each solution set expressed in set-builder notation, select from the column on the right the equivalent interval notation.

1. $\{x \mid x > 5\}$ a) $(-\infty, 5)$

2. $\{x \mid x < 5\}$ b) $[2, 5)$

3. $\{x \mid 2 < x \leq 5\}$ c) $(-\infty, 5]$

4. $\{x \mid x \geq 5\}$ d) $(5, \infty)$

5. $\{x \mid x \leq 5\}$ e) $(2, 5]$

6. $\{x \mid 2 \leq x < 5\}$ f) $[5, \infty)$

Inequalities

Determine whether the given number is a solution of the inequality.

7. $x + 4 < -9; -10$

8. $2x - 5 \geq 7; 7$

9. $4x + 6 > 1 - x; -1$

Inequalities and Interval Notation

Write interval notation for the given set or graph.

10. $\{x \mid x \geq -2\}$ 11. $\{x \mid -3 \leq x \leq 5\}$

12.

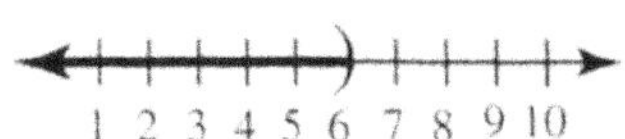

13.

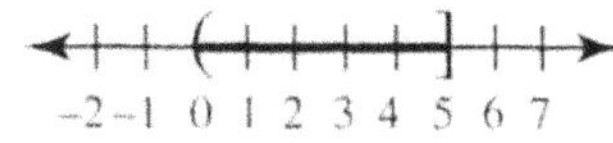

Solving Inequalities

Solve and graph.

14. $x - 7 > -3$

15. $-3x < 18$

16. $2x - 3 \le 5 - 6x$

17. $x + 5 \le 5x + 17$

18. $5(x - 2) + 8 < 2(x + 4) - 3x$

Applications and Problem Solving

Solve.

19. This month Terry is expecting to receive a $300 bonus check in addition to his regular pay rate of $15 per hour. How many hours must Terry work this month in order to have a paycheck that is more than $2850?

20. Jennifer must take three tests in her psychology class, each worth 100 points. She scored an 87 on the first test and a 91 on the second test. In order to make an A in the course Jennifer's test grades must total at least 270. What scores on the third test will give Jennifer an A in the course?

21. Hank is planning to repaint his house and has received estimates from two companies. Company A charges $600 plus $10 per hour. Company B charges $25 per hour. How many hours would the job need to take in order for Company B to cost less?

Intersections of Sets and Conjunctions of Inequalities

ESSENTIALS

Compound inequalities consist of two or more inequalities joined by the word *and* or the word *or*.

The **intersection** of two sets A and B is the set of all members common to A and B. We denote the intersection of sets A and B as $A \cap B$. The intersection of two sets can be illustrated as shown below. The region where the sets overlap is the intersection.

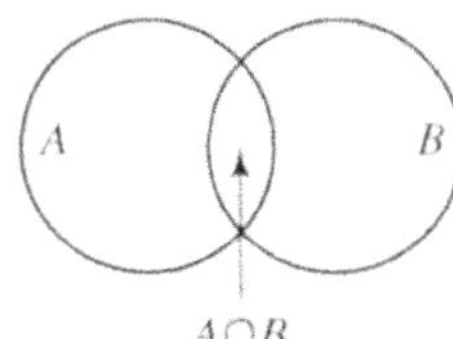

When two or more sentences are joined by the word *and* to make a compound sentence, the new sentence is called a **conjunction** of the sentences.

$a < x$ *and* $x < b$ **can be abbreviated** $a < x < b$;

$b > x$ *and* $x > a$ **can be abbreviated** $b > x > a$.

CAUTION! "$a > x$ *and* $x < b$" cannot be abbreviated as "$a > x < b$".

The word **"and"** corresponds to **"intersection"** and to the symbol "$\cap$". In order for a number to be a solution of a conjunction, it must make each part of the conjunction true.

Sometimes two sets have no elements in common. In such a case, we say that the intersection of the two sets is the **empty set**, denoted $\{\ \}$ or $\varnothing$. Two sets with an empty intersection are said to be **disjoint**. The sets shown below do not overlap, thus they have no elements in common. In this diagram $A \cap B = \varnothing$.

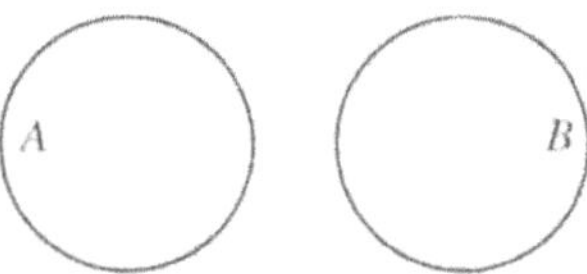

Examples

- The intersection of the sets $\{0, 2, 4, 6, 8\}$ and $\{1, 2, 3, 4, 5\}$ is $\{2, 4\}$.

- The intersection of the sets $\{1, 3, 5, 7, 9\}$ and $\{2, 4, 6, 8, 10\}$ is $\varnothing$.

- The conjunction $3 < x$ *and* $x \leq 7$ can be rewritten as $3 < x \leq 7$.

- Solve $3 < x + 5 \leq 9$.

  $3 < x + 5$ *and* $x + 5 \leq 9$

  $-2 < x$ *and* $x \leq 4$

  The solution set is $\{x \mid -2 < x \leq 4\}$. 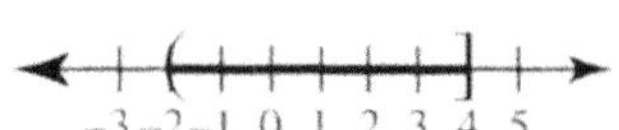

GUIDED LEARNING

🔵 **Textbook**   👤 **Instructor**   ▶ **Video**

EXAMPLE 1	YOUR TURN 1

EXAMPLE 1

Find the intersection:

$\{2, 4, 6, 8, 10\} \cap \{-4, 0, 4, 8, 12\}$.

4 and 8 are the only numbers common to both sets.

The intersection is $\left\{ \boxed{}, \boxed{} \right\}$.

YOUR TURN 1

Find the intersection:

$\{-2, 0, 1, 5, 8\} \cap \{0, 1, 2, 8\}$.

EXAMPLE 2

Solve and graph: $3 < 4y - 1 \le 23$.

Write as the conjunction:

$3 < 4y - 1 \quad \boxed{} \quad 4y - 1 \le 23$.

Solve each inequality separately:

$$3 < 4y - 1 \quad and \quad 4y - 1 \le 23$$

$$\boxed{} < 4y \quad and \quad 4y \le \boxed{} \quad \text{Adding 1}$$

$$\boxed{} < y \quad and \quad y \le \boxed{}. \quad \text{Dividing by 4}$$

The graph is the intersection of these two solution sets.

$\{y \mid 1 < y\}$

$\{y \mid y \le 6\}$

$\{y \mid 1 < y\} \cap \{y \mid y \le 6\} = \left\{ y \mid \boxed{} \right\}$

The solution set is $\{y \mid 1 < y \le 6\}$, or in interval notation, $(1, 6]$.

YOUR TURN 2

Solve and graph: $-4 \le 8y + 4 < 36$.

EXAMPLE 3	YOUR TURN 3
Solve and graph: $17 < 2x + 5$ *and* $4x - 3 < 13$.	Solve and graph: $2 < 3x + 5$ *and* $4 - 4x < -12$.

Solve each inequality separately:

$$17 < 2x + 5 \quad and \quad 4x - 3 < 13$$

$$\boxed{} < 2x \quad and \quad 4x < \boxed{}$$

$$\boxed{} < x \quad and \quad x < \boxed{}.$$

The graph is the intersection of these two solution sets.

$\{x \mid 6 < x\}$ 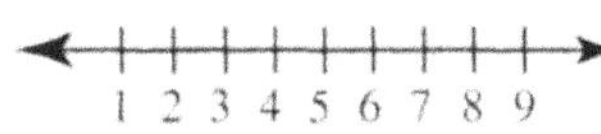

$\{x \mid x < 4\}$

$$\{x \mid 6 < x\} \cap \{x \mid x < 4\} = \boxed{}$$

The solution set is $\varnothing$ because the two sets have no numbers in common.

YOUR NOTES   Write your questions and additional notes.

Unions of Sets and Disjunctions of Inequalities

ESSENTIALS

The **union** of two sets A and B is the collection of all elements belonging to A and/or B. We denote the union of sets A and B by $A \cup \boldsymbol{B}$.

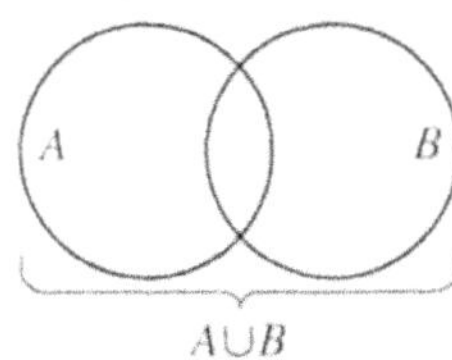

When two or more sentences are joined by the word *or* to make a compound sentence, the new sentence is called a **disjunction** of the sentences.

The word **"or"** corresponds to **"union"** and to the symbol "$\cup$". In order for a number to be in the solution set of a disjunction, it must be in *at least one* of the solution sets of the individual sentences.

Examples

- The union of the sets $\{2, 4, 6\}$ and $\{4, 8, 12\}$ is $\{2, 4, 6, 8, 12\}$.

- The disjunction $3 < x$ *or* $x \leq -7$ can be rewritten as $(-\infty, -7] \cup (3, \infty)$.

GUIDED LEARNING   📖 **Textbook**   👤 **Instructor**   ▶ **Video**

EXAMPLE 1	YOUR TURN 1
Find the union: $\{1, 2\} \cup \{2, 4, 6\}$.	Find the union: $\{0, 3\} \cup \{1, 3, 9\}$.

The numbers in either or both sets are 1, 2, 4, and 6, so the union is $\left\{ \boxed{} \right\}$.

EXAMPLE 2	YOUR TURN 2
Solve and graph: $2x - 7 > 3$ *or* $5 - x \geq 2$.	Solve and graph: $3x + 1 < 7$ *or* $3 - x < -1$.

Solve each inequality separately:

$$2x - 7 > 3 \quad or \quad 5 - x \geq 2$$
$$2x > \boxed{} \quad or \quad -x \geq \boxed{}$$
$$x > \boxed{} \quad or \quad x \leq \boxed{}.$$

(continued) | (continued)

The graph is the union of these two solution sets.

$\{x \mid x > 5\}$ 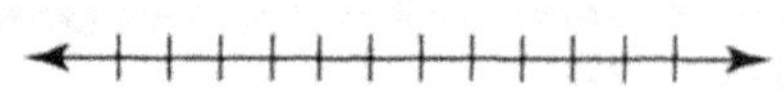

$\{x \mid x \le 3\}$ 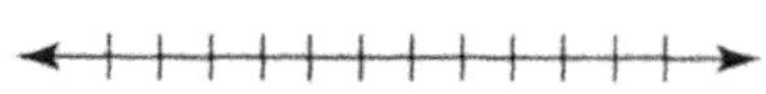

$\{x \mid x > 5\} \cup \{x \mid x \le 3\}$

The solution set is written $\{x \mid x \le 3 \ or \ x > 5\}$, or in interval notation, $(-\infty, 3] \cup (5, \infty)$.

EXAMPLE 3	YOUR TURN 3

EXAMPLE 3

Solve and graph: $2x - 15 > -7 \ or \ 3x + 4 < 25$.

Solve each inequality separately:

$$2x - 15 > -7 \quad or \quad 3x + 4 < 25$$

$$2x > \boxed{} \quad or \quad 3x < \boxed{}$$

$$x > \boxed{} \quad or \quad x < \boxed{}.$$

The graph is the union of these two solution sets.

$\{x \mid x > 4\}$ 

$\{x \mid x < 7\}$ 

$\{x \mid x > 4\} \cup \{x \mid x < 7\}$

The solution set is $\{x \mid x \text{ is a real number}\}$, or in interval notation, $(-\infty, \infty)$.

YOUR TURN 3

Solve and graph: $x - 4 \le 4 \ or \ x - 3 \le 2$.

YOUR NOTES   Write your questions and additional notes.

Applications and Problem Solving

 Textbook   Instructor   Video

GUIDED LEARNING

EXAMPLE 1	YOUR TURN 1

EXAMPLE 1

In order to get a B in her biology class the average of Ruth's midterm exam and final exam must be at least 80 but can be at most 89. If Ruth got a 78 on her midterm what scores could she get on her final exam to get a B in biology?

1., 2. Familiarize and Translate.

Let f = Ruth's score on the final exam.

Ruth's average could be calculated as follows:

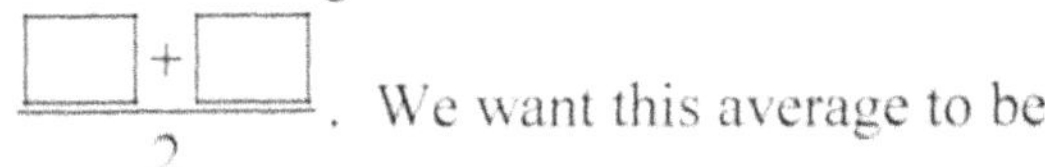

$\dfrac{\boxed{}+\boxed{}}{2}$. We want this average to be

greater than or equal to 80 *and* less than or equal to 89. This can be written as the conjunction

$$\boxed{} \le \frac{78+f}{2} \quad and \quad \frac{78+f}{2} \le \boxed{}.$$

3. Solve. We solve each inequality separately.

$$80 \le \frac{78+f}{2} \quad and \quad \frac{78+f}{2} \le 89$$

$$\boxed{} \le 78+f \quad and \quad 78+f \le \boxed{}$$

$$\boxed{} \le f \quad and \quad f \le \boxed{}.$$

4. Check. Suppose Ruth's final exam score was 83. Her average would be $\dfrac{78+83}{2} = \dfrac{161}{2} = 80.5$.

Similarly it can be shown that if Ruth scored a 99 on the final exam her average would be 88.5. In either scenario her average is at least 80 and at most 89.

5. State. Ruth's final exam score must be at least 82 and at most 100 which can be written as the solution set $\{f \mid 82 \le f \le 100\}$.

YOUR TURN 1

Rob's cookbook says that chicken is fully cooked when it reaches Celsius temperatures C such that $75° \le C \le 80°$. Use the formula $C = \dfrac{5}{9}(F - 32)$ to find an inequality for the corresponding Fahrenheit temperatures.

YOUR NOTES   Write your questions and additional notes.

Practice Exercises

Readiness Check

Determine whether each statement is true or false.

1. $\{1, 2, 3, 4, 5\} \cap \{2, 4, 6, 8, 10\} = \varnothing$

2. $\{1, 2, 3, 4, 5\} \cup \{6\} = \{1, 2, 3, 4, 5, 6\}$

3. The compound inequality $3 < x$ *or* $x < 7$ can be expressed as $3 < x < 7$.

4. The compound inequality $3 > x$ *and* $x < 7$ can be expressed as $3 > x < 7$.

5. The solution set of $x < -2$ *or* $x > 6$ can be written as $(-\infty, -2) \cup (6, \infty)$.

6. The solution set of $x > -2$ *and* $x < 6$ can be written as $(-2, 6)$.

Intersections of Sets and Conjunctions of Inequalities

Find the intersection.

7. $\{1, 2, 3\} \cap \{a, b, c\}$

8. $\{a, b, c, d\} \cap \{c, a, t\}$

Solve and graph.

9. $5 \le x + 2$ *and* $x + 2 < 9$

10. $-3 < 3x + 6 < 12$

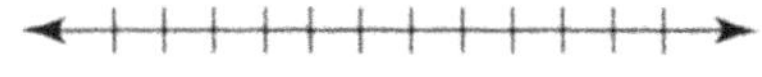

11. $x + 4 \ge -1$ *and* $5 - x < 8$

12. $-2x + 5 > 9$ *and* $9 \le 3x - 6$

Unions of Sets and Disjunctions of Inequalities

Find the union.

13. $\{1, 2, 3\} \cup \{a, b, c\}$

14. $\{1, 2, 3\} \cup \varnothing$

Solve and graph.

15. $7 \le 2x + 1$ *or* $2x + 1 \le 1$

16. $2 - x > 3$ *or* $x + 4 \ge 8$

17. $5x - 3 \ge 7$ *or* $-x + 7 < 2$

18. $x + 8 < 16$ *or* $2x - 3 > 7$

Applications and Problem Solving

Solve.

19. One of the strangest weather days occurred in Browning, Montana, on January 24, 1916. The low temperature for the day was $-49°C$ and the high temperature for the day was $7°C$. Use the inequality $-49° \le C \le 7°$ along with the fact that $C = \dfrac{5}{9}(F - 32)$ to find such an inequality for the corresponding Fahrenheit temperatures.

20. In order to get a grade of A in his psychology class, the average of Rudy's three exams must be greater than or equal to 90. If Rudy got grades of 88 and 96 on his first two exams, what scores could he make on the third exam to get an A in psychology? (Suppose it is possible for him to make higher than 100 on the third exam.)

Properties of Absolute Value

ESSENTIALS

The **absolute value** of a number is the number's distance from zero on the number line.

The **absolute value** of x, denoted $|x|$, is defined as follows:

$$x \geq 0 \rightarrow |x| = x, \qquad x < 0 \rightarrow |x| = -x.$$

Properties of Absolute Value

a) $|ab| = |a| \cdot |b|$, for any real numbers a and b.

   (The absolute value of a product is the product of the absolute values.)

b) $\left|\dfrac{a}{b}\right| = \dfrac{|a|}{|b|}$, for any real numbers a and b and $b \neq 0$.

   (The absolute value of a quotient is the quotient of the absolute values.)

c) $|-a| = |a|$, for any real number a.

   (The absolute value of the opposite of a number is the same as the absolute value of the number.)

Examples

- $|5| = 5$

- $|-5| = 5$

- $|7x| = |7| \cdot |x| = 7|x|$

- $\left|\dfrac{-8}{x}\right| = \dfrac{|-8|}{|x|} = \dfrac{8}{|x|}$

	🔵 **Textbook**	👤 **Instructor**	▶ **Video**

GUIDED LEARNING

EXAMPLE 1	YOUR TURN 1
Simplify, leaving as little as possible inside the absolute-value signs. $$\left|7x^2\right| = \boxed{} \cdot \left|x^2\right| = 7x^2$$ Since x^2 can never be negative for any real number x we can omit the absolute-value signs.	Simplify, leaving as little as possible inside the absolute-value signs. $$\left|5x^4\right|$$

EXAMPLE 2	YOUR TURN 2
Simplify, leaving as little as possible inside the absolute-value signs. $$\left|\frac{6x^3}{2x^2}\right| = \left|\boxed{} \cdot x\right| = 3\boxed{}$$	Simplify, leaving as little as possible inside the absolute-value signs. $$\left|\frac{16y^6}{4y^5}\right|$$
EXAMPLE 3	YOUR TURN 3
Simplify, leaving as little as possible inside the absolute-value signs. $$\left|\frac{5y^7}{-10y^5}\right| = \left|\frac{y^2}{\boxed{}}\right| = \frac{\left|y^2\right|}{\boxed{}} = \frac{\boxed{}}{\boxed{}}$$	Simplify, leaving as little as possible inside the absolute-value signs. $$\left|\frac{6x^9}{-18x^7}\right|$$

YOUR NOTES   Write your questions and additional notes.

Distance on the Number Line

ESSENTIALS

For any real numbers a and b, the **distance** between them is $|a-b|$.

Examples

- The distance between -2 and 5 on the number line is $|-2-5| = |-7| = 7$.

- The distance between -13 and -9 on the number line is $|-13-(-9)| = |-4| = 4$.

- The distance between x and 3 on the number line is $|x-3|$ or $|3-x|$.

GUIDED LEARNING	Textbook	Instructor	Video

EXAMPLE 1	YOUR TURN 1
Find the distance between 6 and -2 on the number line.	Find the distance between 12 and -9 on the number line.

$$\left|\boxed{}-(-2)\right| = \left|\boxed{}\right| = 8 \quad \text{or} \quad \left|\boxed{}-6\right| = \left|\boxed{}\right| = 8$$

6 and -2 are 8 units apart on the number line.

EXAMPLE 2	YOUR TURN 2
Find the distance between -10 and -6 on the number line.	Find the distance between -4 and -15 on the number line.

$$\left|\boxed{}-(-6)\right| = \left|\boxed{}\right| = 4 \quad \text{or}$$

$$\left|\boxed{}-(-10)\right| = \left|\boxed{}\right| = 4$$

-10 and -6 are 4 units apart on the number line.

YOUR NOTES   Write your questions and additional notes.

Equations with Absolute Value

ESSENTIALS

The Absolute-Value Principle

For any positive number p and any algebraic expression X:

a) The solution of $|X| = p$ is those numbers that satisfy $X = -p$ or $X = p$.

b) The equation $|X| = 0$ is equivalent to the equation $X = 0$.

c) The equation $|X| = -p$ has no solution.

Examples

- The equation $|x| = 6$ has the solution set $\{-6, 6\}$.

- Solve $|2x - 8| = 0$.

$$2x - 8 = 0 \qquad \text{Using the second part of the absolute-value principle}$$
$$2x = 8 \qquad \text{Adding 8}$$
$$x = 4 \qquad \text{Dividing by 2}$$

  The solution set is $\{4\}$.

- Solve $|x - 2| = 7$.

$$|x - 2| = 7$$
$$x - 2 = -7 \quad or \quad x - 2 = 7 \qquad \text{Using the first part of the absolute-value principle}$$
$$x = -5 \quad or \quad x = 9 \qquad \text{Adding 2}$$

  The solution set is $\{-5, 9\}$.

- The equation $|x + 5| = -3$ has no solution. The solution set is the empty set, $\varnothing$.

GUIDED LEARNING	Textbook	Instructor	Video

EXAMPLE 1	YOUR TURN 1
Solve: $|4x| = -8$.	Solve: $|x - 3| + 5 = 2$.
Since absolute value is always nonnegative, this equation has no solution. The solution set is $\boxed{}$.	

EXAMPLE 2	YOUR TURN 2
Solve: $\lvert x+7\rvert - 4 = 8$.	Solve: $\lvert x\rvert - 6 = 4$.

We first use the addition principle for equations to get $\lvert x+7\rvert$ by itself.

$$\lvert x+7\rvert - 4 = 8$$

$$\lvert x+7\rvert = \boxed{} \qquad \text{Adding 4}$$

$$x+7 = \boxed{} \quad or \quad x+7 = 12 \qquad \text{Absolute-value principle}$$

$$x = \boxed{} \quad or \qquad x = 5 \qquad \text{Subtracting 7}$$

The solutions are $\boxed{}$ and 5. The solution set is $\left\{\boxed{}, 5\right\}$.

EXAMPLE 3	YOUR TURN 3
Solve: $\lvert 5-3y\rvert = 7$.	Solve: $\lvert y-1\rvert = 3$.

$$\lvert 5-3y\rvert = 7$$

$$5-3y = \boxed{} \quad or \quad 5-3y = 7 \qquad \text{Absolute-value principle}$$

$$-3y = \boxed{} \quad or \qquad -3y = 2 \qquad \text{Subtracting 5}$$

$$y = \boxed{} \quad or \qquad y = -\frac{2}{3} \qquad \text{Dividing by } -3$$

The solutions are $-\dfrac{2}{3}$ and $\boxed{}$. The solution set is $\left\{-\dfrac{2}{3}, \boxed{}\right\}$.

YOUR NOTES   Write your questions and additional notes.

Equations with Two Absolute-Value Expressions

ESSENTIALS

If $|a| = |b|$, this means that a and b are the same distance from 0. If a and b are the same distance from 0, then either they are the same number or they are opposites.

Example

- Solve $|3x + 7| = |x - 3|$.

$$3x + 7 = x - 3 \quad or \quad 3x + 7 = -(x - 3)$$
$$3x = x - 10 \quad or \quad 3x + 7 = -x + 3$$
$$2x = -10 \quad or \quad 3x = -x - 4$$
$$x = -5 \quad or \quad 4x = -4$$
$$x = -5 \quad or \quad x = -1$$

The solution set is $\{-5, -1\}$.

GUIDED LEARNING   📖 **Textbook**   👤 **Instructor**   ▶ **Video**

EXAMPLE 1	YOUR TURN 1
Solve: $|x| = |2x - 4|$.	Solve: $|3x + 1| = |x + 3|$.

$$|x| = |2x - 4|$$

$x = 2x - 4 \quad or \quad x = \boxed{}$

$-x = -4 \quad or \quad x = \boxed{}$

$x = 4 \quad or \quad 3x = \boxed{}$

$x = 4 \quad or \quad x = \boxed{}$

The solutions are $\boxed{}$ and 4. The solution set is $\left\{ \boxed{}, 4 \right\}$.

EXAMPLE 2	YOUR TURN 2
Solve: $\lvert 5x - 6 \rvert = \lvert 5x + 4 \rvert$.	Solve: $\lvert x + 12 \rvert = \lvert x - 12 \rvert$.

$$\lvert 5x - 6 \rvert = \lvert 5x + 4 \rvert$$

$5x - 6 = 5x + 4 \quad or \quad 5x - 6 = \boxed{}$

$-6 = 4 \qquad or \quad 5x - 6 = \boxed{}$

$-6 = 4 \qquad or \quad 10x - 6 = \boxed{}$

$-6 = 4 \qquad or \qquad 10x = \boxed{}$

$-6 = 4 \qquad or \qquad x = \boxed{}$

The first equation has no solution. The solution of the second equation is $\dfrac{1}{5}$.

The solution set is $\left\{ \dfrac{1}{5} \right\}$.

YOUR NOTES   Write your questions and additional notes.

Inequalities with Absolute Value

ESSENTIALS

Solutions of Absolute-Value Inequalities:

For any positive number p and any algebraic expression X:

a) The solutions of $|X| < p$ are those numbers that satisfy $-p < X < p$.

b) The solutions of $|X| > p$ are those numbers that satisfy $X < -p$ *or* $X > p$.

Examples

- Solve: $|x| \le 2$. Then graph.

    $-2 \le x \le 2$

    The solution set is $\{x \mid -2 \le x \le 2\}$, or $[-2, 2]$:

- Solve: $|x| \ge 2$. Then graph.

    $x \le -2$ *or* $x \ge 2$

    The solution set is $\{x \mid x \le -2 \ or \ x \ge 2\}$, or $(-\infty, -2] \cup [2, \infty)$:

- Solve: $|x - 1| \le 3$. Then graph.

    $-3 \le x - 1 \le 3$

    $-2 \le x \le 4$ Adding 1 to each part of the inequality

    The solution set is $\{x \mid -2 \le x \le 4\}$, or $[-2, 4]$:

	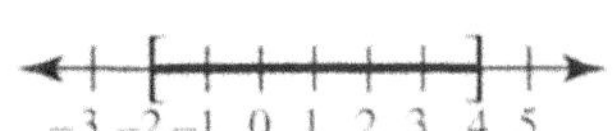 **Textbook**	**Instructor**	**Video**

GUIDED LEARNING

EXAMPLE 1	YOUR TURN 1
Solve: $|x| \le 3$. Then graph.	Solve: $|x| < 2$. Then graph.

Solve: $|x| \le 3$. Then graph.

$|X| \le p$

$|x| \le 3$

$\boxed{} \le x \le \boxed{}$

The solution set is $\left\{ x \middle| \boxed{} \le x \le \boxed{} \right\}$, or

$\left[\boxed{}, \boxed{} \right]$.

The graph is as follows.

YOUR TURN 1

Solve: $|x| < 2$. Then graph.

EXAMPLE 2	YOUR TURN 2
Solve: $\lvert y\rvert > 3$. Then graph. $\lvert X\rvert > p$ $\lvert y\rvert > 3$ $y < \boxed{} \quad or \quad y > \boxed{}$ The solution set is $\left\{ y \,\middle|\, y < \boxed{} \;\; or \;\; y > \boxed{} \right\}$, or $\left(-\infty, \boxed{}\right) \cup \left(\boxed{}, \infty\right)$. The graph is as follows.	Solve: $\lvert y\rvert \ge 1$. Then graph.

```
←—)—+—+—+—+—+—(—→
 -4 -3 -2 -1  0  1  2  3  4
```

EXAMPLE 3	YOUR TURN 3
Solve: $\lvert 2x + 5\rvert \le 3$. Then graph. $\lvert X\rvert \le p$ $\lvert 2x + 5\rvert \le 3$ $\boxed{} \le 2x + 5 \le 3$ $\boxed{} \le 2x \le -2$ $\boxed{} \le x \le -1$ The solution set is $\left\{ x \,\middle|\, \boxed{} \le x \le -1 \right\}$, or $\left[\, \boxed{}, -1 \,\right]$. The graph is as follows.	Solve: $\lvert -3x + 6\rvert < 9$. Then graph.

```
←—+—+—[—+—+—+—]—+—+—+—→
 -6 -5 -4 -3 -2 -1  0  1  2
```

EXAMPLE 4	YOUR TURN 4
Solve: $\left|-x+6\right| \geq 2$. Then graph.	Solve: $\left|x-4\right| \geq 3$. Then graph.

Solve: $\left|-x+6\right| \geq 2$. Then graph.

$$\left|X\right| \geq p$$

$$\left|-x+6\right| \geq 2$$

$$-x+6 \leq \boxed{} \quad or \quad -x+6 \geq 2$$

$$-x \leq \boxed{} \quad or \quad -x \geq -4$$

$$x \geq \boxed{} \quad or \quad x \leq 4$$

The solution set is $\left\{ x \,\middle|\, x \leq \boxed{} \ or \ x \geq \boxed{} \right\}$, or

$\left(-\infty, \boxed{}\right] \cup \left[\boxed{}, \infty\right)$. The graph is as follows.

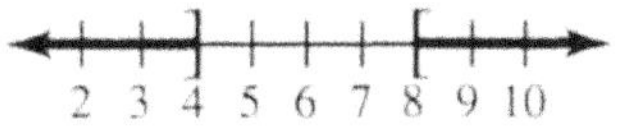

YOUR TURN 4

Solve: $\left|x-4\right| \geq 3$. Then graph.

YOUR NOTES   Write your questions and additional notes.

Practice Exercises

Readiness Check

Determine whether each statement is true or false.

1. Absolute value is never negative.

2. $|4| = -|4|$

3. To find the distance between two numbers on the number line, we simply subtract the two numbers.

4. For any positive number p, the equation $|x| = -p$ has no solution.

5. The inequality $|x| > 4$ has solution set $(-\infty, -4) \cup (4, \infty)$.

Properties of Absolute Value

Simplify, leaving as little as possible inside absolute-value signs.

6. $\left| \dfrac{15x}{5} \right|$

7. $\left| -6x^2 \right|$

Distance on the Number Line

Find the distance between the points on the number line.

8. $-11, 16$

9. $-\dfrac{15}{2}, -\dfrac{5}{2}$

Equations with Absolute Value

Solve.

10. $|x| - 6 = 10$

11. $|-2x + 4| = 16$

12. $|5x+1|+7=8$　　　　　　　**13.** $|7x+2|=0$

Equations with Two Absolute-Value Expressions

Solve.

14. $|2x-1|=|5x+2|$　　　　　　**15.** $|y+8|=|y|$

16. $\left|\dfrac{3}{4}c\right|=|c+1|$　　　　　　**17.** $|x+1|=|x+1|$

Inequalities with Absolute Value

Solve and graph.

18. $|x|\le 4$　　　　　　　　**19.** $|x+4|<3$

20. $|8x|-4>20$　　　　　　　**21.** $|6x+12|\ge 6$

Plotting Ordered Pairs

ESSENTIALS

On a plane, each point is the graph of a number pair. To form the plane, we use two perpendicular number lines called **axes**, which divide the plane into four **quadrants** which are numbered as shown below. They cross at a point called the **origin**. The arrows show the positive directions.

Example

Consider the **ordered pair** $(5, 2)$, shown in the figure at right. The numbers 5 and 2 are called coordinates. In $(5, 2)$, the **first coordinate** (the **abscissa**) is 5 and the

second coordinate (the **ordinate**) is 2. To plot $(5, 2)$, we start at the origin and move *horizontally* to the 5. Then we move up *vertically* 2 units and make a "dot." The point $(-4, 3)$ is also plotted at right. We start at the origin and move horizontally to -4, move up 3 units, and

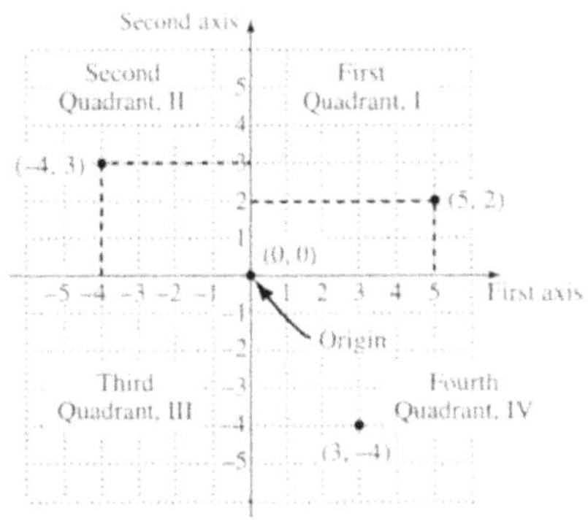

then make a "dot." Note that $(-4, 3)$ and $(3, -4)$ represent different points. The order of the numbers in the pair is important.

GUIDED LEARNING  **Textbook**  **Instructor**  **Video**

EXAMPLE 1	YOUR TURN 1

EXAMPLE 1

Plot the points $(4, -2)$ and $(-3, 0)$.

$(4, -2)$: The first number, 4, is positive.

Starting at the origin, we move ☐

unit(s) to the _____________ (left/right). The second number, -2, is negative. We move

☐ unit(s) _________ (up/down).

$(-3, 0)$: Starting at the origin, we move

☐ unit(s) to the _____________ (left/right).

Then move ☐ unit(s) vertically.

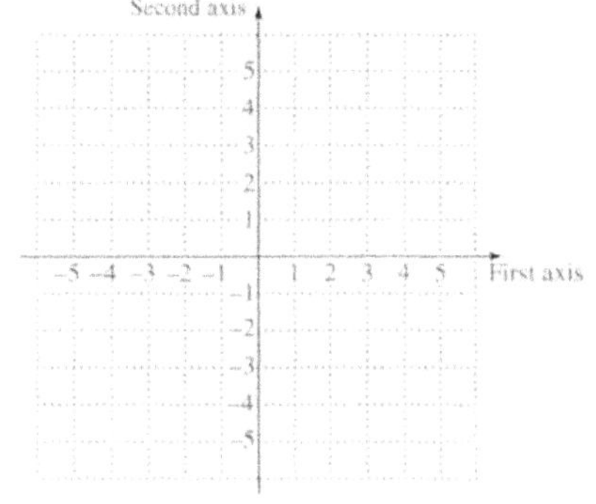

YOUR TURN 1

Plot the points $(-3, -5)$ and $(0, 2)$.

EXAMPLE 2	YOUR TURN 2
In which quadrant, if any, are the points $(-2,6)$, $(0,-1)$, $(5,3)$, and $(-4,-4)$ located? $(-2,6)$: The first coordinate is ________ positive / negative and the second coordinate is ________ . positive / negative The point is in the ________ quadrant. $(0,-1)$: The point is on a(n) ________ and is not in any quadrant. $(5,3)$: Both coordinates are ________ . positive / negative The point is in the ________ quadrant. $(-4,-4)$: Both coordinates are ________ . positive / negative The point is in the ________ quadrant.	In which quadrant, if any, are the points $(-3,-1)$, $(5,-2)$, $(6,0)$, and $(-1,4)$ located?

YOUR NOTES   Write your questions and additional notes.

Solutions of Equations

ESSENTIALS

Unless stated otherwise, to determine whether an ordered pair is a solution of an equation in two variables, we use the first number in the pair to replace the variable that occurs first alphabetically.

In a coordinate plane, the first axis is called the x-axis and the second axis, the y-axis.

Example

- Determine whether each of the following pairs is a solution of $4x - 3y = 12$: $(0, -4)$ and $(1, 2)$.

For $(0, -4)$:

$$\frac{4x - 3y = 12}{\begin{array}{c|c} 4(0) - 3(-4) \ ? \ 12 \\ 0 + 12 \\ 12 \end{array}}$$ TRUE

Thus, $(0, -4)$ *is* a solution.

For $(1, 2)$:

$$\frac{4x - 3y = 12}{\begin{array}{c|c} 4(1) - 3(2) \ ? \ 12 \\ 4 - 6 \\ -2 \end{array}}$$ FALSE

Thus, $(1, 2)$ *is not* a solution

GUIDED LEARNING  **Textbook**    **Instructor**    **Video**

EXAMPLE 1	YOUR TURN 1
Determine whether each of the following pairs is a solution of $3x - 2y = 9$: $(3, 0)$ and $(2, -3)$. For $(3, 0)$: $$\frac{3x - 2y = 9}{\begin{array}{c} 3(\boxed{}) - 2(\boxed{}) \ ? \ 9 \\ 9 - \boxed{} \\ 9 \end{array}}$$ TRUE $(3, 0)$ ______ (is/is not) a solution. For $(2, -3)$: $$\frac{3x - 2y = 9}{\begin{array}{c} 3(\boxed{}) - 2(\boxed{}) \ ? \ 9 \\ 6 + \boxed{} \\ 12 \end{array}}$$ FALSE $(2, -3)$ ______ (is/is not) a solution.	Determine whether each of the following pairs is a solution of $2x - 5y = 10$: $(0, -2)$ and $(5, 1)$.

EXAMPLE 2	YOUR TURN 2

EXAMPLE 2

Show that the pairs $(1,1)$ and $(0,-2)$ are solutions of $y = 3x - 2$. Then graph the two points and use the graph to determine another pair that is a solution.

For $(1,1)$:

$$y = 3x - 2$$

$$1 \;\overset{?}{\;}\; 3\left(\boxed{}\right) - 2$$

$$\boxed{} - 2$$

$$1 \qquad\qquad \text{TRUE}$$

For $(0,-2)$:

$$y = 3x - 2$$

$$-2 \;\overset{?}{\;}\; 3\left(\boxed{}\right) - 2$$

$$\boxed{} - 2$$

$$-2 \qquad\qquad \text{TRUE}$$

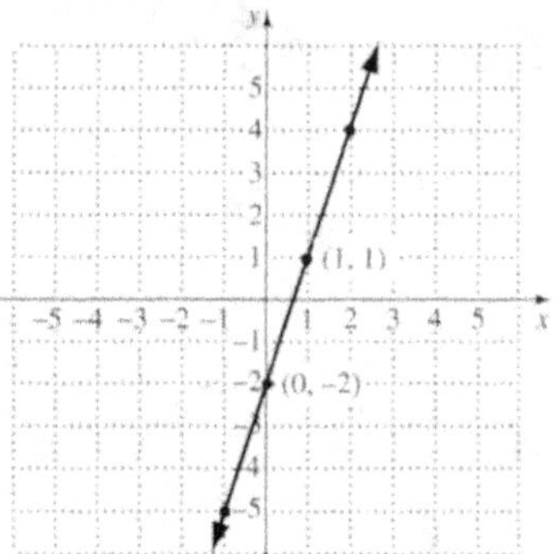

Another solution would be any other point on the line, such as $(2, 4)$ or $\left(-1, \boxed{}\right)$.

These can be checked by substituting the coordinates into $y = 3x - 2$.

YOUR TURN 2

Show that the pairs $(2,-1)$ and $(0,-5)$ are solutions of $y = 2x - 5$. Then graph the two points and use the graph to determine another pair that is a solution.

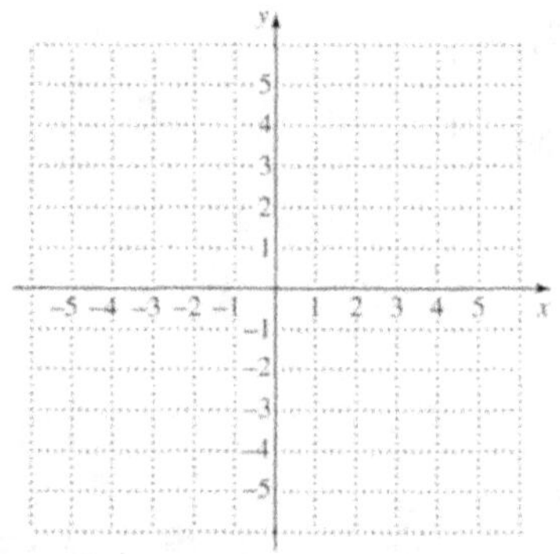

YOUR NOTES   Write your questions and additional notes.

Graphs of Linear Equations

ESSENTIALS

To graph a linear equation:

1. Select a value for one variable and calculate the corresponding value of the other variable. Form an ordered pair using alphabetical order as indicated by the variables.
2. Repeat step (1) to obtain at least two other ordered pairs. Two points are essential to determine a straight line. A third point serves as a check.
3. Plot the ordered pairs and draw a straight line passing through the points.

Example

- Graph: $y = 2x - 2$.

x	$y = 2x - 2$	(x, y)
-1	-4	$(-1, -4)$
0	-2	$(0, -2)$
3	4	$(3, 4)$

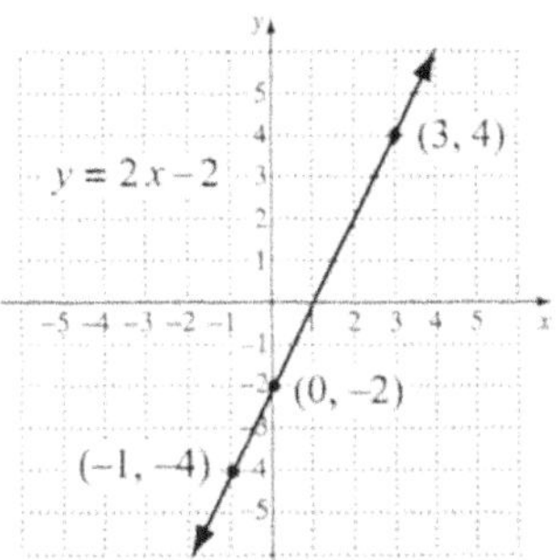

 **Textbook**    **Instructor**    **Video**

GUIDED LEARNING

EXAMPLE 1	YOUR TURN 1

EXAMPLE 1

Graph: $y = -2x - 2$.

x	$y = -2x - 2$	(x, y)
-2	☐	$(-2, \boxed{})$
0	☐	$(0, \boxed{})$
2	☐	$(2, \boxed{})$

Plot the points. We use a ruler or another straightedge to draw a line through the points.

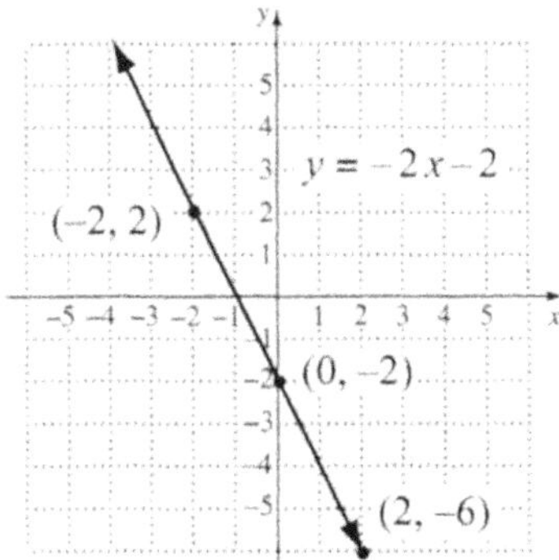

YOUR TURN 1

Graph: $y = -2x - 3$.

x	$y = -2x - 3$	(x, y)

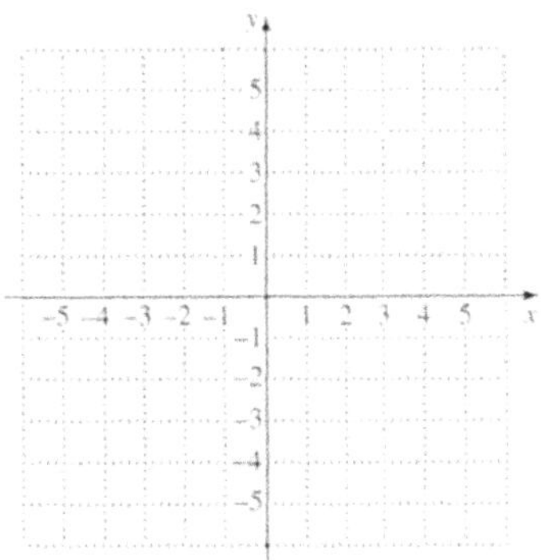

EXAMPLE 2	YOUR TURN 2

EXAMPLE 2

Graph $3x - 6y = 12$.

We first solve for y.

$$3x - 6y = 12$$

$$3x - \boxed{} - 6y = 12 - \boxed{}$$

$$-6y = \boxed{} + 12$$

$$y = \frac{\boxed{}}{\boxed{}} x - 2$$

x	$y = \dfrac{1}{2}x - 2$	(x, y)
0	$\boxed{}$	$\left(0, \boxed{}\right)$
2	$\boxed{}$	$\left(2, \boxed{}\right)$
–2	$\boxed{}$	$\left(-2, \boxed{}\right)$

We then complete and label the graph.

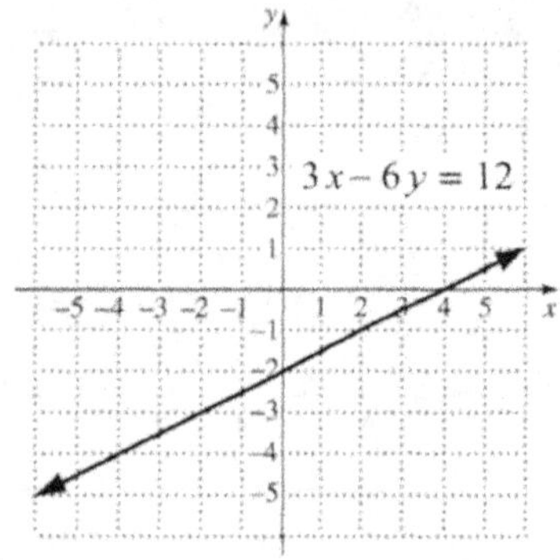

YOUR TURN 2

Graph $2x - 6y = 6$.

x	$y = \boxed{}$	(x, y)

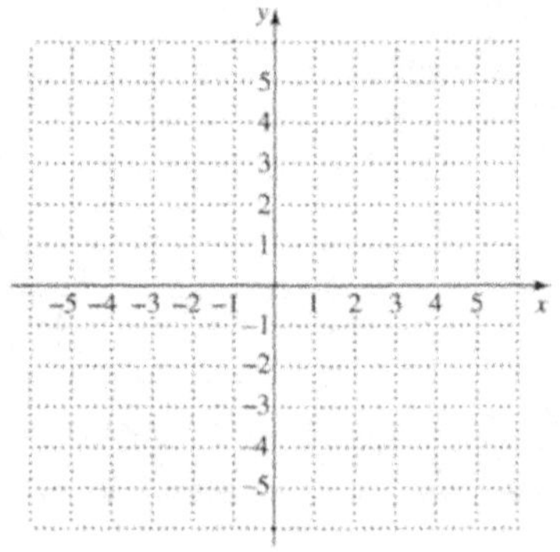

Copyright © 2022 Pearson Education, Inc.

Graphing Nonlinear Equations

ESSENTIALS

The graphs of **nonlinear equations** are not straight lines and they often require us to plot many points to see the general shape of the graph.

Example

- Graph $y = x^2$.

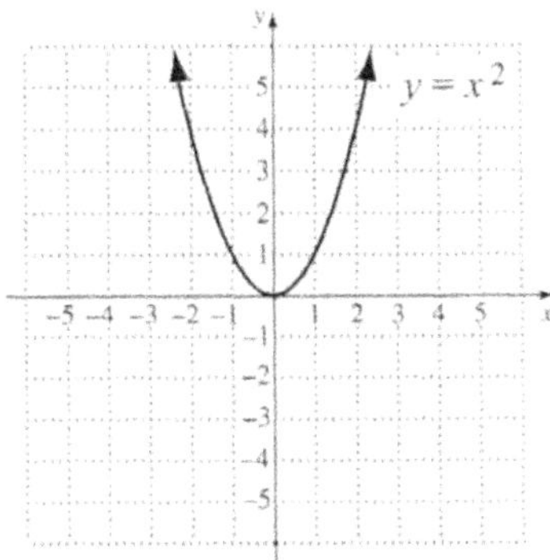

GUIDED LEARNING    **Textbook**    **Instructor**    **Video**

EXAMPLE 1	YOUR TURN 1
Graph: $y = -\lvert x \rvert$.	Graph: $y = 2\lvert x \rvert$.

x	$y = -\lvert x \rvert$	(x, y)
-3	-3	$(-3, -3)$
-2	$\square$	$(-2, -2)$
-1	-1	$\left(\square, \square\right)$
$\square$	0	$(0, 0)$
1	-1	$(1, -1)$
2	$\square$	$\left(2, \square\right)$
3	-3	$(3, -3)$

x	$y = 2\lvert x \rvert$	(x, y)
-3		
-2		
-1		
0		
1		
2		
3		

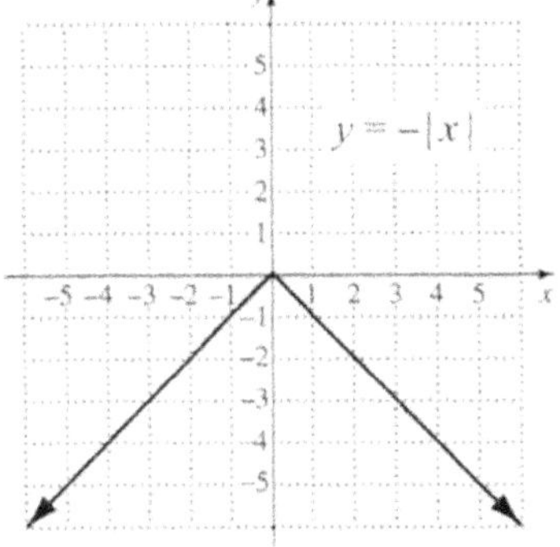

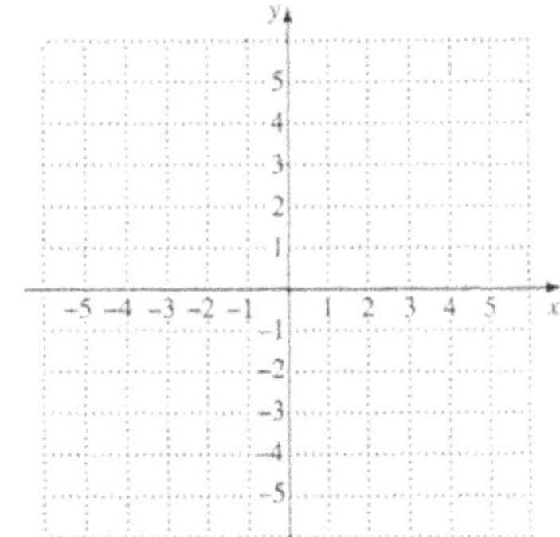

EXAMPLE 2	YOUR TURN 2

EXAMPLE 2

Graph: $y = x^2 + 1$.

x	$y = x^2 + 1$	(x, y)
-3	10	$(-3, 10)$
-2	☐	$(-2, 5)$
-1	2	$(-1, ☐)$
☐	1	$(0, 1)$
1	☐	$(1, ☐)$
2	☐	$(2, ☐)$
3	☐	$(3, ☐)$

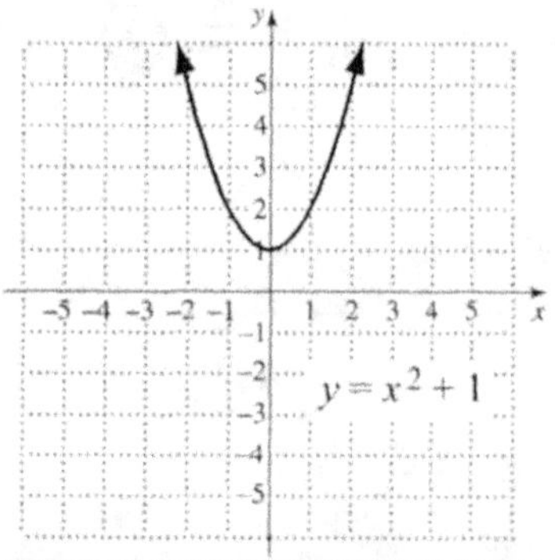

$y = x^2 + 1$

YOUR TURN 2

Graph: $y = -x^2 + 2$.

x	$y = -x^2 + 2$	(x, y)
-2		
-1		
0		
1		
2		

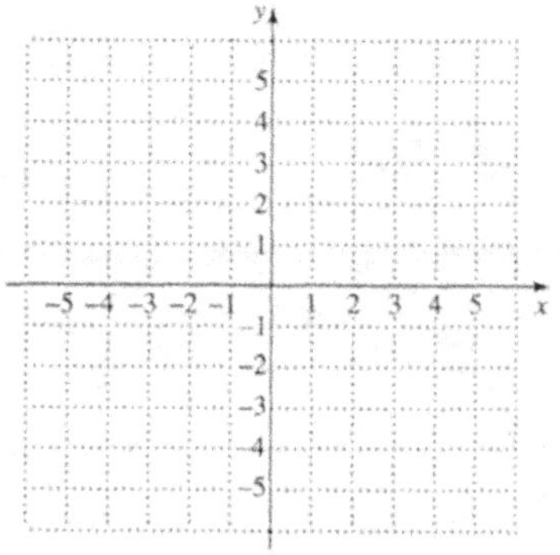

YOUR NOTES   Write your questions and additional notes.

Practice Exercises

Readiness Check

Use the graph of $y = x - 2$ at the right to determine whether each statement is true or false.

1. The point $(-1, -3)$ is in quadrant III.

2. The point $(2, 0)$ is in quadrant I.

3. The ordered pair $(3, 1)$ is a solution of the equation $y = x - 2$.

4. The point $(0, -2)$ is on the y-axis.

5. The graph passes through the origin.

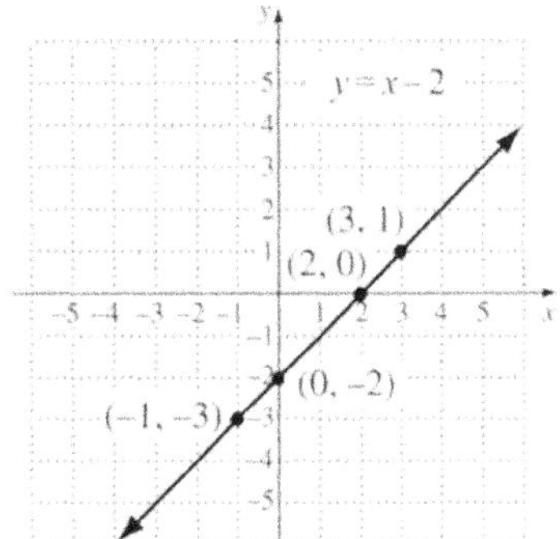

Plotting Ordered Pairs

Plot the following points.

6. $A(2, 5)$, $B(-2, -3)$, $C(-4, -1)$, $D(1, -3)$, $E(0, -4)$, $F(3, 0)$, $G(-2, 1)$, $H(-3, 4)$

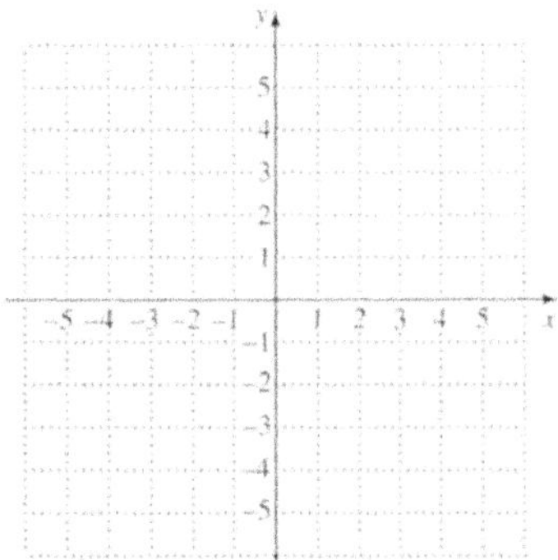

Solutions of Equations

Determine whether the given point is a solution of the equation.

7. $(3, 5)$; $y = 3x - 4$

8. $(-1, -3)$; $2b = 5a - 1$

9. $(4, 3)$; $5x - 3y = 10$

10. An equation and two ordered pairs are given. Show that each pair is a solution of the equation. Then graph the equation and use the graph to determine another solution.

$y = 1 - 2x$; $(-2, 5), (1, -1)$

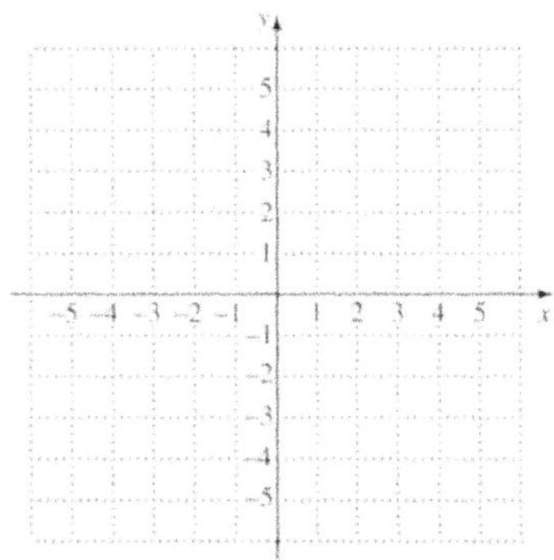

Graphs of Linear Equations

Graph.

11. $y = x - 3$

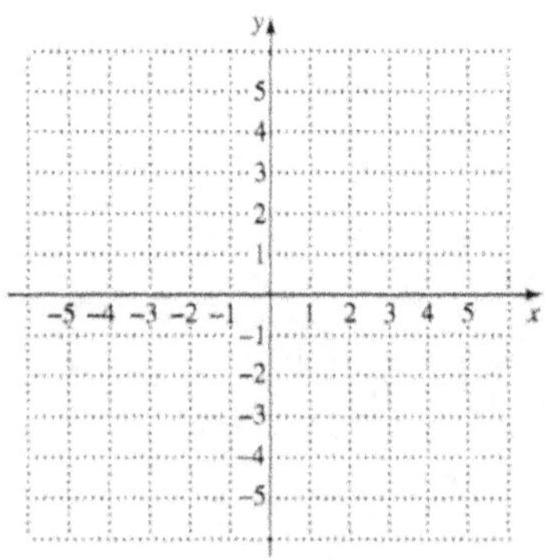

12. $y = \dfrac{1}{2}x + 2$

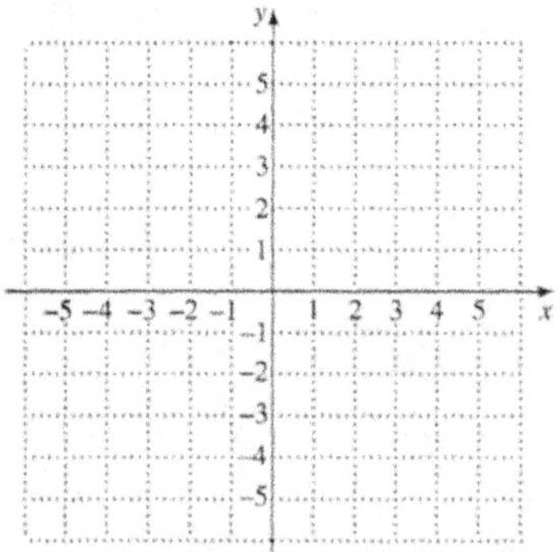

13. $4x + 2y = -2$

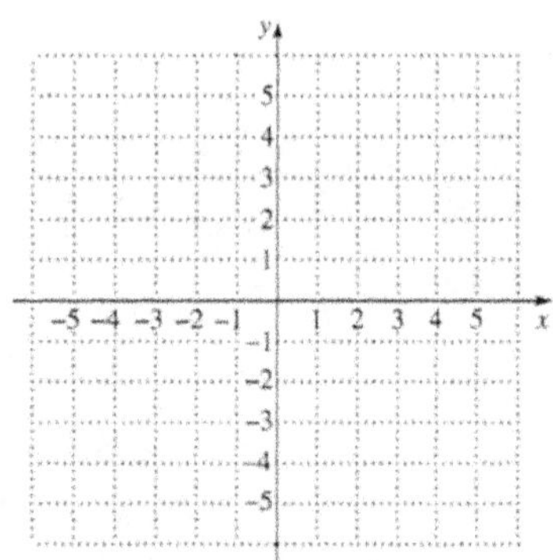

14. $2x - 3y = 6$

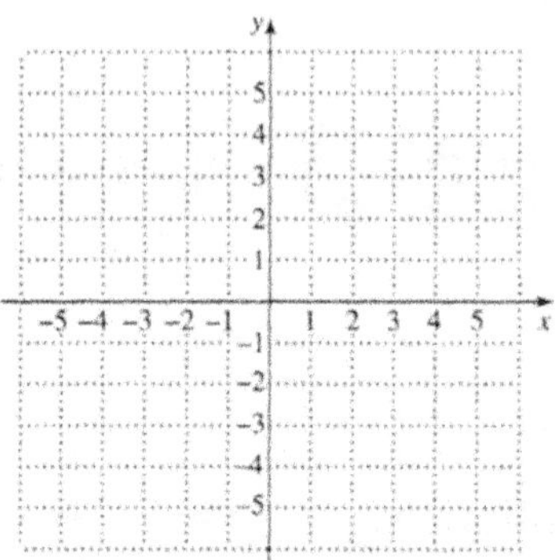

Graphs of Nonlinear Equations

Graph.

15. $y = x^2 - 4$

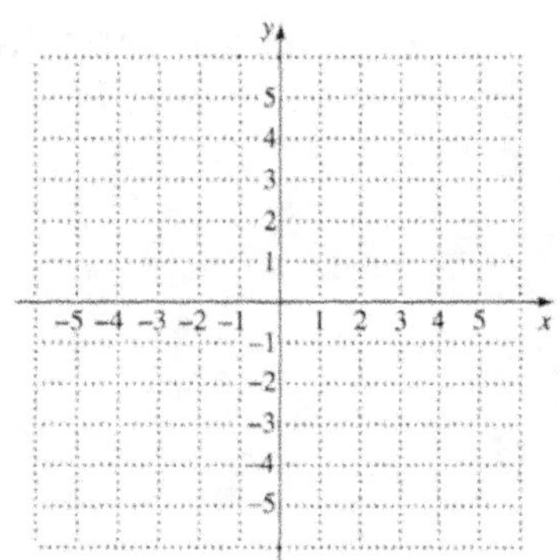

16. $y = |x + 1|$

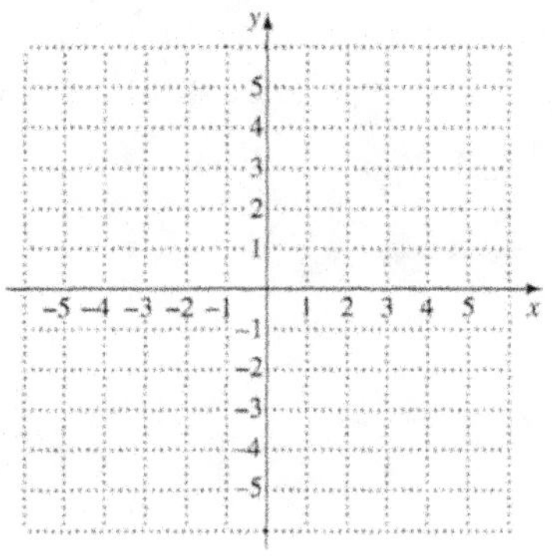

Identifying Functions

ESSENTIALS

A **function** is a correspondence between a first set called the **domain** and a second set called the **range** such that each member of the domain corresponds to *exactly one* member of the range.

Examples

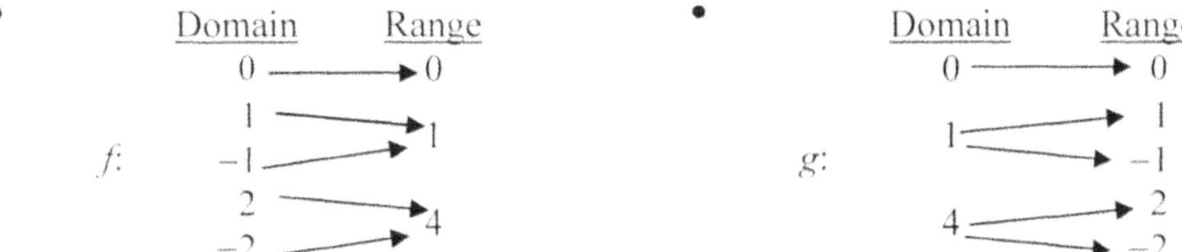

The correspondence *f is* a function because each member of the domain is matched to exactly one member of the range.

The correspondence *g is not* a function because some members of the domain are matched to more than one member of the range.

🔖 **Textbook**	👤 **Instructor**	▶ **Video**

GUIDED LEARNING

EXAMPLE 1	YOUR TURN 1

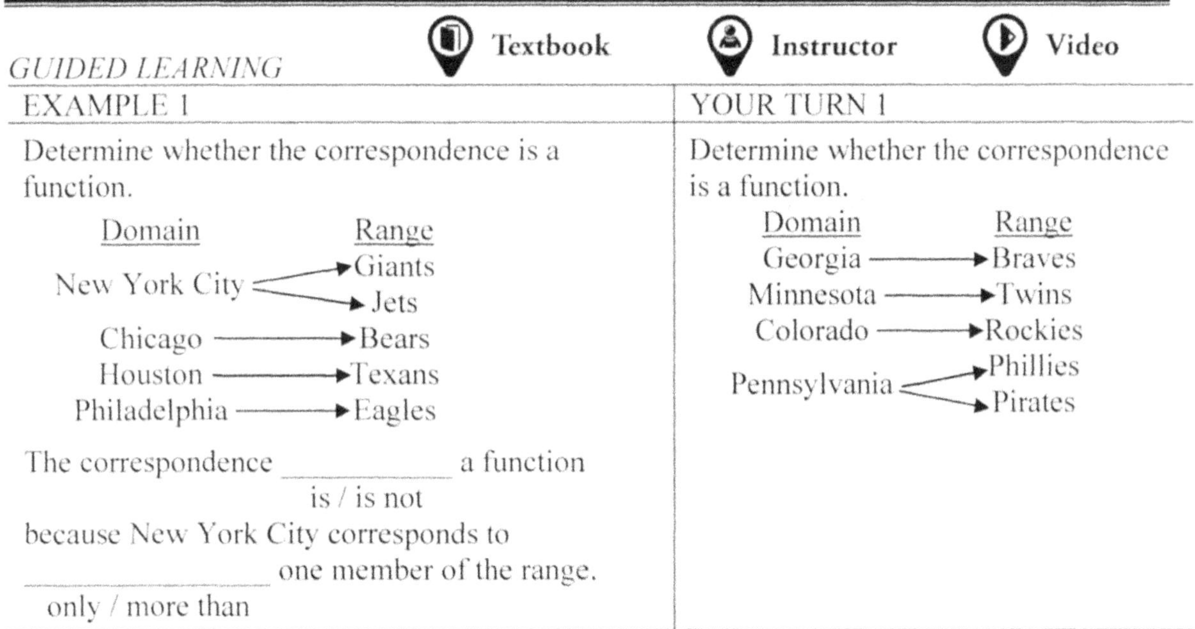

EXAMPLE 1

Determine whether the correspondence is a function.

The correspondence __________ a function
               is / is not

because New York City corresponds to __________ one member of the range.
  only / more than

YOUR TURN 1

Determine whether the correspondence is a function.

EXAMPLE 2	YOUR TURN 2
Determine whether the correspondence is a function.	Determine whether the correspondence is a function.

Domain   Range

-2 -2
-1 -1
0 0
1 1
2 2

The correspondence ___________ a function
is / is not
because each member of the domain is
matched to _______________ one member
only / more than
of the range.

Domain   Range

A ⟶ a
B ⟶ b
C ⟶ c
D ⟶ d

EXAMPLE 3	YOUR TURN 3
Determine whether the correspondence is a function.	Determine whether the correspondence is a function.
Domain: A set of occupied residences	Domain: A set of faculty advisors
Correspondence: A person living in that residence	Correspondence: An advisee assigned to that advisor
Range: A set of people	Range: A set of students

This correspondence ___________ a function
is / is not
because more than one person can live in a
residence.

EXAMPLE 4	YOUR TURN 4
Determine whether the correspondence is a function.	Determine whether the correspondence is a function.
Domain: A set of numbers	Domain: A set of numbers
Correspondence: The cube of the number	Correspondence: The opposite of the number
Range: A set of numbers	Range: A set of numbers

This correspondence is a function because
each number has _______________ one cube.
only / more than

YOUR NOTES   Write your questions and additional notes.

Finding Function Values

ESSENTIALS

For the function f, $f(x)$ is read "f of x" or "f at x" or "the value of f at x" and indicates the output that corresponds to the input x.

Example

- For $f(x) = 5x - 2$, $f(-4) = 5 \cdot (-4) - 2 = -20 - 2 = -22$.

	📖 Textbook	👤 Instructor	▶ Video

GUIDED LEARNING

EXAMPLE 1	YOUR TURN 1
For $g(x) = 3x - 5$, find $g(0)$, $g(-1)$, and $g(a+3)$. $g(0) = 3(\boxed{}) - 5 = 0 - 5 = -5$ $g(-1) = 3(\boxed{}) - 5 = -3 - \boxed{} = \boxed{}$ $g(a+3) = 3(\boxed{}) - 5 = \boxed{} + 9 - 5 = 3a + \boxed{}$	For $g(x) = 3 - 4x$, find $g(0)$, $g(-1)$, and $g(a+2)$.
EXAMPLE 2	YOUR TURN 2
For $f(x) = 13$, find $f(5)$. The function f is a constant function. Every input has the output 13. $f(5) = \boxed{}$	For $f(x) = 7$, find $f(-1)$.
EXAMPLE 3	YOUR TURN 3
For $F(x) = x^2 - x + 1$, find $F(5)$ and $F(-5)$. $F(5) = (\boxed{})^2 - (\boxed{}) + 1$ $\quad = \boxed{} - 5 + 1$ $\quad = \boxed{}$ $F(-5) = (\boxed{})^2 - (\boxed{}) + 1$ $\quad = \boxed{} + 5 + 1$ $\quad = \boxed{}$	For $h(x) = x^2 - 3x$, find $h(1)$ and $h(-1)$.

EXAMPLE 4	YOUR TURN 4
Find the volume of a sphere with radius 5 cm. Use $V(r) = \dfrac{4}{3}\pi r^3$ and 3.14 for π. $V\left(\boxed{}\right) = \dfrac{4}{3}\pi \cdot \left(\boxed{}\right)^3$ $ = \dfrac{4}{3}\pi \cdot \boxed{}\ \text{cm}^3$ $ = \boxed{}\ \text{cm}^3 \approx \boxed{}\ \text{cm}^3$	Find the surface area of a sphere with radius 3 cm. Use $S(r) = 4\pi r^2$ and 3.14 for π.

YOUR NOTES   Write your questions and additional notes.

Graphs of Functions

ESSENTIALS

To graph a function, we find ordered pairs (x, y) or $(x, f(x))$, plot them, and sketch a graph through the points.

Example

- Graph: $f(x) = |x + 2|$.

$f(-3) = |-3 + 2| = |-1| = 1$
$f(-2) = |-2 + 2| = |0| = 0$
$f(-1) = |-1 + 2| = |1| = 1$
$f(0) = |0 + 2| = |2| = 2$
$f(1) = |1 + 2| = |3| = 3$
$f(2) = |2 + 2| = |4| = 4$

x	$f(x)$
-3	1
-2	0
-1	1
0	2
1	3
2	4

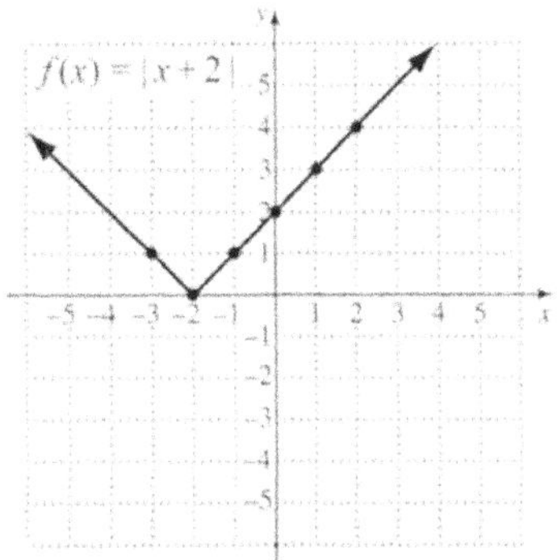

	Textbook		Instructor		Video

EXAMPLE 1	YOUR TURN 1

Graph: $f(x) = 2x - 3$.

Graph: $f(x) = 5 - x$.

First, calculate some function values.

$f(-1) = 2\left(\boxed{}\right) - 3 = \boxed{}$ 

$f(0) = 2\left(\boxed{}\right) - 3 = \boxed{}$ 

$f(1) = 2\left(\boxed{}\right) - 3 = \boxed{}$ 

$f(2) = 2(2) - 3 = 1$

$f(3) = 2(3) - 3 = 3$

x	$f(x)$
-1	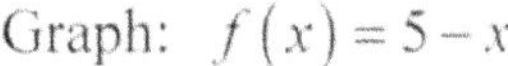
0	
1	
2	1
3	3

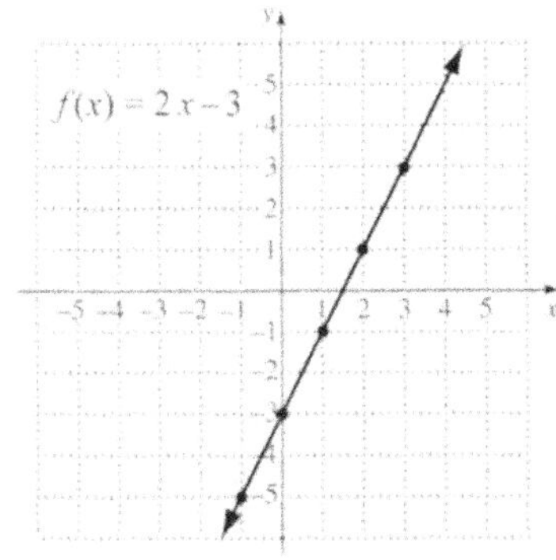

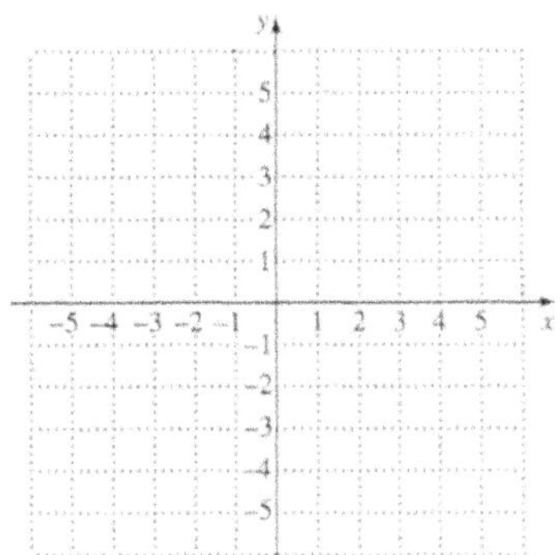

EXAMPLE 2	YOUR TURN 2

EXAMPLE 2

Graph: $g(x) = x^2 - 4x - 5$.

First, calculate some function values.

$g(-2) = (-2)^2 - 4(-2) - 5$

$\quad = 4 + 8 - 5 = 7$

$g(-1) = (-1)^2 - 4(-1) - 5$

$\quad = 1 + 4 - 5 = \boxed{}$

$g(0) = (0)^2 - 4(0) - 5 = \boxed{}$

$g(1) = (1)^2 - 4(1) - 5 = -8$

$g(2) = (2)^2 - 4(2) - 5 = \boxed{}$

$g(3) = (3)^2 - 4(3) - 5 = -8$

$g(4) = (4)^2 - 4(4) - 5 = \boxed{}$

$g(5) = (5)^2 - 4(5) - 5 = \boxed{}$

x	$f(x)$
-2	7
-1	$\boxed{}$
0	$\boxed{}$
1	-8
2	$\boxed{}$
3	-8
4	$\boxed{}$
5	$\boxed{}$

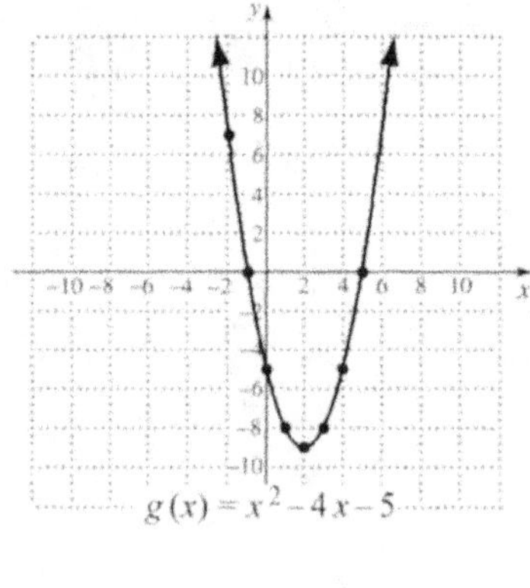

YOUR TURN 2

Graph: $g(x) = x^2 - 3$.

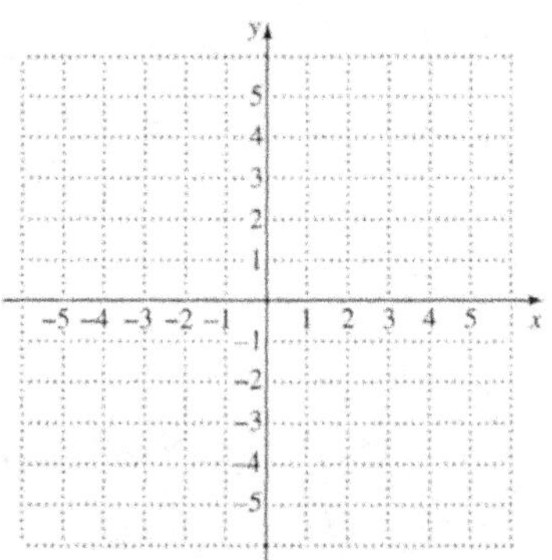

YOUR NOTES   Write your questions and additional notes.

The Vertical-Line Test

The Vertical-Line Test

If it is possible for a vertical line to cross a graph more than once, then the graph is not the graph of a function.

Examples

- A function

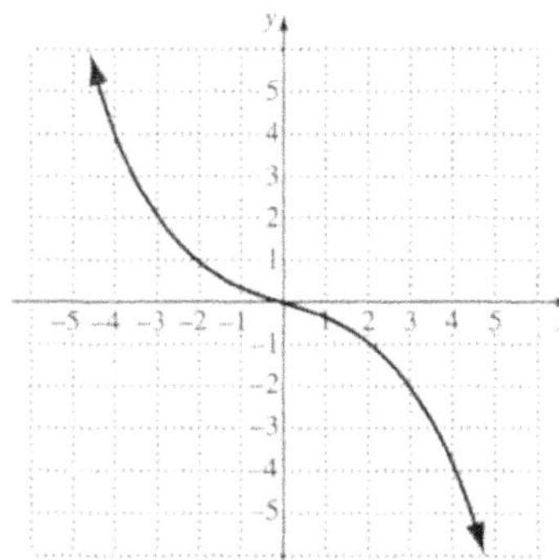

- Not a function

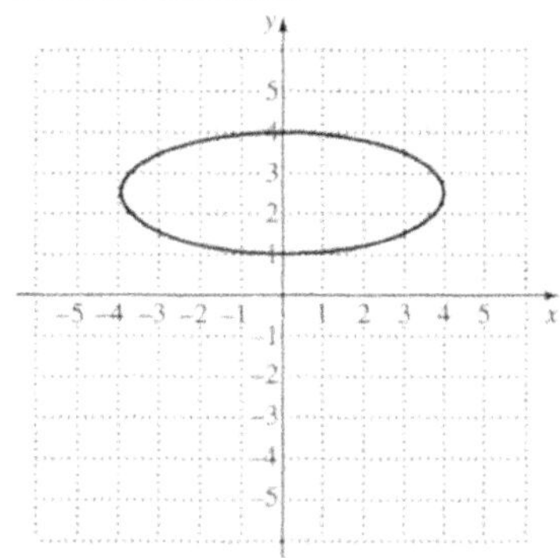

GUIDED LEARNING

 **Textbook**    **Instructor**    **Video**

EXAMPLE 1	YOUR TURN 1
Determine whether the following graph is that of a function.	Determine whether the following graph is that of a function.

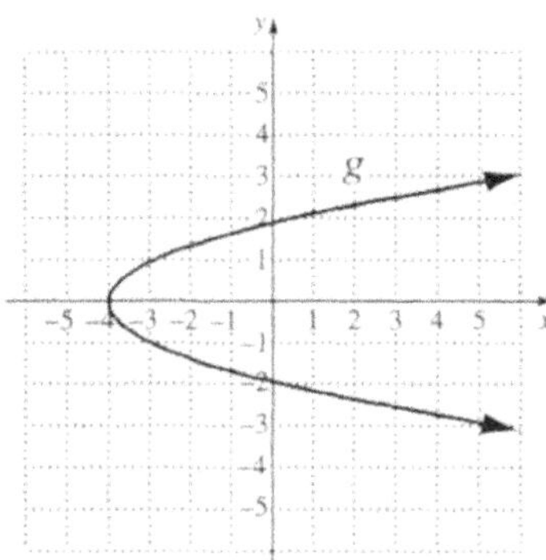

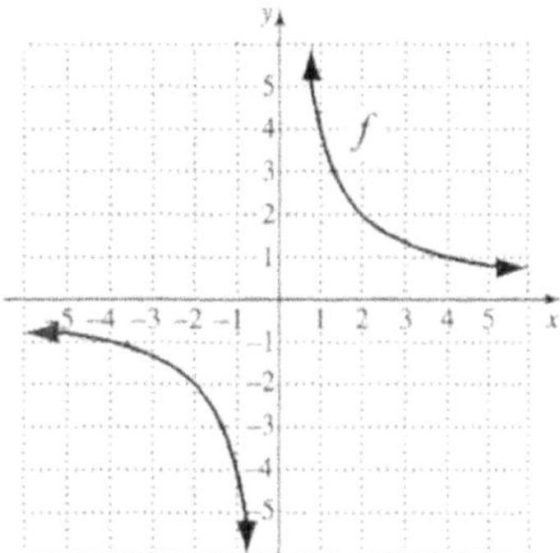

The graph _____________ pass the vertical-line
       does / does not
test. It __________ a function.
    is / is not

YOUR NOTES   Write your questions and additional notes.

Applications of Functions and Their Graphs

ESSENTIALS

Example

- The graph shown represents the average number of tweets per day sent by Twitter users.

  The number of tweets is a function f of the year x.

  To estimate the average number of tweets per day in 2011, locate 2011 on the horizontal axis and move directly up to the graph. Then move across to the vertical axis.

  The average number of tweets per day was about 250 million in 2011. That is, $f(2011) \approx 250,000,000$.

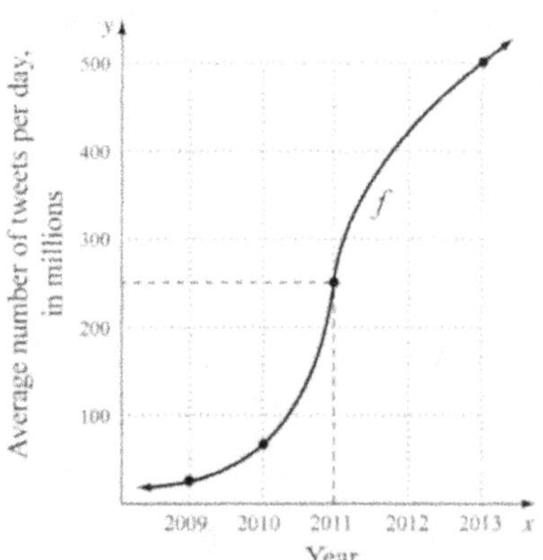

GUIDED LEARNING  **Textbook**  **Instructor**  **Video**

EXAMPLE 1	YOUR TURN 1
The following graph represents the number of for-profit hospitals in the United States from 1975 through 2010. The number of hospitals is a function f of the year x.	Use the graph in Example 1 to estimate the number of for-profit hospitals in 2010. That is, find $f(2010)$.

EXAMPLE 1

The following graph represents the number of for-profit hospitals in the United States from 1975 through 2010. The number of hospitals is a function f of the year x.

Use the graph to estimate the number of for-profit hospitals in 2000. That is, find $f(2000)$.

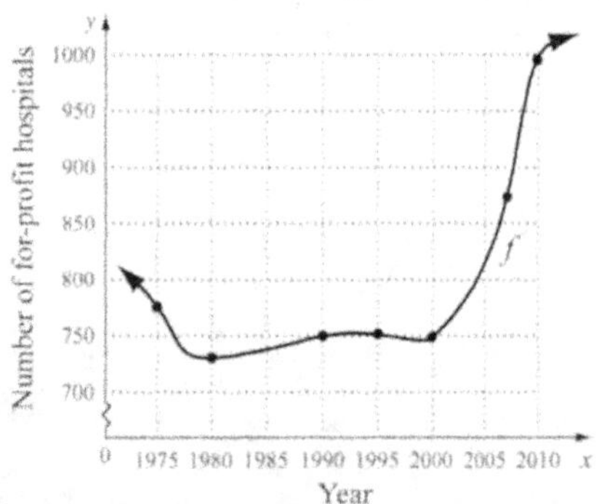

Locate 2000 on the horizontal axis and move up to the graph. Then move across to the vertical axis. In 2000, there were approximately

[____] for-profit hospitals.

YOUR NOTES   Write your questions and additional notes.

Practice Exercises

Readiness Check

Choose the word from the following list that best completes each sentence. Words may be used more than once or not at all.

function	relation	domain	range
horizontal	vertical	input	output

1. A(n) ______________ is a correspondence such that each member of the ______________ is paired with exactly one member of the ______________.

2. The ______________-line test is used to determine if a graph represents a(n) ______________.

3. For the function given by $f(x) = -8$, every input has the same ______________.

Identifying Functions

Determine whether each correspondence is a function.

4.
   Domain   Range
   T-shirt ——→ $12.99
   Shorts ——→ $21.99
   Sandals ——→
   Hat ——→ $16.99

5. *Domain*   *Correspondence*   *Range*

    A set of artists   The songs the artists have recorded   A set of songs

Finding Function Values

6. Find $f(3)$, for $f(x) = 5x - 1$.

7. Find $g(-3)$, for $g(x) = 2x^2 - x - 7$.

8. Find $g(a - 2)$, for $g(x) = 4x - 3$.

9. Find $h(8)$, for $h(x) = 5$.

10. Find the surface area of a cube with a side of 5 cm. Use $A(s) = 6s^2$.

Graphs of Functions

Graph each function.

11. $g(x) = 3x - 1$

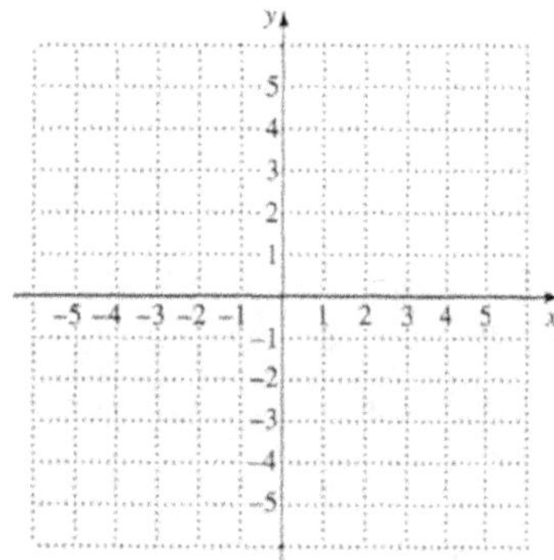

12. $f(x) = x^2 + 1$

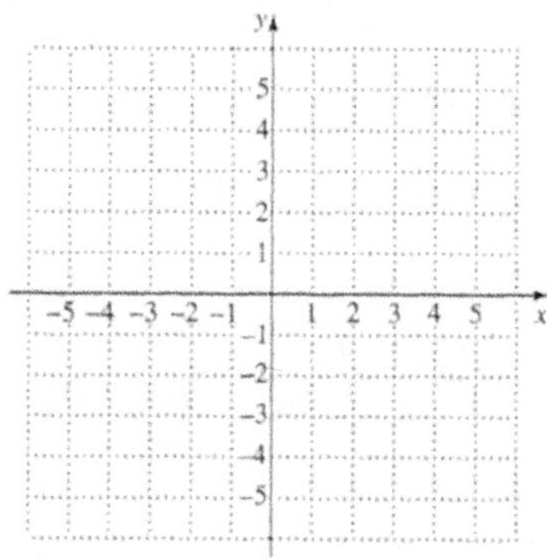

The Vertical-Line Test

Determine whether each of the following is the graph of a function.

13.

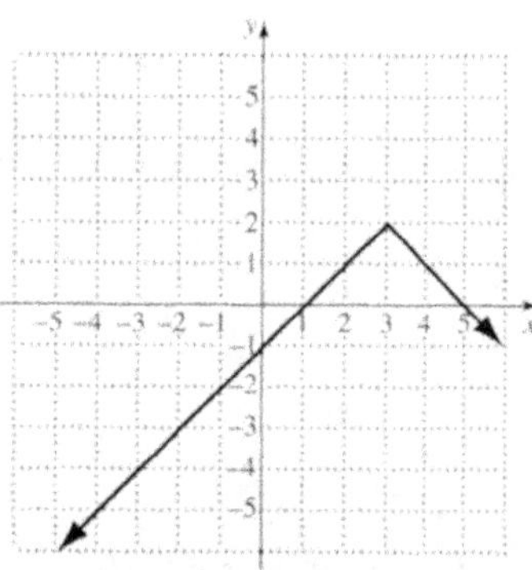

14.

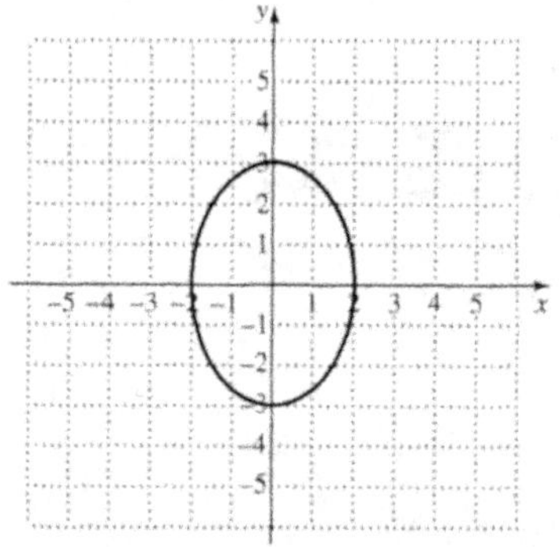

Applications of Functions and Their Graphs

15. The following graph represents the number of hospitals in the United States with 500 beds or more. The number of hospitals is a function g of the year x. Use the graph to estimate the number of hospitals with 500 beds or more in 2000. That is, find $g(2000)$.

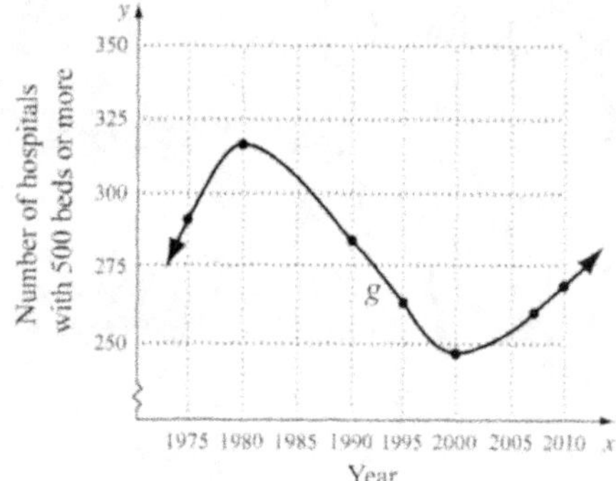

Finding Domain and Range

ESSENTIALS

A **relation** is a set of ordered pairs.

A **function** is a relation in which no two different pairs share a common first coordinate.

The **domain** of a function is the set of all first coordinates.

The **range** of a function is the set of all second coordinates.

Example

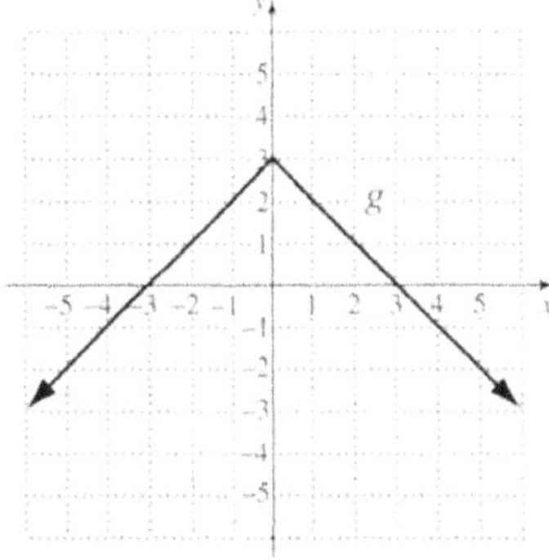

- For the function g shown:

    $g(1) = 2.$

    The domain of g is the set of all real numbers.

    If $g(x) = 1$, then $x = -2$ *or* $x = 2.$

    The range of g is $\{y \mid y \leq 3\}.$

GUIDED LEARNING

 **Textbook**    **Instructor**    **Video**

EXAMPLE 1	YOUR TURN 1

Determine the domain and the range of the function f whose graph is shown below.

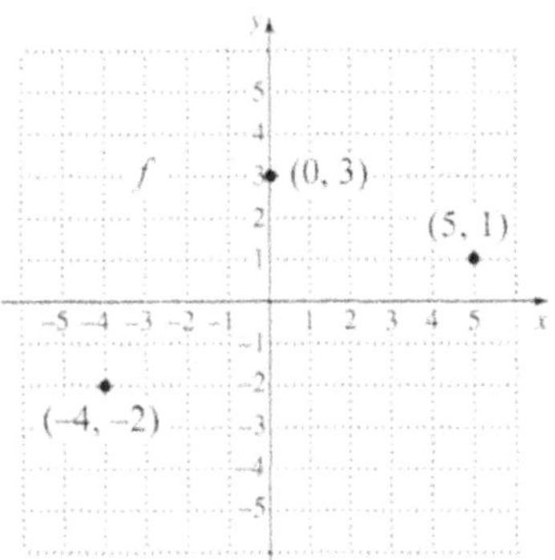

The function f can be written

$$f = \{(0, 3), (5, 1), (-4, -2)\}.$$

The domain is the set of all first coordinates:

$\left\{ 0, 5, \boxed{} \right\}.$

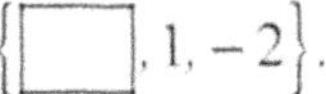

The range is the set of all second coordinates:

$\left\{ \boxed{}, 1, -2 \right\}.$

Determine the domain and the range of the function f whose graph is shown below.

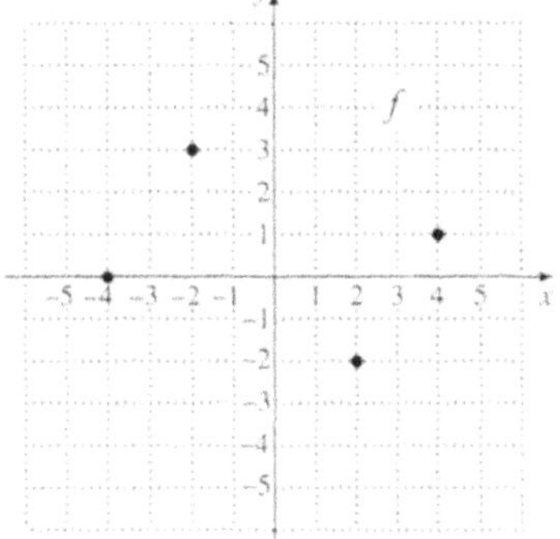

EXAMPLE 2	YOUR TURN 2
Determine the domain and the range of the function.	Determine the domain and the range of the function.

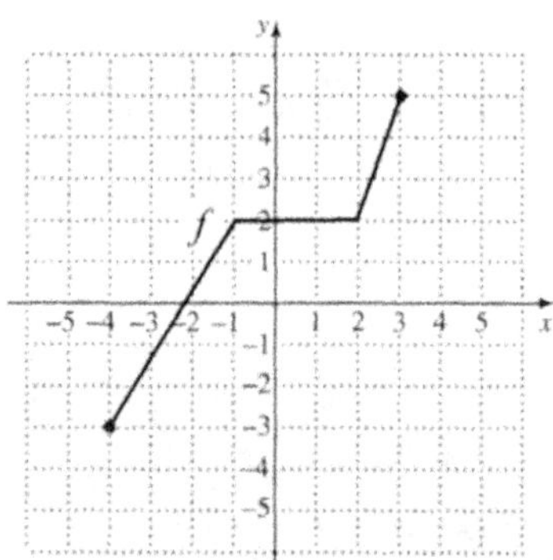

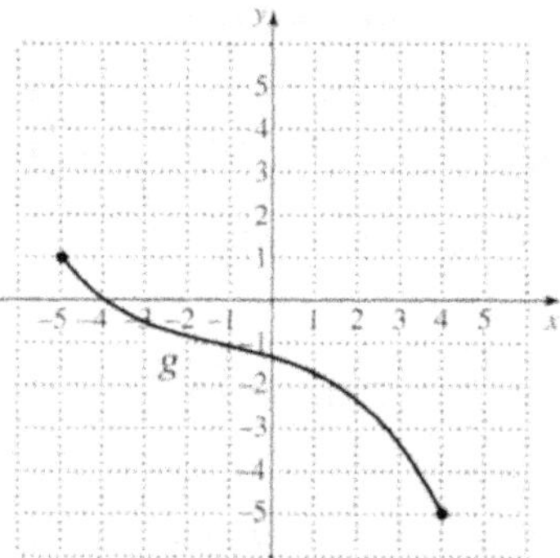

The ☐ is $\{x \mid -4 \le x \le \boxed{}\}$.

The range is $\{y \mid \boxed{} \le y \le \boxed{}\}$.

EXAMPLE 3	YOUR TURN 3
For the graph of the function, determine (a) $f(1)$; (b) the domain; (c) any x-values for which $f(x) = 2$; and (d) the range.	For the graph of the function, determine (a) $f(-2)$; (b) the domain; (c) any x-values for which $f(x) = 2$; and (d) the range.

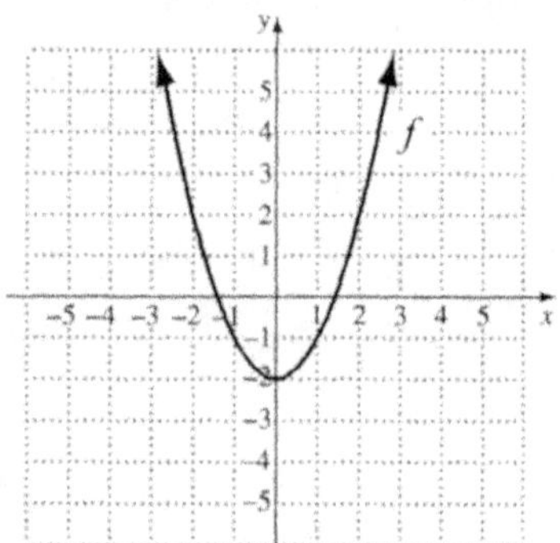

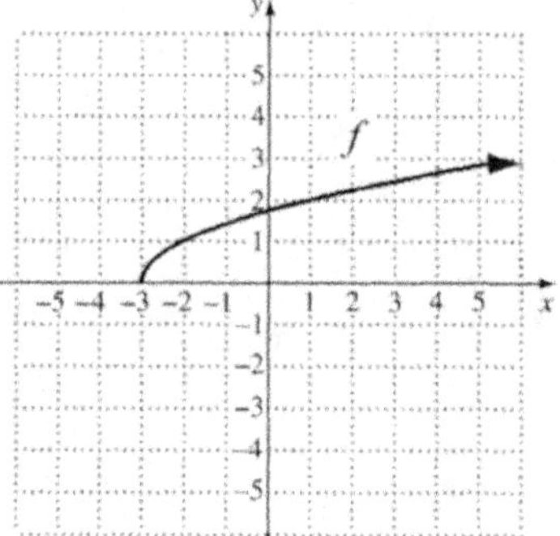

(a) $f(1) = -1$

(b) Domain = ☐

(c) $f\left(\boxed{}\right) = 2$ and $f\left(\boxed{}\right) = 2$

(d) Range = $\{y \mid y \ge \boxed{}\}$

EXAMPLE 4	YOUR TURN 4
Find the domain of $f(x) = \dfrac{3-x}{x+2}$.	Find the domain of $f(x) = \dfrac{5+x}{x-4}$.

EXAMPLE 4 (continued)

Find the values for x for which we cannot compute $f(x)$. These are values that make the _________________ equal to 0.

numerator / denominator

$x + 2 = 0$

$x = \boxed{}$

The domain is

$\left\{ x \middle| x \text{ is a real number } and \ x \neq \boxed{} \right\}.$

EXAMPLE 5	YOUR TURN 5
Find the domain of $f(x) = x^2 - x - 5$.	Find the domain of $f(x) = 3x - 1$.

EXAMPLE 5 (continued)

The domain is the set of all possible inputs. Since we can calculate $x^2 - x - 5$ for any real number, the domain of f is _________________ _________________________________.

YOUR NOTES   Write your questions and additional notes.

Practice Exercises

Readiness Check

Choose the word from the following list that best completes each sentence.

function   relation   domain   range

1. A _______________ is a set of ordered pairs in which no two different pairs share a common first coordinate.

2. Every function is also a _______________.

3. The _______________ of a function is the set of all first coordinates.

4. The _______________ of a function is the set of all second coordinates.

Finding Domain and Range

For the graph of the function shown, determine **(a)** $f(2)$; **(b)** *the domain;* **(c)** *any x-values for which* $f(x) = 1$; *and* **(d)** *the range.*

5.
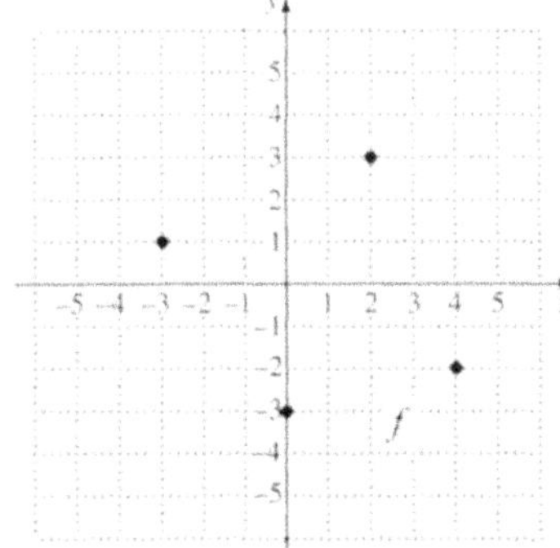

6.
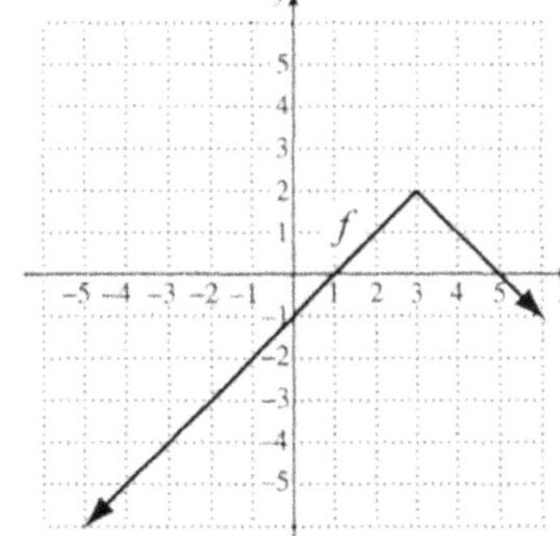

7.

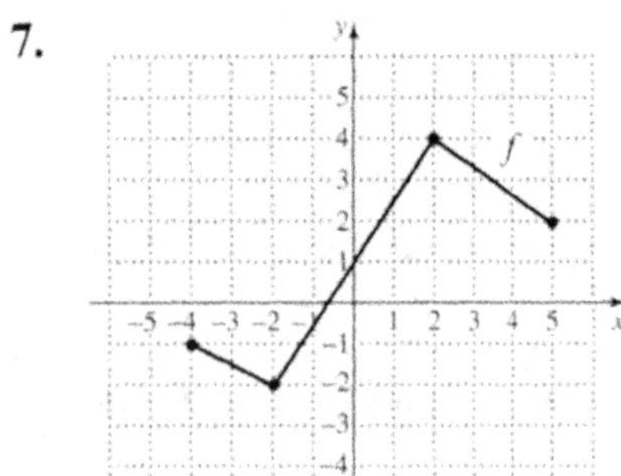

Find the domain of each function.

8. $f(x) = 5 - 7x$

9. $g(x) = \dfrac{4x}{x-5}$

10. $f(x) = \dfrac{8x-1}{2x+3}$

11. $g(x) = |x| + 7$

12. $f(x) = x^2 - x - 6$

The Sum, Difference, Product, or Quotient of Two Functions

ESSENTIALS

The Algebra of Functions

If f and g are functions and x is in the domain of both functions, then:

1. $(f+g)(x)=f(x)+g(x)$;

2. $(f-g)(x)=f(x)-g(x)$;

3. $(f \cdot g)(x)=f(x) \cdot g(x)$;

4. $(f/g)(x)=f(x)/g(x)$, provided $g(x) \neq 0$.

Example

- For $f(x)=2x^2-3$ and $g(x)=x+1$,

a) $\begin{aligned}(f+g)(x) &= f(x)+g(x) \\ &= 2x^2-3+x+1 \\ &= 2x^2+x-2;\end{aligned}$

b) $\begin{aligned}(f-g)(-2) &= f(-2)-g(-2) \\ &= 2(-2)^2-3-(-2+1) \\ &= 8-3-(-1) \\ &= 8-3+1=6\end{aligned}$

c) $\begin{aligned}(f \cdot g)(1) &= f(1) \cdot g(1) \\ &= \left[2(1)^2-3\right] \cdot (1+1) \\ &= (2-3) \cdot 2 \\ &= -1 \cdot 2 \\ &= -2;\end{aligned}$

d) $\begin{aligned}(f/g)(x) &= f(x)/g(x) \\ &= \frac{2x^2-3}{x+1}, \text{ provided } x \neq -1.\end{aligned}$

| | Textbook | Instructor | Video |

GUIDED LEARNING

EXAMPLE 1	YOUR TURN 1
For $f(x)=-x+4$ and $g(x)=x^2+2$, find $(f+g)(x)$. $(f+g)(x)=f(x)+g(x)$ $\qquad = -x+4+\boxed{}+2$ $\qquad = x^2-x+\boxed{}$	For $f(x)=3-x$ and $g(x)=3x^2+1$, find $(f+g)(x)$.

EXAMPLE 2	YOUR TURN 2
For $F(x) = 4 - x^2$ and $G(x) = 2x$, find $(F - G)(-1)$.	For $F(x) = x^2 - 6$ and $G(x) = -3x$, find $(F - G)(-2)$.

For $F(x) = 4 - x^2$ and $G(x) = 2x$, find $(F - G)(-1)$.

$F(-1) = 4 - (-1)^2 = 4 - 1 = \boxed{}$ and

$G(-1) = 2(-1) = \boxed{}$, so we have

$(F - G)(-1) = 3 - (-2) = 3 + 2 = \boxed{}$.

EXAMPLE 3	YOUR TURN 3
For $f(x) = x + 5$ and $g(x) = 2x + 3$, find $(f \cdot g)(a)$.	For $f(x) = 2x - 1$ and $g(x) = x + 4$, find $(f \cdot g)(t)$.

For $f(x) = x + 5$ and $g(x) = 2x + 3$, find $(f \cdot g)(a)$.

$(f \cdot g)(a) = f(a) \cdot g(a)$

$\qquad = \left(\boxed{} + 5\right)\left(2a + \boxed{}\right)$

$\qquad = 2a^2 + 3a + 10a + \boxed{}$

$\qquad = 2a^2 + \boxed{} + 15$

EXAMPLE 4	YOUR TURN 4
For $F(x) = x^2 - 2x$ and $G(x) = x + 3$, find $(F/G)(4)$.	For $F(x) = 2 - x^2$ and $G(x) = x - 1$, find $(F/G)(3)$.

For $F(x) = x^2 - 2x$ and $G(x) = x + 3$, find $(F/G)(4)$.

$(F/G)(x) = F(x)/G(x) = \dfrac{x^2 - 2x}{x + \boxed{}}$

We assume $x \neq -3$. Then

$(F/G)(4) = F(4)/G(4)$

$\qquad = \dfrac{4^2 - 2 \cdot \boxed{}}{4 + \boxed{}}$

$\qquad = \dfrac{16 - \boxed{}}{7} = \dfrac{8}{7}$.

YOUR NOTES   Write your questions and additional notes.

Determining Domain

ESSENTIALS

To find the domain of the sum, the difference, the product, or the quotient of two functions f and g:

1. Find the domain of f and the domain of g.

2. The functions $f + g$, $f - g$, and $f \cdot g$ have the same domain. It is the intersection of the domains of f and g, or, in other words, the set of all values common to the domains of f and g.

3. Find any values of x for which $g(x) = 0$.

4. The domain of f/g is the set found in step (2) (the set of all values common to the domains of f and g) *excluding* any values of x found in step (3).

Example

- Given $f(x) = \dfrac{2}{x-4}$ and $g(x) = x + 7$, find the domains of $f + g$, $f - g$, $f \cdot g$, and f/g.

  The domain of f is $\{x \mid x \text{ is a real number } and \ x \neq 4\}$.

  The domain of g is $\{x \mid x \text{ is a real number}\}$.

  Then the domains of $f + g$, $f - g$, and $f \cdot g$ are the set of all elements common to the domains of f and g, or all real numbers except 4.

  The domain of $f + g =$ the domain of $f - g =$ the domain of $f \cdot g$

  $$= \{x \mid x \text{ is a real number } and \ x \neq 4\}.$$

  The domain of f/g must exclude values of x for which $g(x) = 0$.

  $$g(x) = 0$$
  $$x + 7 = 0$$
  $$x = -7$$

  The domain of f/g is the domain of $f + g$, $f - g$, and $f \cdot g$, but also excludes -7.

  The domain of $f/g = \{x \mid x \text{ is a real number } and \ x \neq 4 \ and \ x \neq -7\}$.

GUIDED LEARNING   🔖 **Textbook**   👤 **Instructor**   ▶ **Video**

EXAMPLE 1	YOUR TURN 1
For $f(x)=2x^3$ and $g(x)=\dfrac{4}{x+3}$, determine the domains of $f+g$, $f-g$, and $f \cdot g$. The domain of f is $\{x \mid x \text{ is a real number}\}$. The domain of g is $\{x \mid x \text{ is a real number } and\ x \neq \boxed{}\}$. The domains of $f+g$, $f-g$, and $f \cdot g$ are the set of all elements common to the domains of f and g, or all real numbers except $\boxed{}$. The domain of $f+g =$ the domain of $f-g =$ the domin of $f \cdot g = \{x \mid x \text{ is a real number } and\ x \neq \boxed{}\}$.	For $f(x)=x^2-1$ and $g(x)=\dfrac{8}{x}$, determine the domains of $f+g$, $f-g$, and $f \cdot g$.
EXAMPLE 2	YOUR TURN 2
For $f(x)=\dfrac{2x}{x-5}$ and $g(x)=6-x$, determine the domain of f/g. The domain of f is $\{x \mid x \text{ is a real number } and\ x \neq \boxed{}\}$. The domain of g is $\{x \mid x \text{ is a real number}\}$. We see that the domain of f/g must exclude $\boxed{}$. Because we cannot divide by 0, the domain of f/g must also exclude x-values for which $g(x)=0$. $g(x)=0$ $6-x=0$ $\boxed{}=x$ Thus, the domain of $f/g =$ $\{x \mid x \text{ is a real number } and\ x \neq 5 \ and\ x \neq \boxed{}\}$.	For $f(x)=\dfrac{4}{x-2}$ and $g(x)=x-7$, determine the domain of f/g.

YOUR NOTES   Write your questions and additional notes.

Practice Exercises

Readiness Check

Given that $f(x) = 2x - 1$ and $g(x) = x + 4$, match each expression with an equivalent expression from the column on the right.

1. $(f + g)(x)$ a) $2x^2 + 7x - 4$

2. $(f - g)(x)$ b) $x^2 + 8x + 16$

3. $(f \cdot g)(x)$ c) $3x + 3$

4. $(g \cdot g)(x)$ d) $x - 5$

The Sum, Difference, Product, or Quotient of Two Functions

Let $f(x) = 3 - x$ and $g(x) = x^2 + 1$. Find each of the following.

5. $(f + g)(2)$ **6.** $(f - g)(-1)$

7. $(f \cdot g)(x)$ **8.** $(f/g)(-3)$

9. $(g - f)(4)$ **10.** $(f \cdot f)(x)$

11. $(g/f)(a)$ **12.** $(f + f)(10)$

Determining Domain

For each pair of functions f and g, determine the domain of the sum, the difference, and the product of the two functions.

13. $f(x) = 2x^2,$

$g(x) = x - 3$

14. $f(x) = 2x + 5,$

$g(x) = 4 - x$

15. $f(x) = \dfrac{1}{x},$

$g(x) = 4 - x^2$

16. $f(x) = 4x^3,$

$g(x) = \dfrac{3}{x - 1}$

For each pair of functions f and g, determine the domain of f/g.

17. $f(x) = -x^3,$

$g(x) = x + 2$

18. $f(x) = 3x + 4,$

$g(x) = 2x - 6$

19. $f(x) = \dfrac{2}{x + 4},$

$g(x) = 8 - x$

20. $f(x) = \dfrac{3}{4x - 4},$

$g(x) = 3x - 9$

The Constant b: The y-Intercept

ESSENTIALS

The y-intercept of the graph of $f(x) = mx + b$ is the point $(0, b)$, or simply b.

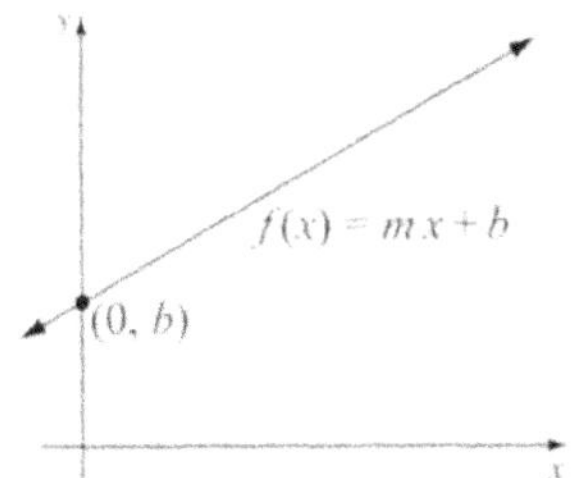

Example

- For $y = 3x + 5$, the y-intercept is $(0, 5)$.

📖 **Textbook**	👤 **Instructor**	▶ **Video**

GUIDED LEARNING

EXAMPLE 1	YOUR TURN 1
Find the y-intercept: $y = 8x + 1$. $y = mx + b$ $y = 8x + 1$ The y-intercept is $\left(\boxed{}, \boxed{}\right)$.	Find the y-intercept: $y = 2x + 13$.
EXAMPLE 2	YOUR TURN 2
Find the y-intercept: $f(x) = 2x - \dfrac{1}{3}$. $f(x) = mx + b$ $f(x) = 2x - \dfrac{1}{3}$ $f(x) = 2x + \boxed{}$ The y-intercept is $\left(\boxed{}, \boxed{}\right)$.	Find the y-intercept: $f(x) = \dfrac{1}{2}x - 9.5$.

EXAMPLE 3	YOUR TURN 3
Find the y-intercept: $2x - 3y = 3$.	Find the y-intercept: $4x + 2y = 6$.

Find the y-intercept: $2x - 3y = 3$.

First solve for y to write the equation in the form $y = mx + b$.

$$2x - 3y = 3$$

$$-3y = 3 - \boxed{}$$

$$\frac{-3y}{-3} = \frac{3 - 2x}{-3}$$

$$y = \boxed{} + \frac{2}{3}x$$

$$y = \frac{2}{3}x - \boxed{}$$

The y-intercept is $\left(\boxed{}, \boxed{}\right)$.

YOUR NOTES   Write your questions and additional notes.

The Constant m: Slope

ESSENTIALS

The **slope** of a line containing points (x_1, y_1) and (x_2, y_2) is given by

$$m = \frac{\text{rise}}{\text{run}} = \frac{\text{change in } y}{\text{change in } x} = \frac{y_2 - y_1}{x_2 - x_1} = \frac{y_1 - y_2}{x_1 - x_2}.$$

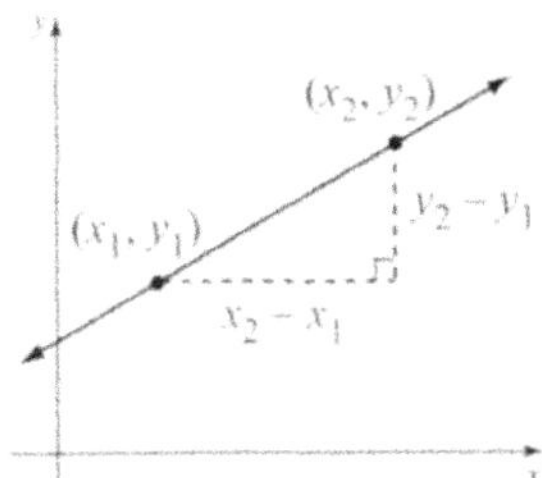

When m is negative, the graph slants down from left to right.
When m is positive, the graph slants up from left to right.
When m is 0, the graph is horizontal.

The **slope** of the line $y = mx + b$ is m.

Slope-intercept equation: $y = mx + b$, where m is the slope and $(0, b)$ is the y-intercept.

Examples

- The slope of the line containing the points $(9, -1)$ and $(8, -7)$ is

$$\frac{\text{rise}}{\text{run}} = \frac{-7 - (-1)}{8 - 9} = \frac{-7 + 1}{8 - 9} = \frac{-6}{-1} = 6.$$

- For $y = 3x + 5$, the slope is 3 and the y-intercept is $(0, 5)$.

	📖 **Textbook**	👤 **Instructor**	▶ **Video**

GUIDED LEARNING

EXAMPLE 1	YOUR TURN 1
Find the slope of the line containing the points $(7, -5)$ and $(3, 2)$. $m = \dfrac{\text{rise}}{\text{run}} = \dfrac{2 - (\boxed{})}{3 - \boxed{}}$ $= \dfrac{2 + \boxed{}}{-4} = \dfrac{\boxed{}}{-4} = -\dfrac{7}{4}$	Find the slope of the line containing the points $(-1, 3)$ and $(4, 5)$.

EXAMPLE 2	YOUR TURN 2
Determine the slope of the line given by $f(x) = 6x - 3$.	Determine the slope of the line given by $f(x) = 9x - 7$.

$$f(x) = mx + b$$
$$f(x) = 6x - 3$$

The slope is $\boxed{}$.

EXAMPLE 3	YOUR TURN 3
Determine the slope of the line given by $2x - y = 3$.	Determine the slope of the line given by $x + 2y = 8$.

First solve for y to write the equation in the form $y = mx + b$.

$$2x - y = 3$$
$$-y = -2x + 3$$
$$\frac{-y}{-1} = \frac{-2x + 3}{\boxed{}}$$
$$y = \boxed{} - 3$$

The slope is $\boxed{}$.

YOUR NOTES   Write your questions and additional notes.

   Copyright © 2022 Pearson Education, Inc.

Applications

ESSENTIALS

Slope can be used to represent *rate of change.*

Example

- A plane ascends from 10,000 ft to 15,000 ft in 5 minutes. Its average rate of ascent is
  $$\frac{15{,}000 \text{ ft} - 10{,}000 \text{ ft}}{5 \text{ min}} = \frac{5000 \text{ ft}}{5 \text{ min}}, \text{ or } 1000 \text{ ft/min.}$$

GUIDED LEARNING   📖 **Textbook**   👤 **Instructor**   ▶ **Video**

EXAMPLE 1	YOUR TURN 1

A road rises 3 ft for every horizontal distance of 100 ft. Find the grade of the road.

$$\text{Grade} = \frac{\text{vertical change}}{\text{horizontal change}} = \frac{\boxed{} \text{ ft}}{\boxed{} \text{ ft}} = 0.03$$

Expressed as a percent, the grade is $\boxed{}$ %.

By 7 p.m., Joe had typed 4 pages of his paper. At 8:30 p.m., he had completed 10 pages. Calculate his typing rate in minutes per page.

EXAMPLE 2	YOUR TURN 2

Find the rate of change for the graph. Remember to use appropriate units.

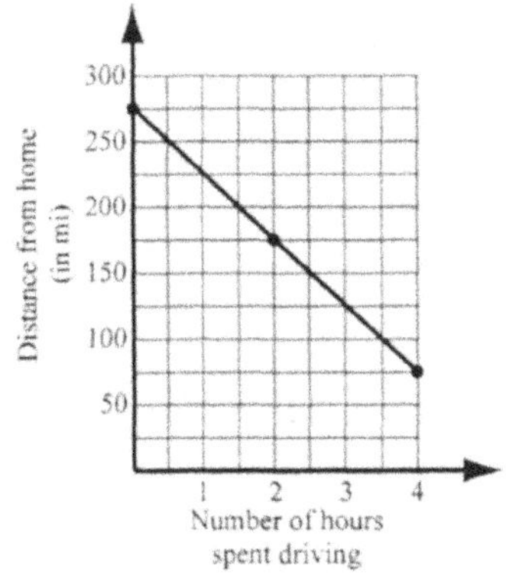

$$\text{Rate of change} = \frac{175 \text{ mi} - 75 \text{ mi}}{2 \text{ hr} - 4 \text{ hr}} = \frac{\boxed{}}{\boxed{}}$$

$$= -50 \text{ mi per hr, or } -50 \text{ mph}$$

Find the rate of change for the graph. Remember to use appropriate units.

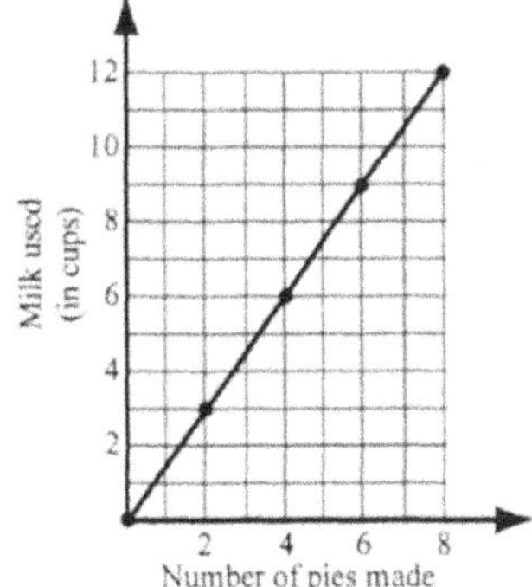

YOUR NOTES   Write your questions and additional notes.

Practice Exercises

Readiness Check

Choose the word or phrase from the following list that best completes each sentence.

y-intercept	x-intercept	slope	up
slope-intercept form		standard form	down

1. The _____________ of the line given by the equation $y = mx + b$ is $(0, b)$.

2. _____________ is a ratio that indicates how a change in vertical direction of a graph corresponds to a change in horizontal direction.

3. Lines with negative slope slant _____________ from left to right.

4. A linear function in the form $f(x) = mx + b$ is written in _____________.

The Constant *b*: The *y*-Intercept

Find the y-intercept of each equation.

5. $f(x) = 2x - \dfrac{1}{2}$

6. $3x - y = 1$

The Constant *m*: Slope

7. Find the slope of the line shown.

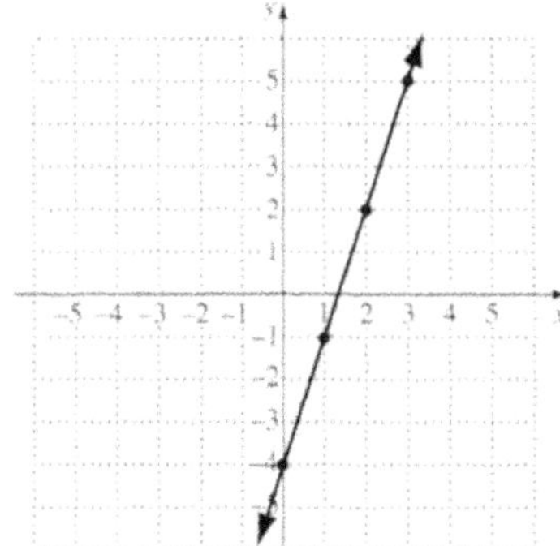

Find the slope of the line containing the given pair of points.

8. $(5, 6)$ and $(1, 2)$

9. $(-3, 2)$ and $(-1, -8)$

The Slope and the *y*-Intercept

Find the slope and the y-intercept of each equation.

10. $f(x) = \dfrac{3}{8}x - 6$

11. $3x - 2y = -6$

12. $x - y = 9$

13. $x + 3y = 2$

Applications

14. At 3:30 p.m., Justin reached highway mile marker 284. At 3:54 p.m., he reached mile marker 312. Assuming a constant rate, find Justin's speed in miles per minute.

15. Find the rate of change.

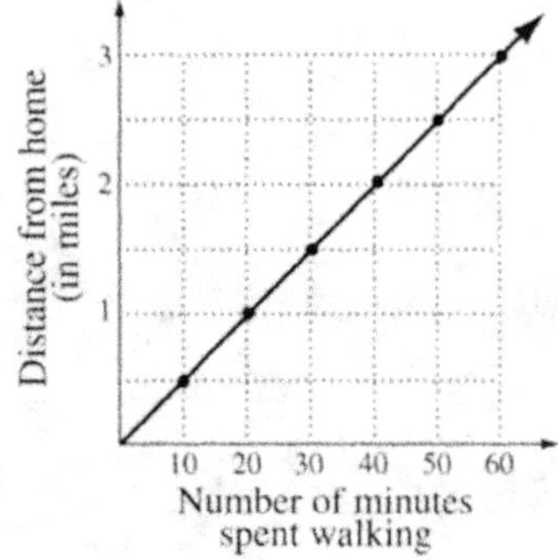

Graphing Using Intercepts

ESSENTIALS

A y-intercept of a graph is a point $(0, b)$. To find b, let $x = 0$ and solve for y.

An x-intercept of a graph is a point $(a, 0)$. To find a, let $y = 0$ and solve for x.

Example

- Find the intercepts of $5x - 4y = 20$.

$$5x - 4 \cdot 0 = 20 \quad \text{Letting } y = 0 \qquad\qquad 5 \cdot 0 - 4y = 20 \quad \text{Letting } x = 0$$
$$5x = 20 \qquad\qquad\qquad\qquad\qquad -4y = 20$$
$$x = 4 \qquad\qquad\qquad\qquad\qquad\quad y = -5$$

The x-intercept is $(4, 0)$. The y-intercept is $(0, -5)$.

 Textbook	 Instructor	 Video

GUIDED LEARNING

EXAMPLE 1	YOUR TURN 1
Find the intercepts of $3x - 4y = -12$ and then graph the line.	Find the intercepts of $-2x + 5y = 10$ and then graph the line.

$$3 \cdot 0 - 4y = -12 \quad \text{Substituting 0 for } x$$
$$\boxed{} = -12$$
$$y = \boxed{}$$

The y-intercept is $\left(\boxed{}, \boxed{}\right)$.

$$3x - 4 \cdot 0 = -12 \quad \text{Substituting 0 for } y$$
$$3x = -12$$
$$x = \boxed{}$$

The x-intercept is $\left(\boxed{}, \boxed{}\right)$.

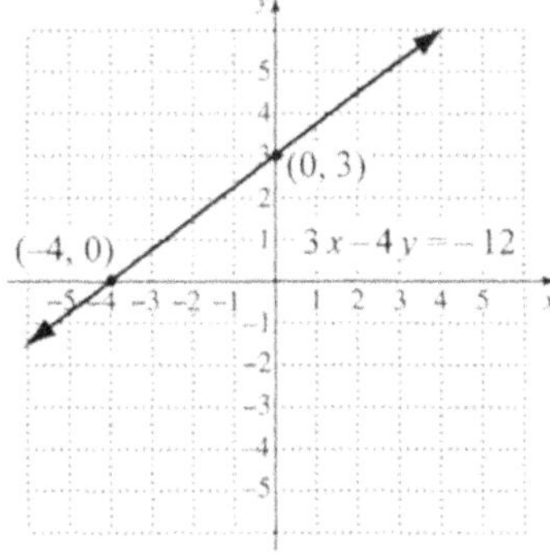

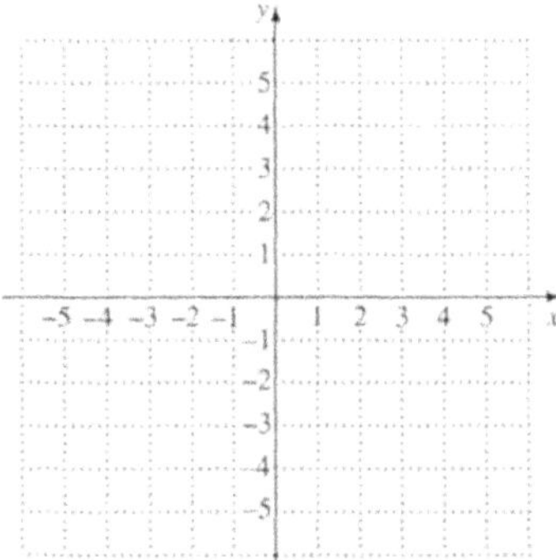

EXAMPLE 2	YOUR TURN 2
Graph $f(x) = 3x - 4$ by using intercepts.	Graph $f(x) = 2x - 3$ by using intercepts.

EXAMPLE 2

Graph $f(x) = 3x - 4$ by using intercepts.

To find the y-intercept, let $x = 0$ and find $f(0)$.

$$f(x) = 3x - 4$$
$$f(0) = 3 \cdot 0 - 4$$
$$= \boxed{}$$

The y-intercept is $\left(0, \boxed{}\right)$.

To find the x-intercept, replace $f(x)$ with 0 and solve for x.

$$f(x) = 3x - 4$$
$$0 = 3x - 4$$
$$4 = \boxed{}$$
$$\boxed{} = x$$

The x-intercept is $\left(\dfrac{4}{3}, 0\right)$.

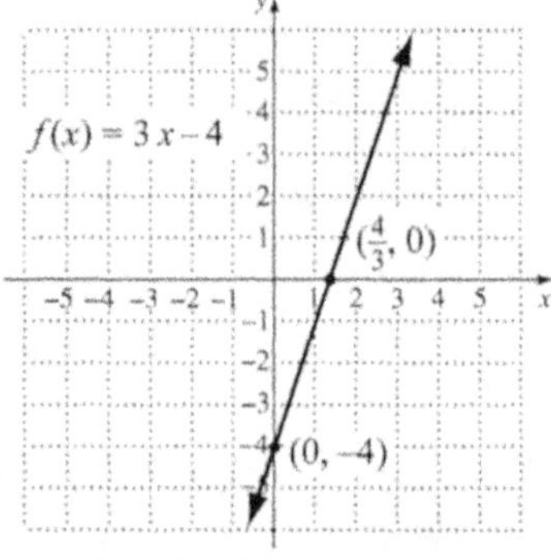

YOUR TURN 2

Graph $f(x) = 2x - 3$ by using intercepts.

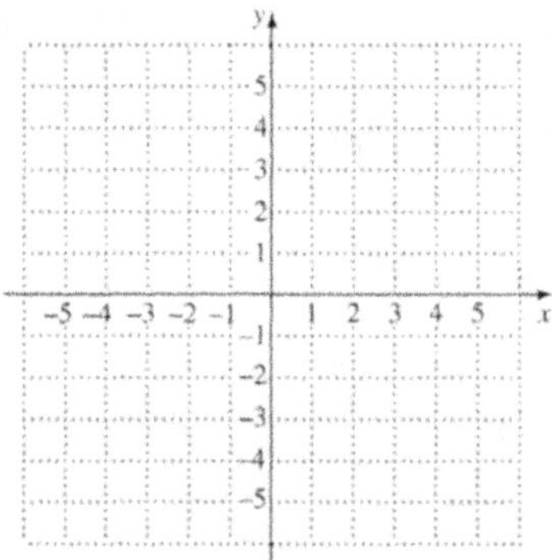

YOUR NOTES   Write your questions and additional notes.

Graphing Using the Slope and the *y*-Intercept

ESSENTIALS

The equation $y = mx + b$ is called the **slope-intercept equation**. The slope is m and the *y*-intercept is $(0, b)$.

Example

- Graph: $f(x) = -\dfrac{3}{4}x - 1$.

  The *y*-intercept is $(0, -1)$.

  The slope is $-\dfrac{3}{4}$, which can be written $\dfrac{-3}{4}$ or $\dfrac{3}{-4}$.

  For $\dfrac{\text{Rise}}{\text{Run}} = \dfrac{-3}{4}$, start at $(0, -1)$ and move down 3 units and to the right 4 units, to the point $(4, -4)$.

  For $\dfrac{\text{Rise}}{\text{Run}} = \dfrac{3}{-4}$, start at $(0, -1)$ and move up 3 units and to the left 4 units, to the point $(-4, 2)$.

  Draw the graph.

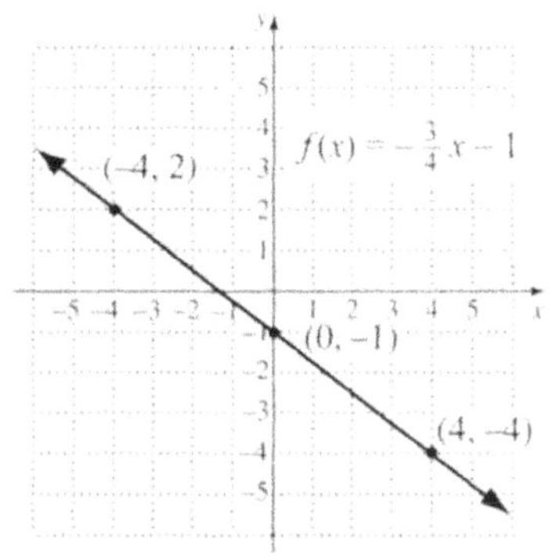

	🔖 **Textbook**	👤 **Instructor**	▶ **Video**

GUIDED LEARNING

EXAMPLE 1	YOUR TURN 1

EXAMPLE 1

Graph: $y = \dfrac{1}{5}x + 3$.

The *y*-intercept is $\left(0, \boxed{}\right)$.

The slope is $\boxed{}$, which can also be written $\dfrac{-1}{-5}$.

For $\dfrac{\text{Rise}}{\text{Run}} = \dfrac{1}{5}$, start at $(0, 3)$ and move up $\boxed{}$ unit and to the right $\boxed{}$ units, to the point $\left(5, \boxed{}\right)$.

YOUR TURN 1

Graph: $y = \dfrac{1}{2}x - 4$.

(continued)

For $\dfrac{\text{Rise}}{\text{Run}} = \dfrac{-1}{-5}$, start at $(0, 3)$ and move down $\boxed{}$ unit

and to the left $\boxed{}$ units, to the point $\left(-5, \boxed{}\right)$.

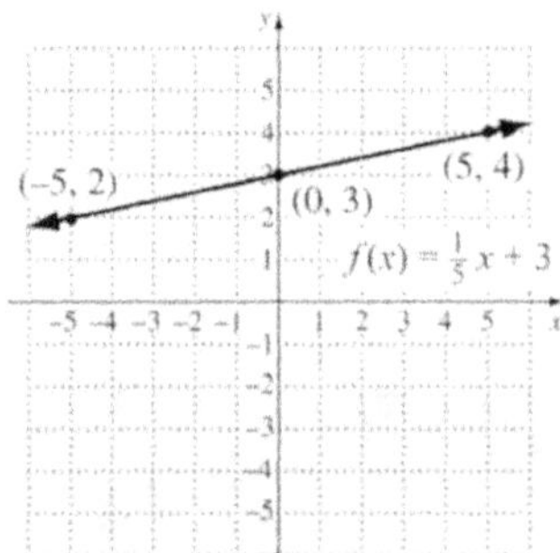

EXAMPLE 2	YOUR TURN 2

Graph: $2x - 3y = 6$.

Graph: $3x - 4y = -12$.

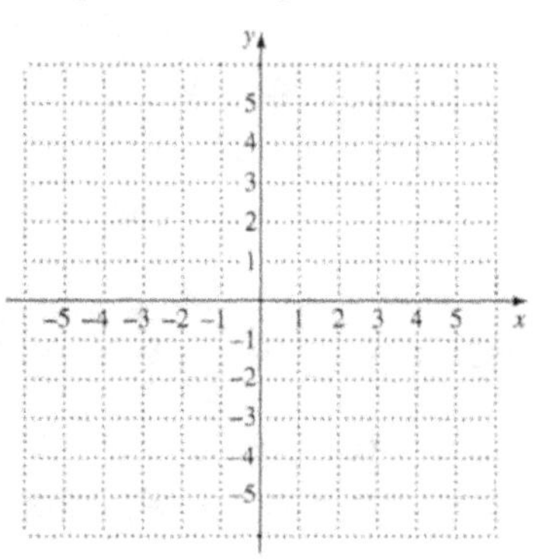

First write the equation in the form $y = mx + b$.

$$2x - 3y = 6$$
$$-3y = -2x + 6$$
$$y = \frac{2}{3}x - 2$$

The y-intercept is $\left(0, \boxed{}\right)$.

The slope is $\boxed{}$, which can also be written $\dfrac{-2}{-3}$.

For $\dfrac{\text{Rise}}{\text{Run}} = \dfrac{2}{3}$, start at $(0, -2)$ and move up $\boxed{}$ units

and to the right $\boxed{}$ units.

For $\dfrac{\text{Rise}}{\text{Run}} = \dfrac{-2}{-3}$, start at $(0, -2)$ and move down $\boxed{}$

units and to the left $\boxed{}$ units.

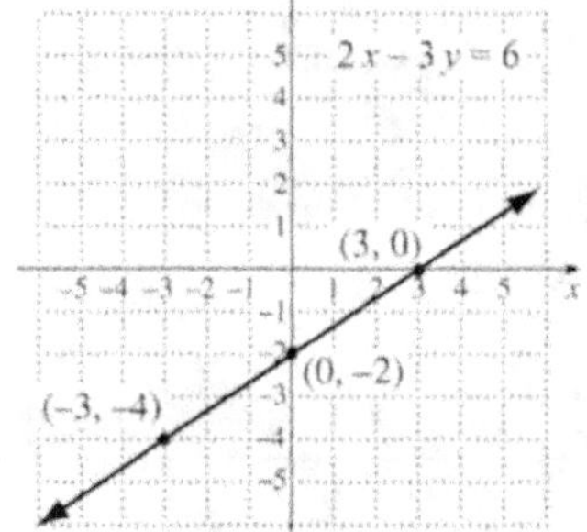

YOUR NOTES   Write your questions and additional notes.

Horizontal Lines and Vertical Lines

ESSENTIALS

The graph of $y = b$ is a **horizontal line** with y-intercept $(0, b)$. Its slope is 0.

The graph of $x = a$ is a **vertical line** with x-intercept $(a, 0)$. Its slope is not defined.

Examples

- Graph $y = 2$. If possible, determine the slope.

  The graph of $y = 2$ is a horizontal line. The slope is 0.

  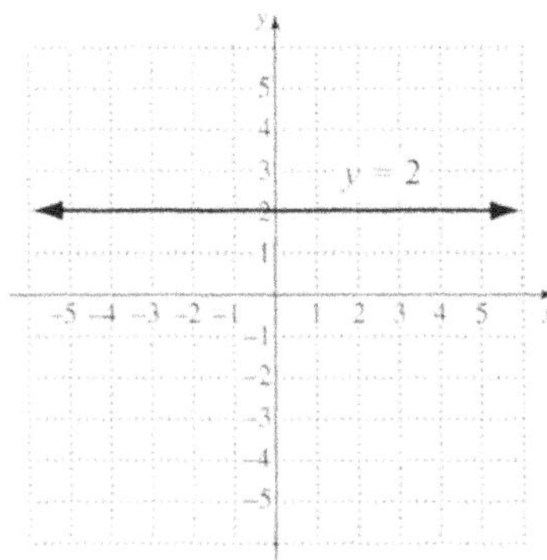

- Graph $x = -3$. If possible, determine the slope.

  The graph of $x = -3$ is a vertical line. The slope is not defined.

  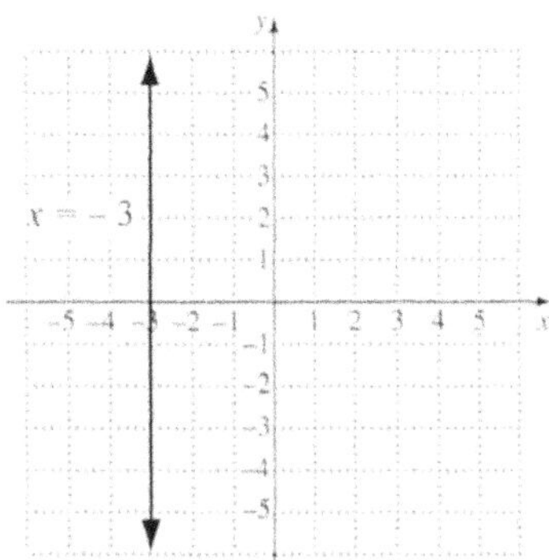

GUIDED LEARNING    **Textbook**    **Instructor**    **Video**

EXAMPLE 1	YOUR TURN 1
Graph: $f(x) = -1$. If possible, determine the slope.	Graph: $f(x) = 3$. If possible, determine the slope.

Graph: $f(x) = -1$. If possible, determine the slope.

The graph is a ________________ line.
                 vertical / horizontal

The slope is ______________.
           0 / not defined

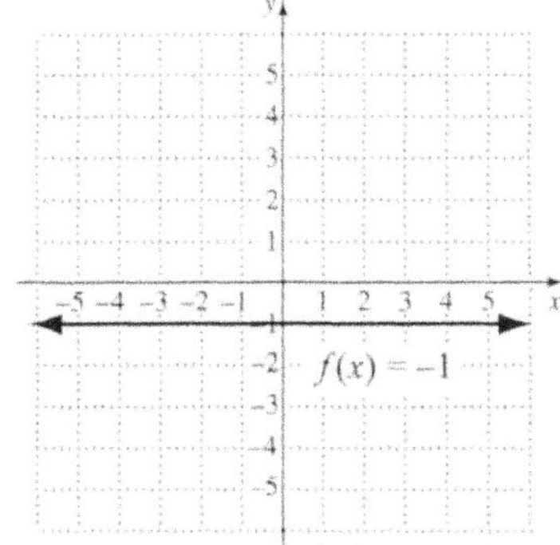

Graph: $f(x) = 3$. If possible, determine the slope.

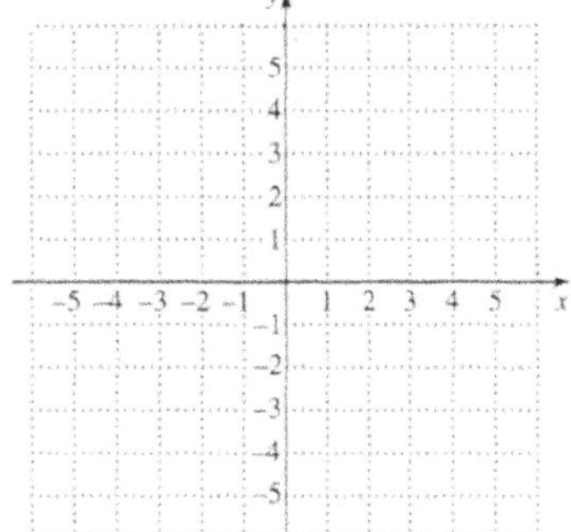

EXAMPLE 2	YOUR TURN 2
Graph: $x = 4$. If possible, determine the slope.	Graph: $x = -2$. If possible, determine the slope.

EXAMPLE 2

Graph: $x = 4$. If possible, determine the slope.

The graph is a ______________________ line.
 vertical / horizontal

The slope is ______________.
 0 / not defined

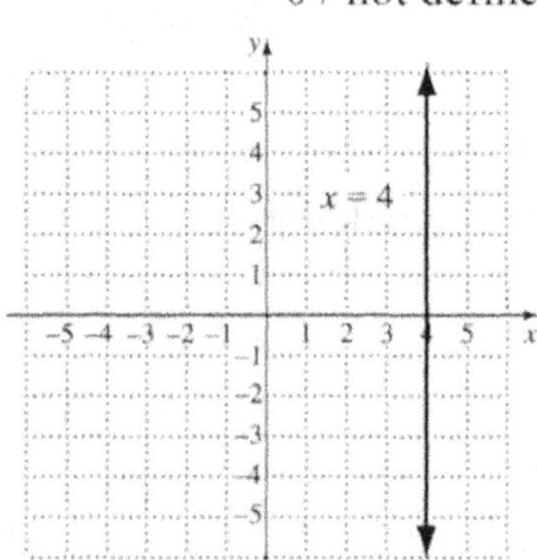

YOUR TURN 2

Graph: $x = -2$. If possible, determine the slope.

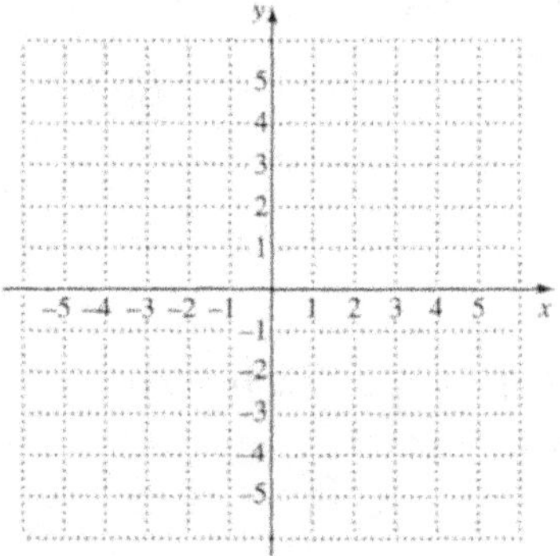

EXAMPLE 3

Graph: $3 + x = 5 - x$. If possible, determine the slope.

The only variable in the equation is x. We write the equation in the form $x = a$.

$$3 + x = 5 - x$$
$$3 + 2x = 5$$
$$2x = 2$$
$$x = 1$$

The graph is a ______________________ line.
 vertical / horizontal

The slope is ______________.
 0 / not defined

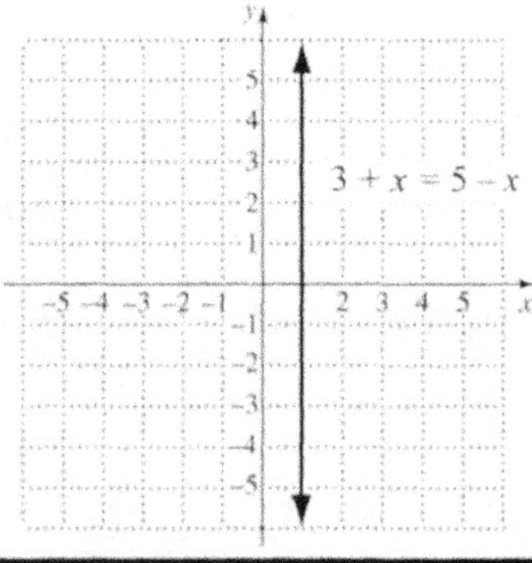

YOUR TURN 3

Graph: $3y - 8 = y + 2$. If possible, determine the slope.

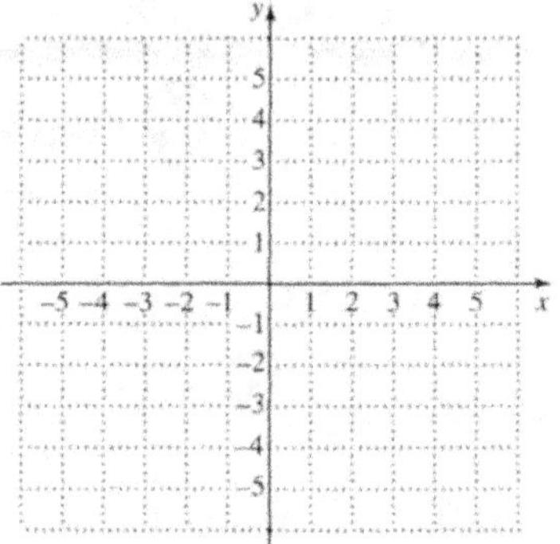

Parallel Lines and Perpendicular Lines

ESSENTIALS

If two lines are vertical, they are parallel.

Two nonvertical lines are parallel if they have the same slope and different y-intercepts.

Two lines are perpendicular if the product of their slopes is -1 or if one line is vertical and the other is horizontal.

Examples

- Determine whether the graphs of $y = -2x - 4$ and $y = -2x + 7$ are parallel.

  The slope of $y = -2x - 4$ is -2. The y-intercept is $(0, -4)$.

  The slope of $y = -2x + 7$ is -2. The y-intercept is $(0, 7)$.

  Since the slopes are the same but the y-intercepts are different, the graphs are parallel.

- Determine whether the graphs of $y = \dfrac{3}{7}x + 5$ and $y = -\dfrac{7}{3}x + 2$ are perpendicular.

  The slope of $y = \dfrac{3}{7}x + 5$ is $\dfrac{3}{7}$. The slope of $y = -\dfrac{7}{3}x + 2$ is $-\dfrac{7}{3}$.

  Since $\dfrac{3}{7}\left(-\dfrac{7}{3}\right) = -1$, the graphs are perpendicular.

GUIDED LEARNING   📍 **Textbook**   👤 **Instructor**   ▶ **Video**

EXAMPLE 1	YOUR TURN 1

Determine whether the graphs of $y = 2x - 5$ and $2y - 4x = 3$ are parallel.
 	
 The slope of $y = 2x - 5$ is ☐ and the y-intercept is ☐.
 	
 We find the slope of $2y - 4x = 3$ by first writing the equation in slope-intercept form.
 	
 $2y - 4x = 3$
 $2y = 4x + 3$
 $y = \boxed{}\,x + \dfrac{3}{2}$
 	
 The slope is ☐ and the y-intercept is $\left(0, \boxed{}\right)$.
 	
 Because the slopes ________ the same and the
                are / are not
 y-intercepts are different, the lines are ________ parallel.
                 are / are not
	Determine whether the graphs of $y = -4x + 2$ and $4y - x = 5$ are parallel.

EXAMPLE 2	YOUR TURN 2
Determine whether the graphs of $3x - y = 7$ and $y = -\dfrac{1}{3}x + 3$ are perpendicular.	Determine whether the graphs of $5x - 6y = 30$ and $5y + 6x = 0$ are perpendicular.

The slope of $y = -\dfrac{1}{3}x + 3$ is $\boxed{}$.

We find the slope of $3x - y = 7$ by first writing the equation in slope-intercept form.

$$3x - y = 7$$
$$-y = -3x + 7$$
$$y = \boxed{}$$

The slope is $\boxed{}$.

We find the product of the slopes:

$$\boxed{}\left(\boxed{}\right) = \boxed{}$$

Since the product of the slopes ________ -1,
$\qquad\qquad\qquad\qquad$ is / is not

the lines ________ perpendicular.
$\qquad$ are / are not

Practice Exercises

Readiness Check

Choose the word or number from the following list that best completes each sentence. Not every choice is used.

horizontal	vertical	perpendicular	parallel
0	-1	rise	run

1. The slope of a _________________ line is undefined.

2. Two lines are _________________ if their slopes are the same and their y-intercepts are different.

3. If the product of the slopes of two lines is _________________, the lines are perpendicular.

4. A slope of $\dfrac{5}{8}$ corresponds to a _________________ of 5 and a _________________ of 8.

Graphing Using Intercepts

Find the intercepts and then graph the line.

5. $2x - 3y = -6$

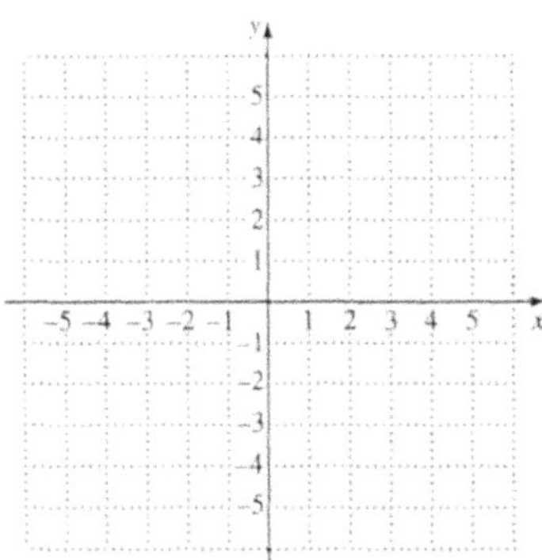

6. $f(x) = x - 4$

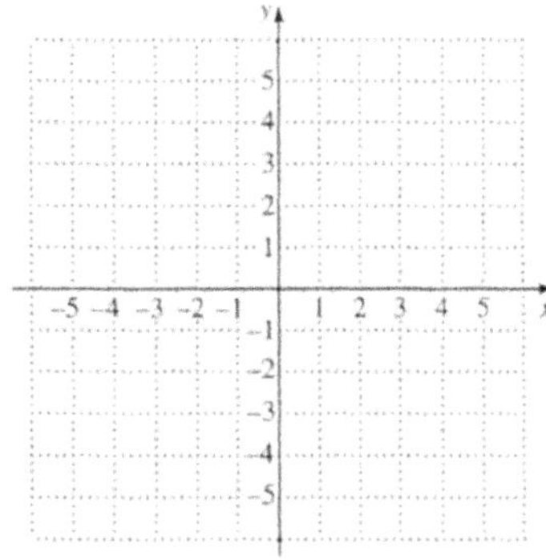

Graphing Using the Slope and the y-Intercept

Graph using the slope and the y-intercept.

7. $y = -\dfrac{2}{3}x + 1$

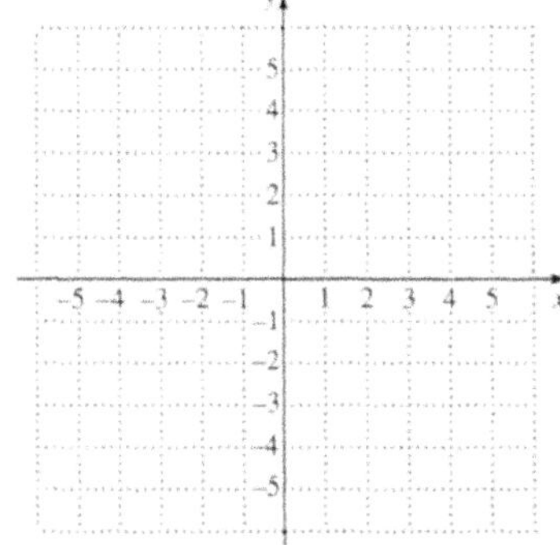

8. $x + 2y = 4$

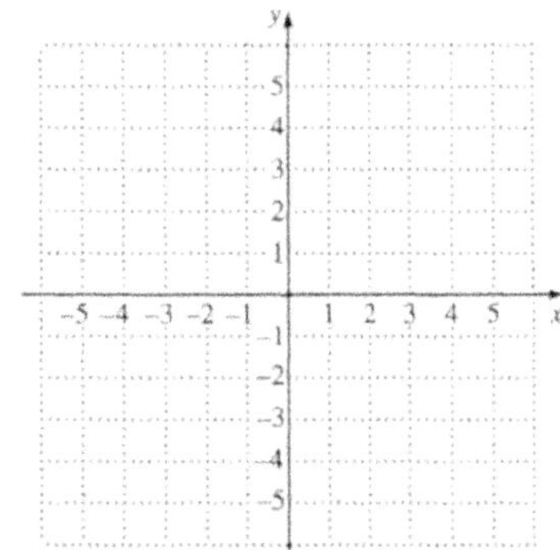

Horizontal Lines and Vertical Lines

Graph and, if possible, determine the slope.

9. $f(x) = -2$

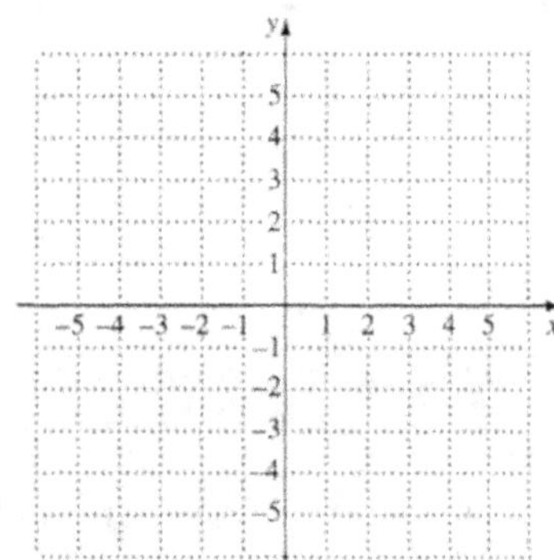

10. $2x = 6$

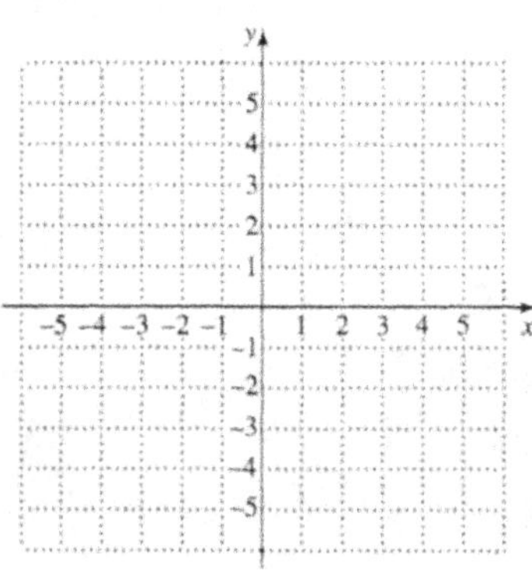

Parallel and Perpendicular Lines

Determine whether the graphs of the given pair of lines are parallel.

11. $10x - 5 = 2y$

$y = 5x + 4$

12. $3x + 4y = -12$

$4y = 3x + 6$

Determine whether the graphs of the given pair of lines are perpendicular.

13. $y = 3x - 1$

$2x + 6y = -5$

14. $x + y = 9$

$x - y = 7$

Finding an Equation of a Line When the Slope and the *y*-Intercept Are Given

ESSENTIALS

Example

- Find an equation for the line with slope $-\dfrac{1}{2}$ and *y*-intercept $(0, 6)$.

  For this line, $m = -\dfrac{1}{2}$ and $b = 6$.

  $y = mx + b$

  $y = -\dfrac{1}{2}x + 6$

GUIDED LEARNING

| | Textbook | Instructor | Video |

EXAMPLE 1	YOUR TURN 1
Find an equation for the line with slope 2 and *y*-intercept $(0, -1)$.	Find an equation for the line with slope -7 and *y*-intercept $(0, 5)$.

For this line, $m = \boxed{}$ and $b = \boxed{}$.

$y = mx + b$

$y = \boxed{}\,x + \left(\boxed{}\right)$

$y = \boxed{}\,x - \boxed{}$

EXAMPLE 2	YOUR TURN 2
Find a linear function $f(x) = mx + b$ whose graph has slope $-\dfrac{2}{3}$ and *y*-intercept $(0, 5)$.	Find a linear function $f(x) = mx + b$ whose graph has slope 4 and *y*-intercept $(0, -10)$.

$f(x) = mx + b$

$f(x) = \boxed{}\,x + \boxed{}$

YOUR NOTES Write your questions and additional notes.

Finding an Equation of a Line When the Slope and a Point Are Given

ESSENTIALS

Point-slope equation of a line: $y - y_1 = m(x - x_1)$

The slope is m, and the line passes through (x_1, y_1).

Example

- Find an equation of the line with slope 2 that passes through $(-3, -9)$.

  Using the point-slope equation:

  $$y - y_1 = m(x - x_1)$$
  $$y - (-9) = 2(x - (-3)) \qquad \text{Substituting 2 for } m, \ -3 \text{ for } x_1, \text{ and } -9 \text{ for } y_1$$
  $$y + 9 = 2(x + 3)$$
  $$y + 9 = 2x + 6$$
  $$y = 2x - 3$$

  Using the slope-intercept equation:

  $$y = mx + b$$
  $$y = 2x + b \qquad \text{Substituting 2 for } m$$
  $$-9 = 2(-3) + b \qquad \text{Substituting } -3 \text{ for } x \text{ and } -9 \text{ for } y$$
  $$-9 = -6 + b$$
  $$-3 = b \qquad \text{Solving for } b$$
  $$y = 2x - 3 \qquad \text{Substituting } -3 \text{ for } b$$

GUIDED LEARNING

 **Textbook**  **Instructor**  **Video**

EXAMPLE 1	YOUR TURN 1
Use the point-slope equation to find an equation of the line with slope -4 that passes through $(-2, 5)$. Here, $m = -4$ and $(x_1, y_1) = (-2, 5)$. $y - y_1 = m(x - x_1)$ $y - \boxed{} = -4\left(x - \left(\boxed{}\right)\right)$ $y - 5 = -4\left(x + \boxed{}\right)$ $y - 5 = -4x - \boxed{}$ $y = -4x - \boxed{}$	Use the point-slope equation to find an equation of the line with slope $\dfrac{2}{3}$ that passes through $(4, -9)$.

EXAMPLE 2	YOUR TURN 2
Use the slope-intercept equation to find an equation of the line with slope 3 that passes through $(1, -7)$.	Use the slope-intercept equation to find an equation of the line with slope -1 that passes through $(-1, 8)$.

First, substitute 3 for m in the slope-intercept equation.

$$y = mx + b$$

$$y = \boxed{}\, x + b$$

Then, substitute 1 for x and -7 for y to find b.

$$y = 3x + b$$

$$\boxed{} = 3\left(\boxed{}\right) + b$$

$$-7 = \boxed{} + b$$

$$\boxed{} = b$$

Finally, substitute -10 for b.

$$y = 3x + \left(\boxed{}\right), \text{ or } y = \boxed{}$$

YOUR NOTES   Write your questions and additional notes.

Finding an Equation of a Line When Two Points Are Given

ESSENTIALS

Example

- Find an equation of the line containing the points $(3, 8)$ and $(-2, 7)$.

$$m = \frac{y_2 - y_1}{x_2 - x_1} = \frac{7-8}{-2-3} = \frac{-1}{-5} = \frac{1}{5}$$

Using the point-slope equation:

We choose to use $(3, 8)$ for (x_1, y_1).

$$y - y_1 = m(x - x_1)$$

$$y - 8 = \frac{1}{5}(x - 3) \qquad \text{Substituting } \frac{1}{5} \text{ for } m, \ 3 \text{ for } x_1, \text{ and } 8 \text{ for } y_1$$

$$y - 8 = \frac{1}{5}x - \frac{3}{5}$$

$$y = \frac{1}{5}x + \frac{37}{5}$$

Using the slope-intercept equation:

$$y = mx + b$$

$$y = \frac{1}{5}x + b \qquad \text{Substituting } \frac{1}{5} \text{ for } m$$

$$8 = \frac{1}{5}(3) + b \qquad \text{Substituting } 3 \text{ for } x \text{ and } 8 \text{ for } y$$

$$8 = \frac{3}{5} + b$$

$$\frac{37}{5} = b \qquad\qquad \text{Solving for } b$$

$$y = \frac{1}{5}x + \frac{37}{5} \qquad \text{Substituting } \frac{37}{5} \text{ for } b$$

GUIDED LEARNING   📍 **Textbook**   👤 **Instructor**   ▶ **Video**

EXAMPLE 1	YOUR TURN 1

EXAMPLE 1

Use the point-slope equation to find an equation of the line containing the points $(-2, 1)$ and $(2, 9)$.

First, we find the slope of the line.

$$m = \frac{y_2 - y_1}{x_2 - x_1} = \frac{9 - 1}{2 - (-2)} = \frac{8}{4} = \boxed{}$$

We choose to use $(2, 9)$ for (x_1, y_1).

$$y - y_1 = m(x - x_1)$$

$$y - \boxed{} = 2\left(x - \boxed{}\right)$$

$$y - 9 = 2x - \boxed{}$$

$$y = 2x + \boxed{}$$

YOUR TURN 1

Use the point-slope equation to find an equation of the line containing the points $(-5, 3)$ and $(3, -1)$.

EXAMPLE 2	YOUR TURN 2

EXAMPLE 2

Use the slope-intercept equation to find an equation of the line containing the points $(4, -1)$ and $(-2, -6)$.

First, we find the slope of the line.

$$m = \frac{y_2 - y_1}{x_2 - x_1} = \frac{-1 - (-6)}{4 - (-2)} = \frac{5}{6}$$

$$y = mx + b$$

$$y = \frac{5}{6}x + b \qquad \text{Substituting } \frac{5}{6} \text{ for } m$$

$$-1 = \frac{5}{6}\left(\boxed{}\right) + b \quad \text{Substituting 4 for } x \text{ and } -1 \text{ for } y$$

$$-1 = \frac{10}{3} + b$$

$$-\frac{13}{3} = b \qquad \text{Solving for } b$$

$$y = \boxed{}\,x + \left(\boxed{}\right), \text{ or } y = \frac{5}{6}x - \frac{13}{3}$$

YOUR TURN 2

Use the slope-intercept equation to find an equation of the line containing the points $(1, 4)$ and $(-2, 7)$.

YOUR NOTES   Write your questions and additional notes.

Finding an Equation of a Line Parallel or Perpendicular to a Given Line Through a Point Not on the Line

ESSENTIALS

Example

- Find an equation of the line containing the point $(-1, 4)$ and parallel to the line $y = 3x - 5$.

  Since the lines are parallel, both lines have slope 3.

  $$y - y_1 = m(x - x_1)$$

  $$y - 4 = 3[x - (-1)], \text{ or } y = 3x + 7$$

GUIDED LEARNING

| 📖 **Textbook** | 👤 **Instructor** | ▶ **Video** |

EXAMPLE 1	YOUR TURN 1
Find an equation of the line containing the point $(8, 11)$ and parallel to the line $3x - 4y = 8$.	Find an equation of the line containing the point $(4, 5)$ and parallel to the line $2x + 5y = -3$.

First, find the equation of the given line in slope-intercept form.

$$3x - 4y = 8$$

$$-4y = -3x + 8$$

$$y = \frac{3}{4}x - 2$$

The slope of the given line is $\boxed{}$. The slope of a

line parallel to it is also $\boxed{}$. We have

$m = \dfrac{3}{4}$ and $(x_1, y_1) = (8, 11)$.

$$y - y_1 = m(x - x_1)$$

$$y - \boxed{} = \boxed{}(x - 8)$$

$$y - 11 = \frac{3}{4}x - \boxed{}$$

$$y = \frac{3}{4}x + \boxed{}$$

EXAMPLE 2	YOUR TURN 2
Find an equation for the line perpendicular to $x = 3y + 5$ and passing through $(4, -2)$. First, find the equation of the given line in slope-intercept form. $$x = 3y + 5$$ $$x - 5 = 3y$$ $$\frac{1}{3}x - \frac{5}{3} = y$$ The slope of the given line is $\boxed{}$. The slope of a line perpendicular to it is the opposite of the reciprocal of $\frac{1}{3}$, or $\boxed{}$. We use the slope-intercept equation. $$y = mx + b$$ $y = -3x + b \qquad$ Substituting -3 for m $-2 = -3\left(\boxed{}\right) + b \quad$ Substituting 4 for x and -2 for y $-2 = -12 + b$ $\boxed{} = b \qquad$ Solving for b Finally, we substitute 10 for b. $$y = -3x + 10$$	Find an equation for the line containing the point $(-5, 1)$ and perpendicular to the line $-2x = 3 - 4y$.

YOUR NOTES   Write your questions and additional notes.

Applications of Linear Functions

ESSENTIALS

Given two points, we can model data with a linear function.

GUIDED LEARNING  **Textbook**  **Instructor** ▶ **Video**

EXAMPLE 1	YOUR TURN 1

EXAMPLE 1

The average monthly revenue of Corp C Enterprise is shown in the table. Use the data from 2010 and 2012 to find a linear function that fits the data. Then use the function to estimate average monthly revenue in 2014.

Year	Average Monthly Revenue
2010	$20,000
2012	21,000

We let $t =$ the number of years since 2000 and $r =$ the average monthly revenue in thousands of dollars. We find a linear function

containing points $\left(10, \boxed{}\right)$ and $\left(\boxed{}, 21\right)$.

$$m = \frac{21 - \boxed{}}{\boxed{} - 10} = \frac{\boxed{}}{\boxed{}}$$

We use m and $\left(10, \boxed{}\right)$ to find an equation of the line.

$$r - \boxed{} = \boxed{}(t - 10)$$

$r = \dfrac{1}{2}t + 15$, or $r(t) = \dfrac{1}{2}t + 15$

To estimate the average monthly revenue in 2014, we find $r(14)$.

$$r(14) = \frac{1}{2}\left(\boxed{}\right) + 15 = \boxed{}$$

Assuming constant growth, average monthly revenue in 2014 is estimated to be $\$\boxed{}$.

YOUR TURN 1

The average monthly expenses of Corp C Enterprise are shown in the table. Use the data for 2009 and 2011 to find a linear function that fits the data. Let $t =$ the number of years since 2000 and $e =$ the average monthly expenses in thousands of dollars. Then use the function to estimate average monthly expenses in 2014.

Year	Average Monthly Expenses
2009	$20,000
2011	19,000

EXAMPLE 2	YOUR TURN 2
Suppose suppliers are willing to sell 100 handmade headbands when the price is \$20 per headband and 60 handmade headbands when the price is \$12 per headband. Find a linear function that expresses the number of headbands suppliers are willing to sell as a function of the price per headband. Use the function to predict how many headbands sellers would be willing to sell if the price were \$15 per headband.	Suppose buyers are willing to buy 100 handmade headbands when the price is \$10 per headband and 70 handmade headbands when the price is \$12 per headband. Find a linear function that expresses the number of headbands buyers are willing to buy as a function of the price per headband. Let $p =$ the price and $h =$ the number of headbands. Use the function to predict how many headbands buyers would be willing to buy if the price were \$15 per headband.

Let $p =$ the price and $h =$ the number of headbands. We find a linear function containing points $\left(20, \boxed{}\right)$ and $\left(\boxed{}, 60\right)$.

$$m = \frac{\boxed{} - 100}{12 - \boxed{}} = \frac{-40}{\boxed{}} = 5$$

We use m and $(20, 100)$ to find an equation of the line.

$$h - \boxed{} = \boxed{}(p - 20)$$
$$h - 100 = 5p - 100$$
$$h = 5p, \text{ or } h(p) = 5p$$

To estimate the number of headbands sellers would be willing to sell if the price were \$15, we find $h(15)$.

$$h(15) = 5 \cdot \boxed{} = \boxed{}$$

The suppliers would be willing to supply $\boxed{}$ headbands.

YOUR NOTES   Write your questions and additional notes.

Practice Exercises

Readiness Check

Pair each item in the first column with the item in the second column that best matches.

1. $y - 4 = -2(x - (-6))$

2. $y = 3x + 22$

3. $y = x - 7$ and $y = x + 5$

4. $y = x - 3$ and $y = 3 - x$

   a) Parallel lines

   b) An equation in slope-intercept form

   c) Perpendicular lines

   d) An equation in point-slope form

Finding an Equation of a Line When the Slope and the y-Intercept Are Given

5. Find an equation of the line with slope -4 and y-intercept $(0, 8)$.

6. Find a linear function $f(x) = mx + b$ whose graph has slope $\dfrac{1}{2}$ and y-intercept $(0, -1)$.

Finding an Equation of a Line When the Slope and a Point Are Given

7. Find an equation of the line with slope 6 and containing the point $(3, 0)$.

8. Find an equation of the line with slope $-\dfrac{1}{2}$ and containing the point $(-2, -6)$.

Finding an Equation of a Line When Two Points Are Given

9. Find an equation of the line containing the points $(5, 6)$ and $(-4, 3)$.

10. Find an equation of the line containing the points $(0, -5)$ and $(-1, -7)$.

Finding an Equation of a Line Parallel or Perpendicular to a Given Line Through a Point Not on the Line

11. Find an equation for the line containing the point $(5, 3)$ and parallel to the line $x - y = -6$.

12. Find an equation for the line containing the point $(10, -1)$ and perpendicular to the line $5x + 4y = -20$.

Applications of Linear Functions

13. Dan joined a fitness club for $150 and pays a monthly fee of $22.95.

   a) Formulate a linear function that models the total cost $C(t)$ of the club membership for t months.

   b) Use the model to determine the total cost after he has been a member of the club for 16 months.

14. In 2005, the number of students participating in an intramural sport at Castlegate Community College was 150. In 2010, the number had risen to 320.

   a) Find a linear function that fits the data. Let $x =$ the number of years after 2005 and $N(x) =$ the number of students participating in an intramural sport.

   b) Use the function of part (a) to estimate the number of students participating in an intramural sport at Castlegate Community College in 2016.

Solving Systems of Equations Graphically

ESSENTIALS

A **system of equations** in two variables is a set of two or more linear equations that are to be solved simultaneously.

A **solution** of a system of two equations in two variables is an ordered pair that makes both equations true. If we graph a system of equations, the point at which the graphs intersect will be the solution of both equations.

Sometimes the equations in a system have graphs that are parallel lines. In such a case, the graphs will never intersect and as such we say the system has **no solution**.

Sometimes the equations in a system have the same graph. In such a case, the system has **infinitely many solutions**.

Consistent Systems and Inconsistent Systems

If a system of equations has at least one solution, then the system is **consistent**.
If a system of equations has no solution, then the system is **inconsistent**.

Dependent Equations and Independent Equations

If for a system of two equations in two variables:
the graphs of the equations are the same line, then the equations are **dependent**.
the graphs of the equations are different lines, then the equations are **independent**.

Examples

- $y = x - 1,$

  $y = -\dfrac{2}{3}x + 4$

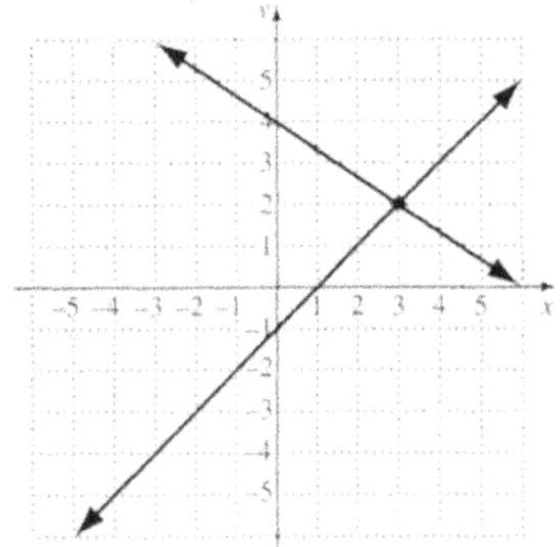

Solution: $(3, 2)$

Consistent system
Independent equations

- $y = \dfrac{1}{2}x - 1,$

  $y = \dfrac{1}{2}x + 2$

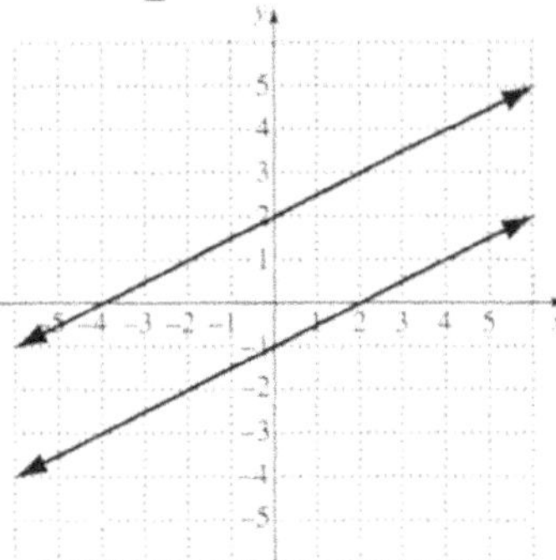

No solution

Inconsistent system
Independent equations

- $y = \dfrac{1}{3}x + 1,$

  $3y = x + 3$

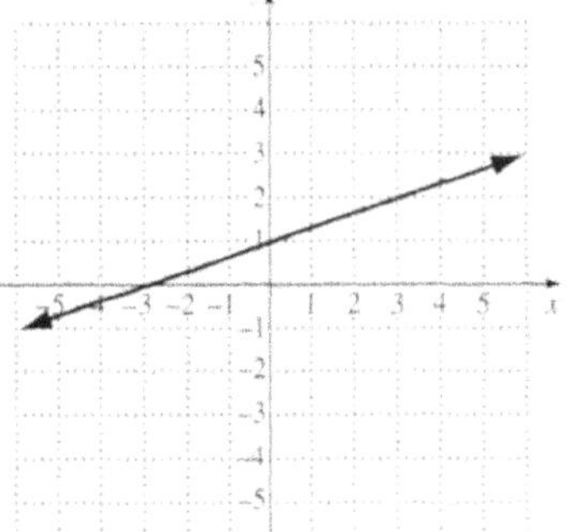

Infinitely many solutions

Consistent system
Dependent equations

GUIDED LEARNING    **Textbook**    **Instructor**    **Video**

EXAMPLE 1	YOUR TURN 1

EXAMPLE 1

Solve this system graphically:
$$2x - y = 3,$$
$$x + 3y = 5.$$

Draw the graph of each equation and find the coordinates of the point of intersection.

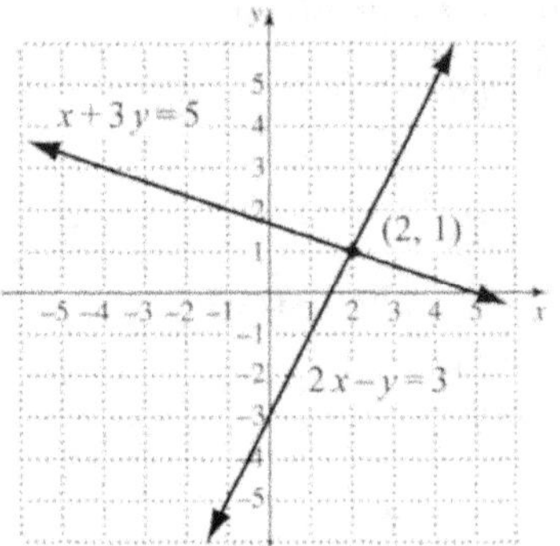

The point of intersection has coordinates that make both equations ______________ .
  true / false

The solution seems to be $\left(\boxed{}, \boxed{} \right)$.

We will check the ordered pair in both equations.
Check:

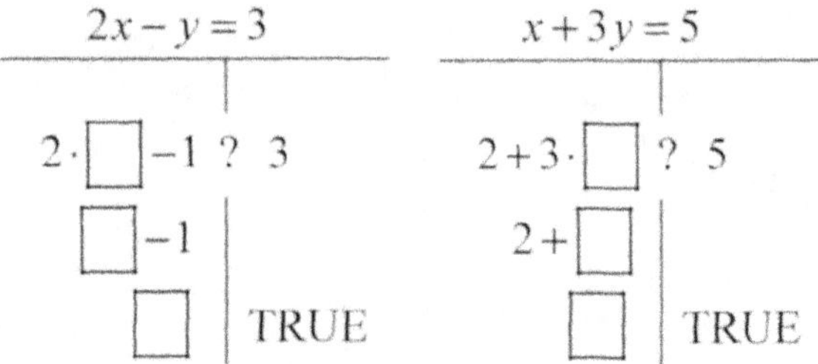

The solution is $\left(\boxed{}, \boxed{} \right)$.

Since this system has at least one solution it is a/an ____________________ system.
  consistent / inconsistent

Since the graphs of the equations are different lines the equations are ____________________ .
  dependent / independent

YOUR TURN 1

Solve this system graphically:
$$x - y = -5,$$
$$x + 2y = 1.$$

Classify the system as consistent or inconsistent and the equations as dependent or independent.

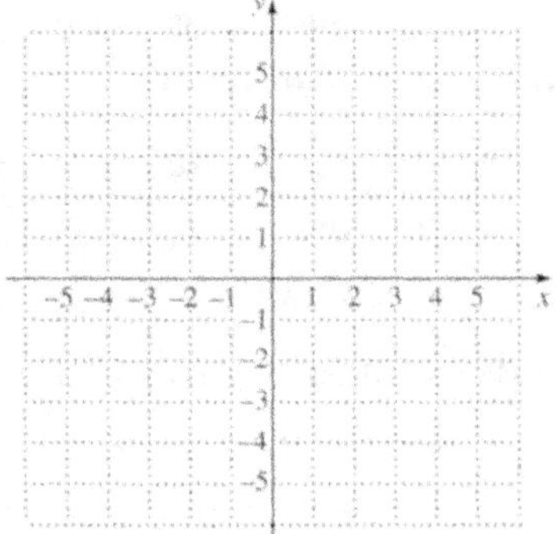

     Copyright © 2022 Pearson Education, Inc.

EXAMPLE 2	YOUR TURN 2
Solve this system graphically:	Solve this system graphically:

Solve this system graphically:

$$f(x) = -2x + 4,$$

$$g(x) = -2x + 2.$$

Graph both equations.

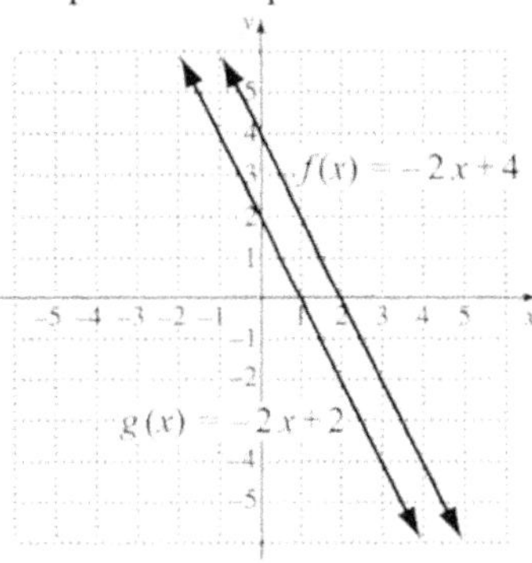

We see that the graphs have the same

___________________, but different
slope / y-intercept

___________________, so the lines are parallel.
slopes / y-intercepts

There is no point at which they cross, so the system has

___________________.
no solution / infinitely many solutions

The solution set is thus the empty set, denoted $\varnothing$, or $\{\ \}$.

Since this system has no solution it is a/an ___________________ system.
   consistent / inconsistent

Since the graphs of the equations are different lines the equations are ___________________.
   dependent / independent

Solve this system graphically:

$$f(x) = x + 2,$$

$$g(x) = x - 2.$$

Classify the system as consistent or inconsistent and the equations as dependent or independent.

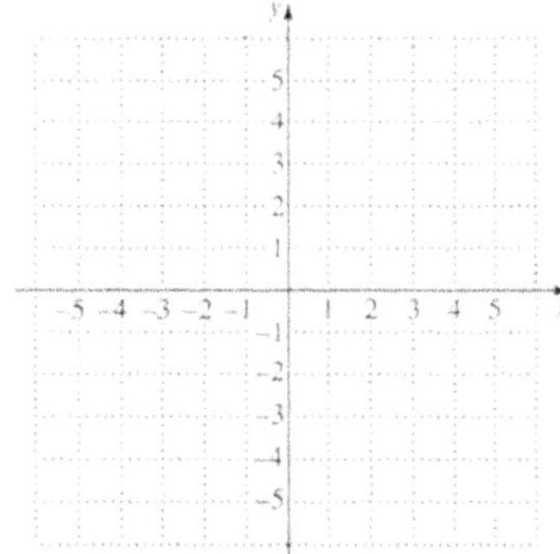

EXAMPLE 3	YOUR TURN 3
Solve this system graphically: $$6x - 3y = 9,$$ $$y = 2x - 3.$$ Graph both equations. 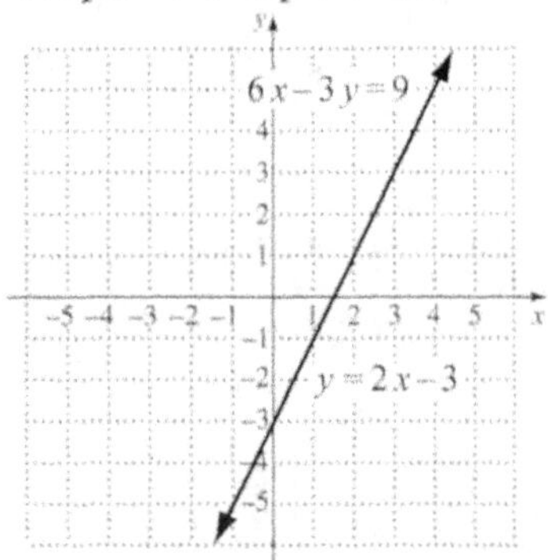 We see that the graphs are _______________. 　　　　　　　　　　　parallel / the same Thus any solution of one of the equations is a solution of the other. The system has a/an _______________ number of solutions. 　finite / infinite Since this system has at least one solution it is a/an _______________ system. 　　consistent / inconsistent Since the graphs of the equations are the same line the equations are _______________. 　　　　　dependent / independent	Solve this system graphically: $$8x + 2y = 6,$$ $$y = -4x + 3.$$ Classify the system as consistent or inconsistent and the equations as dependent or independent. 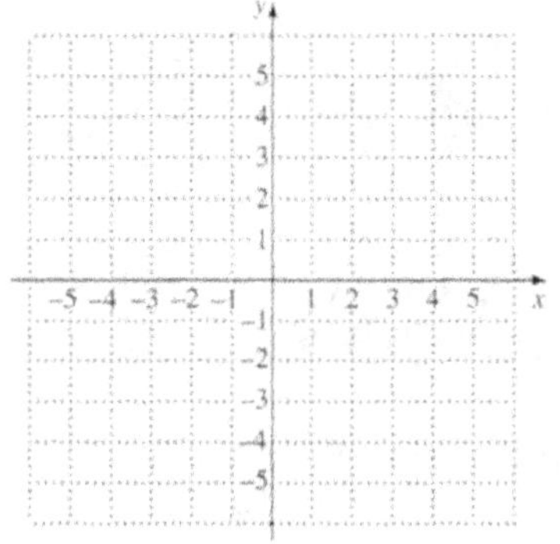

YOUR NOTES　Write your questions and additional notes.

Practice Exercises

Readiness Check

Consider the graphs below when doing Exercises 1-6.

1. Which graph(s) represent(s) a consistent system?

2. Which graph(s) represent(s) an inconsistent system?

3. Which graph(s) represent(s) a system whose equations are dependent?

4. Which graph(s) represent(s) a system whose equations are independent?

5. Which graph(s) represent(s) a system with no solution?

6. Which graph(s) represent(s) a system with a solution at $(2, 4)$?

a)
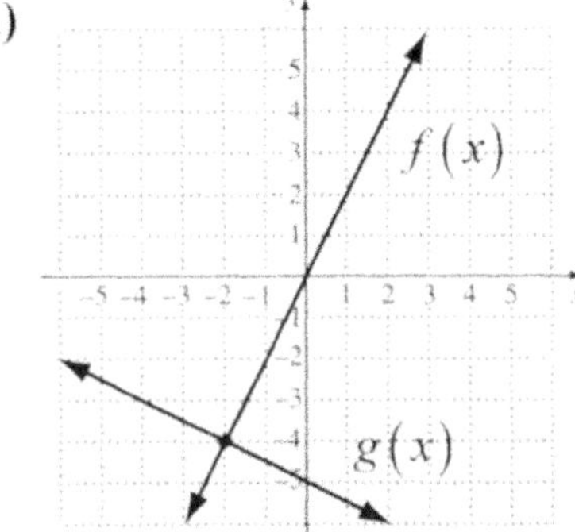

b)
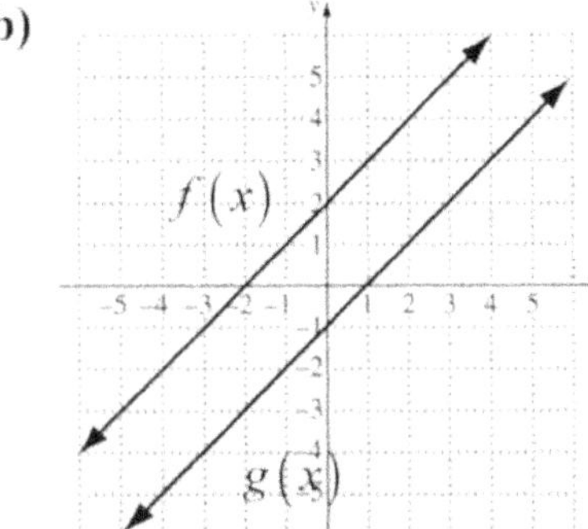

c)
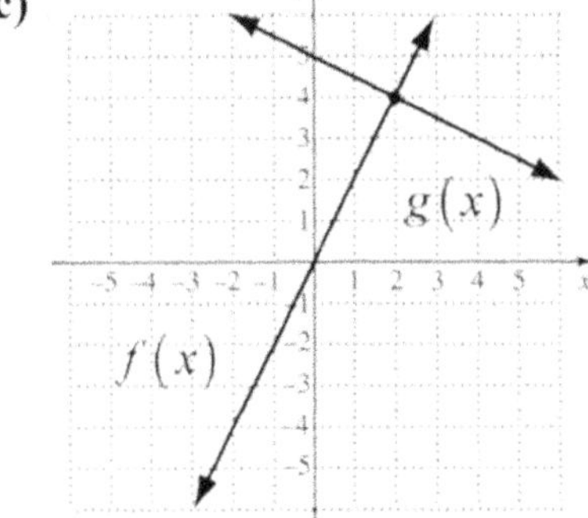

d)
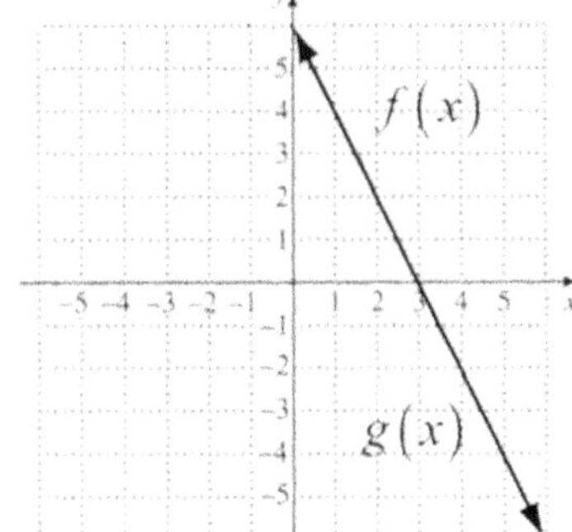

Solving Systems of Equations Graphically

Solve each system of equations graphically. Then classify the system as consistent or inconsistent and the equations as dependent or independent. Complete the check for each.

7. $x - y = 0,$

$x + y = -2$

8. $6x - 2y = 4,$

$y = 3x - 2$

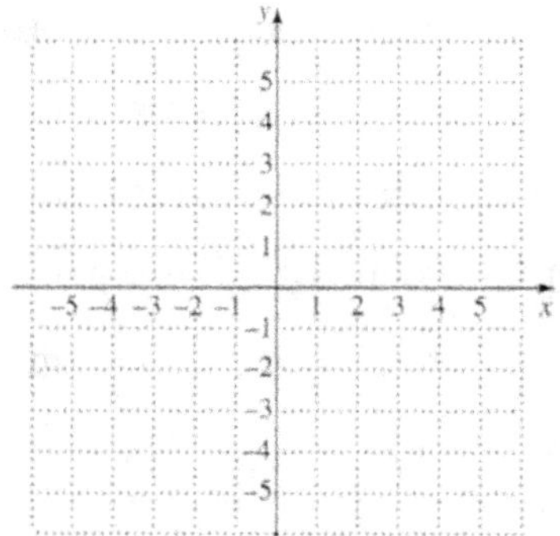

9. $x + 2y = -2,$

$4x - y = 10$

10. $x + y = 5,$

$x + y = 2$

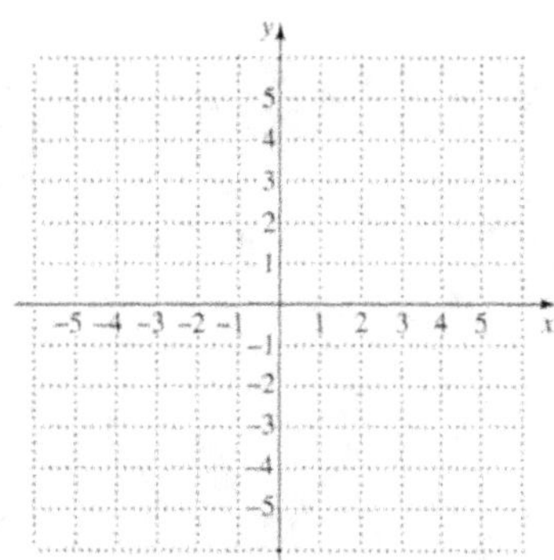

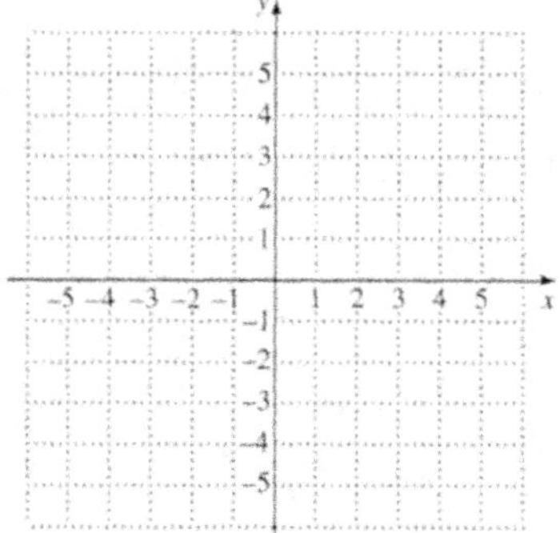

11. $f(x) = -2x + 1,$

$g(x) = 4x - 2$

12. $x = 4,$

$x = -\dfrac{3}{2}$

     Copyright © 2022 Pearson Education, Inc.

The Substitution Method

ESSENTIALS

To use the **substitution method** to solve systems of two equations:

1. If neither equation has a variable alone on one side, solve one equation for one of the variables.
2. Substitute the expression that is equivalent to the variable into the other equation.
3. You should now have an equation that contains only one variable. Solve this equation.
4. Substitute this number in one of the equations to find the value of the other variable.

No matter the order in which you find your answers, coordinates of ordered pairs must be written in alphabetical order.

Special Cases

When solving a system of two linear equations in two variables:

1. If a false equation is obtained, such as $4 = 5$, then the system has no solution.
2. If a true equation is obtained, then the system has an infinite number of solutions.

Example

*   Solve this system by substitution: $4x + y = 11,$ (1)

    $$2x - 3y = 23.\quad (2)$$

    $$\begin{aligned}
    y &= -4x + 11 &&\text{Solving equation (1) for } y \\
    2x - 3(-4x + 11) &= 23 &&\text{Substituting } -4x + 11 \text{ for } y \text{ in Equation (2)} \\
    2x + 12x - 33 &= 23 &&\text{Removing parentheses} \\
    14x - 33 &= 23 &&\text{Collecting like terms} \\
    14x &= 56 &&\text{Adding 33} \\
    x &= 4 &&\text{Dividing by 14}
    \end{aligned}$$

    Now we substitute 4 for x in one of the equations and solve to find y.

    $$y = -4 \cdot (4) + 11 = -16 + 11 = -5$$

    The solution is $(4, -5)$.

<table>
<tr><td> **Textbook**</td><td> **Instructor**</td><td>▶ **Video**</td></tr>
</table>

GUIDED LEARNING

EXAMPLE 1	YOUR TURN 1

EXAMPLE 1

Solve this system:

$$5x + y = 19, \quad (1)$$

$$2x + 8y = 0. \quad (2)$$

First, we solve one equation for one of the variables. Since the coefficient of y is 1 in the first equation, it is the easier one to solve for y:

$$y = -5x + 19. \quad (3)$$

Next, we substitute $-5x + 19$ for y in the second equation and solve for x:

$$2x + 8\left(\boxed{}\right) = 0$$

$$2x - 40x + \boxed{} = 0$$

$$\boxed{} + 152 = 0$$

$$-38x = -152$$

$$x = \boxed{}$$

In order to find y we can return to equation $(1), (2),$ or (3). It is easiest to use equation (3) since we have already solved for y. We substitute 4 for x and solve for y:

$$y = -5 \cdot \boxed{} + 19 = -20 + 19 = \boxed{}$$

We obtain the ordered pair $(4, -1)$.

(continued)

YOUR TURN 1

Solve this system:

$$4x - y = 10,$$

$$x + 2y = -2.$$

Check:

$$\frac{5x+y=19}{5\cdot(4)+(-1)\ ?\ 19} \qquad \frac{2x+8y=0}{2\cdot4+8\cdot(-1)\ ?\ 0}$$

$$\begin{array}{c|c} 20-1 & \\ 19 & \text{TRUE} \end{array} \qquad \begin{array}{c|c} 8+(-8) & \\ 0 & \text{TRUE} \end{array}$$

The solution is $\left(\boxed{},\boxed{}\right)$.

EXAMPLE 2	YOUR TURN 2
Solve this system:	Solve this system:

Solve this system:

$y=-3x-9,$

$9x+3y=12.$

The first equation is already solved for y, so we substitute $-3x-9$ for y in the second equation.

$$9x+3\left(\boxed{}\right)=12$$
$$9x-9x+\left(\boxed{}\right)=12$$
$$\boxed{}=12$$

This is a ________ equation.
$\qquad\qquad$ true / false

This means that the system has

________________________ .

no solution / an infinite number of solutions

Solve this system:

$y=-2x+4,$

$8x+4y=6.$

EXAMPLE 3	YOUR TURN 3
Solve this system: $x - y = 2,$ $3x - 3y = 6.$	Solve this system: $x - 3y = 5,$ $-2x + 6y = -10.$

First, we solve one equation for one of the variables. The coefficient of x is 1 in the first equation so it is the easier one to solve for x.

$x = y + 2.$

Next, we substitute $y + 2$ for x in the second equation and solve for y:

$$3\left(\boxed{}\right) - 3y = 6$$

$$\boxed{} + 6 - 3y = 6$$

$$\boxed{} = 6$$

This is a __________ equation.
         true / false

This means that the system has

__________________________.
no solution / an infinite number of solutions

YOUR NOTES   Write your questions and additional notes.

Solving Applied Problems Involving Two Equations

ESSENTIALS

Many applied problems are easier to solve if we first translate to a system of two equations rather than to a single equation.

GUIDED LEARNING	📖 **Textbook**	👤 **Instructor**	▶ **Video**

EXAMPLE 1	YOUR TURN 1

EXAMPLE 1

The perimeter of a football field (including the end zones) is 1040 feet. The length is 40 feet more than twice the width. Find the length and width.

1. **Familiarize.** We first make a drawing and label it, using l for length and w for width. Also, we use the formula

$$P = 2l + 2w.$$

w = width

l = length

2. **Translate.** We translate as follows:

The perimeter   is   1040 ft

$\downarrow$ $\downarrow$ $\downarrow$

$2l + 2w$ $=$ 1040

We can also write a second equation:

The length   is   40 feet more than twice the width

$\downarrow$ $\downarrow$ $\downarrow$

l $=$ $2w + 40$

We now have a system of equations:

$$2l + 2w = 1040,$$
$$l = 2w + 40.$$

YOUR TURN 1

The perimeter of Mr. McGregor's garden is 110 feet. The length is 5 feet less than twice the width. Find the length and width.

(continued)

3. **Solve.** We substitute $2w+40$ for l in the first equation and solve for w:

$$2\cdot\left(\boxed{}\right)+2w=1040$$

$$4w+\boxed{}+2w=1040$$

$$\boxed{}+80=1040$$

$$6w=960$$

$$w=160$$

Next we substitute 160 for w into either equation and solve for l:

$$l=2\cdot\boxed{}+40$$

$$l=\boxed{}+40$$

$$l=\boxed{}$$

4. **Check.** Consider the dimensions 360 ft and 160 ft. The length is 40 ft more than twice the width: $2\cdot(160)+40=360$. The perimeter is $2\cdot(360)+2\cdot(160)$, or 1040.

5. **State.** The length is 360 ft and the width is 160 ft.

YOUR NOTES   Write your questions and additional notes.

Practice Exercises

Readiness Check

Determine whether each statement is true or false.

1. The substitution method is a graphical method for solving systems of equations.
2. When solving using substitution, if we obtain a true equation, then the system has no solution.
3. When using the substitution method, if neither equation of a system has a variable alone on one side we must solve one equation for one of the variables.
4. When writing the solution of a system of equations, the value that we solved for first is always the first number in the ordered pair.

The Substitution Method

Solve each system of equations by the substitution method.

5. $2x + 5y = 11,$
   $x = 3y$

6. $3x - 2y = -4,$
   $y = 3x - 1$

7. $x - 3y = 4,$
   $-2x + 5y = -5$

8. $2x + y = 5,$
   $4x + 2y = 7$

9. $-3x - 6y = -15,$
$x = -2y + 5$

10. $3x + y = 2,$
$6x - 10y = 7$

11. $5x + 7y = 1,$
$2x + 4y = 4$

12. $3x - 4y = -2,$
$-6x + 8y = 4$

Solving Applied Problems Involving Two Equations

13. The outer perimeter of a picture frame is 32 inches. The length of the frame is 2 inches longer than the width. Find the length and the width.

14. The sum of the measures of **complementary angles** is $90°$. Two complementary angles are such that the measure of one angle is $2°$ more than seven times the measure of the other. Find the measures of the angles.

The Elimination Method

ESSENTIALS

The **elimination method** for solving systems of equations uses the addition principle for equations. Thus, it is also called the addition method. The object of this method is to add the two equations in such a way that one of the variables is eliminated. To eliminate a variable, we sometimes use the multiplication principle to multiply one or both of the equations by a particular number before adding.

To use the elimination method to solve systems of two equations:

1. Write both equations in the form $Ax + By = C$.
2. Clear any decimals or fractions.
3. Choose a variable to eliminate.
4. Make the chosen variable's terms opposites by multiplying one or both equations by appropriate numbers if necessary.
5. Eliminate a variable by adding the equations and then solve for the remaining variable.
6. Substitute in either of the original equations to find the value of the other variable.

Special Cases

When solving a system of two linear equations in two variables:

1. If a false equation is obtained, such as $0 = 5$, then the system has no solution. The system is inconsistent, and the equations are independent.
2. If a true equation is obtained, such as $0 = 0$, then the system has an infinite number of solutions. The system is consistent, and the equations are dependent.

Example

- Solve this system using the elimination method: $3x + y = 8$, (1)

$$2x - y = 7. \quad (2)$$

$$
\begin{aligned}
3x + y &= 8 \quad (1) \\
\underline{2x - y} &= \underline{7} \quad (2) \\
5x + 0y &= 15 \qquad \text{Adding the equations} \\
5x + 0 &= 15 \\
5x &= 15 \\
x &= 3
\end{aligned}
$$

$$3 \cdot (3) + y = 8 \qquad \text{Substituting 3 for } x \text{ in equation (1)}$$

$$
\left.\begin{aligned}
9 + y &= 8 \\
y &= -1
\end{aligned}\right\} \quad \text{Solving for } y
$$

The solution is $(3, -1)$.

GUIDED LEARNING    **Textbook**    **Instructor**   **Video**

EXAMPLE 1	YOUR TURN 1

EXAMPLE 1

Solve this system:

$$2x + y = 2, \quad (1)$$
$$x - y = -5. \quad (2)$$

The y in equation (1) is opposite of $-y$ in equation (2). When added, their sum is zero and the variable y is "eliminated."

We add the two equations:

$$2x + y = 2 \quad (1)$$
$$\underline{x - y = -5 \quad (2)}$$
$$3x + \boxed{} = -3 \qquad \text{Adding}$$
$$3x + 0 = -3$$
$$\boxed{} = -3$$

We eliminated the variable y and now have an equation with one variable, which we solve for x:

$$3x = -3$$
$$x = \boxed{}$$

Next, we substitute -1 for x in either equation and solve for y:

$$2 \cdot \left(\boxed{} \right) + y = 2$$
$$-2 + y = 2$$
$$y = \boxed{}$$

We obtain the ordered pair $\left(\boxed{}, \boxed{} \right)$.

Check:

$2x + y = 2$	$x - y = -5$
$2 \cdot (-1) + 4 \; ? \; 2$	$-1 - 4 \; ? \; -5$
$-2 + 4$	-5 | TRUE
2 | TRUE	

Since $(-1, 4)$ checks, it is the solution.

YOUR TURN 1

Solve this system:

$$4x + 5y = -6,$$
$$-4x - 3y = 2.$$

EXAMPLE 2	YOUR TURN 2
Solve this system: $x + 2y = 4,$ $3x + y = -3.$	Solve this system: $5x + y = 19,$ $x + 4y = 0.$

If we add directly, we will not eliminate a variable. However, note that if the x in the first equation were $-3x$, we could eliminate x.

We multiply the first equation by $\boxed{}$ and then add the equations:

$$\boxed{}x - 6y = \boxed{}$$
$$\underline{3x + \ y = -3}$$
$$\boxed{} - 5y = -15 \qquad \text{Adding}$$

$$\left.\begin{array}{l} -5y = -15 \\ \\ y = \boxed{} \end{array}\right\} \text{Solving for } y$$

Now we substitute 3 for y in one of the equations and solve for x.

$$x + 2 \cdot \boxed{} = 4$$
$$x + \boxed{} = 4$$
$$x = \boxed{}$$

We obtain the ordered pair $\left(\boxed{}, \boxed{}\right)$.

Check:

$x + 2y = 4$	$3x + y = -3$
$-2 + 2 \cdot (3) \ ? \ 4$	$3 \cdot (-2) + 3 \ ? \ -3$
$-2 + 6$	$-6 + 3$
$4 \mid$ TRUE	$-3 \mid$ TRUE

Since $(-2, 3)$ checks, it is the solution.

EXAMPLE 3	YOUR TURN 3

EXAMPLE 3

Solve this system:

$$5x - 4y = 8, \quad (1)$$

$$2x + 3y = -6. \quad (2)$$

We must first multiply in order to make one pair of terms opposites. We decide to do this with the y-terms. We multiply equation (1) by 3 and equation (2) by 4.

The new system is:

$$15x - \boxed{}\,y = \boxed{}$$

$$\boxed{}\,x + 12y = \boxed{}$$

$$23x + \boxed{} = \boxed{} \quad \text{Adding}$$

$$\left.\begin{array}{r} 23x = 0 \\ x = 0 \end{array}\right\} \quad \text{Solving for } x$$

Now substitute 0 for x in one of the equations and solve for y.

$$2\cdot(0) + 3y = -6$$

$$0 + 3y = -6$$

$$3y = -6$$

$$y = \boxed{}$$

We check the ordered pair $(0, -2)$.

Check:

$5x - 4y = 8$	$2x + 3y = -6$
$5\cdot(0) - 4(-2) \;?\; 8$	$2\cdot(0) + 3(-2) \;?\; -6$
$0 + 8$	$0 - 6$
8 | TRUE	-6 | TRUE

Since $(0, -2)$ checks, it is the solution.

YOUR TURN 3

Solve this system:

$$3x - 2y = 15,$$

$$2x - 3y = 15.$$

EXAMPLE 4	YOUR TURN 4

EXAMPLE 4

Solve this system:

$$x - 3y = 5,$$
$$-2x + 6y = -10.$$

We must first multiply in order to make one pair of terms opposites. We decide to do this with the *x*-terms. We multiply equation (1) by 2.

The new system is:

$$2x - \boxed{}\, y = \boxed{}$$
$$\underline{-2x + \quad 6y = -10}$$
$$0x + \boxed{} = \boxed{} \qquad \text{Adding}$$
$$\boxed{} = \boxed{}$$

We have eliminated both variables, and what remains is a _________ equation, $0 = 0$.
  true / false

If an ordered pair is a solution of one of the original equations, then it will be a solution of the other. The system has ________________________.
  no solution / an infinite number of solutions

The system is ________________.
  consistent / inconsistent

The equations are ________________.
  dependent / independent

YOUR TURN 4

Solve this system:

$$3x + y = -9,$$
$$9x + 3y = 12.$$

YOUR NOTES   Write your questions and additional notes.

Solving Applied Problems Using Elimination

ESSENTIALS

Many applied problems are easier to solve if we first translate to a system of two equations rather than to a single equation.

| | 🔖 **Textbook** | 👤 **Instructor** | ▶ **Video** |

GUIDED LEARNING

EXAMPLE 1	YOUR TURN 1

The Lions basketball team scored 69 points on a combination of two-point shots and three-point shots. If they made a total of 30 shots, how many of each kind of shot was made?

1. **Familiarize.** We let $x =$ the number of two-point shots and $y =$ the number of three-point shots. The total points scored from two-point shots alone is $2x$. Similarly, the total points scored from three-point shots alone is $\boxed{}$.

2. **Translate.** We translate to two equations.

$$\underbrace{\text{Total number of points scored}}_{\downarrow} \quad \underset{\downarrow}{\text{is}} \quad \underset{\downarrow}{69}$$

$$2x + 3y \qquad\qquad = \quad 69$$

$$\underbrace{\text{Total number of shots made}}_{\downarrow} \quad \underset{\downarrow}{\text{is}} \quad \underset{\downarrow}{30}$$

$$x + \boxed{} \qquad\qquad = \quad 30$$

We now have a system of equations:
$$2x + 3y = 69, \quad (1)$$
$$x + y = 30. \quad (2)$$

The Tigers basketball team scored 75 points on a combination of two-point shots and three-point shots. If they made a total of 33 shots, how many of each kind of shot was made?

(continued)

3. Solve. First, we multiply by -2 on both sides of equation (2) and add:

$$2x + 3y = 69$$
$$-2x - 2y = \boxed{}$$
$$0x + y = 9 \qquad \text{Adding}$$

$$\left. \begin{array}{l} 0 + y = 9 \\ \quad\ y = 9 \end{array} \right\} \qquad \text{Solving for } y$$

Next we substitute 9 in for y in equation (2) and solve for x:

$$x + \boxed{} = 30$$
$$x = \boxed{}$$

4. Check. If 21 two-point shots and 9 three-point shots were made, there were $2 \cdot 21 + 3 \cdot 9$, or 69, total points scored and $21 + 9$, or 30, shots made. The numbers check in the original problem.

5. State. There were 21 two-point shots made and 9 three-point shots made.

YOUR NOTES   Write your questions and additional notes.

Practice Exercises

Readiness Check

Determine whether each statement is true or false.

1. We call it the elimination method because after the equations are added one variable should have been eliminated.

2. When solving using elimination, if we obtain a false equation, then the equations of the system are dependent.

3. The elimination method is also called the addition method because it uses the addition principle for equations.

4. The first step of the elimination method is to write both equations in the form $Ax + By = C$.

The Elimination Method

Solve each system of equations using the elimination method.

5. $x + 4y = 12,$
   $2x - 4y = 0$

6. $x - y = -4,$
   $-2x + 3y = 9$

7. $2x + 3y = 4,$
   $6x + 9y = 12$

8. $4x - 5y = 7,$
   $5x - 3y = -1$

9. $-3x + 5y = -2,$
$6x - 10y = 6$

10. $7x + 4y = 9,$
$14x + 8y = 18$

11. $2x - 3y = 5,$
$5x + 10y = -12$

12. $x = 2y - 4,$
$6y = 3x + 5$

Solving Applied Problems Using Elimination

13. The sum of two numbers is 4. The larger number minus the smaller number is 26. Find the numbers.

14. A small concert venue sold a total of 375 tickets for a concert. The number of floor seats was 55 more than three times the number of balcony seats. How many of each type of seat was sold?

Total-Value Problems and Mixture Problems

ESSENTIALS

Examples

- A college basketball team scored 83 points in a game. Of those points, 15 points were from free throws. The remaining 68 points were a result of 30 two-point and three-point baskets. How many baskets of each type were made during the game?

  1. **Familiarize.** Let $x =$ the number of two-pointers made and $y =$ the number of three-pointers made.

  2. **Translate.** A total of 30 baskets were made, so $x + y = 30$.

     We also have a second equation.

     Rewording: $\underbrace{\text{The points scored from two-pointers}}$ plus $\underbrace{\text{the points scored from three-pointers}}$ totaled 68.

↓	↓	↓	↓	↓

     Translating: $\qquad x \cdot 2 \qquad + \qquad y \cdot 3 \qquad = \quad 68$

     The system of equations is
     $$x + y = 30, \quad (1)$$
     $$2x + 3y = 68. \quad (2)$$

  3. **Solve.** Multiply equation (1) by -2 and then add the equations.
     $$-2x - 2y = -60 \quad (3)$$
     $$\underline{2x + 3y = 68 \qquad (2)}$$
     $$y = 8$$

     Substituting 8 for y in equation (1), we get $x + 8 = 30$. Thus $x = 22$.

  4. **Check.** 22 two-pointers and 8 three-pointers $= 30$ baskets
     $$22 \cdot 2 + 8 \cdot 3 = 44 + 24 = 68 \text{ points}$$

     The answer checks.

  5. **State.** The team made 22 two-pointers and 8 three-pointers.

- A caterer wants to mix almonds that sell for \$13.25 per pound and walnuts that sell for \$11.25 per pound to make 23 lb of a mixture that sells for \$12.75 per pound. How much of each should be used?

  1. **Familiarize.** Let $a =$ the number of pounds of almonds and $w =$ the number of pounds of walnuts in the mixture.

  2. **Translate.** A 23 lb batch is being made, so $a + w = 23$.

     $\underbrace{\text{The value of the almonds}}$ plus $\underbrace{\text{the value of the walnuts}}$ is $\underbrace{\text{the value of the mixture.}}$

↓	↓	↓	↓	↓

     $\qquad a \cdot 13.25 \qquad + \qquad w \cdot 11.25 \qquad = \qquad 23 \cdot 12.75$

3. **Solve.** When the first equation is solved for a, we have $a = -w + 23$. We then substitute $-w + 23$ for a in the second equation.

$$13.25a + 11.25w = 293.25$$
$$13.25(-w + 23) + 11.25w = 293.25$$
$$-13.25w + 304.75 + 11.25w = 293.25$$
$$-2w + 304.75 = 293.25$$
$$-2w = -11.5$$
$$w = 5.75$$

If $w = 5.75$, then $a = -5.75 + 23 = 17.25$.

4. **Check.** Amount of mixture: $17.25 \text{ lb} + 5.75 \text{ lb} = 23 \text{ lb}$

Value of mixture: $17.25(\$13.25) + 5.75(\$11.25) = \$293.25$.

The numbers check.

5. **State.** The 23 lb mixture should be made by combining 17.25 lb of almonds and 5.75 lb of walnuts.

GUIDED LEARNING   📖 **Textbook**   👤 **Instructor**   ▶ **Video**

EXAMPLE 1	YOUR TURN 1
Matinee ticket prices at Midtown Cinema are $9.50 for adults and $6.50 for children. A total of 90 people purchased tickets for the first showing of a movie. If a total of $681 was collected, how many adults' tickets and how many children's tickets were sold?	A college field hockey team stopped for ice cream on the way home from a game. They ordered 25 cones, some one-scoop at $1.75 and some two-scoop at $2.25. How many of each type of cone was ordered if the total bill was $52.25?

1. **Familiarize.** Let $a =$ the number of adults' tickets and $c =$ the number of children's tickets.

2. **Translate.** 90 people bought tickets, so $a + c = \boxed{}$.

$$\underbrace{\text{Amount collected for adults}} \quad \text{plus} \quad \underbrace{\text{amount collected for children}} \quad \text{totaled } \$681.$$

$$a \cdot \boxed{} \quad + \quad c \cdot \boxed{} \quad = \quad 681$$

We have a system of equations:

$$a + c = 90, \qquad (1)$$
$$9.50a + 6.50c = 681. \qquad (2)$$

(continued)

3. **Solve.** Solve equation (1) for a:

$$a + c = 90$$

$$a = \boxed{} + 90 \quad (3)$$

Next, substitute $\boxed{}$ for a in equation (2):

$$9.50a + 6.50c = 681$$

$$9.50(-c + 90) + 6.50c = 681$$

$$-9.50c + \boxed{} + 6.50c = 681$$

$$\boxed{}\,c + 855 = 681$$

$$-3c = -174$$

$$c = \boxed{}$$

Substitute 58 for c in equation (1) and solve for a.

$$a + c = 90$$

$$a + 58 = 90$$

$$a = \boxed{}$$

4. **Check.** Number of tickets sold:
32 adults + 58 children = 90 tickets

Cost of tickets: $\$9.50(32) + \$6.50(58) = \$681$

The numbers check.

5. **State.** $\boxed{}$ adults' tickets and $\boxed{}$ children's tickets were sold.

EXAMPLE 2	YOUR TURN 2

EXAMPLE 2

Henry wants kelly green paint (18% green pigment). The paint store owner can mix lime green (14.5% pigment) and hunter green (20% pigment) to create kelly green. How much of each color should be mixed to create a 22 gal batch of kelly green?

1. **Familiarize.** Let l = the number of gallons of lime green paint and h = the number of gallons of hunter green paint.

2. **Translate.** Henry needs 22 gal of paint, so

$$l + h = \boxed{} \quad \leftarrow \text{Total amount of paint}$$

To determine the total amount of pigment, we have:

$$0.145l + 0.2h = 0.18 \cdot 22$$

The system of equations is

$$l + h = 22, \quad (1)$$
$$0.145l + 0.2h = 3.96. \quad (2)$$

3. **Solve.** When equation (1) is solved for l, we have $l = \boxed{} + 22$. We then substitute $-h + 22$ for l in equation (2).

$$0.145l + 0.2h = 3.96$$
$$0.145(-h + 22) + 0.2h = 3.96$$
$$-0.145h + \boxed{} + 0.2h = 3.96$$
$$\boxed{} h + 3.19 = 3.96$$
$$0.055h = 0.77$$
$$h = \boxed{}$$

If $h = 14$, then $l = -14 + 22 = 8$.

4. **Check.**

Amount of paint: 8 gal + 14 gal = 22 gal

Amount of pigment: $0.145 \cdot 8 + 0.2 \cdot 14 = 1.16 + 2.8 = 3.96$ gal

The answer checks.

5. **State.** Henry needs $\boxed{}$ gal of lime green and $\boxed{}$ gal of hunter green to make 22 gal of kelly green paint.

YOUR TURN 2

A chemistry professor has a solution that is 60% base and another solution that is 20% base. How much of each should be used to make 120 L of solution that is 52% base?

YOUR NOTES   Write your questions and additional notes.

Motion Problems

ESSENTIALS

Distance, Rate, and Time Equation

If r represents rate, t represents time, and d represents distance, then

$$d = rt.$$

Example

- A small plane flies 2 hr west with a 50-mph tailwind. Returning against the wind takes 3 hr. Find the speed of the plane with no wind.

  1. **Familiarize.** The plane travels the same distance each way. Let d = the distance traveled, in miles. Let r = the speed, in mph, of the plane in still air. Then $r + 50$ = the plane's speed with the wind and $r - 50$ = the plane's speed against the wind.

  2. **Translate.** Construct a table.

	Distance	Rate	Time
With wind	d	$r + 50$	2
Against wind	d	$r - 50$	3

     Using $d = rt$, we have a system of equations.

     $$d = (r + 50)2, \quad (1)$$
     $$d = (r - 50)3 \quad (2)$$

  3. **Solve.** Solve using substitution:

     $$(r + 50)2 = (r - 50)3 \qquad \text{Substituting } (r + 50)2 \text{ for } d \text{ in equation (2)}$$
     $$2r + 100 = 3r - 150$$
     $$-r + 100 = -150$$
     $$-r = -250$$
     $$r = 250.$$

  4. **Check.** When $r = 250$, the speed with the wind is $250 + 50 = 300$ mph and the speed against the wind is $250 - 50 = 200$ mph. The distance with the wind is $300 \cdot 2 = 600$ mi and the distance against the wind is $200 \cdot 3 = 600$ mi. The distances are the same, so the answer checks.

  5. **State.** The speed of the plane with no wind is 250 mph.

| EXAMPLE 1 | YOUR TURN 1 |

EXAMPLE 1

A passenger train leaves Hartford, heading to Atlanta, at a speed of 50 mph. Three hours later a freight train leaves Hartford on a parallel track at 70 mph. How far from Hartford will the freight train catch up to the passenger train?

1. **Familiarize.** The distance traveled when the trains meet is the same. Let d = this distance. Let t = the number of hours the passenger train is running before they meet. Then $t - 3$ = the number of hours the freight train runs before catching up to the passenger train.

2. **Translate.** Construct a table.

	Distance	Rate	Time
Passenger train	d	50	t
Freight train	d	70	$t-3$

Using $d = rt$, we have a system of equations.

$$d = 50t \qquad (1)$$
$$d = 70(t-3) \quad (2)$$

3. **Solve.** Solve using substitution:

$$50t = 70(t-3)$$
$$50t = 70t - \boxed{}$$
$$\boxed{} = -210$$
$$t = \boxed{}$$

Time for the passenger train: 10.5 hr

Time for the freight train: $10.5 - \boxed{} = 7.5$ hr

4. **Check.** Distance passenger train travels at 50 mph: $50(10.5) = 525$ mi. Distance freight train travels at 70 mph: $70(7.5) = 525$ mi. The numbers check.

5. **State.** The freight train catches up to the passenger train $\boxed{}$ mi from Hartford.

YOUR TURN 1

Mary and Susan are training for a triathlon. Mary begins a bike ride at Marker 0 on a bike trail and rides at 6 mph. Two hours later Susan starts at Marker 0 and bikes at 10 mph. How far from Marker 0 will Susan catch up to Mary?

YOUR NOTES   Write your questions and additional notes.

Practice Exercises

Readiness Check

Match each statement with the most appropriate translation.

1. Alice is mixing a 2 lb batch of two kinds of nuts.

2. Tom spent $40 on ice cream cones costing $2 each and milkshakes costing $3 each.

3. A mixture that is 20% acid and another that is 30% acid contains 40 L of acid.

4. Jay spent $20 on candy bars costing $1.25 each and granola bars costing $1.75 each.

a) $2x + 3y = 40$

b) $x + y = 2$

c) $1.25x + 1.75y = 20$

d) $0.2x + 0.3y = 40$

Total-Value Problems and Mixture Problems

Solve. Use the five steps for problem solving.

5. A single day lift ticket at a ski mountain is $75 for adults and $35 for children. A busload of 27 people paid $1305 for lift tickets one day. How many adults and how many children were on the bus?

6. While driving to New York City, Adam stopped and bought gas for $3.89 per gallon and then stopped again and paid $3.49 per gallon. If he purchased a total of 25 gal for $92.85, how many gallons did he buy at each price?

7. A chemist needs to prepare 15 L of a 30% acid solution. How many liters of 20% acid solution and how many liters of 35% acid solution should be mixed to obtain the 30% acid solution?

8. John plans to mix fruit punch costing $1.99 per liter with sparkling water costing $2.19 per liter to make 35 L of punch for a Halloween party. If the punch costs $2.05 per liter, how many liters of each will he use?

Motion Problems

9. Jason drives his boat for 3 hr with a 5-km/h current to reach a campsite. The return trip against the same current took 6 hr. Find the speed of the boat in still water.

10. A small plane leaves Denver flying east at 200 mph. Two hours later, a large jet leaves Denver flying east at 520 mph. How far will the planes be from Denver when the large jet catches up with the small plane?

Solving Systems in Three Variables

ESSENTIALS

Example

- Solve the system of equations:

$$3x - 2y + 6z = 1, \quad (1)$$
$$7x + 3y + z = 4, \quad (2)$$
$$x + y - z = 0. \quad (3)$$

First add equations (2) and (3).

$$
\begin{array}{ll}
7x + 3y + z = 4 & (2) \\
\underline{x + y - z = 0} & (3) \\
8x + 4y \phantom{{}+ z} = 4 & (4)
\end{array}
$$

Select a different pair of equations and eliminate z again. Use equations (1) and (3).

$$
\begin{array}{ll}
3x - 2y + 6z = 1 & (1) \\
x + y - z = 0 & (3) \quad \text{Multiplying both sides by } 6 \rightarrow
\end{array}
\qquad
\begin{array}{l}
3x - 2y + 6z = 1 \\
\underline{6x + 6y - 6z = 0} \\
9x + 4y \phantom{{}+ z} = 1 \quad (5)
\end{array}
$$

Solve the resulting system of equations (4) and (5). Eliminate y first.

$$
\begin{array}{ll}
8x + 4y = 4 & (4) \quad \text{Multiplying both sides by } -1 \rightarrow \\
9x + 4y = 1 & (5)
\end{array}
\qquad
\begin{array}{l}
-8x - 4y = -4 \\
\underline{9x + 4y = 1} \\
x \phantom{{}+ 4y} = -3
\end{array}
$$

We will use equation (4) to find y.

$$
\begin{array}{ll}
8x + 4y = 4 & (4) \\
8(-3) + 4y = 4 & \text{Substituting } -3 \text{ for } x \\
-24 + 4y = 4 & \\
4y = 28 & \\
y = 7 &
\end{array}
$$

Use any of the three original equations to find z. We will use equation (3).

$$
\begin{array}{ll}
x + y - z = 0 & (3) \\
-3 + 7 - z = 0 & \text{Substitute } -3 \text{ for } x \text{ and } 7 \text{ for } y \\
4 - z = 0 & \\
4 = z &
\end{array}
$$

The triple $(-3, 7, 4)$ checks in all three equations. It is the solution.

| | 📖 **Textbook** | 👤 **Instructor** | ▶ **Video** |

GUIDED LEARNING

| EXAMPLE 1 | YOUR TURN 1 |

EXAMPLE 1

Solve the system of equations:

$$5a + 3b \quad\;\; = 4, \quad (1)$$
$$3b - 4c = 4, \quad (2)$$
$$2a \quad\;\; + 2c = 2. \quad (3)$$

Since c is eliminated already from equation 1, eliminate c from equations (2) and (3).

$$3b - 4c = 4 \qquad\qquad\qquad 3b - 4c = 4$$
$$2a \quad + 2c = 2 \quad \text{Multiplying both sides by } 2 \rightarrow 4a \quad + 4c = 4$$
$$\overline{\qquad\qquad\qquad\qquad 4a + 3b \quad\;\; = \boxed{} \;\; (4)}$$

Solve the resulting system of equations (1) and (4).

$$5a + 3b = 4 \qquad (1) \quad \text{Multiplying both sides by } -1 \rightarrow \boxed{} - 3b = -4$$
$$4a + 3b = \boxed{} \quad (4) \qquad\qquad\qquad\qquad\qquad 4a + 3b = 8$$
$$\overline{\qquad\qquad\qquad\qquad\qquad\qquad\qquad -a \quad\;\; = 4}$$
$$a \quad\;\; = -4$$

Use equation (4) to solve for b.

$$4a + 3b = 8 \qquad (4)$$
$$4(-4) + 3b = 8 \qquad \text{Substituting } -4 \text{ for } a$$
$$\boxed{} + 3b = 8$$
$$3b = 24$$
$$b = \boxed{}$$

Use equation (2) or (3) to solve for c.

$$2a + 2c = 2 \qquad (3)$$
$$2(-4) + 2c = 2 \qquad \text{Substituting } -4 \text{ for } a$$
$$\boxed{} + 2c = 2$$
$$2c = 10$$
$$c = \boxed{}$$

The solution is $\boxed{}$.

YOUR TURN 1

Solve the system of equations:

$$3a \quad\;\; + 4c = -2,$$
$$b - 2c = 2,$$
$$a + 2b - \;\; c = 5.$$

YOUR NOTES Write your questions and additional notes.

Practice Exercises

Readiness Check

Each sentence below refers to the following system of equations. Choose the word that best completes each sentence.

$$5x + 3y - z = -1,$$
$$2x + y + 3z = 0,$$
$$x - 2y - 5z = 5$$

1. The system shown is a system of _____________ equations in _____________ variables.

   two / three   two / three

2. Each equation in the system is a _____________ equation in three variables.

   linear / nonlinear

3. The ordered _____________ $(1, -2, 0)$ is a solution of the system because it makes

   pair / triple

   _____________________________________ true.

   all three equations / at least one equation

Solving Systems In Three Variables

Solve.

4. $$x + 2y + 2z = 6,$$
   $$2x - y + 2z = 11,$$
   $$4x + 5y + z = 6$$

5. $$x + 3y - 2z = 0,$$
   $$-x - y + 3z = -1,$$
   $$2x + 5y + z = -5$$

6.
$$2a - b + c = 0,$$
$$4a + b + c = 7,$$
$$-2a + 2b - c = 3$$

7.
$$2w - x - y = 13,$$
$$w + x + 3y = 5,$$
$$w + 4x + 9y = 11$$

8.
$$x - 6y \quad\;\; = 1,$$
$$3x \quad\;\; + 2z = 11,$$
$$y - 7z = 4$$

9.
$$p + r = 7,$$
$$p - 2s = 12,$$
$$2r + 7s = -1$$

Using Systems of Three Equations

ESSENTIALS

Example

- Owen invested $15,000 in three mutual funds. His total earnings in interest for the year were $800. He invested $1000 more in Fund B than in Fund A, and he invested the rest in Fund C. The interest rates for Fund A, Fund B, and Fund C are 3.4%, 5%, and 6% respectively.

  1. **Familiarize.** Let $x =$ the amount invested in Fund A.

     Let $y =$ the amount invested in Fund B.

     Let $z =$ the amount invested in Fund C.

  2. **Translate.** Owen has $15,000 to invest, so

     $$x + y + z = 15,000.$$

     The interest earned is $800. The interest equals the rate times the amount invested, so

     $$0.034x + 0.05y + 0.06z = 800.$$

     There is $1000 more in Fund B than in Fund A, so

     $$y = x + 1000.$$

     We have a system of three equations:

     $$x + y + z = 15,000,$$
     $$0.034x + 0.05y + 0.06z = 800,$$
     $$y = x + 1000.$$

     Rewriting in standard form and clearing decimals, we have:

     $$\begin{aligned} x + y + z &= 15,000, &(1)\\ 34x + 50y + 60z &= 800,000, &(2)\\ -x + y \phantom{{}+ 60z} &= 1000. &(3) \end{aligned}$$

  3. **Solve.** Equation (3) does not have a z-term. We use equations (1) and (2) to write another equation with no z-term. We multiply equation (1) by -60 and then add the equations.

     $$\begin{aligned} -60x - 60y - 60z &= -900,000 \\ 34x + 50y + 60z &= 800,000 \\ \hline -26x - 10y \phantom{{}+ 60z} &= -100,000 \quad (4) \end{aligned}$$

Solve the system of equations (3) and (4). Multiply equation (3) by 10 and then add.

$$-10x + 10y = 10{,}000$$
$$-26x - 10y = -100{,}000$$
$$\overline{-36x = -90{,}000}$$
$$x = 2500$$

Use equation (3) to find y.

$$-x + y = 1000$$
$$-2500 + y = 1000$$
$$y = 3500$$

Use equation (1) to find z.

$$x + y + z = 15{,}000$$
$$2500 + 3500 + z = 15{,}000$$
$$6000 + z = 15{,}000$$
$$z = 9000$$

4. **Check.** Amount of money invested: $\$2500 + \$3500 + \$9000 = \$15{,}000$

   The interest earned: $0.034(\$2500) + 0.05(\$3500) + 0.06(\$9000)$
   $$= \$85 + \$175 + \$540 = \$800$$

   The amount in Fund B, $3500, is $1000 more than the amount in Fund A, $2500.

   The answer checks.

5. **State.** Owen invested $2500 in Fund A, $3500 in Fund B, and $9000 in Fund C.

GUIDED LEARNING	Textbook	Instructor	Video

EXAMPLE 1	YOUR TURN 1

The perimeter of a triangle is 63 in. The longest side is 22 in. longer than the shortest side. The longest side is twice as long as the remaining side. Find the lengths of all three sides.

1. **Familiarize.** Let $x =$ the shortest side, $y =$ the middle side, and $z =$ the longest side.

2. **Translate.** The perimeter is 63 in., so we have $x + y + z = 63$.

The longest side is 22 in. longer than the shortest side, so

$$z = x + 22.$$

The longest side is twice as long as the remaining side, so

$$z = 2y.$$

We have a system of three equations:

$$x + y + z = 63,$$
$$z = x + 22,$$
$$z = 2y.$$

Rewriting in standard form, we have:

$$x + y + z = 63, \quad (1)$$
$$-x + z = 22, \quad (2)$$
$$-2y + z = 0. \quad (3)$$

3. **Solve.** Eliminate x by adding equations (1) and (2).

$$x + y + z = 63$$
$$\underline{-x + z = 22}$$
$$y + \boxed{} = 85 \quad (4)$$

Solve the resulting system of equations (3) and (4).

$$-2y + z = 0$$
$$y + 2z = 85 \quad \text{Multiplying both sides by 2} \rightarrow$$

$$-2y + z = 0$$
$$\underline{2y + 4z = \boxed{}}$$
$$5z = 170$$
$$z = \boxed{}$$

(continued)

YOUR TURN 1: A museum charges $14 admission for adults, $10 for children under 12, and $8 for seniors over 65. During one day, the museum sold 444 tickets and took in $5532. If twice as many adult tickets were sold as the total of children and senior tickets, how many of each type of ticket was sold?

Use equation (4) to find y.

$$y + 2z = 85$$

$y + 2(34) = 85$ Substituting 34 for z

$y + \boxed{} = 85$

$$y = 17$$

Use equation (1) to find x.

$$x + y + z = 63$$

$x + 17 + 34 = 63$ Substituting 17 for y and 34 for z

$x + \boxed{} = 63$

$x = \boxed{}$

4. **Check.** Perimeter: $12 \text{ in.} + 17 \text{ in.} + 34 \text{ in.} = 63 \text{ in.}$

   The longest side, 34 in., is 22 in. longer than the shortest side, 12 in. The longest side is also twice as long as the remaining side: $2 \cdot 17 \text{ in.} = 34 \text{ in.}$

   The answer checks.

5. **State.** The lengths of the sides are $\boxed{}$, $\boxed{}$, and $\boxed{}$.

YOUR NOTES   Write your questions and additional notes.

Practice Exercises

Readiness Check

Match each statement with a translation from the list.

1. The sum of the three numbers is 25.

2. The first number is 25 more than the sum of the other two numbers.

3. The first number minus the second number plus the third is 25.

4. The first number is 25 less than the sum of the other two numbers.

a) $x - y - z = -25$

b) $x + y + z = 25$

c) $x - y + z = 25$

d) $x - y - z = 25$

Using Systems of Three Equations

Solve.

5. The sum of three numbers is 144. The second number is twice the first number. The third number is three times the first. Find the numbers.

6. A college basketball team scored 84 points in a game. The team scored a combination of 2-point field goals, 3-point field goals, and 1-point foul shots. The number of 2-point field goals was 6 more than the number of foul shots and four times the number of 3-point field goals. How many of each type of shot was made?

7. Mary bought a total of 10 gift cards in denominations of $20, $50, and $100. She spent $370. She bought twice as many $20 gift cards as $50 gift cards. How many of each denomination of gift card did she buy?

8. The sum of three numbers is 40. The second number is four times the first number. The sum of the first number and third number is 16. Find the numbers.

Graphs of Linear Inequalities

ESSENTIALS

A **linear inequality in two variables** is an inequality that can be written in the form

$$Ax + By < C,$$

where A, B, and C are real numbers and A and B are not both zero. The symbol $<$ may be replaced with $\leq$, $>$, or $\geq$.

The **solution set** of a linear inequality is the set of all ordered pairs that make it true. A **graph** of an inequality represents its solutions.

The graph of a **linear inequality in two variables** is a region on one side of a line and may include the boundary line. This region is a called a **half-plane**.

To graph an inequality in two variables:

1. Replace the inequality symbol with an equals sign and graph this related equation. If the inequality symbol is $<$ or $>$, draw the line dashed. If the inequality symbol is $\leq$ or $\geq$, draw the line solid.

2. The graph consists of a half-plane on one side of the line and, if the line is solid, the line as well. To determine which half-plane to shade, test a point not on the line in the original inequality. If that point is a solution, shade the half-plane containing that point. If not, shade the opposite half-plane.

Example

- The solutions of the linear inequality $y > 2x + 3$ are given by all of the ordered pairs in the shaded region of the following graph. Note that any points on the dashed line itself are not solutions.

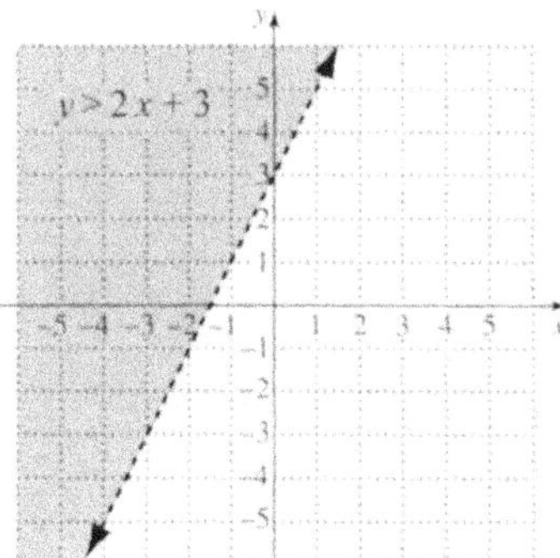

 **Textbook**    **Instructor**   ▶ **Video**

GUIDED LEARNING

EXAMPLE 1	YOUR TURN 1

EXAMPLE 1

Graph $y \geq \dfrac{1}{2}x - 1$.

Since the inequality symbol is "greater than or equal to," graph the related equation

$y = \dfrac{1}{2}x - 1$ using a ______________ line.

solid / dashed

Next, we test a point in either region. We choose $(0, 0)$.

$$y \geq \frac{1}{2}x - 1$$

$$0 \ ? \ \frac{1}{2} \cdot \boxed{} - 1$$

$$\boxed{} - 1$$

$$\boxed{}$$

Since $0 \geq \boxed{}$ is true, the point $(0, 0)$

______________ a solution and we shade the half-

is / is not

plane that ______________ contain $(0, 0)$.

does / does not

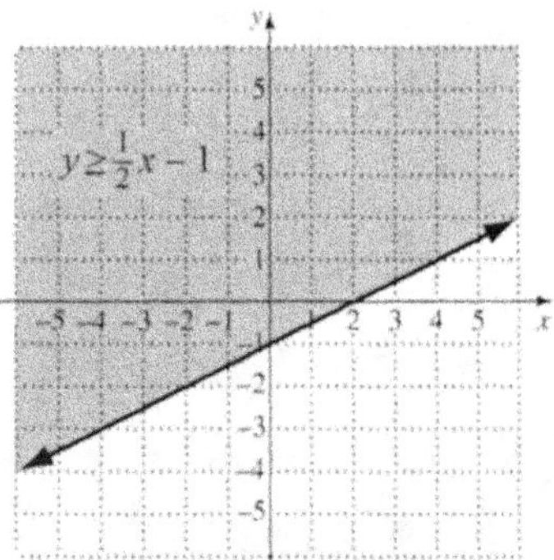

YOUR TURN 1

Graph $y \leq \dfrac{2}{3}x + 1$.

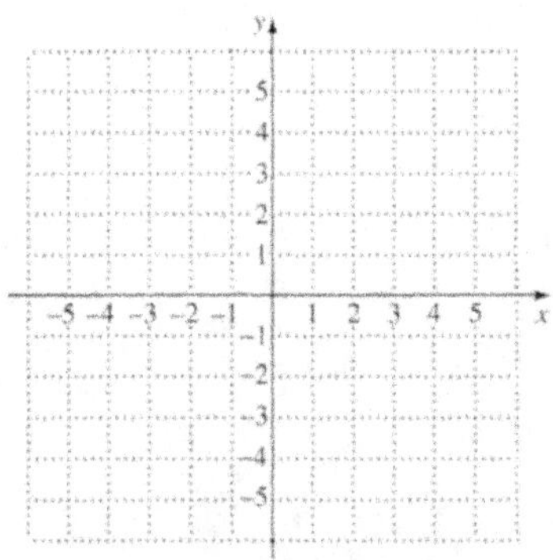

EXAMPLE 2	YOUR TURN 2
Graph $3x - y > 2$.	Graph $3x - 2y < 6$.

EXAMPLE 2

Graph $3x - y > 2$.

Since the inequality symbol is "greater than," graph the related equation $3x - y = 2$ using a __________ line.
solid / dashed

Next, we test the point $(0, 0)$.

$$3x - y > 2$$

$$3 \cdot \boxed{} - 0 \ ? \ 2$$

$$\boxed{} - 0$$

$$\boxed{}$$

Since $\boxed{} > 2$ is __________, the point
true / false

$(0, 0)$ is not a solution and we shade the

half-plane that __________ contain
does / does not

$(0, 0)$.

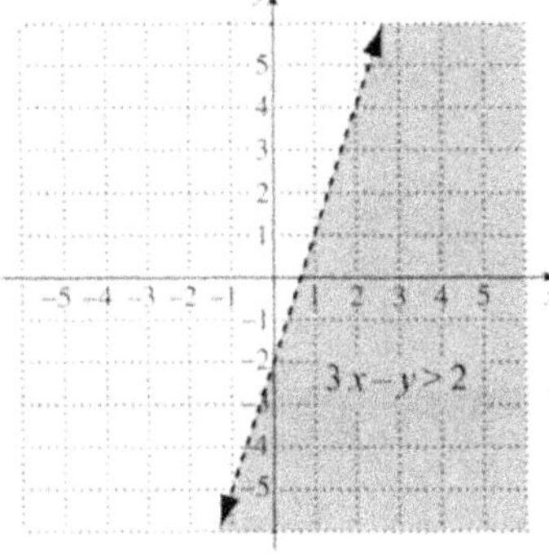

YOUR TURN 2

Graph $3x - 2y < 6$.

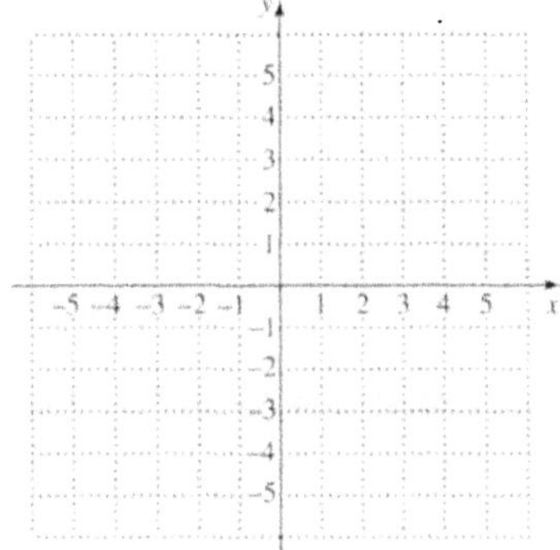

EXAMPLE 3	YOUR TURN 3
Graph $y < -2$ on a plane.	Graph $x \geq 2$ on a plane.

Since the inequality symbol is "less than," graph the related equation $y = -2$ using a __________ line.
solid / dashed

Next, we test the point $(0, 0)$.

$$\frac{y < -2}{\boxed{} \ ? \ -2}$$

Since $\boxed{} < -2$ is __________, the point
true / false

$(0, 0)$ __________ a solution so we shade the
is / is not

half-plane that __________ contain
does / does not

$(0, 0)$.

YOUR NOTES   Write your questions and additional notes.

Systems of Linear Inequalities

ESSENTIALS

A **system of linear inequalities** consists of two or more linear inequalities.

A **solution** of a system of linear inequalities is any ordered pair that is a solution of *each* of the individual inequalities.

In order to **graph** the solution set of a system of linear inequalities we graph each inequality and determine the region that is common to all the solution sets. We draw small arrows on the ends of each line indicating which half-plane contains the solution set.

A system of linear inequalities may have a graph that consists of a polygon and its interior. In many applications it is important to find the vertices of the polygon. This requires solving systems of equations.

Example

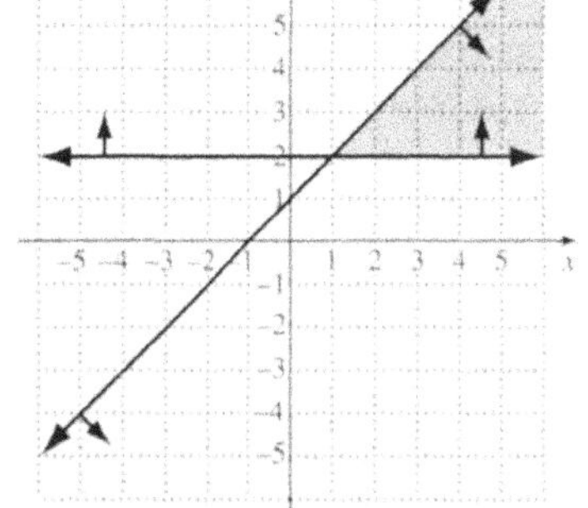

- The solutions of the system

$$y \le x + 1,$$

$$y \ge 2$$

are represented by the shaded region to the right. The vertex is the point of intersection of the two lines. Its coordinates are $(1, 2)$.

 Textbook	 Instructor	 Video

GUIDED LEARNING

EXAMPLE 1	YOUR TURN 1
Graph the solution set of the system. Find the coordinates of any vertices formed. $x + 3y \le 9,$ $2x - 3y \le 0,$ $x \ge -3$ Graph $x + 3y = 9$ using a solid line. Test point $(0,0)$ is a solution of $x + 3y \le 9$, so indicate (with arrows near the ends of the line) the half-plane that ________ does / does not contain $(0, 0)$.	Graph the system. Find the coordinates of any vertices formed. $x - 3y \ge 0,$ $x - y \le 2,$ $y \ge -1$ 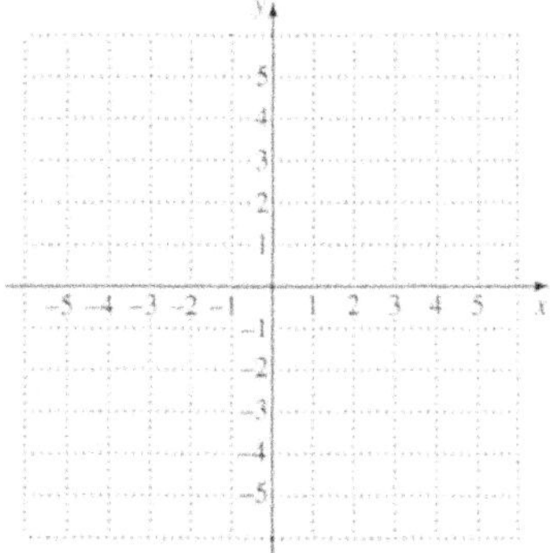
	(continued)

Graph $2x-3y=0$ using a solid line. Test point $(1,2)$ ________ a solution of
<u>is / is not</u>
$2x-3y\le0$, so indicate (with arrows near the ends of the line) the half-plane that contains $(1,2)$.

Graph $x=-3$ using a solid line. Test point $(0,0)$ ________ a solution of $x\ge-3$, so
<u>is / is not</u>
indicate (with arrows near the ends of the line) the half-plane that contains $(0,0)$.

The region of solutions is the triangular region, including segments of the boundary lines, where the half-planes indicated by the arrows near the ends of the lines overlap. Shade this region.

To find the vertices, solve the following systems of equations formed by these lines:

$$x+3y=9, \quad x+3y=9, \quad 2x-3y=0,$$
$$2x-3y=0; \quad\quad x=-3; \quad\quad x=-3.$$

The solutions are
$(3,2)$, $(-3,4)$, and $(-3,-2)$.

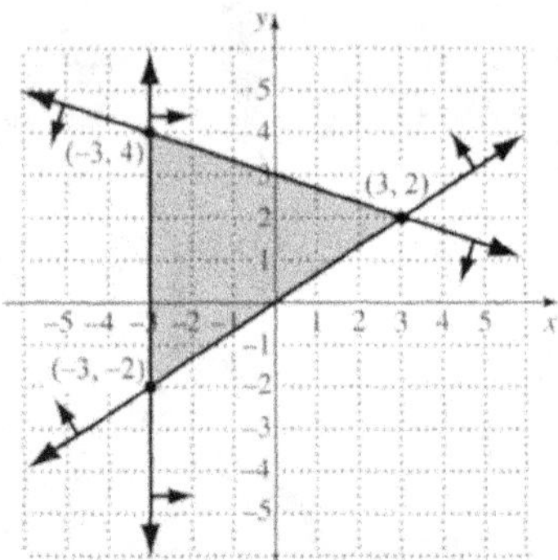

YOUR NOTES   Write your questions and additional notes.

Applications: Linear Programming

ESSENTIALS

Linear programming allows us to find the maximum or minimum value of a linear **objective function** subject to several **constraints** which are expressed as inequalities. The solution set of the system of inequalities composed of the constraints contains all the **feasible solutions**.

It can be shown that the maximum and minimum values of the objective function occur at a vertex of the region of feasible solutions.

Linear Programming Procedure

To find the maximum or minimum value of a linear objective function, subject to constraints:

1. Graph the region of feasible solutions.

2. Determine the coordinates of the vertices of the region.

3. Evaluate the objective function at each vertex. The largest and smallest of those values are the maximum and minimum values of the objective function, respectively.

GUIDED LEARNING   **Textbook**   **Instructor**   **Video**

EXAMPLE 1	YOUR TURN 1
Aspen Carpentry makes bookcases and desks. Each bookcase requires 5 hr of woodworking and 4 hr of finishing. Each desk requires 10 hr of woodworking and 3 hr of finishing. Each month the shop has 600 hr of labor available for woodworking and 240 hr for finishing. The profit on each bookcase is \$75 and on each desk is \$140. How many of each product should be made each month in order to maximize profit? What is the maximum profit?	Fashions by Felicity makes tuxedos and prom dresses and can produce at most 200 of these garments per month. It takes 30 hr to make a tuxedo and 10 hr to make a dress. There is at most 3000 hr of labor available each month. The profit on a tuxedo is \$80 and is \$50 on a dress. How many of each should be made each month in order to maximize the profit? What is the maximum profit?

Let x = the number of bookcases to be produced and sold and let y = the number of desks to be produced and sold.

The profit is given by $P = \boxed{} \cdot x + \boxed{} \cdot y.$

Next we consider the constraints. The total number of hours of woodworking on bookcases and desks must satisfy $5x + 10y \le \boxed{}.$

The total number of hours of finishing on bookcases and desks must satisfy
$4x + \boxed{} y \le 240.$

(continued)

Finally, we have $x \geq 0$ and $y \geq 0$.

Maximize

$$P = 75x + \boxed{}\,y$$

subject to

$$5x + \boxed{}\,y \leq 600,$$
$$4x + 3y \leq \boxed{},$$
$$x \geq 0,$$
$$y \geq 0.$$

We graph the system of inequalities and find the vertices.

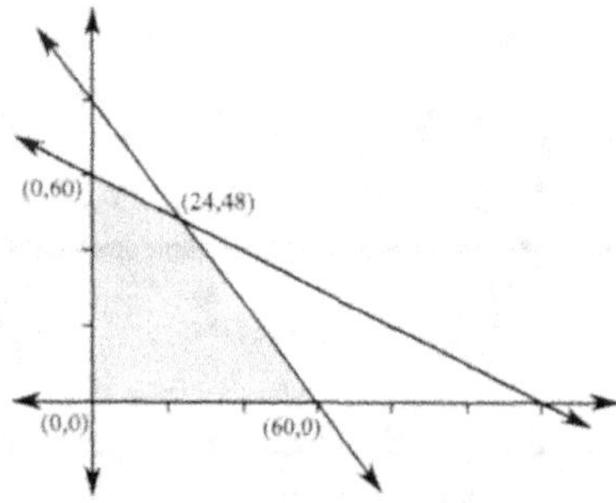

The maximum value will occur at a vertex, so we check the value of P at each one.

Vertex (x, y)	Profit $P = \boxed{}\,x + \boxed{}\,y$
$(0, 0)$	$75 \cdot \boxed{} + 140 \cdot \boxed{} = \boxed{}$
$(60, 0)$	$75 \cdot \boxed{} + 140 \cdot 0 = \boxed{}$
$(24, 48)$	$75 \cdot 24 + 140 \cdot \boxed{} = 8520$
$(0, 60)$	$75 \cdot \boxed{} + 140 \cdot \boxed{} = 8400$

We see that the maximum profit of $\boxed{}$ occurs when 24 bookcases and $\boxed{}$ desks are produced and sold.

YOUR NOTES　　Write your questions and additional notes.

Practice Exercises

Readiness Check

Complete the following sentences.

1. If an ordered pair (x, y) is substituted in an inequality and results in a false statement, then that ordered pair __________ a solution of the inequality.
   is / is not

2. When graphing a linear inequality that contains a $>$ or $<$ symbol, we first graph the related equation using a __________ line.
   solid / dashed

3. When graphing a linear inequality, if the test point (x, y) is a solution, then we shade the half-plane that is on the __________ side of the line as (x, y).
   same / opposite

4. A linear inequality in two variables is graphed on __________.
   a plane / the number line

5. When graphing a system of linear inequalities, we shade the __________ of
   union / intersection
   solution sets of the individual inequalities.

Graphs of Linear Inequalities

Graph each linear inequality on a plane.

6. $y > -\dfrac{1}{3}x + 2$

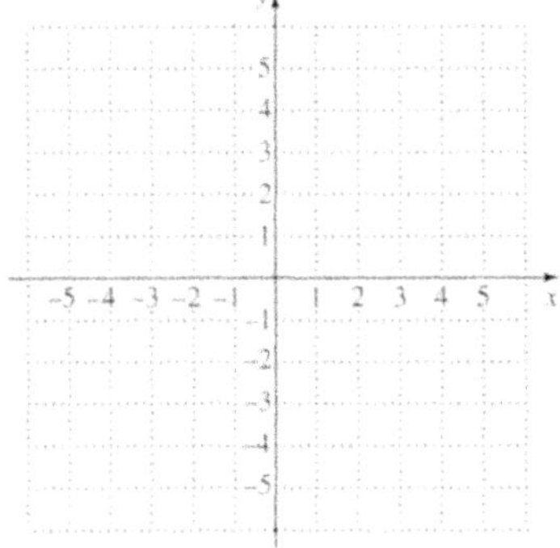

7. $-3x + 2y \le 4$

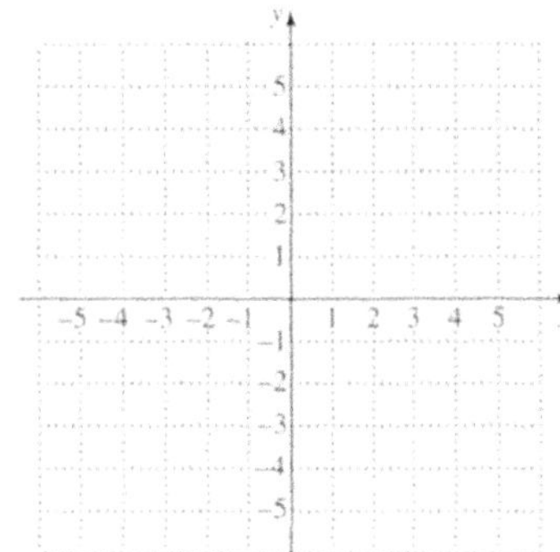

Systems of Linear Inequalities

Graph each system. Find the coordinates of any vertices formed.

8. $y \geq -x - 2,$

  $y \leq 1$

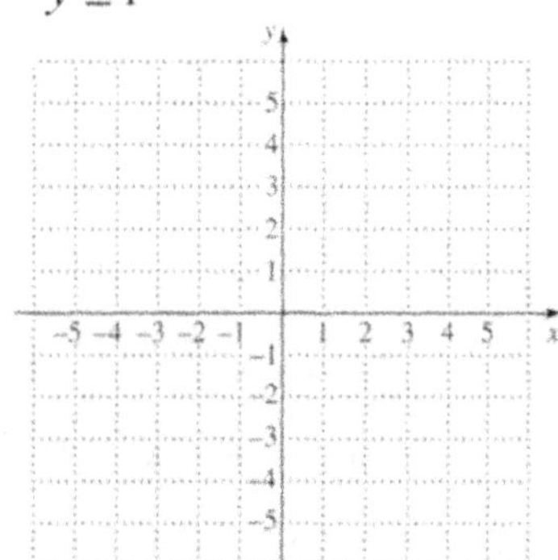

9. $x + y \geq 3,$

  $x - y \leq 1$

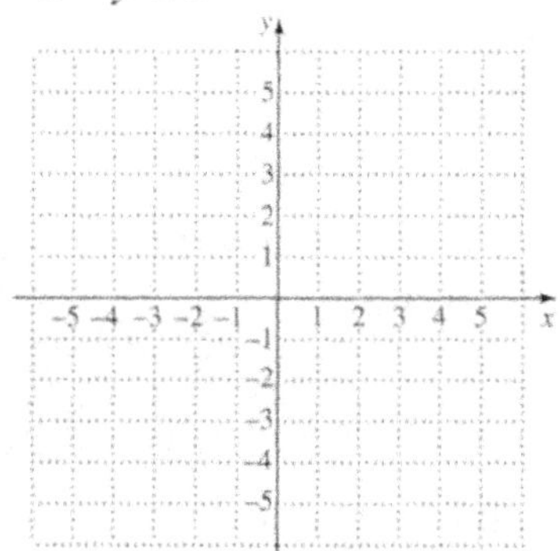

10. $x + y \leq 1,$

  $x + 3y \geq -3,$

  $-x + y \leq 3$

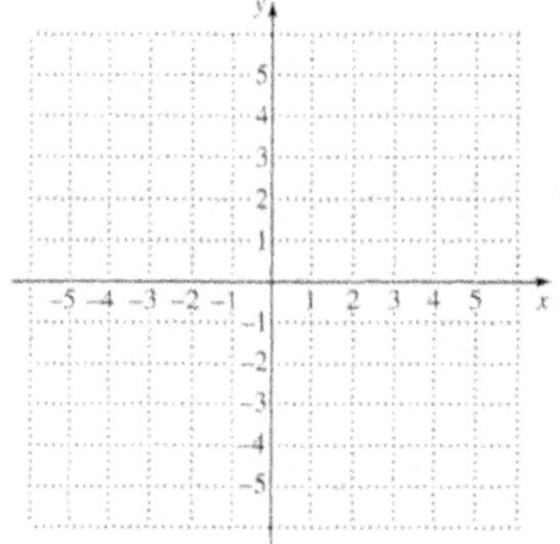

Applications: Linear Programming

11. Corey builds outdoor playhouses. He uses 10 sheets of plywood and 15 two-by-fours for a princess playhouse and 15 sheets of plywood and 45 two-by-fours for each pirate playhouse. He has 60 sheets of plywood and 135 two-by-fours available. The profit on each princess playhouse is $200 and $350 on each pirate playhouse. How many of each should he build in order to maximize the profit? What is the maximum profit?

Copyright © 2022 Pearson Education, Inc.

Polynomial Expressions

ESSENTIALS

A **monomial** is a constant or a constant times some variable or variables raised to powers that are nonnegative integers.

A **polynomial** is a monomial or a combination of sums and/or differences of monomials.

Examples of **polynomials in one variable**: $4a^3$, $2y+1$, x^2+5x-6

Examples of **polynomials in several variables**: $2x^3y^2$, $2x+3y$, $5x+6y^2-4z^3$

The **terms** of a polynomial are separated by $+$ signs.

Names for certain types of polynomials:

Type	Definition: Polynomial of	Examples
Monomial	One term	3, $2x$, $-4y^2$, $7a^4b^8$, $-2x^2y^5z^3$
Binomial	Two terms	$3x-5$, a^2-b^2, $6x^2-7x$, x^2y+4xy^3
Trinomial	Three terms	x^2+6x+9, $2a^2+7ab+3b^2$, $x^2y+3xy-4xy^2$

The **coefficient** is the part of a term that is a constant factor. A **constant term** is a term that contains only a number and no variable.

The **degree of a term** is the sum of the exponents of the variables, if there are variables. The degree of a nonzero constant term is 0. The term 0 has no degree.

The **degree of a polynomial** is the same as the degree of its term of highest degree.

The **leading term** of a polynomial is the term of highest degree. Its coefficient is called the **leading coefficient**.

Descending order: arrangement of terms so that exponents decrease from left to right.

Ascending order: arrangement of terms so that exponents increase from left to right.

Example
- Identify the terms, the degree of each term, and the degree of the polynomial $-6x+4x^3+5-3x^2$. Then identify the leading term, the leading coefficient, and the constant term. Finally, write the polynomial in both descending and ascending order.

Term	$-6x$	$4x^3$	5	$-3x^2$
Degree of term	1	3	0	2
Degree of polynomial		3		
Leading term		$4x^3$		
Leading coefficient		4		
Constant term		5		

Descending order:

$$4x^3-3x^2-6x+5$$

Ascending order:

$$5-6x-3x^2+4x^3$$

GUIDED LEARNING   🔖 **Textbook**   👤 **Instructor**   ▶ **Video**

EXAMPLE 1

Identify the terms, the degree of each term, and the degree of the polynomial
$-6y^5 - 2y - 4y^3 + 12y^9 + 10$. Then identify the leading term, the leading coefficient, and the constant term.

Term	$-6y^5$	$-2y$	$-4y^3$	$12y^9$	10
Degree of term	☐	☐	3	☐	0
Degree of polynomial	☐				
Leading term	$12y^9$				
Leading coefficient	☐				
Constant term	☐				

YOUR TURN 1

Identify the terms, the degree of each term, and the degree of the polynomial
$12y^4 - 7y^7 - 6y^3 - 8$. Then identify the leading term, the leading coefficient, and the constant term.

Term					
Degree of term					
Degree of polynomial					
Leading term					
Leading coefficient					
Constant term					

EXAMPLE 2

Consider $-6y^5 + 12y^9 - 4y^3 - 2y + 10$. Arrange in descending order and then in ascending order.

Descending order: ☐$-6y^5-$☐$-2y+$☐

Ascending order: $10-$☐$-4y^3-$☐$+$☐

YOUR TURN 2

Consider $12y^4 - 7y^7 - 6y^3 - 8$.

Arrange in descending order and then in ascending order.

Descending order:

Ascending order:

YOUR NOTES   Write your questions and additional notes.

Evaluating Polynomial Functions

ESSENTIALS

To **evaluate** a polynomial function, substitute a number for the variable.

Example

- For the polynomial function $P(x) = -2x^4 + 3x - 1$, find $P(4)$.

$$P(x) = -2x^4 + 3x - 1$$

$$P(4) = -2(4)^4 + 3(4) - 1$$

$$= -2(256) + 12 - 1$$

$$= -512 + 12 - 1$$

$$= -501$$

GUIDED LEARNING   📖 **Textbook**   👤 **Instructor**   ▶ **Video**

EXAMPLE 1	YOUR TURN 1
For the polynomial function $f(x) = 3x^2 - 5x + 4$, find $f(-1)$.	For the polynomial function $f(x) = -6x^3 - 4x^2 + 2x - 10$, find $f(2)$.

$$f(x) = 3x^2 - 5x + 4$$

$$f(-1) = 3(-1)^2 - 5(-1) + 4$$

$$= 3(1) - 5(-1) + 4$$

$$= \boxed{} + 5 + 4 = \boxed{}$$

EXAMPLE 2	YOUR TURN 2
When an object is launched upward with an initial velocity of 25 m/sec and from the top of a bridge that is 30 m high, its altitude, in meters, after t seconds is given by the polynomial function $A(t) = 30 + 25t - 4.9t^2$. Find the altitude of the object after 3 seconds.	The polynomial function $$F(x) = -0.00016x^3 + 0.0096x^2 + 0.367x + 10.8$$ can be used to estimate the fuel economy, in miles per gallon (mpg), for a particular vehicle traveling x miles per hour (mph). What is the gas mileage for this vehicle at 40 mph?

We substitute 3 for t.

$$A(t) = 30 + 25t - 4.9t^2$$

$$A(3) = 30 + 25(3) - 4.9(3)^2$$

$$= 30 + 25(3) - 4.9(9)$$

$$= 30 + 75 - \boxed{}$$

$$= \boxed{} \text{ m}$$

YOUR NOTES   Write your questions and additional notes.

Adding Polynomials

Similar, or **like**, **terms** are terms that have the same variable(s) raised to the same power(s).

Polynomials are added by collecting, or combining, like terms.

Example

- Add: $\left(-4x^3 + 5x + 3\right) + \left(6x^3 - 2x^2 - 7\right)$.

$$\left(-4x^3 + 5x + 3\right) + \left(6x^3 - 2x^2 - 7\right) = (-4+6)x^3 - 2x^2 + 5x + (3-7) = 2x^3 - 2x^2 + 5x - 4$$

GUIDED LEARNING ▣ **Textbook**   ▣ **Instructor**   ▶ **Video**

EXAMPLE 1	YOUR TURN 1
Collect like terms: $2x^2 + 12x - 11 - 3x^2 - 15x + 6$. $2x^2 + 12x - 11 - 3x^2 - 15x + 6$ $= (2-3)x^2 + (12-15)x + (-11+6)$ $= \boxed{} - 3x - \boxed{}$	Collect like terms: $-4x^2 + 3x - 1 - 8x^2 - 10x + 6$.
EXAMPLE 2	YOUR TURN 2
Add: $\left(4x^3 - 2xy + y^3\right) + \left(5x^3 + 6xy - 5y^3\right)$. $\left(4x^3 - 2xy + y^3\right) + \left(5x^3 + 6xy - 5y^3\right)$ $= (4+5)x^3 + (-2+6)xy + (1-5)y^3$ $= 9x^3 + \boxed{} - \boxed{}$	Add: $\left(12x^2 y^3 + 3x^2 y - 2x^4 y\right) +$ $\left(5x^2 y^3 - 10x^2 y - x^4 y\right)$.
EXAMPLE 3	YOUR TURN 3
Add using columns: $\left(5y^3 + 6y - 8\right) + \left(-2y^3 + 4y^2 - 4\right)$. In order to use columns to add, we write the polynomials one under the other, listing like terms under one another and leaving spaces for missing terms. $\begin{array}{r} 5y^3 + 6y - 8 \\ -2y^3 + 4y^2 - 4 \\ \hline 3y^3 + \boxed{} + 6y - \boxed{} \end{array}$	Add using columns: $\left(5y^4 + 2y^3 - 5\right) + \left(7y^4 - 3y + 10\right)$.

YOUR NOTES   Write your questions and additional notes.

Subtracting Polynomials

ESSENTIALS

If the sum of two polynomials is 0, the polynomials are **opposites**, or **additive inverses**.

The Opposite of a Polynomial

The *opposite* of a polynomial P can be written as $-P$ or, equivalently, by replacing each term in P with its opposite.

To subtract a polynomial, we add its opposite.

Examples

- Write two equivalent expressions for the opposite of $3x^3 - 2x^2 + 10x - 1$.

  The opposite can be written with parentheses as $-(3x^3 - 2x^2 + 10x - 1)$.

  The opposite can be written without parentheses by replacing each term by its opposite: $-3x^3 + 2x^2 - 10x + 1$.

- Subtract: $(-4x^2 + 3x - 1) - (10x^2 - 6x - 8)$.

$$(-4x^2 + 3x - 1) - (10x^2 - 6x - 8) = (-4x^2 + 3x - 1) + (-10x^2 + 6x + 8)$$
$$= -4x^2 + 3x - 1 - 10x^2 + 6x + 8$$
$$= -14x^2 + 9x + 7$$

	🔲 Textbook	👤 Instructor	▶ Video

GUIDED LEARNING

EXAMPLE 1	YOUR TURN 1
Write two equivalent expressions for the opposite of $3x^2 - 2xy + 4y - 12$.	Write two equivalent expressions for the opposite of $-6xy^2 + 5xy - 16y + 20$.
The opposite can be written with parentheses as $-(3x^2 - 2xy + 4y - 12)$.	
The opposite can be written without parentheses by replacing each term by its opposite: $-3x^2 + \boxed{} - \boxed{} + 12$.	

EXAMPLE 2	YOUR TURN 2
Subtract: $(-7x^3 + 5xy - 1) - (15xy + 7)$.	Subtract: $$(6x^2 + 5x^3y^3) - (8x^2 - 3x^3y^3 + 11).$$
We add the opposite of the polynomial being subtracted.	
$(-7x^3 + 5xy - 1) - (15xy + 7)$	
$= (-7x^3 + 5xy - 1) + (-15xy - 7)$	
$= -7x^3 + 5xy - 1 - 15xy - 7$	
$= \boxed{} - 10xy - \boxed{}$	

EXAMPLE 3	YOUR TURN 3
Subtract: $\left(\dfrac{1}{4}x^2 - \dfrac{5}{8}x + \dfrac{1}{2}\right) - \left(\dfrac{3}{4}x^2 + \dfrac{3}{8}x - \dfrac{1}{2}\right).$	Subtract: $\left(\dfrac{7}{9}x^2 - \dfrac{2}{3}x + \dfrac{5}{6}\right) - \left(-\dfrac{2}{9}x^2 + \dfrac{1}{3}x - \dfrac{1}{6}\right).$

$$\left(\frac{1}{4}x^2 - \frac{5}{8}x + \frac{1}{2}\right) - \left(\frac{3}{4}x^2 + \frac{3}{8}x - \frac{1}{2}\right)$$

$$= \left(\frac{1}{4}x^2 - \frac{5}{8}x + \frac{1}{2}\right) + \left(-\frac{3}{4}x^2 - \frac{3}{8}x + \boxed{}\right)$$

$$= \frac{1}{4}x^2 - \frac{5}{8}x + \frac{1}{2} - \boxed{} - \frac{3}{8}x + \frac{1}{2}$$

$$= -\frac{2}{4}x^2 - \frac{8}{8}x + 1$$

$$= \boxed{} - x + \boxed{}$$

YOUR NOTES   Write your questions and additional notes.

Practice Exercises

Readiness Check

Choose the word, number, or expression that best completes each statement.

1. The expression $4x^2 + 2$ is a $\underline{\hspace{3cm}}$ with a leading coefficient of $\underline{\hspace{1cm}}$.
   binomial / trinomial $\qquad$ 2 / 4
   The degree of the term 2 is $\underline{\hspace{1cm}}$.
   0 / 2

2. The degree of the polynomial $-3xy^2 + 5xyz^2 - 10xy - 4$ is $\underline{\hspace{1cm}}$. The leading
   2 / 4
   coefficient is $\underline{\hspace{1cm}}$. The polynomial has $\underline{\hspace{1cm}}$ terms.
   5 / −3 $\qquad$ 4 / 3

3. The polynomial $-12 + 3x - 15x^2$ is written in $\underline{\hspace{3cm}}$ order.
   ascending / descending

4. To add or subtract polynomials, we add or subtract $\underline{\hspace{3cm}}$.
   like terms / exponents

Polynomial Expressions

Identify the terms, the degree of each term, and the degree of the polynomial. Then identify the leading term, the leading coefficient, and the constant term.

5. $-6x + x^3 + 8 + 3x^4$ $\qquad$ 6. $5 - 12x^4 - 6x^7 + 3x^3 - 2x$ $\qquad$ 7. $-3xy - 6xyz + 2x^2 - 7$

Arrange in descending powers of x.

8. $15 - 3x^4 + 2x^2 - x$ $\qquad\qquad\qquad$ 9. $4x - 3x^2 + 12x^6 - 8x^3 - 10$

Arrange in ascending powers of y.

10. $2y^2 + y^4 - 3y^3 + 9$ $\qquad\qquad\qquad$ 11. $3x^2y - 5xy^4 - 2x^3 + 6y^2$

Evaluating Polynomial Functions

Find $f(5)$ and $f(-1)$ for each polynomial function.

12. $f(x) = 4x^3 - 12x^2 + 20$ $\qquad\qquad$ 13. $f(x) = 4 - 3x^2 + 5x^3$

14. A firm determines that, when it sells x tablet computers, its total revenue, in dollars, is $R(x) = 280x - 0.4x^2$. What is the total revenue from the sale of 62 tablet computers?

Adding Polynomials

Collect like terms to write an equivalent expression.

15. $9x + 3 - 4x + 2x^3 - 4x - 7$

16. $-4x^2 y^2 + 5x^3 - 9x^2 y^2 - 15x^3$

Add.

17. $\left(8xyz - 3x^2 y + 5xy\right) + \left(-2xyz + xy\right)$

18. $\left(3x^2 + 2x\right) + \left(-5x^2 - 8\right) + \left(4x^2 + 7\right)$

Subtracting Polynomials

Write two expressions, one with parentheses and one without, for the opposite of each polynomial.

19. $9x^4 - 3x^3 + 2x^2 - 5$

20. $-12xy - 7x^2 y^2 + xyz - 6$

Subtract.

21. $\left(4x^2 + 5x - 2\right) - \left(2x^2 - 4x + 9\right)$

22. $\left(7a^3 - 4a^2 + 5\right) - \left(-3a^3 - a^2 + 17\right)$

23. $\left(6x^7 yz - x^2 y^2 z^2 + 2x\right) - \left(3x - 2x^7 yz\right)$

24. $\left(-2x^5 - 12\right) - \left(7xy - 15 + 13x^5\right)$

Multiplication of Any Two Polynomials

ESSENTIALS

Multiplying Monomials

To multiply monomials, multiply the coefficients and multiply the variables using the rules for exponents and the commutative and associative laws.

Multiplying Monomials and Binomials

The distributive law is used to multiply polynomials other than monomials.

Multiplying Binomials

To multiply binomials use the distributive law twice, first considering one of the binomials as a single expression and multiplying it by each term of the other binomial.

Multiplying Any Two Polynomials

To find the product of two polynomials P and Q, multiply each term of P by every term of Q and then collect like terms.

Examples

- Multiply: $\left(5x^3y^4\right)\left(-3x^2y^2\right)$.

$$\left(5x^3y^4\right)\left(-3x^2y^2\right) = 5\cdot(-3)\cdot x^3 \cdot x^2 \cdot y^4 \cdot y^2$$
$$= -15x^{3+2}\cdot y^{4+2}$$
$$= -15x^5y^6$$

- Multiply: $4x(5x-2)$.

$$4x(5x-2) = 4x\cdot 5x - 4x\cdot 2 \quad \text{Using the distributive law}$$
$$= 20x^2 - 8x \qquad \text{Multiplying monomials}$$

- Multiply: $\left(x^2-1\right)\left(2x^2+3\right)$.

$$\left(x^2-1\right)\left(2x^2+3\right) = \left(x^2-1\right)2x^2 + \left(x^2-1\right)3 \qquad \text{Using the distributive law}$$
$$= 2x^2\left(x^2-1\right)+3\left(x^2-1\right) \qquad \text{Using a commutative law}$$
$$= 2x^2\cdot x^2 - 2x^2\cdot 1 + 3\cdot x^2 - 3\cdot 1 \qquad \text{Using the distributive law twice}$$
$$= 2x^4 - 2x^2 + 3x^2 - 3 \qquad \text{Multiplying monomials}$$
$$= 2x^4 + x^2 - 3 \qquad \text{Collecting like terms}$$

- Multiply: $\left(x^3-4x^2+5\right)(x+3)$.

$$\left(x^3-4x^2+5\right)(x+3) = \left(x^3-4x^2+5\right)(x) + \left(x^3-4x^2+5\right)(3)$$
$$= x^3(x) - 4x^2(x) + 5(x) + x^3(3) - 4x^2(3) + 5(3)$$
$$= x^4 - 4x^3 + 5x + 3x^3 - 12x^2 + 15$$
$$= x^4 - x^3 - 12x^2 + 5x + 15$$

GUIDED LEARNING	🔖 **Textbook**	👤 **Instructor**	▶ **Video**

EXAMPLE 1	YOUR TURN 1
Multiply and simpify: $\left(-7a^2bc^3\right)\left(3abc^5\right)$. $\left(-7a^2bc^3\right)\left(3abc^5\right) = -7\cdot3\cdot a^2\cdot a\cdot b\cdot b\cdot c^3\cdot c^5$ $ = -21\cdot a^{\square+\square}\cdot b^{1+1}\cdot c^{3+5}$ $ = \boxed{}$	Multiply and simplify: $\left(12a^3b^5c\right)\left(-4a^4bc\right)$.

EXAMPLE 2	YOUR TURN 2
Multiply: $2a^2\left(4a^3-5a^4\right)$. $2a^2\left(4a^3-5a^4\right) = 2a^2\cdot4a^3-\boxed{}\cdot5a^4$ $ = 8a^5-\boxed{}$	Multiply: $9a\left(4a^3-8\right)$.

EXAMPLE 3	YOUR TURN 3
Multiply: $\left(x^2+5\right)\left(x^3-2\right)$. $\left(x^2+5\right)\left(x^3-2\right) = \left(x^2+5\right)x^3+\left(x^2+5\right)(-2)$ $ = x^3\left(x^2+5\right)-2\left(x^2+5\right)$ $ = x^3\cdot x^2+x^3\cdot\boxed{}-2\cdot\boxed{}-2\cdot5$ $ = \boxed{}$	Multiply: $\left(x^4-1\right)\left(3x^5-6\right)$.

EXAMPLE 4	YOUR TURN 4
Multiply: $\left(3x^3+2x-2\right)\left(-2x^2+4x+5\right)$.	Multiply: $\left(2x^3-4x+6\right)\left(-5x^2+x+1\right)$.

$$
\begin{array}{rrrrr}
3x^3 & & +\ 2x & -\ 2 \\
& -\ 2x^2 & +\ 4x & +\ 5 \\
\hline
15x^3 & & +10x & -\ \boxed{} \\
12x^4 & & +\ \boxed{} & -\ 8x \\
\boxed{}x^5 & -\ 4x^3 & +\ 4x^2 \\
\hline
\multicolumn{4}{c}{\boxed{}}
\end{array}
$$

YOUR NOTES   Write your questions and additional notes.

Product of Two Binomials Using the FOIL Method

ESSENTIALS

The FOIL Method

To multiply two binomials $(A+B)$ and $(C+D)$:

1. Multiply **F**irst terms: AC
2. Multiply **O**utside terms: AD
3. Multiply **I**nside terms: BC
4. Multiply **L**ast terms: BD
   $\downarrow$
   FOIL

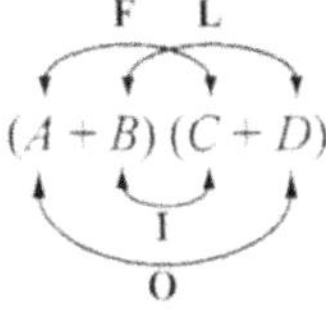

$$(A+B)(C+D) = AC + AD + BC + BD$$

Example

- Multiply: $(x+6)(x-3)$.

$$(x+6)(x-3) = x^2 - 3x + 6x - 18 \qquad \text{FOIL}$$
$$= x^2 + 3x - 18 \qquad \text{Collecting like terms}$$

GUIDED LEARNING   📖 **Textbook**   👤 **Instructor**   ▶ **Video**

EXAMPLE 1	YOUR TURN 1
Multiply: $(2x+1)(-4x+5)$.	Multiply: $(3x-6)(6x-2)$.

$$(2x+1)(-4x+5) = -8x^2 + \boxed{} - 4x + 5$$
$$= \boxed{}$$

EXAMPLE 2	YOUR TURN 2
Multiply: $(m-1)(m+4)(m-7)$.	Multiply: $(m+8)(m+5)(m-4)$.

$$(m-1)(m+4)(m-7)$$
$$= (m^2 + 4m - m - 4)(m-7)$$
$$= (m^2 + \boxed{} - 4)(m-7)$$
$$= (m^2 + 3m - 4)\cdot m + (m^2 + 3m - 4)(-7)$$
$$= \boxed{} + 3m^2 - 4m - \boxed{} - 21m + 28$$
$$= \boxed{}$$

YOUR NOTES   Write your questions and additional notes.

Squares of Binomials

ESSENTIALS

Squaring a Binomial

$$(A+B)^2 = A^2 + 2AB + B^2$$

$$(A-B)^2 = A^2 - 2AB + B^2$$

Trinomials of the form $A^2 + 2AB + B^2$ or $A^2 - 2AB + B^2$ are called **trinomial squares.**

Example

- Multiply: $(x-4)^2$.

$$(A-B)^2 = A^2 - 2\cdot A\cdot B + B^2$$
$$\downarrow \quad \downarrow \qquad \downarrow \qquad \downarrow\ \downarrow \quad \downarrow$$
$$(x-4)^2 = x^2 - 2\cdot x\cdot 4 + 4^2$$
$$= x^2 - 8x + 16$$

	📘 Textbook	👤 Instructor	▶ Video

GUIDED LEARNING

EXAMPLE 1	YOUR TURN 1
Multiply: $(4y+6)^2$. $(4y+6)^2 = (4y)^2 + 2\cdot 4y\cdot 6 + 6^2$ $=\ \boxed{}$	Multiply: $(-3y+5)^2$.
EXAMPLE 2	YOUR TURN 2
Multiply: $\left(\dfrac{1}{4}x - 2\right)^2$. $\left(\dfrac{1}{4}x-2\right)^2 = \left(\dfrac{1}{4}x\right)^2 - 2\cdot\dfrac{1}{4}x\cdot 2 + 2^2$ $=\ \boxed{}$	Multiply: $\left(\dfrac{1}{2}x - 5\right)^2$.

YOUR NOTES   Write your questions and additional notes.

Products of Sums and Differences

The Product of a Sum and a Difference

$$(A+B)(A-B) = A^2 - B^2$$

The product of the sum and the difference of the same two terms is called a **difference of squares**.

Example

- Multiply: $(x+3)(x-3)$.

$$(A+B)(A-B) = A^2 - B^2$$
$$\downarrow \quad \downarrow \quad \downarrow \quad \downarrow \quad \quad \downarrow \quad \quad \downarrow$$
$$(x+3)(x-3) = x^2 - 3^2$$
$$= x^2 - 9$$

GUIDED LEARNING	📖 Textbook	👤 Instructor	▶ Video

EXAMPLE 1	YOUR TURN 1
Multiply: $(2xy+4x)(2xy-4x)$.	Multiply: $(5x^2y+1)(5x^2y-1)$.
$(2xy+4x)(2xy-4x) = (2xy)^2 - (4x)^2$ $= \boxed{}$	

EXAMPLE 2	YOUR TURN 2
Multiply: $(0.3m+2.1n)(0.3m-2.1n)$.	Multiply: $(0.6x+1.2y)(0.6x-1.2y)$.
$(0.3m+2.1n)(0.3m-2.1n) = (0.3m)^2 - (2.1n)^2$ $= \boxed{}$	

YOUR NOTES   Write your questions and additional notes.

Using Function Notation

ESSENTIALS

Example

- Given $f(x) = x^2 + 6x - 7$, find (a) $f(t) - 10$ and (b) $f(t - 10)$.

  a) $f(x) = x^2 + 6x - 7$

  $$f(t) - 10 = (t^2 + 6t - 7) - 10 \quad \text{Evaluating } f(t)$$
  $$= t^2 + 6t - 17 \quad\quad\quad \text{Simplifying}$$

  b) $f(x) = x^2 + 6x - 7$

  $$f(t - 10) = (t - 10)^2 + 6(t - 10) - 7 \quad \text{Substituting } t - 10 \text{ for } x$$
  $$= t^2 - 20t + 100 + 6t - 60 - 7 \quad \text{Multiplying}$$
  $$= t^2 - 14t + 33 \quad\quad\quad\quad\quad \text{Simplifying}$$

GUIDED LEARNING   📘 **Textbook**   👤 **Instructor**   ▶ **Video**

EXAMPLE 1	YOUR TURN 1
Given $f(x) = 2x^2 + 3$, find (a) $f(a) + 1$; (b) $f(a + 1)$.	Given $f(x) = x^2 + 5x$, find (a) $f(a) - 4$; (b) $f(a - 4)$.

a) $f(x) = 2x^2 + 3$

$$f(a) + 1 = (2a^2 + 3) + 1$$
$$= 2a^2 + 4$$

b) $f(x) = 2x^2 + 3$

$$f(a + 1) = 2(a + 1)^2 + 3$$
$$= 2\left(\boxed{}\right) + 3$$
$$= 2a^2 + 4a + \boxed{} + 3$$
$$= \boxed{}$$

EXAMPLE 2	YOUR TURN 2
Given $g(x) = 3x^2y^3 + 2$, find $g(x) \cdot g(x)$. $g(x) \cdot g(x) = \left(3x^2y^3 + 2\right)\left(3x^2y^3 + 2\right)$ $\quad\quad = \left(3x^2y^3 + 2\right)^2$ $\quad\quad = \boxed{} + 2 \cdot \left(3x^2y^3\right) \cdot 2 + 2^2$ $\quad\quad = \boxed{}$	Given $g(x) = 4x^2y^6 - 3$, find $g(x) \cdot g(x)$.

YOUR NOTES   Write your questions and additional notes.

Practice Exercises

Readiness Check

Classify each of the following statements as either true or false.

1. FOIL can be used whenever two binomials are multiplied.

2. Given $f(x) = 2x - 1$, $f(y) + 8 = f(y + 8)$.

3. A trinomial square is of the form $A^2 - B^2$.

4. When multiplying polynomials, the distributive law can always be used.

Multiplying Monomials

Multiply.

5. $2x^5 \cdot 4x^3$

6. $\left(-5x^6 y^7\right)\left(-3x^2 y^2\right)$

7. $\left(-9abc^4\right)\left(3a^2 bc\right)$

Multiplying Monomials and Binomials

Multiply.

8. $6(4 - 2x)$

9. $4a\left(-3a^2 - 7a\right)$

10. $5xy\left(-7x^2 y^3 + 6x^4 y\right)$

Multiplying Any Two Polynomials

Multiply.

11. $(2x + 3)(5x - 1)$

12. $(x + 5)\left(x^2 + x - 2\right)$

13. $(x + 4)\left(3x^2 + x - 5\right)$

The Product of Two Binomials: FOIL

Multiply.

14. $\left(x - \dfrac{1}{6}\right)\left(x - \dfrac{1}{2}\right)$

15. $(4x - 2)(3x + 1)$

16. $(1.3x + 3)(2.4x - 5)$

Squares of Binomials

Multiply.

17. $(x-6)^2$

18. $(2x+3y)^2$

19. $(x^3y-2)^2$

Products of Sums and Differences

Multiply.

20. $(a+10)(a-10)$

21. $(3-6x)(3+6x)$

22. $(-4a+b^2)(4a+b^2)$

Function Notation

23. Let $P(x)=2x-5$ and $Q(x)=4x^2-3x+1$. Find $P(x)\cdot Q(x)$.

24. Given $f(x)=x^2+3x-9$, find $f(t-4)$.

25. Given $f(x)=2x^2+3$, find $f(a+h)$.

26. Given $f(x)=6+4x-x^2$, find $f(a)+7$.

Terms with Common Factors

ESSENTIALS

Factoring

To **factor** a polynomial is to find an equivalent expression that is a product of polynomials. An equivalent expression of this type is called a **factorization** of the polynomial.

A polynomial is **completely factored** if it cannot be factored further.

A polynomial that cannot be factored is a **prime polynomial**.

Example

- Write an expression equivalent to $12x^3 y^2 - 6x^2 y^2 + 4x^4 y$ by factoring out the greatest common factor.

  Look for the greatest common factor of the coefficients of $12x^3 y^2 - 6x^2 y^2 + 4x^4 y$.

  $$12, -6, 4 \quad \rightarrow \quad \text{Greatest common factor} = 2$$

  Look for the greatest common factor of the powers of x.

  $$x^3, x^2, x^4 \quad \rightarrow \quad \text{Greatest common factor} = x^2$$

  Look for the greatest common factor of the powers of y.

  $$y^2, y^2, y \quad \rightarrow \quad \text{Greatest common factor} = y$$

  Thus, $2x^2 y$ is the greatest common factor of the polynomial.

  $$12x^3 y^2 - 6x^2 y^2 + 4x^4 y = 2x^2 y\left(6xy - 3y + 2x^2\right)$$

  Check: $2x^2 y\left(6xy - 3y + 2x^2\right) = 12x^3 y^2 - 6x^2 y^2 + 4x^4 y$

	📖 **Textbook**	👤 **Instructor**	▶ **Video**

GUIDED LEARNING

EXAMPLE 1	YOUR TURN 1
Write an expression equivalent to $10x^6 y^4 + 5x^3 y^3 - 5x^4$ by factoring out the greatest common factor.	Write an expression equivalent to $8xy^2 + 4x^5 y - 2x^3 y$ by factoring out the greatest common factor.
$10x^6 y^4 + 5x^3 y^3 - 5x^4 = 5x^3\left(2x^3 y^4 + \boxed{} - x\right)$	
Check: $5x^3\left(2x^3 y^4 + y^3 - x\right) = 10x^6 y^4 + 5x^3 y^3 - 5x^4$	

EXAMPLE 2	YOUR TURN 2
Write an expression equivalent to $-3x^3 + 6x^2 + 12x$ by factoring out a common factor with a negative coefficient. $$-3x^3 + 6x^2 + 12x = -3x\left(\boxed{} - 2x - 4\right)$$ Check: $-3x\left(x^2 - 2x - 4\right) = -3x^3 + 6x^2 + 12x$	Write an expression equivalent to $-7x^3 + 14x^2 - 21x$ by factoring out a common factor with a negative coefficient.
EXAMPLE 3	YOUR TURN 3
The number of diagonals of a polygon having n sides is given by the polynomial function $$P(n) = \frac{1}{2}n^2 - \frac{3}{2}n.$$ Find an equivalent expression for $P(n)$ by factoring out $\frac{1}{2}n$. $$P(n) = \frac{1}{2}n^2 - \frac{3}{2}n = \boxed{}\left(n - \boxed{}\right)$$ Check: $\frac{1}{2}n(n-3) = \frac{1}{2}n^2 - \frac{3}{2}n$	The surface area of a rectangular prism is represented by the formula $$2lw + 2lh + 2wh.$$ Find an equivalent expression by factoring out 2.

YOUR NOTES   Write your questions and additional notes.

Factoring by Grouping

ESSENTIALS

The largest common factor of a polynomial is sometimes a binomial. In order to identify a common binomial factor in a polynomial with four terms, we must regroup into two groups of two terms each.

Factoring out -1

$$b - a = -1(a - b) = -(a - b)$$

Example

- Factor: $(x+4)(x+2)+(x+4)(x-7)$.

  Here the largest common factor is the binomial $x+4$.

  $$(x+4)(x+2)+(x+4)(x-7) = (x+4)\left[(x+2)+(x-7)\right]$$
  $$= (x+4)(2x-5)$$

GUIDED LEARNING   📖 **Textbook**   👤 **Instructor**   ▶ **Video**

EXAMPLE 1	YOUR TURN 1
Factor: $(2x+1)(x-1)+(2x+1)(x+6)$. The largest common factor is the binomial $2x+1$. $(2x+1)(x-1)+(2x+1)(x+6)$ $= (2x+1)\left[(x-1)+(\boxed{})\right]$ $= (2x+1)\boxed{}$	Factor: $(3x-2)(x+4)+(3x-2)(x-1)$.
EXAMPLE 2	YOUR TURN 2
Factor: $2x^2 + 6x + 5x + 15$. $2x^2 + 6x + 5x + 15$ $= \left(2x^2 + 6x\right) + (5x + 15)$ $= 2x(x+3) + 5\left(\boxed{}\right)$ $= (x+3)\boxed{}$	Factor: $2x^3 - 2x^2 + 3x - 3$.

EXAMPLE 3	YOUR TURN 3
Factor: $6x - 7y + xy - 42$.	Factor: $2a + 4b + ba + 8$.

$6x - 7y + xy - 42$

$= (6x - 42) + (xy - 7y)$

$= 6(x - 7) + y\left(\boxed{}\right)$

$= \boxed{}$

EXAMPLE 4	YOUR TURN 4
Factor: $3x^2 - 3xy + y^2 - xy$.	Factor: $2x^2 - 2xy + 3y^2 - 3xy$.

$3x^2 - 3xy + y^2 - xy$

$= (3x^2 - 3xy) + (y^2 - xy)$

$= 3x(x - y) + y\left(\boxed{}\right)$

$= 3x(x - y) + y(-1)(-y + x)$

$= 3x(x - y) - y(x - y)$

$= \boxed{}$

YOUR NOTES   Write your questions and additional notes.

Practice Exercises

Readiness Check

Choose the word or expression that best completes each statement.

1. The largest common factor of $25x^5 y^3 z^4 + 75x^{10} y^6 z^2$ is ________.
$$25x^{10} y^6 z^4 \ / \ 25x^5 y^3 z^2$$

2. The polynomial $2xy + 3$ is a ________ polynomial.
prime / factorable

3. The largest common factor of the polynomial $(x+7)(x-2)+(x+8)(x-2)$ is

________.
$x + 7 \ / \ x - 2$

4. The expression $x - y$ is equivalent to ________.
$-1(y-x) \ / \ -(x-y)$

Terms with Common Factors

Write an equivalent expression by factoring out the greatest common factor.

5. $12x^2 - 18x + 24$
6. $6a^6 + 9a^5 - 12a^3$

7. $16x^4 y^2 - 8x^5 y^3 + 4x^6 y$
8. $4a^8 - a^5 - a^3$

Write an equivalent expression by factoring out a factor with a negative coefficient.

9. $-9x - 45$
10. $-3x^2 + 9x - 12$

11. $-x^3 - 5x^2 - 6x + 7$
12. $6a^2 - 18ab$

Factoring by Grouping

Write an equivalent expression by factoring.

13. $t(r+2)+s(r+2)$

14. $12a^5 + 8a^3 + 3a^2 + 2$

15. $4x^3 + 3x + 8x^2 + 6$

16. $5x^2 - 5xy + 6y^2 - 6xy$

17. When x hundred DVDs are sold, a company makes a profit of $P(x)$ where $P(x) = x^2 - 4x$, and $P(x)$ is in thousands of dollars. Find an equivalent expression by factoring out a common factor.

18. A company determines that when it sells x units of a product, the total revenue $R(x)$, in dollars, is given by the polynomial function $R(x) = 350x - 0.5x^2$. Find an equivalent expression for $R(x)$ by factoring out a common factor with a negative coefficient.

Factoring Trinomials: $x^2 + bx + c$

To Factor $x^2 + bx + c$

1. If necessary, rewrite the trinomial in descending order.

2. Find a pair of factors that have c as their product and b as their sum.

   - If c is positive, both factors will have the same sign as b.
   - If c is negative, one factor will be positive and the other will be negative. The factor with the larger absolute value will be the factor with the same sign as b.
   - If the sum of the two factors is the opposite of b, changing the signs of both factors will give the desired factors whose sum is b.

3. Check by multiplying.

Examples

- Factor: $x^2 + 5x + 6$.

  There is no common factor. Look for a factorization of 6 in which both factors are positive. Their sum must be 5.

Pairs of factors	Sum of factors
1, 6	7
2, 3	5

  $\leftarrow$ The numbers we need are 2 and 3.

  The factorization is $(x+2)(x+3)$.

  Check: $(x+2)(x+3) = x^2 + 3x + 2x + 6 = x^2 + 5x + 6$.

- Factor: $x^2 + 4x - 12$.

  There is no common factor. Look for a factorization of -12 for which the sum of the factors is 4. The positive factor must have the larger absolute value.

Pairs of factors	Sum of factors
-1, 12	11
-2, 6	4
-3, 4	1

  $\leftarrow$ The numbers we need are -2 and 6.

  The factorization is $(x-2)(x+6)$.

  Check: $(x-2)(x+6) = x^2 + 6x - 2x - 12 = x^2 + 4x - 12$.

GUIDED LEARNING   📖 **Textbook**   👤 **Instructor**   ▶ **Video**

EXAMPLE 1	YOUR TURN 1

Factor: $x^2 - 3x - 10$.

There is no common factor. Look for a factorization of -10 for which the sum of the factors is -3. The negative factor must have the larger absolute value.

Pairs of factors	Sum of factors
1, -10	☐
2, -5	☐

The numbers we need are 2 and -5. The factorization is $(x+2)(x-5)$.

Check: $(x+2)(x-5) = x^2 - 5x + 2x - 10 = x^2 - 3x - 10$.

Factor: $x^2 - 8x - 9$.

EXAMPLE 2	YOUR TURN 2

Factor: $2x^2 + 22x + 36$.

Factor out the largest common factor, 2.

$$2x^2 + 22x + 36 = 2(x^2 + 11x + 18)$$

Look for a factorization of 18 in which both factors are positive and whose sum is ______ .

       $-11 \,/\, 11$

Pairs of factors	Sum of factors
1, 18	19
2, 9	☐
3, 6	☐

Thus, $x^2 + 11x + 18 = (x+2)(x+9)$. The factorization of the original trinomial is $2(x+2)(x+9)$. You can check this by multiplying.

Factor: $3x^2 + 15x + 12$.

YOUR NOTES   Write your questions and additional notes.

Practice Exercises

Readiness Check

Choose from the column on the right the phrase that best completes each statement.

1. To factor $x^2 + 10x + 9$, we look for a factorization of 9 in which __________.
2. To factor $x^2 - 6x - 7$, we look for a factorization of -7 in which __________.
3. To factor $x^2 - 12x + 11$, we look for a factorization of 11 in which __________.
4. To factor $x^2 + 8x - 9$, we look for a factorization of -9 in which __________.

a) both factors are positive.
b) both factors are negative.
c) the positive factor has the greater absolute value.
d) the negative factor has the greater absolute value.

Factoring Trinomials: $x^2 + bx + c$

Factor completely. If a polynomial is not factorable, state this.

5. $x^2 + 7x + 6$

6. $t^2 - 11t - 12$

7. $x^2 - 8x + 15$

8. $x^2 + 4x - 12$

9. $y^2 - 2y + 24$

10. $-48 - 18x + 3x^2$

Factor completely. If a polynomial is not factorable, state this.

11. $y^2 - \dfrac{1}{2}y + \dfrac{1}{16}$

12. $x^7 + 28x^6 + 27x^5$

13. $36 + 5x - x^2$

14. $a^2 + 6ab + 5b^2$

15. $x^4 - 7x^2 - 8$

16. $t^2 + 0.6t + 0.09$

17. $-x^3 + 3x^2 - 2x$

The FOIL Method

ESSENTIALS

To Factor $ax^2 + bx + c$ Using FOIL:

1. Factor out the largest common factor, if one exists. Here we assume that none does.

2. Find two First terms whose product is ax^2.

3. Find two Last terms whose product is c.

4. Repeat steps (2) and (3), if necessary, until a combination is found for which the sum of the Outside and Inside products is bx.

5. Check by multiplying.

Tips:

1. If the largest common factor has been factored out of the original trinomial, then no binomial factor can have a common factor (other than 1 or -1).

2. **a)** If the signs of all the terms are positive, then the signs of all the terms of the binomial factors are positive.

   b) If a and c are positive and b is negative, then the signs of the factors of c are negative.

   c) If a is positive and c is negative, then the factors of c will have opposite signs.

3. Be systematic about your trials. Keep track of those you have tried and those you have not.

4. Changing the signs of the factors of c will change the sign of the middle term.

Example

- Factor using FOIL: $6x^2 + x - 12$.

$$6x^2 + x - 12 = (2x + 3)(3x - 4)$$

GUIDED LEARNING	Textbook	Instructor	Video

EXAMPLE 1	YOUR TURN 1
Factor using FOIL: $3x^2 + 13x - 10$.	Factor using FOIL: $2x^2 + 5x - 12$.
1. There is no common factor other than 1 or -1.	
2. $3x^2$ can be factored as $(3x)(\boxed{})$.	
The factorization must be of the form $(3x +)(x +)$.	
3. There are four pairs of factors of -10, and each can be listed in two ways.	
(continued)	

4.

Pairs of factors	Corresponding trial	Product
$-1, 10$	$(3x-1)(x+10) =$	$3x^2\ \boxed{}\ -10$
$1, -10$	$(3x+1)(x-10) =$	$3x^2\ \boxed{}\ -10$
$-2, \boxed{}$	$(3x-2)(x+5) =$	$3x^2\ \boxed{}\ -10$
$2, -5$	$(3x+2)(x-5) =$	$3x^2\ \boxed{}\ -10$
$10, \boxed{}$	$(3x+10)(x-1) =$	$3x^2\ \boxed{}\ -10$
$-10, 1$	$(3x-10)(x+1) =$	$3x^2\ \boxed{}\ -10$
$5, \boxed{}$	$(3x+5)(x-2) =$	$3x^2\ \boxed{}\ -10$
$-5, 2$	$(3x-5)(x+2) =$	$3x^2\ \boxed{}\ -10$

The factors -2 and 5 give the correct middle term.
The factorization is $(3x-2)(x+5)$.

EXAMPLE 2	YOUR TURN 2

Factor using FOIL: $6+7a-3a^2$.

Factor using FOIL: $18-15a-25a^2$.

1. Write in descending order: $-3a^2 + 7a + 6$. Factor out -1 to make the leading coefficient positive.

$$-3a^2 + 7a + 6 = -1\left(\boxed{}\right)$$

2. The only possible factorization of $3a^2$ is $3 \cdot a \cdot a$. Thus, the factorization must be of the form $(3a+\ \ )(a+\ \ )$.

3. Pairs of factors of -6 are

$$-1, 6 \quad 1, -6 \quad -2, \boxed{} \quad 2, \boxed{}$$
$$6, -1 \quad -6, \boxed{} \quad 3, \boxed{} \quad -3, \boxed{}.$$

4. We conduct trials, stopping when we have the correct factorization.

$$(3a-1)(a+6) = 3a^2\ \boxed{}\ -6$$
$$(3a-2)(a+3) = 3a^2\ \boxed{}\ -6$$
$$(3a+2)(a-3) = 3a^2\ \boxed{}\ -6$$

The factorization of $3a^2 - 7a - 6$ is $(3a+2)(a-3)$.

Thus, the factorization of the original trinomial is $-1(3a+2)(a-3)$. We can also write this as

$(-3a-2)(a-3)$ or $(3a+2)(-a+3)$.

EXAMPLE 3	YOUR TURN 3
Factor using FOIL: $12x^2 - 22xy + 10y^2$.	Factor using FOIL: $8a^4 + 18a^3b + 9a^2b^2$.

EXAMPLE 3

Factor using FOIL: $12x^2 - 22xy + 10y^2$.

1. $12x^2 - 22xy + 10y^2 = 2\left(6x^2 - 11xy + 5y^2\right)$

2. The factorization is of the form
   $\left(2x + \quad\right)\left(3x + \quad\right)$ or $\left(6x + \quad\right)\left(x + \quad\right)$.

3. Pairs of factors of $5y^2$ are
   $\begin{array}{l} y,\ 5y \\ -y,\ -5y \end{array}$ and $\begin{array}{l} 5y,\ y \\ -5y,\ -y. \end{array}$

4. Only the factors of $5y^2$ with negative
   coefficients will give a negative middle term.

   $(2x - y)(3x - 5y) = 6x^2 - \boxed{} + 5y^2$

   $(2x - 5y)(3x - y) = 6x^2 - \boxed{} + 5y^2$

   $(6x - y)(x - 5y) = 6x^2 - \boxed{} + 5y^2$

   $(6x - 5y)(x - y) = 6x^2 - \boxed{} + 5y^2$

Thus, the factorization of $6x^2 - 11xy + 5y^2$ is
$(6x - 5y)(x - y)$. The factorization of the original
trinomial is $2(6x - 5y)(x - y)$.

YOUR NOTES   Write your questions and additional notes.

The *ac*-Method

ESSENTIALS

To Factor $ax^2 + bx + c$ Using the *ac*–Method:

1. Factor out the largest common factor, if one exists. Here we assume that none does.
2. Multiply the leading coefficient a and the constant c.
3. Find a pair of factors of ac whose sum is b.
4. Rewrite the middle term, bx, as a sum or difference using the factors from step (3).
5. Factor by grouping.
6. Check by multiplying.

This method is also referred to as the **grouping method**.

Example

- Factor: $10x^2 - x - 2$.

$$10x^2 - x - 2 = 10x^2 + 4x - 5x - 2 = 2x(5x+2) - 1(5x+2) = (5x+2)(2x-1)$$

GUIDED LEARNING   🔘 **Textbook**   👤 **Instructor**   ▶ **Video**

EXAMPLE 1	YOUR TURN 1
Factor: $6x^2 + 7x - 3$.	Factor: $8x^2 + 2x - 3$.

Factor: $6x^2 + 7x - 3$.

1. There is no common factor (other than 1 or -1).

2. Multiply a and c: $6(-3) = \boxed{}$.

3. Find factors of -18 whose sum is 7.
$$9(-2) = -18 \text{ and } 9 + (-2) = 7.$$

4. Split $7x$ using the results of step (3).
$$7x = 9x - 2x$$

5. Factor by grouping.
$$6x^2 + 7x - 3 = 6x^2 + 9x - 2x - 3$$
$$= 3x(2x+3) - 1(2x+3)$$
$$= (2x+3)(3x-1)$$

The factorization is $(2x+3)(3x-1)$.

EXAMPLE 2	YOUR TURN 2
Factor: $8x^3 + 32x^2 + 30x$.	Factor: $10a^5 - 19a^4 + 6a^3$.

EXAMPLE 2

Factor: $8x^3 + 32x^2 + 30x$.

1. Factor out the largest common factor, $2x$.

$$8x^3 + 32x^2 + 30x = 2x\left(4x^2 + \boxed{} + \boxed{}\right)$$

2. To factor $4x^2 + 16x + 15$, first multiply a and c:

$$4(15) = \boxed{}.$$

3. Find factors of 60 whose sum is 16.

   Since the middle term is positive, we consider only positive factors.

$$6 \cdot 10 = 60 \text{ and } 6 + 10 = 16.$$

4. Split the middle term, $16x$, using the results of step (3).

$$16x = 6x + 10x$$

5. Factor by grouping.

$$4x^2 + 16x + 15 = 4x^2 + 6x + 10x + 15$$
$$= 2x(2x + 3) + 5(2x + 3)$$
$$= (2x + 3)(2x + 5)$$

The factorization of the original trinomial is $2x(2x + 3)(2x + 5)$.

YOUR NOTES   Write your questions and additional notes.

Practice Exercises

Readiness Check

Choose a word from the list below that will make each statement true. Not all words will be used.

common	negative	first
last	prime	positive

1. When factoring a trinomial, first factor out the largest ______________ factor.

2. When factoring a trinomial, $ax^2 + bx + c$, the product of the __________ terms in the binomials is c.

3. When factoring a trinomial $ax^2 + bx + c$, if c is ______________, then the signs in both binomial factors are the same.

4. When factoring a trinomial, $ax^2 + bx + c$, the product of the __________ terms in the binomials is ax^2.

The FOIL Method

Factor completely.

5. $10x^2 + x - 3$

6. $6a^2 - 11a - 10$

7. $15a^3 + 13a^2 + 2a$

8. $t - 1 + 2t^2$

The *ac*-Method

Factor completely.

9. $-20x^2 - 5x + 15$

10. $12p^2 - 38pw + 30w^2$

11. $5x^2 + x - 4$

12. $-18z^3 + 24z^2 + 10z$

13. $4r^2 - 35r + 49$

14. $10t^2 - 29t + 10$

Trinomial Squares

ESSENTIALS

An expression of the form $A^2 + 2AB + B^2$ or $A^2 - 2AB + B^2$ is a **trinomial square**, or a **perfect-square trinomial**.

To Recognize a Trinomial Square

Two terms must be squares, such as A^2 and B^2.

The remaining term must be $2AB$ or its opposite.

Factoring a Trinomial Square

$$A^2 + 2AB + B^2 = (A+B)^2$$

$$A^2 - 2AB + B^2 = (A-B)^2$$

Example

- Factor: $x^2 - 20x + 100$.

  We have a trinomial square of the form $A^2 - 2AB + B^2$, with $A = x$ and $B = 10$.

  $$x^2 - 20x + 100 = (x-10)(x-10)$$

  $$= (x-10)^2$$

GUIDED LEARNING	🔖 Textbook	👤 Instructor	▶ Video

EXAMPLE 1	YOUR TURN 1
Factor: $x^2 + 8x + 16$. Two terms, x^2 and 16, are squares. Twice the product of the square roots is $2 \cdot x \cdot 4$, or $8x$, the remaining term of the trinomial. Thus, we have a trinomial square of the form $A^2 + 2AB + B^2$, with $A = x$ and $B = 4$. $$x^2 + 8x + 16 = \left(x \boxed{} 4\right)^2$$	Factor: $y^2 + 12y + 36$.
EXAMPLE 2	YOUR TURN 2
Factor: $9y^2 - 30y + 25$. We have a trinomial square of the form $A^2 - 2AB + B^2$, with $A = 3y$ and $B = 5$. $$9y^2 - 30y + 25 = \left(3y \boxed{} 5\right)^2$$	Factor: $16y^2 + 56x + 49$.

EXAMPLE 3	YOUR TURN 3
Factor: $-4xy + 4y^2 + x^2$.	Factor: $-6xy + 9y^2 + x^2$.

$$-4xy + 4y^2 + x^2 = x^2 - 4xy + 4y^2$$

We have a trinomial square of the form $A^2 - 2AB + B^2$, with $A = x$ and $B = 2y$.

$$-4xy + 4y^2 + x^2 = x^2 - 4xy + 4y^2 = \left(x - \boxed{}\right)^2$$

EXAMPLE 4	YOUR TURN 4
Factor: $2y^5 + 48y^4 + 288y^3$.	Factor: $12x^4 + 36x^3 + 27x^2$.

Factor out the largest common factor, $2y^3$.

$$2y^5 + 48y^4 + 288y^3 = 2y^3\left(y^2 + \boxed{} + 144\right)$$

$y^2 + 24y + 144$ is a trinomial square of the form $A^2 + 2AB + B^2$, with $A = y$ and $B = 12$.

$$2y^5 + 48y^4 + 288y^3$$
$$= 2y^3\left(y^2 + 24y + 144\right)$$
$$= 2y^3\left(\boxed{}\right)^2$$

Check:
$$2y^3\left(y + 12\right)^2 = 2y^3\left(y + 12\right)\left(y + 12\right)$$
$$= 2y^3\left(y^2 + 12y + 12y + 144\right)$$
$$= 2y^3\left(y^2 + 24y + 144\right)$$
$$= 2y^5 + 48y^4 + 288y^3$$

The factorization is $\boxed{}\left(\boxed{}\right)^2$.

YOUR NOTES   Write your questions and additional notes.

Differences of Squares

ESSENTIALS

An expression of the form $A^2 - B^2$ is a **difference of squares**.

Factoring a Difference of Squares

$$A^2 - B^2 = (A + B)(A - B)$$

Example

- Factor: $x^2 - 16$.

    $A = x$ and $B = 4$

    $$x^2 - 16 = x^2 - 4^2 = (x + 4)(x - 4)$$

GUIDED LEARNING	📖 Textbook	👤 Instructor	▶ Video

EXAMPLE 1	YOUR TURN 1
Factor: $y^2 - 64$.	Factor: $x^2 - 4$.
$y^2 - 64 = y^2 - 8^2 = (y + 8)\left(y - \boxed{}\right)$	
EXAMPLE 2	YOUR TURN 2
Factor: $25t^2 - 16r^2$.	Factor: $49t^2 - 9r^2$.
$25t^2 - 16r^2 = (5t)^2 - (4r)^2 = \left(5t + \boxed{}\right)\left(5t - \boxed{}\right)$	
EXAMPLE 3	YOUR TURN 3
Factor: $16x^8 - 64x^4 y^2$.	Factor: $2x^2 - 162y^2$.
Factor out the largest common factor, $16x^4$.	

$$16x^8 - 64x^4 y^2 = 16x^4\left(x^4 - \boxed{}\right)$$
$$= 16x^4\left[\left(x^2\right)^2 - (2y)^2\right]$$
$$= 16x^4\left(\boxed{} + 2y\right)\left(\boxed{} - 2y\right)$$

YOUR NOTES   Write your questions and additional notes.

More Factoring by Grouping

ESSENTIALS

When factoring by grouping, make sure the result is completely factored. If grouping pairs of terms does not yield a common binomial factor, look to group a trinomial square.

Example

- Factor: $x^3 + 2x^2 - 9x - 18$.

$$x^3 + 2x^2 - 9x - 18 = x^2(x+2) - 9(x+2) \quad \text{Factoring by grouping}$$

$$= (x+2)(x^2 - 9) \quad \text{Factoring out } x+2$$

$$= (x+2)(x+3)(x-3) \quad \text{Factoring } x^2 - 9$$

GUIDED LEARNING   ⬛ **Textbook**   👤 **Instructor**   ▶ **Video**

EXAMPLE 1	YOUR TURN 1
Factor: $a^3 + 2a^2 - 64a - 128$.	Factor: $a^3 - a^2 - 49a + 49$.

$$a^3 + 2a^2 - 64a - 128 = a^2(a+2) - 64(a+2)$$

$$= \left(\boxed{}\right)(a^2 - 64)$$

$$= (a+2)\left(\boxed{}\right)(a-8)$$

EXAMPLE 2	YOUR TURN 2
Factor: $x^2 + 14x + 49 - y^2$.	Factor: $16x^2 + 40x + 25 - y^2$.

$$x^2 + 14x + 49 - y^2 = (x^2 + 14x + 49) - y^2$$

$$= (x+7)^2 - y^2$$

$$= \left((x+7) + \boxed{}\right)\left((x+7) - y\right)$$

$$= \left(\boxed{}\right)(x+7-y)$$

YOUR NOTES   Write your questions and additional notes.

Sums or Differences of Cubes

ESSENTIALS

Factoring a Sum or a Difference of Cubes

$$A^3 + B^3 = (A + B)(A^2 - AB + B^2)$$

$$A^3 - B^3 = (A - B)(A^2 + AB + B^2)$$

Example

- Factor completely: $x^3 - 125$.

  This is a difference of cubes: $x^3 - 125 = x^3 - 5^3$.

  $$A^3 - B^3 = (A - B)(A^2 + AB + B^2)$$
  $$\downarrow \quad \downarrow \quad \downarrow \quad \downarrow \ \downarrow \quad \downarrow \quad \downarrow$$
  $$x^3 - 5^3 = (x - 5)(x^2 + 5x + 25)$$

  Check: $(x - 5)(x^2 + 5x + 25) = x^3 + 5x^2 + 25x - 5x^2 - 25x - 125$
  $$= x^3 - 125.$$

  Thus, $x^3 - 125 = (x - 5)(x^2 + 5x + 25)$.

GUIDED LEARNING   📖 **Textbook**   👤 **Instructor**   ▶ **Video**

EXAMPLE 1	YOUR TURN 1
Factor completely: $y^9 - 64$.	Factor completely: $t^{15} - 27$.

Factor completely: $y^9 - 64$.

This is a difference of cubes: $y^9 - 64 = \left(y^3\right)^3 - \boxed{}^3$.

$$y^9 - 64 = \left(y^3 - 4\right)\left[\left(y^3\right)^2 + 4y^3 + 4^2\right]$$

$$= \left(y^3 - 4\right)\left(\boxed{}\right)$$

Check: $\left(y^3 - 4\right)\left(y^6 + 4y^3 + 16\right)$
$$= y^9 + 4y^6 + 16y^3 - 4y^6 - 16y^3 - 64$$
$$= y^9 - 64.$$

Thus, $y^9 - 64 = \left(y^3 - 4\right)\left(y^6 + 4y^3 + 16\right)$.

EXAMPLE 2	YOUR TURN 2
Factor completely: $8x^3 + 1$.	Factor completely: $27x^3 + 1$.

Factor completely: $8x^3 + 1$.

This is a sum of cubes: $8x^3 + 1 = \left(2x\right)^3 + 1^3$.

$$8x^3 + 1 = \left(2x + 1\right)\left(\boxed{}\right)$$

EXAMPLE 3	YOUR TURN 3
Factor completely: $81x^7 + 24xy^6$.	Factor completely: $128x^{12} + 54y^9$.

$$81x^7 + 24xy^6 = 3x\left(27x^6 + \boxed{}\right)$$
$$= 3x\left[\left(3x^2\right)^3 + \left(2y^2\right)^3\right]$$
$$= 3x\left(3x^2 + 2y^2\right)\left(\boxed{}\right)$$

EXAMPLE 4	YOUR TURN 4
Factor: $a^6 - b^6$.	Factor: $t^6 - 1$.

This is a difference of squares and a difference of cubes. In such cases it is generally better to factor as a difference of squares first.

$$a^6 - b^6$$
$$= \left(a^3\right)^2 - \left(b^3\right)^2$$
$$= \left(a^3 + b^3\right)\left(a^3 - \boxed{}\right)$$
$$= \left(a + \boxed{}\right)\left(a^2 - ab + b^2\right)\left(a - \boxed{}\right)\left(a^2 + ab + b^2\right)$$

YOUR NOTES   Write your questions and additional notes.

Practice Exercises

Readiness Check

Match each polynomial with its classification.

1. $9x^2 - 15$
2. $8 + x^3$
3. $x^2 - 16x + 64$
4. $t^{24} - r^{15}$
5. $16x^2 - 81$

a) Difference of squares
b) Polynomial having a common factor
c) Trinomial square
d) Difference of cubes
e) Sum of cubes

Trinomial Squares

Factor completely.

6. $x^2 + 6x + 9$

7. $t^2 - 18t + 81$

8. $-y^3 - 10y^2 - 25y$

9. $9m^2 + 12mn + 4n^2$

10. $4 + 25x^2 - 20x$

11. $3x^2 + 36x + 108$

Differences of Squares

Factor completely.

12. $x^2 - 144$

13. $25a^2 - 64$

14. $100m^2 - 4$

15. $9x^4 - 16x^2$

16. $8x^2y^4 - 8x^2z^4$

17. $\dfrac{1}{25} - x^2$

More Factoring by Grouping

Factor completely.

18. $\left(c+d\right)^2 - 36$

19. $y^3 + y^2 - 4y - 4$

20. $144z^2 - x^2 + 2xy - y^2$

Sums or Differences of Cubes

Factor completely.

21. $x^3 + 27$

22. $p^3 - 8y^3$

23. $1 - 125a^3$

24. $x^3 y^6 - y^3$

25. $a^3 - \dfrac{1}{8}$

26. $a^6 - 1$

A General Factoring Strategy

ESSENTIALS

A Strategy for Factoring

A. Always factor out the largest common factor (other than 1 or -1).

B. Once the largest common factor has been factored out, *count the number of terms* in the other factor:

Two terms: Try factoring as a difference of squares first. Next, try factoring as a sum or a difference of cubes. Do *not* try to factor a sum of squares, $A^2 + B^2$.

Three terms: If it is a trinomial square, factor as such. If not, try factoring using FOIL or the *ac*-method.

Four terms: Try factoring by grouping and factoring out a common binomial factor. Next, try grouping into a difference of squares, one of which is a trinomial square.

C. Always *factor completely*. If a factor with more than one term can itself be factored further, do so.

D. Write the complete factorization and check by multiplying. If the original polynomial is not factorable, state this.

Example

- Factor: $27x^5 - x^2 y^3$.

A. Factor out the largest common factor: $27x^5 - x^2 y^3 = x^2 \left(27x^3 - y^3\right)$.

B. The factor $27x^3 - y^3$ has two terms and is a difference of cubes.

$$27x^5 - x^2 y^3 = x^2 \left(3x - y\right)\left(9x^2 + 3xy + y^2\right)$$

C. No factor with more than one term can be factored further.

D. Check:
$$x^2 \left(3x - y\right)\left(9x^2 + 3xy + y^2\right) = x^2 \left(27x^3 + 9x^2 y + 3xy^2 - 9x^2 y - 3xy^2 - y^3\right)$$
$$= x^2 \left(27x^3 - y^3\right)$$
$$= 27x^5 - x^2 y^3.$$

	📖 **Textbook**	👤 **Instructor**	▶ **Video**
GUIDED LEARNING			

EXAMPLE 1	YOUR TURN 1

Factor: $12x^2 - 48x + 45$.

A. Factor out the largest common factor:
$$12x^2 - 48x + 45 = \boxed{}\left(4x^2 - 16x + 15\right).$$

B. The trinomial factor is not a perfect square. Try using FOIL or the *ac*-method. We have
$$12x^2 - 48x + 45 = 3\left(\boxed{}\right)(2x - 5).$$

C. We cannot factor further.

D. Check: $3(2x - 3)(2x - 5) = 3\left(4x^2 - 16x + 15\right)$

$$= 12x^2 - 48x + 45.$$

Factor: $10x^2 - 40x - 120$.

EXAMPLE 2	YOUR TURN 2

Factor: $m^2 - 8m + 16 - n^2$.

A. There is no common factor (other than 1 or -1).

B. There are four terms. We try grouping to remove a common binomial factor, but find none. We try grouping as a difference of squares, one of which is a trinomial square.

$$m^2 - 8m + 16 - n^2 = \left(m^2 - 8m + 16\right) - n^2$$

$$= \left(\boxed{}\right)^2 - n^2$$

$$= (m - 4 + n)\left(\boxed{}\right)$$

C. No factor with more than one term can be factored further.

D. Check: $(m - 4 + n)(m - 4 - n)$

$$= m(m - 4 - n) - 4(m - 4 - n) + n(m - 4 - n)$$

$$= m^2 - 4m - mn - 4m + 16 + 4n + mn - 4n - n^2$$

$$= m^2 - 8m + 16 - n^2$$

Factor: $m^2 + 6m + 9 - 25n^2$.

YOUR NOTES   Write your questions and additional notes.

Practice Exercises

Readiness Check

Determine whether each statement is true or false.

1. The largest common factor of the polynomial $36x^3y^2 + 12x^2y - 18x$ is $6x$.

2. The polynomial $25x^2 - 20x + 16$ is a trinomial square.

3. The polynomial $6x^2 + 5x - 6$ is not factorable.

4. The polynomial $x^{15} - 1$ is a difference of cubes.

A General Factoring Strategy

Factor completely.

5. $-7x^3 + 14x^2 + 56x$

6. $18x^2y - 2y$

7. $-2y^2 - 12y - 18$

8. $9m^3 + 18m^2 - 25m - 50$

9. $4x^4 - 32x$

10. $27a^5 - a^2b^3$

Factor completely.

11. $x^2 - 18x + 81 - 36y^2$

12. $8p^5 - 98p^3$

13. $-4t^5 - 12t^2$

14. $24x^3 + 192y^3$

   Copyright © 2022 Pearson Education, Inc.

The Principle of Zero Products

ESSENTIALS

A **polynomial equation** is an equation in which two polynomials are set equal to each other.

A **quadratic equation** is a second-degree polynomial equation in one variable.

The Principle of Zero Products

For any real numbers a and b:

If $ab = 0$, then $a = 0$ or $b = 0$ (or both). If $a = 0$ or $b = 0$, then $ab = 0$.

To Use the Principle of Zero Products

1. Write an equivalent equation with 0 on one side, using the addition principle.
2. Factor the nonzero side of the equation.
3. Set each factor that is not a constant equal to 0.
4. Solve the resulting equations.

Example

- Solve: $x^2 + 7x = 18$.

$$x^2 + 7x = 18$$
$$x^2 + 7x - 18 = 0 \qquad \text{Getting 0 on one side}$$
$$(x+9)(x-2) = 0 \qquad \text{Factoring}$$
$$x + 9 = 0 \quad or \quad x - 2 = 0 \qquad \text{Using the principle of zero products}$$
$$x = -9 \quad or \quad x = 2$$

The solutions are -9 and 2.

GUIDED LEARNING   🔖 **Textbook**   👤 **Instructor**   ▶ **Video**

EXAMPLE 1	YOUR TURN 1
Solve: $5x^2 - 9x - 2 = 0$.	Solve: $3x^2 + 23x + 30 = 0$.

EXAMPLE 1

Solve: $5x^2 - 9x - 2 = 0$.

$$5x^2 - 9x - 2 = 0$$
$$(5x + 1)\left(x - \boxed{}\right) = 0$$
$$\boxed{} = 0 \quad or \quad x - 2 = 0$$
$$5x = -1 \quad or \quad x = \boxed{}$$
$$x = -\frac{1}{5} \quad or \quad x = 2$$

The solutions are $-\dfrac{1}{5}$ and 2.

YOUR TURN 1

Solve: $3x^2 + 23x + 30 = 0$.

EXAMPLE 2	YOUR TURN 2
Let $f(x) = -3x^3 - 147x$ and $g(x) = 42x^2$. Find all x-values for which $f(x) = g(x)$.	Let $f(x) = 2x^2 - 4x$ and $g(x) = -8x + 16$. Find all x-values for which $f(x) = g(x)$.

$$f(x) = g(x)$$
$$-3x^3 - 147x = 42x^2$$
$$-3x^3 - 42x^2 - \boxed{} = 0$$
$$\boxed{}(x^2 + 14x + 49) = 0$$
$$-3x(x+7)(x + \boxed{}) = 0$$
$$-3x = 0 \quad or \quad x + 7 = 0 \quad\quad or \quad x + 7 = 0$$
$$x = 0 \quad or \quad\quad x = \boxed{} \quad or \quad\quad\quad x = -7$$

To check, confirm that $f(0) = g(0) = 0$ and $f(-7) = g(-7) = 2058$.

For $x = 0$ or $x = \boxed{}$, we have $f(x) = g(x)$.

EXAMPLE 3	YOUR TURN 3
Find the domain of F if $F(x) = \dfrac{x}{10x^2 - 90}$.	Find the domain of F if $F(x) = \dfrac{10 - 3x}{-3x^3 + 15x^2}$.

To exclude the values for which the denominator equals 0, we solve:

$$10x^2 - 90 = 0$$
$$\boxed{}(x^2 - 9) = 0$$
$$10(x+3)(x - \boxed{}) = 0$$
$$x + 3 = 0 \quad or \quad x - 3 = 0$$
$$x = \boxed{} \quad or \quad\quad x = 3.$$

The domain of F is $\{x \mid x$ is a real number *and* $x \neq -3$ *and* $x \neq \boxed{}\}$.

YOUR NOTES   Write your questions and additional notes.

Applications and Problem Solving

ESSENTIALS

The **Pythagorean Theorem** relates the lengths of the sides of a right triangle.

A **right triangle** has a $90°$ angle, denoted by the symbol $\ulcorner$ or $\urcorner$.

The longest side of a right triangle, called the **hypotenuse**, is opposite the right angle. The other sides are called the **legs**.

The Pythagorean Theorem

In any right triangle, if a and b are the lengths of the legs and c is the length of the hypotenuse, then $a^2 + b^2 = c^2$.

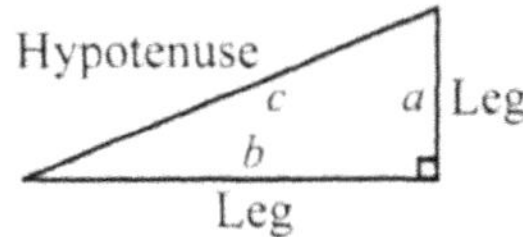

Example

- A wire is stretched from the ground to the top of a pole. The wire is 10 ft long. The height of the pole is 2 ft greater than the distance x from the bottom of the pole to the bottom of the wire. Find the height of the pole and the distance x.

   1. **Familiarize.** The wire, the pole, and the ground form a right triangle. The wire is the hypotenuse. If x is the distance from the bottom of the pole to the bottom of the wire, then $x + 2 =$ the height of the pole.

   2. **Translate.** We use the Pythagorean Theorem.

$$a^2 + b^2 = c^2$$

$$x^2 + (x+2)^2 = 10^2$$

   3. **Solve.** We solve the equation.

$$x^2 + (x+2)^2 = 10^2$$
$$x^2 + x^2 + 4x + 4 = 100$$
$$2x^2 + 4x + 4 = 100$$
$$2x^2 + 4x - 96 = 0$$
$$2(x^2 + 2x - 48) = 0$$
$$2(x+8)(x-6) = 0$$
$$x + 8 = 0 \quad or \quad x - 6 = 0$$
$$x = -8 \ or \quad x = 6$$

   4. **Check.** The length cannot be negative, so we check only 6. If $x = 6$, then $x + 2 = 6 + 2 = 8$. Since $6^2 + 8^2 = 100 = 10^2$, the answer checks.

   5. **State.** The height of the pole is 8 ft, and the distance x is 6 ft.

	🔖 Textbook	👤 Instructor	▶ Video

GUIDED LEARNING

EXAMPLE 1	YOUR TURN 1
A picture frame measures 12 cm by 14 cm, and 48 cm^2 of picture shows. Find the width of the frame.	A picture frame measures 28 in. by 20 in., and 240 in^2 of picture shows. Find the width of the frame.

1. **Familiarize.** Let x represent the width of the frame, in centimeters. Since the frame extends uniformly around the picture, the length of the picture must be $14 - 2x$ and the width must be $12 - 2x$.

2. **Translate.** Translate as follows:

$$\underbrace{\text{Length of picture showing}} \times \underbrace{\text{Width of picture showing}} = \underbrace{\text{Area of picture showing}}$$

$$\downarrow \qquad\qquad \downarrow \qquad\qquad \downarrow$$

$$(14 - 2x) \quad \times \quad (12 - 2x) \quad = \quad \boxed{}$$

3. **Solve.** Solve the equation.

$$(14 - 2x)(12 - 2x) = 48$$
$$168 - 28x - 24x + 4x^2 = 48$$
$$4x^2 - 52x + \boxed{} = 48$$
$$4x^2 - 52x + 120 = 0$$
$$4\left(x^2 - 13x + \boxed{}\right) = 0$$
$$4(x - 3)(x - 10) = 0$$
$$x - 3 = 0 \quad or \quad x - \boxed{} = 0$$
$$x = 3 \quad or \qquad\qquad x = \boxed{}$$

4. **Check.** Note that 10 is not a solution because it would yield negative measurements. If the width of the frame is 3 cm, the length of the picture showing is $14 - 2(3) = 14 - 6 = 8$ cm, the width is $12 - 2(3) = 12 - 6 = 6$ cm, and the area of the picture showing is $8 \cdot 6 = 48$ cm^2. The answer checks.

5. **State.** The width of the frame is $\boxed{}$ cm.

YOUR NOTES   Write your questions and additional notes.

Practice Exercises

Readiness Check

Classify each of the following statements as either true or false.

1. The equation $x^2 + 10x + 25 = 0$ has two unique solutions.

2. The longest side of a right triangle is called the hypotenuse.

3. The Pythagorean Theorem relates the lengths of the sides of any triangle.

4. To solve the equation $x(3x+1) = 2$ we would use the principle of zero products first.

The Principle of Zero Products

Solve.

5. $6x(3x-1) = 0$

6. $x^2 + 10x - 24 = 0$

7. $14t^2 - 35t = 0$

8. $(a+3)(a-3) = 7$

9. $15y^2 - 8 = -14y$

10. $a^3 = 2a^2 + 3a$

11. Let $f(x) = x^2 + 14x + 40$. Find a such that $f(a) = -5$.

12. Let $f(y) = y^3 + 7y^2$ and $g(y) = 4y + 28$. Find all y-values for which $f(y) = g(y)$.

Find the domain of the function f given by each of the following.

13. $f(x) = \dfrac{5}{4x^2 - 64}$

14. $f(x) = \dfrac{10 + x}{3x^3 - 15x^2 + 18x}$

Applications and Problem Solving

Solve.

15. An electronics store determines that the revenue R, in thousands of dollars, from the sale of x hundred televisions is given by

$$R(x) = \frac{4}{25}x^2 + 6x. \text{ If the cost, } C, \text{ in}$$

thousands of dollars, of producing x televisions is given by

$$C(x) = \frac{1}{5}x^2 + 6x - 1, \text{ how many}$$

televisions must be produced and sold to break even (that is, for revenue to equal cost)?

16. Mary's pool is 7 m longer than it is wide and is 17 m diagonally across. How long is Mary's pool?

Rational Expressions and Functions

ESSENTIALS

A **rational expression** consists of the quotient of two polynomials, where the polynomial in the denominator is nonzero.

Example

- Find the domain of f if $f(x) = \dfrac{x-3}{2x+5}$.

  The domain is the set of all replacements for which the rational expression is defined. The rational expression is not defined when the denominator is zero.

  $$2x + 5 = 0 \qquad \text{Setting the denominator equal to 0}$$
  $$2x = -5$$
  $$x = -\frac{5}{2}$$

  We must exclude $-\dfrac{5}{2}$ from the domain.

  The domain of $f = \left\{ x \,\middle|\, x \text{ is a real number } and \ x \neq -\dfrac{5}{2} \right\}$, or $\left(-\infty, -\dfrac{5}{2} \right) \cup \left(-\dfrac{5}{2}, \infty \right)$.

GUIDED LEARNING   📖 **Textbook**   👤 **Instructor**   ▶ **Video**

EXAMPLE 1	YOUR TURN 1
Find all numbers for which the rational expression $\dfrac{3x+5}{x-7}$ is not defined.	Find all numbers for which the rational expression $\dfrac{x}{x+3}$ is not defined.

The rational expression is not defined for replacements that make the denominator 0.

$$x - 7 = \boxed{}$$
$$x = \boxed{}$$

The rational expression is not defined for the replacement $\boxed{}$.

EXAMPLE 2	YOUR TURN 2
Find the domain of f if $f(x) = \dfrac{x^2 - 1}{x^2 + x - 6}$.	Find the domain of g if $g(x) = \dfrac{x^2 - 2x - 10}{x^2 + 3x - 4}$.

Find all replacements that make the denominator 0.

$$\boxed{} = 0$$

$$(x + 3)\left(\boxed{}\right) = 0$$

$$x + 3 = 0 \quad or \quad \boxed{} = 0$$

$$x = -3 \quad or \quad x = \boxed{}$$

The expression is not defined for the replacements -3 and $\boxed{}$.

The domain of

$$f = \left\{ x \mid x \text{ is a real number } and \ x \neq \boxed{} \ and \ x \neq \boxed{} \right\},$$

or $(-\infty, -3) \cup \left(\boxed{}, 2\right) \cup \left(\boxed{}, \infty\right).$

YOUR NOTES   Write your questions and additional notes.

Finding Equivalent Rational Expressions

ESSENTIALS

To multiply rational expressions, multiply numerators and multiply denominators:

$$\frac{A}{B} \cdot \frac{C}{D} = \frac{AC}{BD}.$$

Example

- Multiply to obtain an equivalent expression: $\dfrac{x+1}{x+1} \cdot \dfrac{3x}{x+2}$. Do not simplify.

$$\frac{x+1}{x+1} \cdot \frac{3x}{x+2} = \frac{(x+1)(3x)}{(x+1)(x+2)} \qquad \text{Note: } \frac{x+1}{x+1} = 1$$

GUIDED LEARNING	📖 **Textbook** 👤 **Instructor** ▶ **Video**	

EXAMPLE 1	YOUR TURN 1
Multiply to obtain an equivalent expression: $\dfrac{y-2}{y+3} \cdot \dfrac{y+7}{y+7}$. Do not simplify. Multiply numerators and multiply denominators. Use parentheses if the factors contain more than one term. $\dfrac{y-2}{y+3} \cdot \dfrac{y+7}{y+7} = \dfrac{(y-2)(\boxed{})}{(y+3)(\boxed{})}$	Multiply to obtain an equivalent expression: $\dfrac{3t}{3t} \cdot \dfrac{t-1}{2t+5}$.

YOUR NOTES   Write your questions and additional notes.

Simplifying Rational Expressions

ESSENTIALS

To simplify a rational expression, we remove factors that are equal to 1.

Example

- Simplify by removing a factor equal to 1: $\dfrac{2x^2+6x}{2x^3}$.

$$\frac{2x^2+6x}{2x^3} = \frac{2x(x+3)}{2x \cdot x^2}$$

$$= \frac{2x}{2x} \cdot \frac{x+3}{x^2} \qquad \frac{2x}{2x} = 1$$

$$= \frac{x+3}{x^2}$$

GUIDED LEARNING	📕 **Textbook**	👤 **Instructor**	▶ **Video**

EXAMPLE 1	YOUR TURN 1

Simplify: $\dfrac{5xy}{30x}$.

$\dfrac{5xy}{30x} = \dfrac{5 \cdot x \cdot y}{5 \cdot x \cdot 6}$ Factoring. The largest common factor is $5x$.

$\quad = \dfrac{5x}{5x} \cdot \boxed{}$

$\quad = 1 \cdot \dfrac{y}{6} \qquad \dfrac{5x}{5x} = 1$

$\quad = \dfrac{y}{6} \qquad$ Removing a factor of 1

YOUR TURN 1: Simplify: $\dfrac{12p}{4yp}$.

EXAMPLE 2	YOUR TURN 2

Simplify: $\dfrac{4x-20}{10}$.

$\dfrac{4x-20}{10} = \dfrac{2 \cdot 2(x-5)}{2 \cdot 5} \qquad$ Factoring

$\quad = \boxed{} \cdot \dfrac{2(x-5)}{5}$

$\quad = \dfrac{2(x-5)}{5} \qquad$ Removing a factor of 1

YOUR TURN 2: Simplifying: $\dfrac{x^2+4x}{5x}$.

EXAMPLE 3	YOUR TURN 3
Simplify: $\dfrac{x^2-9}{x^2+2x-3}$. $\dfrac{x^2-9}{x^2+2x-3} = \dfrac{(x-3)(\boxed{})}{(\boxed{})(x+3)}$ Factoring $= \dfrac{x-3}{\boxed{}} \cdot \dfrac{\boxed{}}{x+3}$ $= \dfrac{x-3}{\boxed{}}$ Removing a factor of 1	Simplify: $\dfrac{x^2-2x-8}{x^2-9x+20}$.

EXAMPLE 4	YOUR TURN 4
Simplify: $\dfrac{3a^2-a}{a}$. $\dfrac{3a^2-a}{a} = \dfrac{a\left(\boxed{}\right)}{a\cdot 1}$ $a = a \cdot 1$ $= \dfrac{\cancel{a}(3a-1)}{\cancel{a}\cdot 1}$ Canceling indicates a factor of 1: $\dfrac{a}{a}=1$ $= \dfrac{3a-1}{1} = \boxed{}$	Simplify: $\dfrac{y}{y^2-3y}$.

EXAMPLE 5	YOUR TURN 5
Simplify: $\dfrac{p-7}{7-p}$. $\dfrac{p-7}{7-p} = \dfrac{p-7}{-(p-7)}$ $= \dfrac{\boxed{}(p-7)}{\boxed{}(p-7)}$ $= \dfrac{\boxed{}}{\boxed{}} \cdot \dfrac{p-7}{p-7}$ $= \dfrac{1}{-1} = \boxed{}$ $\dfrac{p-7}{p-7}=1$	Simplify: $\dfrac{y+3}{-y-3}$.

YOUR NOTES   Write your questions and additional notes.

Multiplying and Simplifying

ESSENTIALS

After multiplying rational expressions, we simplify if possible.

Example

- Multiply and simplify: $\dfrac{8}{3x} \cdot \dfrac{5x^2}{2}$.

$$\frac{8}{3x} \cdot \frac{5x^2}{2} = \frac{8 \cdot 5x^2}{3x \cdot 2}$$

$$= \frac{2 \cdot 2 \cdot 2 \cdot 5 \cdot x \cdot x}{3 \cdot x \cdot 2}$$

$$= \frac{20x}{3}$$

	Textbook	**Instructor**	**Video**

GUIDED LEARNING

EXAMPLE 1	YOUR TURN 1
Multiply and simplify: $\dfrac{5y}{2} \cdot \dfrac{y-3}{20}$.	Multiply and simplify: $\dfrac{8x}{3} \cdot \dfrac{x+3}{4x}$.

$$\frac{5y}{2} \cdot \frac{y-3}{20} = \frac{5y(y-3)}{2 \cdot 20} \qquad \text{Multiplying numerators and multiplying denominators}$$

$$= \frac{5 \cdot y \cdot (y-3)}{2 \cdot 4 \cdot 5} \qquad \text{Factoring}$$

$$= \frac{5 \cdot y \cdot (y-3)}{2 \cdot 4 \cdot 5} \qquad \text{Removing a factor of 1: } \frac{5}{5}=1$$

$$= \frac{y(y-3)}{\boxed{}}$$

We leave the answer in factored form.

EXAMPLE 2	YOUR TURN 2
Multiply and simplify: $\dfrac{x^3-5x^2}{x^2+6x+5}\cdot\dfrac{x^2-x-2}{x^3-25x}$.	Multiply and simplify: $\dfrac{x^2-3x-40}{x^2-9}\cdot\dfrac{x^2-2x-3}{x^2-25}$.

$$\frac{x^3-5x^2}{x^2+6x+5}\cdot\frac{x^2-x-2}{x^3-25x}=\frac{\left(x^3-5x^2\right)\left(x^2-x-2\right)}{\left(x^2+6x+5\right)\left(x^3-25x\right)}$$

$$=\frac{x^2\left(\boxed{}\right)(x-2)\left(\boxed{}\right)}{(x+5)\left(\boxed{}\right)x\left(x^2-25\right)}$$

$$=\frac{x\cdot\boxed{}\cdot(x-5)(x-2)(x+1)}{(x+5)(x+1)x(x+5)\left(\boxed{}\right)}$$

$$=\frac{\cancel{x}\cdot x\cdot\cancel{(x-5)}(x-2)\cancel{(x+1)}}{(x+5)\cancel{(x+1)}\cdot\cancel{x}\cdot(x+5)\cancel{(x-5)}}$$

$$=\frac{\boxed{}(x-2)}{\left(\boxed{}\right)^2}$$

YOUR NOTES   Write your questions and additional notes.

Dividing and Simplifying

ESSENTIALS

To divide by a rational expression, multiply by its reciprocal: $\dfrac{A}{B} \div \dfrac{C}{D} = \dfrac{A}{B} \cdot \dfrac{D}{C} = \dfrac{AD}{BC}$.

Example

- Divide: $\dfrac{3x}{5} \div \dfrac{9}{25y}$.

$$\dfrac{3x}{5} \div \dfrac{9}{25y} = \dfrac{3x}{5} \cdot \dfrac{25y}{9} \qquad \text{Multiplying by the reciprocal of the divisor}$$

$$= \dfrac{3x \cdot 25y}{5 \cdot 9}$$

$$= \dfrac{3 \cdot x \cdot 5 \cdot 5 \cdot y}{5 \cdot 3 \cdot 3}$$

$$= \dfrac{\cancel{3} \cdot x \cdot \cancel{5} \cdot 5 \cdot y}{\cancel{5} \cdot \cancel{3} \cdot 3}$$

$$= \dfrac{5xy}{3}$$

GUIDED LEARNING   📖 **Textbook**   🧑‍🏫 **Instructor**   ▶️ **Video**

EXAMPLE 1	YOUR TURN 1
Divide and simplify: $\dfrac{x^2 + 3x}{x-1} \div \dfrac{5x}{x+2}$.	Divide and simplify: $\dfrac{x+2}{x-5} \div \dfrac{5x}{2x-10}$.

$$\dfrac{x^2+3x}{x-1} \div \dfrac{5x}{x+2} = \dfrac{x^2+3x}{x-1} \cdot \dfrac{\boxed{}}{5x}$$

$$= \dfrac{\left(x^2+3x\right)(x+2)}{(x-1) \cdot 5x} \qquad \text{Multiplying}$$

$$= \dfrac{x\boxed{}(x+2)}{(x-1) \cdot 5 \cdot x} \qquad \text{Factoring}$$

$$= \dfrac{\cancel{x}(x+3)(x+2)}{(x-1) \cdot 5 \cdot \cancel{x}} \qquad \dfrac{x}{x} = 1$$

$$= \dfrac{(x+3)(x+2)}{\boxed{}(x-1)} \qquad \text{Simplifying}$$

EXAMPLE 2	YOUR TURN 2
Divide and simplify: $\dfrac{x^2-3x-10}{x+5} \div \dfrac{x^2+2x}{x^2+4x-5}$.	Divide and simplify: $\dfrac{x^2+x-6}{x+1} \div \dfrac{x^2-4}{x^2+4x+3}$.

$$\frac{x^2-3x-10}{x+5} \div \frac{x^2+2x}{x^2+4x-5}$$

$$= \frac{x^2-3x-10}{x+5} \cdot \frac{\boxed{}}{\boxed{}}$$

$$= \frac{(x-5)(x+2)(x+5)(x-1)}{(x+5)x(x+2)} \qquad \text{Factoring}$$

$$= \frac{(x-5)\,\cancel{(x+2)}\,\cancel{(x+5)}\,(x-1)}{\cancel{(x+5)}\,x\,\cancel{(x+2)}}$$

$$= \frac{(x-5)(x-1)}{\boxed{}} \qquad \text{Simplifying}$$

EXAMPLE 3	YOUR TURN 3
Perform the indicated operations and simplify: $\dfrac{a^3-1}{a^2-a-2} \div (a^2-a) \cdot (a+1)^2$. Rewrite the division as multiplication, then multiply.	Perform the indicated operations and simplify: $\dfrac{x^2-y^2}{x^2 y^2} \div (x^3+y^3) \cdot (x^2-xy+y^2)$.

$$\frac{a^3-1}{a^2-a-2} \div (a^2-a) \cdot (a+1)^2$$

$$= \frac{a^3-1}{a^2-a-2} \cdot \frac{1}{\boxed{}} \cdot \frac{(a+1)^2}{1}$$

$$= \frac{(a^3-1)(a+1)^2}{(a^2-a-2)(a^2-a)}$$

$$= \frac{(a-1)\left(\boxed{}+a+1\right)(a+1)(a+1)}{(a+1)\left(\boxed{}\right)a\left(\boxed{}\right)} \qquad \text{Factoring}$$

$$= \frac{\cancel{(a-1)}\,(a^2+a+1)(a+1)\,\cancel{(a+1)}}{\cancel{(a+1)}\,(a-2)\,a\,\cancel{(a-1)}}$$

$$= \frac{(a^2+a+1)\left(\boxed{}\right)}{a\left(\boxed{}\right)}$$

YOUR NOTES   Write your questions and additional notes.

Practice Exercises

Readiness Check

Classify each statement as true or false.

1. _________________ To divide rational expressions, we multiply by the reciprocal of the divisor.

2. _________________ The simplified form of $\dfrac{x+4}{4}$ is x.

3. _________________ A rational expression is not defined for numbers that make the numerator zero.

4. _________________ To simplify rational expressions, we remove factors equal to 1.

Rational Expressions and Functions

5. Find all the numbers for which the rational expression $\dfrac{t^2-4}{3t+2}$ is not defined.

6. Find the domain of f if
$$f(x) = \frac{x-7}{x^2-10x-11}.$$

Finding Equivalent Rational Expressions

7. Multiply to obtain an equivalent expression: $\dfrac{a+3}{a+3} \cdot \dfrac{a-1}{2a+5}$. Do not simplify.

Simplifying Rational Expressions

Simplify.

8. $\dfrac{6a^3}{10a}$

9. $\dfrac{4x-16}{12}$

10. $\dfrac{5x^2 - 5x}{2x - 2}$

11. $\dfrac{t^2 - 3t - 18}{t^2 + 6t + 9}$

Multiplying and Simplifying

Multiply and simplify.

12. $\dfrac{4x + 2}{25y^2} \cdot \dfrac{5y}{2x}$

13. $\dfrac{x^2 - 1}{x^2 + 7x + 12} \cdot \dfrac{x^2 + 3x}{x^2 + 2x + 1}$

Dividing and Simplifying

Divide and simplify.

14. $\dfrac{x - 2}{3} \div \dfrac{4 - 2x}{9}$

15. $\dfrac{y^3 - 8}{y - 5} \div \dfrac{y^2 - 4y + 4}{y^2 - 7y + 10}$

16. $\dfrac{a^2 + 6a}{a^2 + 2a - 3} \div \dfrac{a^2 + 12a + 36}{4a - 4}$

17. Perform the indicated operations and simplify.

$$\dfrac{4x^2 - 25y^2}{2x - y} \div \dfrac{x^2 + 3xy - 10y^2}{5y - 10x} \cdot \dfrac{x^2 - 4xy + 4y^2}{2x + 5y}$$

Finding LCMs by Factoring

ESSENTIALS

To find the LCM (least common multiple) of two or more algebraic expressions, we find their prime factorizations. Then we use each factor the greatest number of times that it occurs in any one prime factorization.

Example

- Find the LCM of $m^2 + 4m + 3$ and $2m^2 + 4m - 6$.

$$m^2 + 4m + 3 = (m+1)(m+3)$$

$$2m^2 + 4m - 6 = 2(m^2 + 2m - 3) = 2(m-1)(m+3)$$

$$\text{LCM} = 2(m+1)(m-1)(m+3)$$

GUIDED LEARNING　　　📖 **Textbook**　　　👤 **Instructor**　　　▶ **Video**

EXAMPLE 1	YOUR TURN 1
Find the LCM of 24 and 50.	Find the LCM of 12 and 30.

$$24 = 2 \cdot 2 \cdot 2 \cdot \boxed{}$$

$$50 = 2 \cdot 5 \cdot \boxed{}$$

$$\text{LCM} = 2 \cdot 2 \cdot 2 \cdot \boxed{} \cdot 5 \cdot \boxed{}$$

$$\text{LCM} = \boxed{}$$

EXAMPLE 2	YOUR TURN 2
Find the LCM of $12a^2b^4$ and $18a^5b^2$.	Find the LCM of $9x^2y$ and $5xyz$.

Write the prime factorizations.

$$12a^2b^4 = \boxed{} \cdot 2 \cdot 3 \cdot \boxed{} \cdot \boxed{} \cdot b \cdot b \cdot b \cdot b$$

$$18a^5b^2 = 2 \cdot 3 \cdot \boxed{} \cdot a \cdot a \cdot a \cdot a \cdot \boxed{} \cdot b \cdot \boxed{}$$

The factors that appear are 2, 3, a, and b.

We use each factor the greatest number of times that it occurs in any one factorization.

$$\text{LCM} = 2 \cdot 2 \cdot 3 \cdot \boxed{} \cdot a \cdot a \cdot a \cdot a \cdot a \cdot b \cdot b \cdot b \cdot \boxed{}$$

$$\text{LCM} = \boxed{}$$

EXAMPLE 3	YOUR TURN 3
Find the LCM of $x^2 + 5x + 6$ and $(x+3)^2$. Write the prime factorizations. $x^2 + 5x + 6 = (\boxed{})(x+3)$ $(x+3)^2 = (\boxed{})(x+3)$ $\text{LCM} = (\boxed{})(x+3)(x+3)$, or $(\boxed{})(\boxed{})^2$	Find the LCM of $2x^2 - 9x - 5$ and $x^2 - 25$.
EXAMPLE 4	YOUR TURN 4
Find the LCM of $x + 2$ and $2x + 6$. Write the prime factorizations. $x + 2 = x + 2$ This expression is prime. $2x + 6 = 2(\boxed{})$ These expressions do not share a common factor other than 1, so the LCM is their product. $\text{LCM} = 2(\boxed{})(\boxed{})$	Find the LCM of $t^2 + t$ and $t - 1$.
EXAMPLE 5	YOUR TURN 5
Find the LCM of $c^2 - 25$, $c^3 + 5c^2$, and $c^2 - 6c + 5$. Write the prime factorizations. $c^2 - 25 = (c+5)(\boxed{})$ $c^3 + 5c^2 = c \cdot c(\boxed{})$ $c^2 - 6c + 5 = (c-1)(\boxed{})$ $\text{LCM} = c^2(c+5)(\boxed{})(\boxed{})$	Find the LCM of $y^6 - y^2$, $2y + 2$, and $y^2 + 2y + 1$.

YOUR NOTES   Write your questions and additional notes.

Adding and Subtracting Rational Expressions

ESSENTIALS

To add or subtract when denominators are the same, add or subtract the numerators and keep the same denominator.

$$\frac{A}{C}+\frac{B}{C}=\frac{A+B}{C} \quad \text{and} \quad \frac{A}{C}-\frac{B}{C}=\frac{A-B}{C}, \text{ where } C \neq 0.$$

To add or subtract rational expressions whose denominators are different,

1. Find the *least common denominator* (LCD) by finding the least common multiple (LCM) of the denominators.

2. Rewrite each of the original rational expressions, as needed, in an equivalent form that has the LCD.

3. Add or subtract the numerators. Write the result over the LCD.

4. Simplify, if possible.

Example

- Add: $\dfrac{5}{4x}+\dfrac{6}{2y}$.

$$\frac{5}{4x}+\frac{6}{2y}=\frac{5}{4x}\cdot\frac{y}{y}+\frac{6}{2y}\cdot\frac{2x}{2x} \qquad \text{The LCD is } 4xy.$$

$$=\frac{5y}{4xy}+\frac{12x}{4xy}$$

$$=\frac{5y+12x}{4xy}$$

GUIDED LEARNING   🔖 **Textbook**   👤 **Instructor**   ▶ **Video**

EXAMPLE 1	YOUR TURN 1
Add: $\dfrac{9+3x}{x}+\dfrac{2}{x}$.	Add: $\dfrac{5+2a}{a}+\dfrac{4}{a}$.

$$\frac{9+3x}{x}+\frac{2}{x}=\frac{9+3x+2}{\boxed{}}$$

$$=\frac{\boxed{}+3x}{\boxed{}}$$

EXAMPLE 2	YOUR TURN 2
Add: $\dfrac{5}{27x}+\dfrac{8}{9x^2}$. $\dfrac{5}{27x}+\dfrac{8}{9x^2}$ $=\dfrac{5}{27x}\cdot\dfrac{x}{x}+\dfrac{8}{9x^2}\cdot\dfrac{\boxed{}}{\boxed{}}$ Multiplying to obtain a common denominator $=\dfrac{5x}{27x^2}+\dfrac{24}{27x^2}$ $=\dfrac{5x+24}{\boxed{}}$	Add: $\dfrac{2}{15y}+\dfrac{3}{5y^3}$.
EXAMPLE 3	YOUR TURN 3
Add: $\dfrac{6ac}{(a+c)(a-c)}+\dfrac{a-2c}{a+c}$. $\dfrac{6ac}{(a+c)(a-c)}+\dfrac{a-2c}{a+c}$ $=\dfrac{6ac}{(a+c)(a-c)}+\dfrac{a-2c}{a+c}\cdot\dfrac{\boxed{}}{\boxed{}}$ $=\dfrac{6ac+(a-2c)(a-c)}{(a+c)(a-c)}$ $=\dfrac{6ac+a^2-ac-2ac+2c^2}{(a+c)(a-c)}$ Multiplying $=\dfrac{\boxed{}}{(a+c)(a-c)}$ Combining like terms $=\dfrac{(a+2c)(a+c)}{(a+c)(a-c)}$ Factoring $=\dfrac{(a+2c)\,\cancel{(a+c)}}{\cancel{(a+c)}\,(a-c)}$ $=\boxed{}$	Add: $\dfrac{2xy}{(x+y)(x+2y)}+\dfrac{x-y}{x+2y}$.

EXAMPLE 4	YOUR TURN 4
Subtract: $\dfrac{5x}{x-7y} - \dfrac{4x-1}{7y-x}$.	Subtract: $\dfrac{3m}{m-5} - \dfrac{2m+1}{5-m}$.

$$\frac{5x}{x-7y} - \frac{4x-1}{7y-x} = \frac{5x}{x-7y} - \frac{-1}{-1}\cdot\frac{4x-1}{7y-x}$$

$$= \frac{5x}{x-7y} - \frac{1-4x}{\boxed{}} = \frac{5x-(1-4x)}{x-7y}$$

$$= \frac{5x-1\,\boxed{}\,4x}{x-7y} = \frac{9x-1}{x-7y}$$

EXAMPLE 5	YOUR TURN 5
Subtract: $\dfrac{3x}{x^2+4x-5} - \dfrac{2x}{x^2+6x+5}$.	Subtract: $\dfrac{2x}{x^2+4x-5} - \dfrac{x}{x^2+7x+10}$.

$$\frac{3x}{x^2+4x-5} - \frac{2x}{x^2+6x+5} = \frac{3x}{(x+5)(x-1)} - \frac{2x}{(x+5)(x+1)}$$

$$\text{LCD} = (x+5)(x-1)(x+1)$$

$$= \frac{3x}{(x+5)(x-1)}\cdot\frac{\boxed{}}{\boxed{}} - \frac{2x}{(x+5)(x+1)}\cdot\frac{x-1}{x-1}$$

$$= \frac{3x^2+\boxed{}}{(x+5)(x-1)(x+1)} - \frac{2x^2-\boxed{}}{(x+5)(x-1)(x+1)}$$

$$= \frac{3x^2+3x-(\boxed{})}{(x+5)(x-1)(x+1)} = \frac{3x^2+3x-\boxed{}+\boxed{}}{(x+5)(x-1)(x+1)}$$

$$= \frac{x^2+\boxed{}}{(x+5)(x-1)(x+1)} = \frac{x(\boxed{})}{(x+5)(x-1)(x+1)}$$

$$= \frac{x\,\cancel{(x+5)}}{\cancel{(x+5)}\,(x-1)(x+1)} = \frac{x}{(x-1)(x+1)}$$

YOUR NOTES   Write your questions and additional notes.

Combined Additions and Subtractions

ESSENTIALS

To simplify a combination of additions and subtractions of rational expressions, find the LCD of all the rational expressions, write all the expressions in an equivalent form that has the LCD, and add or subtract as indicated.

| *GUIDED LEARNING* | 📖 **Textbook** | 👤 **Instructor** | ▶ **Video** |

EXAMPLE 1

Perform the indicated operations and simplify:

$$\frac{6x}{x^2-1}-\frac{2}{x+1}+\frac{1}{1-x}.$$

$$\frac{6x}{x^2-1}-\frac{2}{x+1}+\frac{1}{1-x}$$

$$=\frac{6x}{(x+1)(x-1)}-\frac{2}{x+1}+\frac{-1}{-1}\cdot\frac{1}{1-x}$$

$$=\frac{6x}{(x+1)(x-1)}-\frac{2}{x+1}+\frac{-1}{\boxed{}}\qquad \text{LCD}=(x+1)(x-1)$$

$$=\frac{6x}{(x+1)(x-1)}-\frac{2}{x+1}\cdot\frac{\boxed{}}{\boxed{}}+\frac{-1}{x-1}\cdot\frac{x+1}{x+1}$$

$$=\frac{6x}{(x+1)(x-1)}-\frac{2x-\boxed{}}{(x+1)(x-1)}+\frac{-x-\boxed{}}{(x-1)(x+1)}$$

$$=\frac{6x-\left(\boxed{}\right)+(-x)-1}{(x+1)(x-1)}$$

$$=\frac{6x-2x+\boxed{}+(-x)-1}{(x+1)(x-1)}$$

$$=\frac{3x+\boxed{}}{(x+1)(x-1)}$$

YOUR TURN 1

Perform the indicated operations and simplify:

$$\frac{3x}{x^2-4}+\frac{1}{2-x}+\frac{2}{x-2}.$$

YOUR NOTES   Write your questions and additional notes.

Practice Exercises

Readiness Check

Classify each statement as true or false.

1. _________________ To subtract fractions, they must have a common denominator.

2. _________________ When adding rational expressions, we add numerators and add denominators, and then simplify, if possible.

3. _________________ The LCM of $8x^3$ and $4x$ is $32x^4$.

4. _________________ The LCD of two rational expressions is the LCM of the denominators of the rational expressions.

Finding LCMs by Factoring

Find the LCM.

5. $24, 40$

6. $10x^2y, \ 15x^4y$

7. $y+5, \ y^2-5y$

8. $x^2-1, \ x^2-6x-7$

9. $12a^3-24a^2, \ a^3-4a^2+4a$

10. $6x+18, \ 9x^2-81, \ x^3+6x^2+9x$

Adding and Subtracting Rational Expressions

Add or subtract. Then simplify.

11. $\dfrac{8}{3k}+\dfrac{1}{3k}$

12. $\dfrac{2y-9}{y^2-3y-4}-\dfrac{y+4}{y^2-3y-4}$

13. $\dfrac{x+4}{x+5}+\dfrac{2x-3}{x-6}$

14. $\dfrac{x}{x^2-2x+1}-\dfrac{3}{x^2-6x+5}$

15. $\dfrac{cd}{c^2-d^2}-\dfrac{c+d}{c-d}$

16. $\dfrac{2y}{(y-1)(y-3)}+\dfrac{3}{(y-1)(y+2)}$

Combined Additions and Subtractions

Perform the indicated operations and simplify.

17. $\dfrac{6}{m-2}-\dfrac{m+3}{m^2+m-6}+\dfrac{5}{m+3}$

Divisor a Monomial

ESSENTIALS

To divide a polynomial by a monomial, divide each term of the polynomial by the monomial.

$$\frac{A+B}{C} = \frac{A}{C} + \frac{B}{C}$$

Examples

- $$\frac{4x^2 - 2x + 8}{2x}$$

$$= \frac{4x^2}{2x} - \frac{2x}{2x} + \frac{8}{2x}$$

$$= 2x - 1 + \frac{4}{x}$$

- Divide $8x^3 - 4x^2 + 20x + 40$ by $4x^2$.

$$\left(8x^3 - 4x^2 + 20x + 40\right) \div \left(4x^2\right)$$

$$= \frac{8x^3 - 4x^2 + 20x + 40}{4x^2}$$

$$= \frac{8x^3}{4x^2} - \frac{4x^2}{4x^2} + \frac{20x}{4x^2} + \frac{40}{4x^2}$$

$$= 2x - 1 + \frac{5}{x} + \frac{10}{x^2}$$

GUIDED LEARNING   📖 **Textbook**   👤 **Instructor**   ▶ **Video**

EXAMPLE 1	YOUR TURN 1
Divide $15y^3 - 25y^2 + 5$ by $5y$.	Divide $-48x^4 + 16x^2 - 8$ by $4x$.

Divide each term of the numerator by the monomial, and then perform the three indicated divisions.

$$\left(15y^3 - 25y^2 + 5\right) \div \left(5y\right)$$

$$= \frac{15y^3 - 25y^2 + 5}{5y}$$

$$= \frac{\boxed{}}{5y} - \frac{\boxed{}}{5y} + \frac{\boxed{}}{5y}$$

$$= 3y^2 - 5y + \frac{1}{y}$$

EXAMPLE 2	YOUR TURN 2
Divide: $\left(20x^5y^4 + 4x^3y^5 - 16x^2y^8\right) \div \left(-2x^2y^3\right)$.	Divide:
We first write a single rational expression, then divide each term of the numerator by the monomial, and finally perform the indicated divisions.	$\left(12x^4y^4 - 18x^5y^2 + 30x^2y^5\right) \div \left(-3x^2y\right)$.

$$\left(20x^5y^4 + 4x^3y^5 - 16x^2y^8\right) \div \left(-2x^2y^3\right)$$

$$= \frac{20x^5y^4 + 4x^3y^5 - 16x^2y^8}{-2x^2y^3}$$

$$= \frac{20x^5y^4}{-2x^2y^3} + \frac{4x^3y^5}{-2x^2y^3} - \frac{16x^2y^8}{-2x^2y^3}$$

$$= \boxed{} - 2xy^2 + \boxed{}$$

YOUR NOTES   Write your questions and additional notes.

Divisor Not a Monomial

ESSENTIALS

In division of polynomials when the divisor has more than one term, we use a procedure very similar to long division in arithmetic.

Tips for Dividing Polynomials

1. Arrange the polynomials in descending order.
2. If there are missing terms in the dividend, write them with 0 coefficients or leave space for them.
3. Perform the long division process until the degree of the remainder is less than the degree of the divisor.

Example

- Divide: $\left(x^2 - 5x - 24\right) \div \left(x - 8\right)$.

$$
\begin{array}{r}
x + 3 \\
x - 8 \overline{)\ x^2 - 5x - 24} \\
\underline{x^2 - 8x} \\
3x - 24 \\
\underline{3x - 24} \\
0
\end{array}
$$

The quotient is $x + 3$.

	🔲 **Textbook**	👤 **Instructor**	▶ **Video**

GUIDED LEARNING

EXAMPLE 1	YOUR TURN 1
Divide $x^2 + 8x + 3$ by $x + 2$.	Divide $x^2 + 4x + 1$ by $x + 5$.

EXAMPLE 1

Divide $x^2 + 8x + 3$ by $x + 2$.

$$
\begin{array}{r}
x \\
x + 2 \overline{)\ x^2 + 8x + 3}
\end{array}
$$
Divide: $x^2 / x = x$.

$\underline{x^2 + 2x}$ Multiply: $x\left(x + 2\right) = x^2 + 2x$.

$\boxed{} + 3$ Subtract: $x^2 + 8x - \left(x^2 + 2x\right) = 6x$.

We continue.

(continued)

YOUR TURN 1

Divide $x^2 + 4x + 1$ by $x + 5$.

$$\begin{array}{r}
x + \boxed{} \\
x+2 \overline{\smash{)}\, x^2 + 8x + 3} \\
\underline{x^2 + 2x} \\
\end{array}$$

$6x + \ 3$ Divide: $6x/x = 6.$

$\underline{6x + 12}$ Multiply: $6(x+2) = 6x + 12.$

$\boxed{}$ Subtract: $6x + 3 - (6x + 12) = -9.$

The remainder is -9.

Check: $(x+2)(x+6) + (-9) = x^2 + 8x + 12 + (-9)$

$$= x^2 + 8x + 3$$

We get the dividend, so the answer checks.

The answer is $x + 6$, R -9, or $x + 6 + \dfrac{-9}{x+2}$.

EXAMPLE 2	YOUR TURN 2

Divide: $(n^3 - 6 + 2n^2) \div (n^2 - 2).$

We rewrite the dividend in descending order and write in the missing term, $0n$.

$$\begin{array}{r}
n + 2 \\
n^2 - 2 \overline{\smash{)}\, n^3 + 2n^2 + 0n - 6} \\
\underline{n^3 - 2n} \\
2n^2 + 2n - 6 \\
\underline{2n^2 - 4} \\
2n - \boxed{}
\end{array}$$

The degree of the remainder is less than the degree of the divisor, so we are finished.

The answer is $n + 2 + \dfrac{2n - 2}{\boxed{}}$.

Divide:
$(3y^2 - 4 + y^3) \div (y^2 + 1).$

YOUR NOTES   Write your questions and additional notes.

Synthetic Division

ESSENTIALS

To use synthetic division to divide a polynomial by a binomial, the divisor must be in the form $x - a$.

GUIDED LEARNING

 **Textbook**  **Instructor**  **Video**

EXAMPLE 1	YOUR TURN 1

EXAMPLE 1

Use synthetic division to divide:

$\left(x^3 + 4x^2 + x - 4\right) \div \left(x + 1\right)$.

Note: $x + 1 = x - \left(-1\right)$. We use the format shown below with -1 in the half-box and the coefficients of the dividend, written in descending order, to the right of it. Next we bring down the first 1, multiply it by -1, write the product below the 4, and then add: $4 + \left(-1\right) = 3$. We have

$$\begin{array}{r|rrrr} -1 & 1 & 4 & 1 & -4 \\ & & -1 & & \\ \hline & 1 & 3 & & \end{array}$$

Next we multiply 3 by -1, write the product below the second 1, and add: $1 + \left(-3\right) = -2$. We have

$$\begin{array}{r|rrrr} -1 & 1 & 4 & 1 & -4 \\ & & -1 & -3 & \\ \hline & 1 & 3 & -2 & \end{array}$$

Finally, multiply -2 by -1, write the product below -4 and add: $-4 + 2 = -2$. We have

$$\begin{array}{r|rrr|r} -1 & 1 & 4 & 1 & -4 \\ & & -1 & -3 & \boxed{} \\ \hline & 1 & 3 & -2 & -2 \end{array}$$

The degree of the quotient is one less than the degree of the dividend. The coefficients of the quotient are 1, 3, and -2, and the remainder is -2.

The answer is $x^2 + 3x - 2$, R -2, or

$x^2 + 3x - 2 + \dfrac{-2}{\boxed{}}$.

YOUR TURN 1

Use synthetic division to divide:

$\left(x^3 - 6x^2 + 2x - 5\right) \div \left(x + 2\right)$.

EXAMPLE 2	YOUR TURN 2
Use synthetic division to divide: $(9x - 5 + 4x^3) \div (x - 3)$. Write the dividend in descending order. Include the missing term, $0x^2$: $(4x^3 + 0x^2 + 9x - 5) \div (x - 3)$. We use the procedure outlined in Example 1.	Use synthetic division to divide: $(2x^2 - 1 + 5x^3) \div (x - 2)$.

$$
\begin{array}{r|rrrr}
\underline{3} & 4 & 0 & 9 & -5 \\
& & 12 & 36 & \Box \\
\hline
& 4 & \Box & 45 & 130
\end{array}
$$

The answer is $4x^2 + \Box + 45 + \dfrac{130}{x - 3}$.

YOUR NOTES   Write your questions and additional notes.

Practice Exercises

Readiness Check

Classify each statement as true or false.

1. _________________ When dividing one polynomial by another, we stop dividing when the degree of the remainder is equal to the degree of the divisor.

2. _________________ When dividing a polynomial by a monomial, the quotient has the same number of terms as the dividend.

3. _________________ To check the result after dividing one polynomial by another, we multiply the divisor by the remainder and add the quotient.

4. _________________ When dividing one polynomial by another, if the remainder is zero then the divisor is a factor of the dividend.

Divisor a Monomial

Divide and check.

5. $\dfrac{42x^5 + 14x^3 - 35x^2}{7x^2}$

6. $\dfrac{24m^4n^7 - 15m^3n^3 + 18m^4n^2}{3m^3n^2}$

7. $\left(12y^8 - 3y^2 + 6y\right) \div \left(-3y\right)$

8. $\left(a^4b^4 + 2a^2b^2 - 10ab\right) \div \left(2ab^2\right)$

Divisor Not a Monomial

Divide and check.

9. $\left(x^2 + 2x - 15\right) \div \left(x - 3\right)$

10. $\left(z^2 - 5z - 20\right) \div \left(z + 6\right)$

11. $\left(4x^3 + 8x^2 - 3x + 5\right) \div \left(2x - 1\right)$ **12.** $\left(y^3 - 5 + y\right) \div \left(y + 2\right)$

Synthetic Division

Use synthetic division to divide.

13. $\left(x^3 - 5x^2 + 2x - 1\right) \div \left(x - 3\right)$ **14.** $\left(4x^3 + 3x^2 - 5x - 3\right) \div \left(x + 4\right)$

15. $\left(y^3 - 2y + 8\right) \div \left(y + 1\right)$ **16.** $\left(n^5 - 32\right) \div \left(n + 2\right)$

17. $\left(9x^4 + 3x^3 - 23x^2 + 13x - 2\right) \div \left(x - \dfrac{1}{3}\right)$

Simplifying Complex Rational Expressions

ESSENTIALS

A **complex rational expression** is a rational expression that contains rational expressions within its numerator and/or its denominator.

To simplify a complex rational expression by multiplying by 1,

1. Find the LCM of all the denominators of all the rational expressions within the complex rational expression.

2. Multiply by 1 using $\dfrac{\text{LCM}}{\text{LCM}}$.

3. If possible, simplify.

To simplify a complex rational expression by adding or subtracting in the numerator and the denominator,

1. Add or subtract, as necessary, to get one rational expression in the numerator.

2. Add or subtract, as necessary, to get one rational expression in the denominator.

3. Divide the numerator by the denominator.

4. If possible, simplify.

Example

$$\frac{4-\dfrac{1}{2x}}{\dfrac{4}{x}+\dfrac{1}{x^2}} = \frac{4-\dfrac{1}{2x}}{\dfrac{4}{x}+\dfrac{1}{x^2}} \cdot \frac{2x^2}{2x^2}$$

The LCM of all the denominators is $2x^2$.

Multiplying by 1 using $\dfrac{2x^2}{2x^2}$

$$= \frac{4 \cdot 2x^2 - \dfrac{1}{2x} \cdot 2x^2}{\dfrac{4}{x} \cdot 2x^2 + \dfrac{1}{x^2} \cdot 2x^2}$$

Using the distributive law

$$= \frac{8x^2 - x}{8x + 2}$$

Simplifying in the numerator and denominator

$$= \frac{x(8x-1)}{2(4x+1)}$$

We can't simplify further.

<table>
<tr><td></td><td>⬤ **Textbook**</td><td>⬤ **Instructor**</td><td>⬤ **Video**</td></tr>
</table>

GUIDED LEARNING

EXAMPLE 1	YOUR TURN 1
Simplify: $\dfrac{\dfrac{3}{2x^3}+\dfrac{4}{x^2}}{\dfrac{5}{6x}-\dfrac{1}{x^2}}$.	Simplify: $\dfrac{\dfrac{6}{am^4}+\dfrac{m}{2a^2}}{\dfrac{6}{am^2}+\dfrac{2m}{a}}$.

We use the method of multiplying by 1.

The LCM of all the denominators is $6x^3$. We multiply the rational expression by 1 using the form $\dfrac{6x^3}{6x^3}$.

$$\frac{\dfrac{3}{2x^3}+\dfrac{4}{x^2}}{\dfrac{5}{6x}-\dfrac{1}{x^2}}=\frac{\dfrac{3}{2x^3}+\dfrac{4}{x^2}}{\dfrac{5}{6x}-\dfrac{1}{x^2}}\cdot\frac{\boxed{}}{\boxed{}}$$

$$=\frac{\left(\dfrac{3}{2x^3}+\dfrac{4}{x^2}\right)6x^3}{\left(\dfrac{5}{6x}-\dfrac{1}{x^2}\right)6x^3}$$

$$=\frac{\dfrac{3}{2x^3}\cdot 6x^3+\dfrac{4}{x^2}\cdot 6x^3}{\dfrac{5}{6x}\cdot 6x^3-\dfrac{1}{x^2}\cdot 6x^3}\quad\text{Distributing } 6x^3$$

$$=\frac{\boxed{}+\boxed{}}{\boxed{}-\boxed{}}\quad\text{Simplifying}$$

$$=\frac{3(3+8x)}{x(5x-6)}\quad\text{Factoring}$$

We cannot simplify further.

EXAMPLE 2	YOUR TURN 2

EXAMPLE 2

Simplify: $\dfrac{\dfrac{3}{2x^3}+\dfrac{4}{x^2}}{\dfrac{5}{6x}-\dfrac{1}{x^2}}$.

We use the method of adding or subtracting in the numerator and the denominator.

We multiply to obtain a common denominator in the numerator and in the denominator.

$$\dfrac{\dfrac{3}{2x^3}+\dfrac{4}{x^2}}{\dfrac{5}{6x}-\dfrac{1}{x^2}}=\dfrac{\dfrac{3}{2x^3}+\dfrac{4}{x^2}\cdot\dfrac{2x}{2x}}{\dfrac{5}{6x}\cdot\dfrac{x}{x}-\dfrac{1}{x^2}\cdot\dfrac{6}{6}}$$

$$=\dfrac{\dfrac{3}{2x^3}+\dfrac{8x}{2x^3}}{\dfrac{5x}{6x^2}-\dfrac{6}{6x^2}}$$

$$=\dfrac{\dfrac{\boxed{}}{2x^3}}{\dfrac{\boxed{}}{6x^2}} \qquad \text{Adding in the numerator; subtracting in the denominator}$$

$$=\dfrac{3+8x}{2x^3}\cdot\dfrac{\boxed{}}{\boxed{}} \qquad \text{Multiplying by the reciprocal of the divisor}$$

$$=\dfrac{(3+8x)\cdot 3\cdot \cancel{2}\cdot \cancel{x}}{\cancel{2}\cdot \cancel{x}\cdot x(5x-6)}$$

$$=\dfrac{3(3+8x)}{x(5x-6)} \qquad \dfrac{2x^2}{2x^2}=1$$

YOUR TURN 2

Simplify: $\dfrac{\dfrac{2}{4k}-\dfrac{k}{2}}{\dfrac{6}{k^2}-\dfrac{1}{k}}$.

EXAMPLE 3	YOUR TURN 3
Simplify: $\dfrac{\dfrac{1}{x}-\dfrac{1}{y}}{\dfrac{1}{x^2}-\dfrac{1}{y^2}}$.	Simplify: $\dfrac{\dfrac{1}{m^3}-\dfrac{1}{n^3}}{\dfrac{1}{m^2}-\dfrac{1}{n^2}}$.

$$\dfrac{\dfrac{1}{x}-\dfrac{1}{y}}{\dfrac{1}{x^2}-\dfrac{1}{y^2}}=\dfrac{\dfrac{1}{x}\cdot\dfrac{y}{y}-\dfrac{1}{y}\cdot\boxed{}}{\dfrac{1}{x^2}\cdot\dfrac{y^2}{y^2}-\dfrac{1}{y^2}\cdot\boxed{}}$$

$$=\dfrac{\dfrac{y-x}{xy}}{\dfrac{y^2-x^2}{x^2y^2}}$$ Writing a single rational expression in the numerator and in the denominator

$$=\dfrac{y-x}{xy}\cdot\dfrac{x^2y^2}{y^2-x^2}$$

$$=\dfrac{\cancel{y}-x}{\cancel{x}\cdot\cancel{y}}\cdot\dfrac{\cancel{x}\cdot x\cdot\cancel{y}\cdot y}{(y-x)(y+x)}$$ Factoring

$$=\boxed{}$$

<hr>

YOUR NOTES   Write your questions and additional notes.

Difference Quotients

ESSENTIALS

The expression

$$\frac{f(x+h)-f(x)}{h}$$

is the **difference quotient**, or the **average rate of change**, of a function f.

Examples

- For the function f given by $f(x)=3x-2$, find and simplify the difference quotient.

$$\begin{aligned}
\frac{f(x+h)-f(x)}{h} &= \frac{3(x+h)-2-(3x-2)}{h} \\
&= \frac{3x+3h-2-3x+2}{h} \\
&= \frac{3h}{h}=3
\end{aligned}$$

- For the function f given by $f(x)=\dfrac{3}{x}$, find and simplify the difference quotient.

$$\begin{aligned}
\frac{f(x+h)-f(x)}{h} &= \frac{\dfrac{3}{x+h}-\dfrac{3}{x}}{h} \\
&= \frac{\dfrac{3}{x+h}\cdot\dfrac{x}{x}-\dfrac{3}{x}\cdot\dfrac{x+h}{x+h}}{h} \\
&= \frac{\dfrac{3x}{x(x+h)}-\dfrac{3x+3h}{x(x+h)}}{h}=\frac{\dfrac{3x-(3x+3h)}{x(x+h)}}{h} \\
&= \frac{\dfrac{3x-3x-3h}{x(x+h)}}{h}=\frac{\dfrac{-3h}{x(x+h)}}{h}=\frac{-3\cdot h}{x(x+h)}\cdot\frac{1}{h} \\
&= \frac{-3}{x(x+h)},\ \text{or}\ -\frac{3}{x(x+h)}
\end{aligned}$$

<table>
<tr><td></td><td>📕 **Textbook**</td><td>👤 **Instructor**</td><td>▶ **Video**</td></tr>
</table>

GUIDED LEARNING

EXAMPLE 1	YOUR TURN 1
For the function f given by $f(x) = 2x + 5$, find and simplify the difference quotient. $\dfrac{f(x+h) - f(x)}{h} = \dfrac{2(x+h) + \boxed{} - (2x+5)}{h}$ $= \dfrac{2x + 2h + 5 - 2x - \boxed{}}{h}$ $= \dfrac{\boxed{}}{h} = \boxed{}$	For the function f given by $f(x) = 4x - 3$, find and simplify the difference quotient.

EXAMPLE 2	YOUR TURN 2
For the function f given by $f(x) = x^2 - 4$, find and simplify the difference quotient. $\dfrac{f(x+h) - f(x)}{h} = \dfrac{(x+h)^2 - 4 - \left(\boxed{} - 4\right)}{h}$ $= \dfrac{x^2 + 2xh + h^2 - 4 - x^2 + \boxed{}}{h}$ $= \dfrac{2xh + h^2}{\boxed{}}$ $= \dfrac{h\left(2x + \boxed{}\right)}{h}$ $= \boxed{} + h$	For the function f given by $f(x) = 3 - x^2$, find and simplify the difference quotient.

YOUR NOTES   Write your questions and additional notes.

Practice Exercises

Readiness Check

Choose the word or expression that best completes the sentence.

numerator　　　　　denominator　　　　　LCM　　　　　complex rational expression

1. A ______________________ contains one or more rational expressions within its numerator and/or its denominator.

2. To simplify a complex rational expression, we may begin by multiplying both the numerator and the denominator by the ______________ of all the denominators of all the rational expressions within the complex rational expression.

3. To simplify a complex rational expression, we may begin by obtaining a single rational expression in the ______________ and in the ______________.

4. If we have a single rational expression in the numerator and in the denominator of a complex rational expression, we multiply the rational expression in the ______________ by the reciprocal of the rational expression in the ______________.

Simplifying Complex Rational Expressions

Simplify.

5. $\dfrac{\dfrac{3}{4} - \dfrac{4}{5}}{\dfrac{5}{8} - \dfrac{1}{2}}$

6. $\dfrac{\dfrac{3}{x} + 1}{\dfrac{9}{x^2} - 1}$

7. $\dfrac{\dfrac{2x+4}{x+5}}{\dfrac{3x+6}{x-6}}$

8. $\dfrac{\dfrac{x}{x^2-2x+1}-\dfrac{1}{x-1}}{\dfrac{3}{x^2-6x+5}+\dfrac{x}{x-5}}$

9. $\dfrac{\dfrac{1}{c^2}-\dfrac{1}{d^2}}{\dfrac{1}{c^3}+\dfrac{1}{d^3}}$

10. $\dfrac{\dfrac{2}{x^2-3x-4}-\dfrac{1}{x^2+3x+2}}{\dfrac{2}{x^2-2x-8}-\dfrac{1}{x^2-4}}$

Difference Quotients

For each function f, find and simplify the difference quotient.

11. $f(x)=4x+5$

12. $f(x)=3x^2+1$

13. $f(x)=\dfrac{1}{6x}$

Rational Equations

ESSENTIALS

A **rational equation** is an equation that contains one or more rational expressions.

To solve a rational equation,

1. Clear fractions by multiplying both sides of the equation by the LCM of all the denominators.
2. Solve the resulting equation.
3. Check all possible solutions in the original equation.

Example

- Solve: $\dfrac{1}{x} + \dfrac{3}{4x} = 1$.

  Multiply both sides of the equation by the LCM of the denominators, $4x$.

  $$4x\left(\frac{1}{x} + \frac{3}{4x}\right) = 4x \cdot 1$$

  $$4x \cdot \frac{1}{x} + 4x \cdot \frac{3}{4x} = 4x$$

  $$4 + 3 = 4x$$

  $$7 = 4x$$

  $$\frac{7}{4} = x$$

  Check:

  $$\frac{1}{x} + \frac{3}{4x} = 1$$

  $$\frac{1}{\frac{7}{4}} + \frac{3}{4 \cdot \frac{7}{4}} \;\overset{?}{=}\; 1$$

  $$1 \cdot \frac{4}{7} + \frac{3}{7}$$

  $$\frac{4}{7} + \frac{3}{7}$$

  $$1 \quad \Big| \quad \text{TRUE}$$

  The solution is $\dfrac{7}{4}$.

EXAMPLE 1	YOUR TURN 1

EXAMPLE 1

Solve: $\dfrac{1}{2x} + \dfrac{3}{x^2} = 1$.

$$\frac{1}{2x} + \frac{3}{x^2} = 1$$

$$2x^2\left(\frac{1}{2x} + \frac{3}{x^2}\right) = 2x^2(1) \qquad \text{Multiplying by the LCM}$$

$$2x^2 \cdot \frac{1}{2x} + 2x^2 \cdot \frac{3}{x^2} = 2x^2$$

$$\boxed{} + \boxed{} = 2x^2$$

$$0 = 2x^2 - x - 6$$

$$0 = (2x+3)(x-2)$$

$$x = \boxed{} \quad or \quad x = \boxed{}$$

Check:

For $x = -\dfrac{3}{2}$:

$$\frac{1}{2x} + \frac{3}{x^2} = 1$$

$$\frac{1}{2\left(-\frac{3}{2}\right)} + \frac{3}{\left(-\frac{3}{2}\right)^2} \;?\; 1$$

$$\frac{1}{-3} + \frac{3}{\frac{9}{4}}$$

$$-\frac{3}{9} + \frac{12}{9}$$

$$1 \quad\bigg|\quad \text{TRUE}$$

For $x = 2$:

$$\frac{1}{2x} + \frac{3}{x^2} = 1$$

$$\frac{1}{2 \cdot 2} + \frac{3}{2^2} \;?\; 1$$

$$\frac{1}{4} + \frac{3}{4}$$

$$1 \quad\bigg|\quad \text{TRUE}$$

The solutions are $-\dfrac{3}{2}$ and 2.

YOUR TURN 1

Solve: $\dfrac{2}{x} - \dfrac{1}{x^2} = -3$.

EXAMPLE 2	YOUR TURN 2

Solve: $6 + \dfrac{2x}{x-3} = \dfrac{6}{x-3}$.

We multiply both sides of the equation by the LCM of the denominators, $x-3$, and solve the resulting equation.

$$\boxed{}\left(6 + \frac{2x}{x-3}\right) = \boxed{} \cdot \frac{6}{x-3}$$

$$(x-3)\cdot 6 + (x-3)\cdot \frac{2x}{x-3} = (x-3)\cdot \frac{6}{x-3}$$

$$6x - 18 + 2x = 6$$

$$\boxed{} - 18 = 6$$

$$8x = \boxed{}$$

$$x = 3$$

Check:

$$6 + \frac{2x}{x-3} = \frac{6}{x-3}$$

$$6 + \frac{2\cdot 3}{3-3} \;\overset{?}{\mid}\; \frac{6}{3-3}$$

$$6 + \frac{6}{0} \;\Big|\; \frac{6}{0} \qquad \text{NOT DEFINED}$$

There is no solution.

Solve: $1 + \dfrac{2x}{x+5} = \dfrac{-10}{x+5}$.

EXAMPLE 3	YOUR TURN 3

EXAMPLE 3

Given that $f(x) = \dfrac{x+2}{x-1}$, find all values of x for which $f(x) = \dfrac{1}{2}$.

$$f(x) = \frac{1}{2}$$

$$\frac{x+2}{\boxed{}} = \frac{1}{2}$$

$$2(x-1)\left(\frac{x+2}{x-1}\right) = 2(x-1)\frac{1}{2}$$

$$2(x+2) = \boxed{}$$

$$2x + \boxed{} = x - 1$$

$$x = \boxed{}$$

Check: $f(-5) = \dfrac{-5+2}{-5-1} = \dfrac{-3}{-6} = \dfrac{1}{2}$.

The solution is $\boxed{}$.

YOUR TURN 3

Given that $f(x) = \dfrac{x-2}{x+4}$ find all values of x for which $f(x) = -\dfrac{1}{2}$.

YOUR NOTES   Write your questions and additional notes.

Practice Exercises

Readiness Check

Classify each statement as true or false.

1. If an equation contains a rational expression with a variable in the denominator, then zero cannot be a solution of the equation.

2. Multiplying both sides of a rational equation by the LCM of the denominators might produce an equation that is not equivalent to the original equation.

3. $\dfrac{x+3}{x-2}$ is an example of a rational equation.

4. Numbers that cause a denominator in a rational equation to be zero cannot be solutions of the equation.

Rational Equations

Solve. Don't forget to check!

5. $\dfrac{m}{2} + \dfrac{m}{6} = 5$

6. $\dfrac{n+4}{n-3} = \dfrac{4}{5}$

7. $\dfrac{1}{y} - \dfrac{1}{6y} = \dfrac{8}{9}$

8. $\dfrac{5}{6x+2} + \dfrac{3}{4} = \dfrac{6}{x+3}$

9. $\dfrac{7}{8} - \dfrac{5}{4x} = \dfrac{14}{16}$

10. $\dfrac{k+4}{k-2} = \dfrac{k-3}{k+2}$

11. $\dfrac{x+2}{x-3} = \dfrac{5}{x-3}$

12. $\dfrac{4}{x-5} = \dfrac{x}{x-2}$

13. $\dfrac{a}{a-4} - \dfrac{2}{a} = \dfrac{16}{a^2 - 4a}$

14. $\dfrac{n}{n-2} = \dfrac{5}{n+2} - \dfrac{n-1}{2-n}$

15. For $f(x) = x - \dfrac{18}{x}$, find all values of a for which $f(a) = 3$.

Work Problems

ESSENTIALS

The Work Principle

If $\quad a =$ the time it takes A to do a job,

$\quad\quad\; b =$ the time it takes B to do the same job, and

$\quad\quad\; t =$ the time it takes A and B to do the job working together,

then $\quad \dfrac{t}{a} + \dfrac{t}{b} = 1$.

GUIDED LEARNING

 **Textbook**    **Instructor**    **Video**

EXAMPLE 1	YOUR TURN 1

EXAMPLE 1

The Harrells' pool can be filled in 20 hr using their spring alone and in 15 hr using their well alone. How long will it take to fill the pool if both the spring and the well are used at the same time?

1. **Familiarize.** In the work equation, $\dfrac{t}{a} + \dfrac{t}{b} = 1$, $a = 20$,

   the number of hours required to fill the pool using the spring alone, and $b = 15$, the number of hours required to fill the pool using the well alone.

2. **Translate.**

   $$\dfrac{t}{20} + \dfrac{t}{\boxed{}} = \boxed{}$$

3. **Solve.**

   $$\dfrac{t}{20} + \dfrac{t}{15} = 1$$

   $$\boxed{}\left(\dfrac{t}{20} + \dfrac{t}{15}\right) = \boxed{} \cdot 1 \qquad \text{Multiplying by the LCM of the denominators}$$

   $$60 \cdot \dfrac{t}{20} + 60 \cdot \dfrac{t}{15} = 60$$

   $$\boxed{} + \boxed{} = 60 \qquad \text{Simplifying}$$

   $$7t = 60$$

   $$t = \dfrac{60}{7}, \text{ or } 8\dfrac{4}{7} \text{ hr}$$

4. **Check.** A check confirms the result.

5. **State.** It would take $8\dfrac{4}{7}$ hr to fill the pool using both the spring and the well at the same time.

YOUR TURN 1

It takes Becky 3 hr to clean out a barn. It takes Jon 4 hr to do the same job. How long would it take if they worked together to clean out the barn?

EXAMPLE 2	YOUR TURN 2

EXAMPLE 2

It takes Kori 20 hr longer to paint a townhouse than it takes Chris. Working together, they can do the job in 24 hr. How long would it take each, working alone, to paint a townhouse?

1. **Familiarize.** We let $c =$ the number of hr it takes Chris, working alone, to paint the townhouse. Then $c + 20 =$ the number of hr it would take Kori, working alone, to paint the townhouse.

2. **Translate.** Using the work equation, $\dfrac{t}{a} + \dfrac{t}{b} = 1$, we have

$$\frac{24}{c+20} + \frac{24}{\boxed{}} = 1.$$

3. **Solve.** We multiply both sides by the LCM of the denominators to clear fractions.

$$c(c+20)\left(\frac{24}{c+20} + \frac{24}{c}\right) = c(c+20)\cdot 1$$

$$c\cdot 24 + (c+20)\cdot 24 = c(c+20)$$

$$24c + \boxed{} + \boxed{} = c^2 + 20c \qquad \text{Simplifying}$$

$$0 = c^2 - 28c - 480 \qquad \text{Getting 0 on one side}$$

$$0 = \boxed{}(c - 40)$$

$$c = \boxed{} \quad or \quad c = \boxed{} \qquad \text{Using the principle of zero products}$$

4. **Check.** Since time cannot be negative in this situation, we check only 40. If it takes Chris 40 hr alone, it takes Kori $40 + 20$, or 60 hr alone. In 24 hr, they would have finished $\dfrac{24}{60} + \dfrac{24}{40} = \dfrac{2}{5} + \dfrac{3}{5} = 1$ complete townhouse. The answer checks.

5. **State.** It would take Kori 60 hr to paint the townhouse working alone, and it would take Chris 40 hr.

YOUR TURN 2

Working together, Gavin and Jeff can groom a shelter's dogs in 6 hr. Working alone, it would take Gavin 5 hr longer than it would take Jeff. How long would it take each, working alone, to groom the shelter's dogs?

Applications Involving Proportions

ESSENTIALS

Any rational expression $\dfrac{a}{b}$ represents a **ratio**.

The ratio of two different kinds of measure is called a **rate**.

A **proportion** is an equality of ratios, $\dfrac{A}{B} = \dfrac{C}{D}$, read "*A* is to *B* as *C* is to *D*."

The numbers named in a true proportion are **proportional** to each other.

If $\dfrac{A}{B} = \dfrac{C}{D}$, then the **cross products** AD and BC are equal.

Example

- The dimensions on an architect's drawing are proportional to the dimensions of the building represented. A wall that is 4 cm long in the drawing represents a 15-ft wall in the building. How long is a building wall that is represented by 6 cm in the drawing?

  Let $w =$ the building wall length.

  $$\text{Drawing length} \rightarrow \frac{4}{15} = \frac{6}{w} \leftarrow \text{Drawing length}$$
  $$\text{Building length} \rightarrow \qquad\quad \leftarrow \text{Building length}$$

  $$4w = 15 \cdot 6 \qquad \text{Equating cross products}$$

  $$w = \frac{15 \cdot 6}{4} = 22.5$$

  The building wall is 22.5 ft long.

GUIDED LEARNING | 🔖 **Textbook**   👤 **Instructor**   ▶️ **Video**

EXAMPLE 1	YOUR TURN 1
In city driving, a Subaru Outback can travel 171 mi on 6 gal of gas. How much gas will be required for 228 mi of city driving?	On the highway, a Ford F150 can travel 111 mi on 6 gal of gas. How much gas will be required for 185 mi of highway driving?

1. **Familiarize.** Let $x =$ the number of gallons of gas required to drive 228 mi.

2. **Translate.** We write a proportion.
   $$\text{Miles} \rightarrow \frac{171}{6} = \frac{228}{x} \leftarrow \text{Miles}$$
   $$\text{Gas} \rightarrow \qquad\qquad \leftarrow \text{Gas}$$

(continued)

3. Solve.

$$\frac{171}{6} = \frac{228}{x}$$

$$171x = 6 \cdot 228$$

$$\frac{171x}{171} = \frac{1368}{171}$$

$$x = \boxed{}$$

4. Check. The answer checks.

5. State. The Outback will require $\boxed{}$ gal of gas for 228 mi of city driving.

EXAMPLE 2	YOUR TURN 2

To estimate the number of trout in Jones' Pond, a ranger catches and tags 20 trout and releases them. Later, 30 trout are caught, and 4 of them are tagged. Estimate how many trout are in the pond.

1. Familiarize. Let $t =$ the number of trout in the pond.

2. Translate.

$$\begin{array}{l}\text{Trout tagged} \\ \text{originally} \rightarrow \end{array} \boxed{} = \boxed{} \begin{array}{l}\leftarrow \text{Tagged trout} \\ \text{caught} \end{array}$$

Trout in pond $\rightarrow \;\; t \qquad 30 \;\; \leftarrow$ Trout caught

3. Solve.

$$\frac{20}{t} = \frac{4}{30}$$

$$20 \cdot 30 = 4 \cdot \boxed{}$$

$$\boxed{} = 4t$$

$$\boxed{} = t$$

4. Check. Since $20 \cdot 30 = 600 = 4 \cdot 150$, the answer checks.

5. State. We estimate there are $\boxed{}$ trout in the pond.

YOUR TURN 2 (continued):
To estimate the number of whales in a pod, researchers identify the tail markings of 6 whales. Days later, they identify 10 whales, 2 of which had been identified earlier. Estimate how many whales are in the pod.

YOUR NOTES   Write your questions and additional notes.

Motion Problems

ESSENTIALS

Distance = Rate (or speed) $\times$ Time

$$d = rt; \qquad\qquad r = \frac{d}{t}; \qquad\qquad t = \frac{d}{r}$$

GUIDED LEARNING

 **Textbook**    **Instructor**    **Video**

EXAMPLE 1	YOUR TURN 1

EXAMPLE 1

The speed of the current in Little River is 2 mph. Brent can paddle 12 mi upstream in the same time it takes him to paddle 20 mi downstream. How fast does he paddle in still water?

1. **Familiarize.** Let $b =$ Brent's speed in still water.

	Distance	Speed	Time
Upstream	12	$b-2$	$12/(b-2)$
Downstream	20	$b+2$	$20/(b+2)$

2. **Translate.** The times are the same, so we equate the two expressions in the last column of the table.

3. **Solve.**

$$\frac{12}{b-2} = \frac{20}{b+2}$$

$$(b+2)(b-2)\left(\frac{12}{b-2}\right) = (b+2)(b-2)\left(\frac{20}{b+2}\right)$$

$$\boxed{} \cdot 12 = \boxed{} \cdot 20$$

$$12b + 24 = 20b - 40$$

$$\boxed{} = 8b$$

$$8 = b$$

4. **Check.** The speed upstream is $8-2=6$ mph, so Brent travels 12 mi in $\dfrac{12}{6} = 2$ hr. The speed downstream speed is $8+2=10$ mph, so he travels 20 mi in $\dfrac{20}{10} = 2$ hr. The times are the same, so the answer checks.

5. **State.** Brent paddles 8 mph in still water.

YOUR TURN 1

A moving sidewalk moves at a rate of 1.5 ft/sec. Walking on the moving sidewalk, Karen can travel 240 ft forward in the same time it takes to travel 120 ft in the opposite direction. What is Karen's rate on a nonmoving sidewalk?

EXAMPLE 2	YOUR TURN 2

EXAMPLE 2

Auree drove 100 mi to the ocean. On the way home, she had to travel 10 mph slower due to traffic. Her total driving time was 4.5 hr. How fast did she drive on the way to the ocean?

1. Familiarize. Let $r =$ Auree's speed on the way to the ocean.

	Distance	Speed	Time
To ocean	100	r	$100/r$
Home	100	$r-10$	$100/(r-10)$

2. Translate. The total time was 4.5 hr, or $\dfrac{9}{2}$ hr, so the sum of the two expressions in the last column of the table is $\dfrac{9}{2}$.

$$\frac{100}{r}+\frac{100}{r-10}=\frac{9}{2}$$

3. Solve.

$$\frac{100}{r}+\frac{100}{r-10}=\frac{9}{2}$$

$$2r(r-10)\left(\frac{100}{r}+\frac{100}{r-10}\right)=2r(r-10)\cdot\frac{9}{2} \qquad \text{Clearing fractions}$$

$$2(r-10)\cdot100+2r\cdot100=9r(r-10)$$

$$200r-\boxed{}+200r=9r^2-\boxed{}$$

$$0=9r^2-490r+2000 \qquad \text{Getting 0 on one side}$$

$$0=(9r-40)(r-50)$$

$$r=\boxed{} \quad \text{or} \quad r=\boxed{} \qquad \text{Using the principle of zero products}$$

4. Check. If $r=\dfrac{40}{9}$, then $r-10$, the speed home, is negative, so we reject $\dfrac{40}{9}$. A check confirms that $r=50$ checks.

5. State. Auree's speed to the ocean was 50 mph.

YOUR TURN 2

Seb traveled a total of 320 mi in 5 hr. The first 280 mi of the trip he was on the highway and was able to travel 30 mph faster than the last 40 mi which was on country roads. How fast did he travel on each segment?

YOUR NOTES   Write your questions and additional notes.

Practice Exercises

Readiness Check

Complete each statement using the following words or expressions. Choices may be used more than once or not at all.

rate   distance   work   time   0   1 $\dfrac{A+B}{2}$

1. Distance = ______________ × ______________ .

2. Rate = ______________ ÷ ______________ .

3. Time = ______________ ÷ ______________ .

4. $\dfrac{\text{Time required for A and B to complete the job together}}{\text{Time required for A to complete the job alone}} + \dfrac{\text{Time required for A and B to complete the job together}}{\text{Time required for B to complete the job alone}} =$ ______________ .

Work Problems

5. It takes Clementine 12 hr to cut and piece a full size quilt. It takes Brianna 10 hr to do the same job. How long would it take them working together to cut and piece a full size quilt?

6. Robert can chop and stack the winter firewood in 3 fewer days than it would take Georgia. If they work together, it takes $6\dfrac{2}{3}$ days to cut and stack the wood they need for the winter. How long would it take each to chop and stack the firewood if working alone?

7. Penny can type twice as fast as her younger sister. Working together, it takes them 12 hr to type the annual report for the organization at which they volunteer. How long would it take Penny to type the report working alone?

Applications Involving Proportions

8. A baseball player gets 4 hits in 12 games. At this rate, how many hits will he get in 162 games?

9. To estimate the number of deer in a game preserve, a conservationist catches and tags 18 deer and releases them. Later, 30 deer are caught; 4 of them are tagged. How many deer are in the preserve?

Motion Problems

10. Jackson's running speed is 7 ft/sec slower than Lucia's. Lucia can run 74 ft in the time it takes Jackson to run 60 ft. How fast does Jackson run?

11. A boat moves 15 km/h in still water. It travels 40 km upriver and 40 km downriver in a total time of 6 hours. What is the speed of the current?

Formulas

ESSENTIALS

To solve a rational formula for a given letter, identify the letter, and:

1. Multiply on both sides to clear fractions, if necessary.
2. Multiply to remove parentheses, if necessary.
3. Get all terms with the specified letter alone on one side.
4. Factor out the specified letter if it is in more than one term.
5. Solve for the letter in question, using the multiplication principle.

Example

- Solve $r = \dfrac{d}{t}$ for t.

$$r = \frac{d}{t}$$

$$t \cdot r = t \cdot \frac{d}{t} \quad \text{Clearing fractions}$$

$$tr = d \quad \text{Removing a factor of 1: } \frac{t}{t} = 1$$

$$\frac{tr}{r} = \frac{d}{r} \quad \text{Dividing to isolate } t$$

$$t = \frac{d}{r} \quad \text{Removing a factor of 1: } \frac{r}{r} = 1$$

GUIDED LEARNING 📖 **Textbook**   👤 **Instructor**   ▶ **Video**

EXAMPLE 1	YOUR TURN 1
Solve $\dfrac{1}{m} + \dfrac{1}{x} = \dfrac{1}{z}$ for m.	Solve $\dfrac{2}{k} + \dfrac{1}{y} = 3$ for y.

We multiply by the LCM of the denominators to clear fractions.

$$\boxed{} \cdot \left(\frac{1}{m} + \frac{1}{x} \right) = \boxed{} \cdot \frac{1}{z}$$

$$mxz \cdot \frac{1}{m} + mxz \cdot \frac{1}{x} = mxz \cdot \frac{1}{z} \quad \text{Multiplying to remove parentheses}$$

$$xz + \boxed{} = \boxed{}$$

We need all terms involving m alone on one side.

(continued)

$xz = mx - mz$ Subtracting mz from both sides

$xz = m(x - z)$ Factoring out m

$$\boxed{} = m$$ Dividing both sides by $x - z$

Since m is isolated, we have solved for m.

EXAMPLE 2	YOUR TURN 2

Solve $S = \dfrac{H}{m(t_1 - t_2)}$ for t_1.

Solve $S = \dfrac{H}{m(t_1 - t_2)}$ for t_2.

We multiply by $m(t_1 - t_2)$ to clear the fraction.

$$m(t_1 - t_2) \cdot S = m(t_1 - t_2) \cdot \frac{H}{m(t_1 - t_2)}$$

$$mS(t_1 - t_2) = \boxed{}$$

Since t_1 is inside parentheses, we multiply
to remove the parentheses as the next step in
isolating t_1.

$$mSt_1 - mSt_2 = H$$

$mSt_1 = H + mSt_2$ Adding mSt_2 to both sides

$\dfrac{mSt_1}{mS} = \dfrac{H + mSt_2}{\boxed{}}$ Dividing by mS

$\boxed{} = \dfrac{H + mSt_2}{mS}$ Removing a factor of 1 on the left

YOUR NOTES   Write your questions and additional notes.

Practice Exercises

Readiness Check

For each of Exercises 1–3, indicate which equation is not equivalent to the others.

1.
$$y = kx$$
$$k = \frac{y}{x}$$
$$\frac{y}{k} = x$$
$$yx = k$$

2.
$$ab = ac + c$$
$$b = ac + c - b$$
$$c = \frac{ab}{a+1}$$
$$a(b-c) = c$$

3.
$$\frac{1}{w} + \frac{1}{p} = \frac{1}{t}$$
$$pt + wt = pw$$
$$w + p = t$$
$$p(w-t) = wt$$

Formulas

Solve.

4. $\dfrac{a}{b} = \dfrac{x}{y}$, for b

5. $\dfrac{1}{w} + \dfrac{1}{a} = \dfrac{1}{c}$, for a

6. $x = \dfrac{ab}{b+ac}$, for c

7. $A = \dfrac{(x_1 + x_2)w}{2}$, for x_2

8. $\dfrac{G}{g} = \dfrac{a + A}{A}$, for g

9. $\dfrac{1}{t} = y_1 + y_2$, for t

10. $r = \dfrac{A - P}{Pt}$, for P

Equations of Direct Variation

Direct Variation

If a situation gives rise to a linear function $f(x) = kx$, or $y = kx$, where k is a positive constant, we say that we have **direct variation**, or that y **varies directly as** x, or that y **is directly proportional to** x. The number k is called the **variation constant**, or **constant of proportionality**.

Example

- Find the variation constant and an equation of variation in which y varies directly as x, and $y = 3$ when $x = 8$.

  We know that $(8, 3)$ is a solution of $y = kx$.

  $$y = kx$$

  $$3 = k \cdot 8$$

  $$\frac{3}{8} = k$$

  The variation constant is $\frac{3}{8}$. The equation of variation is $y = \frac{3}{8}x$.

📖 **Textbook**	👤 **Instructor**	▶ **Video**

GUIDED LEARNING

EXAMPLE 1	YOUR TURN 1
Find the variation constant and an equation of variation in which y varies directly as x and $y = 20$ when $x = 4$.	Find the variation constant and an equation of variation in which y varies directly as x and $y = 300$ when $x = 50$.

We know that $(4, 20)$ is a solution of $y = kx$.

$y = kx$ y varies directly as x.

$\boxed{} = k \cdot \boxed{}$ Substituting 20 for y and 4 for x

$\boxed{} = k$ Solving for x

The variation constant is $\boxed{}$. The equation of variation is $y = \boxed{}$.

YOUR NOTES   Write your questions and additional notes.

Applications of Direct Variation

ESSENTIALS

Example

- Mike's pay P varies directly as the number of hours worked H. If Mike works 25 hr, his pay is \$305. Find the pay for 40 hr of work.

$$P = kH$$

$$305 = k \cdot 25$$

$$\frac{305}{25} = k, \text{ or } k = 12.2$$

The equation of variation is $P = 12.2H$. When $H = 40$,

$$P = 12.2H$$

$$P = 12.2(40) \qquad \text{Substituting 40 for } H$$

$$P = 488.$$

If Mike works 40 hr, he will be paid \$488.

GUIDED LEARNING  **Textbook**  **Instructor**  **Video**

EXAMPLE 1	YOUR TURN 1
The number of calories burned, C, varies directly as the time t (in minutes) spent exercising. If Keisha exercises for 25 min she burns 175 calories. How many calories would she burn if she exercised for 45 min?	The amount of simple interest, I, received on a savings account varies directly as the amount of money, m, that is in the account. If \$520 earned \$10.40 in interest in a year, how much interest would be earned with \$650 in the account?

$\qquad C = kt \qquad\qquad C$ varies directly as t.

$\qquad \boxed{} = k \cdot \boxed{} \qquad$ Substituting

$\qquad \boxed{} = k$

The equation of variation is $C = 7t$. When $t = 45$,

$\qquad C = 7t$

$\qquad C = 7\left(\boxed{}\right) \qquad$ Substituting

$\qquad C = \boxed{}$

After 45 min, Keisha will burn $\boxed{}$ calories.

YOUR NOTES   Write your questions and additional notes.

Equations of Inverse Variation

ESSENTIALS

Inverse Variation

If a situation gives rise to a function $f(x) = k/x$, or $y = k/x$, where k is a positive constant, we say that we have **inverse variation**, or that y **varies inversely as** x, or that y **is inversely proportional to** x. The number k is called the **variation constant**, or **constant of proportionality**.

Example

- Find the variation constant and an equation of variation in which y varies inversely as x, and $y = 130$ when $x = 0.6$.

  We know that $(0.6, 130)$ is a solution of $y = k/x$.

  $$y = \frac{k}{x}$$

  $$130 = \frac{k}{0.6}$$

  $$(0.6)130 = k, \text{ or } k = 78$$

  The variation constant is 78. The equation of variation is $y = \dfrac{78}{x}$.

	Textbook		Instructor		Video
GUIDED LEARNING					

EXAMPLE 1	YOUR TURN 1
Find the variation constant and an equation of variation in which y varies inversely as x and $y = 0.4$ when $x = 8$.	Find the variation constant and an equation of variation in which y varies inversely as x and $y = 75$ when $x = 0.1$.
We know that $(8, 0.4)$ is a solution of $y = k/x$.	
$y = \dfrac{k}{x}$ y varies inversely as x.	
$\boxed{} = \dfrac{k}{\boxed{}}$ Substituting 0.4 for y and 8 for x	
$8(0.4) = k$ Solving for k	
$\boxed{} = k$	
The variation constant is $\boxed{}$. The equation of variation is $y = \dfrac{\boxed{}}{x}$.	

EXAMPLE 2	YOUR TURN 2
Find the variation constant and an equation of variation in which y varies inversely as x and $y = 12$ when $x = 15$.	Find the variation constant and an equation of variation in which y varies inversely as x and $y = 32$ when $x = 9$.

We know that $(15, 12)$ is a solution of $y = k/x$.

$$y = \frac{k}{x} \qquad y \text{ varies inversely as } x.$$

$$\boxed{} = \frac{k}{\boxed{}} \qquad \text{Substituting}$$

$$15 \cdot 12 = k \qquad \text{Solving for } k$$

$$\boxed{} = k$$

The variation constant is $\boxed{}$. The equation of variation is $y = \dfrac{\boxed{}}{x}$.

YOUR NOTES   Write your questions and additional notes.

Applications of Inverse Variation

ESSENTIALS

Example

- The cost per person to go on a ski trip for a day is inversely proportional to the number of people going on the trip. If it costs \$90 per person when 30 people go on the trip, how many are going on the ski trip if it costs \$135 per person?

  Let $C =$ the cost per person and let $N =$ the number of people going on the trip.

  Since inverse variation applies, we have $C = \dfrac{k}{N}$.

  Find the equation of variation.

$$C = \frac{k}{N}$$
$$90 = \frac{k}{30}$$
$$30 \cdot 90 = k$$
$$2700 = k$$

The equation of variation is $C = \dfrac{2700}{k}$. When $C = 135$, we have

$$135 = \frac{2700}{N}$$
$$135N = 2700$$
$$N = 20.$$

It would cost \$135 per person if 20 people go on the ski trip.

GUIDED LEARNING   📖 **Textbook**   👤 **Instructor**   ▶ **Video**

EXAMPLE 1	YOUR TURN 1

EXAMPLE 1

The number of minutes that a student should allow for each question on a test is inversely proportional to the number of questions on the test. If a student allows 3 minutes per question on a 20-question test, how many minutes per question should the student allow on a 25-question test? (The times allowed for the two tests are the same.)

Let M = the number of minutes per question and N = the number of questions on the test.

Since inverse variation applies, we have $M = \dfrac{k}{N}$.

Find the equation of variation.

$$M = \frac{k}{N}$$

$$\boxed{} = \frac{k}{\boxed{}} \qquad \text{Substituting 3 for } M \text{ and 20 for } N$$

$$20 \cdot 3 = k$$

$$\boxed{} = k$$

The equation of variation is $M = \dfrac{60}{N}$. When $N = 25$, we have

$$M = \frac{60}{\boxed{}}$$

$$M = \boxed{}.$$

A student should allow 2.4 minutes per question for a 25-question test.

YOUR TURN 1

The number of workers required to paint a house is inversely proportional to the time spent working. It takes 15 hours for 2 people to paint a house. How many workers are needed to paint the house in 6 hours?

YOUR NOTES   Write your questions and additional notes.

Other Kinds of Variation

ESSENTIALS

y varies directly as the nth power of x if there is some positive constant k such that
$y = kx^n$.

y varies inversely as the nth power of x if there is some positive constant k such that
$$y = \frac{k}{x^n}.$$

y varies **jointly** as x and z, if for some positive constant k, $y = kxz$.

Example

- Find an equation of variation in which y varies jointly as w and x and inversely as z, and $y = 100$ when $w = 2$, $x = 5$, and $z = 4$.

$$y = k \cdot \frac{wx}{z}$$

$$100 = k \cdot \frac{2 \cdot 5}{4}$$

$$100 = k \cdot \frac{5}{2}$$

$$40 = k$$

The equation of variation is $y = \dfrac{40wx}{z}$.

GUIDED LEARNING   📖 **Textbook**   👤 **Instructor**   ▶ **Video**

EXAMPLE 1	YOUR TURN 1
Find an equation of variation in which y varies directly as the square of x, and $y = 50$ when $x = 10$.	Find an equation of variation in which y varies directly as the square of x, and $y = 36$ when $x = 3$.

$$y = kx^2$$

$$\boxed{} = k \cdot \boxed{}^2 \qquad \text{Substituting}$$

$$50 = k \cdot \boxed{}$$

$$\boxed{} = k$$

The equation of variation is $y = \boxed{}$.

EXAMPLE 2	YOUR TURN 2
Find an equation of variation in which W varies inversely as the square of d, and $W = 7$ when $d = 2$. $$W = \frac{k}{d^2}$$ $$\boxed{} = \frac{k}{\left(\boxed{}\right)^2} \quad \text{Substituting}$$ $$7 = \frac{k}{\boxed{}}$$ $$\boxed{} = k$$ The equation of variation is $y = \boxed{}$.	Find an equation of variation in which W varies inversely as the square of d, and $W = 8$ when $$d = \frac{1}{2}.$$
EXAMPLE 3	YOUR TURN 3
Find an equation of variation in which y varies jointly as x and z, and $y = 18$ when $x = 3$ and $z = 0.8$. $$y = kxz$$ $$\boxed{} = k \cdot 3(0.8) \quad \text{Substituting}$$ $$18 = k(2.4)$$ $$\boxed{} = k$$ The equation of variation is $y = \boxed{}$.	Find an equation of variation in which y varies jointly as x and z, and $y = 4$ when $x = 2$ and $z = \frac{1}{3}$.

YOUR NOTES   Write your questions and additional notes.

Other Applications of Variation

Example

- The intensity I of a signal varies inversely as the square of the distance d from the transmitter. If the intensity is 20 watts per square meter (W/m^2) at a distance of 3 km, what is the intensity at a distance of 5 km?

$$I = \frac{k}{d^2}$$

$$20 = \frac{k}{3^2}$$

$$20 = \frac{k}{9}$$

$$180 = k$$

The equation of variation is $I = \dfrac{180}{d^2}$.

To find the intensity at a distance of 5 km, substitute 5 for d.

$$I = \frac{180}{d^2} = \frac{180}{5^2} = \frac{180}{25} = 7.2$$

At a distance of 5 km, the intensity of the signal is 7.2 W/m^2.

<table>
<tr><td></td><td>**Textbook**</td><td>**Instructor**</td><td>**Video**</td></tr>
</table>

GUIDED LEARNING

EXAMPLE 1	YOUR TURN 1

The stopping distance d of a car after the brakes have been applied varies directly as the square of the speed r. If a car traveling 50 mph can stop in 250 ft, what is the stopping distance of a car traveling 30 mph?

First find k and the equation of variation.

$$d = kr^2$$

$$\boxed{} = k \cdot \boxed{}^2 \qquad \text{Substituting}$$

$$250 = k \cdot \boxed{}$$

$$\boxed{} = k$$

The equation of variation is $d = \boxed{} \cdot r^2$.

Now find d when $r = 30$ mph.

$$d = 0.1r^2$$

$$= 0.1 \cdot \boxed{}^2 \qquad \text{Substituting}$$

$$= 0.1 \cdot \boxed{}$$

$$= \boxed{}$$

A car traveling 30 mph can stop in $\boxed{}$ ft.

The intensity I of light from a light source varies inversely as the square of the distance r from the source. The intensity of light from a lamp is 80 watts per square meter (W/m^2) at a distance of 4 m from the lamp. What is the intensity at a distance of 10 m from the lamp?

YOUR NOTES   Write your questions and additional notes.

Practice Exercises

Readiness Check

Match each description of variation with the appropriate equation of variation.

1. a varies inversely as b **a)** $a = kb$

2. a varies directly as b **b)** $r = kt$

3. r is directly proportional to t **c)** $r = \dfrac{k}{t}$

4. r is inversely proportional to t **d)** $a = \dfrac{k}{b}$

Equations of Direct Variation

Find the variation constant and an equation of variation in which y varies directly as x and the following are true.

5. $y = 50$ when $x = 10$ 6. $y = 9.6$ when $x = 1.2$

7. $y = 0.2$ when $x = 0.7$ 8. $y = 400$ when $x = 800$

Applications of Direct Variation

9. The distance (in miles) traveled in a car is directly proportional to the average speed (in mph) of the car. If Cody travels 60 mi in a certain amount of time while driving at 55 mph, how many miles can he travel in that same amount of time at 44 mph?

Equations of Inverse Variation

Find the variation constant and an equation of variation in which y varies inversely as x and the following are true.

10. $y = 5$ when $x = 7$

11. $y = 45$ when $x = 0.2$

12. $y = 0.4$ when $x = 25$

13. $y = 22$ when $x = 15$

Applications of Inverse Variation

14. The time required to fill a man-made pond varies inversely as the rate of flow into the pond. A tanker can fill the pond in 4.5 hr at a rate of 800 L/min. How long would it take to fill the pond at a rate of 1200 L/min?

Other Kinds of Variations

Find an equation of variation in which the following are true.

15. y varies inversely as the square of x, and $y = 16$ when $x = 0.2$

16. y varies jointly as x and z and inversely as w, and $y = 60$ when $x = 15$, $z = 20$, and $w = 40$

Other Applications of Variation

17. The amount Q of water emptied by a pipe varies directly as the square of its diameter d. A pipe 2 in. in diameter will empty 300 gal of water over a fixed time period. If we assume the same kind of flow, how many gallons of water are emptied in the same amount of time by a pipe that is 4 in. in diameter?

Square Roots and Square-Root Functions

ESSENTIALS

The number c is a **square root** of a if $c^2 = a$.

Every positive real number has two real-number square roots.
The number 0 has just one square root, 0 itself.
Negative numbers do not have real-number square roots.

The **principal square root** of a nonnegative number is its nonnegative square root. The symbol $\sqrt{a}$ represents the principal square root of a. To name the negative square root of a, we can write $-\sqrt{a}$.

The symbol $\sqrt{}$ is called a **radical**.
An expression written with a radical is called a **radical expression**.
The expression written under the radical is the **radicand**.

Examples

- Find the two square roots of 49.

  The square roots are 7 and -7, because $7^2 = 49$ and $(-7)^2 = 49$.

- Simplify $\sqrt{49}$.

  The radical sign indicates the principal square root, so $\sqrt{49} = 7$.

	📖 **Textbook**	👤 **Instructor**	▶ **Video**

GUIDED LEARNING

EXAMPLE 1	YOUR TURN 1
Simplify $\sqrt{16}$. $\sqrt{16} = \boxed{}$	Simplify $\sqrt{9}$.
EXAMPLE 2	YOUR TURN 2
Simplify $\sqrt{\dfrac{64}{100}}$. $\sqrt{\dfrac{64}{100}} = \dfrac{8}{\boxed{}} = \dfrac{4}{5}$	Simplify $\sqrt{\dfrac{25}{81}}$.
EXAMPLE 3	YOUR TURN 3
Simplify $-\sqrt{144}$. $-\sqrt{144} = -\boxed{}$	Simplify $-\sqrt{121}$.

EXAMPLE 4	YOUR TURN 4
Simplify $\sqrt{0.0081}$. $\sqrt{0.0081} = \boxed{}$	Simplify $\sqrt{0.0025}$.
EXAMPLE 5	YOUR TURN 5
Use a calculator to approximate $\sqrt{74}$ to three decimal places. Using a calculator with a 10-digit readout, we have $\sqrt{74} \approx 8.602325267.$ Rounding to three decimal places, we have $\sqrt{74} \approx \boxed{}.$	Use a calculator to approximate $\sqrt{83}$ to three decimal places.
EXAMPLE 6	YOUR TURN 6
Identify the radicand in $x\sqrt{2x^2 - 7}$. The radicand is the expression under the radical, or $\boxed{}$.	Identify the radicand in $4y^2\sqrt{y+8}$.
EXAMPLE 7	YOUR TURN 7
Find the indicated function value: $f(x) = \sqrt{4x-3}\,;\ f(2).$ $f(2) = \sqrt{4(2)-3}\quad$ Substituting $= \sqrt{\boxed{}} - 3$ $= \boxed{}$	Find the indicated function value: $f(x) = \sqrt{3x+5}\,;\ f(3).$

EXAMPLE 8	YOUR TURN 8

EXAMPLE 8

Find the domain of $f(x) = \sqrt{3x - 15}$.

The domain of f is the set of all x-values for which the radicand is nonnegative.

$3x - 15 \geq \boxed{}$ The radicand is nonnegative.

$3x \geq \boxed{}$

$x \geq \boxed{}$

The domain of $f = \left\{ x \mid x \geq \boxed{} \right\}$, or $\left[\boxed{}, \infty \right)$.

YOUR TURN 8

Find the domain of $g(x) = \sqrt{2x + 5}$.

EXAMPLE 9	YOUR TURN 9

EXAMPLE 9

Graph: $f(x) = \sqrt{x - 3}$.

We choose some values for x and compute $f(x)$.

x	$f(x)$	$(x, f(x))$
3	$\boxed{}$	$\left(3, \boxed{}\right)$
4	1	$\boxed{}$
5	1.4	$(5, 1.4)$
6	1.7	$(6, 1.7)$
7	$\boxed{}$	$\boxed{}$

We then plot the ordered pairs and connect them with a smooth curve.

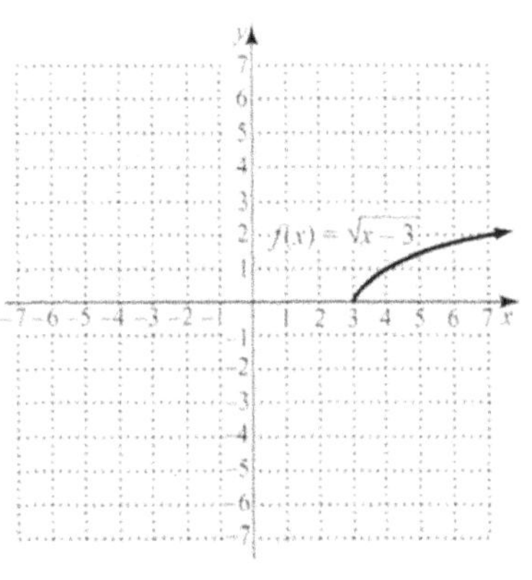

YOUR TURN 9

Graph: $g(x) = 2 + \sqrt{x}$.

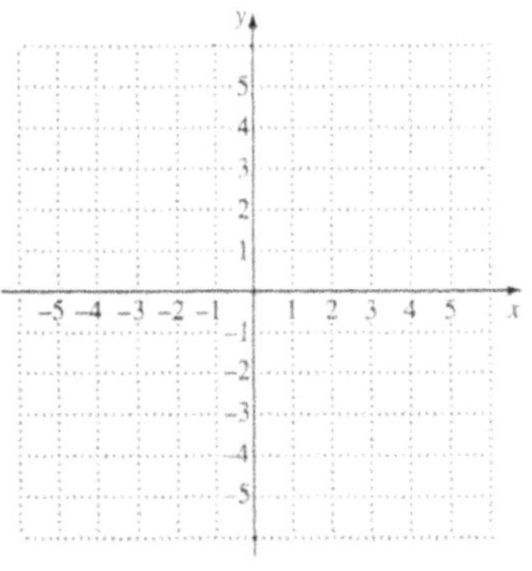

YOUR NOTES   Write your questions and additional notes.

Finding $\sqrt{a^2}$

<hr>

ESSENTIALS

For any real number a, $\sqrt{a^2} = |a|$.

Example

- Simplify $\sqrt{16t^2}$. Assume the variable can represent any real number.

$$\sqrt{16t^2} = \sqrt{(4t)^2} = |4t|, \text{ or } 4|t|$$

<hr>

GUIDED LEARNING	📘 **Textbook** 👤 **Instructor** ▶ **Video**
EXAMPLE 1	**YOUR TURN 1**
Simplify $\sqrt{(-3x)^2}$. Assume that the variable can represent any real number. $$\sqrt{(-3x)^2} = \left|\boxed{}\right|$$ $$= \left|-3\right| \cdot \left|x\right|$$ $$= \boxed{}\,|x|$$	Simplify $\sqrt{(-9y)^2}$. Assume that the variable can represent any real number.
EXAMPLE 2	**YOUR TURN 2**
Simplify $\sqrt{(2x+1)^2}$. Assume that the variable can represent any real number. $$\sqrt{(2x+1)^2} = \boxed{}$$	Simplify $\sqrt{(x+2)^2}$. Assume that the variable can represent any real number.
EXAMPLE 3	**YOUR TURN 3**
Simplify $\sqrt{x^2 + 10x + 25}$. Assume that the expression being squared is nonnegative. $$\sqrt{x^2 + 10x + 25} = \sqrt{\left(\boxed{}\right)} = x + 5$$	Simplify $\sqrt{x^2 + 4x + 4}$. Assume that the expression being squared is nonnegative.

<hr>

YOUR NOTES   Write your questions and additional notes.

Cube Roots

ESSENTIALS

The number c is the **cube root** of a if $c^3 = a$. In symbols, $\sqrt[3]{a}$ denotes the cube root of a.

Example

- Find the indicated function value: $f(x) = \sqrt[3]{x-1}$; $f(9)$.

$$f(9) = \sqrt[3]{9-1}$$
$$= \sqrt[3]{8}$$
$$= 2 \qquad \text{Since } 2 \cdot 2 \cdot 2 = 8$$

GUIDED LEARNING   📖 **Textbook**   👤 **Instructor**   ▶ **Video**

EXAMPLE 1	YOUR TURN 1
Find $\sqrt[3]{64}$. $\sqrt[3]{64} = \boxed{}$ because $\left(\boxed{}\right)^3 = 64$.	Find $\sqrt[3]{-8}$.
EXAMPLE 2	YOUR TURN 2
Simplify $\sqrt[3]{-27x^3}$. $\sqrt[3]{-27x^3} = \boxed{}$ because $\left(\boxed{}\right)^3 = -27x^3$.	Simplify $\sqrt[3]{8x^3}$.
EXAMPLE 3	YOUR TURN 3
Find the indicated function value: $f(x) = \sqrt[3]{x+15}$; $f(-7)$. $f(-7) = \sqrt[3]{\boxed{}} + 15$ $= \sqrt[3]{\boxed{}}$ $= \boxed{}$	Find the indicated function value: $f(x) = \sqrt[3]{2x-3}$; $f(15)$.

YOUR NOTES   Write your questions and additional notes.

Odd and Even *k*th Roots

ESSENTIALS

In the expression $\sqrt[k]{a}$, we call k the **index**. When the index is 2, we do not write it.

If k is an *odd* natural number, then $\sqrt[k]{a^k} = a$.

If k is an *even* natural number, then $\sqrt[k]{a^k} = |a|$.

Examples

- Find $\sqrt[5]{-243}$.

  $\sqrt[5]{-243} = -3$ because $(-3)^5 = -243$.

- Simplify: $\sqrt[4]{x^4}$.

  $\sqrt[4]{x^4} = |x|$

GUIDED LEARNING	📖 **Textbook** 👤 **Instructor** ▶ **Video**
EXAMPLE 1	**YOUR TURN 1**
Find $\sqrt[4]{81}$. $\sqrt[4]{81} = \boxed{}$ because $\left(\boxed{}\right)^4 = 81$.	Find $\sqrt[4]{10{,}000}$.
EXAMPLE 2	**YOUR TURN 2**
Find $-\sqrt[7]{-128}$. $-\sqrt[7]{-128} = -\left(\boxed{}\right) = \boxed{}$	Find $-\sqrt[5]{32}$.
EXAMPLE 3	**YOUR TURN 3**
Find $\sqrt[4]{-16}$. Negative numbers do not have real *k*th roots when k is $\boxed{}$. Thus, $\sqrt[4]{-16}$ does not exist as a real number.	Find $\sqrt[6]{-64}$.
EXAMPLE 4	**YOUR TURN 4**
Simplify: $\sqrt[11]{(x-2)^{11}}$. $\sqrt[11]{(x-2)^{11}} = \boxed{}$	Simplify: $\sqrt[7]{(2x+1)^7}$.
EXAMPLE 5	**YOUR TURN 5**
Simplify: $\sqrt[4]{16x^4}$. Assume that the variable can represent any real number. $\sqrt[4]{16x^4} = |2x|$, or $2\boxed{}$	Simplify: $-\sqrt[4]{81x^4}$. Assume that the variable can represent any real number.

YOUR NOTES   Write your questions and additional notes.

Practice Exercises

Readiness Check

Determine whether each statement is true or false.

1. The index of a square root is 2.

2. The expression $\sqrt[4]{-100}$ does not exist as a real number.

3. The principal square root can be negative.

4. The domain of the function $f(x) = \sqrt{x}$ is $\{x \mid x \geq 0\}$.

Square Roots and Square-Root Functions

For each number, find all of its square roots.

5. 100

6. 256

Simplify.

7. $-\sqrt{\dfrac{100}{81}}$

8. $\sqrt{0.25}$

9. Use a calculator to approximate $\sqrt{153}$ to three decimal places.

10. Identify the radicand in $9x\sqrt{3x-7}$.

For each function, find the specified function value, if it exists.

11. $f(x) = \sqrt{x^2 + 5}$;
    $f(-2)$

12. $f(x)\sqrt{-4x}$;
    $f(5)$

13. $f(x) = \sqrt{(x+3)^2}$;
    $f(-4)$

Finding $\sqrt{a^2}$

Simplify, if possible. Assume variables can represent any real number.

14. $\sqrt{9x^2}$

15. $\sqrt{(-5x)^2}$

16. $\sqrt{y^2 - 12y + 36}$

Cube Roots

Simplify.

17. $\sqrt[3]{125}$

18. $-\sqrt[3]{-1}$

19. $\sqrt[3]{-216x^3}$

20. For $f(x) = \sqrt[3]{5 - 3x}$, find $f(2)$.

Odd and Even *k*th Roots

Find each of the following. Assume that letters can represent any real number.

21. $\sqrt[26]{x^{26}}$

22. $-\sqrt[5]{\dfrac{1}{32}}$

23. $\sqrt[7]{-1}$

24. $\sqrt[9]{a^9}$

25. $\sqrt[6]{(y+8)^6}$

26. $\sqrt[5]{(2x+5)^5}$

Rational Exponents

ESSENTIALS

$$a^{1/n} = \sqrt[n]{a}$$

For any natural numbers m and n $(n \neq 1)$ and any nonnegative real number a,

$a^{m/n}$ means $\sqrt[n]{a^m}$, or $\left(\sqrt[n]{a}\right)^m$.

Example

- Rewrite $64^{5/6}$ without rational exponents, and simplify, if possible.

$$64^{5/6} = \left(\sqrt[6]{64}\right)^5 = 2^5 = 32$$

GUIDED LEARNING   Textbook   Instructor   Video

EXAMPLE 1	YOUR TURN 1
Rewrite without rational exponents, and simplify, if possible: $(-32)^{1/5}$. $$(-32)^{1/5} = \sqrt[5]{-32} = \boxed{}$$	Rewrite without rational exponents, and simplify, if possible: $49^{1/2}$.
EXAMPLE 2	YOUR TURN 2
Rewrite with rational exponents: $\sqrt[7]{5xy}$. $$\sqrt[7]{5xy} = (5xy)^{\boxed{}}$$	Rewrite with rational exponents: $\sqrt[5]{-3x^2y}$.
EXAMPLE 3	YOUR TURN 3
Rewrite without rational exponents, and simplify, if possible: $64^{2/3}$. $64^{2/3}$ means $\left(\sqrt[3]{64}\right)^2$ or, equivalently, $\sqrt[3]{64^2}$. We will use the form $\left(\sqrt[3]{64}\right)^2$. $$64^{2/3} = \left(\sqrt[3]{64}\right)^2 = 4^2 = \boxed{}$$	Rewrite without rational exponents, and simplify, if possible: $81^{3/2}$.

EXAMPLE 4	YOUR TURN 4
Rewrite with rational exponents: $\left(\sqrt[3]{15x}\right)^5$. $$\left(\sqrt[3]{15x}\right)^5 = (15x)^{\boxed{}}$$	Rewrite with rational exponents: $\left(\sqrt[4]{9xy}\right)^5$.

YOUR NOTES   Write your questions and additional notes.

Negative Rational Exponents

ESSENTIALS

For any rational number m/n and any positive real number a,

$$a^{-m/n} \text{ means } \frac{1}{a^{m/n}}.$$

Example

- Rewrite with positive exponents and simplify, if possible: $49^{-1/2}$.

$$49^{-1/2} = \frac{1}{49^{1/2}} = \frac{1}{\sqrt{49}} = \frac{1}{7}$$

GUIDED LEARNING	📖 **Textbook** 👤 **Instructor** ▶ **Video**
EXAMPLE 1	YOUR TURN 1
Rewrite with positive exponents, and simplify, if possible: $81^{-1/2}$. $$81^{-1/2} = \frac{1}{81^{1/2}} = \frac{1}{\sqrt{\Box}} = \frac{1}{\Box}$$	Rewrite with positive exponents, and simplify, if possible: $25^{-1/2}$.
EXAMPLE 2	YOUR TURN 2
Rewrite with positive exponents, and simplify, if possible: $27^{-2/3}$. $$27^{-2/3} = \frac{1}{27^{2/3}} = \frac{1}{\left(\sqrt[3]{\Box}\right)^2} = \frac{1}{\left(\Box\right)^2} = \frac{1}{9}$$	Rewrite with positive exponents, and simplify, if possible: $32^{-3/5}$.
EXAMPLE 3	YOUR TURN 3
Rewrite with positive exponents, and simplify, if possible: $(8xy)^{-4/5}$. $$(8xy)^{-4/5} = \frac{1}{(8xy)^{\Box}}$$	Rewrite with positive exponents, and simplify, if possible: $(11xyz)^{-2/3}$.
EXAMPLE 4	YOUR TURN 4
Rewrite with positive exponents, and simplify, if possible: $3x^{-2/3}y^{2/5}$. $$3x^{-2/3}y^{2/5} = 3 \cdot \frac{1}{x^{\Box}} \cdot y^{2/5} = \frac{\Box}{x^{2/3}}$$	Rewrite with positive exponents, and simplify, if possible: $7x^{-3/2}y^{4/5}$.

EXAMPLE 5	YOUR TURN 5
Rewrite with positive exponents, and simplify, if possible: $\left(\dfrac{5x}{7y}\right)^{-7/3}$. Recall that $\left(\dfrac{a}{b}\right)^{-n} = \left(\dfrac{b}{a}\right)^{n}$. $\left(\dfrac{5x}{7y}\right)^{-7/3} = \left(\dfrac{\boxed{}}{5x}\right)^{7/3}$	Rewrite with positive exponents, and simplify, if possible: $\left(\dfrac{2xy}{5ab}\right)^{-3/5}$.

EXAMPLE 6	YOUR TURN 6
Rewrite with positive exponents, and simplify, if possible: $\dfrac{1}{x^{-5/8}}$. Since $\dfrac{1}{a^{m/n}}$ means $a^{-m/n}$, $\dfrac{1}{x^{-5/8}} = x^{-(-5/8)} = x^{\boxed{}}$	Rewrite with positive exponents, and simplify, if possible: $\dfrac{1}{a^{-2/7}}$.

YOUR NOTES   Write your questions and additional notes.

Laws of Exponents

ESSENTIALS

For any real numbers a and b and any rational exponents m and n:

1. $a^m \cdot a^n = a^{m+n}$

2. $\dfrac{a^m}{a^n} = a^{m-n}$

3. $\left(a^m\right)^n = a^{mn}$

4. $\left(ab\right)^m = a^m b^m$

5. $\left(\dfrac{a}{b}\right)^n = \dfrac{a^n}{b^n}$

Example

- Use the laws of exponents to simplify $2^{1/5} \cdot 2^{2/5}$.

$$2^{1/5} \cdot 2^{2/5} = 2^{1/5 + 2/5} = 2^{3/5}$$

GUIDED LEARNING   📖 **Textbook**   👤 **Instructor**   ▶ **Video**

EXAMPLE 1	YOUR TURN 1
Use the laws of exponents to simplify $x^{1/7} \cdot x^{3/14}$. $x^{1/7} \cdot x^{3/14} = x^{1/7 + 3/14}$ $\phantom{x^{1/7} \cdot x^{3/14}} = x^{2/14 + 3/14}$ $\phantom{x^{1/7} \cdot x^{3/14}} = \boxed{}$	Use the laws of exponents to simplify $x^{1/5} \cdot x^{7/10}$.
EXAMPLE 2	YOUR TURN 2
Use the laws of exponents to simplify $\dfrac{a^{1/6}}{a^{1/2}}$. $\dfrac{a^{1/6}}{a^{1/2}} = a^{1/6 - 1/2}$ $\phantom{\dfrac{a^{1/6}}{a^{1/2}}} = a^{1/6 - 3/6}$ $\phantom{\dfrac{a^{1/6}}{a^{1/2}}} = a^{\boxed{}}$ $\phantom{\dfrac{a^{1/6}}{a^{1/2}}} = a^{-1/3}$, or $\boxed{}$	Use the laws of exponents to simplify $\dfrac{a^{1/4}}{a^{3/4}}$.

EXAMPLE 3	YOUR TURN 3
Use the laws of exponents to simplify $\left(1.2^{3/4}\right)^{1/3}$.	Use the laws of exponents to simplify $\left(3.4^{2/3}\right)^{3/4}$.
$\left(1.2^{3/4}\right)^{1/3} = 1.2^{(3/4)(1/3)}$ $\phantom{\left(1.2^{3/4}\right)^{1/3}} = 1.2^{3/12}$ $\phantom{\left(1.2^{3/4}\right)^{1/3}} = \boxed{}$	

EXAMPLE 4	YOUR TURN 4
Use the laws of exponents to simplify $\left(x^{-1/2}y^{3/4}\right)^{1/3}$.	Use the laws of exponents to simplify $\left(x^{-2/3}y^{1/2}\right)^{1/5}$.
$\left(x^{-1/2}y^{3/4}\right)^{1/3} = x^{(-1/2)(1/3)}y^{(3/4)(1/3)}$ $\phantom{\left(x^{-1/2}y^{3/4}\right)^{1/3}} = x^{-1/6}y^{3/12}$ $\phantom{\left(x^{-1/2}y^{3/4}\right)^{1/3}} = x^{-1/6}y^{1/4}$, or $\dfrac{y^{1/4}}{\boxed{}}$	

YOUR NOTES   Write your questions and additional notes.

Simplifying Radical Expressions

ESSENTIALS

Simplifying Radical Expressions

1. Convert radical expressions to exponential expressions.
2. Use arithmetic and the laws of exponents to simplify.
3. Convert back to radical notation when appropriate.

Example

- Use rational exponents to simplify $\sqrt[7]{(3x)^{14}}$.

$$\sqrt[7]{(3x)^{14}} = (3x)^{14/7}$$
$$= (3x)^{2}$$
$$= 9x^{2}$$

GUIDED LEARNING   📖 **Textbook**   👤 **Instructor**   ▶ **Video**

EXAMPLE 1	YOUR TURN 1
Use rational exponents to simplify $\sqrt[6]{t^{24}}$. $\sqrt[6]{t^{24}} = t^{24/6}$ $= \boxed{}$	Use rational exponents to simplify $\sqrt[5]{t^{15}}$.
EXAMPLE 2	YOUR TURN 2
Use rational exponents to simplify $\sqrt[12]{x^{9}}$. $\sqrt[12]{x^{9}} = x^{9\boxed{}}$ $= x^{3\boxed{}}$ $= \sqrt[\boxed{}]{x^{\boxed{}}}$	Use rational exponents to simplify $\sqrt[10]{x^{4}}$.
EXAMPLE 3	YOUR TURN 3
Use rational exponents to simplify $\sqrt[8]{4}$. $\sqrt[8]{4} = 4^{1\boxed{}}$ $= \left(2^{2}\right)^{1/8}$ $= 2^{1\boxed{}}$ $= \sqrt[\boxed{}]{2}$	Use rational exponents to simplify $\sqrt[4]{9}$.

EXAMPLE 4	YOUR TURN 4
Use rational exponents to simplify $\sqrt[4]{x} \cdot \sqrt[6]{x}$. $$\sqrt[4]{x} \cdot \sqrt[6]{x} = x^{1/4} \cdot x^{1/6}$$ $$= x^{\square/12} \cdot x^{\square/12}$$ $$= x^{\square/12}$$ $$= \sqrt[\square]{x^{\square}}$$	Use rational exponents to simplify $$\sqrt[6]{x} \cdot \sqrt[8]{x}.$$
EXAMPLE 5	YOUR TURN 5
Use rational exponents to simplify $\left(\sqrt[4]{x^2 y^4 z}\right)^{16}$. $$\left(\sqrt[4]{x^2 y^4 z}\right)^{16} = \left(x^2 y^4 z\right)^{16/4}$$ $$= \left(x^2 y^4 z\right)^{4}$$ $$= \left(x^2\right)^{\square}\left(y^4\right)^{\square}(z)^{\square}$$ $$= \boxed{}$$	Use rational exponents to simplify $$\left(\sqrt[3]{x^5 y z^2}\right)^{15}.$$
EXAMPLE 6	YOUR TURN 6
Use rational exponents to write a single radical expression: $\sqrt[3]{\sqrt{x}}$. $$\sqrt[3]{\sqrt{x}} = \sqrt[3]{x^{1/2}}$$ $$= \left(x^{1/2}\right)^{\square}$$ $$= x^{\square}$$ $$= \sqrt[\square]{x}$$	Use rational exponents to write a single radical expression: $\sqrt[5]{\sqrt[3]{x}}$.

YOUR NOTES   Write your questions and additional notes.

Practice Exercises

Readiness Check

Match each expression with its equivalent form.

1. $\left(2^4\right)^5$

2. $2^4 \cdot 2^5$

3. $\dfrac{2^4}{2^5}$

4. $\left(2 \cdot 4\right)^5$

a) 2^{4+5}

b) $2^5 \cdot 4^5$

c) 2^{4-5}

d) $2^{4 \cdot 5}$

Rational Exponents

Rewrite without rational exponents, and simplify, if possible.

5. $x^{2/3}$ **6.** $81^{1/2}$ **7.** $\left(x^2 y^2\right)^{1/3}$ **8.** $4^{5/2}$

Rewrite with rational exponents.

9. $\sqrt[3]{5}$ **10.** $\sqrt[4]{m^3}$ **11.** $\left(\sqrt{4xy}\right)^5$ **12.** $\left(\sqrt[3]{2x^2 y}\right)^8$

Negative Rational Exponents

Write an equivalent expression with positive exponents, and simplify, if possible.

13. $27^{-2/3}$ **14.** $x^2 y^{-1/3} z^4$ **15.** $\dfrac{12a}{c^{-4/7}}$

16. $\left(\dfrac{1}{27}\right)^{-2/3}$ **17.** $5a^{-3/2} b^{-1/5} c^{3/4}$ **18.** $\left(6xyz\right)^{-3/5}$

Laws of Exponents

Use the laws of exponents to simplify. Write the answers with positive exponents.

19. $7^{2/5} \cdot 7^{1/10}$

20. $\left(12^{3/4}\right)^{1/2}$

21. $\dfrac{2.1^{5/6}}{2.1^{-1/2}}$

22. $x^{2/3} \cdot x^{1/5}$

23. $\left(a^{-2/3} b^{1/7}\right)^{1/4}$

24. $\left(5^{-1/3}\right)^{3}$

Simplifying Radical Expressions

Use rational exponents to simplify. Write the answer in radical notation if appropriate.

25. $\sqrt[15]{x^3}$

26. $\left(\sqrt[6]{ab}\right)^{12}$

27. $\sqrt[8]{81}$

Use rational exponents to write a single radical expression.

28. $\sqrt[4]{\sqrt[3]{ab}}$

29. $\left(\sqrt[4]{x^2 y^3}\right)^{16}$

30. $\sqrt{2}\sqrt[4]{2}$

Multiplying and Simplifying Radical Expressions

ESSENTIALS

The Product Rule for Radicals

For any real numbers $\sqrt[k]{a}$ and $\sqrt[k]{b}$,

$$\sqrt[k]{a} \cdot \sqrt[k]{b} = \sqrt[k]{a \cdot b}, \text{ or } a^{1/k} \cdot b^{1/k} = (ab)^{1/k}.$$

Factoring Radical Expressions

For any nonnegative real numbers a and b and any index k,

$$\sqrt[k]{ab} = \sqrt[k]{a} \cdot \sqrt[k]{b}, \text{ or } (ab)^{1/k} = a^{1/k} \cdot b^{1/k}.$$

To simplify a radical expression by factoring:

1. Look for the largest factors of the radicand that are perfect kth powers (where k is the index).
2. Then take the kth root of the resulting factors.
3. A radical expression, with index k, is *simplified* when its radicand has no factors that are perfect kth powers.

Example

- Multiply and simplify: $\sqrt{12}\,\sqrt{8}$.

$$\sqrt{12}\,\sqrt{8} = \sqrt{12 \cdot 8}$$
$$= \sqrt{96}$$
$$= \sqrt{16 \cdot 6}$$
$$= \sqrt{16} \cdot \sqrt{6}$$
$$= 4\sqrt{6}$$

GUIDED LEARNING   📖 **Textbook**   👤 **Instructor**   ▶ **Video**

EXAMPLE 1	YOUR TURN 1
Multiply: $\sqrt{6}\,\sqrt{7}$.	Multiply: $\sqrt{5}\,\sqrt{13}$.
$\sqrt{6}\,\sqrt{7} = \sqrt{6 \cdot \boxed{}}$ $= \sqrt{\boxed{}}$	

EXAMPLE 2	YOUR TURN 2
Multiply: $\sqrt[5]{6x}\,\sqrt[5]{2y^2}$.	Multiply: $\sqrt[3]{4x^2}\,\sqrt[3]{5y}$.
$\sqrt[5]{6x}\,\sqrt[5]{2y^2} = \sqrt[5]{6x \cdot 2y^2}$ $= \boxed{}$	

EXAMPLE 3	YOUR TURN 3
Multiply: $\sqrt[4]{7}\,\sqrt{3}$.	Multiply: $\sqrt[3]{5}\,\sqrt{2}$.

The indices are different, so we must use rational exponents.

$$\sqrt[4]{7}\,\sqrt{3} = 7^{1/4} \cdot 3^{\boxed{}}$$
$$= 7^{1/4} \cdot 3^{\boxed{}/4}$$
$$= \left(7 \cdot 3^2\right)^{\boxed{}}$$
$$= 63^{1/4}$$
$$= \sqrt[\boxed{}]{63}$$

EXAMPLE 4	YOUR TURN 4
Simplify: $\sqrt{84}$.	Simplify: $\sqrt{125}$.
$\sqrt{84} = \sqrt{4 \cdot 21}$ $= \sqrt{\boxed{}} \cdot \sqrt{21}$ $= \boxed{}$	

EXAMPLE 5	YOUR TURN 5
Simplify: $\sqrt{24x^2y^3}$.	Simplify: $\sqrt{48xy^4}$.
$\sqrt{24x^2y^3} = \sqrt{4 \cdot 6 \cdot x^2 \cdot y^2 \cdot y^1}$ $= \sqrt{\boxed{}} \cdot \sqrt{6y}$ $= \boxed{}\sqrt{6y}$	

EXAMPLE 6	YOUR TURN 6
Simplify: $\sqrt[3]{24a^6b^{14}}$.	Simplify: $\sqrt[3]{32a^{10}b^7}$.

$$\sqrt[3]{24a^6b^{14}} = \sqrt[3]{8 \cdot 3 \cdot a^6 \cdot b^{12} \cdot b^2}$$

$$= \sqrt[3]{8}\,\boxed{}\,\sqrt[3]{b^{12}}\,\sqrt[3]{3b^2}$$

$$= 2\,\boxed{}\,b^{12/3}\sqrt[3]{3b^2}$$

$$= \boxed{}$$

EXAMPLE 7	YOUR TURN 7
Multiply and simplify: $\sqrt[3]{18}\,\sqrt[3]{3}$.	Multiply and simplify: $\sqrt[3]{12}\,\sqrt[3]{6}$.

$$\sqrt[3]{18}\,\sqrt[3]{3} = \sqrt[3]{18 \cdot 3}$$

$$= \sqrt[3]{2 \cdot 3 \cdot 3 \cdot \boxed{}}$$

$$= \boxed{}\,\sqrt[3]{2}$$

EXAMPLE 8	YOUR TURN 8
Multiply and simplify: $2\sqrt{6} \cdot 5\sqrt{4}$.	Multiply and simplify: $-3\sqrt{5} \cdot 4\sqrt{8}$.

$$2\sqrt{6} \cdot 5\sqrt{4} = 2 \cdot 5\sqrt{6 \cdot 4}$$

$$= \boxed{}\,\sqrt{24}$$

$$= 10\sqrt{\boxed{} \cdot 6}$$

$$= 10\sqrt{4} \cdot \sqrt{6}$$

$$= 10 \cdot 2\sqrt{6}$$

$$= \boxed{}$$

EXAMPLE 9	YOUR TURN 9
Multiply and simplify: $\sqrt[4]{8x^5y^3}\,\sqrt[4]{6x^2y}$.	Multiply and simplify: $\sqrt[4]{9x^8y}\,\sqrt[4]{9x^3y}$.

$$\sqrt[4]{8x^5y^3}\,\sqrt[4]{6x^2y} = \sqrt[4]{8\cdot 6\cdot x^5\cdot x^2\cdot y^3\cdot y}$$

$$= \sqrt[4]{48x^7\,\boxed{}}$$

$$= \sqrt[4]{16x^4y^4\cdot 3x^3}$$

$$= \sqrt[4]{16x^4y^4}\,\sqrt[4]{3x^3}$$

$$= \boxed{}$$

YOUR NOTES   Write your questions and additional notes.

Dividing and Simplifying Radical Expressions

ESSENTIALS

The Quotient Rule for Radicals

For any nonnegative number a, any positive number b, and any index k,

$$\frac{\sqrt[k]{a}}{\sqrt[k]{b}} = \sqrt[k]{\frac{a}{b}}, \text{ or } \frac{a^{1/k}}{b^{1/k}} = \left(\frac{a}{b}\right)^{1/k}.$$

kth Roots of Quotients

For any nonnegative number a, any positive number b, and any index k,

$$\sqrt[k]{\frac{a}{b}} = \frac{\sqrt[k]{a}}{\sqrt[k]{b}}, \text{ or } \left(\frac{a}{b}\right)^{1/k} = \frac{a^{1/k}}{b^{1/k}}.$$

Example

- Simplify by taking the roots of the numerator and the denominator: $\sqrt{\dfrac{16}{y^2}}$.

$$\sqrt{\frac{16}{y^2}} = \frac{\sqrt{16}}{\sqrt{y^2}} = \frac{4}{y}$$

GUIDED LEARNING   📖 **Textbook**   👤 **Instructor**   ▶ **Video**

EXAMPLE 1	YOUR TURN 1
Divide and simplify: $\dfrac{\sqrt{32}}{\sqrt{2}}$.	Divide and simplify: $\dfrac{\sqrt{50}}{\sqrt{2}}$.
$\dfrac{\sqrt{32}}{\sqrt{2}} = \sqrt{\dfrac{32}{\boxed{}}} = \sqrt{16} = \boxed{}$	
EXAMPLE 2	YOUR TURN 2
Divide and simplify: $\dfrac{6\sqrt[3]{54}}{\sqrt[3]{2}}$.	Divide and simplify: $\dfrac{4\sqrt[3]{192}}{\sqrt[3]{3}}$.
$\dfrac{6\sqrt[3]{54}}{\sqrt[3]{2}} = 6\sqrt[3]{\dfrac{54}{2}}$ $= 6\sqrt[3]{\boxed{}}$ $= 6 \cdot 3$ $= \boxed{}$	

EXAMPLE 3	YOUR TURN 3
Simplify: $\sqrt[3]{\dfrac{27}{8}}$.	Simplify: $\sqrt[4]{\dfrac{16}{81}}$.

$$\sqrt[3]{\frac{27}{8}} = \frac{\sqrt[3]{27}}{\boxed{}} = \frac{3}{\boxed{}}$$

EXAMPLE 4	YOUR TURN 4
Simplify: $\sqrt{\dfrac{50x^6}{y^{10}}}$.	Simplify: $\sqrt{\dfrac{3x^9}{25y^6}}$.

$$\sqrt{\frac{50x^6}{y^{10}}} = \frac{\sqrt{50x^6}}{\sqrt{y^{10}}}$$

$$= \frac{\sqrt{25x^6 \cdot \boxed{}}}{\sqrt{y^{10}}}$$

$$= \frac{5\boxed{}\sqrt{2}}{y^5}$$

EXAMPLE 5	YOUR TURN 5
Divide and simplify: $\dfrac{\sqrt[4]{a^6b^5}}{\sqrt[3]{a^2b}}$.	Divide and simplify: $\dfrac{\sqrt[5]{a^7b^8}}{\sqrt{ab^3}}$.

The indices are different, so we must use rational exponents.

$$\frac{\sqrt[4]{a^6b^5}}{\sqrt[3]{a^2b}} = \frac{\left(a^6b^5\right)^{1/4}}{\left(a^2b\right)^{1/3}}$$

$$= \frac{a^{6/4}b^{\boxed{}}}{a^{2/3}b^{\boxed{}}}$$

$$= a^{6/4-2/3}b^{5/4-1/3}$$

$$= a^{18/12-\boxed{}/12}b^{15/12-\boxed{}/12}$$

$$= a^{\boxed{}/12}b^{\boxed{}/12}$$

$$= \sqrt[12]{a^{\boxed{}}b^{\boxed{}}}$$

YOUR NOTES   Write your questions and additional notes.

Practice Exercises

Readiness Check

Choose the word or expression that best completes each sentence.

1. When simplifying $\sqrt[4]{162x^5}$, the largest perfect fourth power is __________.
   $$81x^4 \ / \ 27x^5$$

2. When multiplying radicals with the same index, multiply the radicands and then __________, if possible.
   $$\text{add} \ / \ \text{simplify}$$

3. When simplifying $\sqrt[3]{54x^6}$, you must find the largest __________.
   $$\text{perfect cube} \ / \ \text{perfect sixth power}$$

4. In order to use the product and quotient rules for radicals, the indices must be

   __________.
   $$\text{different} \ / \ \text{the same}$$

Multiplying and Simplifying Radical Expressions

Simplify. Assume that no radicands were formed by raising negative numbers to even powers.

5. $\sqrt{50}$

6. $\sqrt[3]{40}$

7. $\sqrt{120x^5}$

8. $\sqrt[3]{-24x^9}$

9. $\sqrt{50ab^7}$

10. $\sqrt[4]{32x^{12}y^5}$

Multiply and simplify. Assume that no radicands were formed by raising negative numbers to even powers.

11. $\sqrt{6}\,\sqrt{10}$

12. $\sqrt{7a^3}\,\sqrt{14a^5}$

13. $\sqrt[3]{3a^2}\,\sqrt[3]{6a}$

14. $\sqrt[4]{x^2y^3}\,\sqrt[4]{x^8y^4}$

15. $\sqrt{12x^3y}\,\sqrt[4]{20xy^5}$

Dividing and Simplifying Radical Expressions

Simplify. Assume that all expressions under radicals represent positive numbers.

16. $\dfrac{\sqrt{125}}{\sqrt{5}}$

17. $\dfrac{\sqrt{32x^5y}}{\sqrt{16xy}}$

18. $\dfrac{\sqrt[3]{250x^4y^7}}{\sqrt[3]{2x^2y}}$

19. $\dfrac{\sqrt[4]{x^3y^2}}{\sqrt[3]{xy}}$

Simplify.

20. $\sqrt{\dfrac{16}{25}}$

21. $\sqrt{\dfrac{49}{x^4}}$

22. $\sqrt[4]{\dfrac{3x^4y^8}{z^4}}$

23. $\sqrt[3]{\dfrac{125a}{8b^3}}$

24. $\sqrt[3]{\dfrac{81x^5}{y^6}}$

25. $\sqrt[5]{\dfrac{a^9b^2c^6}{z^{15}}}$

Addition and Subtraction

ESSENTIALS

Like radicals are radical expressions that have the same indices and the same radicands.

Example

- Simplify $2\sqrt{3} + 5\sqrt{3}$ by collecting like radical terms, if possible.

$$2\sqrt{3} + 5\sqrt{3} = (2+5)\sqrt{3} = 7\sqrt{3}$$

GUIDED LEARNING	🔖 **Textbook**	👤 **Instructor**	▶ **Video**

EXAMPLE 1	YOUR TURN 1
Simplify $4\sqrt{6} + 7\sqrt{6}$ by collecting like radical terms. $$4\sqrt{6} + 7\sqrt{6} = \left(4 + \boxed{}\right)\sqrt{6}$$ $$= 11\sqrt{6}$$	Simplify $2\sqrt{3} + 8\sqrt{3}$ by collecting like radical terms.
EXAMPLE 2	YOUR TURN 2
Simplify $10\sqrt[3]{5} - 4\sqrt[3]{3}$ by collecting like radical terms, if possible. $10\sqrt[3]{5} - 4\sqrt[3]{3}$ cannot be simplified because the radicands are different.	Simplify $8\sqrt[3]{2} - 3\sqrt[3]{7}$ by collecting like radical terms, if possible.
EXAMPLE 3	YOUR TURN 3
Simplify $9\sqrt[3]{3x} + 2\sqrt[3]{3x} - 10\sqrt[4]{3x}$ by collecting like radical terms, if possible. $$9\sqrt[3]{3x} + 2\sqrt[3]{3x} - 10\sqrt[4]{3x} = (9+2)\sqrt[3]{3x} - 10\sqrt[4]{3x}$$ $$= \boxed{} - 10\sqrt[4]{3x}$$	Simplify $6\sqrt[4]{2xy} + 3\sqrt[4]{2xy} - 5\sqrt[3]{2xy}$ by collecting like radical terms, if possible.

EXAMPLE 4	YOUR TURN 4
Simplify $3\sqrt{8}-\sqrt{32}$ by collecting like radical terms, if possible.	Simplify $5\sqrt{18}+\sqrt{50}$ by collecting like radical terms, if possible.

$$3\sqrt{8}-\sqrt{32}=3\sqrt{4\cdot 2}-\sqrt{16\cdot 2}$$
$$=\boxed{}\cdot 2\sqrt{2}-4\sqrt{2}\quad\bigg\}\ \text{Simplifying the radicands}$$
$$=6\sqrt{2}-4\sqrt{2}$$
$$=(6-4)\sqrt{2}$$
$$=\boxed{}\sqrt{2}$$

EXAMPLE 5	YOUR TURN 5
Simplify $\sqrt[3]{3x^3y^5}+5\sqrt[3]{3y^2}$ by collecting like radical terms, if possible.	Simplify $\sqrt[3]{5xy^9}+\sqrt[3]{5x}$ by collecting like radical terms, if possible.

$$\sqrt[3]{3x^3y^5}+5\sqrt[3]{3y^2}=\sqrt[3]{x^3y^3\cdot 3y^2}+5\sqrt[3]{3y^2}$$
$$=\sqrt[3]{x^3y^3}\cdot\sqrt[3]{3y^2}+5\sqrt[3]{3y^2}$$
$$=\boxed{}y\sqrt[3]{3y^2}+5\sqrt[3]{3y^2}$$
$$=\left(\boxed{}\right)\sqrt[3]{3y^2}$$

YOUR NOTES　　Write your questions and additional notes.

More Multiplication

ESSENTIALS

Expressions of the form $\sqrt{a} + \sqrt{b}$ and $\sqrt{a} - \sqrt{b}$ are called **conjugates**. Their product is always an expression that contains no radicals.

To multiply radical expressions in which some factors contain more than one term, we use the procedures for multiplying polynomials.

Example

- Multiply: $\left(\sqrt{x} + 2\sqrt{3}\right)\left(\sqrt{x} + 5\sqrt{3}\right)$.

$$\left(\sqrt{x} + 2\sqrt{3}\right)\left(\sqrt{x} + 5\sqrt{3}\right) = \sqrt{x} \cdot \sqrt{x} + \sqrt{x} \cdot 5\sqrt{3} + 2\sqrt{3} \cdot \sqrt{x} + 2\sqrt{3} \cdot 5\sqrt{3} \quad \text{Using FOIL}$$

$$= x + 5\sqrt{3x} + 2\sqrt{3x} + 10 \cdot 3$$

$$= x + 7\sqrt{3x} + 30$$

GUIDED LEARNING	📖 **Textbook**	👤 **Instructor**	▶ **Video**

EXAMPLE 1	YOUR TURN 1
Multiply: $\sqrt{2}\left(x - \sqrt{3}\right)$.	Multiply: $\sqrt{5}\left(\sqrt{3} + y\right)$.
$\sqrt{2}\left(x - \sqrt{3}\right) = \sqrt{2} \cdot \boxed{} - \sqrt{2} \cdot \boxed{}$ $\qquad = x\sqrt{2} - \boxed{}$	
EXAMPLE 2	YOUR TURN 2
Multiply: $\sqrt[3]{x}\left(\sqrt[3]{3x^2} + \sqrt[3]{24x^2}\right)$.	Multiply: $\sqrt[3]{y}\left(\sqrt[3]{4y^2} + \sqrt[3]{256y^2}\right)$.
$\sqrt[3]{x}\left(\sqrt[3]{3x^2} + \sqrt[3]{24x^2}\right) = \sqrt[3]{x} \cdot \sqrt[3]{3x^2} + \sqrt[3]{x} \cdot \sqrt[3]{24x^2}$ $\qquad = \sqrt[3]{3x^3} + \boxed{}$ $\qquad = \sqrt[3]{x^3}\,\sqrt[3]{3} + \sqrt[3]{8x^3}\,\sqrt[3]{3}$ $\qquad = x\sqrt[3]{3} + \boxed{}\,\sqrt[3]{3} \qquad \text{Simplifying}$ $\qquad = \boxed{}\,\sqrt[3]{3}$	

EXAMPLE 3	YOUR TURN 3

Multiply: $\left(2+\sqrt{6}\right)^2$.

$$\left(2+\sqrt{6}\right)^2 = \left(2+\sqrt{6}\right)\left(2+\sqrt{6}\right)$$

$$= (2)^2 + 2\sqrt{6} + 2\sqrt{6} + \left(\sqrt{6}\right)^2 \qquad \text{FOIL}$$

$$= 4 + \boxed{}\sqrt{6} + 6$$

$$= \boxed{}$$

Multiply: $\left(3-\sqrt{7}\right)^2$.

EXAMPLE 4	YOUR TURN 4

Multiply: $\left(2\sqrt{3}+\sqrt{5}\right)\left(\sqrt{3}-4\sqrt{5}\right)$.

$$\left(2\sqrt{3}+\sqrt{5}\right)\left(\sqrt{3}-4\sqrt{5}\right) = 2\left(\sqrt{3}\right)^2 - 8\sqrt{3}\cdot\sqrt{5} + \sqrt{5}\cdot\sqrt{3} - 4\left(\sqrt{5}\right)^2$$

$$= 2\cdot 3 - 8\sqrt{15} + \sqrt{15} - \boxed{}\cdot 5$$

$$= 6 - 8\sqrt{15} + \sqrt{15} - 20$$

$$= -14 - \boxed{}\sqrt{15}$$

Multiply:
$\left(4\sqrt{6}+\sqrt{2}\right)\left(\sqrt{6}-2\sqrt{2}\right)$.

EXAMPLE 5	YOUR TURN 5

Multiply: $\left(3+\sqrt{6}\right)\left(3-\sqrt{6}\right)$.

$$\left(3+\sqrt{6}\right)\left(3-\sqrt{6}\right) = (3)^2 - 3\sqrt{6} + 3\sqrt{6} - \left(\sqrt{6}\right)^2$$

$$= 9 - 3\sqrt{6} + \boxed{} - 6$$

$$= 9 - \boxed{}$$

$$= \boxed{}$$

Multiply:
$\left(5-\sqrt{2}\right)\left(5+\sqrt{2}\right)$.

YOUR NOTES   Write your questions and additional notes.

Practice Exercises

Readiness Check

Determine whether each statement is true or false.

1. The conjugate of $\sqrt{a} + \sqrt{b}$ is $\sqrt{a} - \sqrt{b}$.

2. To add radical expressions, only the indices must be the same.

3. The simplified form of a product of two radical expressions always contains a radical.

4. The product of conjugates does not contain a radical.

Addition and Subtraction

Add or subtract. Then simplify by collecting like radical terms, if possible. Assume that no radicands were formed by raising negative numbers to even powers.

5. $4\sqrt{3} + 2\sqrt{3}$

6. $8\sqrt[3]{5} - 5\sqrt[3]{5}$

7. $2\sqrt{6} - 5\sqrt[4]{7} + 2\sqrt{6} + 9\sqrt[4]{7}$

8. $3\sqrt{20} - \sqrt{500}$

9. $\sqrt{49x^3} + 5\sqrt{x}$

10. $9\sqrt[3]{16} - 2\sqrt[3]{54}$

11. $\sqrt{9x + 36} - \sqrt{x + 4}$

12. $\sqrt[3]{8x^5} + \sqrt[3]{27x^5}$

More Multiplication

Multiply. Assume that no radicands were formed by raising negative numbers to even powers.

13. $\sqrt{5}\left(6-3\sqrt{5}\right)$

14. $\sqrt[3]{3y}\left(\sqrt[3]{9y^2}-\sqrt[3]{18y}\right)$

15. $\left(\sqrt{3}+x\right)^2$

16. $\left(\sqrt{a}+4\right)\left(\sqrt{a}-9\right)$

17. $\left(4\sqrt{5}+3\right)\left(\sqrt{5}-1\right)$

18. $\left(2\sqrt[3]{5}-4\sqrt[3]{7}\right)\left(\sqrt[3]{5}-6\sqrt[3]{7}\right)$

19. $\left(6-\sqrt{3}\right)\left(6+\sqrt{3}\right)$

20. $\left(\sqrt{x}+\sqrt{w}\right)\left(\sqrt{x}-\sqrt{w}\right)$

Rationalizing Denominators

ESSENTIALS

When we **rationalize the denominator** of a radical expression, we find an equivalent expression without a radical in the denominator.

Example

- Rationalize the denominator: $\dfrac{2\sqrt{5}}{3\sqrt{2}}$.

  Multiply by 1 using $\dfrac{\sqrt{2}}{\sqrt{2}}$ so that the radicand in the denominator will be a perfect square.

  $$\frac{2\sqrt{5}}{3\sqrt{2}} = \frac{2\sqrt{5}}{3\sqrt{2}} \cdot \frac{\sqrt{2}}{\sqrt{2}} = \frac{2\sqrt{5}\cdot\sqrt{2}}{3\sqrt{2}\cdot\sqrt{2}} = \frac{2\sqrt{10}}{3\sqrt{4}} = \frac{2\sqrt{10}}{3\cdot 2} = \frac{\sqrt{10}}{3}$$

GUIDED LEARNING   📍 **Textbook**   📍 **Instructor**   📍 **Video**

EXAMPLE 1	YOUR TURN 1
Rationalize the denominator: $\sqrt{\dfrac{5}{3}}$.	Rationalize the denominator: $\sqrt{\dfrac{1}{2}}$.

Multiply by 1 using $\dfrac{\sqrt{3}}{\sqrt{3}}$ so that the

denominator of the radicand will be a perfect square.

$$\sqrt{\frac{5}{3}} = \frac{\sqrt{5}}{\sqrt{3}}\cdot\frac{\sqrt{\boxed{}}}{\sqrt{\boxed{}}} = \frac{\sqrt{5}\cdot\sqrt{3}}{\sqrt{3}\cdot\sqrt{3}} = \frac{\sqrt{15}}{\sqrt{\boxed{}}} = \frac{\sqrt{15}}{\boxed{}}$$

EXAMPLE 2	YOUR TURN 2
Rationalize the denominator: $\sqrt[3]{\dfrac{3}{25}}$.	Rationalize the denominator: $\sqrt[3]{\dfrac{9}{2}}$.

Since $25 = 5\cdot 5$, we need another factor of 5 so that the denominator of the radicand is a perfect cube.

$$\sqrt[3]{\frac{3}{25}} = \frac{\sqrt[3]{3}}{\sqrt[3]{25}}\cdot\frac{\sqrt[3]{\boxed{}}}{\sqrt[3]{\boxed{}}} = \frac{\sqrt[3]{15}}{\sqrt[3]{\boxed{}}} = \frac{\sqrt[3]{15}}{\boxed{}}$$

EXAMPLE 3	YOUR TURN 3
Rationalize the denominator: $\dfrac{\sqrt{3}}{3\sqrt{x^3}}$.	Rationalize the denominator: $\dfrac{\sqrt{5}}{2\sqrt{x}}$.

Multiply by 1 using $\dfrac{\sqrt{x}}{\sqrt{x}}$ so that the radicand

in the denominator is a perfect square.

$$\frac{\sqrt{3}}{3\sqrt{x^3}} = \frac{\sqrt{3}}{3\sqrt{x^3}} \cdot \frac{\sqrt{x}}{\sqrt{x}} = \frac{\boxed{}}{3\sqrt{x^4}} = \boxed{}$$

EXAMPLE 4	YOUR TURN 4
Rationalize the denominator: $\dfrac{x\sqrt[3]{y}}{\sqrt[3]{9y^2z}}$.	Rationalize the denominator: $\dfrac{\sqrt[3]{xz}}{\sqrt[3]{16x^2yz^2}}$.

Multiply by $\dfrac{\sqrt[3]{3yz^2}}{\sqrt[3]{3yz^2}}$ so that the radicand in the

denominator is a perfect cube.

$$\frac{x\sqrt[3]{y}}{\sqrt[3]{9y^2z}} = \frac{x\sqrt[3]{y}}{\sqrt[3]{9y^2z}} \cdot \frac{\sqrt[3]{3yz^2}}{\sqrt[3]{3yz^2}}$$

$$= \frac{x\sqrt[3]{\boxed{}}}{\sqrt[3]{27y^3z^3}}$$

$$= \frac{x\sqrt[3]{3y^2z^2}}{\boxed{}}$$

YOUR NOTES   Write your questions and additional notes.

Rationalizing When There Are Two Terms

ESSENTIALS

Pairs of radical expressions like $\sqrt{a} + \sqrt{b}$ and $\sqrt{a} - \sqrt{b}$ are called **conjugates**. The product of conjugates has no radicals.

Example

- Rationalize the denominator: $\dfrac{4}{6 + \sqrt{5}}$.

$$\frac{4}{6 + \sqrt{5}} = \frac{4}{6 + \sqrt{5}} \cdot \frac{6 - \sqrt{5}}{6 - \sqrt{5}}$$

Multiplying by 1, using the conjugate of $6 + \sqrt{5}$, which is $6 - \sqrt{5}$

$$= \frac{4\left(6 - \sqrt{5}\right)}{\left(6 + \sqrt{5}\right)\left(6 - \sqrt{5}\right)}$$

Multiplying numerators and denominators

$$= \frac{4\left(6 - \sqrt{5}\right)}{6^2 - \left(\sqrt{5}\right)^2}$$

Using $(A + B)(A - B) = A^2 - B^2$

$$= \frac{24 - 4\sqrt{5}}{36 - 5} = \frac{24 - 4\sqrt{5}}{31}$$

GUIDED LEARNING   📕 **Textbook**   👤 **Instructor**   ▶ **Video**

EXAMPLE 1	YOUR TURN 1
Rationalize the denominator: $\dfrac{2}{\sqrt{5} + x}$.	Rationalize the denominator: $\dfrac{4}{x + \sqrt{3}}$.

$$\frac{2}{\sqrt{5} + x} = \frac{2}{\sqrt{5} + x} \cdot \frac{\sqrt{5} - x}{\boxed{}}$$

$$= \frac{2\left(\sqrt{5} - x\right)}{\left(\sqrt{5} + x\right)\left(\boxed{}\right)}$$

$$= \frac{2\sqrt{5} - \boxed{}}{\left(\sqrt{5}\right)^2 - x^2}$$

$$= \frac{2\sqrt{5} - 2x}{\boxed{} - x^2}$$

EXAMPLE 2	YOUR TURN 2
Rationalize the denominator: $\dfrac{3+\sqrt{3}}{\sqrt{7}-\sqrt{3}}$.	Rationalize the denominator: $\dfrac{4-\sqrt{11}}{\sqrt{2}+\sqrt{11}}$.

$$\frac{3+\sqrt{3}}{\sqrt{7}-\sqrt{3}} = \frac{3+\sqrt{3}}{\sqrt{7}-\sqrt{3}} \cdot \frac{\sqrt{7}+\sqrt{3}}{\sqrt{7}+\sqrt{3}}$$

$$= \frac{\left(3+\sqrt{3}\right)\left(\sqrt{7}+\sqrt{3}\right)}{\left(\sqrt{7}-\sqrt{3}\right)\left(\sqrt{7}+\sqrt{3}\right)}$$

$$= \frac{3\sqrt{7}+\boxed{}+\sqrt{3}\sqrt{7}+\left(\sqrt{3}\right)^{2}}{\left(\sqrt{7}\right)^{2}-\left(\sqrt{3}\right)^{2}}$$

$$= \frac{3\sqrt{7}+3\sqrt{3}+\sqrt{\boxed{}}+3}{7-3}$$

$$= \frac{3\sqrt{7}+3\sqrt{3}+\sqrt{21}+3}{\boxed{}}$$

YOUR NOTES   Write your questions and additional notes.

Practice Exercises

Readiness Check

Determine whether each statement is true or false.

1. The conjugate of $a - \sqrt{b}$ is $a + \sqrt{b}$.

2. To rationalize the denominator of $\dfrac{\sqrt[3]{5x}}{\sqrt[3]{3y}}$, we can multiply by 1 using $\dfrac{\sqrt[3]{3y}}{\sqrt[3]{3y}}$.

3. To rationalize the denominator of $\dfrac{2+\sqrt{3}}{5-\sqrt{7}}$, multiply by a form of 1 using $\dfrac{5-\sqrt{7}}{5-\sqrt{7}}$.

4. After a denominator is rationalized, it contains no radical expressions.

Rationalizing Denominators

Rationalize the denominator. Assume that no radicands were formed by raising negative numbers to even powers.

5. $\sqrt{\dfrac{1}{3}}$

6. $\dfrac{\sqrt{6}}{\sqrt{7}}$

7. $\dfrac{2\sqrt{5}}{3\sqrt{6}}$

8. $\sqrt[3]{\dfrac{3}{xy^2}}$

9. $\sqrt{\dfrac{7x}{12}}$

10. $\sqrt[4]{\dfrac{x^2}{16yz^4}}$

Rationalizing When There Are Two Terms

Rationalize the denominator. Assume that no radicands were formed by raising negative numbers to even powers.

11. $\dfrac{2}{5+\sqrt{3}}$

12. $\dfrac{4\sqrt{y}}{2-\sqrt{y}}$

13. $\dfrac{\sqrt{2}+\sqrt{3}}{\sqrt{2}-\sqrt{7}}$

14. $\dfrac{3-2\sqrt{x}}{2+3\sqrt{x}}$

The Principle of Powers

ESSENTIALS

A **radical equation** is an equation in which the variable appears in one or more radicands.

The Principle of Powers

If $a = b$ is true, then $a^n = b^n$ is true for any natural number n.

To solve an equation with one radical term:

1. Isolate the radical term on one side of the equation.
2. Use the principle of powers and then solve the resulting equation.
3. Check any possible solution in the original equation.

When we raise both sides of an equation to an even exponent, it is essential that we check the answer in the original equation.

Example

- Solve: $x = \sqrt{x+3} - 1$.

$$x = \sqrt{x+3} - 1$$

$$x + 1 = \sqrt{x+3} \qquad \text{Isolating the radical}$$

$$(x+1)^2 = \left(\sqrt{x+3}\right)^2 \qquad \text{Using the principle of powers}$$

$$x^2 + 2x + 1 = x + 3$$

$$x^2 + x - 2 = 0$$

$$(x+2)(x-1) = 0$$

$$x + 2 = 0 \quad or \quad x - 1 = 0$$

$$x = -2 \quad or \quad x = 1$$

We check both -2 and 1 in the original equation. The number 1 checks but -2 does not, so the solution is 1.

GUIDED LEARNING   🔖 **Textbook**   👤 **Instructor**   ▶ **Video**

EXAMPLE 1	YOUR TURN 1
Solve: $\sqrt{x} - 5 = 1$.	Solve: $\sqrt{x} + 3 = 8$.

EXAMPLE 1

Solve: $\sqrt{x} - 5 = 1$.

$$\sqrt{x} - 5 = 1$$

$$\sqrt{x} = \boxed{} \qquad \text{Isolating the radical}$$

$$\left(\sqrt{x}\right)^2 = 6^2 \qquad \text{Using the principle of powers}$$

$$x = \boxed{}$$

A check shows that 36 is the solution.

EXAMPLE 2	YOUR TURN 2

EXAMPLE 2

Solve: $\sqrt{2y-1} = y-2$.

$$\sqrt{2y-1} = y-2$$

$$\left(\sqrt{2y-1}\right)^2 = (y-2)^2$$

$$2y-1 = y^2 - \boxed{} + 4$$

$$0 = y^2 - \boxed{} + 5$$

$$0 = (y-1)\left(\boxed{}\right)$$

$$y = 1 \ \ or \ \ y = \boxed{}$$

Check: For 1:

$$\sqrt{2y-1} = y-2$$

$$\sqrt{2(1)-1} \ ? \ 1-2$$

$$\sqrt{2-1} \ \Big| \ -1$$

$$\sqrt{1}$$

$$1 \ \Big| \ \text{FALSE}$$

For 5:

$$\sqrt{2y-1} = y-2$$

$$\sqrt{2(5)-1} \ ? \ 5-2$$

$$\sqrt{10-1} \ \Big| \ 3$$

$$\sqrt{9}$$

$$3 \ \Big| \ \text{TRUE}$$

The number 5 checks and 1 does not, so the solution is $\boxed{}$.

YOUR TURN 2

Solve: $\sqrt{3-x} + 1 = x$.

EXAMPLE 3	YOUR TURN 3

EXAMPLE 3

Solve: $\sqrt[3]{x-5} + 2 = 1$.

$$\sqrt[3]{x-5} + 2 = 1$$

$$\sqrt[3]{x-5} = -1 \qquad \text{Isolating the radical}$$

$$\left(\sqrt[3]{x-5}\right)^3 = \boxed{} \qquad \text{Using the principle of powers}$$

$$x - 5 = -1$$

$$x = \boxed{}$$

The number 4 checks and is the solution.

The solution is $\boxed{}$.

YOUR TURN 3

Solve: $\sqrt[3]{x+4} + 5 = 3$.

YOUR NOTES   Write your questions and additional notes.

Equations with Two Radical Terms

ESSENTIALS

To solve an equation with two or more radical terms:

1. Isolate one of the radical terms.
2. Use the principle of powers.
3. If a radical remains, perform steps (1) and (2) again.
4. Solve the resulting equation.
5. Check possible solutions in the original equation.

Example

- Solve: $\sqrt{x+2} - 2\sqrt{x-1} = 0$.

$$\sqrt{x+2} - 2\sqrt{x-1} = 0$$

$$\sqrt{x+2} = 2\sqrt{x-1} \qquad \text{Isolating a radical}$$

$$\left(\sqrt{x+2}\right)^2 = \left(2\sqrt{x-1}\right)^2 \qquad \text{Using the principle of powers}$$

$$x+2 = 4(x-1)$$

$$x+2 = 4x - 4$$

$$6 = 3x$$

$$2 = x$$

Since 2 checks in the original equation, the solution is 2.

📖 **Textbook**	👤 **Instructor**	▶ **Video**

GUIDED LEARNING

EXAMPLE 1	YOUR TURN 1
Solve: $\sqrt{4y-2} = \sqrt{3y+9}$.	Solve: $\sqrt{2y-7} = \sqrt{8-y}$.

EXAMPLE 1

Solve: $\sqrt{4y-2} = \sqrt{3y+9}$.

$$\sqrt{4y-2} = \sqrt{3y+9}$$

$$\left(\sqrt{4y-2}\right)^2 = \left(\sqrt{3y+9}\right)^2$$

$$4y - \boxed{} = 3y + \boxed{}$$

$$\boxed{} - 2 = 9$$

$$y = \boxed{}$$

A check shows that $\boxed{}$ is the solution.

EXAMPLE 2	YOUR TURN 2
Solve: $\sqrt{a+1} = \sqrt{a-2}+1$.	Solve: $\sqrt{a+8}-\sqrt{a}=2$.

$$\sqrt{a+1} = \sqrt{a-2}+1$$

$$\left(\sqrt{a+1}\right)^2 = \left(\sqrt{a-2}+1\right)^2$$

$$a+1 = a-2+2\sqrt{a-2}+1$$

$$\boxed{} = 2\sqrt{a-2}$$

$$1 = \sqrt{a-2}$$

$$1^2 = \left(\boxed{}\right)^2$$

$$1 = a-2$$

$$\boxed{} = a$$

A check shows that $\boxed{}$ is the solution.

EXAMPLE 3	YOUR TURN 3
Solve: $\sqrt{3x+4} = 1+\sqrt{2x+2}$.	Solve: $\sqrt{3-x} = 1+\sqrt{4-2x}$.

$$\sqrt{3x+4} = 1+\sqrt{2x+2}$$

$$\left(\sqrt{3x+4}\right)^2 = \left(1+\sqrt{2x+2}\right)^2$$

$$3x+\boxed{} = 1+2\sqrt{2x+2}+2x+\boxed{}$$

$$3x+4 = \boxed{}+2\sqrt{2x+2}+2x$$

$$x+\boxed{} = 2\sqrt{2x+2}$$

$$(x+1)^2 = \left(2\sqrt{2x+2}\right)^2$$

$$x^2+2x+1 = \boxed{}(2x+2)$$

$$x^2+2x+1 = 8x+\boxed{}$$

$$x^2-6x-\boxed{} = 0$$

$$(x+1)\left(x-\boxed{}\right) = 0$$

$$x+1=0 \quad or \quad x-7=0$$

$$x = \boxed{} \quad or \quad x = \boxed{}$$

Since both numbers check, the solutions are $\boxed{}$ and $\boxed{}$.

YOUR NOTES   Write your questions and additional notes.

Applications

Example

- At a height of h meters, one can see V kilometers to the horizon, where $V = 3.5\sqrt{h}$. A technician can see 70 km to the horizon from atop a tower. What is the altitude of the technician's eyes?

  1. **Familiarize.** Use the formula $V = 3.5\sqrt{h}$.

  2. **Translate.** Substitute 70 for V in the formula: $70 = 3.5\sqrt{h}$.

  3. **Solve.** Solve the equation for h.

  $$70 = 3.5\sqrt{h}$$
  $$20 = \sqrt{h}$$
  $$(20)^2 = \left(\sqrt{h}\right)^2$$
  $$400 = h$$

  4. **Check.** Let $h = 400$ and calculate V: $V = 3.5\sqrt{400} = 3.5(20) = 70$. The answer checks.

  5. **State.** The altitude of the technician's eyes is 400 m.

GUIDED LEARNING	🔖 Textbook	👤 Instructor	▶ Video

EXAMPLE 1	YOUR TURN 1
At a height of h feet, one can see d miles to the horizon, where $d = \sqrt{1.5h}$. From an airplane window, Anne can see 90 miles to the horizon. What is the altitude of the airplane? 1. **Familiarize.** Use the formula $\boxed{}$. 2. **Translate.** Substitute $\boxed{}$ for d. $\quad d = \sqrt{1.5h}$ $\quad \boxed{} = \sqrt{1.5h}$ (continued)	The formula $r = 2\sqrt{5L}$ can be used to approximate the speed r, in miles per hour, of a car that has left a skid mark of length L, in feet. How far will a car skid at 30 mph?

3. **Solve**. Solve for h.

$$90 = \sqrt{1.5h}$$

$$(90)^2 = \left(\boxed{}\right)^2$$

$$8100 = \boxed{}$$

$$\frac{8100}{\boxed{}} = \frac{1.5h}{\boxed{}}$$

$$\boxed{} = h$$

4. **Check**. Let $h = \boxed{}$ and calculate d:

$$d = \sqrt{1.5h}$$

$$= \sqrt{1.5\left(\boxed{}\right)}$$

$$= \sqrt{\boxed{}}$$

$$= \boxed{}$$

The answer checks.

5. **State**. The altitude of the plane is $\boxed{}$ ft.

YOUR NOTES   Write your questions and additional notes.

Practice Exercises

Readiness Check

Fill in the word that best completes each sentence.

To solve an equation with two or more radical terms:

1. _______________ one of the radical terms.

2. Use the principle of _______________ .

3. If a _______________ remains, perform steps (1) and (2) again.

4. _______________ the resulting equation.

5. Check possible solutions in the _______________ equation.

The Principle of Powers

Solve.

6. $\sqrt{x} - 6 = 5$

7. $\sqrt[3]{2a + 5} + 3 = 0$

8. $\sqrt{8x + 1} = 5$

9. $3\sqrt{x - 1} = x + 1$

10. $\sqrt{2y + 7} - y = 2$

11. $\sqrt[4]{y + 16} + 4 = 0$

Equations with Two Radical Terms

Solve.

12. $\sqrt{2m-1} = \sqrt{1-2m}$

13. $\sqrt{x-12} = 2 - \sqrt{x}$

14. $\sqrt{2x-9} - \sqrt{x-1} = -1$

15. $\sqrt{x+1} + \sqrt{x+6} = 5$

16. $\sqrt[3]{6x-5} + 3 = -2$

17. $\sqrt{x+5} = \sqrt{2x-5}$

Applications

18. The speed v, in meters per second, of a wave on the surface of the ocean can be approximated by the formula $v = 3.1\sqrt{d}$, where d is the depth of the water, in meters. A wave is traveling 12.4 m/sec. What is the water depth?

Applications

ESSENTIALS

The Pythagorean Theorem

In any right triangle, if a and b are the lengths of the legs and c is the length of the hypotenuse, then $a^2 + b^2 = c^2$.

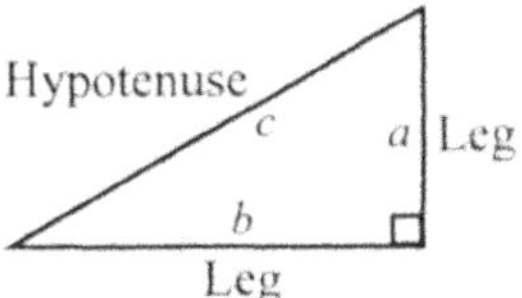

Example

- One leg of a right triangle is 6 m and the hypotenuse is 12 m. Find the length of the other leg. Give an exact answer and an approximation to three decimal places.

$$a^2 + b^2 = c^2$$
$$6^2 + b^2 = 12^2$$
$$36 + b^2 = 144$$
$$b^2 = 108$$
$$b = \sqrt{108} \qquad \text{The length must be positive.}$$
$$b = \sqrt{36 \cdot 3}$$
$$b = 6\sqrt{3} \approx 10.392$$

The exact answer is $6\sqrt{3}$ m, and the approximation is 10.392 m.

GUIDED LEARNING   📖 **Textbook**   👤 **Instructor**   ▶ **Video**

EXAMPLE 1	YOUR TURN 1
In the right triangle shown, $a = \sqrt{2}$ and $c = 4$. Find b. Give an exact answer and an approximation to three decimal places.	In the right triangle shown, $b = 3$ and $c = \sqrt{11}$. Find a. Give an exact answer and an approximation to three decimal places.

$$a^2 + b^2 = c^2$$
$$\left(\boxed{}\right)^2 + b^2 = 4^2$$
$$\boxed{} + b^2 = 16$$
$$b^2 = \boxed{}$$
$$b = \sqrt{14} \approx \boxed{}$$

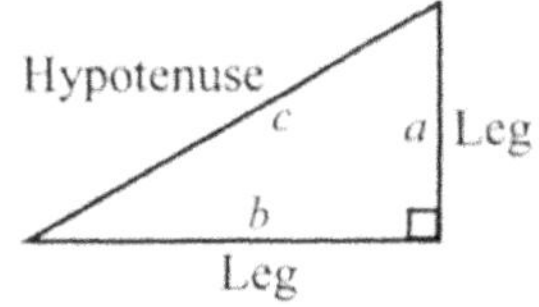

EXAMPLE 2	YOUR TURN 2
How long is a guy wire reaching from the top of a 36-ft pole to a point on the ground 12 ft from the pole? Give an exact answer and an approximation to three decimal places. Let c = the length (in feet) of the guy wire.	A ladder is leaning against a house. The bottom of the ladder is 3 ft from the house. The top of the ladder touches the house at a point 20 ft above the ground. How long is the ladder? Give an exact answer and an approximation to three decimal places.

$$12^2 + 36^2 = c^2$$
$$144 + 1296 = c^2$$
$$1440 = c^2$$
$$\sqrt{\boxed{}} = c$$

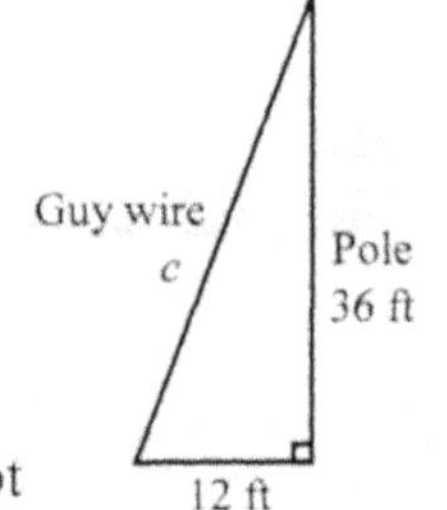

(Use the positive square root because length is not negative.)

$$\sqrt{1440}, \text{ or } 12\sqrt{\boxed{}} = c \quad \text{Exact answer}$$
$$\boxed{} \approx c \quad \text{Approximation}$$

The length of the guy wire is _______ ft, or about _______ ft.

YOUR NOTES   Write your questions and additional notes.

Practice Exercises

Readiness Check

Determine whether each statement is true or false.

1. The longest side of a right triangle is the hypotenuse.

2. The Pythagorean theorem is true for any triangle.

3. When solving for a length, we do not consider negative numbers since length cannot be negative.

4. If the lengths of the legs of a right triangle are rational numbers, then the length of the hypotenuse is an irrational number.

Applications

In Exercises 5-10, give an exact answer and, where appropriate, an approximation to three decimal places.

5. In the right triangle shown, $a = 6$ and $b = 7$. Find c.

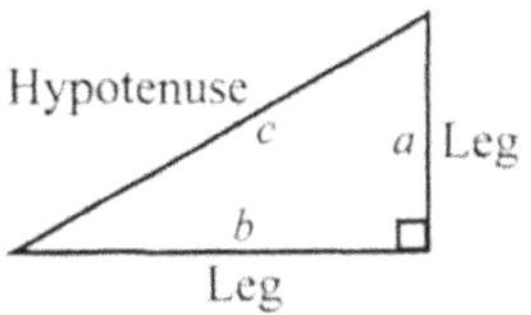

6. In the right triangle shown, $a = 6$ and $c = 10$. Find b.

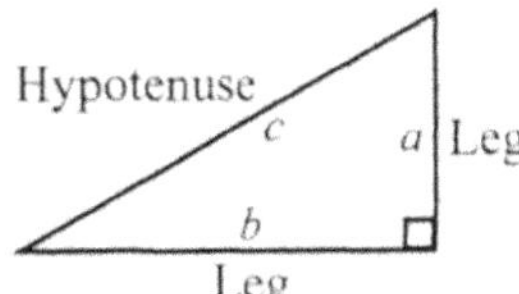

7. A right triangle's hypotenuse is 10 cm, and one leg is $3\sqrt{2}$ cm. Find the length of the other leg.

8. One leg in a right triangle is 2 m, and the hypotenuse measures $\sqrt{6}$ m. Find the length of the other leg.

9. A zipline ride extends 400 ft along the ground in a wooded area. The riders start the ride from a platform 30 ft high. How long is the zipline?

10. Mary walks through a rectangular park on her way to work. If the park is 50 m long and 45 m wide, how far does Mary walk when she walks across the park diagonally?

Increasing, Decreasing, and Constant Functions

ESSENTIALS

A function f is said to be **increasing** on an *open* interval I, if for all a and b in that interval, $a < b$ implies $f(a) < f(b)$.

A function f is said to be **decreasing** on an *open* interval I, if for all a and b in that interval, $a < b$ implies $f(a) > f(b)$.

A function f is said to be **constant** on an *open* interval I, if for all a and b in that interval, $f(a) = f(b)$.

In other words, on a given interval, if the graph of a function rises from left to right, it is said to be **increasing** on that interval. If the graph drops from left to right, it is said to be **decreasing**. If the function values stay the same on the interval, the function is said to be **constant**.

Example

- Determine the intervals on which the function is
  (a) increasing;   (b) decreasing;   (c) constant.

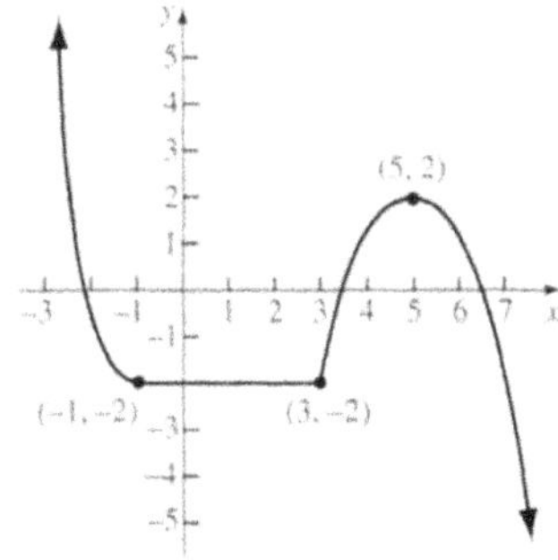

a) As x-values increase from 3 to 5, y-values increase from -2 to 2. Thus the function is increasing on $(3, 5)$.

b) As x-values increase from $-\infty$ to -1, y-values decrease. Also, as x-values increase from 5 to ∞, y-values decrease. Thus the function is decreasing on $(-\infty, -1)$ and $(5, \infty)$.

c) As x-values increase from -1 to 3, the y-values stay the same, -2. The function is constant on $(-1, 3)$.

<table>
<tr><td colspan="2">GUIDED LEARNING</td><td>🔖 Textbook</td><td>👤 Instructor</td><td>▶ Video</td></tr>
</table>

EXAMPLE 1	YOUR TURN 1

EXAMPLE 1

Determine the intervals on which the function is (a) increasing; (b) decreasing; (c) constant.

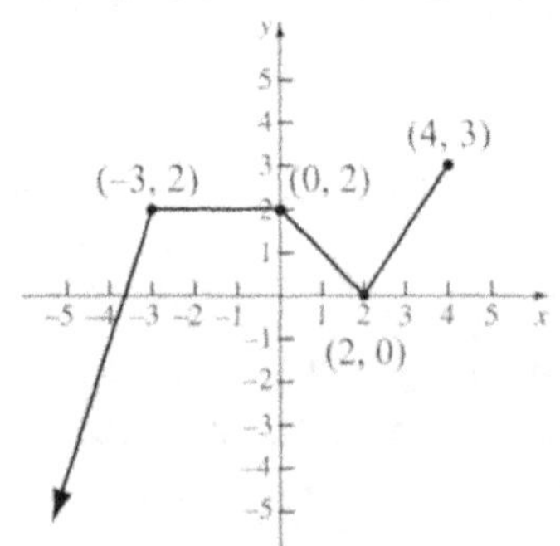

a) As x-values increase from ☐ to -3, y-values increase. Also, as x-values increase from ☐ to 4, y-values increase. Thus the function is increasing on $\left(-\infty, \boxed{}\right)$ and $\left(\boxed{}, 4\right)$.

b) As x-values increase from 0 to 2, y-values ______________ . Thus the function
   increase / decrease
   is decreasing on $\left(0, \boxed{}\right)$.

c) As x-values increase from -3 to ☐, the y-values stay the same, 2. Thus the function is constant on $\left(\boxed{}, 0\right)$.

YOUR TURN 1

Determine the intervals on which the function is (a) increasing; (b) decreasing; (c) constant.

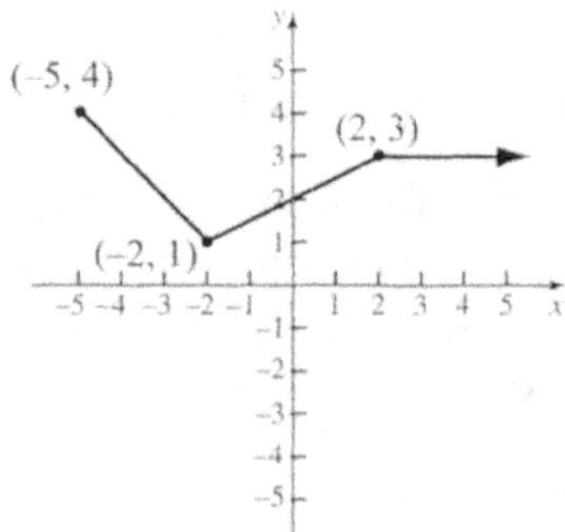

YOUR NOTES   Write your questions and additional notes.

Relative Maximum and Minimum Values

ESSENTIALS

Suppose that f is a function for which $f(c)$ exists for some c in the domain of f. Then:

$f(c)$ is a **relative maximum** (plural, **maxima**) if there exists an *open* interval I containing c such that $f(c) > f(x)$, for all x in I where $x \neq c$; and

$f(c)$ is a **relative minimum** (plural **minima**) if there exists an *open* interval I containing c such that $f(c) < f(x)$, for all x in I where $x \neq c$.

Simply stated, $f(c)$ is a *relative maximum* if $(c, f(c))$ is the highest point in some *open* interval, and $f(c)$ is a *relative minimum* if $(c, f(c))$ is the lowest point in some *open* interval.

Example

- Using the graph shown below, determine any relative maxima or minima of the function $f(x) = x^3 - 2x^2 - 4.5x + 3$ and the intervals on which the function is increasing or decreasing.

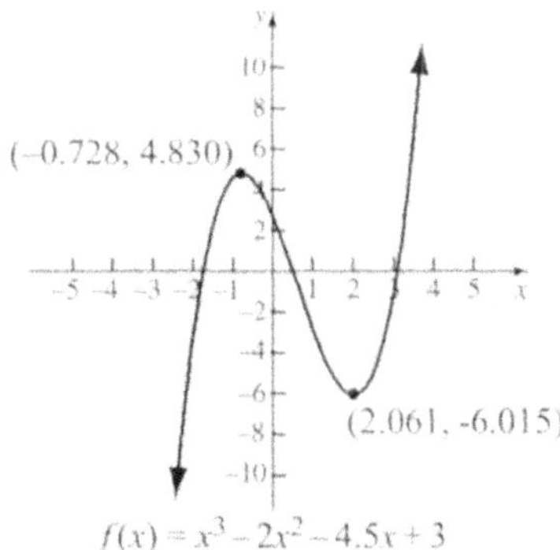

We see that the relative maximum value of the function is 4.830. It occurs when $x = -0.728$. We also see that the relative minimum is -6.015 at $x = 2.061$.

The graph starts rising, or increasing, from the left and stops increasing at the relative maximum. Then it decreases to the relative minimum and then rises again. Thus the function is increasing on the intervals $(-\infty, -0.728)$ and $(2.061, \infty)$ and is decreasing on the interval $(-0.728, 2.061)$.

 **Textbook**    **Instructor**    **Video**

GUIDED LEARNING

EXAMPLE 1	YOUR TURN 1

EXAMPLE 1

Using the graph, determine any relative maxima or minima of the function $f(x) = -2x^3 - 6x^2 - 2.5x + 2$ and the intervals on which the function is increasing or decreasing.

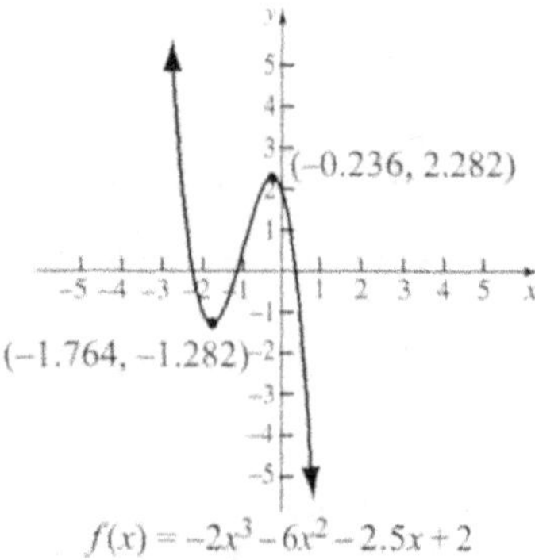

$$f(x) = -2x^3 - 6x^2 - 2.5x + 2$$

We see that the relative maximum value of the function is [　　] at $x =$ [　　]. We also see that the relative minimum value is [　　] at

$x =$ [　　].

The graph starts falling or decreasing from the left and stops decreasing at the relative

____
maximum / minimum

Then it increases to the relative

______________ and then falls again. Then
maximum / minimum

the function is increasing on the interval

$\left(-1.764, \boxed{}\right)$ and is decreasing on

$\left(-\infty, \boxed{}\right)$ and $\left(-0.236, \boxed{}\right)$.

YOUR TURN 1

Using the graph, determine any relative maxima or minima of the function $f(x) = 2x^3 + 4x^2 - 1.5x + 1$ and the intervals on which the function is increasing or decreasing.

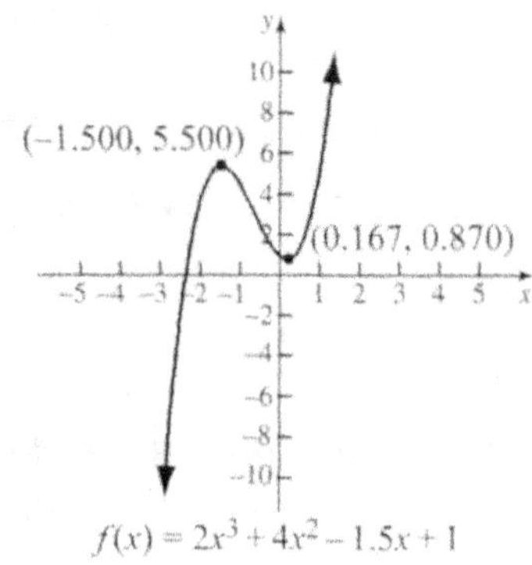

$$f(x) = 2x^3 + 4x^2 - 1.5x + 1$$

YOUR NOTES   Write your questions and additional notes.

Applications of Functions

ESSENTIALS

Many real-world situations can be modeled by functions.

Example

Jenna's restaurant supply store has 20 ft of dividers with which to set off a rectangular area for the storage of overstock. If a corner of the store is used for the storage area, the partition needs to form only two sides of a rectangle.

a) Express the floor area of the storage space as a function of the length of the partition.

Let $x =$ the length of the partition, in feet.

$$A(x) = x(20 - x)$$
$$A(x) = 20x - x^2$$

b) Find the domain of the function.

$$A(x) = 20x - x^2$$

The domain could be $(-\infty, \infty)$ but, since only 20 ft of dividers are available, a more realistic domain is $(0, 20)$.

c) Using the graph shown below, determine the dimensions that maximize the area of the floor.

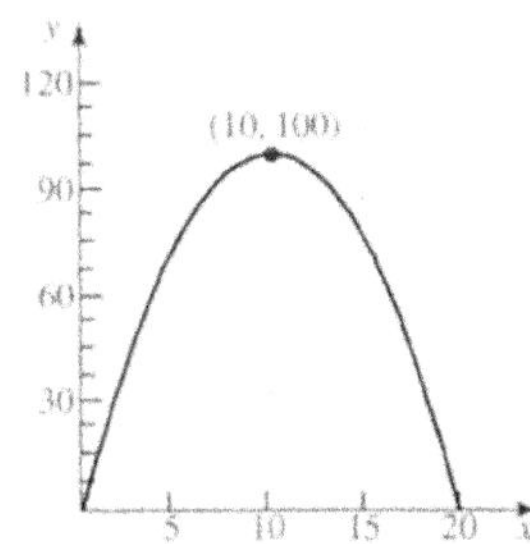

Relative maximum is 100 when $x = 10$.

Maximum area $= 100$ ft^2 when $x = 10$ ft.

Length: 10 ft

Width: $20 - 10 = 10$ ft

Dimensions: 10 ft $\times$ 10 ft

 **Textbook**    **Instructor**   **Video**

GUIDED LEARNING

EXAMPLE 1	YOUR TURN 1

EXAMPLE 1

A dentist has 24 ft of dividers with which to set off a rectangular area for a play space in her waiting room. If a corner of the waiting room is used for the play space, the partition needs to form only two sides of the rectangle.

a) Express the floor area of the play space as a function of the length of the partition.

b) Find the domain of the function.

c) Using the graph shown below, determine the dimensions that maximize the floor area.

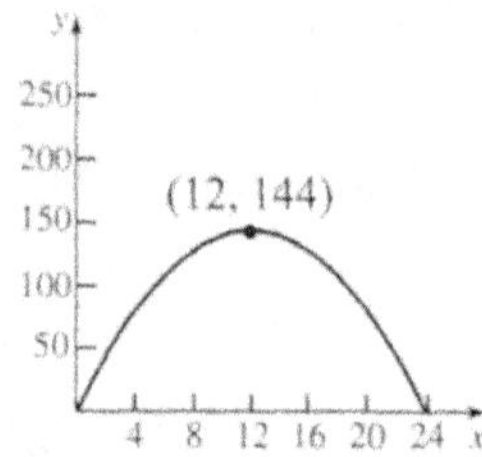

a) Let $x =$ the length of the rectangle, in feet. Then $\boxed{} - x =$ the width.

$$A(x) = x(24 - x) = 24x - \boxed{}$$

b) Because the rectangle's length and width must be positive and only 24 ft of dividers are available, we restrict the domain of A to $\{x \mid 0 < x < 24\}$, or $(0, 24)$.

c) From the graph we see that the maximum value of the area appears to be 144 when $x = 12$. Thus the dimensions that maximize the area are

Length $= x = 12$ ft and

Width $= 24 - x = 24 - \boxed{} = 12$ ft.

YOUR TURN 1

Joseph has 36 ft of fencing with which to enclose a garden plot in his yard. If an already fenced corner of the yard is used for the plot, the new fencing needs to form only two sides of the plot.

a) Express the area of the garden plot as a function of the length of the plot.

b) Find the domain of the function.

c) Using the graph shown below, determine the dimensions that maximize the area of the plot.

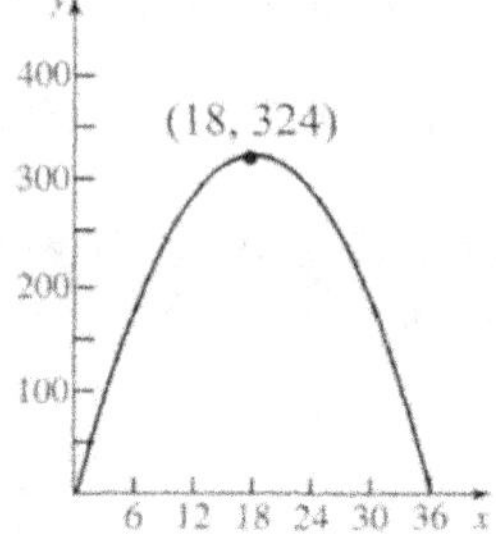

YOUR NOTES   Write your questions and additional notes.

Functions Defined Piecewise

ESSENTIALS

Some functions are defined **piecewise** using different output formulas for different pieces, or parts, of the domain.

A piecewise function with importance in calculus and computer programming is the **greatest integer function** denoted $f(x) = [\![x]\!]$, or $f(x) = \text{int}(x)$, and defined as follows:

$$f(x) = [\![x]\!] = \text{the greatest integer less than or equal to } x.$$

Example

- For the function defined as

$$f(x) = \begin{cases} x+1, & \text{for } x < -2, \\ 5, & \text{for } -2 \le x \le 3, \\ x^2, & \text{for } x > 3, \end{cases}$$

find $f(-5)$, $f(0)$, $f(3)$, and $f(10)$.

Since $-5 < -2$, we use the formula $f(x) = x+1$:

$$f(-5) = -5+1 = -4.$$

Since $-2 \le 0 \le 3$, we use the formula $f(x) = 5$:

$$f(0) = 5.$$

Since $-2 \le 3 \le 3$, we use the formula $f(x) = 5$ again:

$$f(3) = 5.$$

Since $10 > 3$, we use the formula $f(x) = x^2$:

$$f(10) = 10^2 = 100.$$

GUIDED LEARNING

EXAMPLE 1	YOUR TURN 1

EXAMPLE 1

For the function defined as

$$g(x) = \begin{cases} x^2 - 1, & \text{for } x \leq -1, \\ -x, & \text{for } -1 \leq x \leq 3 \\ 6, & \text{for } x > 3, \end{cases}$$

find $g(-2)$, $g(1)$, $g(3)$, and $g(4)$.

Since $-2 \leq -1$, we use the formula $g(x) = x^2 - 1$:

$$g(-2) = (-2)^2 - 1 = 4 - 1 = \boxed{}.$$

Since $-1 < 1 \leq 3$, we use the formula $g(x) = -x$:

$$g(1) = -1.$$

Since $-1 < 3 \leq 3$, we use the formula $g(x) = -x$:

$$g(3) = \boxed{}.$$

Since $4 > 3$, we use the formula $g(x) = 6$:

$$g(4) = \boxed{}.$$

YOUR TURN 1

For the function defined as

$$f(x) = \begin{cases} 2x + 3, & \text{for } x < -2, \\ 1, & \text{for } -2 \leq x \leq 1, \\ x^2, & \text{for } x > 1, \end{cases}$$

find $f(-5)$, $f(-2)$, $f(0)$, and $f(3)$.

EXAMPLE 2	YOUR TURN 2

EXAMPLE 2

Graph the function defined as

$$g(x) = \begin{cases} 2x, & \text{for } x < -1, \\ x-3, & \text{for } x \geq -1. \end{cases}$$

We create the graph in two pieces, or parts. First we graph $g(x) = 2x$ only for inputs less than -1. We find some ordered pairs that are solutions of this piece of the function.

x $(x < -1)$	$g(x) = 2x$
-3	-6
-2	
$-\dfrac{3}{2}$	-3

We graph $g(x) = x - 3$ only for inputs greater than or equal to $\boxed{}$. We find some ordered pairs that are solutions of this piece of the function.

x $(x \geq -1)$	$g(x) = x - 3$
-1	
1	-2
4	

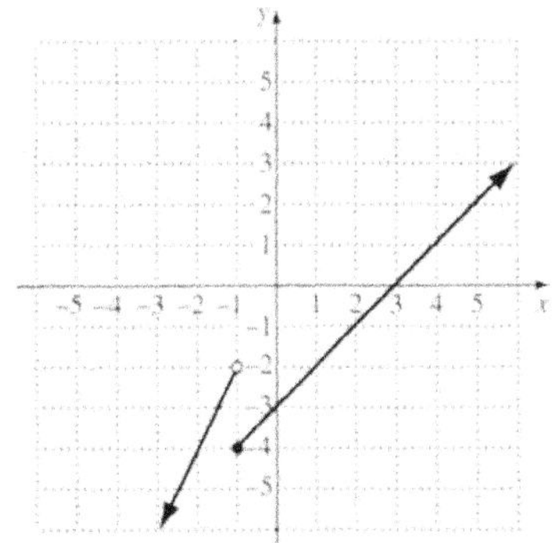

YOUR TURN 2

Graph the function defined as

$$f(x) = \begin{cases} -x+2, & \text{for } x \leq 3, \\ x-5, & \text{for } x > 3. \end{cases}$$

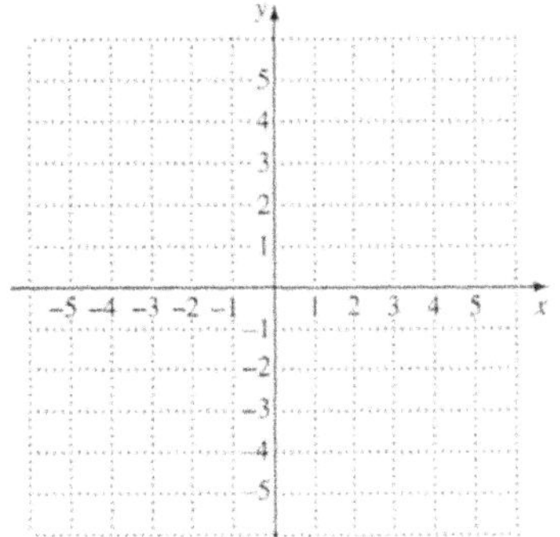

EXAMPLE 3	YOUR TURN 3

EXAMPLE 3

Graph the greatest integer function $f(x) = [\![x]\!]$.

The greatest integer function can be thought of as a piecewise function with an infinite number of statements.

$$f(x) = \begin{cases} \vdots \\ -3, & \text{for } -3 \le x < -2, \\ -2, & \text{for } -2 \le x < -1, \\ \boxed{}, & \text{for } -1 \le x < 0, \\ 0, & \text{for } 0 \le x < 1, \\ \boxed{}, & \text{for } 1 \le x < 2, \\ 2, & \text{for } 2 \le x < 3, \\ \boxed{}, & \text{for } 3 \le x < 4, \\ \vdots \end{cases}$$

YOUR TURN 3

Graph $f(x) = [\![x]\!] - 1$.

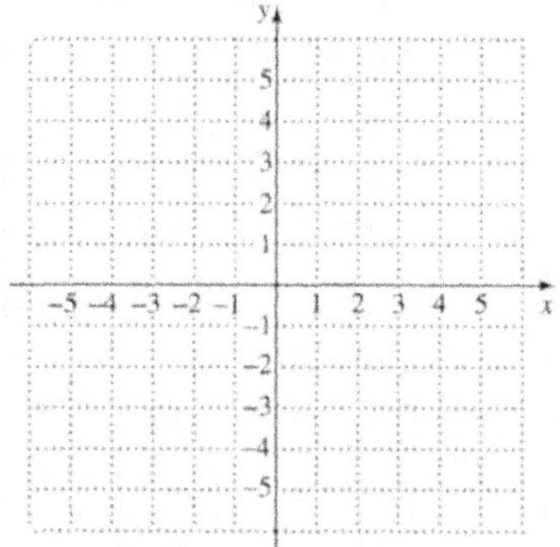

YOUR NOTES   Write your questions and additional notes.

Practice Exercises

Readiness Check

Determine whether each statement is true or false.

1. If the graph of a function drops from left to right on an interval, the function is increasing on the interval.

2. For a function f, $f(c)$ is a relative maximum if $(c, f(c))$ is this highest point in some open interval.

3. The greatest integer function is a piecewise function.

4. For the greatest integer function $f(x) = [\![x]\!]$, $f(-5.3) = -5$.

Increasing, Decreasing, and Constant Functions

5. Determine the intervals on which the function is

    (a) increasing; (b) decreasing; (c) constant.

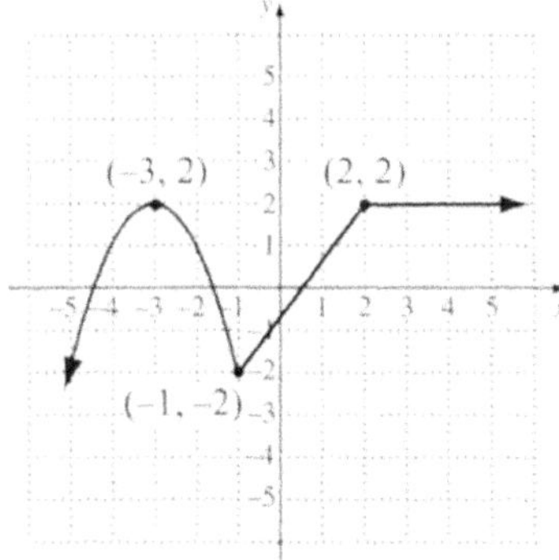

Relative Maximum and Minimum Values

6. Using the graph, determine any maxima or minima of the function $f(x) = -x^3 - 1.1x^2 + 5x + 2$ and the intervals on which the function is increasing or decreasing.

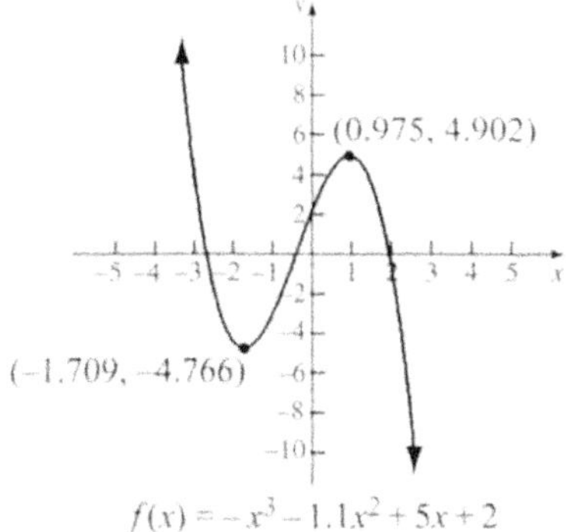

$f(x) = -x^3 - 1.1x^2 + 5x + 2$

Applications of Functions

7. Mason is designing a triangular pennant so that the height of the triangle, in inches, is 6 more than twice the base. Express the area of the pennant as a function of the base.

8. A rancher has 360 yd of fencing with which to enclose two adjacent rectangular corrals. A river forms one side of the corrals. Suppose the width of each corral is x yards.

   a) Express the total area of the two corrals as a function of x.

   b) Find the domain of the function.

   c) Using the graph, determine the dimensions that yield the maximum area.

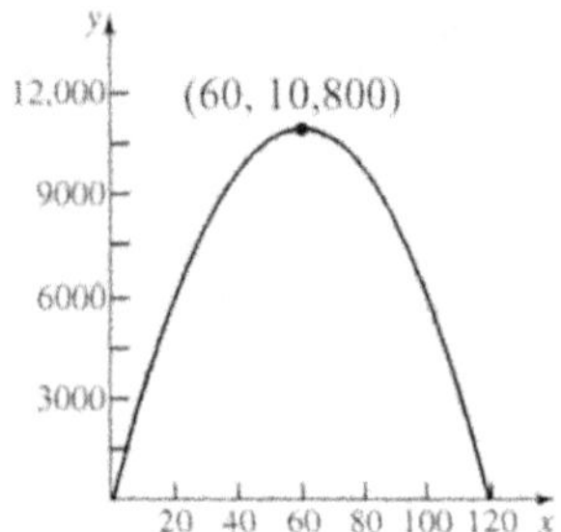

Functions Defined Piecewise

Simplify.

9. For the function defined as

$$f(x) = \begin{cases} 2x - 3, & \text{for } x \le -3, \\ 1, & \text{for } -3 < x < 0, \\ x^2 + 2, & \text{for } x \ge 0, \end{cases}$$

   find $f(-3), f(-1), f(0),$ and $f(4)$.

Graph each of the following.

10. $g(x) = \begin{cases} -x, & \text{for } x \le -4, \\ x+1, & \text{for } x > -4 \end{cases}$

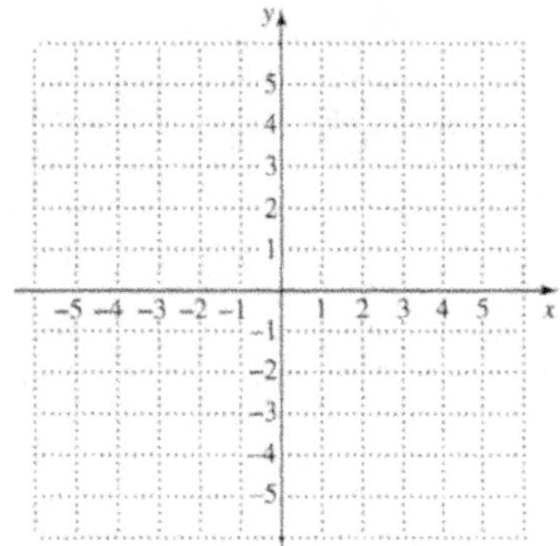

11. $f(x) = 2[\![x]\!]$

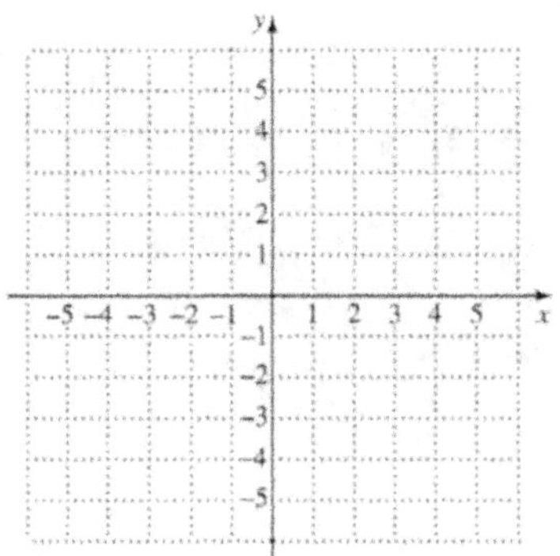

Section 7.1 Symmetry

> **Algebraic Tests of Symmetry**
>
> *x-axis*: If replacing y with $-y$ produces an equivalent equation, then the graph is *symmetric with respect to the x-axis*.
>
> *y-axis*: If replacing x with $-x$ produces an equivalent equation, then the graph is *symmetric with respect to the y-axis*.
>
> *Origin*: If replacing x with $-x$ and y with $-y$ produces an equivalent equation, then the graph is *symmetric with respect to the origin*.

Example 1 Test $y = x^2 + 2$ for symmetry with respect to the x-axis, the y-axis, and the origin.

x-axis:

$$y = x^2 + 2$$
$$\downarrow$$
$$-y = x^2 + 2 \qquad \text{Replacing } y \text{ with } -y$$
$$y = -x^2 - \boxed{} \qquad \text{Not equivalent to } y = x^2 + 2$$

The graph _____________ symmetric with respect to the x-axis.
is / is not

y-axis:

$$y = x^2 + 2$$
$$\downarrow$$
$$y = \left(-x\right)^2 + 2 \qquad \text{Replacing } x \text{ with } -x$$
$$y = \boxed{} + 2 \qquad \text{Obtaining the original equation}$$

The graph _____________ symmetric with respect to the y-axis.
is / is not

Origin:

$$y = x^2 + 2$$
$$\downarrow \quad \downarrow$$
$$-y = \left(-x\right)^2 + 2 \qquad \text{Replacing } y \text{ with } -y \text{ and } x \text{ with } -x$$
$$\boxed{} = x^2 + 2$$
$$\boxed{} = -x^2 - 2 \qquad \text{Not equivalent to } y = x^2 + 2$$

The graph _____________ symmetric with respect to the origin.
is / is not

Your Turn 1   Test $3y^2 - x^3 = 2$ for symmetry with respect to the x-axis, the y-axis, and the origin.

Example 2   Test $x^2 + y^4 = 5$ for symmetry with respect to the x-axis, the y-axis, and the origin.

x-axis:

$$x^2 + y^4 = 5$$

$$x^2 + (-y)^4 = 5 \qquad \text{Replacing } y \text{ with } -y$$

$$x^2 + \boxed{} = 5 \qquad \text{Obtaining the original equation}$$

The graph $\underline{\hspace{2cm}}$ symmetric with respect to the x-axis.
$$\text{is / is not}$$

y-axis:

$$x^2 + y^4 = 5$$

$$(-x)^2 + y^4 = 5 \qquad \text{Replacing } x \text{ with } -x$$

$$x^2 + y^4 = \boxed{} \qquad \text{Obtaining the original equation}$$

The graph $\underline{\hspace{2cm}}$ symmetric with respect to the y-axis.
$$\text{is / is not}$$

Origin:

$$x^2 + y^4 = 5$$

$$(-x)^2 + (-y)^4 = 5 \qquad \text{Replacing } x \text{ with } -x \text{ and } y \text{ with } -y$$

$$\boxed{} + y^4 = 5 \qquad \text{Obtaining the original equation}$$

The graph $\underline{\hspace{2cm}}$ symmetric with respect to the origin.
$$\text{is / is not}$$

Your Turn 2 Test $y^2 + 3 = 7x^4$ for symmetry with respect to the x-axis, the y-axis, and the origin.

Even Functions and Odd Functions

If the graph of a function f is symmetric with respect to the y-axis, we say that it is an **even function**. That is, for each x in the domain of f, $f(x) = f(-x)$.

If the graph of a function f is symmetric with respect to the origin, we say that it is an **odd function**. That is, for each x in the domain of f, $f(-x) = -f(x)$.

Example 3

a) Determine whether $f(x) = 5x^7 - 6x^3 - 2x$ is even, odd, or neither.

$$f(-x) = 5(-x)^7 - 6(-x)^3 - 2(-x)$$
$$= -5x^7 + 6x^3 + 2x$$

$$f(-x) \neq f(x)$$

Thus, $f(x)$ ___________ even.
$\quad\quad\quad\quad\quad$ is / is not

$$-f(x) = -\left(5x^7 - 6x^3 - 2x\right)$$
$$= -5x^7 + 6x^3 + 2x$$

$$f(-x) = -f(x)$$

Thus, $f(x)$ ___________ odd.
$\quad\quad\quad\quad\quad$ is / is not

(continued)

b) Determine if $h(x) = 5x^6 - 3x^2 - 7$ is even, odd, or neither.

$$h(-x) = 5(-x)^6 - 3(-x)^2 - 7$$
$$= 5x^6 - 3x^2 - \boxed{}$$

$$h(-x) = h(x)$$

Thus, $h(x)$ is an _____________ function.
even / odd

The function cannot be both even and odd, so we stop here.

Your Turn 3

a) Determine whether $g(x) = x^5 + 3x^4 - x + 2$ is even, odd, or neither.

b) Determine if $f(x) = 2x^4 + 5x^2 - 3$ is even, odd, or neither.

Practice Exercises

Determine whether the graph is symmetric with respect to the x-axis, the y-axis, and the origin.

1. $y = 4x^2 - x^3$

2. $x^4 + 7y^2 = 100$

3. $5y = 4x^2 - x$

4. $xy = 24$

5. $2x = |y|$

6. $3x^3 = y^3 - 5$

Determine whether the function is even, odd, or neither even nor odd.

7. $f(x) = -2x^2 - x^4 + 2$

8. $g(x) = x^2 - 4x^3 + 15$

9. $f(x) = 3x + \dfrac{1}{3x}$

10. $h(x) = x^5 - 3x$

11. $g(x) = 4\sqrt{x^2 + 1}$

12. $h(x) = x^2 - |x|$

Section 7.2 Transformations

Vertical Translation and Horizontal Translation

For $b > 0$:

  the graph of $y = f(x) + b$ is the graph of $y = f(x)$ shifted *up* b units;

  the graph of $y = f(x) - b$ is the graph of $y = f(x)$ shifted *down* b units.

For $d > 0$:

  the graph of $y = f(x - d)$ is the graph of $y = f(x)$ shifted *to the right* d units;

  the graph of $y = f(x + d)$ is the graph of $y = f(x)$ shifted *to the left* d units.

Example 1 Graph each of the following. Before doing so, describe how each graph can be obtained from one of the basic graphs shown earlier in this section.

a) $g(x) = x^2 - 6$

  We compare this with the graph of $f(x) = x^2$.

  $g(x) = f(x) - 6$, so the graph of $g(x)$ is the graph of $f(x)$ translated

  _____________ 6 units.
   up / down

  We compare some points on the graphs of *f* and *g*.

f	g
$(-3, 9)$	$(-3, 3)$
$(0, 0)$	$\left(0, \boxed{}\right)$
$(2, 4)$	$\left(2, \boxed{}\right)$

  The *y*-coordinate of a point on the graph of $g(x)$ is 6 _____________ than the
                                                      more / less

  corresponding point on the graph of $f(x)$.

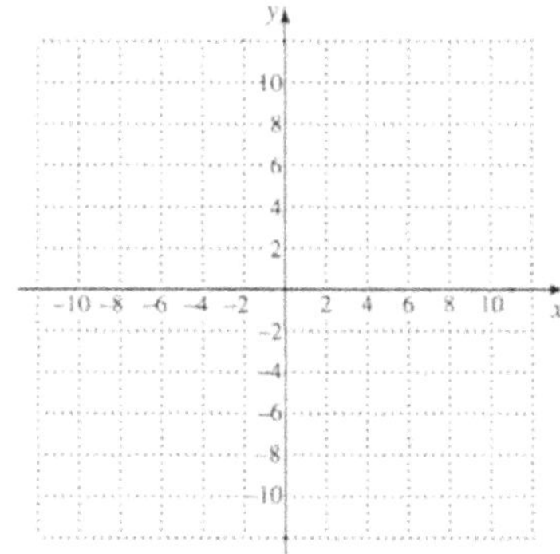

(continued)

b) $g(x) = |x - 4|$

We compare this with the graph of $f(x) = |x|$.

$g(x) = f(x - 4)$, so the graph of $g(x)$ is the graph of $f(x)$ shifted

__________ 4 units.
right / left

We compare some points on the graphs of f and g.

f	g
$(-4, 4)$	$(0, 4)$
$(0, 0)$	$\left(\boxed{}, 0\right)$
$(6, 6)$	$\left(10, \boxed{}\right)$

The x-coordinate of a point on the graph of $g(x)$ is 4 __________ than the
more / less

corresponding point on the graph of $f(x)$.

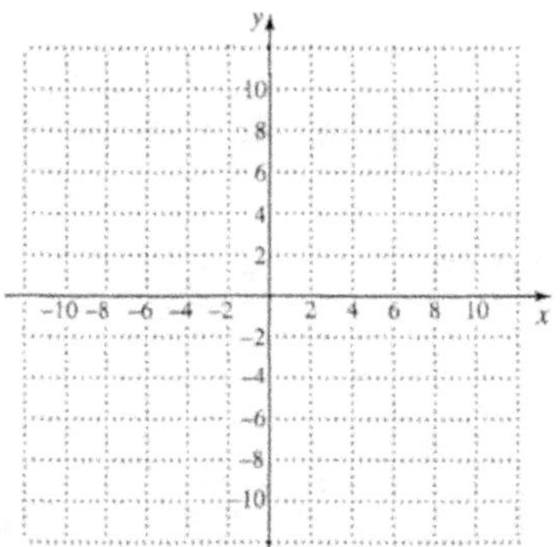

c) $g(x) = \sqrt{x + 2}$

We compare this with the graph of $f(x) = \sqrt{x}$.

$g(x) = f(x + 2)$, so the graph of $g(x)$ is the graph of $f(x)$ shifted

__________ 2 units, as we can see from the graphs of the functions.
right / left

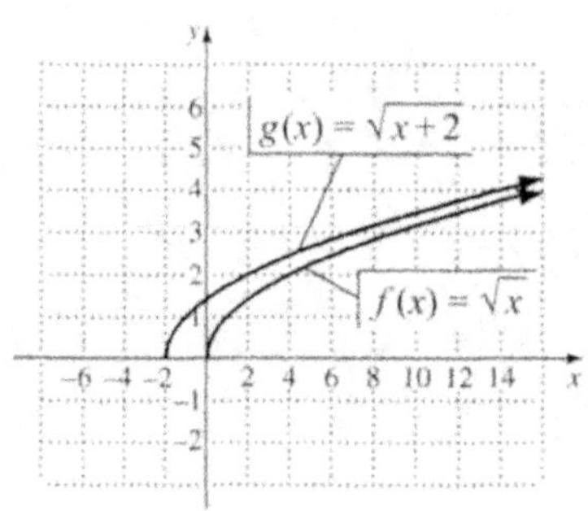

(continued)

d) $h(x) = \sqrt{x+2} - 3$

We compare this with the graph of $f(x) = \sqrt{x}$.

In part (c) we saw that the graph of $g(x) = \sqrt{x+2}$ is the graph of $f(x)$ shifted left

◻ units.

$h(x) = g(x) - 3$, so the graph of $h(x)$ is the graph of $g(x)$ shifted _______ 3 units.

up / down

Together, the graph of $h(x)$ is the graph of $f(x) = \sqrt{x}$ shifted left ◻ units and down

◻ units.

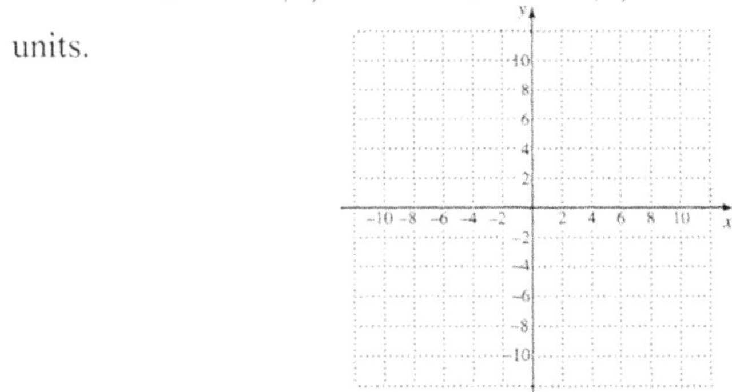

✎ Your Turn 1   Graph each of the following. Before doing so, describe how each graph can be obtained from one of the basic graphs.

a) $g(x) = x^2 + 4$

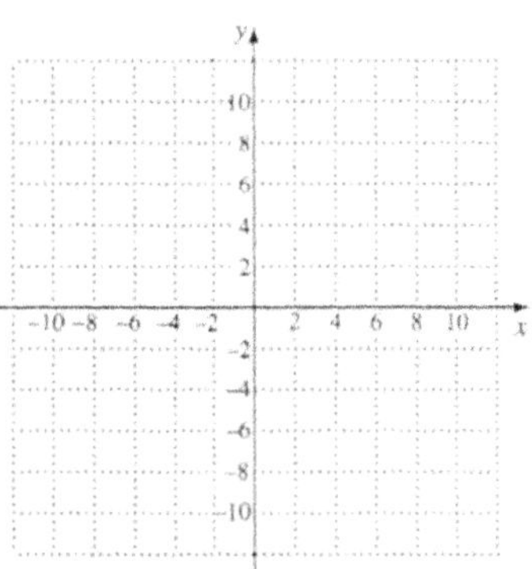

(continued)

b) $g(x) = |x+3|$

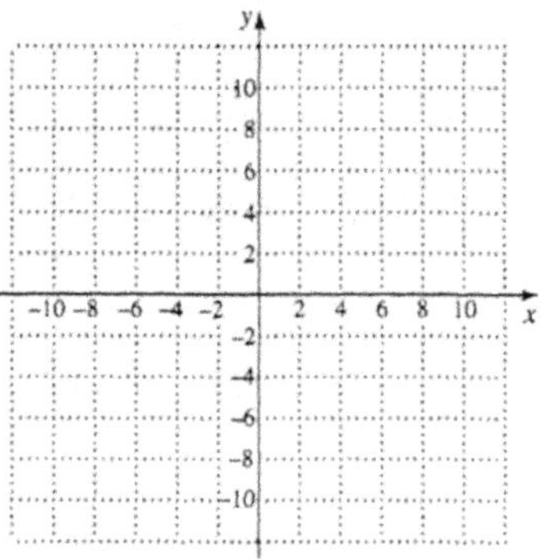

c) $g(x) = \sqrt{x-4}$

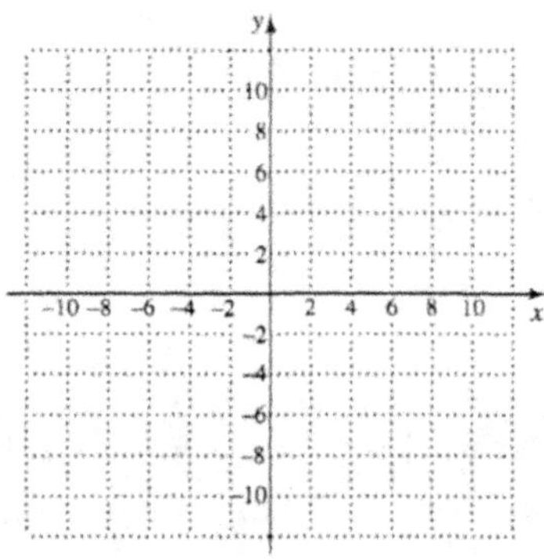

d) $h(x) = \sqrt{x-2} + 5$

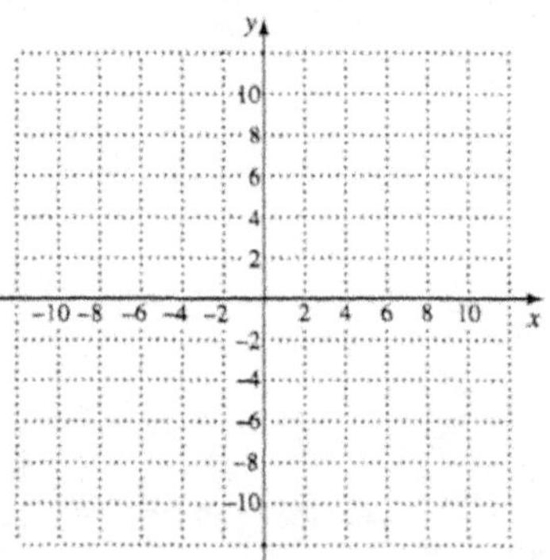

Reflections

The graph of $y = -f(x)$ is the **reflection** of the graph of $y = f(x)$ across the x-axis.

The graph of $y = f(-x)$ is the **reflection** of the graph of $y = f(x)$ across the y-axis.

If a point (x, y) is on the graph of $y = f(x)$, then $(x, -y)$ is on the graph of $y = -f(x)$, and $(-x, y)$ is on the graph of $y = f(-x)$.

Example 2 Graph each of the following. Before doing so, describe how each graph can be obtained from the graph of $f(x) = x^3 - 4x^2$.

a) $g(x) = (-x)^3 - 4(-x)^2$

$f(-x) = (-x)^3 - 4(-x)^2 = g(x)$, so $g(x)$ is a reflection of $f(x)$ across the

________ axis.
    x- / y-

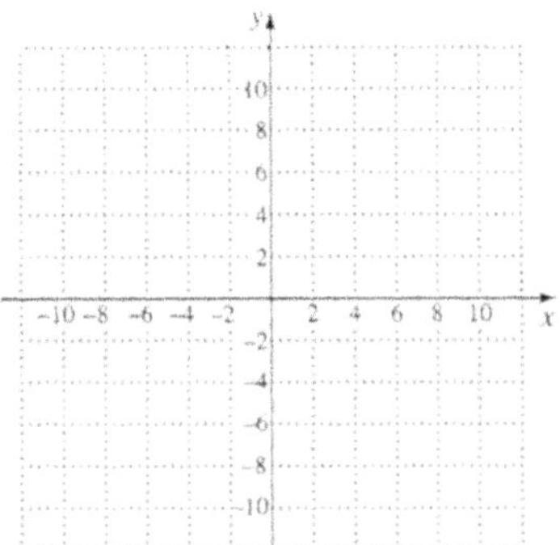

If (x, y) is on the graph of f, then $\left(\boxed{}, y\right)$ is on the graph of g. For example, if $(2, -8)$ is on the graph of f, then $(-2, -8)$ is on the graph of g.

b) $h(x) = 4x^2 - x^3$

$-f(x) = -(x^3 - 4x^2) = -x^3 + 4x^2 = 4x^2 - x^3 = h(x)$, so the graph of $h(x)$ is the

reflection of the graph of $f(x)$ across the ________ axis.
                                    x- / y-

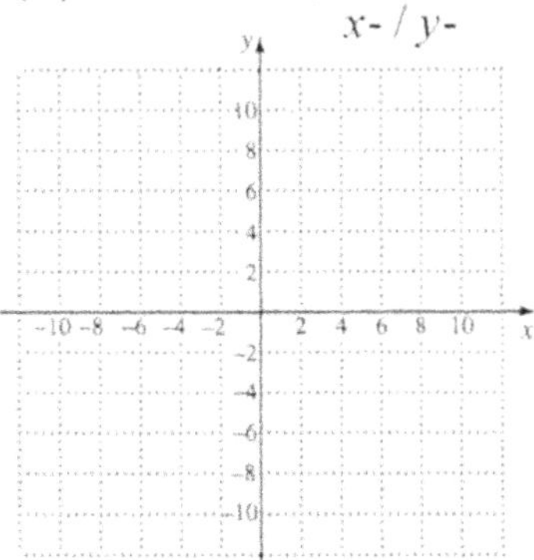

If (x, y) is on the graph of f, then $\left(x, \boxed{}\right)$ is on the graph of h. For example, if $(2, -8)$ is on the graph of f, then $(2, 8)$ is on the graph of h.

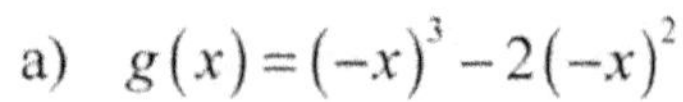

Your Turn 2   Graph each of the following. Before doing so, describe how each graph can be obtained from the graph of $f(x) = x^3 - 2x^2$.

a) $g(x) = (-x)^3 - 2(-x)^2$

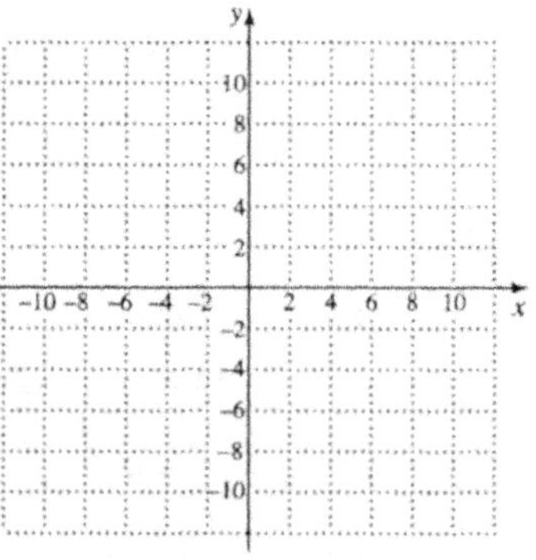

b) $h(x) = 2x^2 - x^3$

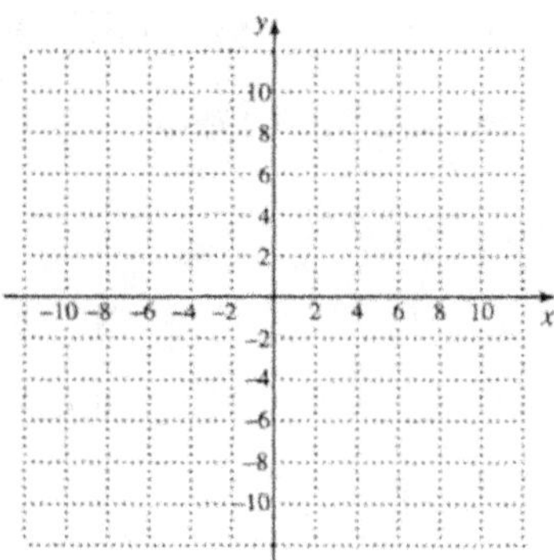

Vertical Stretching and Shrinking and Horizontal Stretching and Shrinking

The graph of $y = a f(x)$ can be obtained from the graph of $y = f(x)$ by

   stretching vertically for $|a| > 1$, or

   shrinking vertically for $0 < |a| < 1$.

For $a < 0$, the graph is also reflected across the x-axis. (The y-coordinates of the graph of $y = a f(x)$ can be obtained by multiplying the y-coordinates of $y = f(x)$ by a.)

The graph of $y = f(cx)$ can be obtained from the graph of $y = f(x)$ by

   shrinking horizontally for $|c| > 1$, or

   stretching horizontally for $0 < |c| < 1$.

For $c < 0$, the graph is also reflected across the y-axis. (The x-coordinates of the graph of $y = f(cx)$ can be obtained by dividing the x-coordinates of the graph of $y = f(x)$ by c.)

Example 3 Shown is a graph of $y = f(x)$ for some function f. No formula for f is given.

Graph each of the following.

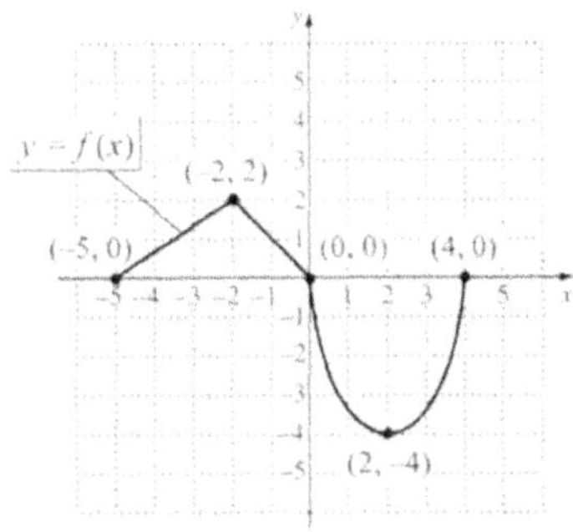

a) $g(x) = 2f(x)$

The graph of $y = a f(x)$ is a vertical stretching or shrinking of the graph of $y = f(x)$. Here $|a| = |2| > 1$, so this is a stretching. Each y-value on $f(x)$ will be multiplied by 2 to obtain the graph of $g(x)$.

Points on f	Corresponding points on g
$(-5, 0)$	$(-5, 0)$
$(-2, 2)$	$\left(-2, \boxed{}\right)$
$(0, 0)$	$(0, 0)$
$(2, -4)$	$\left(2, \boxed{}\right)$
$(4, 0)$	$(4, 0)$

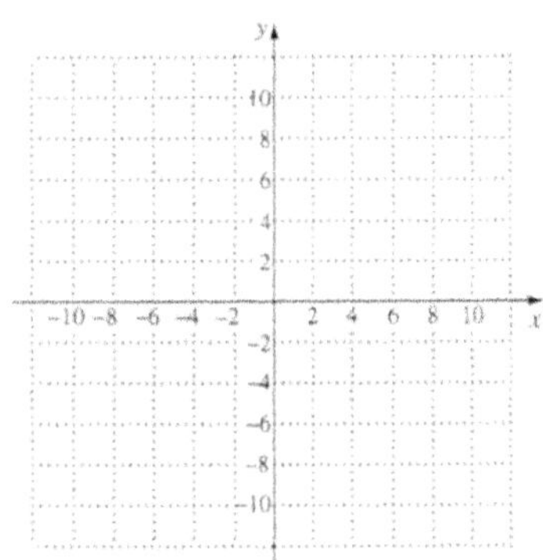

(continued)

b) $h(x) = \dfrac{1}{2} f(x)$

The graph of $y = a f(x)$ is a vertical stretching or shrinking of the graph of $y = f(x)$.

Here $|a| = \left|\dfrac{1}{2}\right| < 1$, so this is a shrinking. Each y-value on $f(x)$ will be multiplied by $\dfrac{1}{2}$ to obtain the graph of $h(x)$.

Points on f	Corresponding points on h
$(-5, 0)$	$\left(-5,\ \boxed{}\ \right)$
$(-2, 2)$	$(-2, 1)$
$(0, 0)$	$\left(0,\ \boxed{}\ \right)$
$(2, -4)$	$\left(2,\ \boxed{}\ \right)$
$(4, 0)$	$(4, 0)$

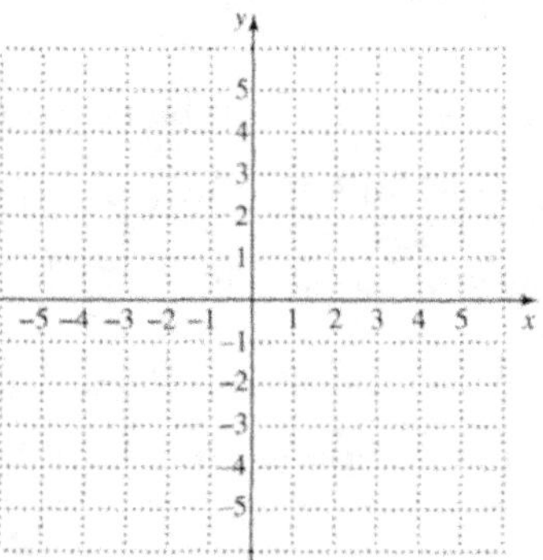

c) $r(x) = f(2x)$

The graph of $y = f(cx)$ is a horizontal stretching or shrinking of the graph of $y = f(x)$.
Here $|c| = |2| > 1$, so this is a shrinking. Each x-coordinate on $f(x)$ will be divided by 2 to obtain the graph of r.

Points on f	Corresponding points on r
$(-5, 0)$	$(-2.5, 0)$
$(-2, 2)$	$\left(\boxed{},\ 2\right)$
$(0, 0)$	$\left(\boxed{},\ 0\right)$
$(2, -4)$	$(1, -4)$
$(4, 0)$	$\left(\boxed{},\ 0\right)$

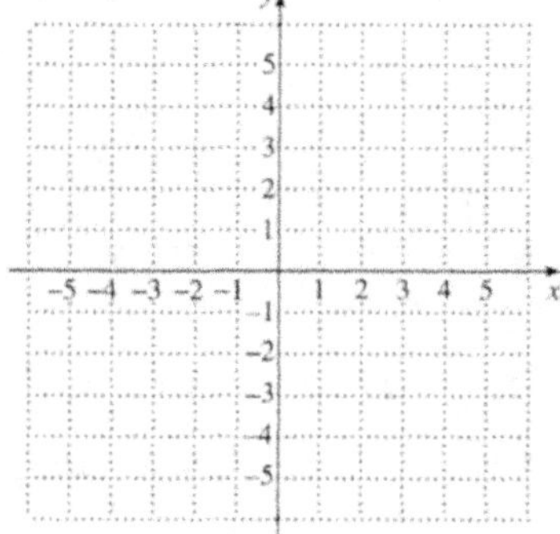

(continued)

d) $s(x) = f\left(\dfrac{1}{2}x\right)$

The graph of $y = f(cx)$ is a horizontal stretching or shrinking of the graph of $y = f(x)$.

Here $|c| = \left|\dfrac{1}{2}\right| < 1$, so this is a stretching. Each x-coordinate on $f(x)$ will be multiplied by 2 to obtain the graph of s.

Points on f	Corresponding points on s
$(-5, 0)$	$(\boxed{}, 0)$
$(-2, 2)$	$(-4, 2)$
$(0, 0)$	$(\boxed{}, 0)$
$(2, -4)$	$(\boxed{}, -4)$
$(4, 0)$	$(8, 0)$

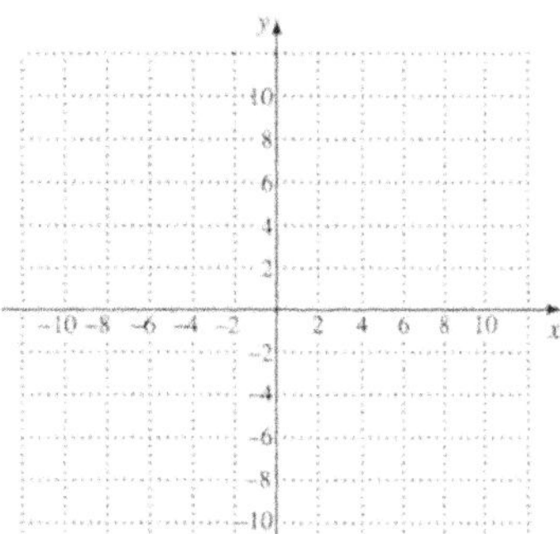

e) $t(x) = f\left(-\dfrac{1}{2}x\right)$

The graph of $y = f(cx)$ is a horizontal stretching or shrinking of the graph of $y = f(x)$.

Here $c = -\dfrac{1}{2}$. Because $c < 0$, we will also have the reflection of f across the y-axis. We have $|c| = \left|-\dfrac{1}{2}\right| = \boxed{} < 1$, so this is a horizontal stretching along with the reflection across the y-axis. We will multiply each x-coordinate of f by 2 and then change its sign.

Points on f	Corresponding points on t
$(-5, 0)$	$(10, 0)$
$(-2, 2)$	$(\boxed{}, 2)$
$(0, 0)$	$(0, 0)$
$(2, -4)$	$(\boxed{}, -4)$
$(4, 0)$	$(\boxed{}, 0)$

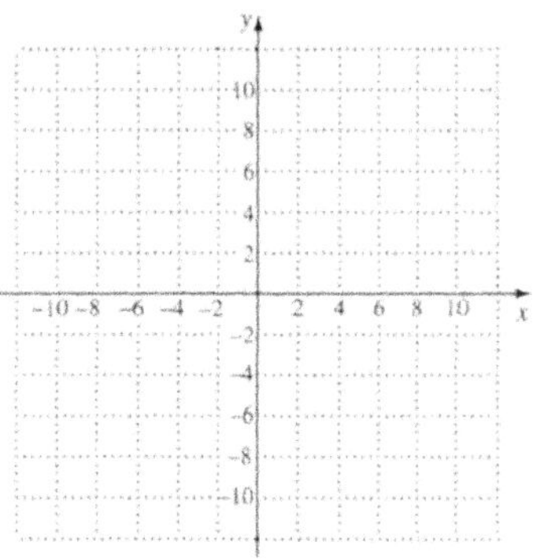

Your Turn 3   Shown is a graph of $y = f(x)$ for some function f. No formula for f is given. Graph each of the following.

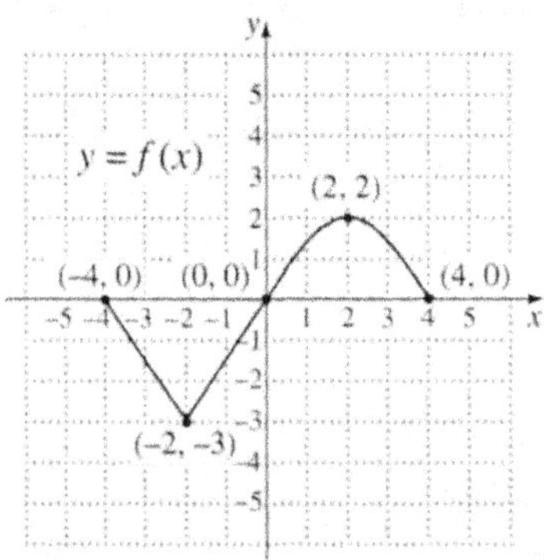

a) $h(x) = 2f(x)$

Points on f	Points on h
(,)	(,)
(,)	(,)
(,)	(,)
(,)	(,)
(,)	(,)

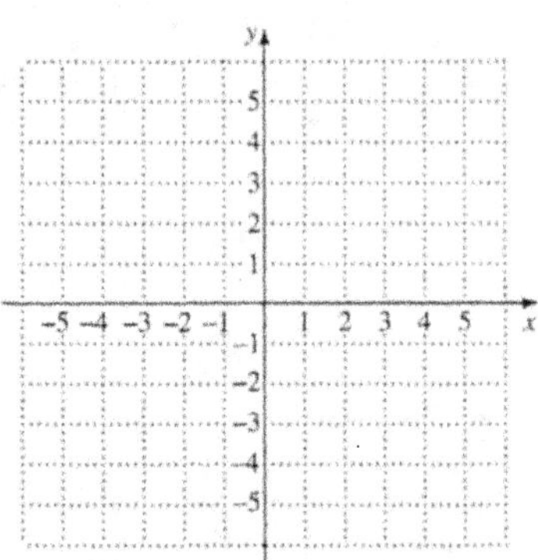

b) $s(x) = \dfrac{1}{2} f(x)$

Points on f	Points on s
(,)	(,)
(,)	(,)
(,)	(,)
(,)	(,)
(,)	(,)

(continued)

c) $t(x) = f(2x)$

Points on f	Points on t
(.)	(.)
(.)	(.)
(.)	(.)
(.)	(.)
(.)	(.)

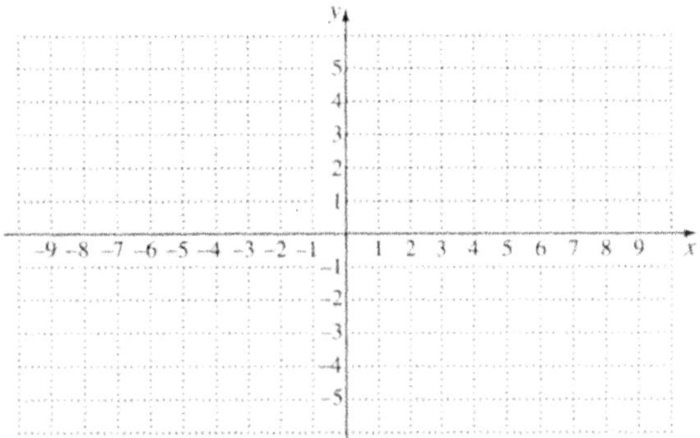

d) $g(x) = f\left(\dfrac{1}{2}x\right)$

Points on f	Points on g
(,)	(,)
(,)	(,)
(,)	(,)
(,)	(,)
(,)	(,)

e) $r(x) = f\left(-\dfrac{1}{2}x\right)$

Points on f	Points on r
(.)	(.)
(.)	(.)
(.)	(.)
(.)	(.)
(.)	(.)

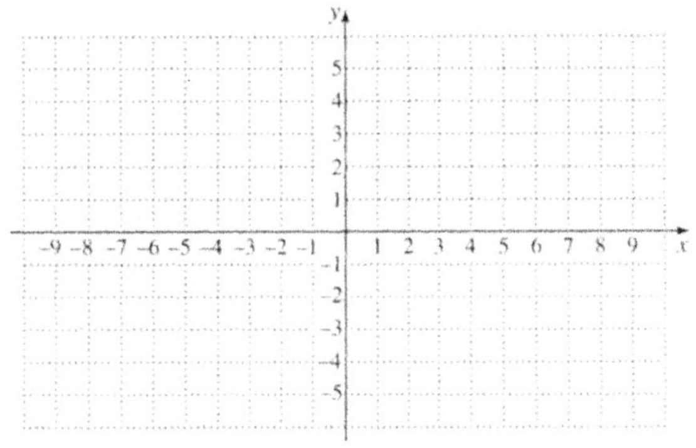

Example 4   Use the graph of $y = f(x)$ shown below to graph $y = -2f(x-3)+1$.

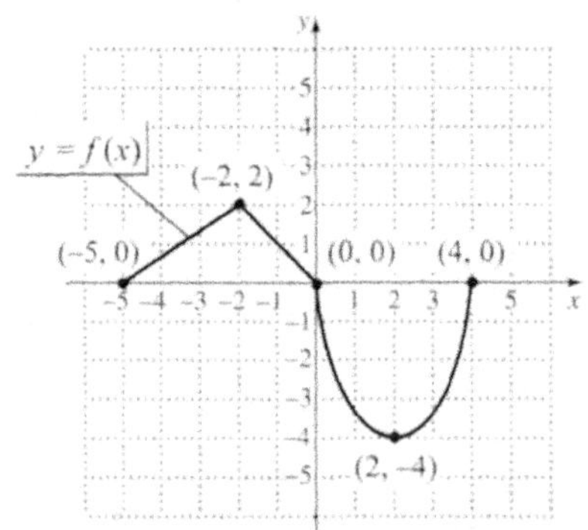

First consider $y = f(x-3)$. This is a shift of the given graph $\boxed{}$ units to

the $\underline{}$.
$\quad$ right / left

Points on $y = f(x)$	Points on $y = f(x-3)$
$(-5, 0)$	$(-2, 0)$
$(-2, 2)$	$\left(\boxed{}, 2\right)$
$(0, 0)$	$(3, 0)$
$(2, -4)$	$\left(\boxed{}, -4\right)$
$(4, 0)$	$(7, 0)$

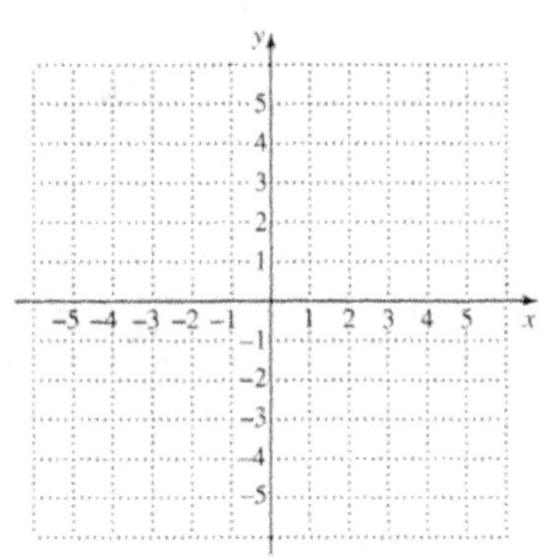

Now sketch the graph of $y = f(x-3)$.

Next consider $y = 2f(x-3)$. This will stretch the graph of $y = f(x-3)$ vertically by a factor of 2.

Points on $y = f(x-3)$	Points on $y = 2f(x-3)$
$(-2, 0)$	$(-2, 0)$
$(3, 0)$	$(3, 0)$
$(7, 0)$	$\left(7, \boxed{}\right)$
$(1, 2)$	$\left(1, \boxed{}\right)$
$(5, -4)$	$(5, -8)$

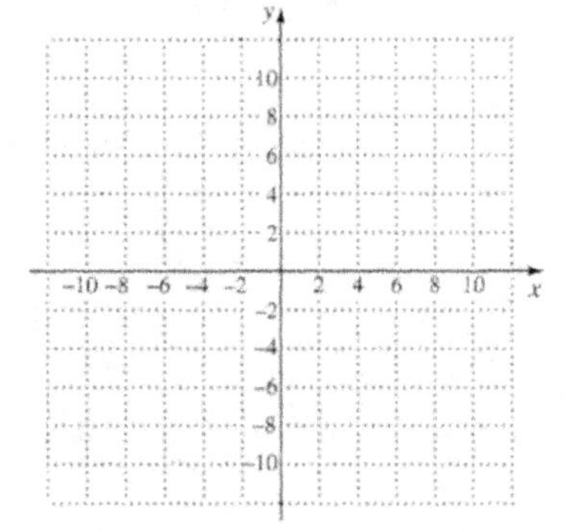

Now sketch the graph of $y = 2f(x-3)$.

(continued)

Next consider $y = -2f(x-3)$. This is the reflection of the graph of $y = 2f(x-3)$ across the _________ axis.
_x- / y-_

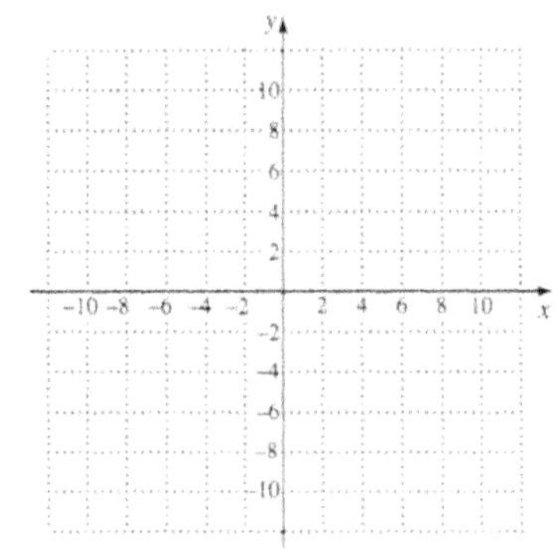

Points on $y = 2f(x-3)$	Points on $y = -2f(x-3)$
$(-2, 0)$	$(-2, 0)$
$(3, 0)$	$\left(3, \boxed{}\right)$
$(7, 0)$	$(7, 0)$
$(1, 4)$	$\left(1, \boxed{}\right)$
$(5, -8)$	$\left(5, \boxed{}\right)$

Now sketch the graph of $y = -2f(x-3)$.

Finally, consider $y = -2f(x-3)+1$. This is a shift of the graph of $y = -2f(x-3)$ _________ 1 unit.
up / down

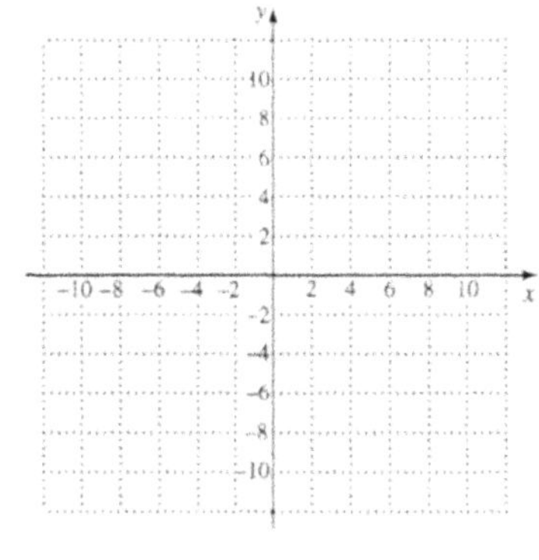

Points on $y = -2f(x-3)$	Points on $y = -2f(x-3)+1$
$(-2, 0)$	$(-2, 1)$
$(1, -4)$	$\left(1, \boxed{}\right)$
$(3, 0)$	$\left(3, \boxed{}\right)$
$(5, 8)$	$(5, 9)$
$(7, 0)$	$\left(7, \boxed{}\right)$

Now we can sketch the graph of $y = -2f(x-3)+1$.

Your Turn 4 Use the graph of $y = f(x)$ shown below to graph $y = -\dfrac{1}{2}f(x+3)-1$.

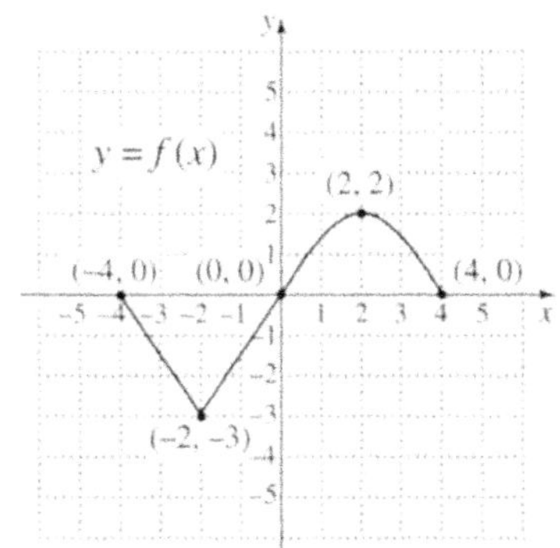

(contiued)

First consider $y = f(x+3)$.

$y = f(x)$	$y = f(x+3)$
(,)	(,)
(,)	(,)
(,)	(,)
(,)	(,)
(,)	(,)

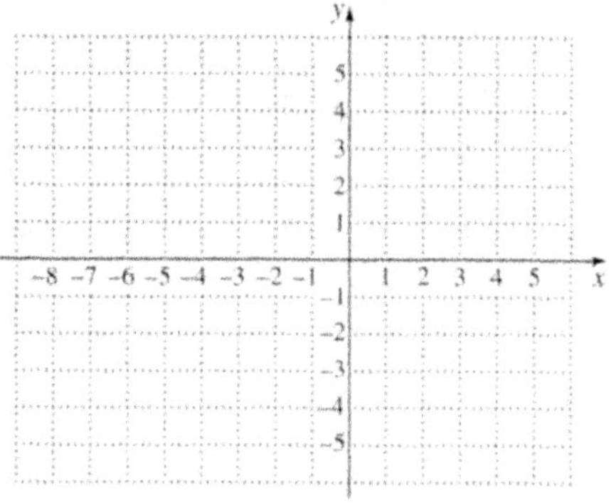

Next consider $y = \dfrac{1}{2} f(x+3)$.

$y = f(x+3)$	$y = \dfrac{1}{2} f(x+3)$
(,)	(,)
(,)	(,)
(,)	(,)
(,)	(,)
(,)	(,)

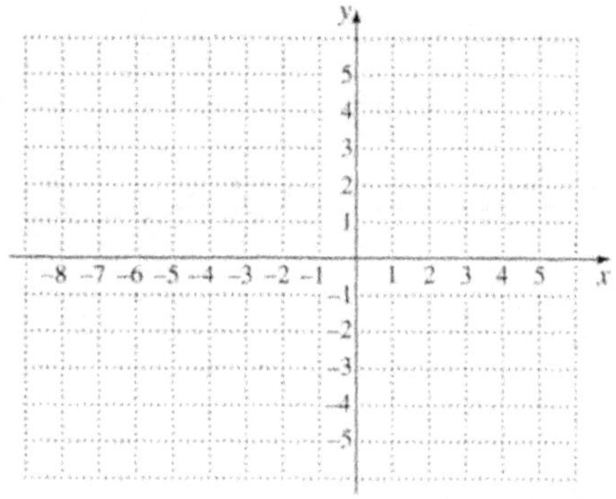

Next consider $y = -\dfrac{1}{2} f(x+3)$.

$y = \dfrac{1}{2} f(x+3)$	$y = -\dfrac{1}{2} f(x+3)$
(,)	(,)
(,)	(,)
(,)	(,)
(,)	(,)
(,)	(,)

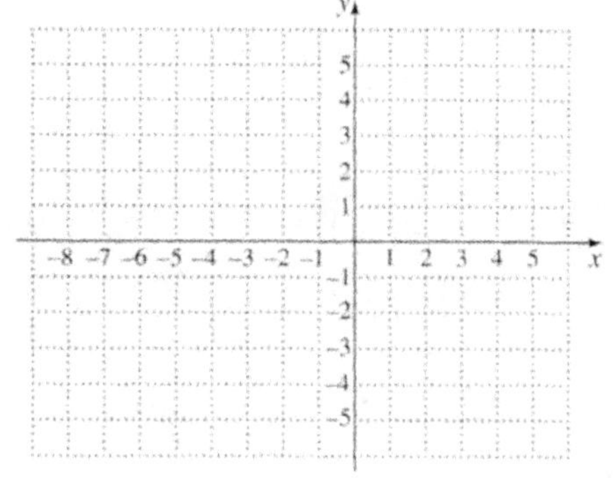

Finally, consider $y = -\dfrac{1}{2} f(x+3) - 1$.

$y = -\dfrac{1}{2} f(x+3)$	$y = -\dfrac{1}{2} f(x+3) - 1$
(,)	(,)
(,)	(,)
(,)	(,)
(,)	(,)
(,)	(,)

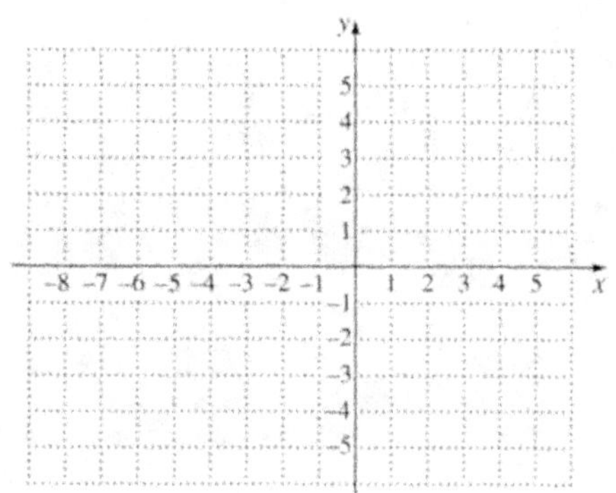

Practice Exercises

Graph each of the following. Before doing so, describe how each graph can be obtained from a basic graph.

1. $g(x) = x^2 - 2$

The graph of $g(x) = x^2 - 2$
is the graph of $f(x) = x^2$
shifted _________ ☐ units.
    up / down

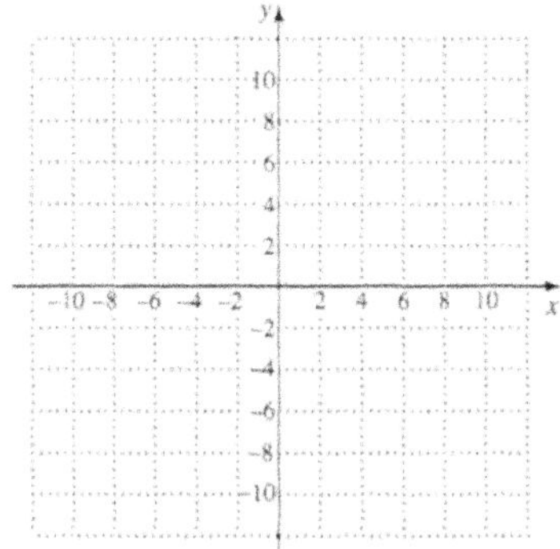

2. $g(x) = |x + 4|$

The graph of $g(x) = |x + 4|$
is the graph of $f(x) = |x|$
shifted _________ ☐ units.
    left / right

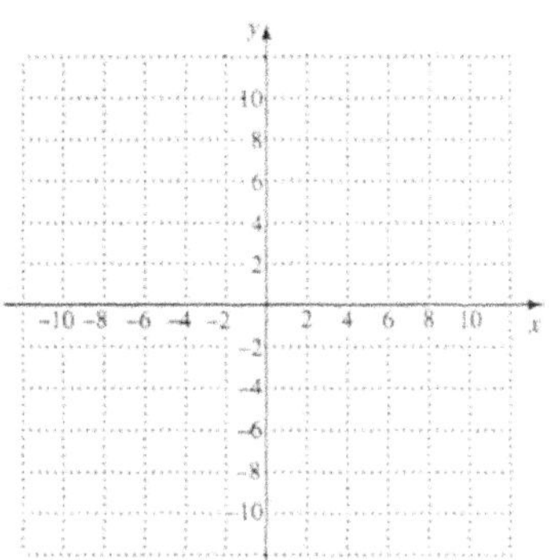

3. $g(x) = \sqrt{x + 4}$

The graph of $g(x) = \sqrt{x + 4}$
is the graph of $f(x) = \sqrt{x}$
shifted _________ ☐ units.
    left / right

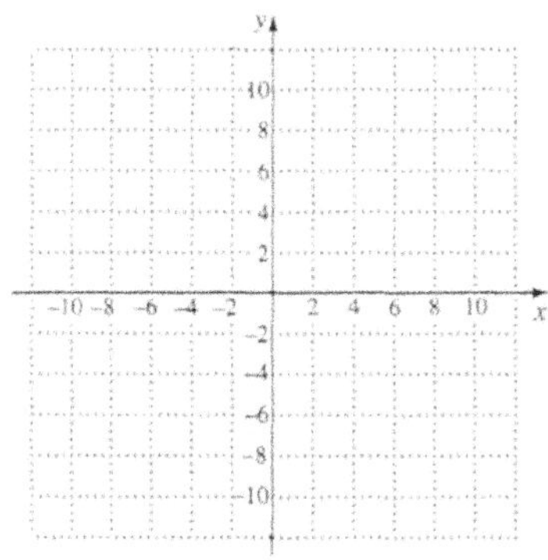

4. $g(x) = \sqrt{x - 2} + 3$

The graph of $g(x) = \sqrt{x - 2} + 3$
is the graph of $f(x) = \sqrt{x}$
shifted _________ ☐ units
    left / right
and _________ ☐ units.
    up / down

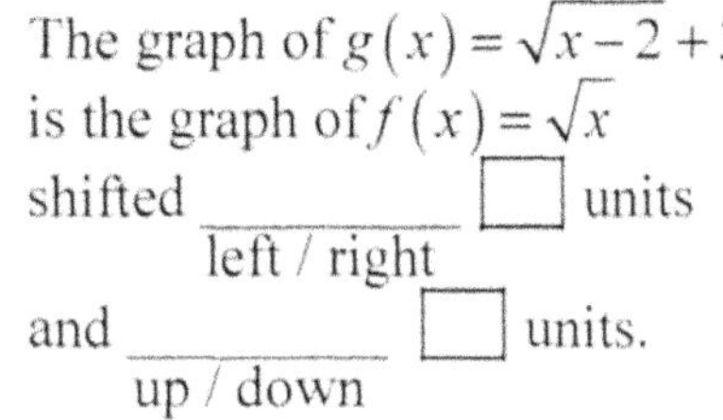

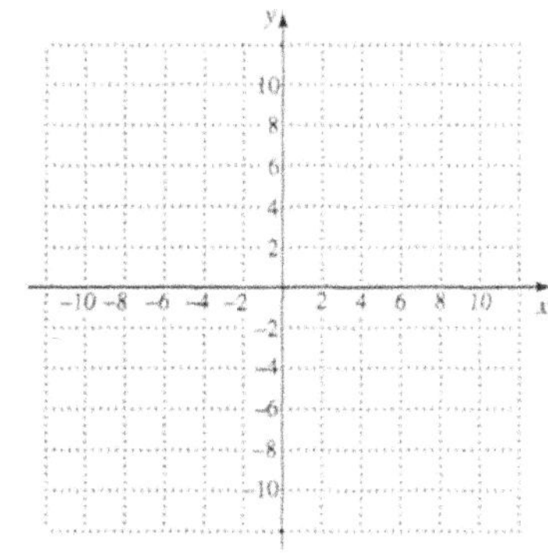

Graph each of the following. Before doing so, describe how each graph can be obtained from the graph of $f(x) = x^3 - x^2$.

5. $g(x) = (-x)^3 - (-x)^2$

(Hint: find $f(-x)$ and compare $g(x)$ with $f(-x)$.)The graph of $g(x)$ is the reflection of $f(x)$ across the ______ axis.

$x\text{-} / y\text{-}$

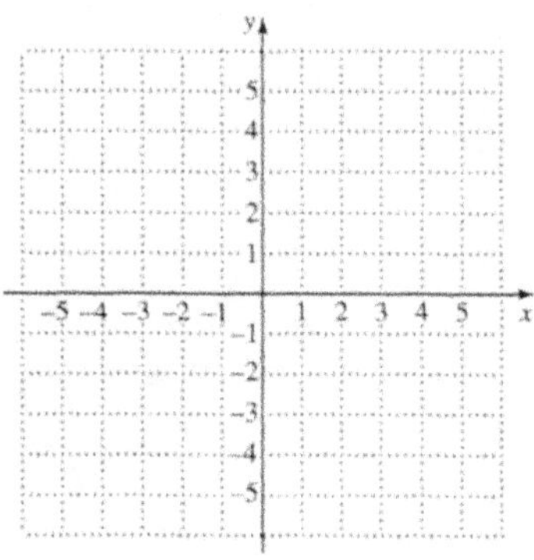

6. $h(x) = x^2 - x^3$

(Hint: find $-f(x)$ and compare $h(x)$ with $-f(x)$.) The graph of $h(x)$ is the reflection of $f(x)$ across the ______ axis.

$x\text{-} / y\text{-}$

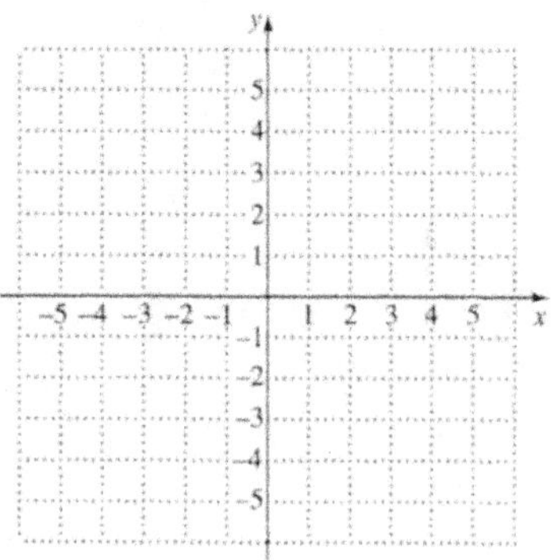

7. Shown is a graph of $y = f(x)$ for some function f. No formula for f is given. Graph each of the following.

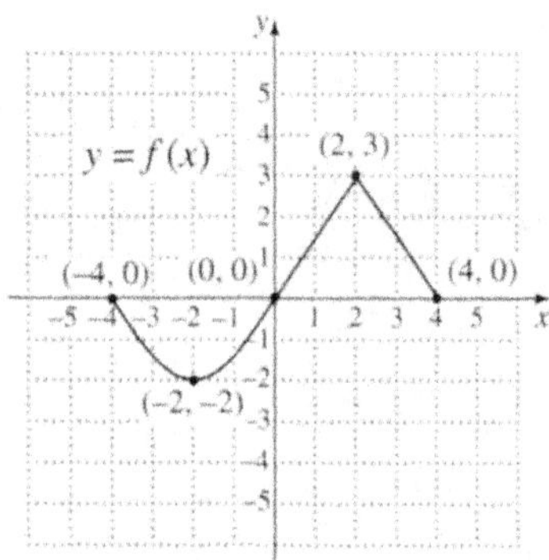

a) $g(x) = 2f(x)$

b) $h(x) = \dfrac{1}{2} f(x)$

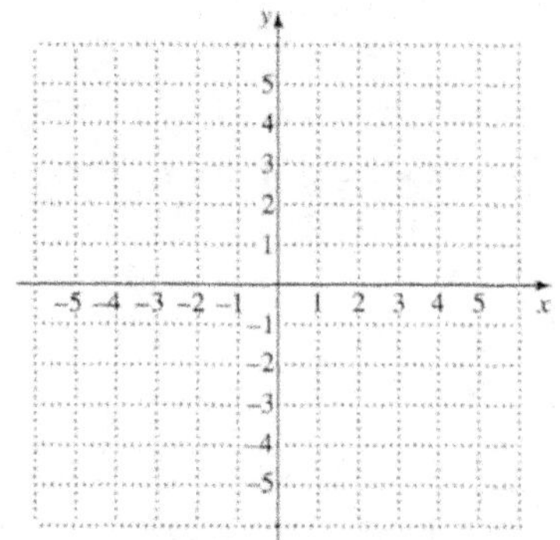

c) $r(x) = f(2x)$

d) $s(x) = f\left(-\dfrac{1}{2}x\right)$

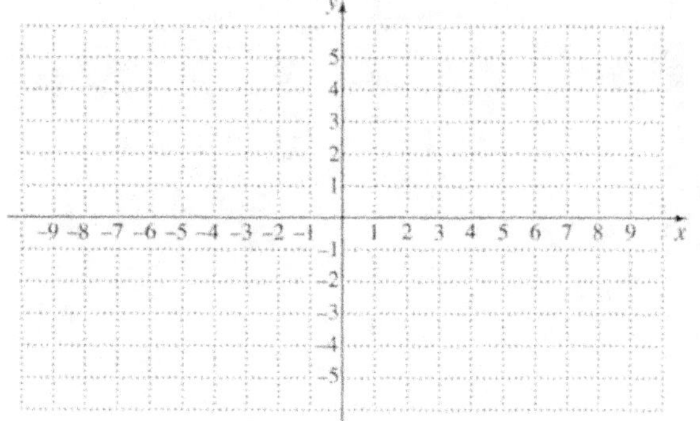

Section 7.3 The Complex Numbers

The Number i

The number i is defined such that

$$i = \sqrt{-1} \text{ and } i^2 = -1.$$

Example 1 Express each number in terms of i.

a) $\sqrt{-7} = \sqrt{-1 \cdot 7}$

$\quad\quad = \boxed{} \cdot \sqrt{7}$

$\quad\quad = \boxed{}\sqrt{7}, \text{ or } \sqrt{7}\,i$

b) $\sqrt{-16} = \sqrt{-1 \cdot 16}$

$\quad\quad = \boxed{} \cdot \sqrt{16}$

$\quad\quad = i \cdot \boxed{}$

$\quad\quad = 4i$

c) $-\sqrt{-13} = -\sqrt{-1 \cdot 13}$

$\quad\quad = -\sqrt{-1} \cdot \sqrt{13}$

$\quad\quad = -\boxed{}\sqrt{13}, \text{ or } -\sqrt{13}\,i$

d) $-\sqrt{-64} = -\sqrt{-1 \cdot 64}$

$\quad\quad = -\boxed{} \cdot \sqrt{64}$

$\quad\quad = -i \cdot \boxed{}$

$\quad\quad = -8i$

e) $\sqrt{-48} = \sqrt{-1 \cdot 48}$

$\quad\quad = \sqrt{-1} \cdot \sqrt{48}$

$\quad\quad = \boxed{} \cdot \sqrt{48}$

$\quad\quad = i \cdot \sqrt{16 \cdot 3}$

$\quad\quad = i \cdot \boxed{} \cdot \sqrt{3}$

$\quad\quad = i \cdot 4 \cdot \sqrt{3}$

$\quad\quad = \boxed{}$

Your Turn 1 Express each number in terms of i.

a) $\sqrt{-3}$

b) $\sqrt{-36}$

c) $-\sqrt{-41}$

d) $-\sqrt{-100}$

e) $\sqrt{-84}$

Complex Numbers

A **complex number** is a number of the form $a + bi$, where a and b are real numbers. The number a is said to be the **real part** of $a + bi$ and the number b is said to be the **imaginary part** of $a + bi$.

Example 2   Add or subtract and simplify each of the following.

a) $(8 + 6i) + (3 + 2i)$

$$(8 + 6i) + (3 + 2i) = \left(8 + \boxed{}\right) + \left(6i + \boxed{}\right)$$
$$= 11 + \boxed{}$$

b) $(4 + 5i) - (6 - 3i)$

$$(4 + 5i) - (6 - 3i) = (4 - 6) + \left(5i + \boxed{}\right)$$
$$= \boxed{} + 8i$$

Your Turn 2   Add or subtract and simplify each of the following.

a) $(7 + 3i) + (2 + i)$

b) $(8 + 7i) - (4 - 5i)$

Example 3   Multiply and simplify each of the following.

a) $\sqrt{-16} \cdot \sqrt{-25} = \boxed{} \cdot \sqrt{16} \cdot \boxed{} \cdot \sqrt{25}$
$$= \boxed{} \cdot 4 \cdot i \cdot \boxed{}$$
$$= 20i^2 \qquad\qquad\qquad i = \sqrt{-1}$$
$$= 20\left(\boxed{}\right) \qquad\qquad i^2 = -1$$
$$= -20$$

(continued)

b) $(1+2i)(1+3i) = 1+3i+2i+\boxed{}$

$= 1+3i+2i+6\left(\boxed{}\right)$

$= 1+\boxed{}-6$

$= -5+5i$

c) $(3-7i)^2 = 9-\boxed{}+49i^2$

$= 9-42i+49(-1)$

$= 9-42i-49$

$= \boxed{}-42i$

✏ **Your Turn 3**   Multiply and simplify each of the following.

a) $\sqrt{-4}\sqrt{-49}$

b) $(3+4i)(5-2i)$

c) $(2-5i)^2$

Example 4   Simplify each of the following.

a) $i^{37} = \boxed{}\cdot i$

$= \left(i^2\right)^{\boxed{}}\cdot i$

$= \left(\boxed{}\right)^{18}\cdot i$

$= 1\cdot i$

$= i$

b) $i^{58} = \left(i^2\right)^{\boxed{}}$

$= (-1)^{29}$

$= \boxed{}$

c) $i^{75} = \boxed{}\cdot i$

$= \left(i^2\right)^{\boxed{}}\cdot i$

$= (-1)^{37}\cdot i$

$= -1\cdot i$

$= \boxed{}$

d) $i^{80} = \left(i^2\right)^{\boxed{}}$

$= (-1)^{40}$

$= \boxed{}$

✏ **Your Turn 4**   Simplify each of the following.

a) i^{21}

b) i^{72}

c) i^{35}

d) i^{18}

> **Conjugate of a Complex Number**
>
> The **conjugate** of a complex number $a+bi$ is $a-bi$. The numbers $a+bi$ and $a-bi$ are **complex conjugates**.

Example 5 Multiply each of the following.

a) $(5+7i)(5-7i)$ $(A+B)(A-B)=A^2-B^2$

$$=5^2-\left(\boxed{}\right)^2$$

$$=25-49i^2$$

$$=25-49\left(\boxed{}\right)$$

$$=25+49$$

$$=\boxed{}$$

b) $(8i)(-8i)=-64i^2$

$$=-64\left(\boxed{}\right)$$

$$=64$$

Your Turn 5 Multiply each of the following.

a) $(3+10i)(3-10i)$

b) $(6i)(-6i)$

Example 6 Divide $2-5i$ by $1-6i$.

$$\frac{2-5i}{1-6i}=\frac{2-5i}{1-6i}\cdot\frac{\boxed{}}{\boxed{}}=\frac{(2-5i)(1+6i)}{(1-6i)(1+6i)}$$

$$=\frac{2+12i-5i-\boxed{}}{1-\left(\boxed{}\right)^2}$$

$$=\frac{2+12i-5i+\boxed{}}{1-36i^2}=\frac{\boxed{}+7i}{1+36}$$

$$=\frac{32+7i}{\boxed{}}=\frac{32}{37}+\frac{7}{37}i$$

Your Turn 6 Divide $3-2i$ by $4+5i$.

Practice Exercises

Express the number in terms of i.

1. $\sqrt{-29}$

2. $\sqrt{-81}$

3. $\sqrt{-124}$

4. $-\sqrt{-25}$

5. $\sqrt{-175}$

6. $-\sqrt{-37}$

Add or subtract and simplify.

7. $(2+i)+(10+6i)$

8. $(3-5i)-(1+2i)$

9. $(9+8i)-(10-7i)$

10. $(5-3i)+(4+9i)$

Multiply and simplify.

11. $\sqrt{-81}\cdot\sqrt{-100}$

12. $(6-3i)(5+4i)$

13. $(7-5i)(8+i)$

14. $(4-9i)^2$

Simplify.

15. i^{31}

16. i^{22}

17. i^{92}

18. i^{61}

Multiply and simplify.

19. $(-12i)(12i)$

20. $(3+2i)(3-2i)$

21. $(1-10i)(1+10i)$

22. $(5i)(-5i)$

Divide and simplify.

23. $\dfrac{4-3i}{1+2i}$

24. $\dfrac{4}{-6+i}$

25. $\dfrac{i}{7-i}$

26. $\dfrac{8-i}{2-6i}$

Section 7.4 Quadratic Equations, Functions, Zeros, and Models

Quadratic Equations

A **quadratic equation** is an equation that can be written in the form

$ax^2 + bx + c = 0$, $a \neq 0$, where a, b, and c are real numbers.

Quadratic Functions

A **quadratic function** f is a function that can be written in the form

$f(x) = ax^2 + bx + c$, $a \neq 0$, where a, b, and c are real numbers.

The **zeros** of a quadratic function $f(x) = ax^2 + bx + c$ are the *solutions* of the associated quadratic equation $ax^2 + bx + c = 0$.

Equation-Solving Principles

The Principle of Zero Products: If $ab = 0$ is true, then $a = 0$ or $b = 0$, and if $a = 0$ *or* $b = 0$, then $ab = 0$.

The Principle of Square Roots: If $x^2 = k$, then $x = \sqrt{k}$ *or* $x = -\sqrt{k}$.

Example 1 Solve: $2x^2 - x = 3$.

$$2x^2 - x = 3$$
$$2x^2 - x - 3 = 0$$
$$(2x - 3)\left(\boxed{}\right) = 0$$

$2x - 3 = 0 \qquad or \quad x + 1 = 0$

$\quad 2x = 3 \qquad or \qquad x = -1$

$\quad\; x = \boxed{} \quad or \qquad x = -1$

The solutions are $\boxed{}$

and $\boxed{}$.

Check:

For $x = -1$:

$$2x^2 - x = 3$$
$$2\left(\boxed{}\right)^2 - \left(\boxed{}\right) \;?\; 3$$
$$2 + 1$$
$$3 \;\big|\; 3 \qquad \text{TRUE}$$

For $x = \dfrac{3}{2}$:

$$2x^2 - x = 3$$
$$2\left(\dfrac{3}{2}\right)^2 - \dfrac{3}{2} \;?\; 3$$
$$2\left(\dfrac{9}{4}\right) - \dfrac{3}{2}$$
$$\boxed{} - \dfrac{3}{2}$$
$$3 \;\big|\; 3 \qquad \text{TRUE}$$

Your Turn 1 Solve: $3x^2 - 10x = 8$.

Example 2 Solve: $2x^2 - 10 = 0$.

$2x^2 - 10 = 0$

$2x^2 = 10$

$x^2 = 5$

$x = \boxed{}$, or $\sqrt{5}$ and $-\sqrt{5}$

The solutions are $\sqrt{5}$ and $-\sqrt{5}$, or $\boxed{}$.

Check:

$$2x^2 - 10 = 0$$

$2\left(\boxed{}\right)^2 - 10 \; \overset{?}{\;} \; 0$

$10 - 10$

$0 \;\vert\; 0 \quad$ TRUE

Your Turn 2 Solve: $5x^2 - 35 = 0$.

Example 3 Find the zeros of $f(x) = x^2 - 6x - 10$ by completing the square.

$x^2 - 6x - 10 = 0$

$x^2 - 6x \quad = 10$

$x^2 - 6x + \boxed{} = 10 + \boxed{} \qquad \frac{1}{2}\left(\boxed{}\right) = -3; \; (-3)^2 = \boxed{}$

$\left(\boxed{}\right)^2 = 19$

$x - 3 = \boxed{}$

$x = 3 \pm \sqrt{19}$

$x = 3 + \sqrt{19} \quad or \quad x = 3 - \sqrt{19}$

$x \approx 7.359 \quad or \quad x \approx -1.359$

Zeros of $f(x) = x^2 - 6x - 10$ are $3 + \sqrt{19}$ and $\boxed{}$,

or approximately $\boxed{}$ and -1.359.

Your Turn 3   Find the zeros of $f(x) = x^2 - 8x - 13$.

Example 4   Solve: $2x^2 - 1 = 3x$.

$$2x^2 - 1 = 3x$$

$$2x^2 - \boxed{} - 1 = 0$$

$$2x^2 - 3x = 1$$

$$x^2 - \frac{3}{2}x = \boxed{}$$

$$x^2 - \frac{3}{2}x + \boxed{} = \frac{1}{2} + \boxed{} \qquad \text{Completing the square}$$

$$\left(\boxed{}\right)^2 = \frac{17}{16}$$

$$x - \frac{3}{4} = \pm\sqrt{\frac{17}{16}} = \pm\frac{\sqrt{17}}{\sqrt{16}} = \pm\frac{\sqrt{17}}{\boxed{}}$$

$$x = \frac{3}{4} \pm \frac{\sqrt{17}}{4}$$

$$x = \frac{3 \pm \sqrt{17}}{4}$$

The solutions are $\dfrac{3 + \sqrt{17}}{4}$ and $\dfrac{3 - \sqrt{17}}{4}$, or $\boxed{}$.

Your Turn 4   Solve: $2x^2 + 1 = 5x$.

The Quadratic Formula

The solutions of $ax^2 + bx + c = 0$, $a \neq 0$, are given by

$$x = \frac{-b \pm \sqrt{b^2 - 4ac}}{2a}.$$

Example 5   Solve $3x^2 + 2x = 7$. Find exact solutions and approximate solutions rounded to three decimal places.

$$3x^2 + 2x = 7$$

$$3x^2 + 2x - 7 = 0$$

$$a = 3 \quad b = \boxed{} \quad c = \boxed{}$$

$$x = \frac{-\boxed{} \pm \sqrt{2^2 - 4 \cdot 3 \cdot \left(\boxed{}\right)}}{2(3)} \qquad \text{Quadratic formula: } x = \frac{-b \pm \sqrt{b^2 - 4ac}}{2a}$$

$$= \frac{-2 \pm \sqrt{4 + 84}}{6} = \frac{-2 \pm \sqrt{\boxed{}}}{6}$$

$$= \frac{-2 \pm \sqrt{4 \cdot \boxed{}}}{6} = \frac{-2 \pm 2\sqrt{22}}{6}$$

$$= \frac{2\left(-1 \pm \sqrt{22}\right)}{2 \cdot 3} = \frac{\boxed{} \pm \sqrt{22}}{3}$$

Exact solutions: $\dfrac{-1 - \sqrt{22}}{3}$ and $\dfrac{-1 + \sqrt{22}}{3}$

Approximate solutions: -1.897 and $\boxed{}$

Your Turn 5   Solve: $4x^2 + 3x = 5$. Find exact solutions and approximate solutions rounded to three decimal places.

Example 6 Solve: $x^2 + 5x + 8 = 0$.

$a = \boxed{}$ $b = 5$ $c = 8$

$$x = \frac{-\boxed{} \pm \sqrt{5^2 - 4 \cdot \boxed{} \cdot \boxed{}}}{2 \cdot 1}$$

Quadratic formula: $x = \dfrac{-b \pm \sqrt{b^2 - 4ac}}{2a}$

$$x = \frac{-5 \pm \sqrt{\boxed{}}}{2}$$

$$x = \frac{-5 \pm \sqrt{7}\,\boxed{}}{2}$$

$$x = -\frac{5}{2} + \frac{\sqrt{7}}{2}i \ \text{ or } \ x = -\frac{5}{2} - \frac{\sqrt{7}}{2}i$$

Your Turn 6 Solve: $x^2 + 3x + 5 = 0$.

Discriminant

For $ax^2 + bx + c = 0$, where a, b, and c are real numbers, $a \neq 0$:

$\quad b^2 - 4ac = 0 \rightarrow$ One real-number solution;

$\quad b^2 - 4ac > 0 \rightarrow$ Two different real-number soluions;

$\quad b^2 - 4ac < 0 \rightarrow$ Two different imaginary-number solutions,
$\qquad\qquad\qquad\qquad$ complex conjugates.

Example 7 Solve: $x^4 - 5x^2 + 4 = 0$.

Let $u = \boxed{}$.

$u^2 - 5\boxed{} + 4 = 0$

$(u - 1)\left(\boxed{}\right) = 0$

$u - 1 = 0 \quad or \quad \boxed{} = 0$

$\quad u = 1 \quad or \qquad u = 4$

$x^2 = \boxed{} \quad or \quad x^2 = 4$

$x = \pm 1 \quad or \quad x = \boxed{}$

The solutions are -1, 1, -2, and 2.

Your Turn 7 Solve: $x^4 - 10x^2 + 9 = 0$.

Example 8 Solve: $t^{2/3} - 2t^{1/3} - 3 = 0$.

Let $u = t^{1/3}$.

$$\left(t^{1/3}\right)^{\square} - 2t^{1/3} - 3 = 0$$

$$u^2 - 2\boxed{} - 3 = 0$$

$$\left(\boxed{}\right)(u - 3) = 0$$

$$u + 1 = 0 \qquad or \qquad u - 3 = 0$$

$$u = \boxed{} \qquad or \qquad u = 3$$

$$\boxed{} = -1 \qquad or \qquad t^{1/3} = 3$$

$$\left(t^{1/3}\right)^3 = (-1)^3 \qquad or \qquad \left(t^{1/3}\right)^3 = 3^3$$

$$\boxed{} = -1 \qquad\qquad t = 27$$

The solutions are $\boxed{}$ and $\boxed{}$.

Your Turn 8 Solve: $t^{2/3} - 2t^{1/3} - 8 = 0$.

Example 9 *Museums in China.* The number of museums in China increased from approximately 2000 in the year 2000 to over 3500 by the end of 2012. In 2012, a record 451 new museums opened. For comparison, in the United States, only 20–40 new museums were opened per year from 2000 to 2008. The function

$$h(x) = 30.992x^2 + 4.108x + 2294.594$$

can be used to estimate the number of museums in China, x years after 2005.

a) Estimate the number of museums that will be in China in 2017 if the number of new museums that open per year continues at the same rate.

$$h(12) = 30.992\left(\boxed{}\right)^2 + 4.108\left(\boxed{}\right) + 2294.594$$

$$h(12) \approx 6807$$

In the year 2017, there are approximately $\boxed{}$ museums in China.

b) In what year was the number of museums in China 2600?

$$\boxed{} = 30.992x^2 + 4.108x + 2294.594$$

$$0 = 30.992x^2 + 4.108x - \boxed{}$$

$$a = 30.992 \qquad b = 4.108 \qquad c = -305.406$$

$$x = \frac{-b \pm \sqrt{b^2 - 4ac}}{2a}$$

$$x = \frac{-\boxed{} \pm \sqrt{(4.108)^2 - 4(30.992)(-305.406)}}{2(30.992)}$$

$$x = \frac{-4.108 \pm \sqrt{37,877.44667}}{61.984}$$

$$x = \boxed{} \quad or \quad x = \boxed{}$$

We are looking for a year after 2005, so we use the positive solution 3.074. Thus, there were about 2600 museums in China 3 years after 2005, or in $\boxed{}$.

Your Turn 9 Small businesses are firms employing less than 500 people. The number of small businesses in the United States increased from approximately 4.9 million in 2010 to approximately 20.9 million by the end of 2018 (Source: Small Business Administration; U.S. Census Bureau). The function

$$g(x) = 0.370x^2 - 1.384x + 5.534$$

can be used to estimate the number of small businesses in the United States, in millions, x years after 2010.

(continued)

a)   Estimate the number of small businesses in the United States in 2021 if the number of new small businesses that open per year continues at the same rate.

b)   In what year was the number of small businesses approximately 8.5 million?

Example 10   *Sales of New Homes.* Sales of new homes have increased in recent years. The function

$$h(x) = 22.1x^2 - 72.2x + 371.9$$

can be used to estimate the sales of new homes, in thousands, in the United States, where x is the number of years after 2009 (*Source*: IHS Global Insight). In what year were the number of sales of new homes about 563,400, or 563.4 thousands?

$$\boxed{} = 22.1x^2 - 72.2x + 371.9$$

$$0 = 22.1x^2 - 72.2x - \boxed{}$$

$$a = 22.1 \quad b = -72.2 \quad c = -191.5$$

$$x = \frac{-\left(\boxed{}\right) \pm \sqrt{(-72.2)^2 - 4(22.1)(-191.5)}}{2(22.1)} \qquad \text{Quadratic formula: } x = \frac{-b \pm \sqrt{b^2 - 4ac}}{2a}$$

$$x = \frac{72.2 \pm \sqrt{22,141.44}}{44.2}$$

$$x = \boxed{} \quad \text{or} \quad x = \boxed{}$$

We are looking for a year after 2009, so we use the positive solution. Thus, there were about 563,400 sales of new homes 5 years after 2009, or in $\boxed{}$.

Your Turn 10   The function $h(x) = 22.1x^2 - 72.2x + 371.9$ can be used to estimate the sales of new homes, in thousands, in the United States, where x is the number of years after 2009 (Source: IHS Global Insight). In what year were the number of sales of new homes about 734,300, or 734.3 thousands?

Example 11   *Train Speeds*. Two trains leave a station at the same time. One train travels due west, and the other travels due south. The train traveling west travels 20 km/h faster than the train traveling south. After 2 hr, the trains are 200 km apart. Find the speed of each train.

We use $d = rt$ to find the distance each train traveled in 2 hr.

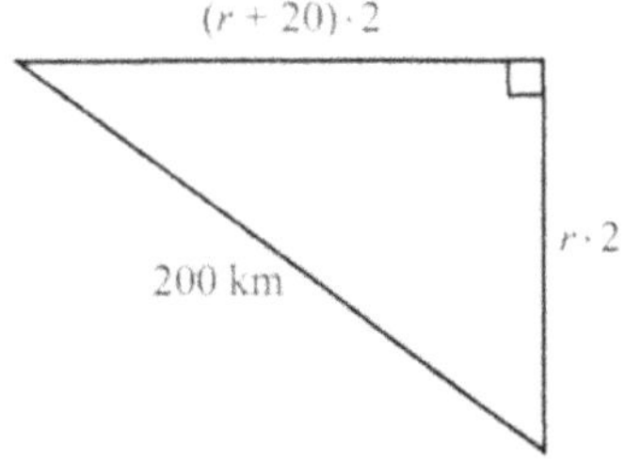

We use the Pythagorean theorem.

$$\left[2(r+20)\right]^2 + \left(\boxed{}\right)^2 = 200^2$$

$$\frac{\boxed{}}{4} + \frac{4r^2}{4} = \frac{40,000}{4}$$

$$(r+20)^2 + r^2 = 10,000$$

$$r^2 + 40r + \boxed{} + r^2 = 10,000$$

$$\frac{2r^2}{2} + \frac{40r}{2} - \frac{9600}{2} = 0$$

$$r^2 + 20r - 4800 = 0$$

$$\left(\boxed{}\right)(r - 60) = 0$$

$$r + 80 = 0 \quad or \quad r - 60 = 0$$

$$r = -80 \quad or \quad r = \boxed{}$$

The negative value of r doesn't make sense in the problem. We check 60.

(continued)

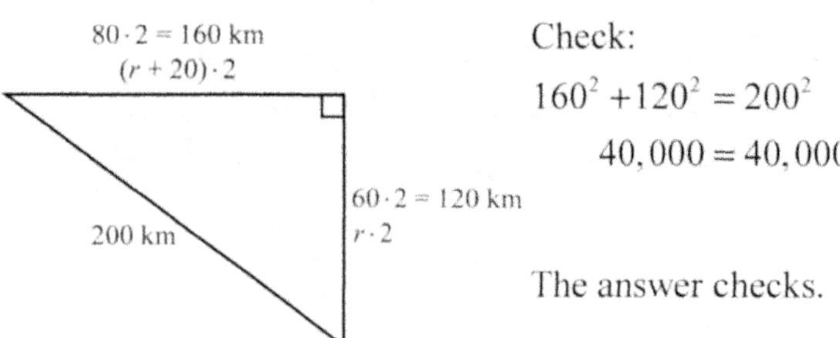

Check:

$$160^2 + 120^2 = 200^2$$

$$40,000 = 40,000$$

The answer checks.

The speed of the train heading south is ☐ km/h, and the speed of the train heading west is ☐ km/h.

Your Turn 11   Two trains leave a station at the same time. One train travels due west, and the other travels due south. The train traveling west travels 25 km/h faster than the train traveling south. After 3 hr, the trains are 375 km apart. Find the speed of each train. We use $d = rt$ to find the distance each train traveled in 3 hr.

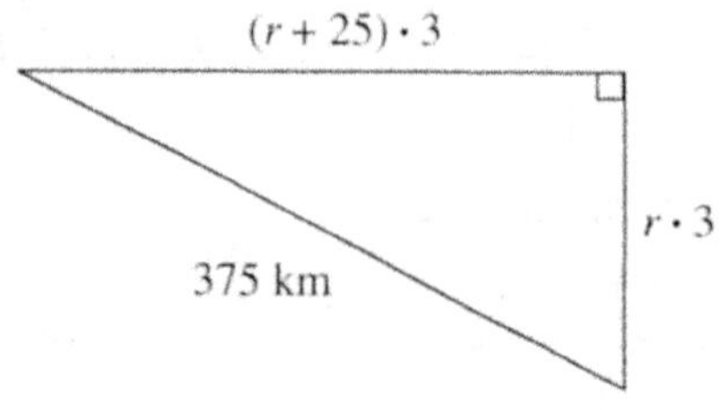

Practice Exercises

Solve using the principle of zero products.

1. $3x^2 + 2x - 5 = 0$

2. $4x^2 + 11x = -6$

Solve using the principle of square roots.

3. $8x^2 - 88 = 0$

4. $3x^2 - 51 = 0$

Find the zeros of the function by completing the square.

5. $f(x) = x^2 - 10x - 6$

6. $f(x) = 4x^2 + 20x - 5$

Solve using the quadratic formula.

7. $5x^2 - 9x = -3$

8. $x^2 + 3x + 11 = 0$

Find the zeros of the function. Give exact solutions and approximate solutions rounded to three decimal places when possible.

9. $f(x) = x^2 - 2x - 9$

10. $f(x) = 3x^2 + 9x + 10$

Solve.

11. $x^4 - 13x^2 + 36 = 0$

12. $z^{2/3} + 3z^{1/3} - 10 = 0$

13. The number of small businesses in the United States increased from approximately 4.9 million in 2010 to approximately 20.9 million by the end of 2018 (Source: Small Business Administration; U.S. Census Bureau). The function $g(x) = 0.370x^2 - 1.384x + 5.534$ can be used to estimate the number of small businesses in the United States, in millions, x years after 2010.

a) Estimate the number of small businesses in the United States in 2023 if the number of new small businesses that open per year continues at the same rate.

b) In what year was the number of small businesses approximately 14.1 million?

14. Two trains leave a station at the same time. One train travels due west, and the other travels due south. The train traveling west travels 15 mph faster than the train traveling south. After 2 hr, the trains are 150 mi apart. Find the speed of each train. We use $d = rt$ to find the distance each train traveled in 2 hr.

Section 7.5 Analyzing Graphs of Quadratic Functions

Graphing Quadratic Functions

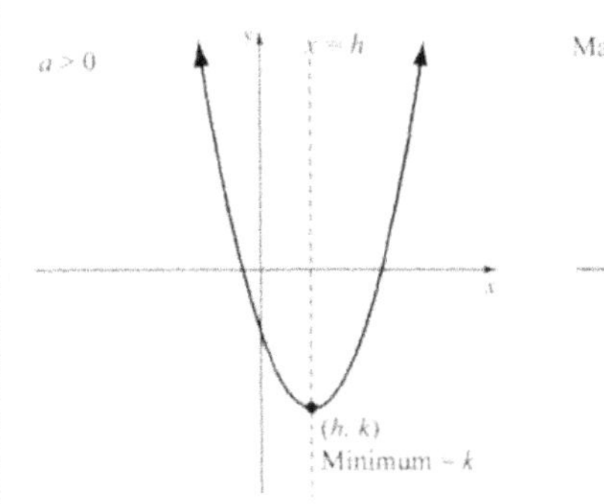
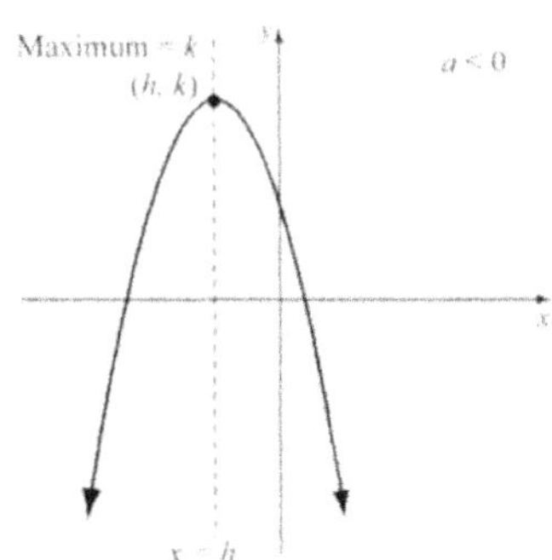

The graph of the function

$f(x) = a(x-h)^2 + k$ is a

parabola that

- opens up if $a > 0$ and down if $a < 0$;
- has (h, k) as the vertex;
- has $x = h$ as the axis of symmetry;
- has k as a minimum value (output) if $a > 0$;
- has k as a maximum value if $a < 0$.

Example 1 Find the vertex, the axis of symmetry, and the maximum or minimum value of $f(x) = x^2 + 10x + 23$. Then graph the function.

$$f(x) = a(x-h)^2 + k$$

$$f(x) = \left(x^2 + 10x + \boxed{}\right) - \boxed{} + 23$$

$$= (x+5)^2 - 2$$

$$= \left[x - \left(\boxed{}\right)\right]^2 + \left(\boxed{}\right)$$

Vertex: $(-5, -2)$

Axis of symmetry: $x = \boxed{}$

Minimum: -2

Some points on the graph:

$(-5, -2)$

$(-4, -1)$

$(-2, 7)$

$(-7, 2)$

$(-8, 7)$

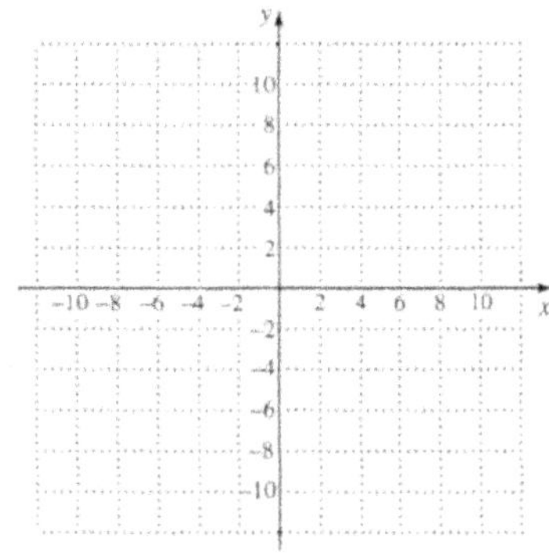

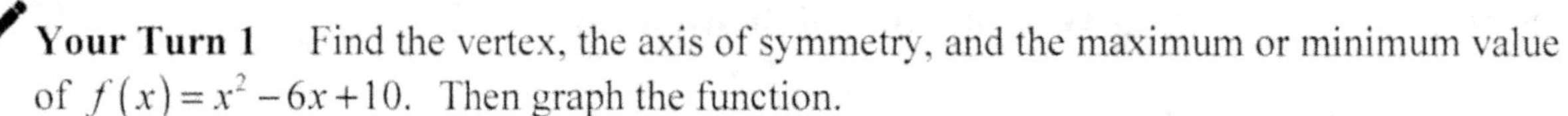

Your Turn 1 Find the vertex, the axis of symmetry, and the maximum or minimum value of $f(x) = x^2 - 6x + 10$. Then graph the function.

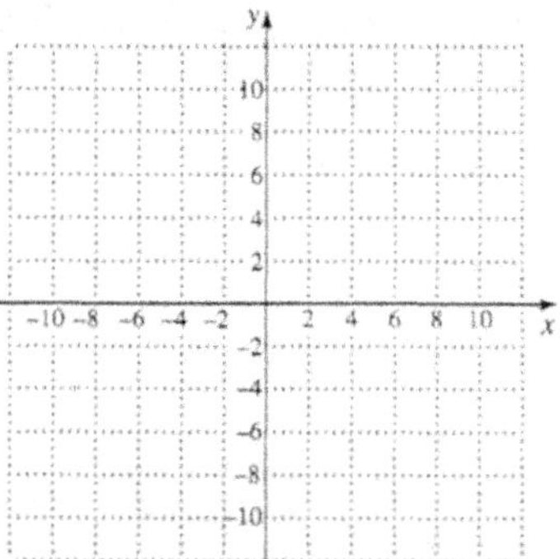

Example 2 Find the vertex, the axis of symmetry, and the maximum or minimum value of $g(x) = \dfrac{x^2}{2} - 4x + 8$. Then graph the function.

$$g(x) = a(x-h)^2 + k$$

$$g(x) = \frac{x^2}{2} - 4x + 8 = \frac{1}{2}\left(x^2 - 8x\right) + 8$$

$$= \frac{1}{2}\left(x^2 - 8x + \boxed{} - \boxed{}\right) + 8$$

$$= \frac{1}{2}\left(x^2 - 8x + 16\right) - \frac{1}{2}\cdot 16 + 8$$

$$= \frac{1}{2}\left(\boxed{}\right)^2 + \boxed{}, \text{ or } \frac{1}{2}(x-4)^2$$

Vertex: $(4, 0)$

Axis of symmetry: $x = \boxed{}$

Minimum: $\boxed{}$

Some points on the graph:

$(4, 0)$

$(0, 8)$

$(2, 2)$

$(6, 2)$

$(8, 8)$

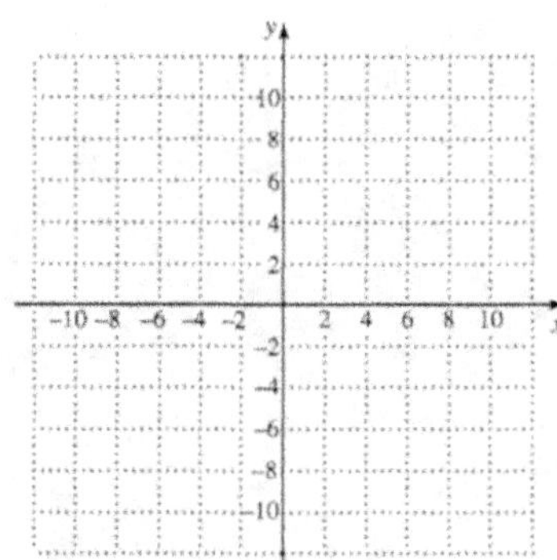

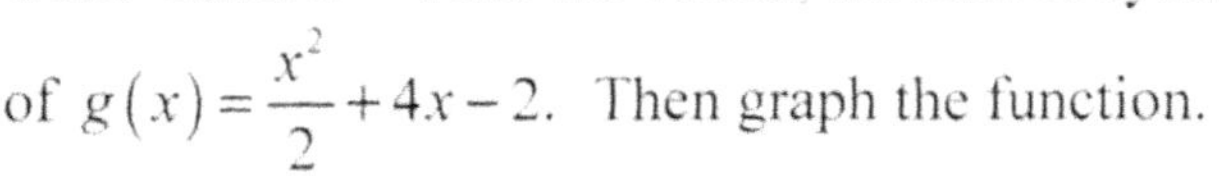 **Your Turn 2**   Find the vertex, the axis of symmetry, and the maximum or minimum value of $g(x) = \dfrac{x^2}{2} + 4x - 2$. Then graph the function.

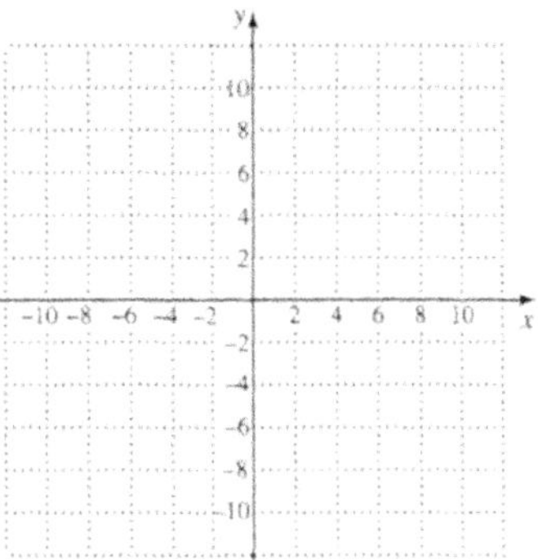

Example 3   Find the vertex, the axis of symmetry, and the maximum or minimum value of $f(x) = -2x^2 + 10x - \dfrac{23}{2}$. Then graph the function.

$$f(x) = -2x^2 + 10x - \frac{23}{2}$$

$$= -2\left(\boxed{}\right) - \frac{23}{2}$$

$$= -2\left(x^2 - 5x + \frac{25}{4} - \frac{25}{4}\right) - \frac{23}{2} \qquad\qquad \left(-\frac{5}{2}\right)^2 = \frac{25}{4}$$

$$= -2\left(x^2 - 5x + \frac{25}{4}\right) + \boxed{} - \frac{23}{2}$$

$$f(x) = -2\left(\boxed{}\right)^2 + \boxed{}$$

Vertex: $\left(\dfrac{5}{2}, 1\right)$

Axis of symmetry: $x = \boxed{}$

Maximum value is $\boxed{}$ when $x = \dfrac{5}{2}$.

Some points on the graph:

$$\left(\frac{5}{2}, 1\right)$$

$$\left(2, \frac{1}{2}\right)$$

$$\left(3, \frac{1}{2}\right)$$

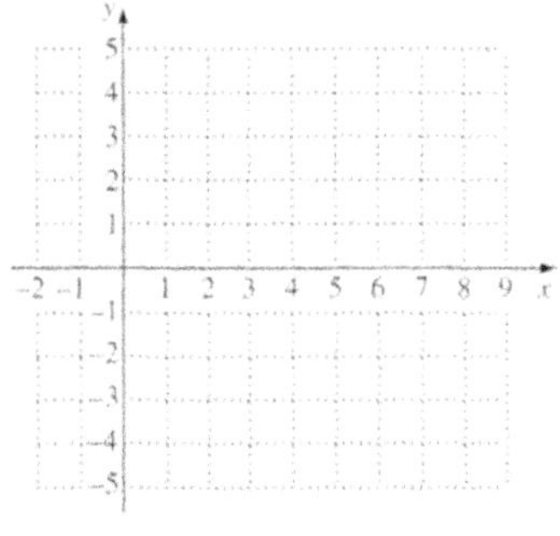

Your Turn 3   Find the vertex, the axis of symmetry, and the maximum or minimum value of

$h(x) = -2x^2 + 6x - \dfrac{3}{2}$.   Then graph the function.

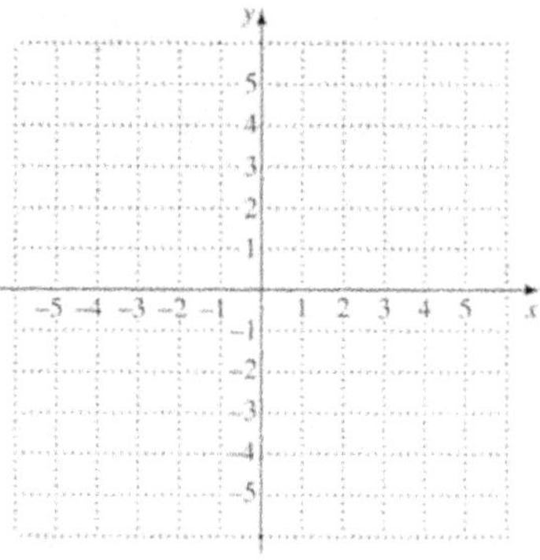

The Vertex of a Parabola

The vertex of the graph of $f(x) = ax^2 + bx + c$ is

$$\left(-\frac{b}{2a},\, f\left(-\frac{b}{2a}\right)\right).$$

We calculate the x-coordinate. Then we substitute to find the y-coordinate.

Example 4   For the function $f(x) = -x^2 + 14x - 47$:

a) Find the vertex.

$$a = -1 \qquad b = 14 \qquad c = -47$$

$$\text{Vertex:} \left(\frac{-14}{2(-1)},\quad \right) = (7,\quad)$$

$$f(7) = -(7)^2 + 14 \cdot 7 - 47$$
$$= -49 + 98 - 47 = \boxed{}$$

The vertex is $(7, 2)$.

b) Determine whether there is a maximum or a minimum value and find that value.

Since the coefficient of x^2 is negative, the function has a ________________ value at

maximum / minimum

the vertex.

Maximum value: $\boxed{}$

c) Find the range.

Vertex: $(7, 2)$

Maximum value: 2

Range: $\left(-\infty, \boxed{}\right]$

(continued)

d) On what intervals is the function increasing? decreasing?

Increasing as we approach the vertex from the left: $\left(-\infty, \boxed{}\right)$

Decreasing as we move away from the vertex on the right: $\left(\boxed{}, \infty\right)$

Your Turn 4 For the function $f(x) = -x^2 + 12x - 35$:

a) Find the vertex.

b) Determine whether there is a maximum or a minimum value and find that value.

c) Find the range.

d) On what intervals is the function increasing? decreasing?

Example 5 *Maximizing Area.* A landscaper has enough stone to enclose a rectangular koi pond next to an existing garden wall of the Englemans' house with 24 ft of stone wall. If the garden wall forms one side of the rectangle, what is the maximum area that the landscaper can enclose? What dimensions of the koi pond will yield this area?

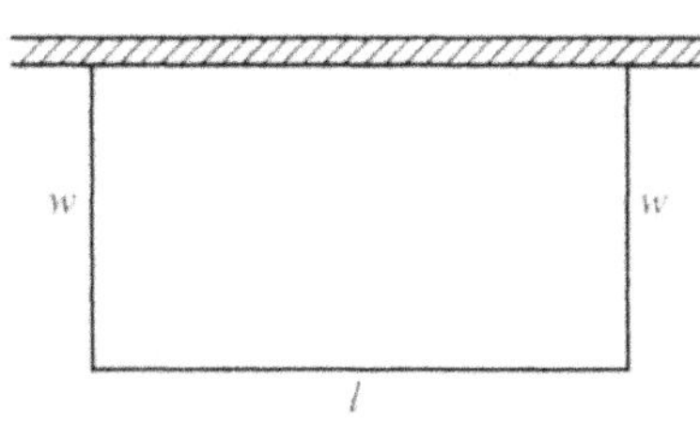

$2w + l = \boxed{}$

$l = 24 - 2w$

$A = l \cdot w$

$A = \left(\boxed{}\right) \cdot w$

$A = 24w - 2w^2$

$A = -2w^2 + 24w \qquad A(w) = aw^2 + bw + c$

Maximum occurs at the vertex:

$w = \dfrac{-b}{2a} = \dfrac{\boxed{}}{2(-2)} = \dfrac{-24}{-4} = \boxed{}$

$l = 24 - 2w = 24 - 2\left(\boxed{}\right) = 24 - 12 = \boxed{}$

$A = 12\ \text{ft} \cdot 6\ \text{ft} = 72\ \text{ft}^2$

The maximum possible area is $\boxed{}$ ft² when the koi pond is $\boxed{}$ ft wide and $\boxed{}$ ft long.

Your Turn 5   A landscaper has enough stone to enclose a rectangular pond next to an existing garden wall with 40 ft of stone wall. If the garden wall forms one side of the rectangle, what is the maximum area that the landscaper can enclose? What dimensions of the pond will yield this area?

Example 6   *Height of a Rocket.* A model rocket is launched with an initial velocity of 100 ft/sec from the top of a hill that is 20 ft high. Its height, in feet, t seconds after it has been launched is given by the function $s(t) = -16t^2 + 100t + 20$. Determine the time at which the rocket reaches its maximum height and find the maximum height.

1. & 2.   Familiarize and Translate.

$$s(t) = -16t^2 + 100t + 20$$

3.   Carry out.

$$a = -16 \qquad b = 100 \qquad c = 20$$

$$t = -\frac{b}{2a} = -\frac{\boxed{}}{2(-16)} = -\frac{100}{-32}$$

$$= \boxed{}$$

$$s(3.125) = -16\left(\boxed{}\right)^2 + 100\left(\boxed{}\right) + 20$$

$$= \boxed{}$$

4.   Check. We can check the answer by completing the square on the function. If we did that, we would get

$$s(t) = -16(t - 3.125)^2 + 176.25.$$

This confirms that the vertex is $(3.125, 176.25)$.

5.   State. The rocket reaches a maximum height of $\boxed{}$ ft $\boxed{}$ sec after it has been launched.

Your Turn 6   A model rocket is launched with an initial velocity of 80 ft/sec from the top of a hill that is 15 ft high. Its height, in feet, t seconds after it has been launched is given by the function $s(t) = -16t^2 + 80t + 15$. Determine the time at which the rocket reaches its maximum height and find the maximum height.

Example 7   Jared drops a screwdriver from the top of an elevator shaft. Exactly 5 sec later, he hears the sound of the screwdriver hitting the bottom of the shaft. The speed of sound is 1100 ft/sec. How tall is the elevator shaft?

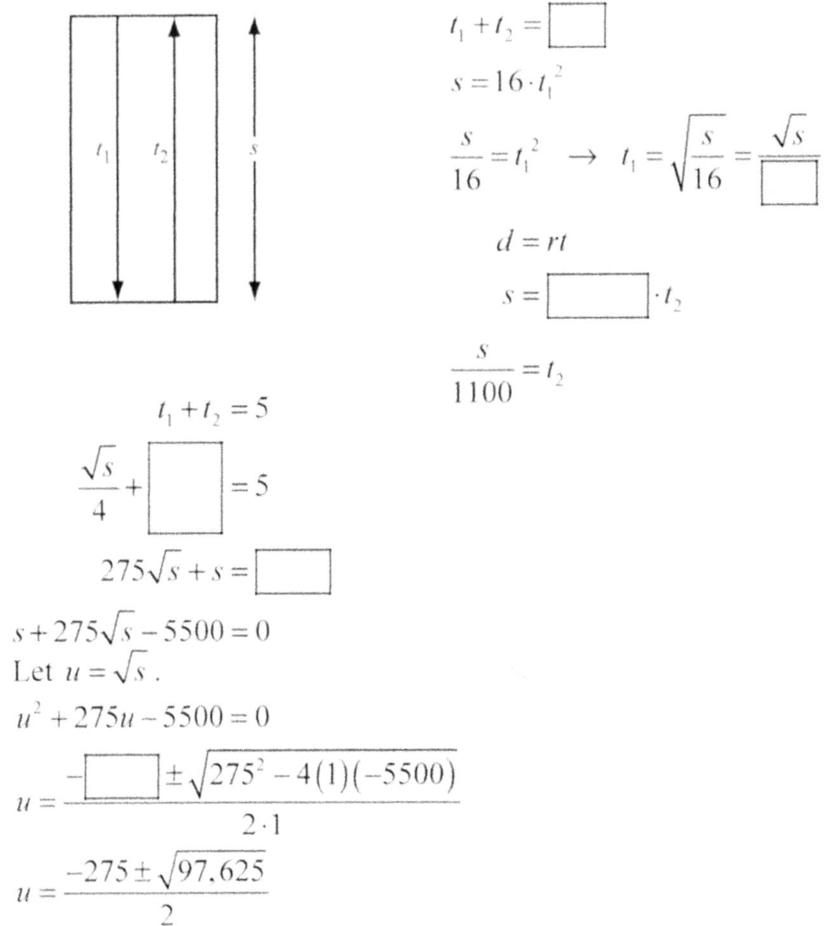

$$t_1 + t_2 = \boxed{}$$

$$s = 16 \cdot t_1^{\,2}$$

$$\frac{s}{16} = t_1^{\,2} \quad \rightarrow \quad t_1 = \sqrt{\frac{s}{16}} = \frac{\sqrt{s}}{\boxed{}}$$

$$d = rt$$

$$s = \boxed{} \cdot t_2$$

$$\frac{s}{1100} = t_2$$

$$t_1 + t_2 = 5$$

$$\frac{\sqrt{s}}{4} + \boxed{} = 5$$

$$275\sqrt{s} + s = \boxed{}$$

$$s + 275\sqrt{s} - 5500 = 0$$

Let $u = \sqrt{s}$.

$$u^2 + 275u - 5500 = 0$$

$$u = \frac{-\boxed{} \pm \sqrt{275^2 - 4(1)(-5500)}}{2 \cdot 1}$$

$$u = \frac{-275 \pm \sqrt{97{,}625}}{2}$$

(continued)

$$u = \frac{-275 + \sqrt{97,625}}{2}$$ We want only the positive solution.

$$u \approx \boxed{}$$

$$\sqrt{s} \approx 18.725$$

$$s \approx 350.6 \text{ ft}$$

The height of the elevator shaft is about $\boxed{}$ ft.

Your Turn 7   Lexi drops a screwdriver from the top of an elevator shaft. Exactly 4 sec later, she hears the sound of the screwdriver hitting the bottom of the shaft. The speed of sound is 1100 ft/sec. How tall is the elevator shaft?

Practice Exercises

Find the vertex, the axis of symmetry, and the maximum or minimum value of the function. Then graph the function.

1. $f(x) = x^2 + 8x + 15$

Vertex: (,)

Axis of symmetry:

_________________ value:
 maximum / minimum

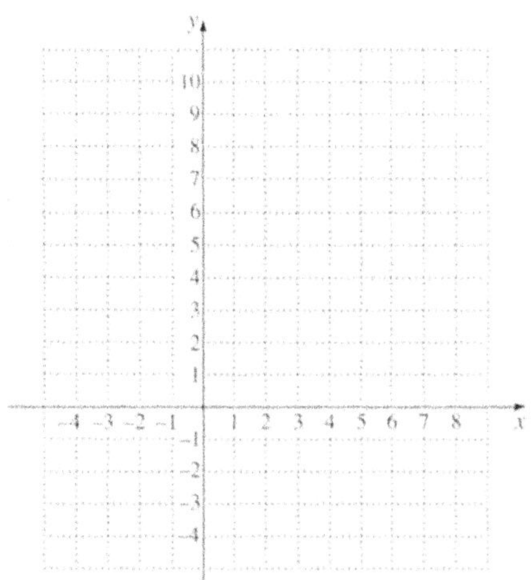

2. $g(x) = -2x^2 + 14x - \dfrac{31}{2}$

Vertex: (,)

Axis of symmetry:

_________________ value:
 maximum / minimum

For the function:

a) Find the vertex.

b) Determine whether there is a maximum or a minimum value and find that value.

c) Find the range.

d) On what intervals is the function increasing? decreasing?

3. $h(x) = -x^2 - 2x + 11$

a) Vertex: (,)

b) _________________ value:
  maximum / minimum

c) Range:

d) Increasing:

  Decreasing:

4. $f(x) = \dfrac{1}{2}x^2 - 5x + \dfrac{7}{2}$

a) Vertex: (,)

b) _________________ value:
  maximum / minimum

c) Range:

d) Increasing:

  Decreasing:

5. An agriculture class at the community college decides to enclose a rectangular plot for an experiment, using a side of the building as one side of the rectangle. What is the maximum area that the class can enclose using 92 feet of fence? What should the dimensions of the plot be in order to yield this area?

6. A model rocket is launched with an initial velocity of 60 ft/sec from the top of a hill that is 30 ft high. Its height, in feet, t seconds after it has been launched is given by the function $s(t) = -16t^2 + 60t + 30$. Determine the time at which the rocket reaches its maximum height and find the maximum height.

7. Harry drops his phone from the top of a storage bin. Exactly 2 sec later, he hears the sound of the phone hitting the bottom of the bin. The speed of sound is 1100 ft/sec. How tall is the bin? (Hint: See the formulas $s = 16t^2$ and $\text{distance} = \text{rate} \times \text{time}$ in Example 7.)

Section 8.1 Polynomial Functions and Models

Polynomial Function

A **polynomial function** P is given by

$$P(x) = a_n x^n + a_{n-1} x^{n-1} + a_{n-2} x^{n-2} + \cdots + a_1 x + a_0,$$

where the coefficients $a_n, a_{n-1}, \ldots, a_1, a_0$ are real numbers and the exponents are whole numbers.

The Leading-Term Test

If $a_n x^n$ is the leading term of a polynomial function, then the behavior of the graph as $x \to \infty$ or as $x \to -\infty$ can be described in one of the four following ways:

n	$a_n > 0$	$a_n < 0$
Even		
Odd		

The 〰 portion of the graph is not determined by this test.

Example 1 Use the leading-term test to match the functions below with one of the graphs on the next page.

a) $f(x) = 3x^4 - 2x^3 + 3$

Exponent: <u>even</u> / odd (circle one)
Coefficient: <u>positive</u> / negative (circle one)
Graph: <u>A, B, C, or D</u> (circle one)

b) $f(x) = -5x^3 - x^2 + 4x + 2$

Exponent: even / <u>odd</u> (circle one)
Coefficient: positive / <u>negative</u> (circle one)
Graph: <u>A, B, C, D</u> (circle one)

c) $f(x) = x^5 + \dfrac{1}{4}x + 1$

Exponent: even / <u>odd</u> (circle one)
Coefficient: <u>positive</u> / negative (circle one)
Graph: <u>A, B, C, D</u> (circle one)

d) $f(x) = -x^6 + x^5 - 4x^3$

Exponent: <u>even</u> / odd (circle one)
Coefficient: positive / <u>negative</u> (circle one)
Graph: <u>A, B, C, D</u> (circle one)

(continued)

A. 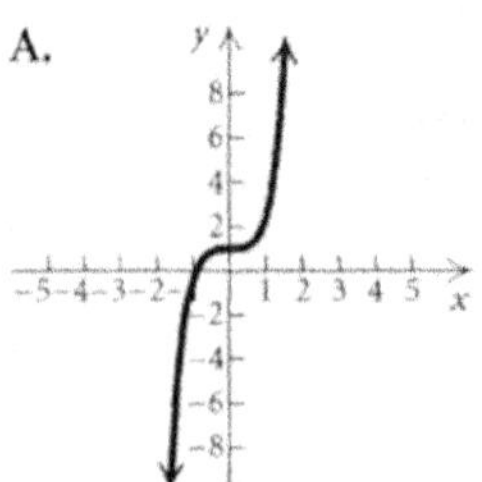 **B.** 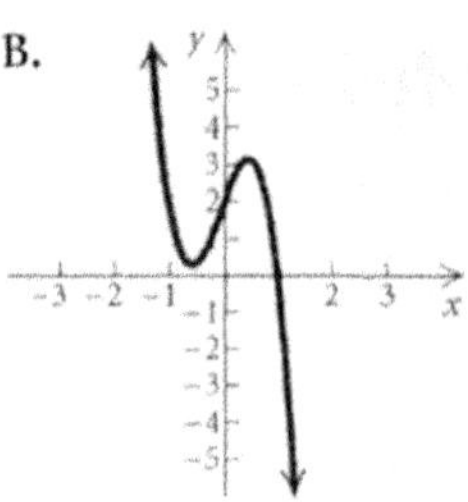 **C.** 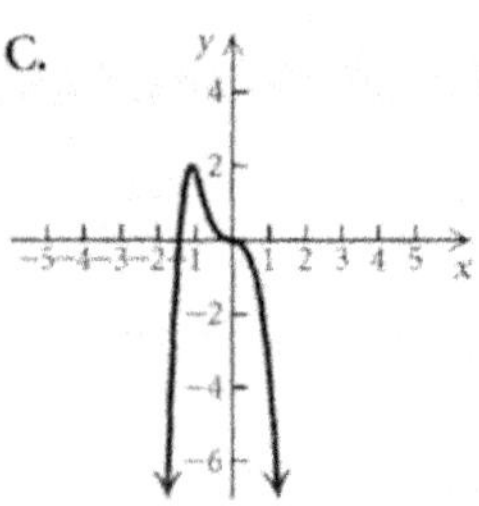 **D.** 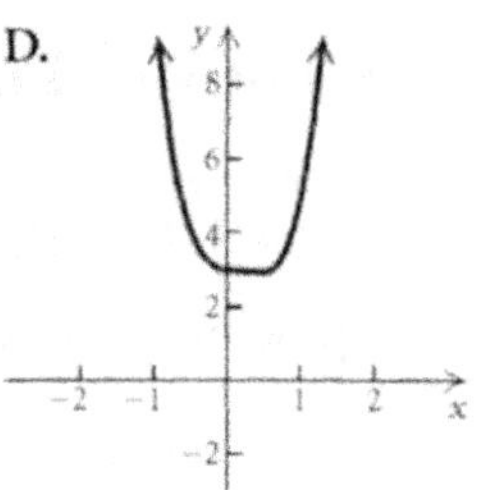

Your Turn 1 Use the leading-term test to match the functions below with one of the graphs below.

a) $f(x) = -3x^3 + 2x^2 - x + 1$

b) $f(x) = 2x^2 - x + 1$

c) $f(x) = -x^4 + 2x^3 - x + 3$

d) $f(x) = x^3 + 3x^2 - 2$

A. 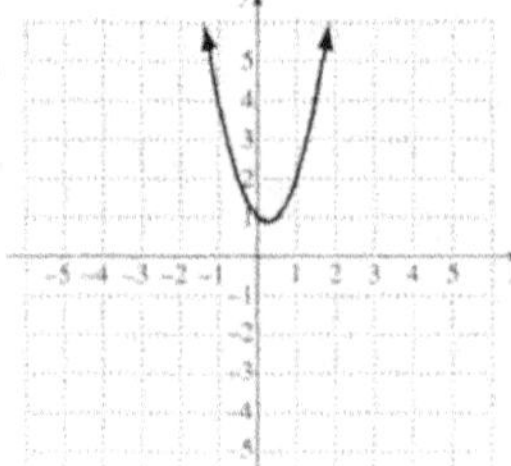 **B.** 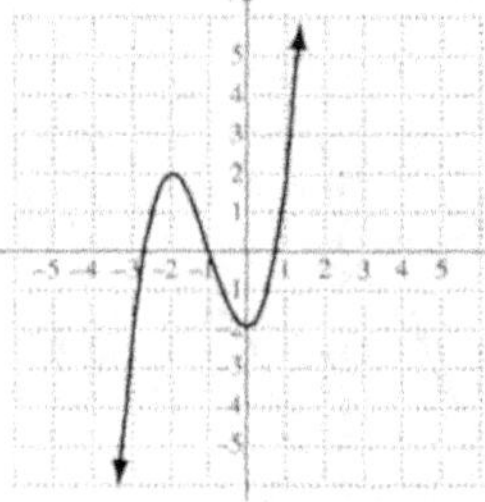 **C.** 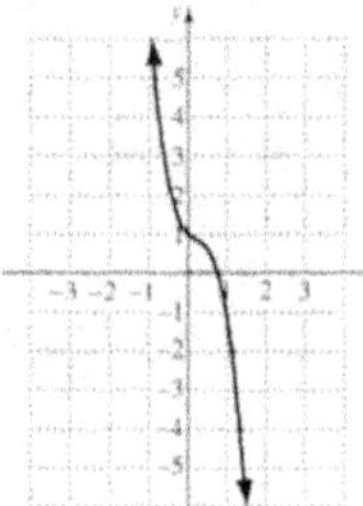 **D.** 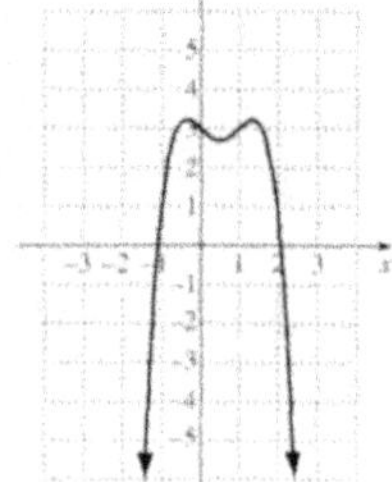

Example 2 Consider $P(x) = x^3 + x^2 - 17x + 15$. Determine whether each of the numbers 2 and -5 is a zero of $P(x)$.

If $P(c) = 0$, then c is a zero of the polynomial.

Let us first evaluate $P(2)$.

$P(2) = 2^3 + 2^2 - 17 \cdot 2 + 15$

$P(2) = 8 + \boxed{} - 34 + 15$

$P(2) = \boxed{}$

Since $P(2) \neq 0$, 2 <u>is / is not</u> (circle one) a zero of this function.

(continued)

Let us now evaluate $P(-5)$.

$$P(-5) = (-5)^3 + (-5)^2 - 17 \cdot (-5) + 15$$

$$P(-5) = -125 + 25 + 85 + 15$$

$$P(-5) = 0$$

Since $P(-5) = 0$, -5 <u>is / is not</u> (circle one) a zero of this function.

Your Turn 2 Consider $P(x) = x^3 - x^2 - 10x - 8$. Determine whether each of the numbers -2 and 3 is a zero of $P(x)$.

Example 3 Find the zeros of

$$f(x) = 5(x-2)(x-2)(x-2)(x+1)$$

$$= 5(x-2)^3(x+1).$$

According to the principle of zero products, either
$$x - 2 = 0 \quad or \quad x + 1 = 0$$
$$x = 2 \quad or \quad x = -1 \quad \text{So the zeros of } f(x) \text{ are } \boxed{} \text{ and } \boxed{}.$$

Your Turn 3 Find the zeros of

$$f(x) = 3(x-4)(x-4)(x-4)(x+2)$$

$$= 3(x-4)^3(x+2).$$

Example 4   Find the zeros of
$$g(x) = -(x-1)(x-1)(x+2)(x+2)$$
$$= -(x-1)^2(x+2)^2.$$

To find the zeros, we set $g(x) = 0$, and use the principle of zero products.

$$0 = -(x-1)^2(x+2)^2$$
$$x-1 = 0 \quad or \quad x+2 = 0$$
$$x = 1 \quad or \qquad x = -2 \qquad \text{The zeros are } \boxed{} \text{ and } \boxed{}.$$

Your Turn 4   Find the zeros of
$$g(x) = -2(x-5)(x-5)(x+3)(x+3)$$
$$= -2(x-5)^2(x+3)^2.$$

Example 5   Find the zeros of $f(x) = x^3 - 2x^2 - 9x + 18$.

$$f(x) = x^3 - 2x^2 - 9x + 18$$
$$= x^2\left(x - \boxed{}\right) - 9\left(\boxed{} - 2\right)$$
$$= (x-2)(x^2 - 9)$$
$$= (x-2)(x+3)\left(x - \boxed{}\right)$$

To find the zeros, we solve the equation $f(x) = 0$ using the principle of zero products.

$$0 = (x-2)(x+3)(x-3)$$
$$x-2 = 0 \quad or \quad x+3 = 0 \quad or \quad x-3 = 0$$
$$x = 2 \quad or \qquad x = -3 \quad or \qquad x = 3$$

The solutions of $f(x) = 0$ are 2, $\boxed{}$, and 3.

The zeros of $f(x)$ are $\boxed{}$, -3, and $\boxed{}$.

Your Turn 5   Find the zeros of $f(x) = x^3 + x^2 - 4x - 4$.

Example 6   Find the zeros of $f(x) = x^4 + 4x^2 - 45$.

To find the zeros, we solve the equation $f(x) = 0$.

$$0 = x^4 + 4x^2 - 45$$
$$0 = \left(x^2 - \boxed{}\right)\left(x^2 + 9\right)$$

Using the principle of zero products, this becomes

$$x^2 - 5 = 0 \quad or \quad x^2 + 9 = 0$$
$$x^2 = 5 \quad or \quad x^2 = -9$$
$$x = \pm\sqrt{5} \quad or \quad x = \pm\sqrt{-9}$$
$$x = \pm 3i$$

The solutions of $f(x) = 0$ are $-\sqrt{5}$, $\sqrt{5}$, $\boxed{}$, and $3i$.

The zeros of $f(x)$ are $-\sqrt{5}$, $\boxed{}$, $-3i$, and $\boxed{}$.

Your Turn 6   Find the zeros of $g(x) = x^4 - 2x^2 - 24$.

Example 7 The polynomial function

$$M(t) = 0.5t^4 + 3.45t^3 - 96.65t^2 + 347.7t$$

can be used to estimate the number of milligrams of the pain relief medication ibuprofen in the bloodstream t hours after 400 mg of the medication has been taken. Find the number of milligrams in the bloodstream at $t = 0$, 0.5, 1, 1.5, and so on, up to 6 hr. Round the function values to the nearest tenth.

Using a calculator, we compute the function values:

$M(0) = 0$, $\qquad\qquad$ $M(3.5) = 255.9$,

$M(0.5) = 150.2$, $\qquad$ $M(4) = \boxed{}$,

$M(1) = 255$, $\qquad\qquad$ $M(4.5) = 126.9$,

$M(1.5) = 318.3$, $\qquad$ $M(5) = 66$,

$M(2) = \boxed{}$, $\qquad\quad$ $M(5.5) = 20.2$,

$M(2.5) = 338.6$, $\qquad$ $M(6) = 0$.

$M(3) = 306.9$,

Your Turn 7 A model rocket is launched with an initial velocity of 80 ft/sec from the top of a hill that is 35 ft high. Its height, in feet, t seconds after it has been launched is given by the function $S(t) = -16t^2 + 80t + 35$. Find the height of the rocket 2 seconds after it is launched.

Practice Exercises

1. Use the leading-term test to match the functions below with one of the graphs below.

   a) $f(x) = -2x^2 + 3x + 2$

   b) $f(x) = x^3 - 2x^2 + 3x$

   c) $f(x) = -3x^3 - x + 1$

   d) $f(x) = x^6 + 3x^4 - x - 3$

   A. 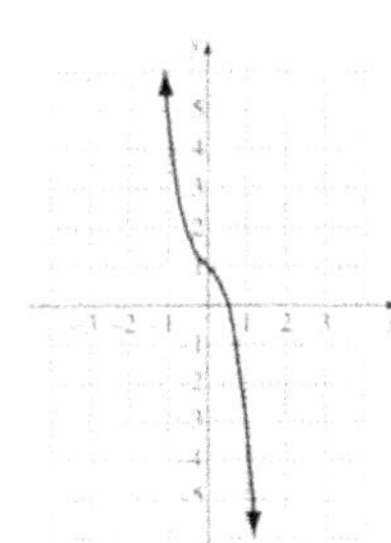   **B.** 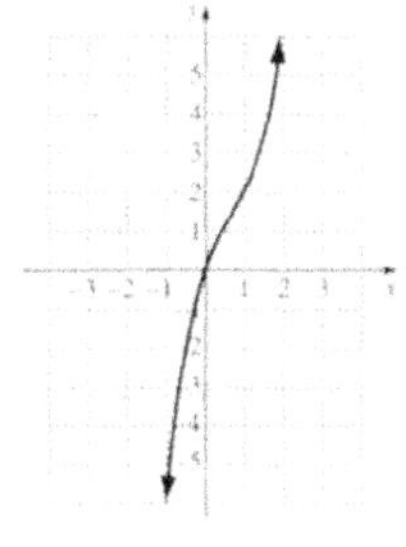   **C.** 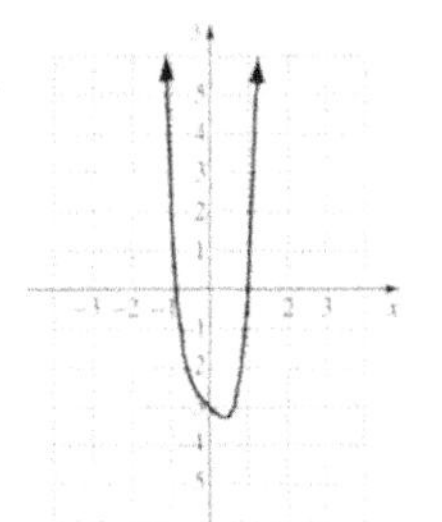   **D.** 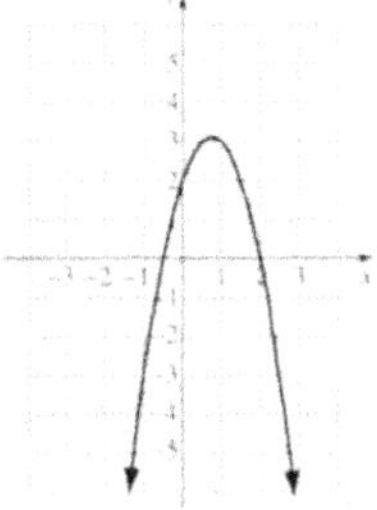

2. Use substitution to determine if each of the numbers -1 and -2 is a zero of
   $f(x) = x^3 + 7x^2 + 4x - 12$.

3. Use substitution to determine if each of the numbers -1 and 1 is a zero of
   $f(x) = x^3 + 2x^2 - 5x - 6$.

Find the zeros of the polynomial function.

4. $f(x) = 2(x-3)^2(x+4)$

5. $f(x) = -3(x+1)^3(x-6)$

6. $f(x) = (x-1)(x+5)(x+3)$

7. $f(x) = -(x-2)(x-2)(x+8)$

8. $f(x) = x^3 + 4x^2 - x - 4$

9. $f(x) = 2x^3 + 2x^2 - 32x - 32$

10. $f(x) = x^4 - 2x^2 - 3$

11. $f(x) = x^4 - 4x^2 - 12$

12. Gentamicin is an antibiotic used by veterinarians. The concentration in micrograms per milliliter (mcg/mL) of Gentamicin in a horse's bloodstream t hours after injection can be approximated by the polynomial function

$$C(t) = -0.005t^4 + 0.003t^3 + 0.35t^2 + 0.5t.$$

Find the amount of Gentamicin in the bloodstream at $t = 2$ and at $t = 4$. Round to the nearest tenth.

Section 8.2 Graphing Polynomial Functions

To Graph a Polynomial Function:

1. Use the leading-term test to determine the end behavior.

2. Find the zeros of the function by solving $f(x) = 0$. Any real zeros are the first coordinates of the x-intercepts.

3. Use the x-intercepts (zeros) to divide the x-axis into intervals and choose a test point in each interval to determine the sign of all function values in that interval.

4. Find $f(0)$. This gives the y-intercept of the function.

5. If necessary, find additional function values to determine the general shape of the graph and then draw the graph.

6. As a partial check, use the facts that the graph has at most n x-intercepts and at most $n-1$ turning points. Multiplicity of zeros can also be considered in order to check where the graph crosses or is tangent to the x-axis.

Example 1   Graph the polynomial function $h(x) = -2x^4 + 3x^3$.

1. Use the leading-term test.
    Degree: 4, even; coefficient: $-2 < 0$
    As $x \to \infty$ and as $x \to -\infty$, $h(x) \to \boxed{}$.

2. Find the zeros. Solve $h(x) = 0$.

$$-2x^4 + 3x^3 = 0$$

$$-x^3(2x - 3) = 0$$

$$\boxed{} = 0 \quad or \quad 2x - 3 = 0$$

$$x = 0 \quad or \qquad x = \boxed{}$$

3.

Choose a test value from each interval and find $h(x)$.

Interval	$(-\infty, 0)$	$\left(0, \dfrac{3}{2}\right)$	$\left(\dfrac{3}{2}, \infty\right)$
Test value	-1	1	2
Function value, $h(x)$	-5	1	-8
Sign of $h(x)$	$-$	$+$	$-$
Location of points on graph	Above/Below x-axis (circle one)	Above/Below x-axis (circle one)	Above/Below x-axis (circle one)

(continued)

Three points on the graph are $\left(-1,\boxed{}\right)$, $(1,1)$, and $\left(\boxed{},-8\right)$.

4. To determine the y-intercept, find $h(0)$.

$$h(x) = -2x^4 + 3x^3$$

$$h(0) = -2\cdot 0^4 + 3\cdot 0^3 = \boxed{}$$

y-intercept: $(0, 0)$

5. Find additional function values and draw the graph.

x	$h(x)$
-1.5	-20.25
-0.5	-0.5
0.5	$\boxed{}$
2.5	-31.25

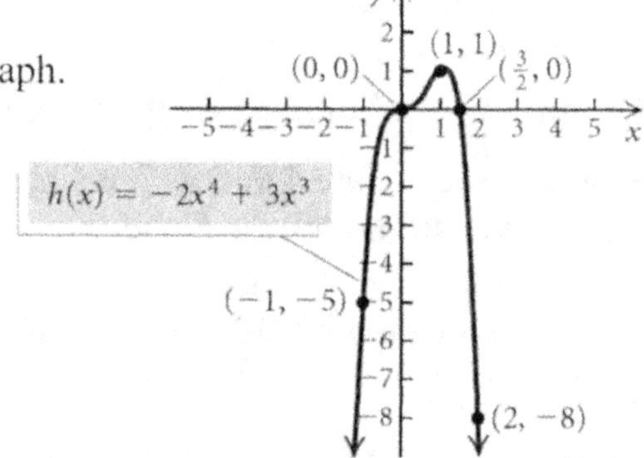

Your Turn 1 Graph the polynomial function $f(x) = x^4 - 2x^3$.

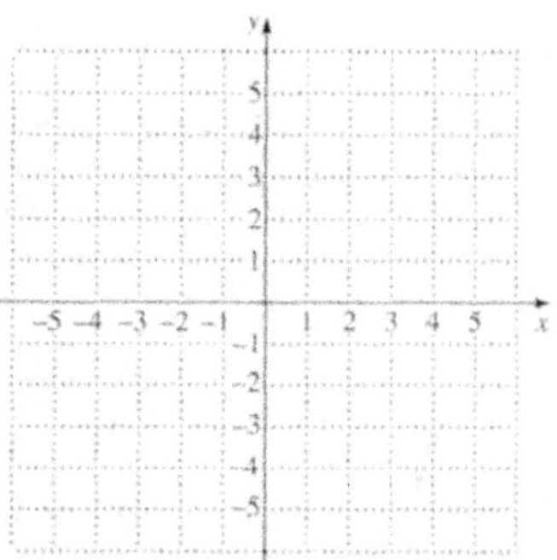

Example 2 Graph the polynomial function $f(x) = 2x^3 + x^2 - 8x - 4$.

1. Use the leading-term test.
   Degree: 3, odd; coefficient: $2 > 0$
   As $x \to \infty$, $f(x) \to \infty$; as $x \to -\infty$, $f(x) \to -\infty$.

2. Find the zeros. Solve $f(x) = 0$.

$$2x^3 + x^2 - 8x - 4 = 0$$

$$x^2(2x+1) - 4(2x+1) = 0$$

$$(2x+1)\left(x^2 - \boxed{}\right) = 0$$

$$(2x+1)(x+2)(x-2) = 0$$

$$\boxed{} = 0 \quad or \quad x+2 = 0 \quad or \quad x-2 = 0$$

$$x = -\frac{1}{2} \quad or \quad x = \boxed{} \quad or \quad x = 2$$

(continued)

3.

Choose a test value from each interval and find $f(x)$.

Interval	$(-\infty, -2)$	$\left(\boxed{}, -\dfrac{1}{2}\right)$	$\left(-\dfrac{1}{2}, 2\right)$	$\left(\boxed{}, \infty\right)$
Test value	-3	-1	1	3
Function value, $f(x)$	-25	3	-9	35
Sign of $f(x)$	$-$	$+$	$-$	$+$
Location of points on graph	Above/Below x-axis (circle one)	Above/Below x-axis (circle one)	Above/Below x-axis (circle one)	Above/Below x-axis (circle one)

Four points on the graph are $(-3, -25)$, $\left(-1, \boxed{}\right)$, $(1, -9)$, and $\left(\boxed{}, 35\right)$.

4. To determine the y-intercept, find $f(0)$.

$$f(x) = 2x^3 + x^2 - 8x - 4$$

$$f(0) = 2 \cdot 0^3 + 0^2 - 8 \cdot 0 - 4 = \boxed{}$$

y-intercept: $\left(0, \boxed{}\right)$

5. Find additional function values and draw the graph.

x	$f(x)$
-2.5	-9
-1.5	3.5
0.5	-7.5
1.5	-7

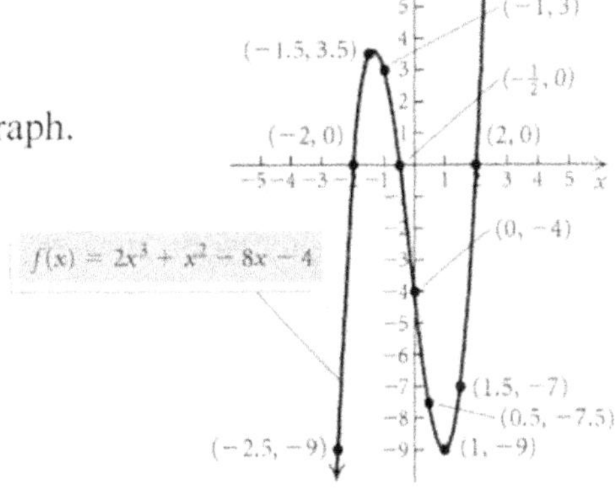

Your Turn 2 Graph the polynomial function $f(x) = x^3 + 2x^2 - x - 2$.

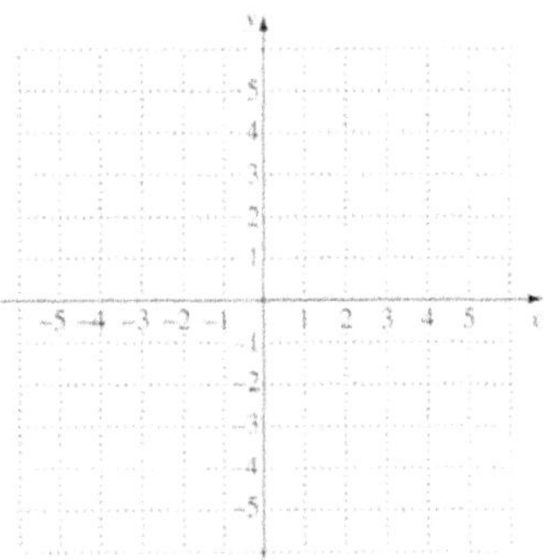

Example 3 Graph the polynomial function $g(x) = x^4 - 7x^3 + 12x^2 + 4x - 16$
$$= (x+1)(x-2)^2(x-4).$$

1. Use the leading-term test.
   Degree: 4, even; coefficient: $1 > 0$
   As $x \to \infty$ and as $x \to -\infty$, $g(x) \to \infty$.

2. Find the zeros. Solve $g(x) = 0$.

$$(x+1)(x-2)^2(x-4) = 0$$

$x + 1 = 0$ *or* $(x-2)^2 = 0$ *or* $x - 4 = 0$

$x = -1$ *or* $\qquad x = 2$ *or* $\qquad x = 4$

3.

Choose a test value from each interval and find $g(x)$.

Interval	$(-\infty, -1)$	$\left(\boxed{}, 2\right)$	$(2, 4)$	$\left(\boxed{}, \infty\right)$
Test value	-1.25	1	3	4.25
Function value, $g(x)$	≈ 13.9	-6	-4	≈ 6.6
Sign of $g(x)$	$+$	$-$	$-$	$+$
Location of points on graph	Above/Below x-axis (circle one)	Above/Below x-axis (circle one)	Above/Below x-axis (circle one)	Above/Below x-axis (circle one)

4. To determine the y-intercept, find $g(0)$.

$$g(0) = 0^4 - 7\cdot 0^3 + 12\cdot 0^2 + 4\cdot 0 - 16 = \boxed{}$$

y-intercept: $\left(\boxed{}, -16\right)$

5. Find additional function values and draw the graph.

x	$g(x)$
-0.5	-14.1
0.5	-11.8
1.5	-1.6
2.5	-1.3
3.5	-5.1

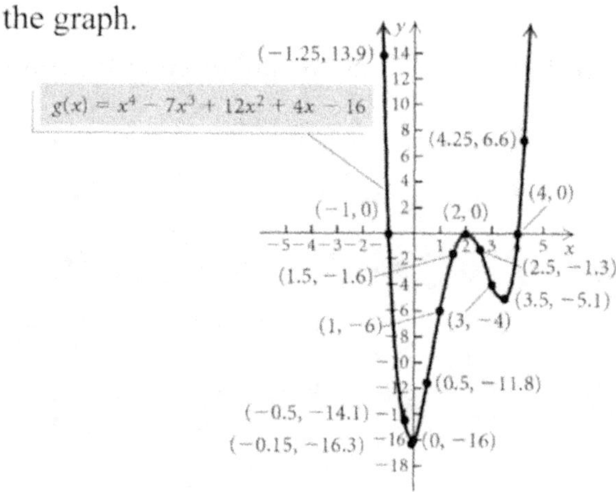

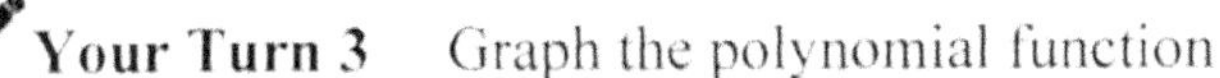

Your Turn 3   Graph the polynomial function

$$g(x) = x^4 - 4x^3 + 2x^2 + 4x - 3$$

$$= (x-1)^2 (x+1)(x-3).$$

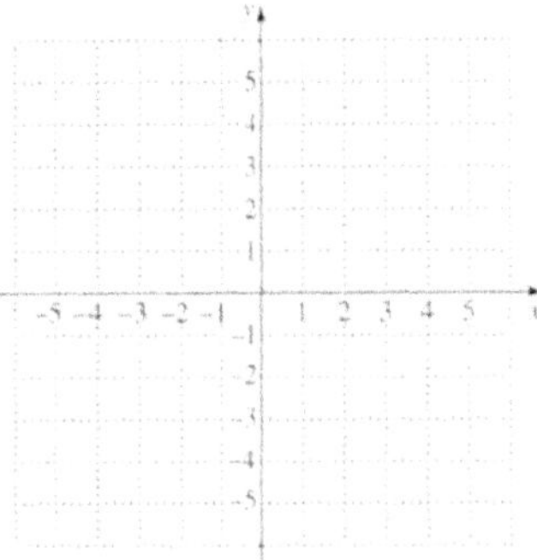

The Intermediate Value Theorem

For any polynomial function $P(x)$ with real coefficients, suppose that for $a \neq b$,

$P(a)$ and $P(b)$ are of opposite signs. Then the function has a real zero between a and b.

The intermediate value theorem *cannot* be used to determine whether there is a real zero between a and b when $P(a)$ and $P(b)$ have the *same* sign.

Example 4

a) Using the intermediate value theorem, determine, if possible, whether $f(x) = x^3 + x^2 - 6x$ has a real zero between -4 and -2.

$$f(-4) = (-4)^3 + (-4)^2 - 6(-4) = \boxed{}$$

$$f(-2) = (-2)^3 + (-2)^2 - 6(-2) = \boxed{}$$

Because $f(-4)$ and $f(-2)$ have opposite signs, $f(x)$ <u>has / does not have</u> (circle one) a zero between -4 and -2.

b) Using the intermediate value theorem, determine, if possible, whether $f(x) = x^3 + x^2 - 6x$ has a real zero between -1 and 3.

$$f(-1) = (-1)^3 + (-1)^2 - 6(-1) = \boxed{}$$

$$f(3) = (3)^3 + (3)^2 - 6(3) = \boxed{}$$

Because $f(-1)$ and $f(3)$ have the same sign, it <u>is / is not</u> (circle one) possible to determine with the intermediate value theorem if there is a zero between -1 and 3.

(continued)

c) Using the intermediate value theorem, determine, if possible, whether $g(x) = \frac{1}{3}x^4 - x^3$ has a real zero between $-\frac{1}{2}$ and $\frac{1}{2}$.

$$g\left(-\frac{1}{2}\right) = \frac{1}{3}\left(-\frac{1}{2}\right)^4 - \left(-\frac{1}{2}\right)^3 = \boxed{}$$

$$g\left(\frac{1}{2}\right) = \frac{1}{3}\left(\frac{1}{2}\right)^4 - \left(\frac{1}{2}\right)^3 = \boxed{}$$

Because $g\left(-\frac{1}{2}\right)$ and $g\left(\frac{1}{2}\right)$ have opposite signs, $g(x)$ <u>has / does not have</u> (circle one) a zero between $-\frac{1}{2}$ and $\frac{1}{2}$.

d) Using the intermediate value theorem, determine, if possible, whether $g(x) = \frac{1}{3}x^4 - x^3$ has a real zero between 1 and 2.

$$g(1) = \frac{1}{3}(1)^4 - (1)^3 = \boxed{}$$

$$g(2) = \frac{1}{3}(2)^4 - (2)^3 = \boxed{}$$

Because $g(1)$ and $g(2)$ have the same sign, it <u>is / is not</u> (circle one) possible to determine with the intermediate value theorem if there is a zero between 1 and 2.

Your Turn 4 Use the intermediate value theorem to determine, if possible, whether the function f has a real zero between a and b.

a) $f(x) = x^3 - 4x^2 + 4x$;

  $a = 3, \quad b = 4$

b) $f(x) = x^3 - 4x^2 + 4x$;

  $a = -1, \quad b = 1$

c) $f(x) = 2x^4 - 9x^2$;

  $a = 2, \quad b = 3$

d) $f(x) = 2x^4 - 9x^2$;

  $a = 1, \quad b = 2$

Practice Exercises

Graph the polynomial function.

1. $g(x) = x^3 + x^2 - x - 1$

2. $f(x) = -3x^4 + 5x^3$

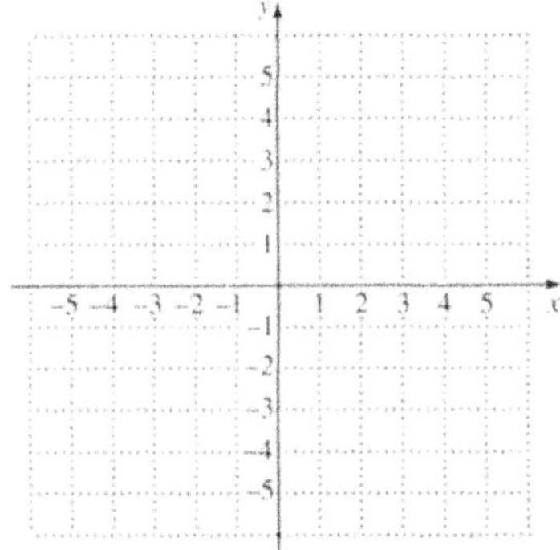

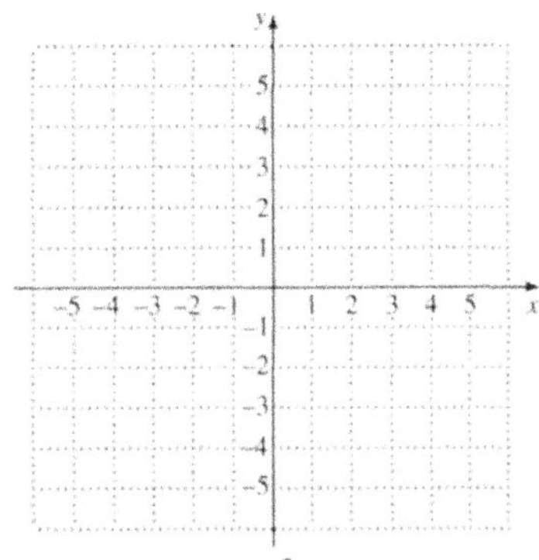

3. $f(x) = x^4 - 4x^2$

4. $h(x) = (1-x)^3$

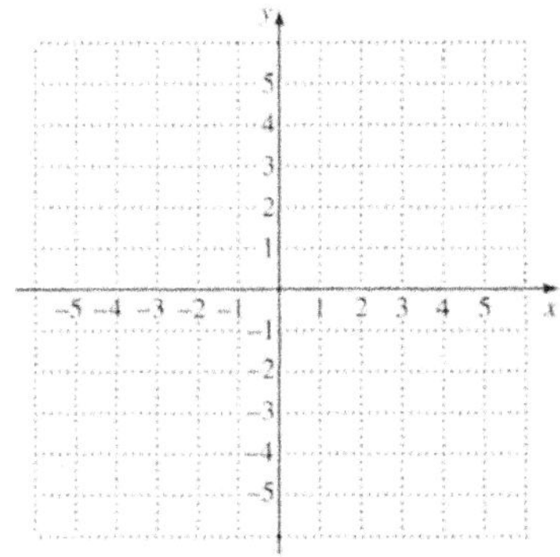

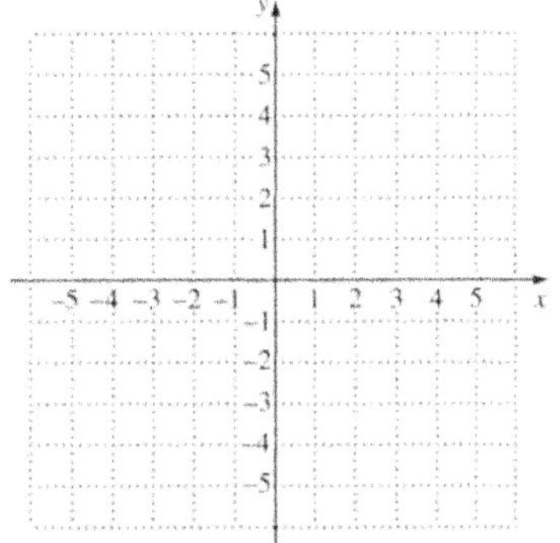

Use the intermediate value theorem to determine, if possible, whether the function f has a real zero between a and b.

5. $f(x) = x^3 - 3x^2 + 2x - 1$;

   $a = 2, \quad b = 3$

6. $f(x) = x^3 - 3x^2 + 2x - 1$;

   $a = -1, \quad b = 1$

7. $f(x) = x^4 - x^2 - 4$;

   $a = -1, \quad b = 0$

8. $f(x) = x^4 - x^2 - 4$;

   $a = 1, \quad b = 2$

9. $f(x) = x^3 + 2x^2 - 3$;

   $a = 0, \quad b = 2$

10. $f(x) = x^3 + 2x^2 - 3$;

   $a = -2, \quad b = -1$

Section 8.3 Polynomial Division; The Remainder Theorem and the Factor Theorem

Example 1 Divide to determine whether $x+1$ and $x-3$ are factors of x^3+2x^2-5x-6.

First Divisor: $x+1$

$$
\begin{array}{r}
\boxed{} \\
x+1 \enclose{longdiv}{x^3 \;+\; 2x^2 \;-\; 5x \;-\; 6} \\
x^3 \;+\; \boxed{} \\
x^2 \;-\; 5x \\
x^2 \;+\; x \\
\boxed{} \;-\; 6 \\
-6x \;-\; 6 \\
\boxed{} \quad \leftarrow \text{Remainder}
\end{array}
$$

Since the remainder is 0, we know that $x+1$ <u>is</u> / <u>is not</u> (circle one) a factor of x^3+2x^2-5x-6.

Second Divisor: $x-3$

$$
\begin{array}{r}
\boxed{} \\
x-3 \enclose{longdiv}{x^3 \;+\; 2x^2 \;-\; 5x \;-\; 6} \\
x^3 \;-\; 3x^2 \\
5x^2 \;-\; 5x \\
\boxed{} \;-\; 15x \\
10x \;-\; \boxed{} \\
10x \;-\; 30 \\
\boxed{} \quad \leftarrow \text{Remainder}
\end{array}
$$

Since the remainder is 24, we know that $x-3$ <u>is</u> / <u>is not</u> (circle one) a factor of x^3+2x^2-5x-6.

Your Turn 1   Divide to determine whether $x+2$ and $x+1$ are factors of x^3-2x^2-5x+6.

Example 2   Use synthetic division to find the quotient and the remainder:
$$\left(2x^3+7x^2-5\right)\div(x+3).$$

First, rewrite the dividend including the missing x-term, and write the divisor in the form $(x-c)$:
$$\left(2x^3+7x^2+0x-5\right)\div\left(x-(-3)\right)$$

Now perform synthetic division:

$$
\begin{array}{r|rrrr}
-3 & 2 & 7 & 0 & -5 \\
   &   & -6 & \square & 9 \\
\hline
   & 2 & \square & -3 & \square
\end{array}
$$

The quotient is $2x^2+x-3$. The remainder is $\boxed{}$.

Your Turn 2   Use synthetic division to find the quotient and the remainder:
$$\left(x^3+3x^2-5x+4\right)\div(x+1).$$

> **The Remainder Theorem**
>
> If a number c is substituted for x in the polynomial $f(x)$, then the result $f(c)$ is the remainder that would be obtained by dividing $f(x)$ by $x-c$.
>
> That is, if $f(x) = (x-c) \cdot Q(x) + R$, then $f(c) = R$.

Example 3 Given that $f(x) = 2x^5 - 3x^4 + x^3 - 2x^2 + x - 8$, find $f(10)$.

$f(10)$ is the remainder when $f(x)$ is divided by $x - 10$.

We can use synthetic division to find $f(10)$.

$$
\begin{array}{r|rrrrrr}
10 & 2 & -3 & 1 & -2 & 1 & -8 \\
   &   & 20 & \boxed{} & 1710 & 17{,}080 & 170{,}810 \\
\hline
   & 2 & 17 & 171 & \boxed{} & 17{,}081 & \boxed{}
\end{array}
$$

Thus, $f(10) = \boxed{}$.

Your Turn 3 Given that $f(x) = 2x^4 - x^3 + 3x^2 - 2x - 7$, find $f(3)$.

Example 4 Determine whether 5 is a zero of $g(x)$, where $g(x) = x^4 - 26x^2 + 25$.

If 5 is a zero of $g(x)$, then $g(5) = 0$. If we divide $g(x)$ by $x - 5$ and get a remainder of 0, then $g(5) = 0$ and 5 is a zero of the function.

We use synthetic division to find $g(5)$, remembering to write zeros for the missing terms in $g(x)$.

$$
\begin{array}{r|rrrrr}
5 & 1 & 0 & -26 & 0 & 25 \\
  &   & 5 & \boxed{} & -5 & -25 \\
\hline
  & \boxed{} & 5 & -1 & -5 & \boxed{}
\end{array}
$$

$g(5) = \boxed{}$, so 5 is a zero of $g(x)$.

Your Turn 4   Determine whether -1 is a zero of $g(x)$, where $g(x) = x^4 - 2x^2 - 8$.

Example 5   Determine whether i is a zero of $f(x)$, where $f(x) = x^3 - 3x^2 + x - 3$.

If i is a zero of $f(x)$, then $f(i) = 0$. If we divide $f(x)$ by $x - i$ and get a remainder of 0, then $f(i) = 0$ and i is a zero of the function.

We use synthetic division to find $f(i)$.

$$
\begin{array}{r|rrrr}
i & 1 & -3 & 1 & -3 \\
  &   & \boxed{} & -1-3i & 3 \\
\hline
  & 1 & -3+i & \boxed{} & \boxed{}
\end{array}
$$

$f(i) = \boxed{}$, so i is a zero of $f(x)$.

Your Turn 5   Determine whether $2i$ is a zero of $h(x)$, where $h(x) = x^3 - x^2 + 4x - 4$.

> **The Factor Theorem**
>
> For a polynomial $f(x)$, if $f(c) = 0$, then $x - c$ is a factor of $f(x)$.

Example 6 Let $f(x) = x^3 - 3x^2 - 6x + 8$. Factor $f(x)$ and solve the equation $f(x) = 0$.

According to the factor theorem, if $f(c) = 0$, then $x - c$ is a factor of $f(x)$. At this point, we make some arbitrary choices for c and determine if they are zeros.

First try $c = -1$. This means we are dividing by $x + 1$.

$$
\begin{array}{r|rrrr}
-1 & 1 & -3 & -6 & 8 \\
   &   & -1 & \Box & \Box \\
\hline
   & 1 & -4 & -2 & \Box
\end{array}
$$

$10 \neq 0$, so -1 is not a zero of $f(x)$. That means that $x + 1$ is not a factor of $f(x)$.

Now try $c = 1$. This tests whether $x - 1$ is a factor of $f(x)$.

$$
\begin{array}{r|rrrr}
1 & 1 & -3 & -6 & 8 \\
  &   & 1 & -2 & -8 \\
\hline
  & 1 & \Box & -8 & \Box
\end{array}
$$

This tells us that one of the factors of $f(x)$ is $x - 1$. Another factor is the quotient that we found when dividing. We have

$$
\begin{aligned}
f(x) &= (x-1)(x^2 - 2x - 8) \\
     &= (x-1)(x-4)(x+2).
\end{aligned}
$$

Finally, we solve the equation $f(x) = 0$.

$$(x-1)(x-4)(x+2) = 0$$

$$x - 1 = 0 \quad \text{or} \quad x - 4 = 0 \quad \text{or} \quad x + 2 = 0$$

$$x = \Box \quad \text{or} \quad x = \Box \quad \text{or} \quad x = \Box$$

The solutions are $\Box$, $\Box$, and $\Box$.

Your Turn 6   Let $f(x) = x^3 + 4x^2 + x - 6$. Factor $f(x)$ and solve the equation $f(x) = 0$.

Practice Exercises

1. Use division to determine if each of the following is a factor of
   $f(x) = x^3 - x^2 - 10x - 8$.
   a) $x + 2$ b) $x - 1$ c) $x + 1$

Use synthetic division to find the quotient and the remainder.

2. $\left(x^4 + 2x^3 - 7x^2 - 8x + 10\right) \div \left(x + 3\right)$

3. $\left(3x^3 - 4x + 1\right) \div \left(x - 2\right)$

Use synthetic division to find the function values.

4. $f(x) = x^3 - 2x^2 + 3x + 4$;
   find $f(1)$ and $f(-2)$.

5. $f(x) = x^4 - 2x^2 + 3x - 1$;
   find $f(2)$ and $f(-1)$.

Use synthetic division to determine if the given numbers are zeros of the given polynomial functions.

6. $-1, 1;\ f(x) = 3x^3 - 7x^2 - 18x - 8$

7. $2, -3;\ f(x) = 2x^3 + 7x^2 + 2x - 3$

8. $i, 3i;\ f(x) = x^3 - x^2 + 9x - 9$

Factor the polynomial function $f(x)$. Then solve $f(x) = 0$.

9. $f(x) = x^3 + 3x^2 - 6x - 8$ **10.** $f(x) = x^3 - 13x - 12$

Section 8.4   Theorems about Zeros of Polynomial Functions

In this section, we will need to use the following relationship between the factors of $f(x)$ and the zeros of $f(x)$.

The Factor Theorem

For a polynomial $f(x)$, if $f(c) = 0$, then $x - c$ is a factor of $f(x)$.

The Fundamental Theorem of Algebra

Every polynomial function of degree n, with $n \geq 1$, has at least one zero in the set of complex numbers.

We will also need to know one of the results of the fundamental theorem of algebra:

Every polynomial function f of degree n, with $n \geq 1$, can be factored into n linear factors (not necessarily unique); that is,

$$f(x) = a_n (x - c_1)(x - c_2) \cdots (x - c_n).$$

Example 1   Find a polynomial function of degree 3, having the zeros 1, $3i$, and $-3i$.

If these are the zeros, then the factors must be

$(x - 1)$, $(x - 3i)$, $\left(x - (-3i)\right)$, or

$(x - 1)$, $(x - 3i)$, $(x + 3i)$.   Writing $\left(x - (-3i)\right)$ as $(x + 3i)$

So the function must have the form

$f(x) = a_n (x - 1)(x - 3i)(x + 3i).$

Let $a_n = 1$.

$f(x) = (x - 1)(x - 3i)(x + 3i)$

$\quad = (x - 1)(x^2 + 9)$

$\quad = \boxed{} + 9x - x^2 - 9$

$\quad = x^3 - \boxed{} + 9x - 9$

Your Turn 1   Find a polynomial function of degree 3 having the zeros 3, $2i$, and $-2i$.

Example 2   Find a polynomial function of degree 5 with -1 as a zero of multiplicity 3, 4 as a zero of multiplicity 1, and 0 as a zero of multiplicity 1.

If these are the zeros, then the function in factored form must be:

$$f(x) = a_n \left(x + \boxed{} \right)^3 \left(x - \boxed{} \right)(x - 0)$$

Let $a_n = 1$.

$$f(x) = (x+1)^3 (x-4)(x-0)$$
$$= x^5 - \boxed{} - 9x^3 - \boxed{} - 4x$$

Your Turn 2   Find a polynomial function of degree 4 with 1 as a zero of multiplicity 2, -2 as a zero of multiplicity 1, and 0 as a zero of multiplicity 1.

Nonreal Zeros: $a + bi$ and $a - bi$, $b \neq 0$

If a complex number $a + bi$, $b \neq 0$, is a zero of a polynomial function $f(x)$ with *real* coefficients, then its conjugate, $a - bi$, is also a zero. For example, if $2 + 7i$ is a zero of a polynomial function $f(x)$, with real coefficients, then its conjugate, $2 - 7i$, is also a zero. (Nonreal zeros occur in conjugate pairs.)

Irrational Zeros: $a + c\sqrt{b}$ and $a - c\sqrt{b}$, b **Is Not a Perfect Square**

If $a + c\sqrt{b}$, where a, b, and c are rational and b is not a perfect square, is a zero of a polynomial function $f(x)$ with *rational* coefficients, then its conjugate, $a - c\sqrt{b}$, is also a zero. For example, if $-3 + 5\sqrt{2}$ is a zero of a polynomial function $f(x)$ with rational coefficients, then its conjugate, $-3 - 5\sqrt{2}$, is also a zero. (Irrational zeros occur in conjugate pairs.)

Example 3 Suppose that a polynomial function of degree 6 with rational coefficients has $-2 + 5i$, $-2i$, and $1 - \sqrt{3}$ as three of its zeros. Find the other zeros.

Since the function has rational coefficients, the other zeros are the conjugates of the given zeros. They are

$$-2 - 5i, \quad \boxed{}, \quad \text{and } 1 + \sqrt{3}.$$

Your Turn 3 Suppose that a polynomial function of degree 5 with rational coefficients has $-4, 1 - 2i$, and $\sqrt{6}$ as three of its zeros. Find the other zeros.

Example 4 Find a polynomial function of lowest degree with rational coefficients that has $-\sqrt{3}$, and $1+i$ as two of its zeros.

Two other zeros must be ⬚ and ⬚. (We will not include additional zeros, because we want the function of lowest degree.)

$$f(x) = \left[x-\left(-\sqrt{3}\right)\right]\left[x-\left(\boxed{}\right)\right]\left[x-(1+i)\right]\left[x-\left(\boxed{}\right)\right]$$

$$= \left(x+\sqrt{3}\right)\left(x-\sqrt{3}\right)\left((x-1)-i\right)\left((x-1)+\boxed{}\right)$$

$$= \left(x^2 - \boxed{}\right)\left((x-1)^2 - i^2\right)$$

$$= \left(x^2 - 3\right)\left(x^2 - 2x + 1^2 + 1\right)$$

$$= \left(x^2 - 3\right)\left(x^2 - 2x + 2\right)$$

$$= x^4 - 2x^3 - \boxed{} + 6x - \boxed{}$$

✎ **Your Turn 4** Find a polynomial function of lowest degree with rational coefficients that has $\sqrt{2}$ and $3-i$ as two of its zeros.

The Rational Zeros Theorem

Let $P(x) = a_n x^n + a_{n-1}x^{n-1} + \cdots + a_1 x + a_0$, where all the coefficients are integers. Consider a rational number denoted by p/q, where p and q are relatively prime (having no common factor besides -1 and 1). If p/q is a zero of $P(x)$, then p is a factor of a_0 and q is a factor of a_n.

Example 5   Given $f(x) = 3x^4 - 11x^3 + 10x - 4$:

a) Find the rational zeros and then the other zeros; that is, solve $f(x) = 0$.

b) Factor $f(x)$ into linear factors.

First, notice the degree of the polynomial is 4, so there are at most 4 distinct zeros.

a) The rational zeros theorem says that if a rational number p/q is a zero of $f(x)$, then p must be a factor of -4 and q must be a factor of 3.

$$\frac{Possibilities\ for\ p}{Possibilities\ for\ q} : \frac{\pm 1,\ \pm 2,\ \pm 4}{\pm 1,\ \pm 3}$$

$$Possibilities\ for\ \frac{p}{q} : 1, -1, 2, -2, 4, -4, \frac{1}{3}, -\frac{1}{3}, \frac{2}{3}, -\frac{2}{3}, \frac{4}{3}, -\frac{4}{3}$$

To find which are zeros, we use synthetic division.

Try 1:

1⌋	3	-11	0	10	-4
		3	☐	-8	2
	☐	-8	-8	2	☐

Since $f(1) = $ ☐, 1 is not a zero.

Try -1:

-1⌋	3	-11	0	10	-4
		☐	14	-14	☐
	3	-14	☐	-4	0

Since $f(-1) = $ ☐, -1 is a zero.

Using the results of the synthetic division, we can rewrite $f(x)$ as

$$f(x) = (x+1)(3x^3 - 14x^2 + 14x - 4).$$

Now consider $3x^3 - 14x^2 + 14x - 4$. Continue using synthetic division to find other zeros.

Try -1 again since it might have multiplicity 2.

-1⌋	3	-14	14	-4
		☐	17	☐
	3	-17	☐	-35

Since $f(-1) = -35$, -1 does not have multiplicity 2.   (continued)

Synthetic division shows that 2, -2, 4, and -4 are not zeros.

Now try $\dfrac{2}{3}$:

$$
\begin{array}{r|rrrr}
2/3 & 3 & -14 & 14 & -4 \\
    &   & 2 & \Box & 4 \\
\hline
    & \Box & -12 & 6 & \Box
\end{array}
$$

Since $f\!\left(\dfrac{2}{3}\right) = \Box$, $\dfrac{2}{3}$ is a zero of $f(x)$ and $\left(x - \dfrac{2}{3}\right)$ is a factor of $f(x)$.

We can rewrite $f(x)$ as

$$f(x) = (x+1)\left(x - \frac{2}{3}\right)\left(3x^2 - 12x + 6\right)$$

$$= (x+1)\left(x - \frac{2}{3}\right)3\left(x^2 - 4x + 2\right)$$

and the last factor can be used to find the two remaining zeros by solving the equation $x^2 - 4x + 2 = 0$ using the quadratic formula.

$$x = \frac{-b \pm \sqrt{b^2 - 4ac}}{2a}$$

$$x = \frac{-(-4) \pm \sqrt{(-4)^2 - 4 \cdot 1 \cdot 2}}{2 \cdot 1}$$

$$x = \frac{4 \pm \sqrt{\Box}}{2} = \frac{4 \pm 2\sqrt{2}}{2} = \frac{2\left(2 \pm \sqrt{2}\right)}{2} = 2 \pm \sqrt{2}$$

So the rational zeros are $\Box$, $\dfrac{2}{3}$, $2 + \sqrt{2}$, $2 - \sqrt{2}$.

b) The complete factorization of $f(x)$ is

$$f(x) = 3(x+1)\left(x - \frac{2}{3}\right)\left[x - \left(2 - \sqrt{2}\right)\right]\left[x - \left(2 + \sqrt{2}\right)\right], \text{ or}$$

$$f(x) = (x+1)\left(3x - \Box\right)\left[x - \left(2 - \sqrt{2}\right)\right]\left[x - \left(2 + \sqrt{2}\right)\right].$$

Your Turn 5   Given $f(x) = 2x^4 + 3x^3 - 7x^2 + 2$:

a) Find the rational zeros and then the other zeros: that is, solve $f(x) = 0$.

b) Factor $f(x)$ into linear factors.

Example 6   Given $f(x) = 2x^5 - x^4 - 4x^3 + 2x^2 - 30x + 15$:

a) Find the rational zeros and then the other zeros; that is, solve $f(x) = 0$.

b) Factor $f(x)$ into linear factors.

First, notice the degree of the polynomial is 5, so there are at most 5 distinct zeros.

a) The rational zeros theorem says that if a rational number p/q is a zero of $f(x)$, then p must be a factor of 15 and q must be a factor of 2.

$$\frac{\textit{Possibilities for } p}{\textit{Possibilities for } q} : \frac{\pm 1,\ \pm 3,\ \pm 5,\ \pm 15}{\pm 1,\ \pm 2}$$

$$\textit{Possibilities for } \frac{p}{q} : 1, -1, 3, -3, 5, -5, 15, -15, \frac{1}{2}, -\frac{1}{2}, \frac{3}{2}, -\frac{3}{2}, \frac{5}{2}, -\frac{5}{2}, \frac{15}{2}, -\frac{15}{2}$$

(continued)

To find which are zeros, we use synthetic division.

Try 1:

$$\begin{array}{r|rrrrrr} 1 & 2 & -1 & -4 & 2 & -30 & 15 \\ & & 2 & 1 & -3 & -1 & -31 \\ \hline & 2 & 1 & -3 & -1 & -31 & \boxed{} \end{array}$$

Since $f(1) = \boxed{}$, 1 is not a zero.

Try -1:

$$\begin{array}{r|rrrrrr} -1 & 2 & -1 & -4 & 2 & -30 & 15 \\ & & -2 & 3 & 1 & -3 & 33 \\ \hline & 2 & -3 & -1 & 3 & -33 & \boxed{} \end{array}$$

Since $f(-1) = \boxed{}$, -1 is not a zero.

Try $\dfrac{1}{2}$:

$$\begin{array}{r|rrrrrr} 1/2 & 2 & -1 & -4 & 2 & -30 & 15 \\ & & 1 & 0 & -2 & 0 & -15 \\ \hline & 2 & \boxed{} & -4 & 0 & -30 & \boxed{} \end{array}$$

Since $f\left(\dfrac{1}{2}\right) = \boxed{}$, $\dfrac{1}{2}$ is a zero of $f(x)$ and $\left(x - \dfrac{1}{2}\right)$ is a factor of $f(x)$.

We can rewrite $f(x)$ as

$$f(x) = \left(x - \frac{1}{2}\right)\left(2x^4 - 4x^2 - 30\right)$$

$$= \left(x - \frac{1}{2}\right)2\left(x^4 - 2x^2 - 15\right) = \left(x - \frac{1}{2}\right)2\left(x^2 - 5\right)\left(x^2 + 3\right).$$

We now solve the equation $f(x) = 0$ using the principle of zero products.

$$\left(x - \frac{1}{2}\right)2\left(x^2 - 5\right)\left(x^2 + 3\right) = 0$$

$$x - \frac{1}{2} = 0 \quad\quad \textit{or} \quad x^2 - 5 = 0 \quad\quad \textit{or} \quad x^2 + 3 = 0$$

$$x = \frac{1}{2} \quad\quad \textit{or} \quad\quad x^2 = 5 \quad\quad \textit{or} \quad\quad x^2 = -3$$

$$x = \boxed{} \quad \textit{or} \quad\quad x = \boxed{} \quad \textit{or} \quad\quad x = \boxed{}$$

The rational zero is $\dfrac{1}{2}$. The other zeros are $\sqrt{5}$, $-\sqrt{5}$, $\sqrt{3}i$, and $-\sqrt{3}i$.

(continued)

b) The complete factorization of $f(x)$ is

$$f(x) = 2\left(x - \boxed{}\right)\left(x + \sqrt{5}\right)\left(x - \sqrt{5}\right)\left(x + \sqrt{3}i\right)\left(x - \boxed{}\right), \text{ or}$$

$$f(x) = (2x - 1)\left(x + \sqrt{5}\right)\left(x - \boxed{}\right)\left(x + \sqrt{3}i\right)\left(x - \sqrt{3}i\right).$$

Your Turn 6   Given $f(x) = 3x^5 - x^4 + 9x^3 - 3x^2 - 12x + 4$:

a) Find the rational zeros and the other zeros; that is, solve $f(x) = 0$.

b) Factor $f(x)$ into linear factors.

Descartes' Rule of Signs

Let $P(x)$, written in descending order or ascending order, be a polynomial function with real coefficients and a nonzero constant term. The number of positive real zeros of $P(x)$ is either:

1. The same as the number of variations of sign in $P(x)$, or

2. Less than the number of variations of sign in $P(x)$ by a positive even integer.

The number of negative real zeros of $P(x)$ is either:

3. The same as the number of variations of sign in $P(-x)$, or

4. Less than the number of variations of sign in $P(-x)$ by a positive even integer.

A zero of multiplicity m must be counted m times.

Example 7 Determine the number of variations of sign in the polynomial function $P(x) = 2x^5 - 3x^2 + x + 4$.

There are ☐ variations of sign.

Your Turn 7 Determine the number of variations of sign in the polynomial function $P(x) = 2x^4 + 3x^3 - x^2 + 4x - 5$.

Example 8 Given the function $P(x) = 2x^5 - 5x^2 - 3x + 6$, what does Descartes' rule of signs tell you about the number of positive real zeros and the number of negative real zeros?

There are ☐ variations of sign in $P(x)$.

Thus, the number of positive real zeros is ☐ or $2 - 2 = $ ☐.

Now find $P(-x)$:

$$P(-x) = 2(-x)^5 - 5(-x)^2 - 3(-x) + 6$$

$$= -2x^5 \ \boxed{} \ 5x^2 + 3x + 6$$

There is ☐ variation of sign in $P(-x)$.

Thus, there is ☐ negative real zero.

Total Number of Zeros	5	
Positive Real	2	0
Negative Real	1	1
Nonreal	2	4

Your Turn 8   Given the function $P(x) = 2x^5 + x^3 - 3x^2 + 5$, what does Descartes' rule of signs tell you about the number of positive real zeros and the number of negative real zeros?

Example 9   Given the function $P(x) = 5x^4 - 3x^3 + 7x^2 - 12x + 4$, what does Descartes' rule of signs tell you about the number of positive real zeros and the number of negative real zeros?

There are ☐ variations of sign in $P(x)$.

Thus $P(x)$ has ☐ or $4 - 2 = $ ☐ or $4 - 4 = $ ☐ positive real zeros.

Now find $P(-x)$:

$$P(-x) = 5(-x)^4 - 3(-x)^3 + 7(-x)^2 - 12(-x) + 4$$
$$= 5x^4 + 3x^3 + 7x^2 + 12x + 4$$

There are ☐ variations of sign in $P(-x)$.

Thus, there are ☐ negative real zeros.

Total Number of Zeros	4		
Positive Real	4	2	0
Negative Real	0	0	0
Nonreal	0	2	4

Your Turn 9   Given the function $P(x) = 3x^4 + 2x^3 - 5x^2 - x - 3$, what does Descartes' rule of signs tell you about the number of positive real zeros and the number of negative real zeros?

Example 10   Given the function $P(x) = 6x^6 - 2x^2 - 5x$, what does Descartes' rule of signs tell you about the number of positive real zeros and the number of negative real zeros?

$$P(x) = 6x^6 - 2x^2 - 5x = x\left(6x^5 - 2x - 5\right)$$

Note that $x = 0$ is a zero of this function.

Let $Q(x) = 6x^5 - 2x - 5$.

There is $\boxed{}$ variation of sign in $Q(x)$.

Thus there is $\boxed{}$ positive real zero.

Now find $Q(-x)$:

$$Q(-x) = 6(-x)^5 - 2(-x) - 5 = -6x^5 + 2x - 5$$

There are $\boxed{}$ variations of sign in $Q(-x)$.

Thus, there are $\boxed{}$ or $2 - 2 = \boxed{}$ negative real zeros.

Total Number of Zeros		6	
0 as a Zero		1	1
Positive Real		1	1
Negative Real		2	0
Nonreal		2	4

Your Turn 10   Given the function $P(x) = 3x^4 - x^3 - 6x^2 + x$, what does Descartes' rule of signs tell you about the number of positive real zeros and the number of negative real zeros?

Practice Exercises

Find a polynomial of degree 3 with the given numbers as zeros.

1. $-3, i, -i$

2. $2, -\sqrt{5}, \sqrt{5}$

3. Find a polynomial function of degree 4 with 2 as a zero of multiplicity 2, -1 as a zero of multiplicity 1, and -3 as a zero of multiplicity 1.

Suppose that a polynomial function of degree 4 with rational coefficients has the given numbers as zeros. Find the other zeros.

4. $2, \sqrt{5}, -3$

5. $1 - 2i, 3 + \sqrt{2}$

6. Find a polynomial function of lowest degree with rational coefficients that has 1 and $3 + i$ as two of its zeros.

For each polynomial function, (a) find the rational zeros and then the other zeros; that is, solve $f(x) = 0$; and (b) factor $f(x)$ into linear factors.

7. $f(x) = x^3 - 2x^2 + x - 2$

8. $f(x) = 2x^4 - x^3 - 7x^2 + 2x + 6$

What does Descartes' rule of signs tell you about the number of positive real zeros and the number of negative real zeros of the function?

9. $P(x) = 2x^4 - 3x^2 + 4x - 1$

10. $Q(x) = x^5 + x^4 - 2x^3 - 3x^2 + 2x + 4$

11. $g(x) = 2x^6 + 4x^3 + 2x + 6$

Section 8.5 Rational Functions

Rational Function

A **rational function** is a function f that is a quotient of two polynomials. That is,

$$f(x) = \frac{p(x)}{q(x)}$$

where $p(x)$ and $q(x)$ are polynomials and where $q(x)$ is not the zero polynomial. The domain of f consists of all inputs x for which $q(x) \neq 0$.

Example 1 Consider $f(x) = \dfrac{1}{x-3}$. Find the domain and graph f.

The domain is all the x-values which do not make the denominator zero.

$x - 3 = 0$ when $x = \boxed{}$.

Domain: $\left\{ x \mid x \neq \boxed{} \right\}$, or $(-\infty, 3) \cup (3, \infty)$

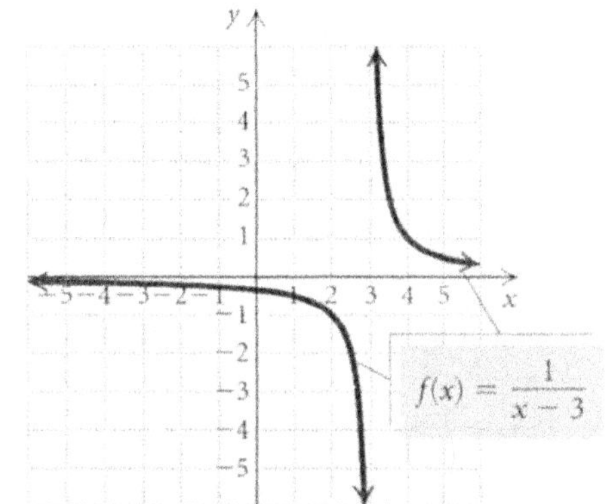

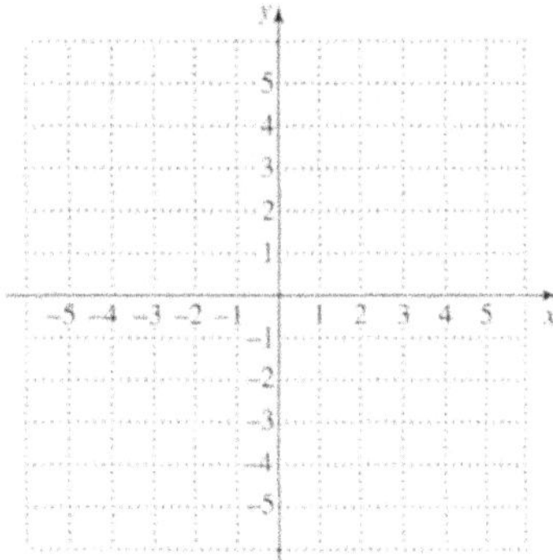

Your Turn 1 Consider $f(x) = \dfrac{1}{x+2}$. Find the domain and graph f.

Example 2 Determine the domain of each of the functions.

Function	Domain
$f(x) = \dfrac{1}{x}$	$\left\{ x \mid x \neq \boxed{} \right\}$, or $(-\infty, 0) \cup (0, \infty)$
$f(x) = \dfrac{1}{x^2}$	$\left\{ x \mid x \neq 0 \right\}$, or $\left(-\infty, \boxed{}\right) \cup (0, \infty)$
$f(x) = \dfrac{x-3}{x^2+x-2} = \dfrac{x-3}{(x+2)(x-1)}$	$\left\{ x \mid x \neq -2 \text{ and } x \neq \boxed{} \right\}$, or $(-\infty, -2) \cup (-2, 1) \cup (1, \infty)$
$f(x) = \dfrac{2x+5}{2x-6} = \dfrac{2x+5}{2(x-3)}$	$\left\{ x \mid x \neq \boxed{} \right\}$, or $(-\infty, 3) \cup (3, \infty)$
$f(x) = \dfrac{x^2+2x-3}{x^2-x-2} = \dfrac{x^2+2x-3}{(x+1)(x-2)}$	$\left\{ x \mid x \neq -1 \text{ and } x \neq 2 \right\}$, or $(-\infty, -1) \cup (-1, 2) \cup (2, \infty)$
$f(x) = \dfrac{-x^2}{x+1}$	$\left\{ x \mid x \neq -1 \right\}$, or $\left(-\infty, -1\right) \cup \left(\boxed{}, \infty\right)$

Your Turn 2 Determine the domain of each of the functions.

Function	Domain
$f(x) = -\dfrac{2}{x}$	
$f(x) = \dfrac{3}{x^2}$	
$f(x) = \dfrac{x-1}{x^2-4} = \dfrac{x-1}{(x+2)(x-2)}$	
$f(x) = \dfrac{3x-1}{4x+8} = \dfrac{3x-1}{4(x+2)}$	
$f(x) = \dfrac{x^2+x-1}{x^2-3x-10} = \dfrac{x^2+x-1}{(x+2)(x-5)}$	
$f(x) = \dfrac{3x^2}{x-4}$	

Determining Vertical Asymptotes

> For a rational function $f(x) = p(x)/q(x)$, where $p(x)$ and $q(x)$ are polynomials with *no common factors other than constants,* if a is a zero of the denominator, then the line $x = a$ is a vertical asymptote for the graph of the function.

Example 3   Determine the vertical asymptotes for the graph of each of the following functions.

a) $f(x) = \dfrac{2x - 11}{x^2 + 2x - 8}$

First, factor the denominator.

$$f(x) = \dfrac{2x - 11}{(x + 4)\left(x - \boxed{}\right)}$$

The numerator and the denominator have no common factors.

Zeros of the denominator: $-4, \boxed{}$

Vertical asymptotes: $x = \boxed{}$ and $x = 2$

b) $h(x) = \dfrac{x^2 - 4x}{x^3 - x}$

First, factor the numerator and the denominator.

$$h(x) = \dfrac{x\left(x - \boxed{}\right)}{x\left(x^2 - 1\right)} = \dfrac{x\left(x - \boxed{}\right)}{x(x + 1)(x - 1)}$$

Zeros of the numerator: $0, \boxed{}$

Zeros of the denominator: $0, -1,$ and $\boxed{}$

Vertical asymptotes: $x = -1$ and $x = 1$

c) $g(x) = \dfrac{x - 2}{x^3 - 5x}$

First, factor the denominator: $g(x) = \dfrac{x - 2}{x\left(x^2 - 5\right)}.$

The numerator and the denominator have no common factors.

Solving $x\left(x^2 - 5\right) = 0,$ we have $x = \boxed{}$ or $x = \pm\sqrt{5}.$

Zeros of the denominator: $0, -\sqrt{5},$ and $\boxed{}$

Vertical asymptotes: $x = \boxed{}, x = -\sqrt{5},$ and $x = \sqrt{5}$

Your Turn 3 Determine the vertical asymptotes for the graph of each of the following functions.

a) $g(x) = \dfrac{3x-4}{x^2-4x-5}$

b) $f(x) = \dfrac{3x+12}{x^3-4x}$

c) $h(x) = \dfrac{2x+1}{x^3-3x}$

Example 4 Find the horizontal asymptote: $f(x) = \dfrac{-7x^4-10x^2+1}{11x^4+x-2}$.

When the numerator and the denominator of a rational function have the same degree, the horizontal asymptote is given by $y = a/b$ where a and b are the leading coefficients of the numerator and the denominator, respectively.

Thus, the horizontal asymptote is $y = \dfrac{-7}{\boxed{}} = -\dfrac{\boxed{}}{11}$, or $y = -0.\overline{63}$.

Your Turn 4 Find the horizontal asymptote: $f(x) = \dfrac{3x^3-2x+5}{5x^3+x^2-1}$.

Example 5 Find the horizontal asymptote: $f(x) = \dfrac{2x+3}{x^3 - 2x^2 + 4}$.

Let's call the numerator $p(x)$ and the denominator $q(x)$. Notice $p(x)$ has degree 1 and $q(x)$ has degree 3. As x increases, a degree 3 polynomial will increase much faster than a degree 1 polynomial.

So, as $x \to \infty$, $\dfrac{p(x)}{q(x)} \to 0$, and as $x \to -\infty$, $\dfrac{p(x)}{q(x)} \to 0$.

Thus, the horizontal asymptote is $y = \boxed{}$.

This is the case anytime the degree of the numerator is less than the degree of the denominator.

Your Turn 5 Find the horizontal asymptote: $f(x) = \dfrac{x-1}{2x^2 - 5x + 4}$.

Determining a Horizontal Asymptote

- When the numerator and the denominator of a rational function have the same degree, the line $y = a/b$ is the horizontal asymptote, where a and b are the leading coefficients of the numerator and the denominator, respectively.

- When the degree of the numerator of a rational function is less than the degree of the denominator, the x-axis, or $y = 0$, is the horizontal asymptote.

- When the degree of the numerator of a rational function is greater than the degree of the denominator, there is no horizontal asymptote.

Example 6 Graph $g(x) = \dfrac{2x^2 + 1}{x^2}$. Include and label all asymptotes.

The denominator is zero when $x = \boxed{}$, so the vertical asymptote is $x = 0$.

The degree of the numerator is the same as the degree of the denominator, so the ratio of the leading coefficients gives us the horizontal asymptote: $y = \boxed{}$.

Find and plot some points on the graph. We have:

$(-2, 2.25)$, $(-1, 3)$, $(-0.5, 6)$,

$(0.5, 6)$, $(1.5, 2.\overline{4})$, $(2, 2.25)$

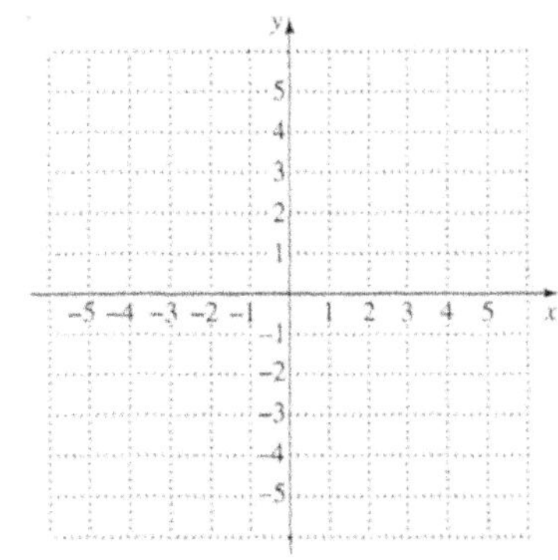

Your Turn 6   Graph $h(x) = \dfrac{x^2 + 1}{x^2}$. Include and label all asymptotes.

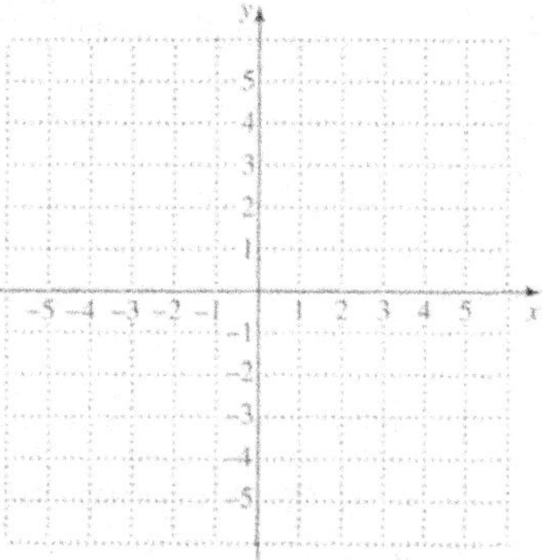

Sometimes a line that is neither horizontal nor vertical is an asymptote. Such a line is called an **oblique asymptote**, or a **slant asymptote**.

Example 7   Find all the asymptotes of $f(x) = \dfrac{2x^2 - 3x - 1}{x - 2}$.

The denominator is zero when $x = \boxed{}$, so the vertical asymptote is $x = 2$.

There are no horizontal asymptotes, because the degree of the numerator is greater than that of the denominator.

When the degree of the numerator is 1 more than the degree of the denominator, we divide to find an equivalent expression.

$$f(x) = \frac{2x^2 - 3x - 1}{x - 2} = 2x + 1 + \frac{1}{x - 2}$$

As $x \to \infty$ or as $x \to -\infty$, $\dfrac{1}{x - 2} \to 0$, so $f(x) \to 2x + 1$.

The line $y = \boxed{}$ is an oblique asymptote.

Your Turn 7   Find all the asymptotes of $f(x) = \dfrac{3x^2 - x + 1}{x + 1}$.

   Copyright © 2022 Pearson Education, Inc.

Example 8 Graph: $f(x) = \dfrac{2x+3}{3x^2+7x-6}$.

First, find the zeros of the denominator.

$$3x^2 + 7x - 6 = 0$$

$$(3x-2)(x+3) = 0$$

$$3x - 2 = 0 \quad or \quad x + 3 = 0$$

$$x = \frac{2}{3} \quad or \quad x = \boxed{}$$

Neither zero is a zero of the numerator, so the vertical asymptotes are $x = \dfrac{2}{3}$ and $x = -3$.

The degree of the numerator is less than the degree of the denominator, so the horizontal asymptote is $y = \boxed{}$.

The zero of the numerator tells us the x-intercept.

$$2x + 3 = 0$$

$$2x = -3$$

$$x = \boxed{}$$

Thus the x-intercept is $\left(-\dfrac{3}{2}, 0\right)$.

To find the y-intercept, find $f(0)$.

$$f(0) = \frac{2 \cdot 0 + 3}{3 \cdot 0^2 + 7 \cdot 0 - 6}$$

$$= \frac{3}{-6} = -\frac{1}{2}$$

The y-intercept is $\left(0, -\dfrac{1}{2}\right)$.

Finally, choose some other x-values and find their corresponding function values to find enough points to determine the shape of the graph.

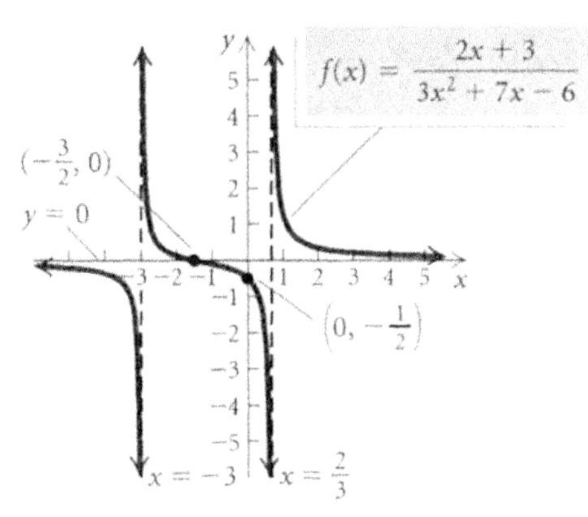

Your Turn 8 Graph: $f(x) = \dfrac{x+2}{2x^2 - x - 1}$.

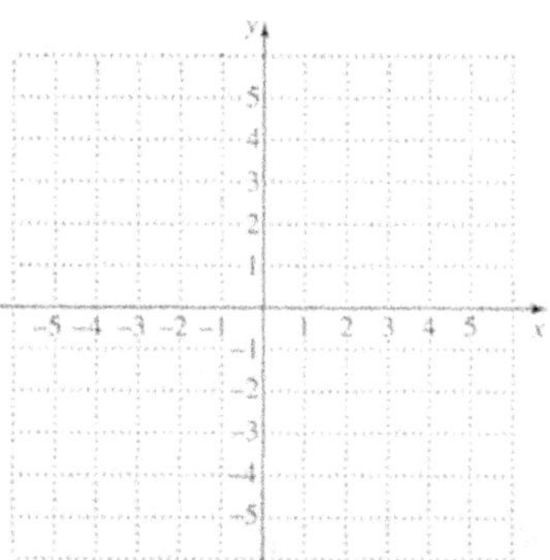

Example 9 Graph: $g(x) = \dfrac{x^2 - 1}{x^2 + x - 6}$.

First, find the zeros of the denominator.

$$x^2 + x - 6 = 0$$
$$(x+3)(x-2) = 0$$
$$x = -3 \quad or \quad x = 2$$

Neither zero is a zero of the numerator, so the vertical asymptotes are $x = -3$ and $x = \boxed{}$.

The degree of the numerator is the same as the degree of the denominator, so the horizontal asymptote is $y = \dfrac{1}{1} = 1$.

The zeros of the numerator tell us the x-intercepts.

$$x^2 - 1 = 0$$
$$(x+1)(x-1) = 0$$
$$x = -1 \quad or \quad x = \boxed{}$$

The x-intercepts are $(-1, 0)$ and $(1, 0)$.

To find the y-intercept, find $g(0)$.

$$g(0) = \frac{0^2 - 1}{0^2 + 0 - 6} = \frac{-1}{-6} = \frac{1}{6}$$

The y-intercept is $\left(0, \dfrac{1}{6}\right)$.

Finally, choose some other x-values and find their corresponding function values to find enough points to determine the shape of the graph.

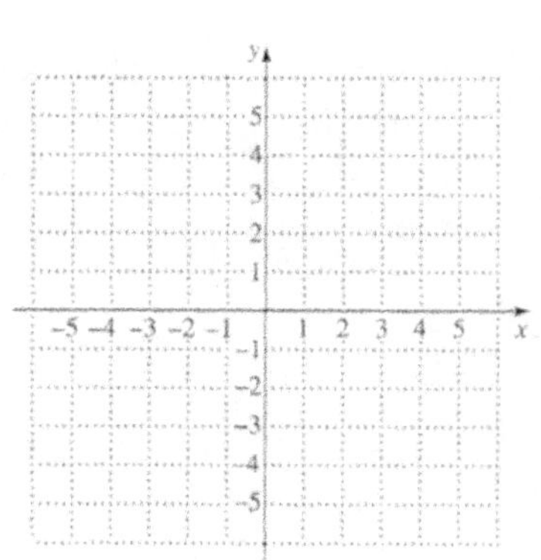

Your Turn 9   Graph: $f(x) = \dfrac{x^2 - 2}{x^2 - x - 6}$.

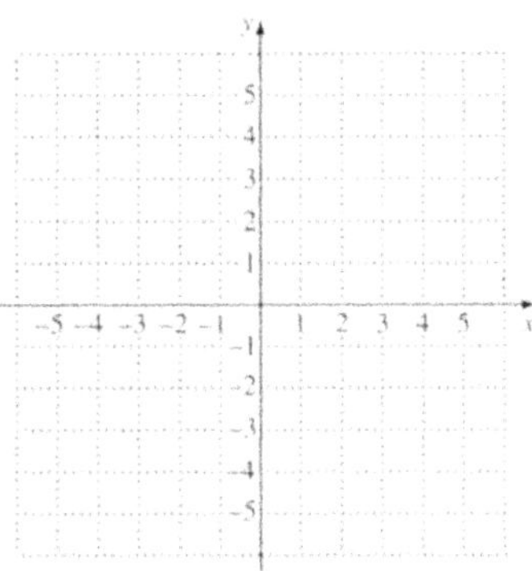

Example 10   Graph: $g(x) = \dfrac{x - 2}{x^2 - x - 2}$.

Factoring the denominator, we have

$$g(x) = \frac{x - 2}{(x - 2)(x + 1)}.$$

The denominator is equal to 0 when $x = 2$ or $x = -1$.

Domain: $\{x \mid x \neq -1 \text{ and } x \neq 2\}$

2 is also a zero of the numerator, so the only vertical asymptote is $x = -1$.

The degree of the numerator is less than the degree of the denominator, so the horizontal asymptote is $y = \boxed{}$.

The function has no zeros because the only zero of the numerator, 2, is not in the domain of the function.

$$g(x) = \frac{x - 2}{x^2 - x - 2} = \frac{x - 2}{(x - 2)(x + 1)} = \frac{1}{x + 1}, \ x \neq 2$$

The graph will be the graph of

$g(x) = \dfrac{1}{x + 1}$ with a hole at $x = 2$.

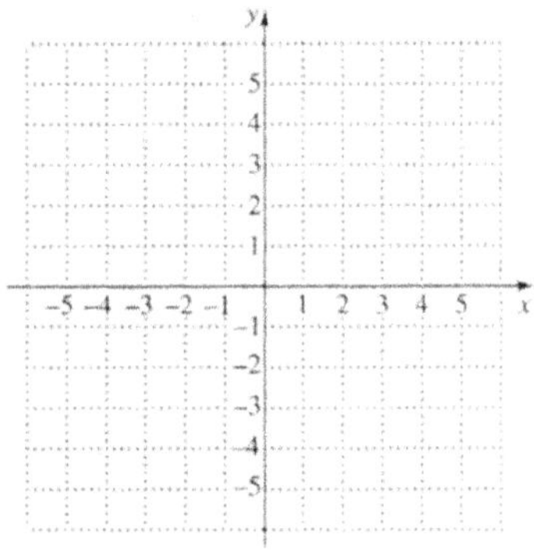

Your Turn 10   Graph: $g(x) = \dfrac{x-3}{x^2 - 2x - 3}$.

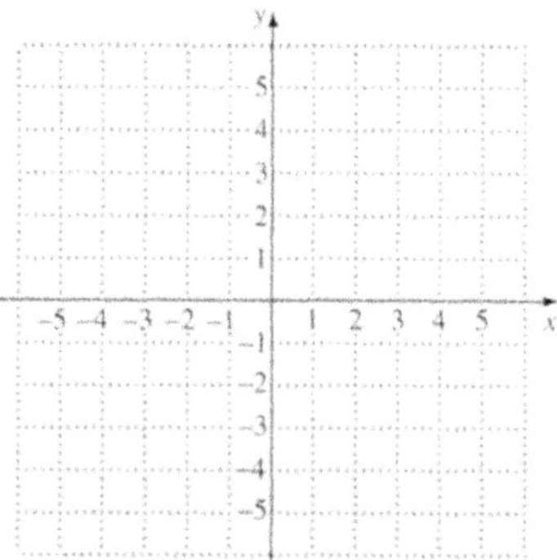

Example 11   Graph: $f(x) = \dfrac{-2x^2 - x + 15}{x^2 - x - 12}$.

Factoring: $f(x) = \dfrac{(-2x+5)(x+3)}{(x-4)(x+3)}$.

Domain: $\left\{x \mid x \neq \boxed{} \text{ and } x \neq \boxed{} \right\}$, or $(-\infty, -3) \cup (-3, 4) \cup (4, \infty)$

The only zero of the denominator that is not a zero of the numerator is 4, so the only vertical asymptote is $x = 4$.

The degree of the numerator is equal to the degree of the denominator, so the horizontal asymptote is $y = \dfrac{-2}{1} = -2.$

The zeros of the numerator are $\dfrac{5}{2}$ and -3, but -3 is not in the domain of the function, so the only x-intercept is $\left(\dfrac{5}{2}, 0 \right)$.

To find the y-intercept, find $f(0)$.

$$f(0) = \dfrac{15}{-12} = -\dfrac{5}{4}$$

The y-intercept is $\left(0, -\dfrac{5}{4} \right)$.

Simplifying: $f(x) = \dfrac{(-2x+5)(x+3)}{(x-4)(x+3)} = \dfrac{-2x+5}{x-4}, \ x \neq -3 \text{ and } x \neq 4$

The graph will be the graph of $f(x) = \dfrac{-2x+5}{x-4}$ with a hole at $x = -3$.

Determine the second coordinate of the hole:

$$f(-3) = \dfrac{-2(-3)+5}{-3-4} = \dfrac{11}{\boxed{}} = -\dfrac{11}{7}$$

(continued)

The hole is at $\left(-3,\ \boxed{}\right)$, indicated by an open circle on the graph.

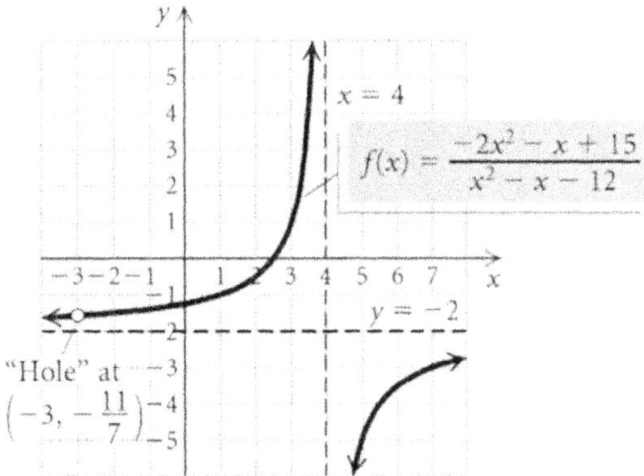

✐ **Your Turn 11** Graph: $f(x) = \dfrac{x^2 + x - 2}{x^2 + 3x + 2}$.

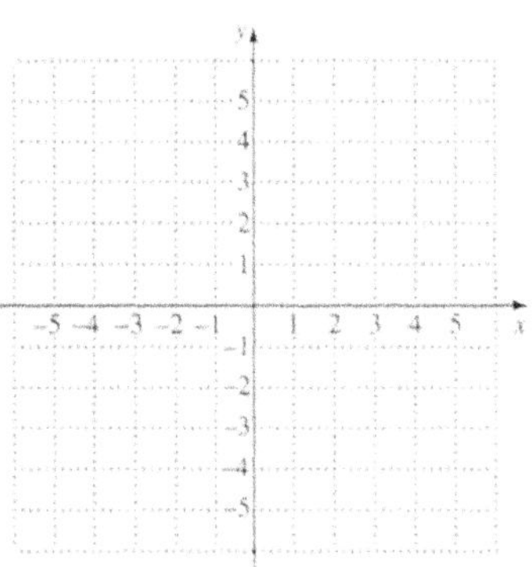

Example 12 *Temperature During an Illness.* A person's temperature T, in degrees Fahrenheit, during an illness is given by the function

$$T(t) = \frac{4t}{t^2 + 1} + 98.6,$$

where time t is given in hours since the onset of the illness.

a) Find the temperature at $t = 0,\ 1,\ 2,\ 5,\ 12,$ and 24.

$$T(0) = 98.6$$
$$T(1) = 100.6$$
$$T(2) = \boxed{}$$
$$T(5) = 99.369$$
$$T(12) = 98.931$$
$$T(24) = \boxed{}$$

(continued)

b) Find the horizontal asymptote of the graph of $T(t)$. Complete $T(t) \rightarrow \boxed{}$ as $t \rightarrow \infty$.

$$T(t) = \frac{4t}{t^2+1} + 98.6$$

$$= \frac{4t + 98.6t^2 + 98.6}{t^2+1} \qquad \text{Writing with a common denominator}$$

$$= \frac{98.6t^2 + 4t + 98.6}{t^2+1}$$

Since the degree of the numerator is the same as the degree of the denominator, the horizontal asymptote is the ratio of the leading coefficients: $y = \dfrac{98.6}{1}$, or $y = \boxed{}$.

So, $T(t) \rightarrow \boxed{}$ as $t \rightarrow \infty$.

c) Give the meaning of the answer to part (b) in terms of the application.
As time goes on, a person's temperature returns to normal, $98.6°\,F$.

Your Turn 12 A medical researcher determines that the concentration in the body of a certain dosage of a new medication, in parts per million, after time t in hours is given by the function

$$C(t) = \frac{0.5t + 3000}{10t + 100}, \, t \geq 1.$$

a) Find the concentration at $t = 1$, 3, and 4. Round to the nearest tenth.

b) Find the horizontal asymptote of the graph of the function and complete the following:
$$C(t) \rightarrow \boxed{} \text{ as } t \rightarrow \infty.$$

c) Give the meaning of the answer to part (b) in terms of the application.

Practice Exercises

Determine the domain of the function.

1. $f(x) = \dfrac{x}{x-3}$

2. $g(x) = \dfrac{x+2}{x^2+3x-18}$

Determine the vertical asymptotes of the graph of the function.

3. $h(x) = \dfrac{2-x}{x^2-6x+5}$

4. $f(x) = \dfrac{2x+3}{2x^2+x-6}$

Determine the horizontal asymptotes of the graph of the function.

5. $g(x) = \dfrac{3x^2+5}{x^2-4x+4}$

6. $f(x) = \dfrac{x+8}{x^3+4}$

Determine the oblique asymptotes of the graph of the function.

7. $f(x) = \dfrac{2x^2+5x-1}{x+2}$

8. $g(x) = \dfrac{x^3-4x}{x^2+1}$

Graph the function.

9. $f(x) = \dfrac{5}{x-1}$

10. $g(x) = \dfrac{x-1}{x^2 + x - 6}$

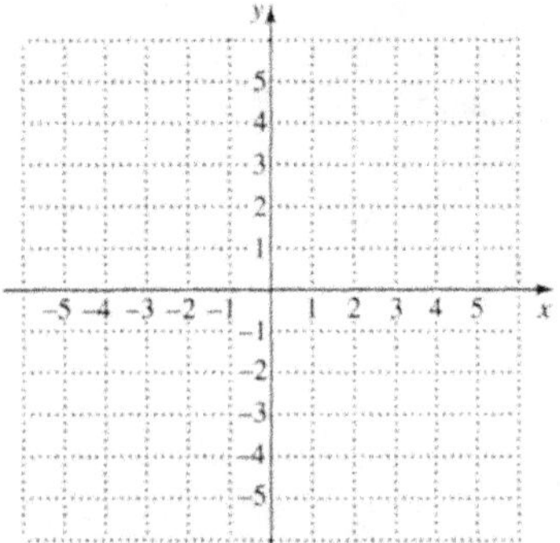

11. $h(x) = \dfrac{x^2 + 6x + 8}{x^2 + 3x - 4}$

12. $f(x) = \dfrac{x^2 - 2}{2x^2 + x - 1}$

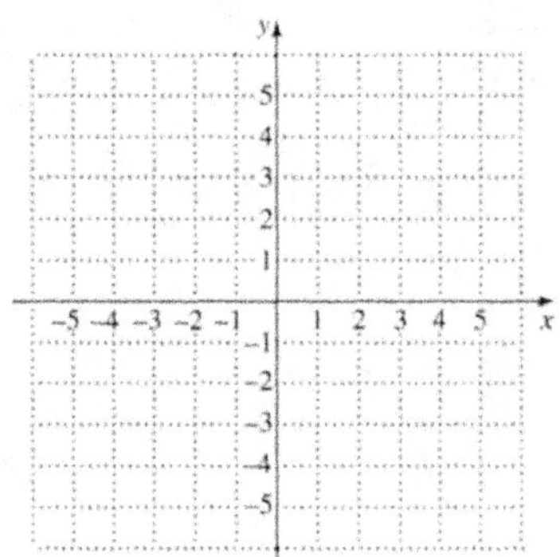

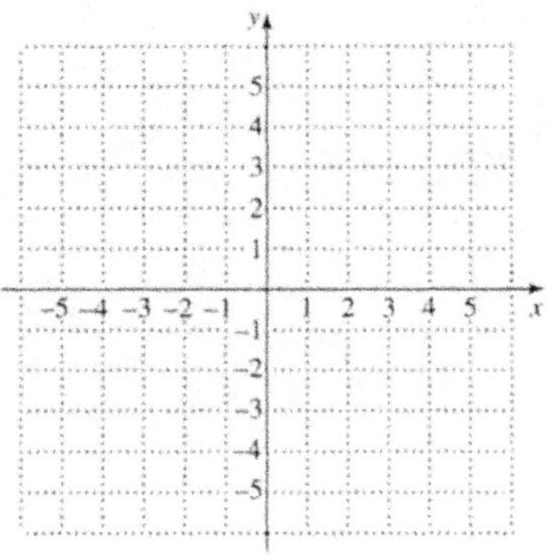

Section 8.6   Polynomial Inequalities and Rational Inequalities

Example 1   Solve: $x^2 - 4x - 5 > 0$.

The related equation is $x^2 - 4x - 5 = 0$.

Graph $f(x) = x^2 - 4x - 5$.

Find the x-intercepts by solving the related equation.

$$x^2 - 4x - 5 = 0$$

$$(x - 5)(x + 1) = 0$$

$$x - 5 = 0 \quad or \quad x + 1 = 0$$

$$x = \boxed{} \quad or \quad x = \boxed{}$$

x-intercepts: $(5, 0)$ and $(-1, 0)$

The intercepts divide the x-axis into three intervals.

Interval	$(-\infty, -1)$	$(-1, 5)$	$(5, \infty)$
Test value	$f(-2) = 7$	$f(0) = -5$	$f(7) = 16$
Sign of $f(x)$	$+$	$-$	$+$

The solution set is $\left(-\infty, \boxed{}\right) \cup \left(\boxed{}, \infty\right)$, or $\left\{x \mid x < -1 \ or \ x \boxed{} 5\right\}$.

Your Turn 1   Solve: $x^2 + x - 6 > 0$.

Example 2   Solve: $x^2 + 3x - 5 \le x + 3$.

$$x^2 + 3x - 5 - x - 3 \le 0$$
$$x^2 + 2x - 8 \le 0$$

Graph $f(x) = x^2 + 2x - 8$.

Find the x-intercepts by solving the related equation.

$$x^2 + 2x - 8 = 0$$
$$\left(x + \boxed{}\right)\left(x - \boxed{}\right) = 0$$
$$x + 4 = 0 \quad or \quad x - 2 = 0$$
$$x = -4 \quad or \quad x = 2$$

x-intercepts: $\left(\boxed{}, 0\right), \left(\boxed{}, 0\right)$

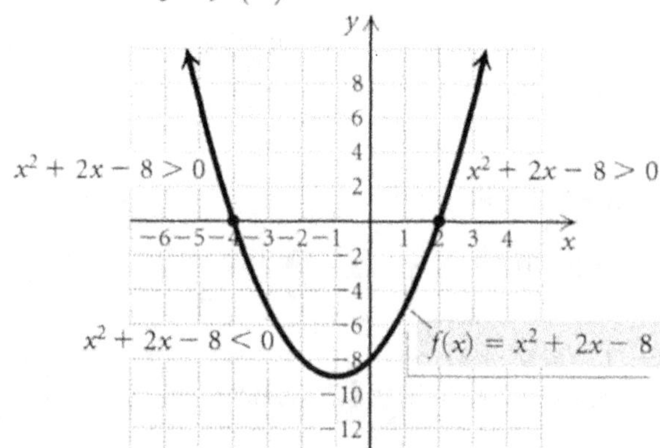

The solution set is $\left[-4, \boxed{}\right]$, or $\left\{x \mid \boxed{} \le x \le 2\right\}$.

Your Turn 2   Solve: $x^2 + 4x - 4 < x + 6$.

Example 3   Solve: $x^3 - x > 0$.

Find the zeros of the related equation.

$$x^3 - x = 0$$
$$x\left(x^2 - \boxed{}\right) = 0$$
$$x(x + 1)(x - 1) = 0$$
$$x = 0 \quad or \quad x + 1 = 0 \quad or \quad \boxed{} = 0$$
$$x = 0 \quad or \quad x = \boxed{} \quad or \quad x = 1$$

x-intercepts: $(0, 0), (-1, 0), \left(\boxed{}, 0\right)$

(continued)

Interval	$(-\infty, -1)$	$(-1, 0)$	$(0, 1)$	$(1, \infty)$
Test value	$f(-2) = -6$	$f(-0.5) = 0.375$	$f(0.5) = -0.375$	$f(2) = 6$
Sign of $f(x)$	$-$	$+$	$-$	$+$

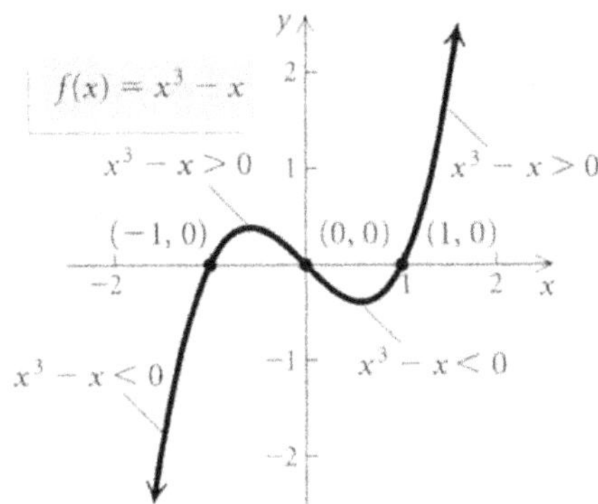

The solution set is $\left(-1, \boxed{}\right) \cup \left(\boxed{}, \infty\right)$, or $\left\{x \,\middle|\, -1 < x < 0 \;or\; x > \boxed{}\right\}$.

Your Turn 3 Solve: $x^3 - 4x \le 0$.

Example 4 Solve: $3x^4 + 10x \le 11x^3 + 4$.

$$3x^4 - 11x^3 + 10x - 4 \le 0$$

Find the zeros of the related equation $3x^4 - 11x^3 + 10x - 4 = 0$.

This equation was solved in Example 5 in Section 8.4. The exact solutions are

$$-1, \; 2 - \sqrt{2}, \; \frac{2}{3}, \; \text{and} \; 2 + \sqrt{2}.$$

The approximate solutions are

$$-1, \; 0.586, \; 0.667, \; \text{and} \; 3.414.$$

(continued)

Interval	Test Value	Sign of $f(x)$
$(-\infty,\ -1)$	$f(-2)=112$	$+$
$\left(-1,\ 2-\sqrt{2}\right)$	$f(0)=-4$	$-$
$\left(2-\sqrt{2},\ \dfrac{2}{3}\right)$	$f(0.6)=0.0128$	$+$
$\left(\dfrac{2}{3},\ 2+\sqrt{2}\right)$	$f(1)=-2$	$-$
$\left(2+\sqrt{2},\ \infty\right)$	$f(4)=100$	$+$

The solution set is $\left[\boxed{},\, 2-\sqrt{2}\right]\cup\left[\dfrac{2}{3},\, 2+\sqrt{2}\right]$, or

$$\left\{x\,\middle|\,-1\le x\le 2-\sqrt{2}\ \text{or}\ \dfrac{2}{3}\le x\le \boxed{}\right\}.$$

Your Turn 4   Solve: $2x^{4}-x^{3}-10x^{2}+x+2\ge 0$
$$(x+2)(2x-1)\left(x^{2}-2x-1\right)\ge 0.$$

Example 5 Solve: $\dfrac{3x}{x+6} < 0$.

Related function:

$$f(x) = \dfrac{3x}{x+6}.$$

To find the critical values, first find the values of x for which the function is not defined.

$$x + 6 = 0 \text{ when } x = \boxed{}.$$

The values of x for which the related function equals 0 are also critical values.

$$f(x) = 0$$

$$\dfrac{3x}{x+6} = 0$$

$$\left(x + \boxed{}\right) \cdot \dfrac{3x}{x+6} = 0 \cdot \left(x + \boxed{}\right)$$

$$3x = 0$$

$$x = \boxed{}$$

Interval	$(-\infty, -6)$	$(-6, 0)$	$(0, \infty)$
Test value	$f(-8) = 12$	$f(-2) = -\dfrac{3}{2}$	$f(3) = 1$
Sign of $f(x)$	+	−	+

The solution set is $\left(-6, \boxed{}\right)$, or $\left\{ x \mid \boxed{} < x < 0 \right\}$.

Your Turn 5 Solve: $\dfrac{2x}{x-3} < 0$.

Example 6   Solve: $\dfrac{x+1}{2x-4} \le 1$.

$$\dfrac{x+1}{2x-4} - 1 \le 0$$

Related function:

$$f(x) = \dfrac{x+1}{2x-4} - 1.$$

Critical values are the values of x for which $f(x)$ is not defined or 0. $f(x)$ is not defined when $2x - 4 = 0$, or $x = 2$.

Now solve $f(x) = 0$.

$$\dfrac{x+1}{2x-4} - 1 = 0$$

$$\dfrac{x+1}{2x-4} = 1$$

$$x + 1 = 2x - 4$$

$$1 = x - 4$$

$$\boxed{} = x$$

The critical values are 2 and 5.

Interval	Test Value	Sign of $f(x)$
$(-\infty,\ 2)$	$f(0) = -\dfrac{5}{4}$	$\boxed{}$
$(2,\ 5)$	$f(3) = 1$	$\boxed{}$
$(5,\ \infty)$	$f(6) = -\dfrac{1}{8}$	$\boxed{}$

The critical value 2 is not in the solution set. The critical value 5 is in the solution set.

The solution set is $\left(-\infty,\ \boxed{}\right) \cup \left[\boxed{},\ \infty\right)$.

Your Turn 6 Solve: $\dfrac{x+2}{3x-6} \geq 1$.

Example 7 Solve: $\dfrac{x-3}{x+4} \geq \dfrac{x+2}{x-5}$.

$$\frac{x-3}{x+4} - \frac{x+2}{x-5} \geq 0$$

Related function: $f(x) = \dfrac{x-3}{x+4} - \dfrac{x+2}{x-5}$.

Find the critical values.

$f(x)$ is not defined for $x = -4$ and $x = \boxed{}$.

Now solve $f(x) = 0$.

$$\frac{x-3}{x+4} - \frac{x+2}{x-5} = 0$$

$$(x+4)\cdot(x-5)\cdot\left[\frac{x-3}{x+4} - \frac{x+2}{x-5}\right] = 0\cdot(x+4)\cdot(x-5)$$

$$(x-5)\cdot\left(x - \boxed{}\right) - (x+4)\cdot\left(x + \boxed{}\right) = 0$$

$$x^2 - 8x + 15 - \left(x^2 + \boxed{}\,x + 8\right) = 0$$

$$x^2 - 8x + 15 - x^2 - 6x - 8 = 0$$

$$-14x + 7 = 0$$

$$-14x = -7$$

$$x = \frac{7}{14} = \boxed{}$$

(continued)

Critical values: $\boxed{}$, $\dfrac{1}{2}$, and $\boxed{}$

Interval	Test Value	Sign of $f(x)$
$(-\infty,\ -4)$	$f(-5)=7.7$	$+$
$\left(-4,\ \dfrac{1}{2}\right)$	$f(-2)=-2.5$	$\boxed{}$
$\left(\dfrac{1}{2},\ 5\right)$	$f(3)=2.5$	$\boxed{}$
$(5,\ \infty)$	$f(6)=-7.7$	$-$

The critical values -4 and 5 are not in the domain. The critical value $\dfrac{1}{2}$ is.

The solution set is $\left(-\infty, \boxed{}\right) \cup \left[\dfrac{1}{2}, 5\right)$, or $\left\{x \middle| x < \boxed{} \ or \ \dfrac{1}{2} \le x < \boxed{}\right\}$.

Your Turn 7   Solve: $\dfrac{x+1}{x-2} \le \dfrac{x+3}{x-4}$.

Practice Exercises

Solve.

1. $x^2 + 2x - 15 \le 0$

2. $x^2 + 2x > 2x + 1$

3. $x^2 + 3x - 6 \ge 2x + 6$

4. $x^2 - 3x + 2 < 0$

5. $x^3 - 9x < 0$

6. $x^3 + x^2 \ge 4x + 4$

7. $\dfrac{4x}{x-2} \le 0$

8. $\dfrac{x}{x+4} \ge 0$

9. $\dfrac{x-3}{4x+4} > 1$

10. $\dfrac{2x-6}{x+1} < 3$

11. $\dfrac{x-1}{x+3} \ge \dfrac{2x+1}{2x}$

12. $\dfrac{x+3}{x-4} > \dfrac{x+1}{x+2}$

Section 9.1 The Composition of Functions

Composition of Functions

The **composite function** $f \circ g$, the **composition** of f and g, is defined as

$$(f \circ g)(x) = f(g(x)),$$

where x is in the domain of g and $g(x)$ is in the domain of f.

Example 1 Given that $f(x) = 2x - 5$ and $g(x) = x^2 - 3x + 8$, find each of the following.

a) $(f \circ g)(x)$ and $(g \circ f)(x)$

$$(f \circ g)(x) = f(g(x)) = f\left(x^2 - 3x + \boxed{}\right)$$

$$= 2\left(x^2 - 3x + 8\right) - 5 = 2x^2 - 6x + \boxed{} - 5$$

$$= 2x^2 - 6x + 11$$

$$(g \circ f)(x) = g(f(x)) = g\left(2x - \boxed{}\right)$$

$$= (2x - 5)^2 - 3(2x - 5) + 8$$

$$= 4x^2 - \boxed{}x + 25 - 6x + 15 + 8$$

$$= 4x^2 - 26x + \boxed{}$$

b) $(f \circ g)(7)$ and $(g \circ f)(7)$

$$(f \circ g)(7) = f\left(7^2 - 3 \cdot \boxed{} + 8\right) = f(36)$$

$$= 2 \cdot \boxed{} - 5 = 67$$

$$(g \circ f)(7) = g(2 \cdot 7 - 5) = g\left(\boxed{}\right)$$

$$= 9^2 - 3 \cdot 9 + \boxed{} = 62$$

We could also use the results from part (a) to find $(f \circ g)(7)$ and $(g \circ f)(7)$.

$$(f \circ g)(x) = 2x^2 - 6x + 11$$

$$(f \circ g)(7) = 2 \cdot 7^2 - 6 \cdot \boxed{} + 11 = 67$$

$$(g \circ f)(x) = 4x^2 - 26x + 48$$

$$(g \circ f)(7) = 4 \cdot 7^2 - 26 \cdot 7 + \boxed{} = 62$$

(continued)

c) $(g \circ g)(1)$

$$(g \circ g)(1) = g(g(1)) = g\left(1^2 - 3 \cdot \boxed{} + 8\right)$$
$$= g(6)$$
$$= \boxed{}^2 - 3 \cdot 6 + 8 = 26$$

d) $(f \circ f)(x)$

$$(f \circ f)(x) = f(f(x)) = f(2x - 5)$$
$$= 2\left(2x - \boxed{}\right) - 5$$
$$= \boxed{}x - 10 - 5$$
$$= 4x - 15$$

Your Turn 1   Given that $f(x) = 3x + 2$ and $g(x) = x^2 - 2x + 7$, find each of the following.

a) $(f \circ g)(x)$ and $(g \circ f)(x)$

b) $(f \circ g)(2)$ and $(g \circ f)(2)$

(continued)

c) $(g \circ g)(1)$

d) $(f \circ f)(x)$

Example 2

a) Given that $f(x) = \sqrt{x}$ and $g(x) = x - 3$, find $f \circ g$ and $g \circ f$.

$$(f \circ g)(x) = f(g(x)) = \sqrt{x - 3}$$

$$(g \circ f)(x) = g(f(x)) = \boxed{} - 3$$

b) Given that $f(x) = \sqrt{x}$ and $g(x) = x - 3$, find the domain of $f \circ g$ and the domain of $g \circ f$.

$$(f \circ g)(x) = \sqrt{x - 3}$$

Domain of $f(x) = \left[\boxed{}, \infty\right)$

Domain of $g(x) = (-\infty, \infty)$

Domain of $f \circ g = \{x \mid x \geq 3\}$, or $\left[\boxed{}, \infty\right)$ because $g(x) \geq 0$, or $x - 3 \geq 0$, when $x \geq 3$.

We could also get this by considering the domain of $(f \circ g)(x) = \sqrt{x - 3}$.

$$(g \circ f)(x) = \sqrt{x} - 3$$

Domain of $g \circ f$ = Domain of $f(x) = [0, \infty)$

We could also find this by considering the domain of $(g \circ f)(x) = \sqrt{x} - 3$.

Your Turn 2

a) Given that $f(x) = x + 4$ and $g(x) = \sqrt{x}$, find $f \circ g$ and $g \circ f$.

b) Given that $f(x) = x + 4$ and $g(x) = \sqrt{x}$, find the domain of $f \circ g$ and the domain of $g \circ f$.

Example 3 Given that $f(x) = \dfrac{1}{x-2}$ and $g(x) = \dfrac{5}{x}$, find $f \circ g$ and $g \circ f$ and the domain of each.

$$(f \circ g)(x) = f\big(g(x)\big) = \dfrac{1}{\dfrac{5}{x} - 2}$$

$$= \dfrac{1}{\dfrac{5-2x}{x}} = 1 \cdot \dfrac{x}{5-2x}$$

$$= \dfrac{x}{5-2x}$$

$$(g \circ f)(x) = g\big(f(x)\big) = \dfrac{5}{\dfrac{1}{x-2}}$$

$$= 5 \cdot \dfrac{x-2}{1}$$

$$= 5(x-2)$$

(continued)

Domain of g: $\{x \mid x \neq 0\}$

Domain of f: $\{x \mid x \neq 2\}$

In addition to 0, values of x for which $g(x) = 2$ must be excluded from the domain of $f \circ g$. We have

$$\frac{5}{x} = 2$$

$$5 = 2x$$

$$\frac{5}{2} = x.$$

Both $\dfrac{5}{2}$ and $\boxed{}$ must be excluded from the domain of $f \circ g$.

Domain of $f \circ g$: $\left(-\infty, \boxed{}\right) \cup \left(0, \dfrac{5}{2}\right) \cup \left(\boxed{}, \infty\right)$

In addition to 2, values of x for which $f(x) = 0$ must be excluded from the domain of $g \circ f$. We have

$$\frac{1}{x-2} = 0$$

$$\boxed{} = 0 \qquad \text{False}$$

Thus, only 2 must be excluded.

Domain of $g \circ f$: $\left(-\infty, \boxed{}\right) \cup (2, \infty)$

Your Turn 3   Given that $f(x) = \dfrac{1}{x-5}$ and $g(x) = \dfrac{3}{x}$, find $f \circ g$ and $g \circ f$ and the domain of each.

$f \circ g =$

$g \circ f =$

(continued)

Domain of $f \circ g$:

Domain of $g \circ f$:

Example 4 If $h(x) = (2x-3)^5$, find $f(x)$ and $g(x)$ such that $h(x) = (f \circ g)(x)$.

For $f(x) = x^5$ and $g(x) = 2x - \boxed{}$,

$$h(x) = (f \circ g)(x) = f(g(x))$$
$$= f(2x-3)$$
$$= \left(\boxed{}\right)^5.$$

These functions give us $h(x)$, but there are other choices for $f(x)$ and $g(x)$.

For $f(x) = (x+7)^5$ and $g(x) = 2x - 10$,

$$h(x) = (f \circ g)(x) = f(g(x))$$
$$= f\left(2x - \boxed{}\right)$$
$$= (2x - 10 + 7)^5$$
$$= \left(2x - \boxed{}\right)^5.$$

This is also a correct choice. There are others as well.

Your Turn 4   If $h(x) = (3x-4)^7$, find $f(x)$ and $g(x)$ such that $h(x) = (f \circ g)(x)$.

Example 5   If $h(x) = \dfrac{1}{(x+3)^3}$, find $f(x)$ and $g(x)$ such that $h(x) = (f \circ g)(x)$.

For $f(x) = \dfrac{1}{x^3}$ and $g(x) = x + \boxed{}$,

$$h(x) = (f \circ g)(x) = f(g(x))$$
$$= f\left(\boxed{}\right)$$
$$= \frac{1}{(x+3)^3}.$$

These functions give us $h(x)$, but there are other correct choices for $f(x)$ and $g(x)$.

For $f(x) = \dfrac{1}{x}$ and $g(x) = (x+3)^3$,

$$h(x) = (f \circ g)(x) = f(g(x))$$
$$= f\left((x+3)^3\right)$$
$$= \frac{1}{\boxed{}}.$$

This is also a correct choice. There are others as well.

Your Turn 5   If $h(x) = \dfrac{1}{(x-4)^5}$, find $f(x)$ and $g(x)$ such that $h(x) = (f \circ g)(x)$.

Practice Exercises

1. Given that $f(x) = 4x - 1$ and $g(x) = x^2 - 3x + 2$, find each of the following.

   a) $(f \circ g)(x)$ and $(g \circ f)(x)$ b) $(f \circ g)(3)$ and $(g \circ f)(3)$

   c) $(g \circ g)(-2)$ d) $(f \circ f)(x)$

2. Given that $f(x) = \sqrt{x + 1}$ and $g(x) = 2x - 3$

   a) find $f \circ g$ and $g \circ f$,

   b) find the domain of $f \circ g$ and the domain of $g \circ f$.

3. Given that $f(x) = \dfrac{4}{x}$ and $g(x) = \dfrac{1}{3x+5}$,

   a) find $f \circ g$ and $g \circ f$,

   b) find the domain of $f \circ g$ and the domain of $g \circ f$.

4. If $h(x) = \dfrac{1}{(x+7)^5}$, find $f(x)$ and $g(x)$ such that $h(x) = (f \circ g)(x)$.

5. If $h(x) = \sqrt[3]{x^2 - x + 3}$, find $f(x)$ and $g(x)$ such that $h(x) = (f \circ g)(x)$.

Section 9.2 Inverse Functions

Inverse Relation

Interchanging the first and second coordinates of each ordered pair in a relation produces the **inverse relation**.

Example 1 Consider the relation g given by

$$g = \{(2, 4), (-1, 3), (-2, 0)\}.$$

Graph the relation in blue. Find the inverse relation and graph it in red.

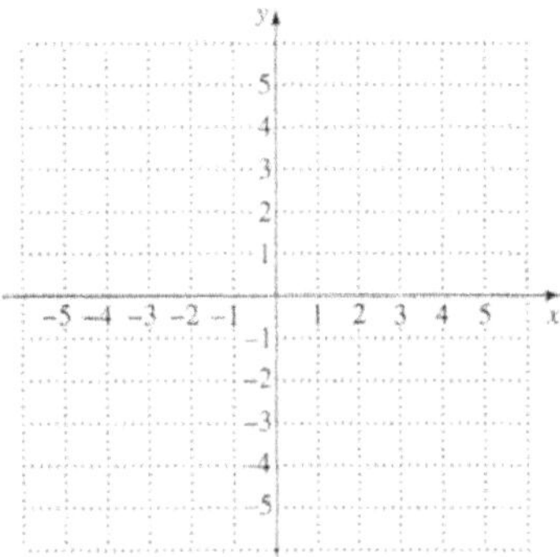

Inverse of $g = \{ \boxed{} \}$

Your Turn 1 Consider the relation h given by

$$h = \{(3, -4), (-1, -2), (5, 1), (0, 4)\}.$$

Graph the relation in blue. Find the inverse relation and graph it in red.

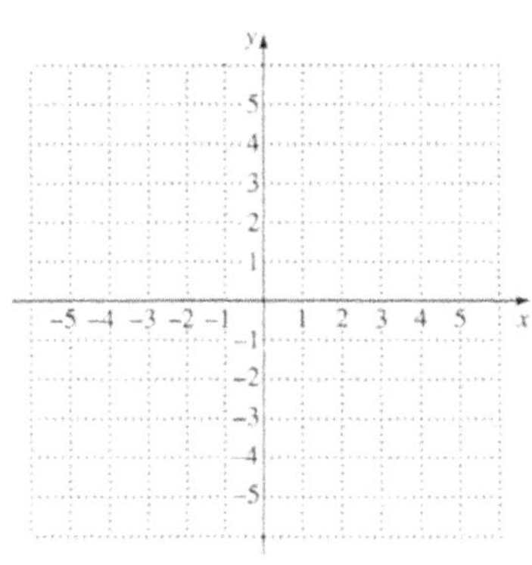

Inverse of $h = \{ \}$

Inverse Relation

If a relation is defined by an equation, interchanging the variables produces an equation of the inverse relation.

Example 2 Find an equation for the inverse of the relation: $y = x^2 - 5x.$

The equation of the inverse relation is $\boxed{}$.

Your Turn 2   Find an equation for the inverse of the relation $y = x^2 - 2x + 3$.

One-to-One Functions

A function f is **one-to-one** if different inputs have different outputs – that is,

$\quad$ if $a \neq b$, then $f(a) \neq f(b)$.

Or a function is **one-to-one** if when the outputs are the same, the inputs are the same – that is,

$\quad$ if $f(a) = f(b)$, then $a = b$.

If the inverse of a function f is also a function, it is named f^{-1} (read "f-inverse").

One-to-One Functions and Inverses
- If a function f is one-to-one, then its inverse f^{-1} is a function.
- The domain of a one-to-one function f is the range of the inverse f^{-1}.
- The range of a one-to-one function f is the domain of the inverse f^{-1}.
- A function that is increasing over its entire domain or is decreasing over its entire domain is a one-to-one function.

Example 3   Given the function f described by $f(x) = 2x - 3$, prove that f is one-to-one (that is, it has an inverse that is a function).

Given $f(x) = 2x - 3$.

If $f(a) = f(b)$, then $\boxed{}$.

$\quad f(a) = \boxed{} \qquad\qquad f(b) = \boxed{}$

$\quad 2a - 3 = 2b - 3$

$\quad \boxed{} = \boxed{}$

$\quad \boxed{} = \boxed{}$

Thus, if $f(a) = f(b)$, then $a = b$.

This tells us that $f(x)$ is one-to-one.

Your Turn 3 Given the function f described by $f(x) = 7x + 10$, prove that f is one-to-one (that is, it has an inverse that is a function).

Example 4 Given the function g described by $g(x) = x^2$, prove that g is not one-to-one.

To show that g is not a one-to-one function, we need to find two numbers such that $a \neq b$ and $g(a) = g(b)$.

Try the values $x = 3$ and $x = -3$.

$$g\left(\boxed{}\right) = 3^2 = \boxed{}$$

$$g\left(\boxed{}\right) = (-3)^2 = \boxed{}$$

Thus, g _________ one-to-one.
 is / is not

Your Turn 4 Given the function $f(x) = x^4 + 1$, prove that f is not one-to-one.

Horizontal-Line Test

If it is possible for a horizontal line to intersect the graph of a function more than once, then the function is *not* one-to-one and its inverse is *not* a function.

Example 5 From the graphs shown, determine whether each function is one-to-one and thus has an inverse that is a function.

a)

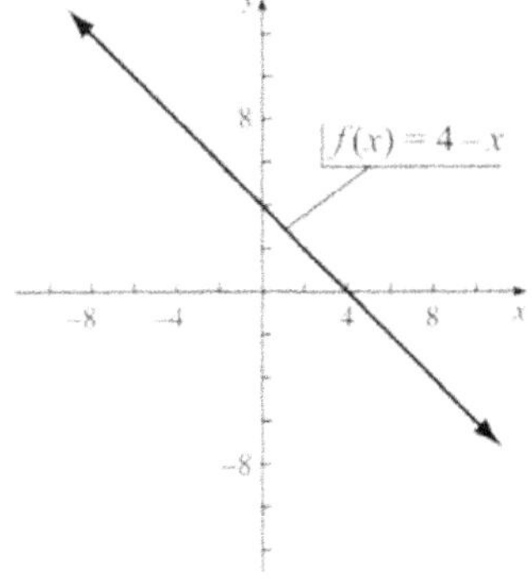

b)

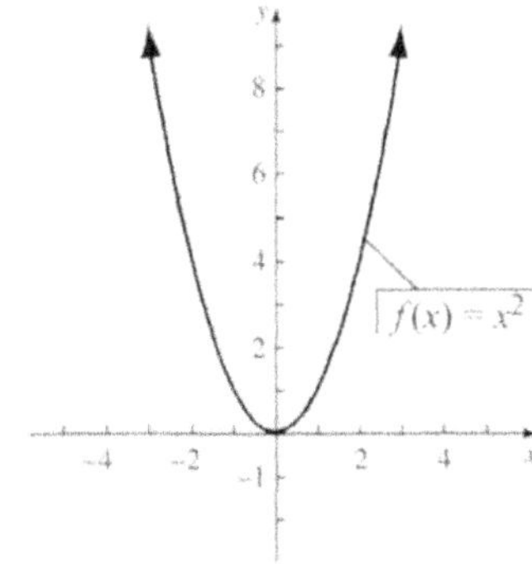

one-to-one / not one-to-one $\qquad\qquad$ one-to-one / not one-to-one

(continued)

c)

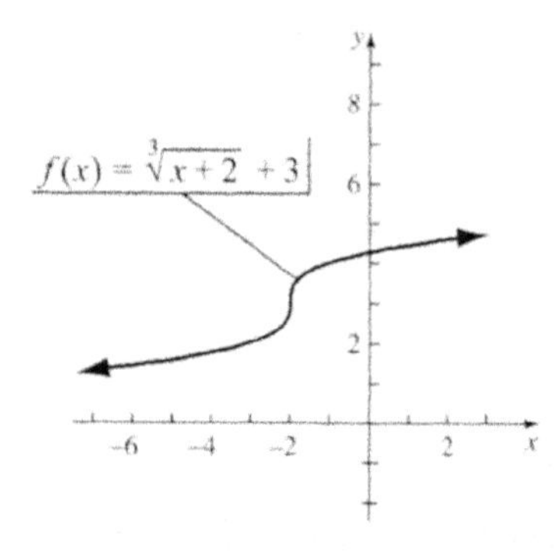

____
one-to-one / not one-to-one

d)

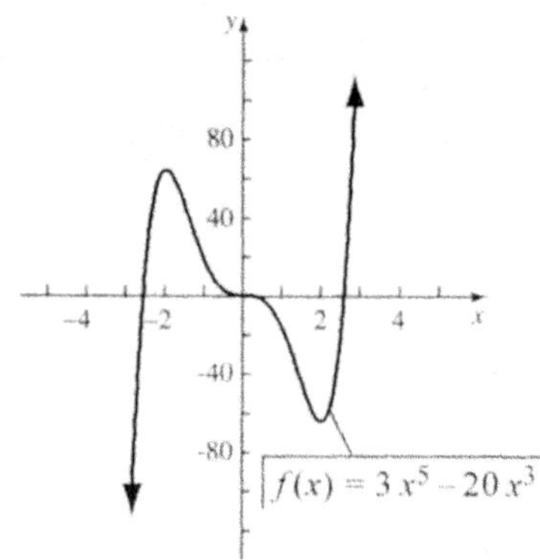

____
one-to-one / not one-to-one

✐ **Your Turn 5** From the graphs shown, determine whether each function is one-to-one and thus has an inverse.

a)

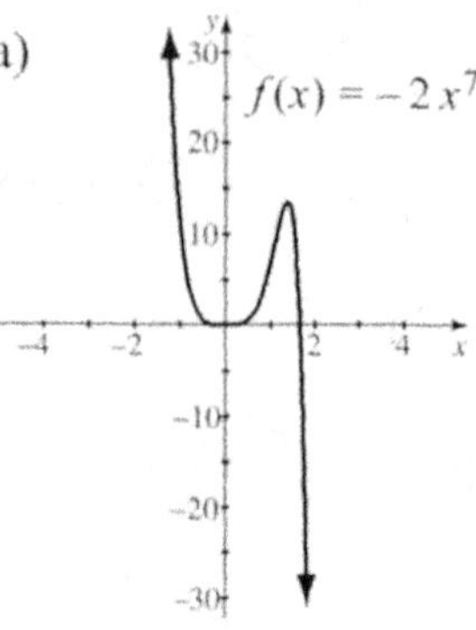

____
one-to-one / not one-to-one

b)

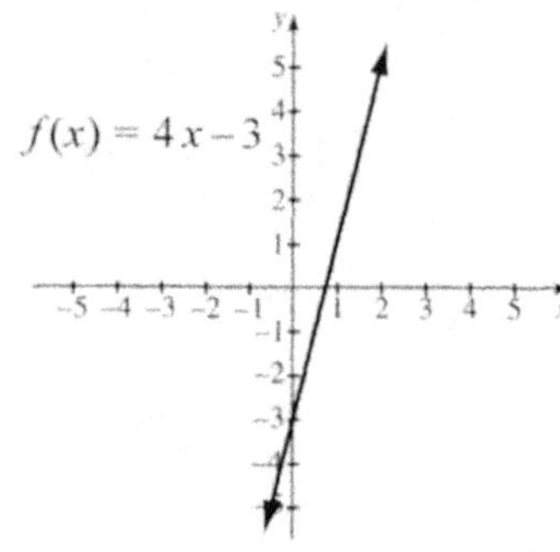

____
one-to-one / not one-to-one

c)

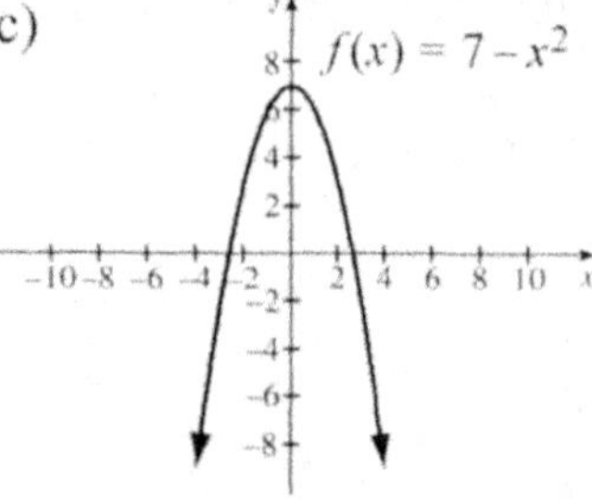

____
one-to-one / not one-to-one

d)

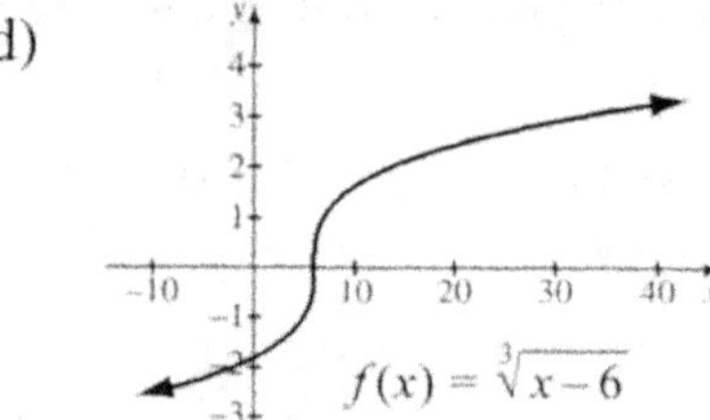

____
one-to-one / not one-to-one

Obtaining a Formula for an Inverse

If a function f is one-to-one, a formula for its inverse can generally be found as follows.

1. Replace $f(x)$ with y.

2. Interchange x and y.

3. Solve for y.

4. Replace y with $f^{-1}(x)$.

Example 6 Determine whether the function $f(x) = 2x - 3$ is one-to-one, and if it is, find a formula for $f^{-1}(x)$.

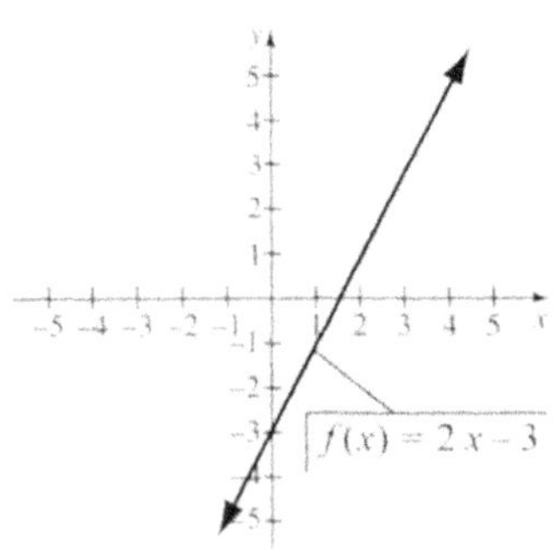

The graph passes the horizontal-line test, so it is one-to-one. Find the inverse.

1. Replace $f(x)$ with y:

   $$\boxed{}$$

2. Interchange x and y.

   $$\boxed{}$$

3. Solve for y:

   $$x + 3 = 2y$$

   $$\boxed{} = y$$

4. Replace y with $f^{-1}(x)$:

   $$f^{-1}(x) = \boxed{}$$

Your Turn 6 Determine whether the function $f(x) = 5 - 2x$ is one-to-one, and if it is, find a formula for $f^{-1}(x)$.

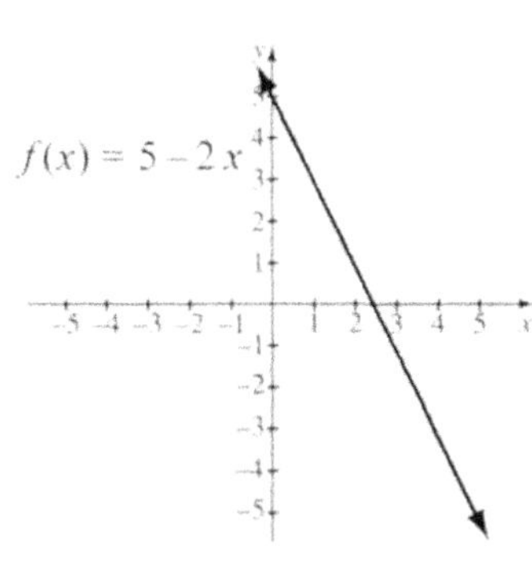

Example 7   Graph $f(x) = 2x - 3$ and $f^{-1}(x) = \dfrac{x+3}{2}$ using the same set of axes. Then compare the two graphs.

x	$f(x) = 2x - 3$
-1	-5
0	
2	
3	

x	$f^{-1}(x) = \dfrac{x+3}{2}$
-5	-1
-3	
1	
3	

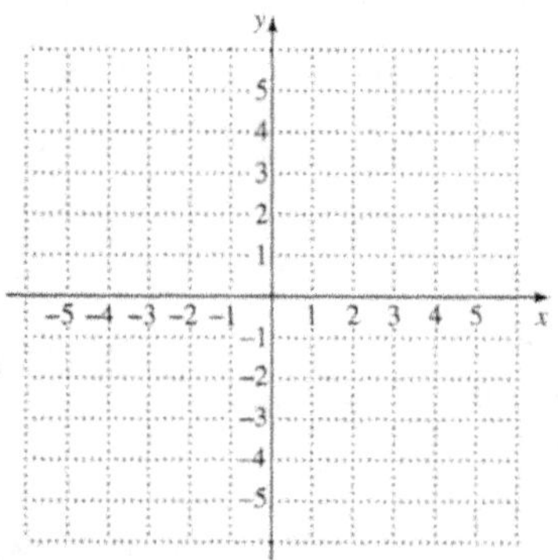

The graphs are reflections of each other across the line $y = x$.

The graph of f^{-1} is a reflection of the graph of f across the line $y = x$.

Your Turn 7   Graph $f(x) = 1 - 2x$ and $f^{-1}(x) = \dfrac{1-x}{2}$ using the same set of axes. Then compare the two graphs.

x	$f(x) = 1 - 2x$

x	$f^{-1}(x) = \dfrac{1-x}{2}$

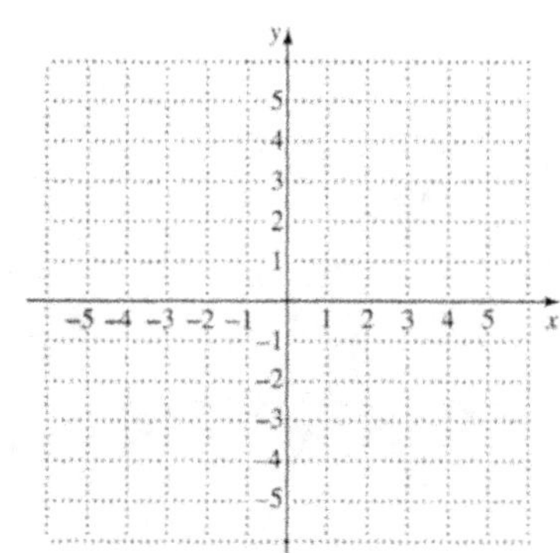

Example 8   Consider $g(x) = x^3 + 2$.

   a) Determine whether the function is one-to-one.

   b) If it is one-to-one, find a formula for its inverse.

   c) Graph the function and its inverse.

a) $g(x)$ passes the horizontal-line test, so it is one-to-one.

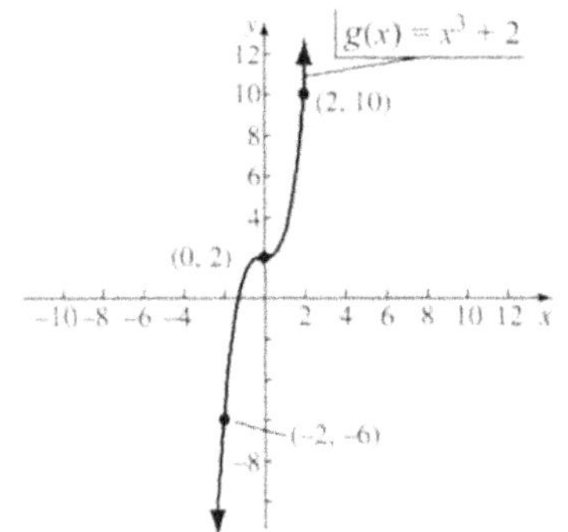

b) **1.** Replace $g(x)$ with y.

$$y = x^3 + 2$$

   2. Interchange x and y.

   3. Solve for y.

$$x = y^3 + 2$$

$$\boxed{} = y^3$$

$$\boxed{} = y$$

c)
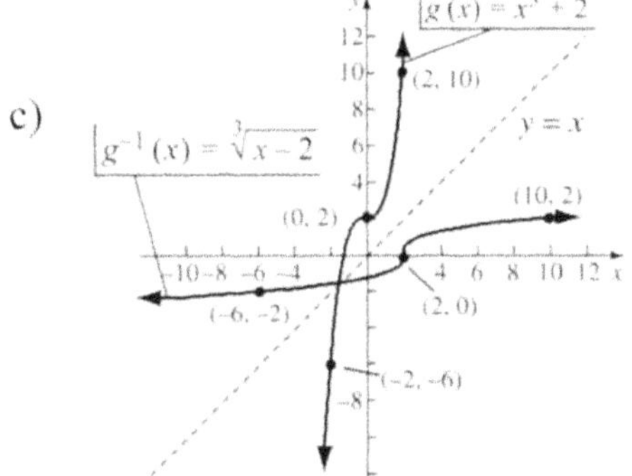

   4. Replace y with $g^{-1}(x)$

$$g^{-1}(x) = \boxed{}$$

✏ **Your Turn 8**   Consider $h(x) = 3 - x^3$

   a) Determine whether the function is one-to-one.

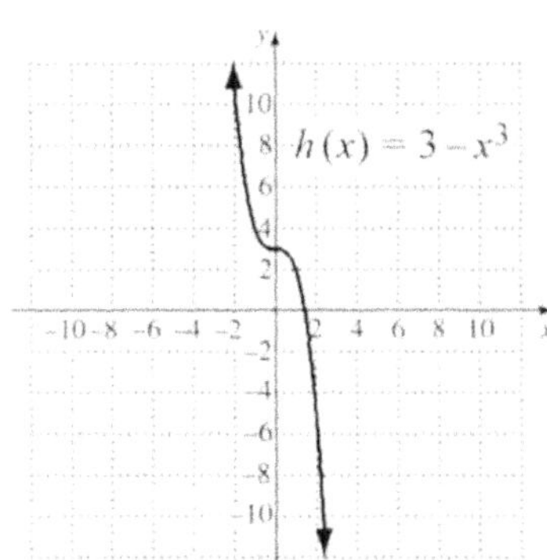

   b) If it is one-to-one, find a formula for its inverse.

(continued)

c) Graph the function and its inverse.

x	$h(x) = 3 - x^3$		x	$h^{-1}(x) =$

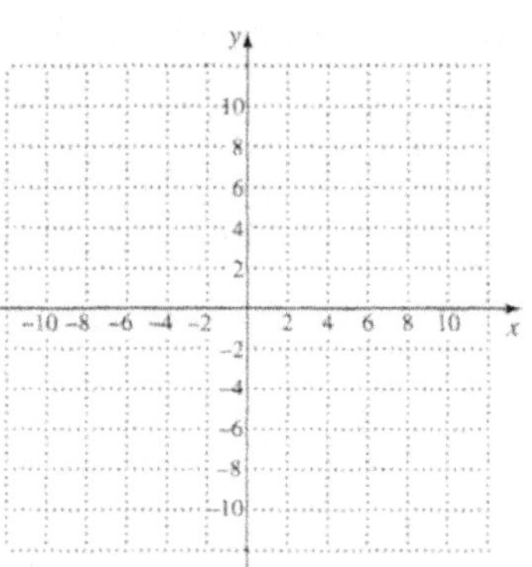

If a function f is one-to-one, then f^{-1} is the unique function such that each of the following holds:

$$\left(f^{-1} \circ f\right)(x) = f^{-1}\left(f(x)\right) = x, \quad \text{for each } x \text{ in the domain of } f, \text{ and}$$

$$\left(f \circ f^{-1}\right)(x) = f\left(f^{-1}(x)\right) = x, \quad \text{for each } x \text{ in the domain of } f^{-1}.$$

Example 9 Given that $f(x) = 5x + 8$, use composition of functions to show that

$$f^{-1}(x) = \frac{x-8}{5}.$$

$$\left(f^{-1} \circ f\right)(x) = f^{-1}\left(f(x)\right) = f^{-1}\left(\boxed{}\right) = \frac{\left(\boxed{} - 8\right)}{5} = \frac{5x}{5} = \boxed{}$$

$$\left(f \circ f^{-1}\right)(x) = f\left(f^{-1}(x)\right) = f\left(\boxed{}\right) = 5\left(\boxed{}\right) + 8 = x - 8 + 8 = \boxed{}$$

These two compositions show us that $f^{-1}(x) = \dfrac{x-8}{5}$ is the inverse of $f(x) = 5x + 8$.

Your Turn 9 Given that $g(x) = 3x - 4$, use composition of functions to show that

$$g^{-1}(x) = \frac{x+4}{3}.$$

Practice Exercises

1. Find the inverse relation of $f = \{(-6, 0), (2, 3), (5, -4)\}$.

   Inverse of $f = \{ \qquad\qquad\qquad \}$

2. Find an equation for the inverse of the relation:

   $y = 2x^2 + 4x - 7.$

3. Given the function g described by $g(x) = -2x + 7$, prove that g is one-to-one (that is, it has an inverse that is a function).

4. Given the function f described by $f(x) = 3 - x^2$, prove that f is not one-to-one.

From the graph shown, determine whether the function is one-to-one and thus has an inverse.

5. 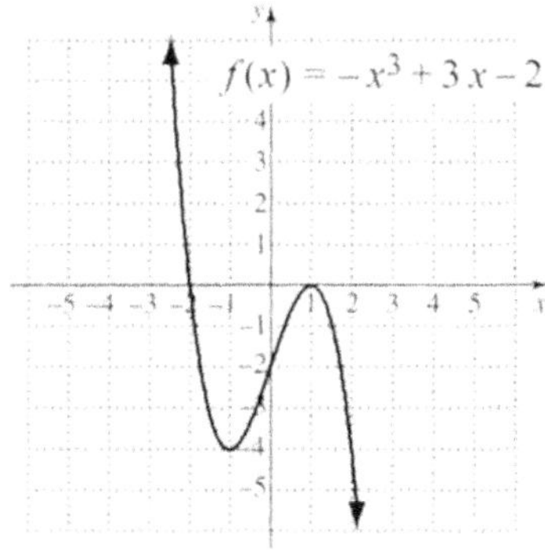

   one-to-one / not one-to-one

6. 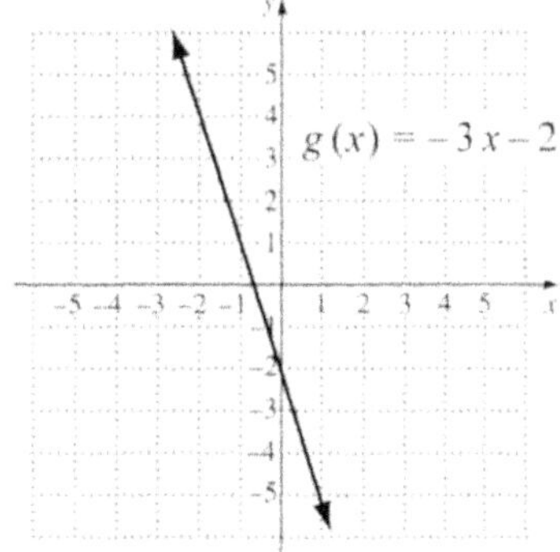

   one-to-one / not one-to-one

For each of the following one-to-one functions, find a formula for its inverse and then graph the function and its inverse using the same set of axes.

7. $g(x) = 3x - 1$

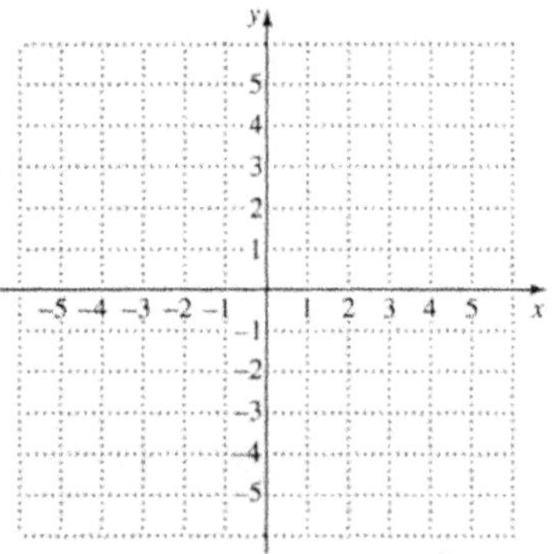

8. $f(x) = \dfrac{1}{2}x^3 + 1$

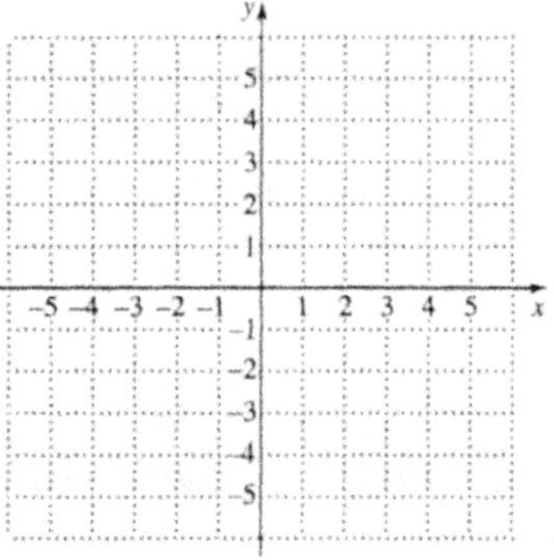

9. $h(x) = \dfrac{-4}{x}$

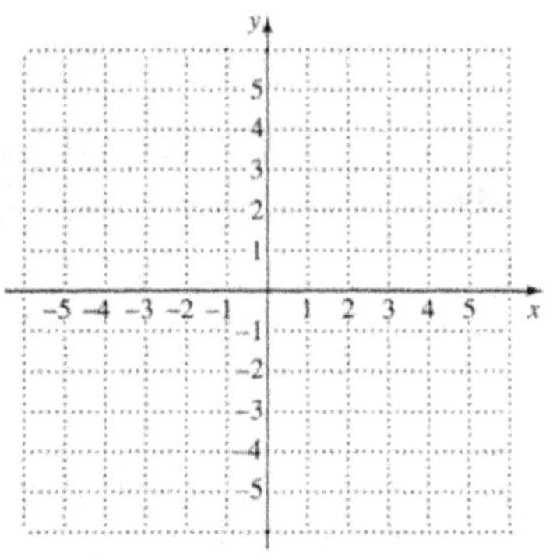

10.   Given that $h(x) = 5 - 6x$, use composition of functions to show that $h^{-1} = \dfrac{5 - x}{6}$.

Section 9.3 Exponential Functions and Graphs

Exponential Function

The function $f(x) = a^x$, where x is a real number, $a > 0$ and $a \neq 1$, is called the **exponential function, base a.**

Example 1 Graph the exponential function $y = f(x) = 2^x$.

x	$y = f(x) = 2^x$	(x, y)
0	1	$(0, 1)$
1	☐	$(1, 2)$
2	4	$\left(2, \Box\right)$
3	8	$(3, 8)$
−1	$\dfrac{1}{2}$	$\left(-1, \dfrac{1}{2}\right)$
−2	☐	$\left(-2, \dfrac{1}{4}\right)$
−3	$\dfrac{1}{8}$	$\left(\Box, \dfrac{1}{8}\right)$

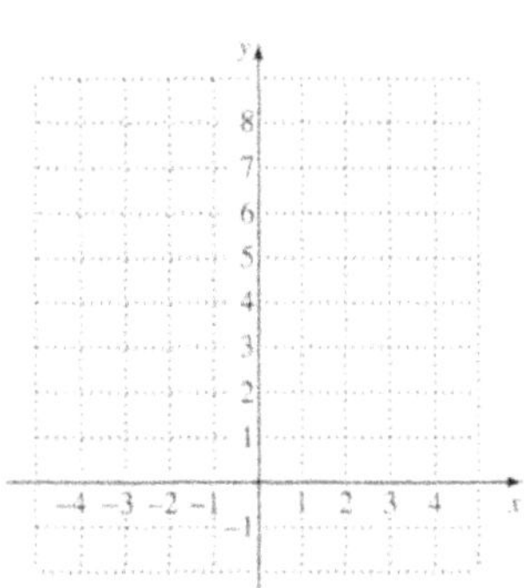

Your Turn 1 Graph the exponential function $y = f(x) = 3^x$.

x	$y = f(x) = 3^x$	(x, y)

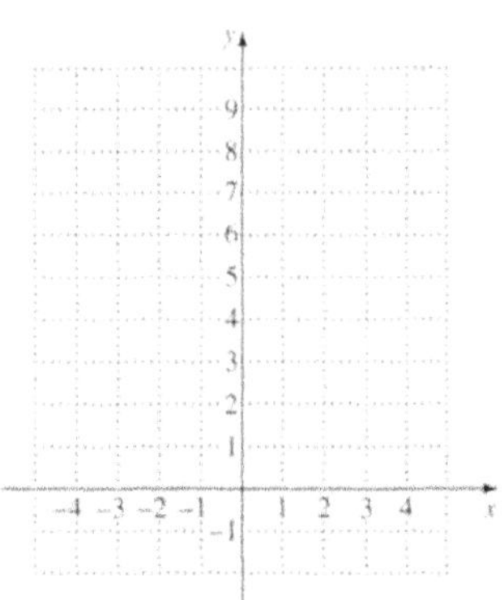

Example 2 Graph the exponential function $y = f(x) = \left(\dfrac{1}{2}\right)^x$.

Points of $g(x) = 2^x$	Points of $f(x) = \left(\dfrac{1}{2}\right)^x = 2^{-x}$
$(0, 1)$	$(0, 1)$
$(1, 2)$	$(-1, 2)$
$(2, 4)$	$(\boxed{}, 4)$
$(3, 8)$	$(-3, 8)$
$\left(-1, \dfrac{1}{2}\right)$	$\left(\boxed{}, \dfrac{1}{2}\right)$
$\left(-2, \dfrac{1}{4}\right)$	$\left(2, \dfrac{1}{4}\right)$
$\left(-3, \dfrac{1}{8}\right)$	$\left(3, \dfrac{1}{8}\right)$

$$y = f(x) = \left(\dfrac{1}{2}\right)^x = 2^{-x}$$

$$y = g(x) = 2^x$$

Your Turn 2 Graph the exponential function $y = f(x) = \left(\dfrac{1}{3}\right)^x$.

Points of $g(x) = 3^x$	Points of $f(x) = \left(\dfrac{1}{3}\right)^x = 3^{-x}$

Example 3   Graph each of the following. Before doing so, describe how each graph can be obtained from the graph of $f(x) = 2^x$.

a) $f(x) = 2^{x-2}$ b) $f(x) = 2^x - 4$ c) $f(x) = 5 - 0.5^x$

a) Graph $f(x) = 2^{x-2}$.

The graph of $f(x) = 2^{x-2}$ is the graph of $f(x) = 2^x$ shifted ___________ 2 units.

left / right

x	$f(x) = 2^{x-2}$
-1	$\dfrac{1}{8}$
0	
1	$\dfrac{1}{2}$
2	
3	2
4	4

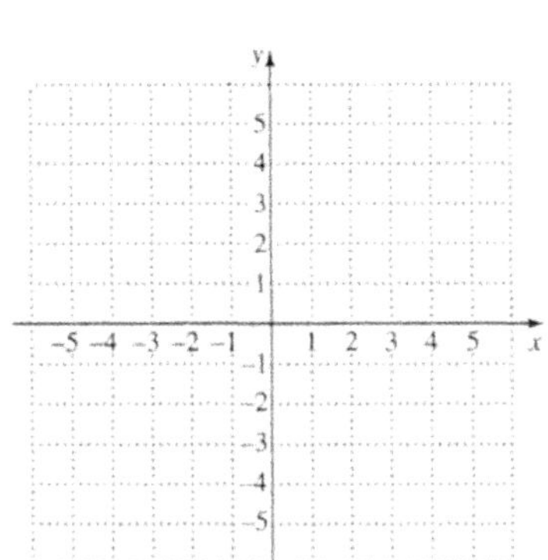

b) Graph $f(x) = 2^x - 4$.

Next, we consider the graph of $f(x) = 2^x - 4$, which is the graph of $f(x) = 2^x$ shifted ___________ 4 units.

up / down

x	$f(x) = 2^x - 4$
-2	$-3\dfrac{3}{4}$
-1	
0	-3
1	
2	0
3	4

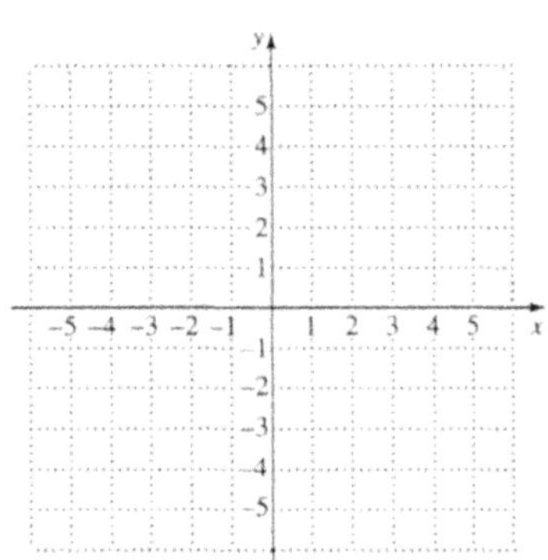

(continued)

c) Graph $f(x) = 5 - 0.5^x$.

$$f(x) = 5 - 0.5^x = 5 - \left(\frac{1}{2}\right)^x$$

$$= 5 - \left(2^{-1}\right)^x$$

$$= 5 - 2^{-x}$$

The graph of $f(x) = 5 - 0.5^x$ is the graph of $f(x) = 2^x$ reflected across the

__________ , then reflected across the __________ , and finally shifted
   x-axis / *y*-axis   *x*-axis / *y*-axis

__________ 5 units.
up / down

x	$f(x) = 5 - 0.5^x$
-3	-3
-2	
-1	3
0	
1	$4\frac{1}{2}$
2	$4\frac{3}{4}$

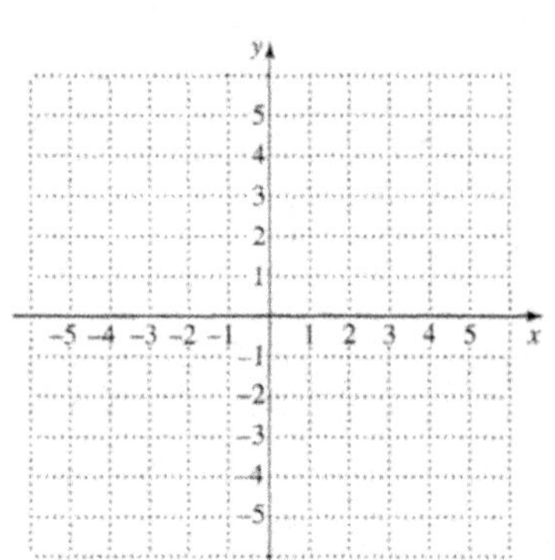

Your Turn 3   Graph each of the following. Before doing so, describe how each graph can be obtained from $f(x) = 3^x$.

a) Graph $f(x) = 3^{x+2}$.

The graph of $f(x) = 3^{x+2}$ is the graph of $f(x) = 3^x$ shifted __________ 2 units.
                                                  left / right

x	$f(x) = 3^{x+2}$

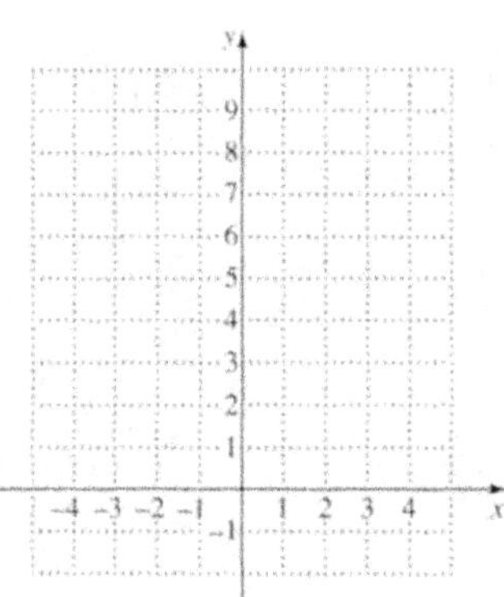

(continued)

b) Graph $f(x) = 3^x - 5$.

The graph of $f(x) = 3^x - 5$ is the graph of $f(x) = 3^x$ shifted _________ 5 units.
$$\underline{\text{up / down}}$$

x	$f(x) = 3^x - 5$

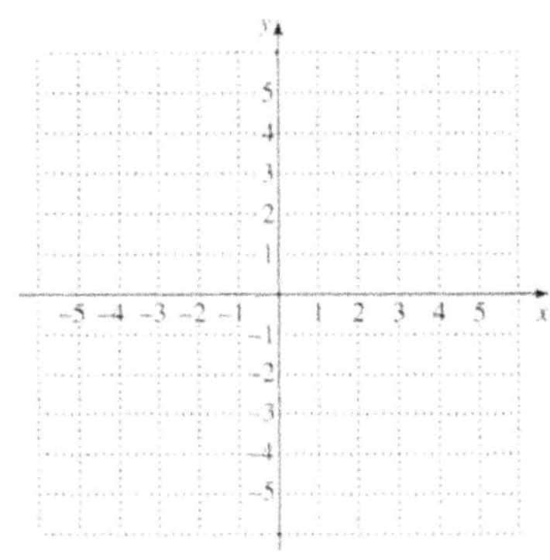

c) Graph $f(x) = 2 - \left(\dfrac{1}{3}\right)^x = 2 - 3^{-x}$.

The graph of $f(x) = 2 - \left(\dfrac{1}{3}\right)^x$ is the graph of $f(x) = 3^x$ reflected across

the _________ , then reflected across the _________ , and finally
$\quad\underline{x\text{-axis} / y\text{-axis}}\qquad\qquad\qquad\underline{x\text{-axis} / y\text{-axis}}$

shifted _________ 2 units.
$$\underline{\text{up / down}}$$

x	$f(x) = 2 - 3^{-x}$

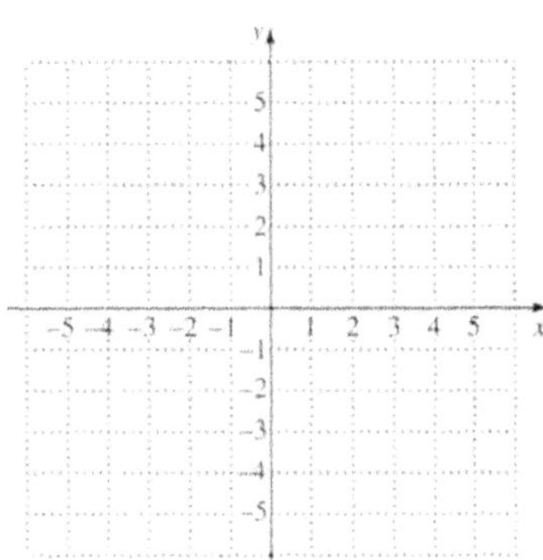

Example 4 *Compound Interest.* The amount of money A to which a principal P will grow after t years at interest rate r (in decimal form), compounded n times per year, is given by the formula

$$A = P\left(1 + \frac{r}{n}\right)^{nt}.$$

Suppose that \$100,000 is invested at 6.5% interest, compounded semiannually.

a) Find a function for the amount to which the investment grows after t years.

$$A = P\left(1 + \frac{r}{n}\right)^{nt}$$

$$P = 100{,}000 \qquad r = \boxed{} \qquad n = 2$$

$$A(t) = 100{,}000\left(1 + \frac{0.065}{2}\right)^{2t}$$

$$A(t) = 100{,}000\left(\boxed{}\right)^{2t}$$

b) Find the amount of money in the account at $t = 0, 4, 8,$ and 10 years.

$$A(0) = 100{,}000(1.0325)^{2\cdot\boxed{}} = \boxed{}$$

$$A(4) = 100{,}000(1.0325)^{2\cdot4} \approx \$129{,}157.75$$

$$A(8) = 100{,}000(1.0325)^{2\cdot\boxed{}} \approx \boxed{}$$

$$A(10) = 100{,}000(1.0325)^{2\cdot10} \approx \$189{,}583.79$$

c) Graph the function.

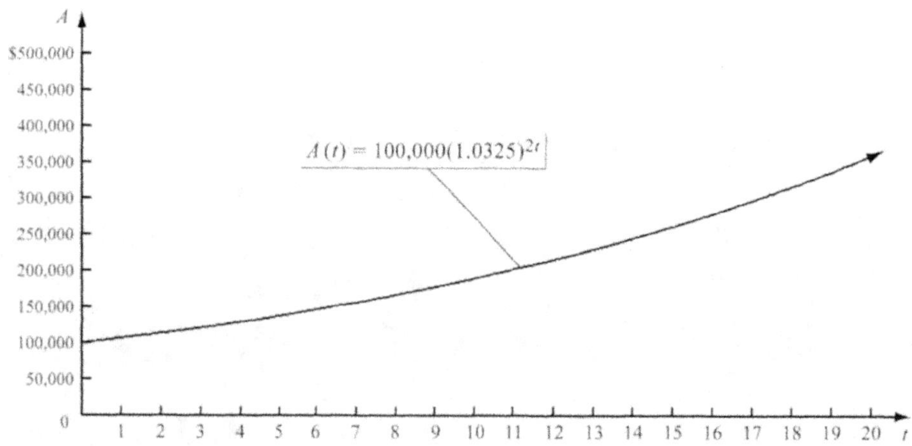

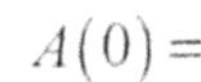**Your Turn 4**   The amount of money A to which a principal P will grow after t years at interest rate r (in decimal form), compounded n times per year, is given by the formula

$$A = P\left(1 + \frac{r}{n}\right)^{nt}.$$

Suppose that \$200,000 is invested at 5% interest, compounded quarterly.

a) Find a function for the amount to which the investment grows after t years.

b) Find the amount of money in the account at $t = 0$, 3, 6, 8, and 10.

$A(0) =$

$A(3) =$

$A(6) =$

$A(8) =$

$A(10) =$

c) Graph the function

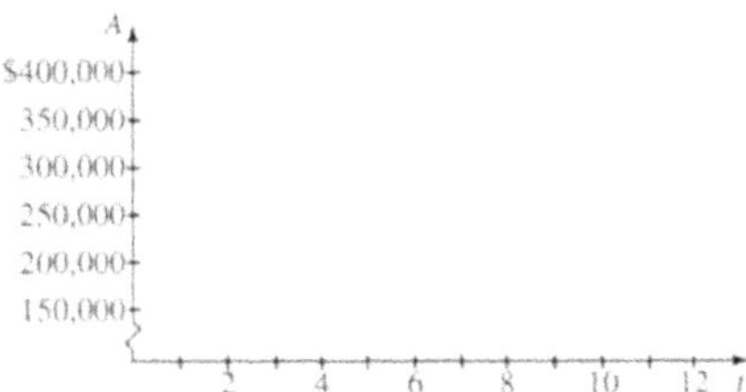

$e = 2.7182818284\ldots$

Example 5   Find each value of e^x, to four decimal places, using the e^x key on a calculator.

a) $e^3 = e^{\wedge}(3) \approx \boxed{}$

b) $e^{-0.23} = e^{\wedge}(-0.23) \approx \boxed{}$

c) $e^0 = e^{\wedge}(0) = \boxed{}$

Your Turn 5   Find each value of e^x, to four decimal places, using the e^x key on a calculator.

a) e^4

b) $e^{-0.25}$

c) e^0

Example 6 Graph $f(x)=e^x$ and $g(x)=e^{-x}$.

x	$f(x)=e^x$	$g(x)=e^{-x}$
-2	0.135	7.389
-1	0.368	
0		1
1	2.718	0.368
2	7.389	0.135

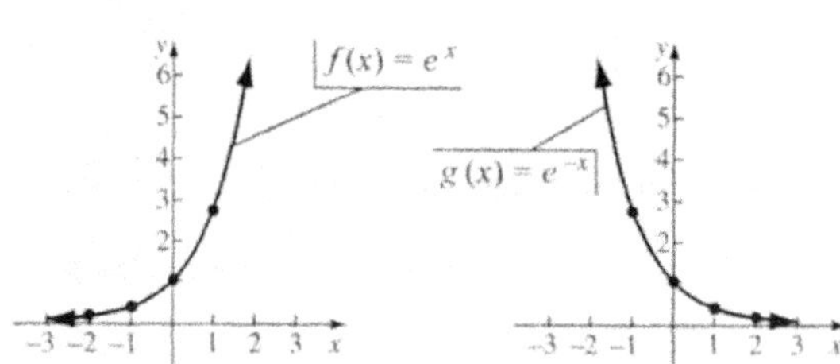

Your Turn 6 Graph $f(x)=e^x$ and $g(x)=e^{-x}$.

x	$f(x)=e^x$	$g(x)=e^{-x}$

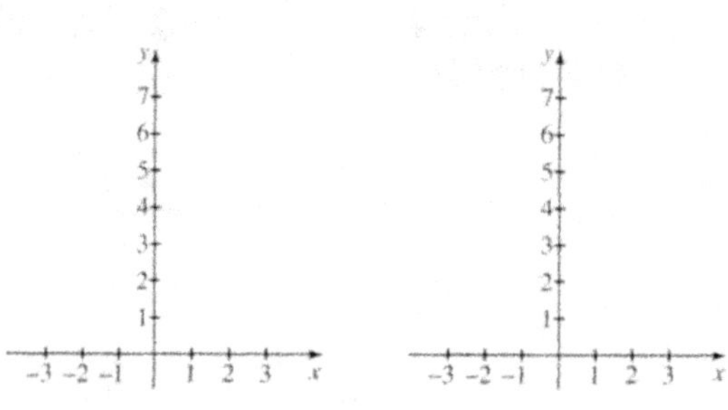

Example 7 Graph each of the following. Before doing so, describe how each graph can be obtained from the graph of $y=e^x$.

a) $f(x)=e^{x+3}$ b) $f(x)=e^{-0.5x}$ c) $f(x)=1-e^{-2x}$

a) Graph $f(x)=e^{x+3}$.

The graph of $f(x)=e^{x+3}$ is the graph of $f(x)=e^x$ shifted _____________ 3 units.
left / right

x	$f(x)=e^{x+3}$
-5	0.135
-4	0.368
-3	
-2	
-1	7.389
0	20.086
1	54.598

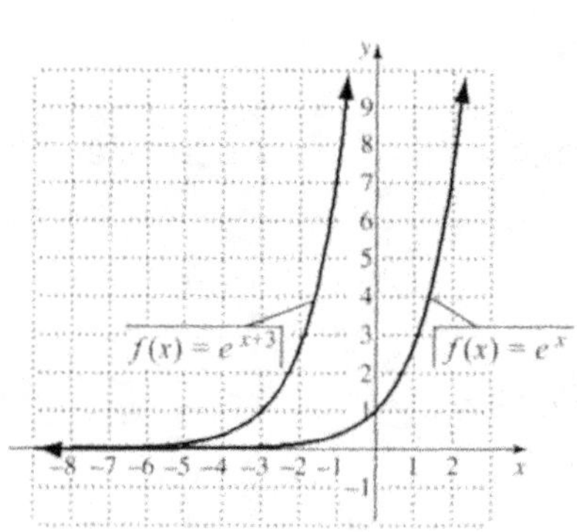

(continued)

Copyright © 2022 Pearson Education, Inc.

b) Graph $f(x) = e^{-0.5x}$.

The graph of $f(x) = e^{-0.5x}$ is the graph of $f(x) = e^x$ stretched

___________________________ and reflected across the ________________.
  horizontally / vertically x-axis / y-axis

x	$f(x) = e^{-0.5x}$
-3	4.482
-2	
-1	1.649
0	
1	0.607
2	0.368

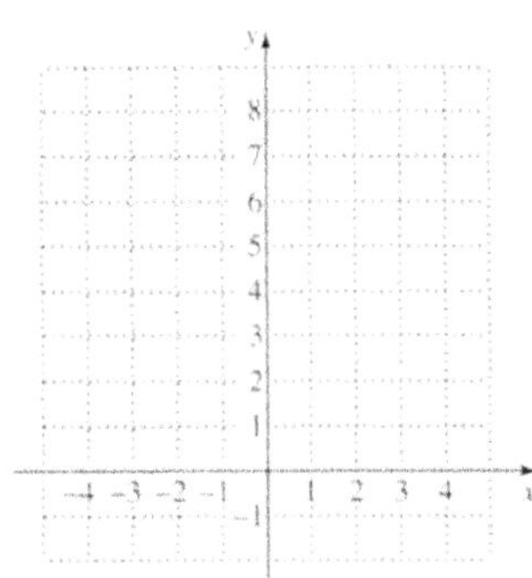

c) Graph $f(x) = 1 - e^{-2x}$.

The graph of $f(x) = 1 - e^{-2x}$ is the graph of $f(x) = e^x$ shrunk ___________________.
                                                        horizontally / vertically

reflected across the ______________ and also reflected across the ______________,
                     x-axis / y-axis x-axis / y-axis

and finally shifted ____________ 1 unit.
                   up / down

x	$f(x) = 1 - e^{-2x}$
-1	-6.389
0	
1	0.865
2	0.982
3	

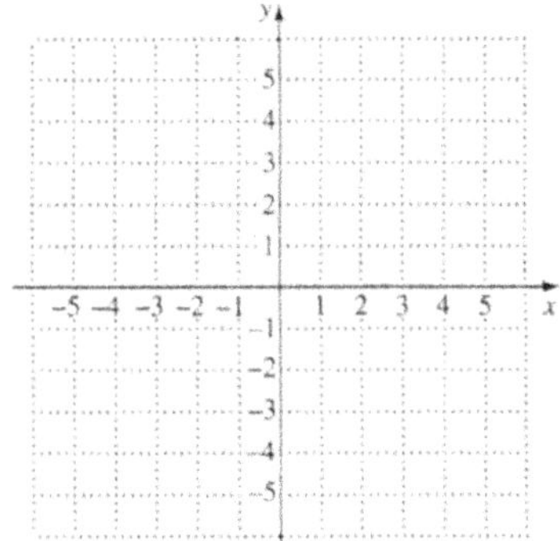

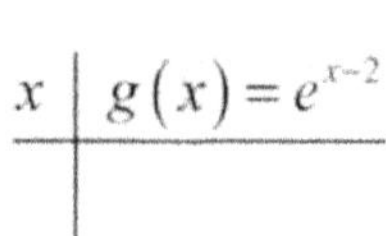

Your Turn 7   Graph each of the following. Before doing so, describe how each graph can be obtained from the graph of $f(x) = e^x$.

a) Graph $g(x) = e^{x-2}$.

The graph of $g(x) = e^{x-2}$ is the graph of $f(x) = e^x$ shifted __________ 2 units.
$$\text{left / right}$$

x	$g(x) = e^{x-2}$

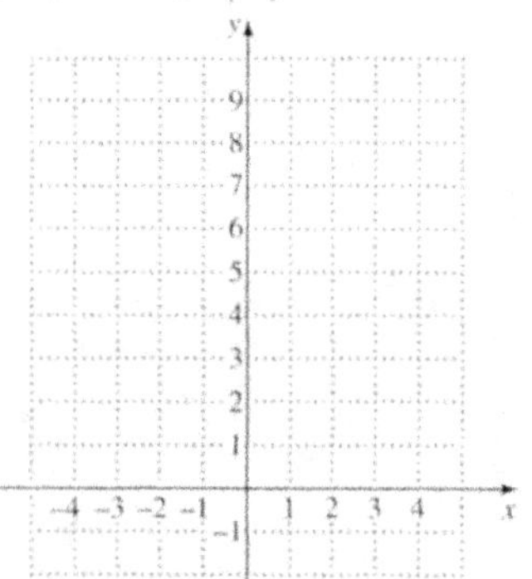

b) Graph $h(x) = e^{-2x}$.

The graph of $h(x) = e^{-2x}$ is the graph of $f(x) = e^x$ __________ horizontally
$$\text{shrunk / stretched}$$

and reflected across the __________ .
$$x\text{-axis} / y\text{-axis}$$

x	$h(x) = e^{-2x}$

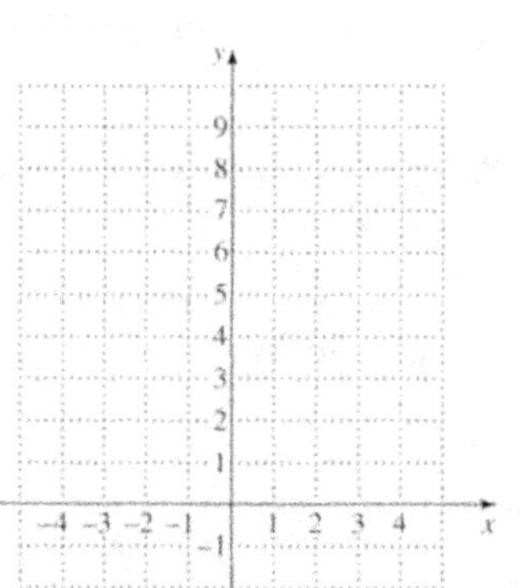

c) Graph $t(x) = 3 - \dfrac{1}{2} e^x$.

The graph of $t(x) = 3 - \dfrac{1}{2} e^x$ is the graph of $f(x) = e^x$ shrunk

__________ and reflected across the __________ and finally
$$\text{horizontally / vertically} \qquad x\text{-axis} / y\text{-axis}$$

shifted __________ 3 units.
$$\text{up / down}$$

x	$t(x) = 3 - \dfrac{1}{2} e^x$

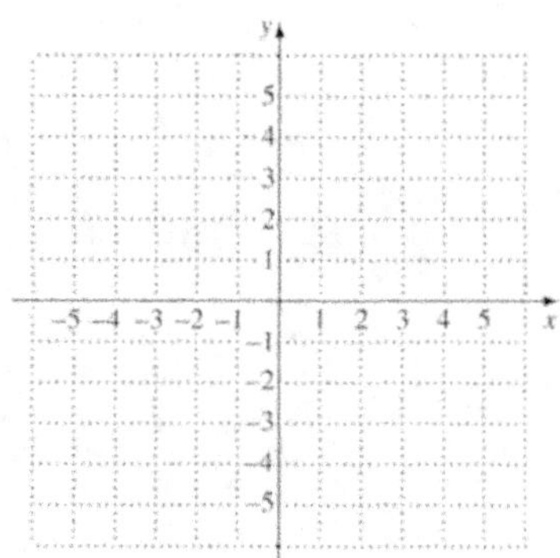

Practice Exercises

Find each value of e^x, to four decimal places, using the e^x key on a calculator.

1. e^0 **2.** e^{-6} **3.** e^3 **4.** $e^{-0.9}$

Graph each of the following. Describe how each graph can be obtained from the graph of an exponential function.

5. $f(x) = 3 - 2^x$

The graph of $f(x) = 3 - 2^x$ is
the graph of $f(x) = 2^x$ reflected
across the _______________ and
        x-axis / *y*-axis
then shifted _______________ 3 units.
        up / down

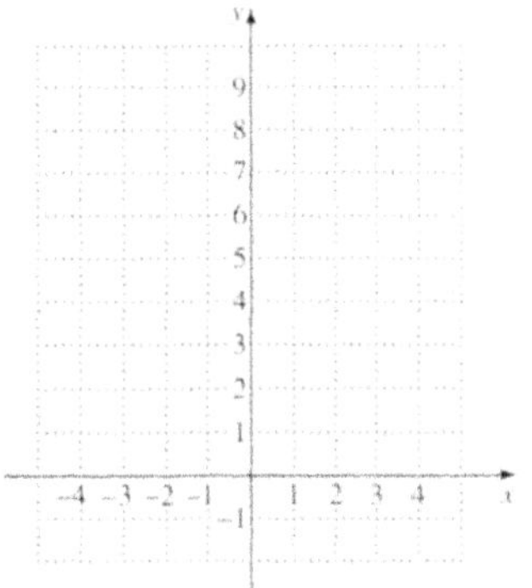

6. $f(x) = 4e^{0.5x}$

The graph of $f(x) = 4e^{0.5x}$ is
the graph of $f(x) = e^x$
_______________ horizontally
  shrunk / stretched
and _______________ vertically.
    shrunk / stretched

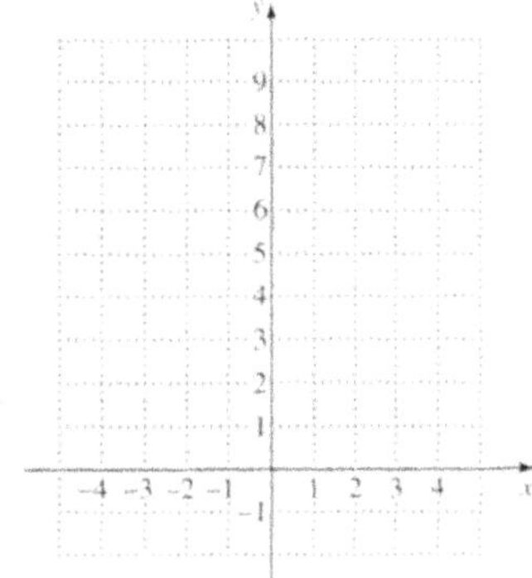

7. $f(x) = 2^{x+3} - 4$

The graph of $f(x) = 2^{x+3} - 4$ is
the graph of $f(x) = 2^x$ shifted
_______________ 3 units and shifted
  left / right
_______________ 4 units.
  up / down

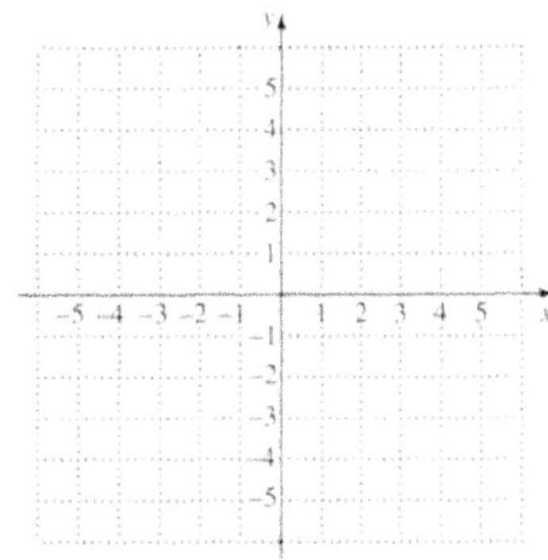

8. $f(x) = \left(\dfrac{1}{2}\right)^{2x}$

The graph of $f(x) = \left(\dfrac{1}{2}\right)^{2x}$ is

the graph of $f(x) = 2^x$ shrunk

___________________ and
$\underline{\text{horizontally / vertically}}$

then reflected across

the ___________.
$\underline{\text{x-axis / y-axis}}$

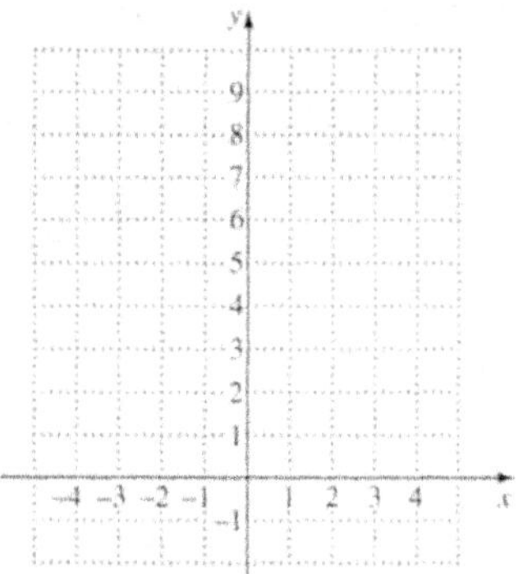

9. $f(x) = -e^{x-4} - 1$

The graph of $f(x) = -e^{x-4} - 1$

is the graph of $f(x) = e^x$ shifted

___________ 4 units and reflected
$\underline{\text{left / right}}$

across the ___________ and
$\underline{\text{x-axis / y-axis}}$

finally shifted ___________ 1 unit.
$\underline{\text{up / down}}$

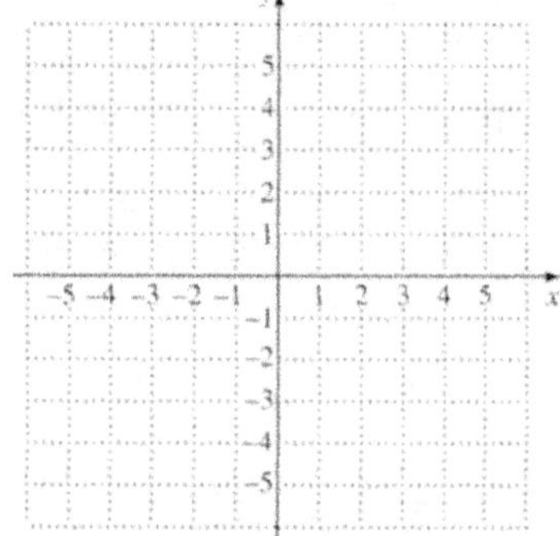

10. Suppose that \$150,000 is invested at 4% compounded semiannually.

a) Find a function for the amount to which the investment grows after t years.

b) Find the amount of money in the account at $t = 0,\ 8,\ 15,$ and 25.

$A(0) =$ $A(8) =$ $A(15) =$ $A(25) =$

c) Graph the function.

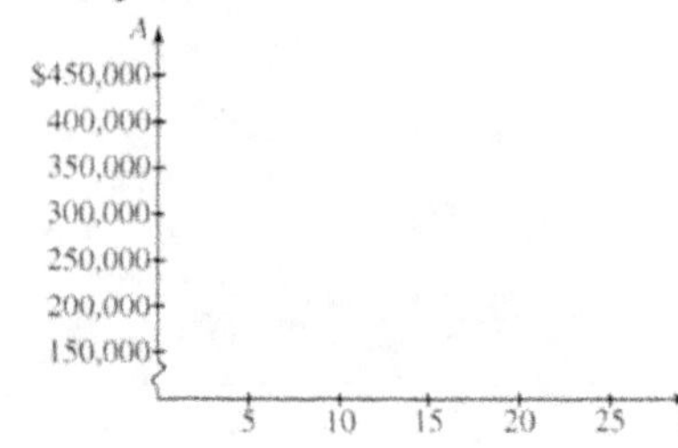

Section 9.4 Logarithmic Functions and Graphs

Example 1 Graph $x = 2^y$.

$x\ (x=2^y)$	y	(x, y)
1	0	$(1, 0)$
	1	
4	2	$(4, 2)$
	-1	
	-2	
$\dfrac{1}{8}$	-3	$\left(\dfrac{1}{8}, -3\right)$

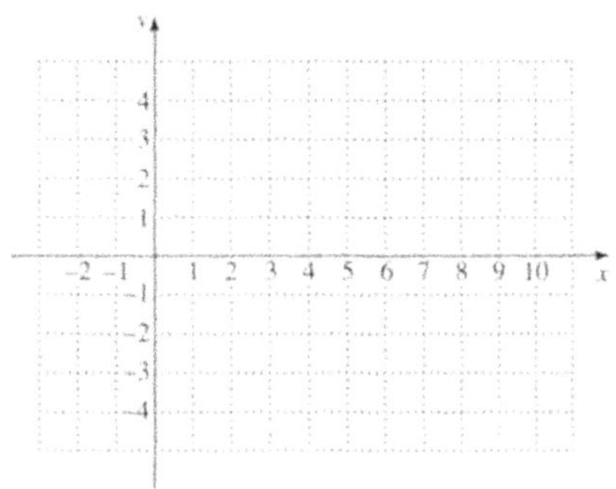

Your Turn 1 Graph $x = 3^y$.

$x\ (x=3^y)$	y	(x, y)

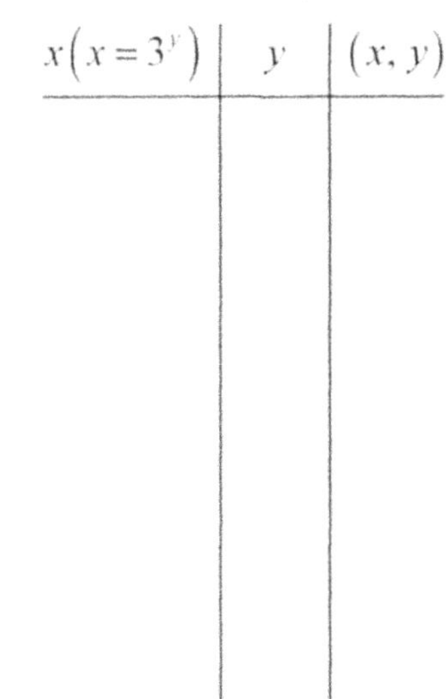

Logarithmic Function, Base a

We define $y = \log_a x$ as that number y such that $x = a^y$, where $x > 0$ and a is a positive constant other than 1.

Example 2

a) Find $\log_{10} 10{,}000$.

$$10^{\boxed{?}} = 10{,}000$$

$$\log_{10} 10{,}000 = \boxed{}$$

b) Find $\log_{10} 0.01$.

$$10^{\boxed{?}} = \frac{1}{100} = 10^{\boxed{}}$$

$$\log_{10} 0.01 = \boxed{}$$

c) Find $\log_2 8$.

$$2^{\boxed{?}} = 8$$

$$\log_2 8 = \boxed{}$$

d) Find $\log_9 3$.

$$9^{\boxed{?}} = 3 = 9^{1/2}$$

$$\log_9 3 = \boxed{}$$

e) Find $\log_6 1$.

$$6^{\boxed{?}} = 1$$

$$\log_6 1 = \boxed{}$$

f) Find $\log_8 8$.

$$8^{\boxed{?}} = 8$$

$$\log_8 8 = \boxed{}$$

Your Turn 2

a) Find $\log_{10} 100{,}000$.

b) Find $\log_{10} 0.001$.

c) Find $\log_{25} 5$.

d) Find $\log_3 81$.

e) Find $\log_{23} 23$.

f) Find $\log_9 1$.

$\log_a 1 = 0$ and $\log_a a = 1$, for any logarithmic base a.

$\log_a x = y \longleftrightarrow x = a^y$ A logarithm is an exponent!

Example 3

Convert each of the following to a logarithmic equation.

a) $16 = 2^x$ $\log_{\square} 16 = \boxed{}$

b) $10^{-3} = 0.001$ $\log_{\square} 0.001 = \boxed{}$

c) $e^t = 70$ $\log_{\square} 70 = \boxed{}$

Your Turn 3   Convert each of the following to a logarithmic equation.

a) $27 = 3^x$ b) $10^{-6} = 0.000001$ c) $e^y = 15$

Example 4

Convert each of the following to an exponential equation.

a) $\log_2 32 = 5$ $2^{\square} = \boxed{}$

b) $\log_a Q = 8$ $a^{\square} = \boxed{}$

c) $x = \log_t M$ $t^{\square} = \boxed{}$

Your Turn 4   Convert each of the following to an exponential equation.

a) $\log_4 64 = 3$ b) $a = \log_c B$ c) $\log_e W = 45$

Common Logarithms (Base 10): $\log M = \log_{10} M$

Natural Logarithms (Base e): $\ln M = \log_e M$

Example 5   Find each of the following common logarithms on a calculator. If you are using a graphing calculator, set the calculator in REAL mode. Round to four decimal places.

a) $\log(645{,}778) \approx$ ⬜

b) $\log(0.0000239) \approx$ ⬜

c) $\log(-3)$

The domain of the log function is $(0, \infty)$; -3 is not in this domain.

$\log(-3)$ _______________ exist.
$\qquad\qquad$ does / does not

Your Turn 5   Find each of the following common logarithms on a calculator. Round to four decimal places.

a) $\log 0.000167 \approx$

b) $\log 23{,}072 \approx$

c) $\log(-6)$

Example 6   Find each of the following natural logarithms on a calculator. If you are using a graphing calculator, set the calculator in REAL mode. Round to four decimal places.

a) $\ln(645{,}778) \approx$ ⬜

b) $\ln(0.0000239) \approx$ ⬜

c) $\ln(-5)$

The domain of the ln function is $(0, \infty)$; -5 is not in this domain.

$\ln(-5)$ _______________ exist.
$\qquad\qquad$ does / does not

d) $\ln(e) = 1$ $\qquad\qquad\qquad$ $e^{?} = e$
$\qquad\qquad\qquad\qquad\qquad$ $e^{\square} = e$

e) $\ln(1) = 0$ $\qquad\qquad\qquad$ $e^{?} = 1$
$\qquad\qquad\qquad\qquad\qquad$ $e^{\square} = 1$

> $\ln 1 = 0$ and $\ln e = 1$, for the logarithmic base e.

Your Turn 6   Find each of the following natural logarithms on a calculator. Round to four decimal places.

a) $\ln 0.000167 \approx$

b) $\ln 23{,}072 \approx$

c) $\ln(-6)$

d) $\ln 1 =$

e) $\ln e =$

The Change-of-Base Formula

For any logarithmic bases a and b, and any positive number M,
$$\log_b M = \frac{\log_a M}{\log_a b}.$$

Example 7   Find $\log_5 8$ using common logarithms.

Change-of-base formula:

$$\log_b M = \frac{\log_a M}{\log_a b}$$

Let $a = \boxed{}$, $b = \boxed{}$, and $M = 8$.

$$\log_5 8 = \frac{\log_{10} \boxed{}}{\log_{10} 5} \approx \boxed{}$$

✏ **Your Turn 7**   Find $\log_3 7$ using common logarithms.

Example 8   Find $\log_5 8$ using natural logarithms.

Change-of-base formula:

$$\log_b M = \frac{\log_a M}{\log_a b}$$

Let $a = \boxed{}$, $b = 5$, and $M = \boxed{}$.

$$\log_5 8 = \frac{\log_{\boxed{}} 8}{\log_{\boxed{}} 5} = \frac{\ln 8}{\ln 5} \approx \boxed{}$$

✏ **Your Turn 8**   Find $\log_3 7$ using natural logarithms.

Example 9   Graph $y = f(x) = \log_5 x$.

Think: $x = \boxed{}$

x, or 5^y	y
1	0
$\boxed{}$	1
$\boxed{}$	-1
$\dfrac{1}{25}$	-2

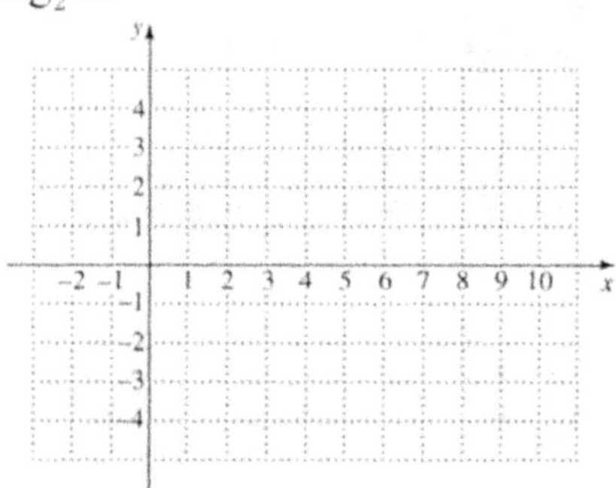

Your Turn 9   Graph $y = f(x) = \log_2 x$.

x, or 2^y	y

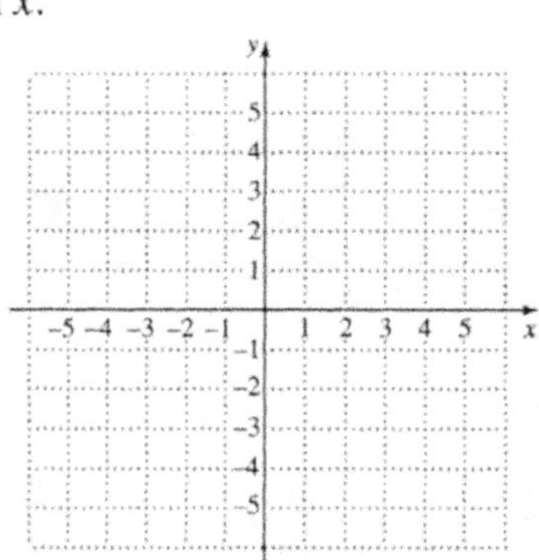

Example 10   Graph $g(x) = \ln x$.

x	$g(x) = \ln x$
0.5	-0.7
1	$\boxed{}$
2	$\boxed{}$
3	$\boxed{}$
4	1.4
5	1.6

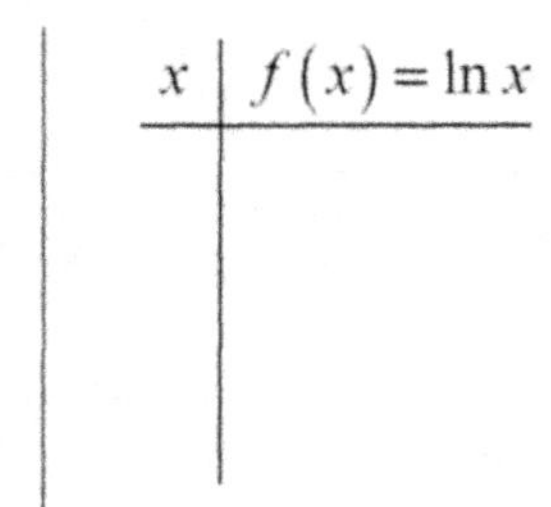

Your Turn 10   Graph $f(x) = \ln x$.

x	$f(x) = \ln x$

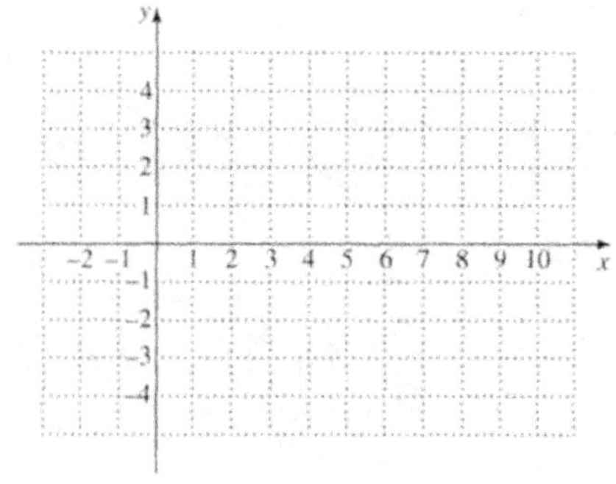

Example 11 Graph each of the following. Before doing so, describe how each graph can be obtained from the graph of $y = \ln x$. Give the domain and the vertical asymptote of each function.

a) $f(x) = \ln(x+3)$ b) $f(x) = 3 - \dfrac{1}{2}\ln x$ c) $f(x) = \left|\ln(x-1)\right|$

a) Graph $f(x) = \ln(x+3)$.

The graph of $f(x) = \ln(x+3)$ is a shift of the graph of $y = \ln x$ $\underline{\hspace{2cm}}$ 3 units.
 left / right

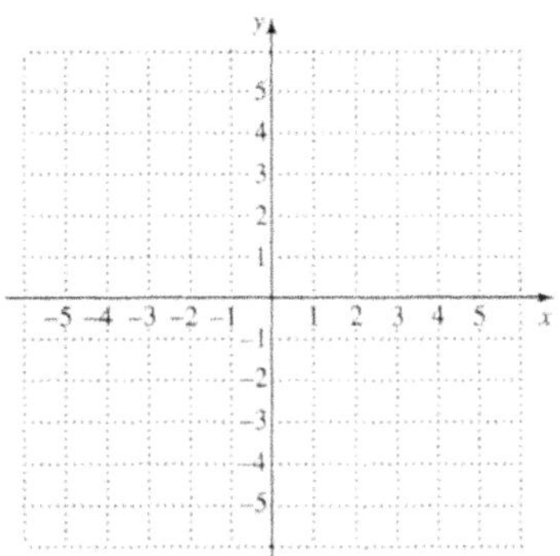

x	$f(x)$
-2.9	-2.303
-2	0
0	1.099
2	1.609
4	1.946

Domain: $\left(\boxed{}, \infty\right)$

Vertical asymptote: $\boxed{}$

b) Graph $f(x) = 3 - \dfrac{1}{2}\ln x$.

The graph of $f(x) = 3 - \dfrac{1}{2}\ln x$ is a $\underline{\hspace{4cm}}$ shrinking of the graph of
 vertical / horizontal

$y = \ln x$, followed by a reflection across the $\underline{\hspace{3cm}}$, and then a translation
 x-axis / y-axis

$\underline{\hspace{3cm}}$ 3 units.
up / down

x	$f(x)$
0.1	4.151
1	3
3	2.451
6	2.104
9	1.901

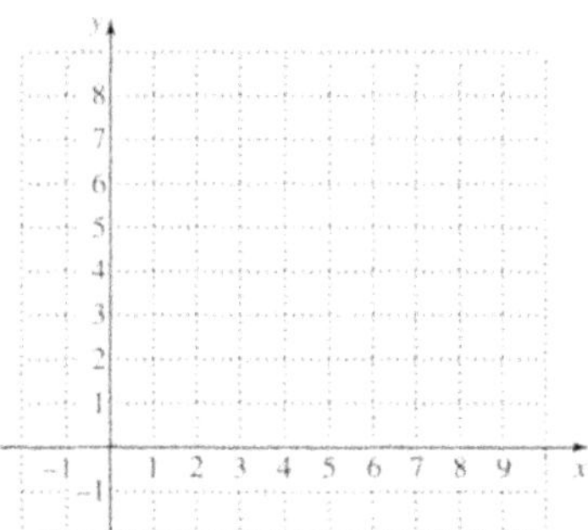

Domain: $\left(\boxed{}, \infty\right)$

Vertical asymptote: $\boxed{}$

(continued)

c) Graph $f(x) = |\ln(x-1)|$.

The graph of $f(x) = |\ln(x-1)|$ is a shift of the graph of $y = \ln x$ __________ 1 unit
left / right

followed by a reflection of negative outputs across the __________ .
x-axis / y-axis

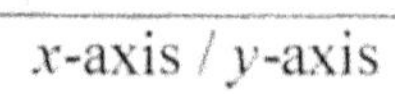

Domain: $\left(\boxed{}, \infty \right)$

Vertical asymptote: $\boxed{}$

x	$f(x)$
1.1	2.303
2	0
4	1.099
6	1.609
8	1.946

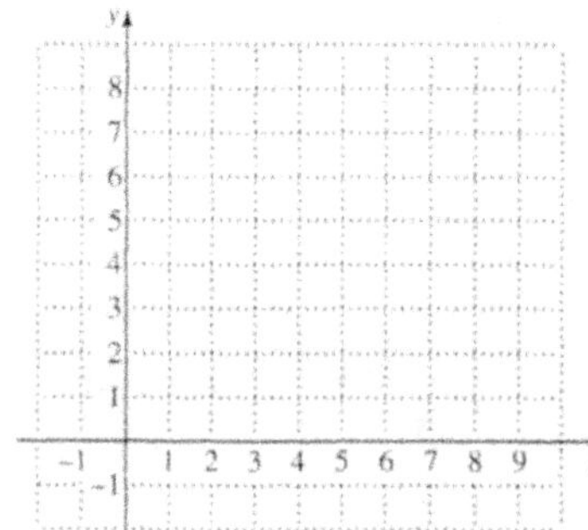

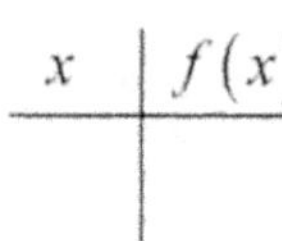**Your Turn 11** Graph each of the following. Before doing so, describe how each graph can be obtained from the graph of $y = \ln x$. Give the domain and the vertical asymptote of each function.

a) Graph $f(x) = \ln(x-4)$.

The graph of $f(x) = \ln(x-4)$ is a shift of the graph of $y = \ln x$ __________ 4 units.
left / right

x	$f(x)$

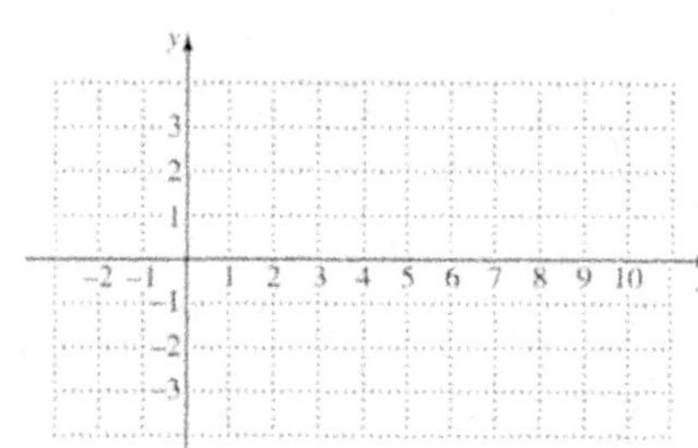

Domain: (   )

Asymptote:

b) Graph $f(x) = 2 - \dfrac{1}{2}\ln x$

The graph of $f(x) = 2 - \dfrac{1}{2}\ln x$ is a __________ shrinking of the graph of
vertical / horizontal

$y = \ln x$, followed by a reflection across the __________ , and then a translation
x-axis / y-axis

__________ 2 units.
up / down

x	$f(x)$

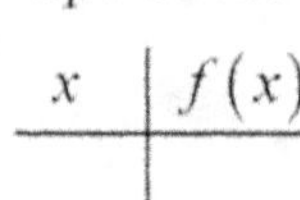

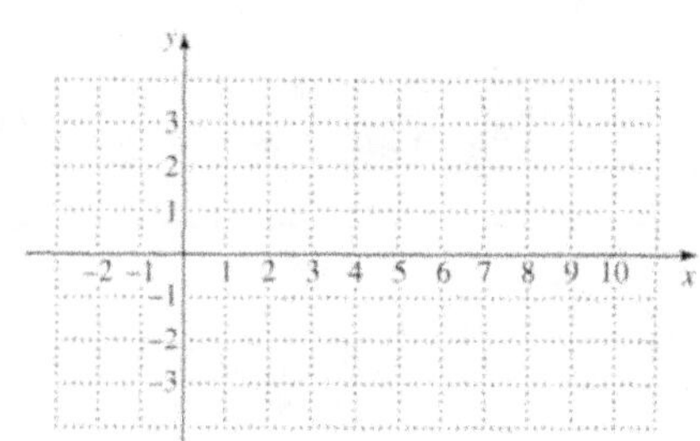

Domain: (   )

Asymptote:

(continued)

c) Graph $f(x) = \left| \ln(x+3) \right|$.

The graph of $f(x) = \left| \ln(x+3) \right|$. is a shift of the graph of $y = \ln x$ ___________ 3 units
$\overline{\text{left / right}}$

followed by a reflection of negative outputs across the ___________.
$\overline{x\text{-axis} / y\text{-axis}}$

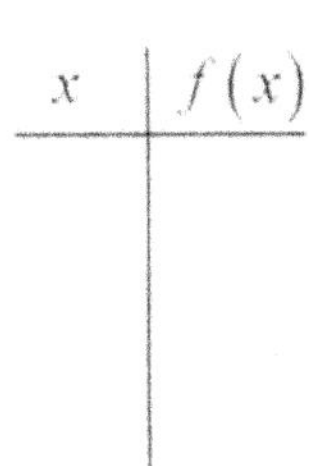

x	$f(x)$

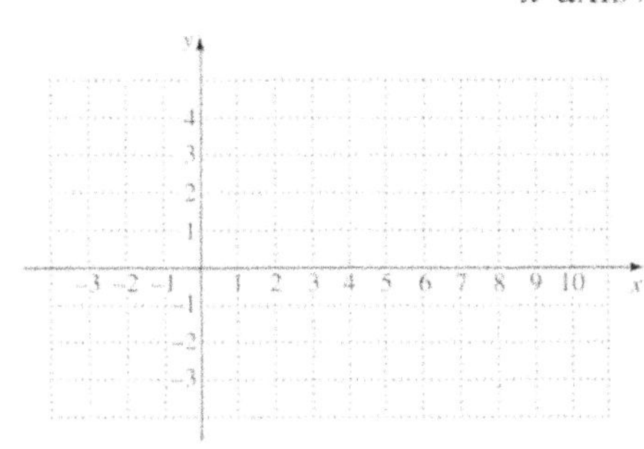

Domain: (   )

Asymptote:

Example 12   In a study by psychologists Bornstein and Bornstein, it was found that the average walking speed w, in feet per second, of a person living in a city of population P, in thousands, is given by the function

$$w(P) = 0.37 \ln P + 0.05.$$

a) The population of Billings, Montana, is 106,954. Find the average walking speed of people living in Billings.

$$106,954 = 106.954 \text{ thousands}$$

$$w(106.954) = 0.37 \ln\left(\boxed{} \right) + 0.05$$

$$\approx \boxed{} \text{ ft/sec}$$

b) The population of Chicago, Illinois, is 2,714,856. Find the average walking speed of people living in Chicago.

$$2,714,856 = 2714.856 \text{ thousands}$$

$$w(2714.856) = 0.37 \ln\left(\boxed{} \right) + 0.05$$

$$\approx \boxed{} \text{ ft/sec}$$

Your Turn 12   The average walking speed w, in feet per second, of a person living in a city of population P, in thousands, is given by the function $w(P) = 0.37 \ln P + 0.05$.

a) The population of Los Angeles, California, is 4,015,940. Find the average walking speed of people living in Los Angeles.

(continued)

b) The population of Flagstaff, Arizona, is 75,752. Find the average walking speed of people living in Flagstaff.

Example 13   Measured on the Richter scale, the magnitude R of an earthquake of intensity I is defined as

$$R = \log \frac{I}{I_0},$$

where I_0 is a minimum intensity used for comparison. We can think of I_0 as a threshold intensity that is the weakest earthquake that can be recorded on a seismograph. If one earthquake is 10 times as intense as another, its magnitude on the Richter scale is 1 greater than that of the other. If one earthquake is 100 times as intense as another, its magnitude on the Richter scale is 2 higher, and so on. Thus an earthquake whose magnitude is 7 on the Richter scale is 10 times as intense as an earthquake whose magnitude is 6. Earthquake intensities can be interpreted as multiples of the minimum intensity I_0.

The undersea Tohoku earthquake and tsunami, near the northeast coast of Honshu, Japan, on March 1, 2011, had an intensity of $10^{9.0} \cdot I_0$ (*Source*: earthquake.usgs.gov). They caused extensive loss of life and severe structural damage to buildings, railways, and roads. What was the magnitude on the Richter scale?

$$R = \log \frac{I}{I_0} \qquad\qquad I = 10^{9.0} \cdot I_0$$

$$R = \log \frac{\boxed{}}{I_0}$$

$$R = \log \boxed{}$$

$$R = 9.0$$

The magnitude of the earthquake was $\boxed{}$ on the Richter scale.

Your Turn 13   The earthquake near Port-au-Prince, Haiti, on January 12, 2010, had an intensity of $10^{7.0} \cdot I_0$. What was the magnitude on the Richter scale?

Practice Exercises

Find each of the following. Do not use a calculator.

1. $\log_2 128$

2. $\ln e^{-4}$

3. $\ln 1$

2. $\log_5 \sqrt[3]{5}$

5. $\log 1{,}000{,}000$

6. $\log_{36} 6$

Convert to a logarithmic equation.

7. $y^a = 7$

8. $4^{-3} = \dfrac{1}{64}$

Convert to an exponential equation.

9. $\log 0.0001 = -4$

10. $\log_c Q = s$

Find each of the following using a calculator. Round to four decimal places.

11. $\ln 40.6$

12. $\log(-1000)$

13. $\log 0.036$

14. $\ln 217$

Find the logarithm using common logarithms and the change-of-base formula. Round to four decimal places.

15. $\log_5 16$

16. $\log_3 100$

Find the logarithm using natural logarithms and the change-of-base formula. Round to four decimal places.

17. $\log_5 16$

18. $\log_3 100$

For each of the following functions, describe how the graph can be obtained from the graph of a basic logarithmic function. Then graph the function. Give the domain and an equation of the asymptote.

19. $f(x) = -\log(x+1)$

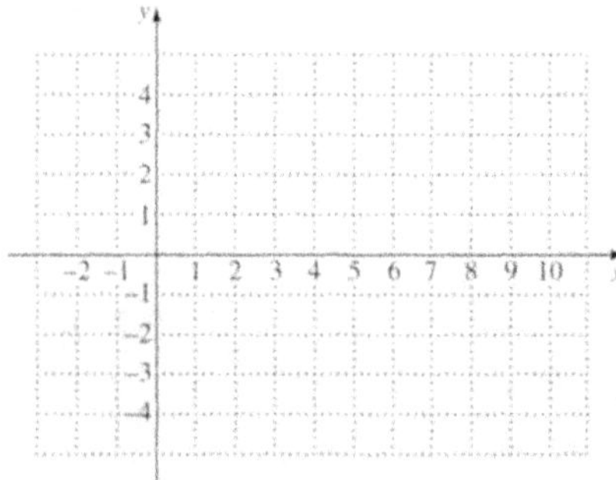

20. $f(x) = \ln 4x - 1$

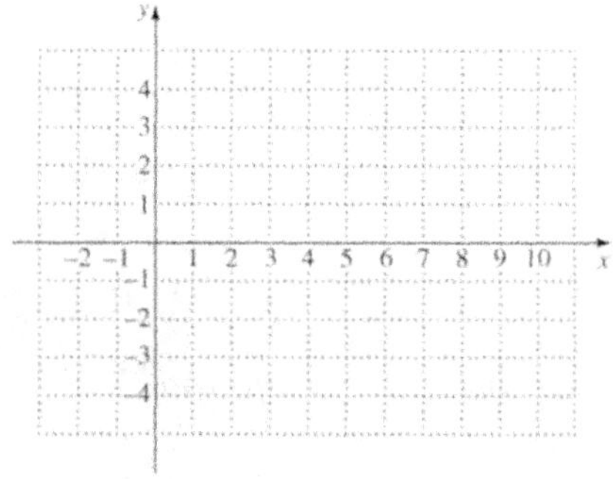

21. $f(x) = |\ln(x-3)|$

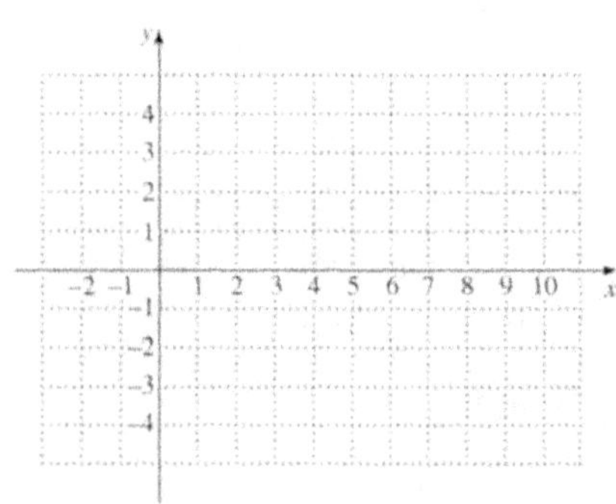

22. $f(x) = \dfrac{1}{2}\log(x-2) + 3$

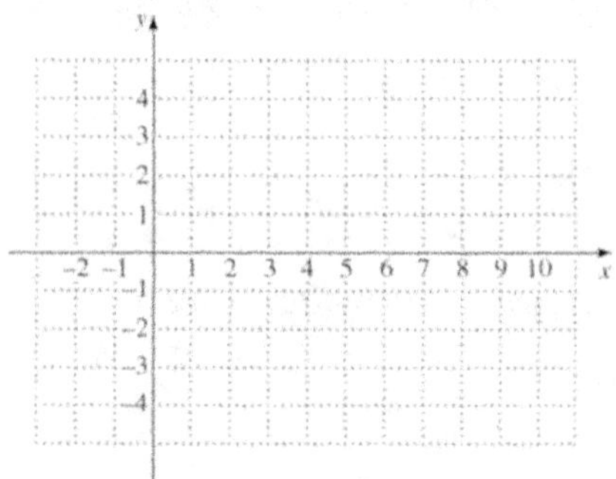

23. The average walking speed w, in feet per second, of a person living in a city of population P, in thousands, is given by the function $w(P) = 0.37 \ln P + 0.05$. The population of Boston, Massachusetts, is 694,583. Find the average walking speed of people in Boston.

24. The earthquake in Sumatra, Indonesia, on March 28, 2005, had an intensity of $10^{8.6} \cdot I_0$. What was the magnitude on the Richter scale?

Section 9.5 Properties of Logarithmic Functions

Properties of Logarithms

For any positive numbers M and N, any logarithmic base a, and any real number p:

Product Rule: $\log_a MN = \log_a M + \log_a N$.

Power Rule: $\log_a M^p = p \log_a M$.

Quotient Rule: $\log_a \dfrac{M}{N} = \log_a M - \log_a N$.

Example 1   Express as a sum of logarithms.

$\log_3 (9 \cdot 27) = \boxed{} + \boxed{}$ Using the product rule

Your Turn 1   Express as a sum of logarithms.

$\log_4 (16 \cdot 64) =$

Example 2   Express as a single logarithm.

$\log_2 p^3 + \log_2 q = \log_2 \left(\boxed{} \right)$ Using the product rule

Your Turn 2   Express as a single logarithm.

$\log_6 t + \log_6 s^2 =$

Example 3

a) Express $\log_a 11^{-3}$ as a product.

Power Rule: $\log_a M^p = p \log_a M$

$M = 11$ and $p = -3$

$\log_a 11^{-3} = \boxed{} \log_a \boxed{}$

(continued)

b) Express $\log_a \sqrt[4]{7}$ as a product.

First rewrite $\sqrt[4]{7}$ with an exponent using $\sqrt[n]{x} = x^{\frac{1}{n}}$.

$$\log_a \sqrt[4]{7} = \log_a 7^{\boxed{}}$$

$$= \boxed{} \log_a 7 \qquad \text{Using the power rule}$$

c) Express $\ln x^6$ as a product.

Using the power rule:

$$\ln x^6 = \boxed{} \ln \boxed{}$$

Your Turn 3   Express as a product.

a) $\log_a 9^5$
 b) $\log_a \sqrt[3]{26}$
 c) $\ln x^{-2}$

Example 4   Express as a difference of logarithms.

$$\log_t \frac{8}{w} = \log_t \boxed{} - \log_t \boxed{} \qquad \text{Using the quotient rule}$$

Your Turn 4   Express as a difference of logarithms.

$$\log_c \frac{a}{9}$$

Example 5   Express as a single logarithm.

$$\log_b 64 - \log_b 16 = \boxed{} \qquad \text{Using the quotient rule}$$

$$= \boxed{}$$

Your Turn 5   Express as a single logarithm.

$$\log_a 81 - \log_a 9$$

Example 6

a) Express $\log_a \dfrac{x^2 y^5}{z^4}$ in terms of sums and differences of logarithms.

$$\log_a \frac{x^2 y^5}{z^4} = \log_a \left(\boxed{} \right) - \log_a \boxed{}$$

$$= \log_{\boxed{}} x^2 + \log_{\boxed{}} y^5 - \log_a z^4$$

$$= \boxed{}\, \log_a x + \boxed{}\, \log_a y - \boxed{}\, \log_a z$$

b) Express $\log_a \sqrt[3]{\dfrac{a^2 b}{c^5}}$ in terms of sums and differences of logarithms.

$$\log_a \sqrt[3]{\frac{a^2 b}{c^5}} = \log_a \left(\frac{a^2 b}{c^5} \right)^{\boxed{}}$$

$$= \boxed{}\, \log_a \left(\frac{a^2 b}{c^5} \right)$$

$$= \frac{1}{3}\left(\log_a \left(\boxed{} \right) - \log_a \boxed{} \right)$$

$$= \frac{1}{3}\left(\log_a \boxed{} + \log_a \boxed{} - \log_a c^5 \right)$$

$$= \frac{1}{3}\left(\boxed{}\, \log_a a + \log_a b - \boxed{}\, \log_a c \right)$$

$$= \boxed{} + \boxed{}\, \log_a b - \boxed{}\, \log_a c$$

c) Express $\log_b \dfrac{a y^5}{m^3 n^4}$ in terms of sums and differences of logarithms.

$$\log_b \frac{a y^5}{m^3 n^4} = \log_b \left(a y^5 \right) - \log_{\boxed{}} \left(m^3 n^4 \right)$$

$$= \log_b \boxed{} + \log_b \boxed{} - \left(\log_{\boxed{}} m^3 + \log_{\boxed{}} n^4 \right)$$

$$= \log_b a + \log_b y^5 \boxed{} \log_b m^3 \boxed{} \log_b n^4$$

$$= \log_b a + \boxed{}\, \log_b y - \boxed{}\, \log_b m - \boxed{}\, \log_b n$$

Your Turn 6   Express in terms of sums and differences of logarithms.

a) $\log_a \dfrac{y^4 z^3}{x^6}$

b) $\log_c \sqrt[3]{\dfrac{c^4 a^5}{b^2}}$

c) $\log_b \dfrac{c^3 m^2}{a^2 n^5}$

Example 7   Express as a single logarithm.

$$5\log_b x - \log_b y + \frac{1}{4}\log_b z$$

$$= \log_b x^{\square} - \log_b y + \log_b z^{\square}$$

$$= \log_b \frac{\rule{1cm}{0.4pt}}{\rule{1cm}{0.4pt}} + \log_b z^{\frac{1}{4}}$$

$$= \log_b \frac{\rule{1cm}{0.4pt}}{\rule{1cm}{0.4pt}}$$

Your Turn 7   Express as a single logarithm.

$$\frac{1}{3}\log_a x + 4\log_a y - \log_a z$$

Example 8   Express as a single logarithm.

$$\ln\left(3x+1\right) - \ln\left(3x^2 - 5x - 2\right)$$

$$= \ln\frac{\boxed{}}{3x^2 - 5x - 2} = \ln\frac{3x+1}{\left(3x+1\right)\left(\boxed{}\right)}$$

$$= \ln\frac{1}{\boxed{}}$$

Your Turn 8   Express as a single logarithm.

$$\ln\left(y+5\right) - \ln\left(2y^2 + 3y - 35\right)$$

Example 9

a)   Given that $\log_a 2 \approx 0.301$ and $\log_a 3 \approx 0.477$, find $\log_a 6$, if possible.

$$\log_a 6 = \log_a \left(2\cdot 3\right)$$

$$= \log_a \boxed{} + \log_a \boxed{}$$

$$\approx 0.301 + \boxed{}$$

$$\approx \boxed{}$$

(continued)

b) Given that $\log_a 2 \approx 0.301$ and $\log_a 3 \approx 0.477$, find $\log_a \dfrac{2}{3}$, if possible.

$$\log_a \frac{2}{3} = \log_a 2 - \log_a \boxed{}$$

$$\approx \boxed{} - 0.477$$

$$\approx \boxed{}$$

c) Given that $\log_a 2 \approx 0.301$ and $\log_a 3 \approx 0.477$, find $\log_a 81$, if possible.

$$\log_a 81 = \log_a \boxed{}$$

$$= \boxed{} \log_a 3$$

$$\approx 4\left(\boxed{}\right)$$

$$\approx 1.908$$

d) Given that $\log_a 2 \approx 0.301$ and $\log_a 3 \approx 0.477$, find $\log_a \dfrac{1}{4}$, if possible.

$$\log_a \frac{1}{4} = \log_a 1 - \log_a \boxed{}$$

$$= \boxed{} - \log_a 2^2$$

$$= -2 \log_a \boxed{}$$

$$\approx -2\left(\boxed{}\right)$$

$$\approx -0.602$$

e) Given that $\log_a 2 \approx 0.301$ and $\log_a 3 \approx 0.477$, find $\log_a 5$, if possible.

$$\log_a 5 = \log_a (2+3) \neq \log_a 2 + \log_a 3$$

Finding $\log_a 5$ using the given logarithms _____________ possible.
$$\text{is / is not}$$

f) Given that $\log_a 2 \approx 0.301$ and $\log_a 3 \approx 0.477$, find $\dfrac{\log_a 3}{\log_a 2}$, if possible.

$$\frac{\log_a 3}{\log_a 2} \approx \frac{\boxed{}}{\boxed{}} \approx 1.585$$

Your Turn 9 Given that $\log_a 3 \approx 0.477$ and $\log_a 8 \approx 0.903$, find each of the following, if possible. Round the answer to the nearest thousandth.

a) $\log_a 24$

b) $\log_a \dfrac{3}{8}$

c) $\log_a 64$

d) $\log_a \dfrac{1}{9}$

e) $\log_a 11$

f) $\dfrac{\log_a 8}{\log_a 3}$

The Logarithm of a Base to a Power

For any base a and any positive real number x,

$$\log_a a^x = x.$$

Example 10

a) Simplify.

$$\log_a a^8 = \boxed{}$$

b) Simplify.

$$\ln e^{-t} = \log_{\boxed{}} e^{-t}$$

$$= \boxed{}$$

c) Simplify.

$$\log 10^{3k} = \log_{\boxed{}} 10^{3k}$$

$$= \boxed{}$$

Your Turn 10   Simplify.

a) $\log 10^{9t}$

b) $\log_c c^{10}$

c) $\ln e^{-2k}$

A Base to a Logarithmic Power

For any base a and any positive real number x,

$$a^{\log_a x} = x.$$

Example 11

a) Simplify.

$4^{\log_4 k} = \boxed{}$

b) Simplify.

$e^{\ln 5} = e^{\boxed{}} = \boxed{}$

c) Simplify.

$10^{\log 7t} = 10^{\boxed{}} = \boxed{}$

Your Turn 11   Simplify.

a) $e^{\ln 3}$

b) $6^{\log_6 b}$

c) $10^{\log 3y}$

Practice Exercises

1. Express $\log_5 (25 \cdot 125)$ as a sum of logarithms.

2. Express as a product.

   a) $\log_a 3^4$ b) $\ln y^{-5}$ c) $\log_t \sqrt[5]{40}$

3. Express $\log_c \dfrac{a}{7}$ as a difference of logarithms.

Express in terms of sums and differences of logarithms.

4. $\log_b 4x^3 yz^2$

5. $\ln \dfrac{7}{2w^4 z}$

6. $\log \sqrt{s^5 z}$

7. $\ln \sqrt[3]{\dfrac{a^2}{b^5 c^6}}$

Express as a single logarithm and, if possible, simplify.

8. $\log_a 40 + \log_a 0.4$

9. $\dfrac{1}{4}\log_t w - 8\log_t x$

10. $\ln\left(2x^2 - 9x - 5\right) - \ln\left(2x + 1\right)$

11. $\log 100{,}000 - \log 100$

12. $6\log_c y + \dfrac{1}{3}\log_c x - \log_c b$

13. $\ln\dfrac{1}{x-6} + \ln\left(x^2 - 36\right)$

Given that $\log_a 2 \approx 0.301$ *and* $\log_a 9 \approx 0.954$, *find each of the following, if possible. Round to the nearest thousandth.*

14. $\log_a \dfrac{2}{9}$

15. $\dfrac{\log_a 9}{\log_a 2}$

16. $\log_a 11$

17. $\log_a 4$

18. $\log_a 18$

19. $\log_a \dfrac{1}{81}$

Simplify.

20. $8^{\log_8 k}$

21. $\ln e^{-a}$

22. $\log 10^{5c}$

23. $e^{\ln 13}$

Section 9.6 Solving Exponential Equations and Logarithmic Equations

---

Base-Exponent Property

For any $a > 0$, $a \neq 1$,

$$a^x = a^y \longleftrightarrow x = y.$$

Example 1   Solve $2^{3x-7} = 32$.

$2^{3x-7} = 32$

$2^{3x-7} = \boxed{}$

$3x - 7 = 5$

$3x = \boxed{}$

$x = \boxed{}$

Check:

$$
\begin{array}{c|c}
\multicolumn{2}{c}{2^{3x-7} = 32} \\
\hline
2^{3\boxed{}-7} \ ? \ 32 & \\
2^{\boxed{}} & \\
32 & 32 \quad \text{TRUE}
\end{array}
$$

The solution is $\boxed{}$.

Your Turn 1   Solve $3^{2x+21} = 27$.

Property of Logarithmic Equality

For any $M > 0$, $N > 0$, $a > 0$, and $a \neq 1$,

$$\log_a M = \log_a N \longleftrightarrow M = N.$$

Example 2   Solve $3^x = 20$.

$3^x = 20$

$\log 3^x = \log 20$

$\boxed{} \log 3 = \log 20$

$x = \dfrac{\log 20}{\boxed{}}$

$x \approx 2.7268$

Check: $3^{2.7268} \approx \boxed{}$

The solution is about $\boxed{}$.

Your Turn 2   Solve $5^x = 45$.

Example 3   Solve $100e^{0.08t} = 2500$.

$$100e^{0.08t} = 2500$$

$$e^{0.08t} = \boxed{}$$

$$\ln e^{0.08t} = \ln 25$$

$$\boxed{} = \ln 25$$

$$t = \frac{\ln 25}{\boxed{}}$$

$$t \approx 40.2$$

The solution is about $\boxed{}$.

Your Turn 3   Solve $400e^{0.03t} = 1600$.

Example 4   Solve $4^{x+3} = 3^{-x}$.

$$4^{x+3} = 3^{-x}$$

$$\log 4^{x+3} = \log 3^{-x}$$

$$\left(\boxed{}\right)\log 4 = \boxed{}\log 3$$

$$x \log 4 + \boxed{}\log 4 = -x \log 3$$

$$x \log 4 + \boxed{} = -3 \log 4$$

$$\boxed{}\left(\log 4 + \log 3\right) = -3 \log 4$$

$$x = \frac{-3 \log 4}{\boxed{}}$$

$$x \approx -1.6737$$

The solution is about $\boxed{}$.

Your Turn 4   Solve $7^{x-2} = 4^{-x}$.

Example 5   Solve $e^x + e^{-x} - 6 = 0$.

$$e^x + e^{-x} - 6 = 0$$

$$e^x + \frac{1}{\boxed{}} - 6 = 0$$

$$\boxed{}\left(e^x + \frac{1}{e^x} - 6\right) = \boxed{} \cdot 0$$

$$e^{2x} + 1 - 6e^x = 0$$

$$e^{2x} - 6e^x + 1 = 0 \qquad e^{2x} = \left(e^x\right)^2$$

$$\left(\boxed{}\right)^2 - 6e^x + 1 = 0$$

$$u^2 - 6 \cdot \boxed{} + 1 = 0 \qquad \text{Let } u = e^x.$$

$$a = 1 \quad b = -6 \quad c = 1$$

$$u = \frac{-b \pm \sqrt{b^2 - 4ac}}{2a}$$

$$= \frac{-\left(\boxed{}\right) \pm \sqrt{(-6)^2 - 4 \cdot 1 \cdot 1}}{2 \cdot 1}$$

$$= \frac{6 \pm \sqrt{36 - 4}}{2}$$

$$= \frac{6 \pm \sqrt{\boxed{}}}{2} = \frac{6 \pm \boxed{} \cdot \sqrt{2}}{2} = 3 \pm \boxed{}$$

$$e^x = 3 \pm 2\sqrt{2}$$

$$\ln e^x = \boxed{}\left(3 \pm 2\sqrt{2}\right)$$

$$\boxed{} \cdot \ln e = \ln\left(3 \pm 2\sqrt{2}\right)$$

$$x = \ln\left(3 \pm 2\sqrt{2}\right)$$

The solutions are approximately $\boxed{}$ and $\boxed{}$.

Your Turn 5   Solve $e^x + e^{-x} - 10 = 0$.

Example 6   Solve $\log_3 x = -2$.

$$\log_3 x = -2$$

$$3^{\square} = x$$

$$\frac{1}{3^2} = x$$

$$\boxed{} = x$$

Check:

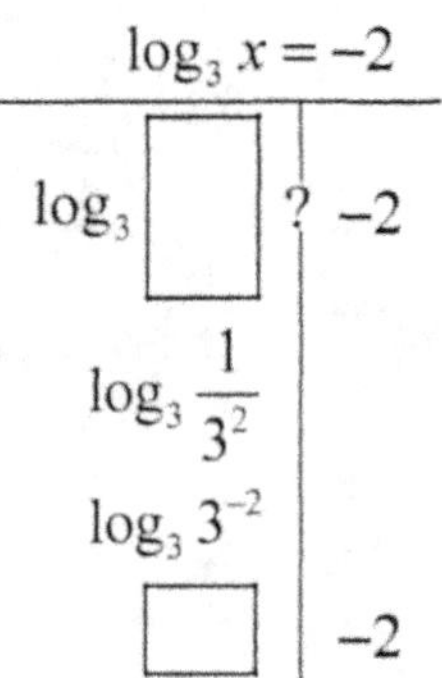

The solution is $\dfrac{1}{9}$.

Your Turn 6   Solve $\log_2 x = -6$.

Example 7   Solve $\log x + \log(x+3) = 1$.

$$\log x + \log(x+3) = 1$$

$$\log_{10}\left[x\left(\boxed{}\right)\right] = 1$$

$$x(x+3) = \boxed{}^{1}$$

$$x^2 + \boxed{} = 10$$

$$x^2 + 3x - 10 = 0$$

$$(x-2)\left(\boxed{}\right) = 0$$

$$x - 2 = 0 \qquad or \quad x + 5 = 0$$

$$x = \boxed{} \quad or \qquad x = \boxed{}$$

Check: For 2:

$$\log x + \log(x+3) = 1$$

$$\log\boxed{} + \log\left(\boxed{} + 3\right) \ ? \ 1$$

$$\log 2 + \log 5$$

$$\log(2 \cdot 5)$$

$$\log 10$$

$$1 \mid 1 \quad \text{TRUE}$$

Check: For -5:

$$\log x + \log(x+3) = 1$$

$$\log\left(\boxed{}\right) + \log(-5+3) \ ? \ 1$$

We see that -5 ___________ be a
$\qquad\qquad\quad$ can / cannot
solution because the logarithmic
function is not defined for negative
values of x.

The solution is $\boxed{}$.

Your Turn 7   Solve $\log x + \log(x+21) = 2$.

Example 8   Solve $\log_3(2x-1) - \log_3(x-4) = 2$.

$$\log_3(2x-1) - \log_3(x-4) = 2$$

$$\log_3 \frac{2x-1}{\boxed{}} = 2$$

$$\frac{2x-1}{x-4} = \boxed{}^{\,2}$$

$$\frac{2x-1}{x-4} = 9$$

$$(x-4)\cdot\frac{2x-1}{x-4} = (x-4)\cdot 9$$

$$2x-1 = 9x-36$$

$$35 = 7x$$

$$5 = x$$

Check:

$$\log_3(2x-1) - \log_3(x-4) = 2$$

$$\log_3(2\cdot 5 - 1) - \log_3(5-4) \ \overset{?}{\underset{}{}}\ 2$$

$$\log_3 \boxed{} - \log_3 \boxed{}$$

$$2 - \boxed{}$$

$$2 \ \big|\ 2 \qquad \text{TRUE}$$

The solution is $\boxed{}$.

Your Turn 8   Solve $\log_2(3x+4)-\log_2(x-3)=3$.

Example 9   Solve $\ln(4x+6)-\ln(x+5)=\ln x$.

$$\ln(4x+6)-\ln(x+5)=\ln x$$

$$\ln\frac{\boxed{}}{x+5}=\ln x$$

$$\frac{4x+6}{x+5}=\boxed{}$$

$$(x+5)\cdot\frac{4x+6}{x+5}=(x+5)\cdot x$$

$$4x+6=\boxed{}+5x$$

$$0=x^2+x-6$$

$$0=(x+3)\left(\boxed{}\right)$$

$$x+3=0 \quad or \quad x-2=0$$

$$x=\boxed{} \quad or \quad x=\boxed{}$$

The number -3 ______________ a solution because $4(-3)+6=-6$ and $\ln(-6)$ is not a real
 is / is not

number. The value $\boxed{}$ checks and is the solution.

Your Turn 9   Solve $\ln(5x+4)-\ln(x+2)=\ln x$.

Practice Exercises

Solve the exponential equation.

1. $2^{x+9} = 64$

2. $4^x = 15$

3. $5^{x-2} = 6^{-x}$

4. $20e^{0.5t} = 2000$

Solve the logarithmic equation.

5. $\log_2 x = 5$

6. $\log(x+9) + \log x = 1$

7. $\log_3(x+3) - \log_3(x-1) = 2$

8. $\ln x = -3$

Solve.

9. $\log_5 x = -3$

10. $5^{2x+3} = 625$

11. $9^{4x-3} = 2^x$

12. $\log_3(x-1) + \log_3(x+1) = 2$

13. $e^x + e^{-x} - 3 = 0$

14. $\ln(8x+28) - \ln(x+5) = \ln x$

Section 9.7   Applications and Models: Growth and Decay; Compound Interest

The **population growth function** is

$$P(t) = P_0 e^{kt}, \; k > 0,$$

where P_0 is the population at time 0, $P(t)$ is the population after time t, and k is the **exponential growth rate**.

Example 1   In 2013, the population of Ghana, located on the west coast of Africa, was about 25.2 million, and the exponential growth rate was 2.19% per year (*Source: CIA World Factbook, 2014*).

a) Find the exponential growth function.

$P(t) = 25.2e^{\boxed{}t}$, where t = number of years after 2013

b) Estimate the population in 2018.

$P\left(\boxed{}\right) = 25.2e^{0.0219\left(\boxed{}\right)} \approx 28.1$ million

c) After how long will the population be double what it was in 2013?

$$P(T) = 25.2e^{0.0219T}$$

$$\boxed{} = 25.2e^{0.0219T}$$

$$\boxed{} = e^{0.0219T}$$

$$\ln 2 = \ln e^{0.0219T}$$

$$\ln 2 = \boxed{}$$

$$\frac{\ln 2}{\boxed{}} = T$$

$$31.7 \approx T$$

The population of Ghana will be double what it was in 2013 about $\boxed{}$ years after 2013.

(continued)

d) At this growth rate, when will the population be 40 million?

$$\boxed{} = 25.2e^{0.0219t}$$

$$\frac{40}{25.2} = e^{0.0219t}$$

$$\ln\left(\frac{40}{25.2}\right) = \boxed{}$$

$$\frac{\ln\left(\dfrac{40}{25.2}\right)}{\boxed{}} = t$$

$$21 \approx t$$

The population of Ghana will be 40 million about $\boxed{}$ years after 2013.

Your Turn 1 In 2020, the population of the United States was about 331.0 million, and the exponential growth rate was 0.59% per year (*Source: United Nations Population Division*).

a) Find the exponential growth function.

b) Estimate the population in 2025.

c) After how long will the population be double what it was in 2020?

d) At this growth rate, when will the population be 420 million?

Interest Compounded Continuously

When an amount P_0 is invested in a savings account at interest rate r compounded continuously, the amount $P(t)$ in the account after t years is given by the function

$$P(t) = P_0 e^{kt}, \; k > 0.$$

Example 2 Suppose that $2000 is invested at interest rate k, compounded continuously, and grows to $2504.65 in 5 years.

a) What is the interest rate?

$$P(t) = 2000e^{kt}$$

$$\boxed{} = 2000e^{k\,\boxed{}}$$

$$\frac{2504.65}{2000} = e^{5k}$$

$$\ln\frac{2504.65}{2000} = \ln e^{5k}$$

$$\ln\frac{2504.65}{2000} = \boxed{}$$

$$\frac{\ln\dfrac{2504.65}{2000}}{\boxed{}} = k$$

$$0.045 \approx k$$

The interest rate is about $\boxed{}$.

b) Find the exponential growth function.

$$P(t) = P_0 e^{kt}$$

$$P(t) = \boxed{}\, e^{\boxed{}\,t}$$

c) What will the balance be after 10 years?

$$P(t) = 2000e^{0.045t}$$

$$P(10) = 2000e^{0.045(\boxed{})}$$

$$= 2000e^{\boxed{}}$$

$$\approx 3136.62$$

The balance after 10 years is $\boxed{}$.

(continued)

d) After how long will the $2000 have doubled?

$$\boxed{} = 2000e^{0.045T}$$

$$\boxed{} = e^{0.045T}$$

$$\ln 2 = \ln e^{0.045T}$$

$$\ln 2 = \boxed{}$$

$$\frac{\ln 2}{\boxed{}} = T$$

$$15.4 \approx T$$

The investment of $2000 will double in about $\boxed{}$ years.

Growth Rate and Doubling Time

The **growth rate k** and the **doubling time T** are related by

$$kT = \ln 2, \ \text{ or } k = \frac{\ln 2}{T}, \ \text{ or } T = \frac{\ln 2}{k}.$$

Your Turn 2   Suppose that $10,000 is invested at interest rate k, compounded continuously, and grows to $12,917.53 in 8 years.

a) What is the interest rate?

b) Find the exponential growth function.

(continued)

c) What is the balance after 10 years?

d) After how long will the $10,000 have doubled?

Example 3   The population of the Philippines is now doubling every 37.7 years. What is the exponential growth rate?

$$k = \frac{\ln 2}{T}$$

$$k = \frac{\ln 2}{\boxed{}} \approx \boxed{} \approx 1.84\%$$

The growth rate of the population of the Philippines is about $\boxed{}$ per year.

Your Turn 3   The population of Vietnam is now doubling every 76.2 years. What is the exponential growth rate?

Example 4 A lake is stocked with 400 fish of a new variety. The size of the lake, the availability of food, and the number of other fish restrict the growth of that type of fish in the lake to a limiting value of 2500. The population gets closer and closer to this limiting value, but never reaches it. The population of fish in the lake after time t, in months, is given by the function

$$P(t) = \frac{2500}{1 + 5.25e^{-0.32t}}.$$

Note that this function increases toward a limiting value of 2500. The graph has $y = 2500$ as a horizontal asymptote. Find the population after 0, 1, 5, 10, 15, and 20 months.

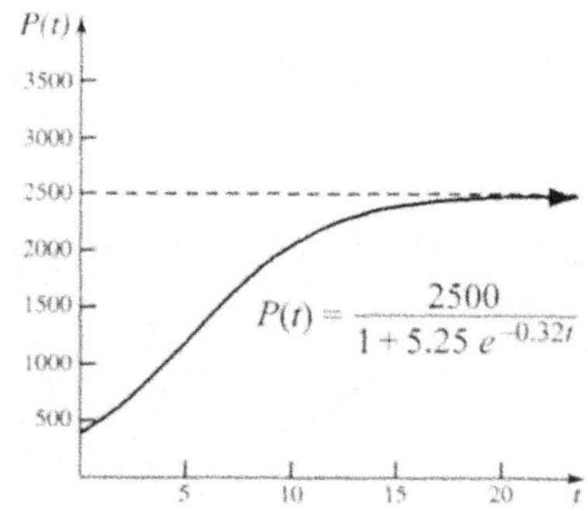

$$P(0) = \frac{2500}{1 + 5.25e^{-0.32(\boxed{})}} = \boxed{}$$

$$P(1) \approx \boxed{}$$

$$P(5) \approx \boxed{}$$

$$P(10) \approx \boxed{}$$

$$P(15) \approx \boxed{}$$

$$P(20) \approx \boxed{}$$

Your Turn 4 A lake is stocked with 325 fish of a new variety. The size of the lake, the availability of food, and the number of other fish restrict the growth of that type of fish in the lake to a limiting value of 1900. The population gets closer and closer to this limiting value, but never reaches it. The population of fish in the lake after time t, in months, is given by the function

$$P(t) = \frac{1900}{1 + 4.85e^{-0.32t}}.$$

Find the population after 0, 4, 12, and 18 months.

The **exponential decay function** is

$$P(t) = P_0 e^{-kt}, \quad k > 0,$$

where P_0 is the population at time t, and k is the **decay rate**.

Decay Rate and Half-Life

The decay rate k and the half-life T are related by

$$kT = \ln 2, \text{ or } k = \frac{\ln 2}{T}, \text{ or } T = \frac{\ln 2}{k}.$$

Example 5 The radioactive element carbon-14 has a half-life of 5750 years. The percentage of carbon-14 present in the remains of organic matter can be used to determine the age of that organic matter. Archaeologists discovered that the linen wrapping from one of the Dead Sea Scrolls had lost 22.3% of its carbon-14 at the time it was found. How old was the linen wrapping?

$$P(t) = P_0 e^{-kt}, \quad k > 0$$

Find k:

$$k = \frac{\ln 2}{T} = \frac{\ln 2}{\boxed{}} \approx 0.00012$$

$$P(t) = P_0 e^{\boxed{} t}$$

$$\boxed{} = P_0 e^{-0.00012t} \qquad \text{(77.7\% of the carbon-14 remains.)}$$

$$0.777 P_0 = P_0 e^{-0.00012t}$$

$$0.777 = \boxed{}$$

$$\ln 0.777 = \ln e^{-0.00012t}$$

$$\ln 0.777 = \boxed{}$$

$$\frac{\ln 0.777}{\boxed{}} = t$$

$$2103 \approx t$$

The linen wrapping on the Dead Sea Scrolls was about $\boxed{}$ years old.

Your Turn 5   A contractor plans to use wood beams from an old barn in a community center and wants to know the age of the beams. A scientist determines that the beams had lost 1.8% of their carbon-14. How old are the beams? (The decay rate of carbon-14 is about 0.00012.)

Practice Exercises

1. In 2020, the population of Mexico was about 128.9 million, and the exponential growth rate was 1.06% per year.

   a) Find the exponential growth function.

   b) Estimate the population in 2025.

   c) After how long will the population be double what it was in 2020?

2. Suppose that $22,000 is invested at interest rate k, compounded continuously, and grows to $26,024.61 in 6 years.

   a) What is the interest rate?

   b) Find the exponential growth function.

   c) What is the balance after 15 years?

   d) After how long will the $22,000 have doubled?

3. The population of India is now doubling in 70 years. What is the exponential growth rate?

4. A lake is stocked with 440 fish of a new variety. The size of the lake, the availability of food, and the number of other fish restrict the growth of that type of fish in the lake to a limiting value of 3200. The population gets closer and closer to this limiting value, but never reaches it. The population of fish in the lake after time t, in months, is given by the function

$$P(t) = \frac{3200}{1 + 6.28e^{-0.41t}}.$$

Find the population after 0, 5, 10, and 15 months.

5. Artwork can be dated by using carbon-14 dating of the organic material used to create the piece. A museum determined that the material used in a painting had lost 6% of its carbon-14. How old is the painting? (*Source: scientificarttests.com/carbon-dating.html*)

Student Activities

Correlation Guide

The **Student Activities** in this Notebook accompany *College Algebra with Intermediate Algebra* by Beecher/Penna/Johnson/Bittinger. They are located in Tools for Success within MyLab Math. The following table correlates the Student Activities with sections in the text.

Student Activity		Section in the Text	
1	American Football	R.2	Operations with Real Numbers
2	Finding the Magic Number	R.4	Introduction to Algebraic Expressions
3	Pets in the United States	1.3	Applications and Problem Solving
4	Mobile Data	2.4	The Algebra of Functions
5	Linear Regression on a Graphing Utility: TV Viewership and Frozen Entrée Sales	2.7	Finding Equations of Lines; Applications
6	Going Beyond High School	2.7	Finding Equations of Lines; Applications
7	Fruit Juice Consumption	3.4	Systems of Equations in Two Variables
8	Waterfalls	3.7	Systems of Inequalities and Linear Programming
9	Construction	3.7	Systems of Inequalities and Linear Programming
10	Visualizing Factoring	4.4	Factoring Trinomials: $x^2 + bx + c$
11	Matching Factorizations	4.5	Factoring Trinomials: $ax^2 + bx + c,\ a \neq 1$
12	Data and Downloading	5.8	Variation and Applications
13	Firefighting Formulas	6.3	Simplifying Radical Expressions
14	Music Downloads	7.4	Analyzing Graphs of Quadratic Functions
15	Let's Go to the Movies	7.4	Analyzing Graphs of Quadratic Functions
16	Teaching about Zeros: Creating a Video	8.4	Theorems about Zeros of Polynomial Functions
17	Earthquake Magnitude	9.4	Logarithmic Functions and Graphs
18	A Cosmic Path	11.2	The Circle and the Ellipse
19	Bargaining for a Used Car	12.3	Geometric Sequences and Series
20	Interest Rates on Credit Cards	Left to the instructor	
21	Home Loans	Left to the instructor	

Activity 1: **American Football**

Focus: Order on the number line, addition with and without the number line, absolute value

American football is played on a field that is 100 yards long, not counting the end zones, and 53 1/3 yards wide. The yardage on a football field is marked as shown in the diagram. Each team's objective is to move the ball, by running with it or passing it, into the other team's end zone.

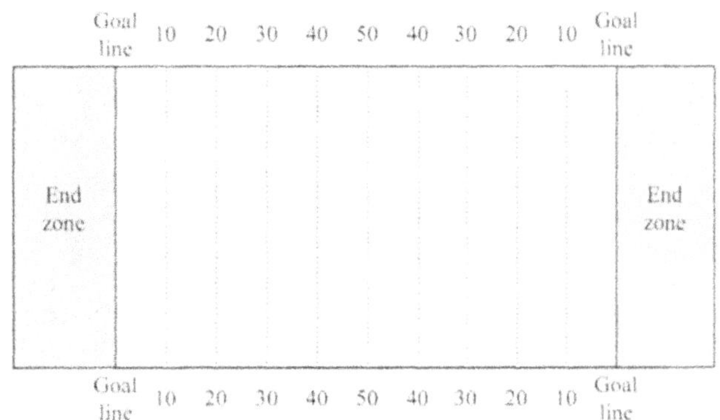

1. Draw a number line to model the length of the football field. Place 0 at the center of the field.

The football team with possession of the ball has 4 tries, called downs, to gain 10 yards. If it does not gain 10 yards after 4 downs, the other team wins possession of the ball.

On one possession, starting at the center of the field (location 0), a team heading in the positive direction on the number line made 4 plays in the following order:

    4-yard loss   6-yard gain   3-yard loss   12-yard gain

2. Use a number line to show where the ball was at the end of each play.

3. Where was the ball after the end of the 4 plays?

4. Did the team reach the goal of gaining 10 yards?

5. Write the final position of the ball as a sum.

Suppose the order of the plays was different:

    4-yard loss   6-yard gain   12-yard gain   3-yard loss

6. Write the final position of the ball as a sum.

7. How did the final position of the ball compare to the earlier result?

8. As soon as a team gains 10 yards, it earns 4 more downs to gain 10 more yards. Does this make any practical difference in the result of either of the previous sets of plays?

9. The distance the ball was moved on each play, not considering the direction of the move, is the **absolute value** of each number. For the game of football, is the direction of each play important, or is just the absolute value of each play important?

When a football team loses possession of the ball without scoring, it is called a turnover. There are two types of turnovers: turnovers due to fumbles or interceptions, and turnovers due to not advancing the ball 10 yards on 4 downs. Teams are ranked according to the following statistics:

- *Turnovers gained*: the number of times the team gains possession of the ball due to fumbles made by the other team plus the number of interceptions it makes

- *Turnovers lost*: the number of times the teams loses possession of the ball due to fumbles plus the number of interceptions made by the other team

- *Turnover margin*: the difference between turnovers gained and turnovers lost

10. If g = turnovers gained, l = turnovers lost, and m = turnover margin, write an equation that can be used to calculate turnover margin.

11. Is a positive turnover margin good or bad? Which turnover margin is better: -8 or 1?

12. Turnover margins for several NFL football teams during a recent season are listed. Draw a number line and graph the turnover margins, ranking the teams from best to worst.

Dallas Cowboys	6
Oakland Raiders	-15
Green Bay Packers	14
Pittsburgh Steelers	0
New York Giants	-2

Activity 2: Finding the Magic Number

Focus: Evaluating polynomials in several variables

Materials: A coin for each person

Note: This activity is designed for 3-member groups.

Can you determine the requirements for your baseball team to clinch first place?

A team's *magic number* is the combined number of wins by that team and losses by the second-place team that guarantee the leading team a first-place finish. For example, if the Cubs' magic number is 3 over the Reds, any combination of Cubs wins and Reds losses that totals 3 will guarantee a first-place finish for the Cubs. A team's magic number is computed using the polynomial

$$G - P - L + 1,$$

where G is the length of the season, in games, P is the number of games that the leading team has played, and L is the total number of games that the second-place team has lost minus the total number of games that the leading team has lost.

1. The standings below are from a fictitious league. Each group should calculate the Jaguars' magic number with respect to the Catamounts as well as the Jaguars' magic number with respect to the Wildcats. (Assume that the schedule is 162 games long.)

	W	L
Jaguars	92	64
Catamounts	90	66
Wildcats	89	66

2. Each group member should play the role of one of the teams, using coin tosses to simulate the remaining games. If a group member correctly predicts the side (heads or tails) that comes up, the coin toss represents a win for that team. Should the other side appear, the toss represents a loss. Assume that these games are against other (unlisted) teams in the league. Each group member should perform three coin tosses and then update the standings.

3. Recalculate the two magic numbers found in Question 1, using the updated standings from Question 2.

4. Slowly – one coin toss at a time – play out the remainder of the season. Record all wins and losses, update the standings, and recalculate the magic numbers each time all three group members have completed a round of coin tosses.

5. Examine the work in part (4) and explain why a magic number of 0 indicates that a team has been eliminated from contention.

Activity 3: Pets in the United States

Focus: Percents, Formulas, Applications

Data: American Veterinary Medical Association

Although dogs and cats are the most popular pets in the United States, many American households contain other animals as well. The following table lists the number of households with at least one of five types of pets and the total pet population.

	Number of Households (in millions) in 2012	Number of Households (in millions) in 2017	Total Pet Population (in millions) in 2012	Total Pet Population (in millions) in 2017
Birds	3.671	3.509	8.300	7.538
Horses	1.780	0.893	4.856	1.914
Fish	7.738	10.475	57.750	76.323
Rabbits	1.408	1.534	3.210	2.244
Poultry	1.020	1.397	12.591	15.367

1. Familiarize yourself with the data. What information is given for each type of animal? The data is given "in millions." What does that mean? For example, what was the total population of pet fish in 2017? How many households had pet horses in 2017?

2. For which pets did the number of households increase from 2012 to 2017? For which pets did the total pet population increase from 20012 to 2017? Why are these lists not the same?

3. If we consider only households containing a specific pet, then the formula $p = ah$ gives the pet population p given the average number of pets per household a and the number of households h that contain at least one pet. Solve the formula for a. For which pets listed in the table was the average number of pets per household greater than 2 in 2017?

4. Complete the following table. Use a + sign to indicate an increase and a − sign to indicate a decrease. Round percent change to the nearest tenth of a percent and number of households to the nearest one. (Recall that we find percent increase or decrease by computing $\dfrac{\text{amount of increase or decrease}}{\text{original amount}}$. For example, if a population increases from 200 to 225, the amount of increase is $225 - 200$, or 25, and the percent increase is $\dfrac{25}{200}$, or 0.125, or 12.5%.)

	Number of Households (in millions) in 2012	Number of Households (in millions) in 2017	Change in the Number of Households (in millions)	Percent Change in the Number of Households
Birds	3.671	3.509	−0.162	−4.4%
Horses	1.780	0.893		
Fish	7.738	10.475	+2.737	+35.4%
Rabbits	1.408	1.534		
Reptiles	2.966		+0.703	
Ferrets		0.326	−0.008	
Livestock			−0.167	−25.3%

5. For which pet listed in the first table did the number of households increase the most from 2012 to 2017? For which pet was the percent increase in the number of households the greatest? Why are these not the same?

6. Estimate the number of households with birds as pets in 2022 in two ways. First, assume that the change in the number of households from 2017 to 2022 will be the same as the change from 2012 to 2017. Second, assume that the percent change in the number of households from 2017 to 2022 will be the same as the change from 2012 to 2017. Which method resulted in the smaller decrease? Why?

7. The total population of birds kept as household pets in 2012 was about 5 times the population of lizards kept as household pets. How many lizards were kept as household pets in 2012?

8. In 2017 there were 48,255,000 households with dogs as pets. The total population of pet dogs was 488,000 more than the number of pet fish. What was the average number of dogs per household in 2017?

9. Write a paragraph for a news source about pet ownership in the United States using some of the information from this activity. Include what you feel are the most interesting and important numbers you calculated and explain their relevance.

Activity 4: Mobile Data

Focus: Algebra of functions; percent increase

Data: Cisco Visual Networking Index: Forecast and Methodology, 2014–2019.

Consumer use of mobile devices to send and receive data and to access the internet is growing rapidly. The growth, however, does not follow the same pattern in all regions of the world. The table below lists projections for the average monthly mobile data and internet traffic, in petabytes.
(A petabyte is 1×10^{15} bytes.)

Region	2014	2015	2016	2017	2018	2019	Amount of Increase 2014 to 2019	Percent Increase 2014 to 2019
Asia Pacific, A	977	1622	2616	4114	6245	9459	8482	868%
North America, N	563	849	1287	1897	2704	3798	3235	575%
Central and Eastern Europe, C	242	464	832	1409	2231	3488	3246	
Middle East and Africa, M	199	383	690	1194	1927	3051		
Western Europe, W	341	504	760	1137	1653	2392		
Latin America, L	201	354	581	915	1380	2032		
Total, T	2523	4176						

1. Explore the data. The regions are listed in order of their projected average monthly mobile data usage in 2019. Is this order the same as it was in 2014? If not, how is the order different?

2. Complete the table, filling in the Total row and the Amount of Increase and Percent Increase columns. Which region saw the greatest amount of increase? Which saw the greatest percent increase?

3. In the table, a function name is given after each region name. We can refer to entries in the table using those function names. For example, $A(2014)$ represents the average monthly mobile data usage in the Asia Pacific region in 2014, so $A(2014) = 977$. What is the value of each of the following, and what quantity does it represent?

(a) $W(2019)$

(b) $(C + W)(2014)$

(c) $(N - L)(2019)$

(d) $(A + N + C + M + W + L)(2019)$

4. By stacking the regional monthly mobile data for each year, we can show regional and total data in the same graph. The data for 2014 to 2017 are graphed below. Complete the graph through 2019. If you have colored pencils available, shade the regions between each pair of lines. What does the height of each shaded region represent?

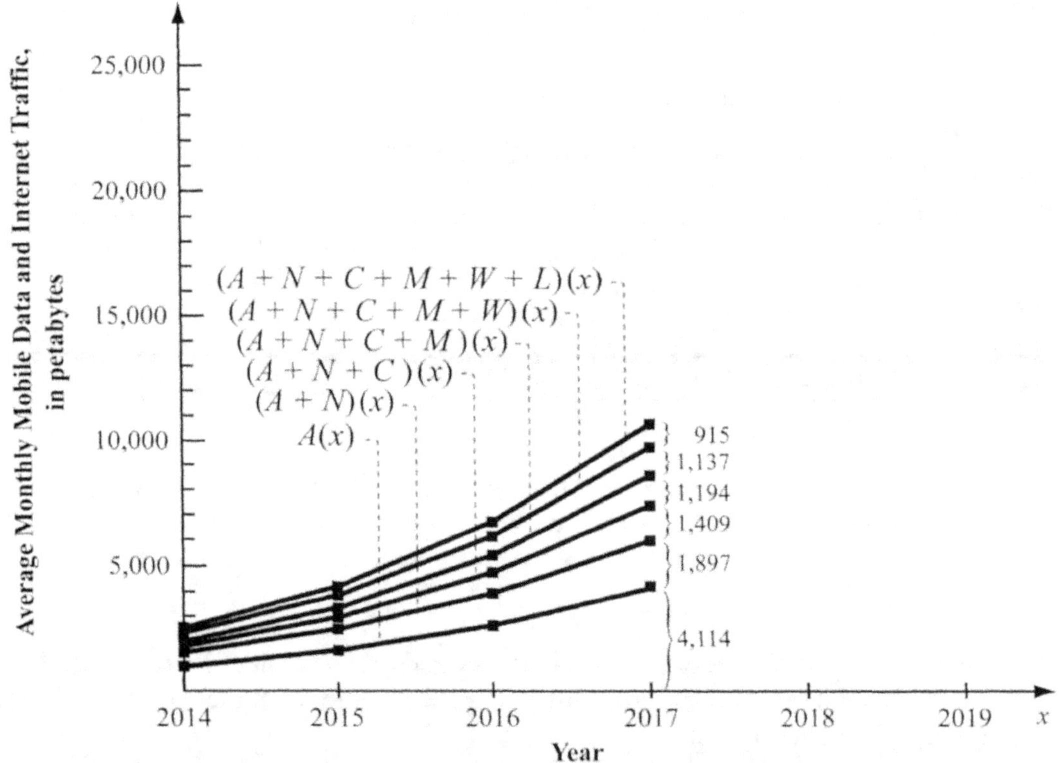

5. Why is mobile data usage predicted to grow at different rates across the regions? List two or three factors that might affect the growth of mobile data usage, and explain how they might differ across these geographic regions.

Activity 5: Linear Regression on a Graphing Utility

Focus: Linear regression on a graphing utility

Data: Nielsen

Due to the rise of streaming services such as Netflix, Amazon, and Hulu as well as a resurgence in video games, TV viewing among people ages 18 to 34 has declined at a fast pace in recent years. The following table gives the percentage of viewers in this age range watching television, including DVR playback, in an average minute in the period from September 24 to October 28 for the years 2014 through 2018.

Year, x	Percentage of Viewers, y
2014, 0	26.4
2015, 1	24.2
2016, 2	22.3
2017, 3	19.8
2018, 4	16.8

We will use a graphing utility to fit a linear equation $y = mx + b$ to these data, where x is the number of years after 2014. (The keystrokes are for a TI-84 Plus calculator.) First we enter the data from the table on the STAT list screen. We will enter the x-values in L1 and the y-values in L2.

We begin by clearing any existing lists. To clear L1, move the cursor to the heading L1. Press CLEAR and then the down arrow. Do this for each of the other lists that contain entries.

Now position the cursor at the first blank space in L1 and enter the x-values. To enter 0, press 0 ENTER. Continue entering the x-values 1 through 4, each followed by ENTER. Press the right arrow to move to the top of the L2 column. Type the y-values in succession, each followed by ENTER.

To select linear regression from the STAT CALC menu and to store the equation as Y1 on the Y = screen, press STAT right arrow 4 VARS right arrow 1 1 ENTER. The display will show the coefficients a and b of the regression line $y = ax + b$. (The values of r and r^2, which are indications of how well the equation fits the data, are also shown).

```
LinReg
 y=ax+b
 a=-2.36
 b=26.62
 r2=.9924447613
 r=-.9962152183
```

1. What does *a* represent in the regression equation? What does *b* represent?

2. Will the graph slant up from left to right or down? How do you know? Graph the equation in the window with $Xmin = 0$, $Xmax = 15$, $Yscl = 1$, $Ymin = 0$, $Ymax = 30$, and $Yscl = 5$ to confirm your answer.

3. Estimate the percentage of viewers ages 18 to 34 watching television in the given time period in 2019 and in 2020.

4. Predict when there will be no TV viewers age 18 to 34 in the given time period. Does this answer seem reasonable? Why or why not?

5. Data on the sales of frozen entrees, in billions of dollars, are given in the following table. Use a graphing utility to fit a linear equation to the data. Let $x =$ the number of years after 2015.

Year, x	Sales of Frozen Entrees, y
2015, 0	$17.6
2016, 1	17.8
2017, 2	18.0
2018, 3	19.0

6. Estimate the sales of frozen entrees in 2020 and in 2021.

7. Research the reasons for this rise in sales in recent years.

8. Do you think this trend will continue? Does your research provide any insight into this?

Activity 6: Going Beyond High School

Focus: Circle, bar, and line graphs, percent, rate of change, equations of lines, graphing lines

Data: National Center for Education Statistics

Patterns of college attendance have changed significantly in the past few decades. The table below gives percentages of high school completers enrolling in college, for several demographics, for selected years between 1985 and 2017.

Year	Percent of Recent High School Completers Enrolled in a Two-Year College	Percent of Recent High School Completers Enrolled in a Four-Year College	Percent of Recent Male High School Completers Enrolled in College	Percent of Recent Female High School Completers Enrolled in College
1985	19.6	38.1	58.6	56.8
1990	20.1	40.0	58.0	62.2
1995	21.5	40.4	62.6	61.3
2000	21.4	41.9	59.9	66.2
2005	24.0	44.6	66.5	70.4
2010	26.7	41.4	62.8	74.0
2015	25.2	44.0	65.8	72.5
2017	22.6	44.2	61.1	71.7

1. Examine the data in the table. What do the numbers represent? Why do you think that the column heads reference "high school completers" rather than "high school graduates"?

2. There were 2,870,000 high school completers in 2017. How many enrolled in a college? How many enrolled in a two-year college? How many enrolled in a four-year college? How many did not enroll in college?

3. For 2017, make a graph to illustrate the percent of high school completers enrolling in two-year colleges, four-year colleges, and not in any college. What type of graph will best illustrate this data?

4. For 2017, make a graph to illustrate the number of high school completers enrolling in two-year colleges, four-year colleges, and not in any college, given that there were 2,870,000 high school completers in 2017. What type of graph will best illustrate this data?

5. Using one pair of axes, draw a graph to illustrate the percent of high school completers who enrolled in a two-year college and a graph to illustrate the percent of high school completers who enrolled in a four-year college for years from 1985 to 2017. What type of graph will best illustrate this data? What trends can you see from the graph?

6. Using data from the two years 1985 and 2017, what is the rate of change for the percent of male high school completers who enrolled in college? Using data from the same years, what is the rate of change for the percent of female high school completers who enrolled in college? What do these rates mean? Do you think that these rates are good indicators of changes in college enrollment for high school completers between 1985 and 2017?

7. Use data from the two years 1985 and 2017 to write an equation that could be used to model the percent of recent male high school completers who enrolled in college. Using data from the same years, write an equation that could be used to model the percent of recent female high school completers who enrolled in college. Using one pair of axes, graph both equations. Describe what these graphs tell you about these two categories of students.

8. Use regression and all the data from 1985 through 2017 to write an equation to model the percent of recent male high school completers who enrolled in college. Using regression and data from the same years, write an equation to model the percent of recent female high school completers who enrolled in college. Graph both equations on one pair of axes. Describe what these graphs tell you about these two categories of students. How do these equations differ from those found in Question 7?

9. Examine the data in the table shown below. For what categories in the table before Question 1 are percentages given here? How do these categories relate to the categories in that table?

Year	Percent of Recent Male High School Completers Enrolled in a Two-Year College	Percent of Recent Male High School Completers Enrolled in a Four-Year College	Percent of Recent Female High School Completers Enrolled in a Two-Year College	Percent of Recent Female High School Completers Enrolled in a Four-Year College
2014	21.2	42.8	28.0	44.6
2015	24.3	41.5	24.2	46.4
2016	25.3	42.2	22.3	49.6
2017	23.9	37.2	21.4	50.3

10. Using one pair of axes, draw a line graph for each of the four categories of students in the table in Question 9, using a different color for each category, if possible. Describe the patterns you see. What events in the United States in those years may have affected the college decisions of high school completers?

11. In 2018, a high school guidance counselor wants to know what her ninth grade students will be doing after graduation. Using data and equations from this activity, write a paragraph describing the overall picture of the types of colleges high school completers enroll in and what she can expect her students to do.

Activity 7: Fruit Juice Consumption

Focus: Systems of equations and graphing

Data: U. S. Department of Agriculture

The table below lists the per capita consumption of various fruit juices in the United States, in gallons per person, for several years.

Year	Orange Juice	Apple Juice and Cider	Grape Juice	Grapefruit Juice	Pineapple Juice
1985	4.81	1.55	0.23	0.61	0.34
1990	3.25	1.74	0.28	0.91	0.50
1995	4.73	1.59	0.45	0.59	0.38
2000	5.54	1.80	0.34	0.53	0.30
2005	4.76	1.87	0.51	0.23	0.26
2010	3.45	2.21	0.37	0.22	0.23
2013	4.01	1.78	0.43	0.20	0.21

1. Explore the data. For which juices does per capita consumption appear to be increasing? Decreasing? Neither?

2. Graph each set of data using a line graph. Which patterns appear to be linear?

3. Focus on grape juice, grapefruit juice, and pineapple juice. Fit a linear equation to each set of data, and graph the lines along with the data.

4. Do the slopes of the lines support your initial analysis of whether per capita consumption is increasing or decreasing?

5. Examine the points of intersection of the three lines. What does each point tell you about the consumption of these juices?

6. Now focus on orange juice and apple juice and cider. Fit a linear equation to each set of data, and graph the lines along with the data. What is the predicted per capita consumption of orange juice in 2015? What is the predicted per capita consumption of apple juice and cider in 2015? In what year will per capita orange juice consumption be equal to per capita apple juice and cider consumption?

7. Which of the lines that you fit for the five juices appear to be a good fit for the data?

The previous data looked at long-term trends of per capita consumption of several fruit juices. The table below lists per capita consumption for the same juices, in gallons per person, for every year from 2000 to 2010.

Year	Orange Juice	Apple Juice and Cider	Grape Juice	Grapefruit Juice	Pineapple Juice
2000	5.54	1.80	0.34	0.53	0.30
2001	5.15	1.79	0.33	0.54	0.31
2002	5.02	1.80	0.37	0.47	0.32
2003	4.09	1.95	0.40	0.40	0.34
2004	4.94	2.13	0.38	0.38	0.27
2005	4.76	1.87	0.51	0.23	0.26
2006	4.39	2.22	0.44	0.20	0.27
2007	4.13	2.28	0.56	0.29	0.22
2008	3.79	2.11	0.45	0.30	0.27
2009	3.92	2.09	0.38	0.26	0.27
2010	3.45	2.21	0.37	0.22	0.23

8. Explore the data. Do any of the trends seem different for this decade than they did for the years from 1985 to 2013?

9. Graph each set of data. Which patterns appear to be linear?

10. Now focus on orange juice and apple juice and cider. Fit a linear equation to each
    set of data, and graph the lines along with the data. What is the predicted per capita
    consumption of orange juice in 2015? What is the predicted per capita consumption of
    apple juice and cider in 2015? In what year will per capita consumption of orange juice
    be equal to per capita consumption of apple juice and cider?

11. Were your predictions the same for Questions 6 and 10? If they were different, which
    prediction do you think is the most reliable? Find the actual values for 2015 for both
    orange juice and apple juice and cider consumption. How close were your estimates?
    What factors, other than historic trends, might a producer of juice consider when trying
    to predict demand?

Activity 8: **Waterfalls**

Focus: Linear inequalities

1. The Western New York Waterfall Society has compiled classifications of waterfalls by their shape. Write and graph an inequality that describes each of the types of waterfalls.

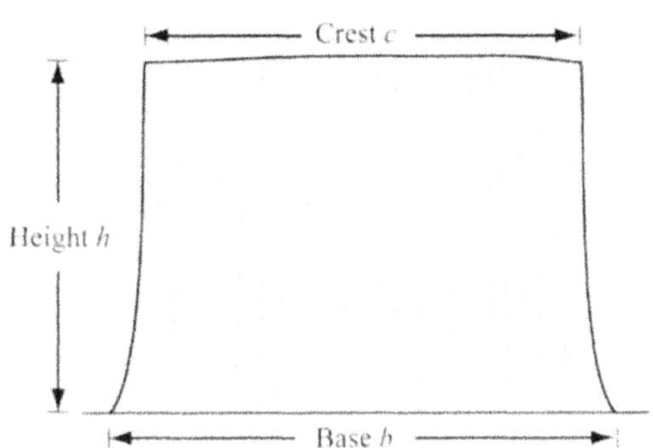

Fan: The crest width c is less than or equal to one-third the base width b.

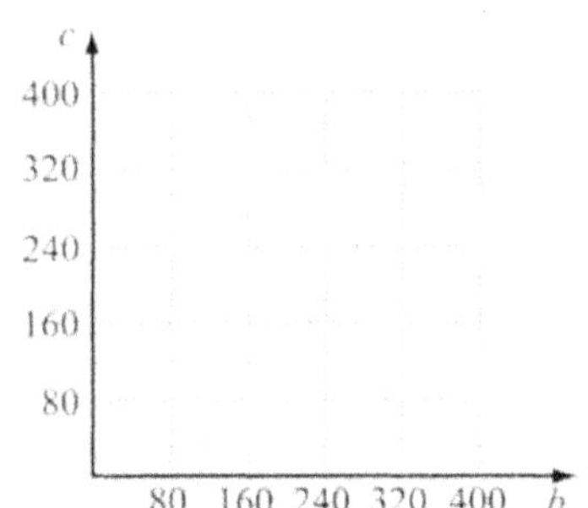

Funnel: The crest c is greater than three times the base width b.

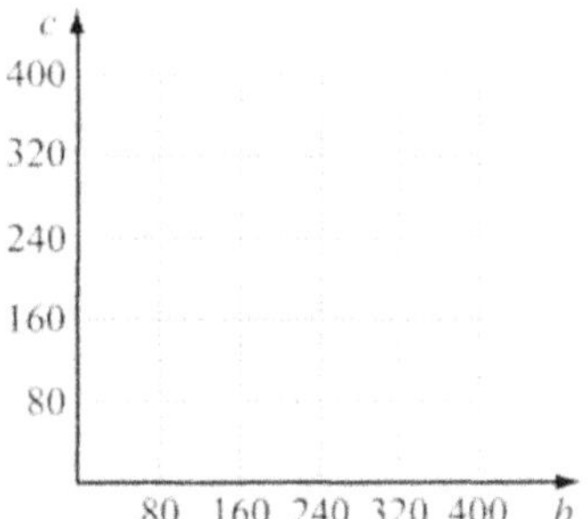

Curtain: The crest width c is greater than one and one-half times the height h.

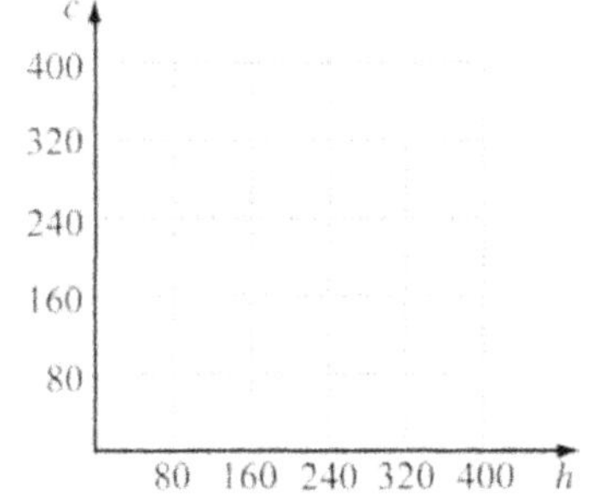

Ribbon: The crest width c is less than or equal to one-half the height h.

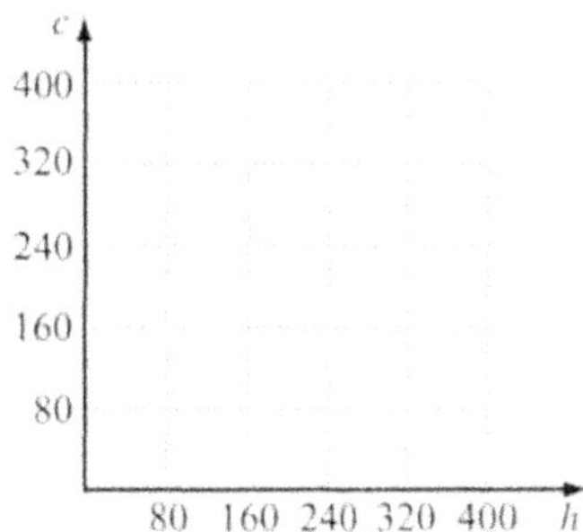

2. Each of the falls described below can be plotted as a point on a graph. Locate each of the waterfalls described below on the graph of each inequality above. What is the most appropriate classification for each waterfall?

(a) Angeline Falls
    King County, Washington
    Height: 400 ft
    Crest width: 100 ft
    Base width: 100 ft

(b) Brotherhood Falls
    Skagit County, Washington
    Height: 66 ft
    Crest width: 5 ft
    Base width: 40 ft

(c) Lower Lewis River Falls
    Skamania County, Washington
    Height: 43 ft
    Crest width: 200 ft
    Base width: 200 ft

Activity 9: Construction

Focus: Linear inequalities; systems of linear inequalities

Guardrails for worker safety at construction sites are required to be 42 inches high. In some cases, it is impossible to build guardrails to this height. According to the Oregon Occupational Health and Safety Administration, for these sites it is acceptable to build a shorter barrier that meets the following two requirements:
- The barrier must be a minimum of 30 inches high.
- The sum of the height and depth of the barrier must be at least 48 inches.

1. Which of the following barriers meets these requirements?

(a)

(b)

(c)

(d)

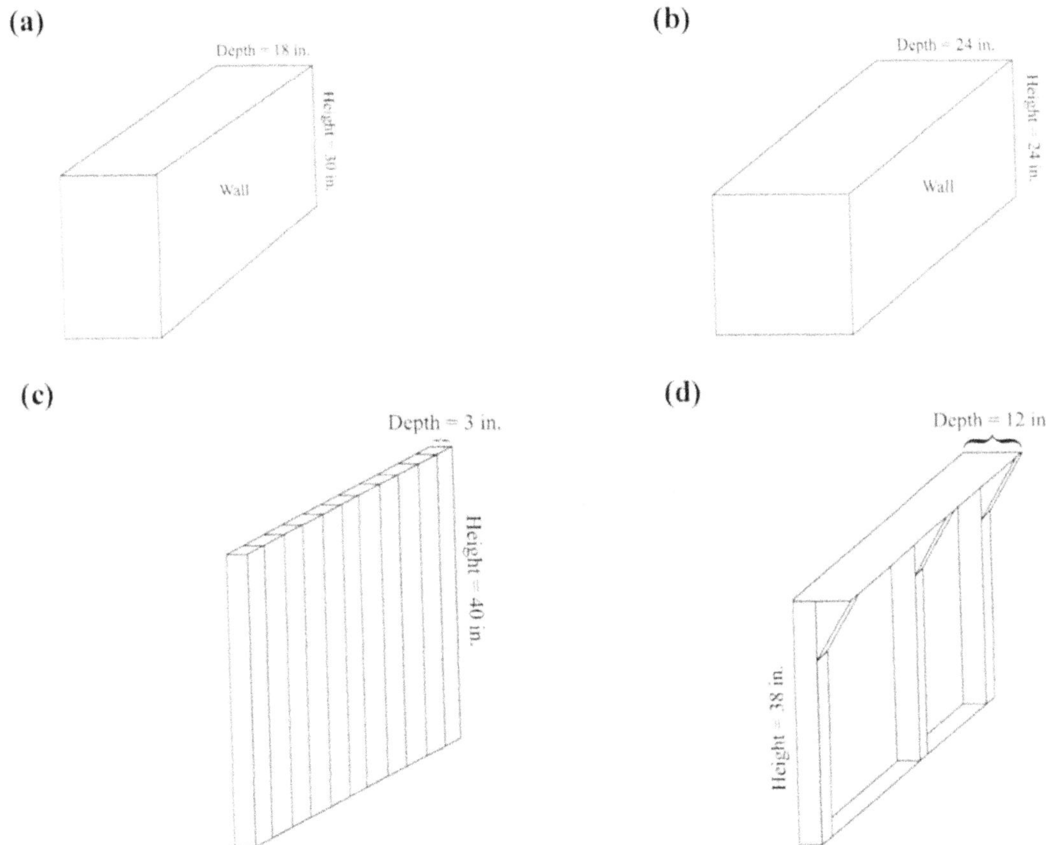

2. Write a linear inequality that models the requirement that the barrier must be a minimum of 30 inches high. Let h represent the height of the barrier. Graph the inequality on the grid below.

3. Write a linear inequality that models the requirement that the sum of the height and depth of the barrier must be at least 48 inches. Let h represent the height of the barrier and d represent the depth of the barrier. Graph the inequality on the same grid, using a different shading.

You have now graphed a system of inequalities. The overlap in the shading shows the solution set of the system.

4. Plot each of the barriers described above as a point on the grid. How does this graph support your answer to question 1?

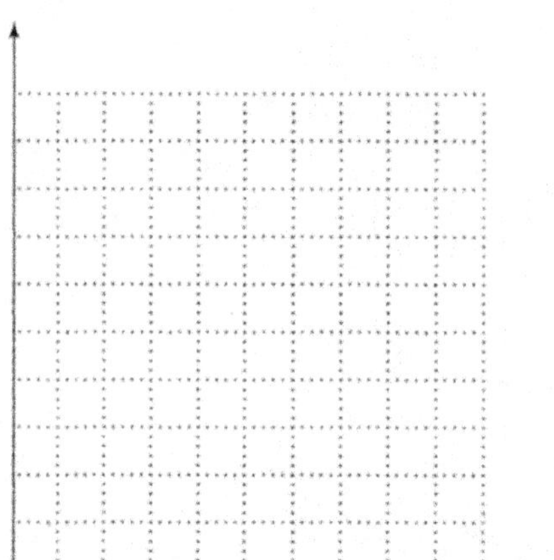

     Copyright © 2022 Pearson Education, Inc.

Activity 10: **Visualizing Factoring**

Focus: Factoring trinomials

Materials: Graph paper and scissors

Factoring a polynomial like $x^2 + 5x + 6$ can be thought of as determining the length and the width of a rectangle that has area $x^2 + 5x + 6$.

1. a) To factor $x^2 + 11x + 10$ geometrically, first cut out shapes like those below to represent x^2, $11x$, and 10. This can be done by either tracing the figures below or by selecting a value for x, say 4, and using the squares on the graph paper to cut out the following:

    x^2: Using the value selected for x, cut out a square that is x units on each side.

    $11x$: Using the value selected for x, cut out a rectangle that is 1 unit wide and x units long. Repeat this to form 11 such strips.

    10: Cut out two rectangles with whole-number dimensions and an area of 10. One should be 2 units by 5 units and the other 1 unit by 10 units.

    b) Then attempt to use one of the two rectangles with area 10, along with all the other shapes, to piece together one large rectangle. Only one of the rectangles with area 10 will work.

    c) From the large rectangle formed in part (b), use the length and the width to determine the factorization of $x^2 + 11x + 10$. Where do the dimensions of the rectangle representing 10 appear in the factorization?

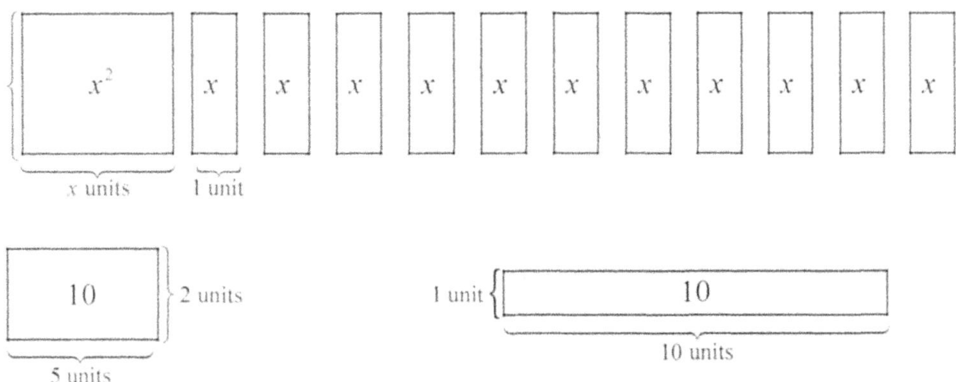

2. Repeat step (1) above, but this time use the other rectangle with area 10, and use only 7 of the 11 strips, along with the x^2-shape. Piece together the shapes to form one large rectangle. What factorization do the dimensions of this rectangle suggest?

3. Cut out rectangles with area 12 use the above approach to factor $x^2 + 8x + 12$. What dimensions should be used for the rectangle with area 12?

Activity 11: Matching Factorizations

Focus: Factoring polynomials

Materials: One set of cards, copied and cut out from the attached master, for each student or group of students.

Note: You will need a large surface such as a table for this activity.

The top portion of each of the cards in your set contains a polynomial, and the bottom portion contains a factorization. Shuffle the cards. Then choose one card and match the polynomial at the top of that card to the factorization at the bottom of a second card. Then match the polynomial at the top of the second card to the factorization at the bottom of a third card. Continue to match polynomials with their factorizations, forming a circle as you go. If your work is correct, the cards should form a complete circle when you are finished.

$x^2 - x - 6$	$x^2 + 5x + 6$	$2x^2 + 5x + 3$
$(x+2)(x+3)$	$(2x+3)(x+1)$	$(x-2)(x-3)$
$x^2 - 5x + 6$	$6x^2 + x - 1$	$x^2 + x - 6$
$(2x+1)(3x-1)$	$(x-2)(x+3)$	$(2x+3)(3x-2)$

$6x^2+5x-6$	$6x^2-5x-1$	$6x^2-5x-6$	$2x^2-5x+3$
$(6x+1)(x-1)$	$(2x-3)(3x+2)$	$(2x-3)(x-1)$	$(2x-1)(3x-1)$
$6x^2-5x+1$	$6x^2+13x+6$	$6x^2+5x-1$	$2x^2+x-3$
$(2x+3)(3x+2)$	$(6x-1)(x+1)$	$(2x+3)(x-1)$	$(2x-1)(3x+1)$
$6x^2-x-1$	$6x^2-7x+1$		
$(6x-1)(x-1)$	$(x+2)(x-3)$		

Activity 12: Data and Downloading

Focus: Direct variation, inverse variation

Although download speed and data plan limits continue to increase, so do the size of the files consumers download. The table below lists a typical file size for several types of files, along with the time it takes to download one of those files and the number of that type of file that could be downloaded within a 5 GB data limit. The units used are related as follows.

1 MB (megabyte) = 1000 KB (kilobyte)
1 GB (gigabyte) = 1000 MB = 1,000,000 KB

File Type	File Size	Time It Takes to Download a File at 10 Mbps (megabytes per second) Download Speed	Maximum Number of Downloads in One Month with 5 GB Monthly Plan Limit
Email without attachments	5 KB (0.005 MB)	0.0005 sec	1,000,000
Word document	100 KB (0.1 MB)	0.01 sec	50,000
High-resolution image	5 MB	0.5 sec	1000
2-minute YouTube video	10 MB	1 sec	500
3-hour movie	1000 MB	100 sec	5
1-hour HDTV show	2500 MB	250 sec	2

1. Explore the data. As the file size increases, does the time it takes to download that type of file increase or decrease? As the file size increases, does the number of downloads possible of that type of file increase or decrease?

2. Divide the time it takes, in seconds, to download each type of file by the file size, in MB. What is the common ratio? Considering the units of this ratio, what does the ratio represent?

3. Multiply each of the file sizes, in MB, by the corresponding maximum number of downloads possible of that type of file. What is the common product? What does it represent?

4. Graph the data. On one grid, place file size on the horizontal axis and download time on the vertical axis. On a second grid, place file size on the horizontal axis and number of downloads on the vertical axis.

5. Which two quantities appear to vary directly? Which two quantities appear to vary inversely?

6. Find an equation of variation that describes how download time t varies with file size x. How is the constant in this equation related to the common ratio found in Question 2?

7. Find an equation of variation that describes how the number of downloads n varies with file size x. How is the constant in this equation related to the common product found in Question 3?

8. A 2-minute MP3 music file is about 2 MB. Use one of the equations from the table to determine how long it would take to download a 2-minute song. Use one of the equations above to determine how many 2-minute songs could be downloaded within a 5 GB data limit.

9. Write a paragraph comparing direct variation and inverse variation.

Activity 13: Firefighting Formulas

Topics: Evaluating radical expressions, graphing radical equations, simplifying
   radical expressions

Data for Formulas: Manchaca Fire Rescue (mvfd.org) and
                Truckee Meadow Fire Protection District (tmfire.us)

In order to effectively fight fires, firefighters rely on accurate estimations of amount and
velocity of water flow. Accuracy and safety also depend on water pressure in the hose and
the nozzle reaction force. The following table lists water flow for selected nozzle pressures
for a $1\frac{3}{4}$-inch diameter smooth-bore nozzle tip.

Nozzle pressure, NP, in pounds per square inch	Water flow, Q, in gallons per minute	Nozzle reaction force, in pounds
20	410	
40	575	
60	705	
80	815	
100	910	
120	1000	
140	1075	

1. Explore the data. As nozzle pressure increases, does water flow increase or decrease?
    Does the relationship appear to be linear?

2. Graph the data.

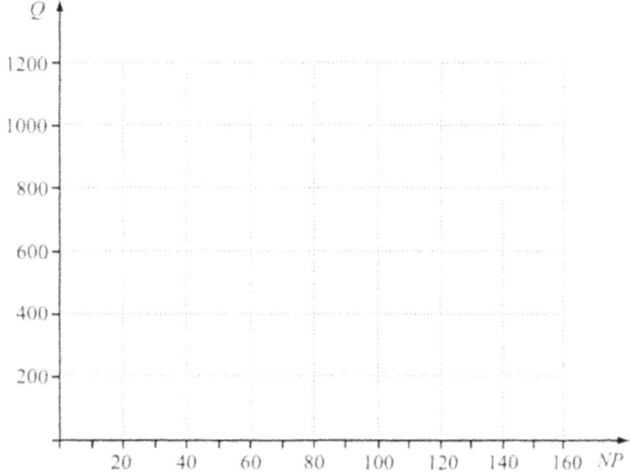

Firefighters use a number of formulas to calculate nozzle reaction, water flow, and volume. Several of these formulas are listed below.

$$Q = 29.72D^2\sqrt{NP} \qquad\qquad \text{Equation 1}$$
$$NR = 1.57D^2 \times NP \text{ (for a solid stream)} \qquad \text{Equation 2}$$
$$V = 12.1\sqrt{NP} \qquad\qquad \text{Equation 3}$$

where

Q = water flow, in gallons per minute
V = water velocity, in feet per second
D = nozzle diameter, in inches
NR = nozzle reaction force, in pounds
NP = nozzle pressure, in pounds per square inch

3. Graph the equation $Q = 29.72D^2\sqrt{NP}$, where $D = 1.75$, on the same graph as the data given in the table. Does the equation appear to be a good fit?

4. Equation 2 gives nozzle reaction in terms of diameter and nozzle pressure. Use Equation 1 and Equation 2 to find a formula that gives nozzle reaction in terms of water flow and nozzle pressure. (*Hint:* Solve Equation 1 for D^2.)

5. Suppose, for a $1\frac{1}{2}$-inch nozzle tip, the nozzle pressure is 60 pounds per square inch. Find the water flow, the nozzle reaction force, and the water velocity.

6. Use Equation 2 to fill in the nozzle reaction force for a $1\frac{3}{4}$-inch nozzle tip in the third column of the table on the preceding page.

7. As nozzle pressure increases, both water flow and nozzle reaction force increase. An increase in water flow means more water can be used to extinguish a fire. Is an increase in nozzle reaction force helpful or harmful for firefighters? Research nozzle reaction force, and explain what it is and how it affects firefighting.

Activity 14: Music Downloads

Focus: Linear and quadratic models

Data: Recording Industry Association of America

Sales of physical recorded music, such as CDs, declined as digital sales increased. In time, digital music sales also began to decrease. The table below lists sales of albums and single tracks, in millions of units, for years between 2008 and 2014.

Year	Number of Digital Albums Sold, in millions	Number of Digital Single Tracks Sold, in millions
2008	428	1070
2009	379	1159
2010	326	1172
2011	331	1271
2012	316	1336
2013	289	1259
2014	257	1108

1. Explore the data. How many digital albums were sold in 2008? In 2014? How many digital single tracks were sold in 2008? In 2014?

2. Looking at only the data from 2008 and 2014, did the number of digital albums sold increase or decrease? Looking at only the data from 2008 and 2014, did the number of digital single tracks increase or decrease?

3. Let x represent the number of years after 2008, and graph the number of digital albums sold for all years from 2008 to 2014. How would you describe the trend of the data? What type of function would you use to model the data?

4. Let x represent the number of years after 2008, and graph the number of digital single tracks sold for all years from 2008 to 2014. How would you describe the trend of the data? What type of function would you use to model the data?

5. Choose two points from the digital album data and fit a linear function to the data. Graph this line with the data. How well does the function model the data? Is there a pair of points that you think would have resulted in a better model?

6. Choose three points from the digital single tracks data and fit a quadratic function to the data. Graph this parabola with the data. How well does the function model the data? Is there another set of three points that you think would have resulted in a better model?

7. Using all the points and a graphing utility, fit a linear function to the digital album data. Graph this line with the data. How does the fit of this function compare with the function you found in Question 5?

8. Using all the points and a graphing utility, fit a quadratic function to the digital single tracks data. Graph this line with the data. How does the fit of this function compare with the function you found in Question 6?

9. If the trends continue, will the number of single tracks sold ever equal the number of digital albums sold? If so, when? How might these trends influence the way artists record and market their music?

10. In what year were the sales of single tracks at a maximum? What do you think caused the sales of single tracks to decline after that year? In order to answer this question, you may need additional information. What other data would you like to collect, and why?

Activity 15: Let's Go to the Movies

Focus: Modeling with polynomial functions

Data: National Association of Theatre Owners

The following table shows the number of annual movie admissions in the United States and Canada for several years from 1988 to 2018.

Year, x	Number of Movie Admissions in the United States and Canada, in billions
1988, 0	1.08
1991, 3	1.14
1994, 6	1.24
1997, 9	1.354
2000, 12	1.383
2003, 15	1.521
2006, 18	1.401
2009, 21	1.414
2012, 24	1.36
2015, 27	1.32
2018, 30	1.301

1. What trends do you see in the data?

2. Graph the data on the grid below or on a graphing calculator. Choose a scale that shows the trends. Would you use a linear function or a quadratic function to model the data?

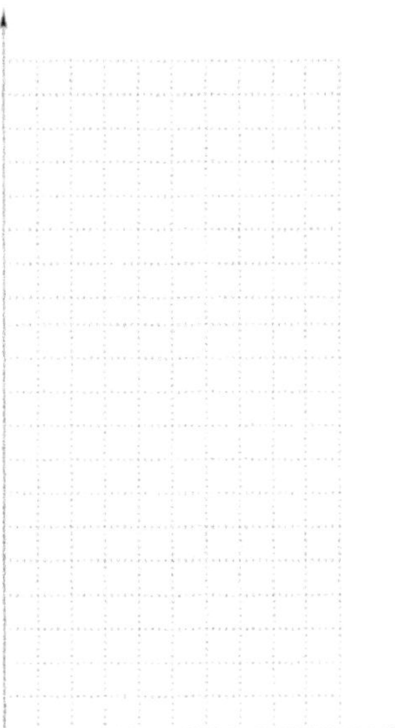

3. Fit a quadratic function to the data, either algebraically by choosing three points or using a graphing utility and quadratic regression. Label this function E.

4. Now look only at the years from 1988 to 2003. If this were all the data you had, would you use a linear function or a quadratic function to model the data? Fit a function to this first part of the data set either algebraically or with regression using a graphing utility. Label this function F.

5. Now look only at the years from 1997 to 2012. If this were all the data you had, would you use a linear function or a quadratic function to model the data? Fit a function to this second part of the data set either algebraically or with regression using a graphing utility. Let $x =$ the number of years after 1988. Label this function S.

6. The number of movie admissions in 2014 was 1.27 billion. Would any of your three functions E, F, or S have predicted this number? Which one was the closest?

7. Research to find the number of movie admissions in the United States and Canada in 2010. Would any of your three functions E, F, or S have predicted this number? Which one was the closest?

8. Write a paragraph describing the results of your analysis. Include your thoughts about what might have caused the overall increasing and decreasing trends in movie admissions that you observed.

Activity 16: Teaching about Zeros; Creating a Video

Focus: Theorems about zeros of polynomial functions

Materials: Phone or computer for recording video

Group size: 2-4

One of the best ways to know that you understand a concept is to be able to explain it to someone else. Choose one of the following questions, or use the one assigned to you by your instructor, and use your textbook or other resources to thoroughly examine theorems that answer the question. Then create a video that answers the question, teaching the concepts in such a way that your classmates will also understand the question and the answer. Your video should be less than five minutes in length and should contain at least the following:

- Description and examples of the types of polynomial functions you are discussing
- Clear answers to the question using language that is both accurate and understandable
- At least one worked out example; more may be necessary if your result has several parts
- At least one exercise for viewers to try on their own

When all videos are complete, play yours for the class and check the answers to the exercises you requested your classmates to solve.

1. What do you know about complex number solutions of polynomial functions with real coefficients?

2. What do you know about irrational zeros of polynomial functions with rational coefficients?

3. What do you know about rational zeros of polynomial functions with integer coefficients?

4. What can you tell about the number of positive real zeros and the number of negative real zeros from the coefficients of a polynomial function?

- Use a phone to record yourself teaching at a blackboard or whiteboard.

- Use a smart pen to record your explanation as you write it on paper or on a tablet.

- Prepare a slide presentation on your computer and record audio as you work through the presentation.

- Prepare a presentation on your computer and make a screencast.

- Use educational software to make a video.

Activity 17: Earthquake Magnitude

Focus: Applications of logarithms

One way of describing the size of an earthquake is by estimating its *seismic moment* or the force required to produce the waves recorded on a seismograph. Rather than actually giving the seismic moment M_0 (measured in dyne·cm), scientists often instead list an earthquake's *moment magnitude* M_w, where

$$M_w = \frac{2}{3}\log(M_0) - 10.7 .$$

The following table lists the moment magnitude for significant earthquakes in Chile in 2014 and 2015.

Date	Moment Magnitude	Seismic Moment
March 16, 2014	7.0	
April 1, 2014	8.2	
April 1, 2014	7.5	
April 1, 2014	7.0	
April 2, 2014	7.7	
August 23, 2014	6.4	
October 8, 2014	7.0	
March 18, 2015	6.3	
March 23, 2015	6.4	
September 16, 2015	8.3	

1. Explore the data. Which was the strongest earthquake listed? The weakest? Can you tell from the moment magnitude how much greater the strongest earthquake was than the weakest?

2. Solve the equation $M_w = \frac{2}{3}\log(M_0) - 10.7$ for M_0. Use this new equation to find the seismic moment for each earthquake in the table.

3. We can divide the seismic moments of two earthquakes to find out how many times stronger one earthquake was than another. How many times stronger was the September 16, 2015, earthquake than the March 18, 2015, earthquake?

4. The September 16, 2015, earthquake was 2 units of moment magnitude greater than the March 18, 2015, earthquake. Will the proportional moment magnitude of any two earthquakes that are 2 units apart be the same as the proportional seismic moment of these two earthquakes?

5. Suppose two earthquakes are one unit of moment magnitude apart. Use the equation $M_W = \dfrac{2}{3}\log(M_0) - 10.7$ to determine how many times stronger the stronger earthquake is than the weaker earthquake.

6. Calculate the seismic moment of earthquakes with moment magnitude 1, 2, 3, 4, 5, 6, 7, 8, and 9. Which would be easier to illustrate on a graph: all 9 moment magnitudes or all 9 seismic moments?

7. Why might scientists use moment magnitudes instead of seismic moments to describe the strength of an earthquake?

Activity 18: A Cosmic Path

Focus: Ellipses

On May 4, 2007, Comet 17P/Holmes was at the point closest to the sun in its orbit. Comet 17P is traveling in an elliptical orbit with the sun as one focus, and one orbit takes about 6.88 years. One astronomical unit (AU) is 93,000,000 mi. One group member should do the following calculations in AU and the other in millions of miles.

Data: Harvard-Smithsonian Center for Astrophysics

1. At its *perihelion*, a comet with an elliptical orbit is at the point in its orbit closest to the sun. At its *aphelion*, the comet is at the point farthest from the sun. The perihelion distance for Comet 17P is 2.053218 AU, and the aphelion distance is 5.183610 AU. Use these distances to find a. (See the following diagram.)

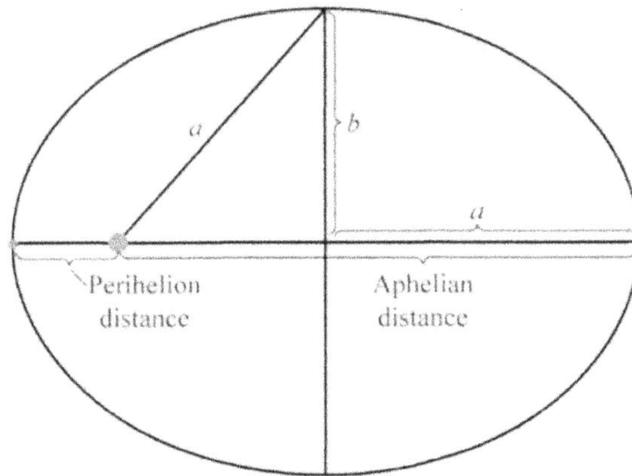

2. Using the figure above, express b^2 as a function of a. Then find b using the value found for a in part (1).

3. One formula for approximating the perimeter of an ellipse is

$$P = \pi\left(3a + 3b - \sqrt{(3a+b)(a+3b)}\right),$$

developed by the Indian mathematician S. Ramanujan in 1914. How far does comet 17P travel in one orbit?

4. What is the speed of the comet? Find the answer in AU per year and in miles per hour.

5. Which calculations – AU or mi – were easier to use? Why?

Activity 19: Bargaining for a Used Car

Focus: Geometric series

Time: 30 minutes

Group size: 2

Materials: Calculators

1. One group member (the "seller") is asking \$5500 for a car. The second member (the "buyer") offers \$1500. The seller splits the difference (\$5500 - \$1500 = \$4000, and \$4000 ÷ 2 = \$2000) and lowers the price to \$3500 (\$5500 − \$2000 = \$3500). The buyer splits the difference again (\$3500 - \$1500 = \$2000, and \$2000 ÷ 2 = \$1000) and gives a counter offer of \$2500 (\$1500 + \$1000 = \$2500). Continue in this manner until you are able to agree on the car's selling price to the nearest penny.

2. Check several guesses to find what the buyer's initial offer should be in order to achieve a purchase price of \$2500 or less.

3. The seller's price in the bargaining above can be modeled recursively by the sequence

$$a_1 = 5500, \qquad a_n = a_{n-1} - \frac{d}{2^{2n-3}},$$

   where d is the difference between the initial price and the first offer. Use this recursively defined sequence to solve parts (1) and (2) above.

4. The first four terms in the sequence in part (3) can be written as

$$a_1, \quad a_1 - \frac{d}{2}, \quad a_1 - \frac{d}{2} - \frac{d}{8}, \quad a_1 - \frac{d}{2} - \frac{d}{8} - \frac{d}{32}.$$

Use the formula for the limit of an infinite geometric series to find a simple algebraic formula for the eventual sale price, P, when the bargaining process from above is followed. Verify the formula by using it to solve parts (1) and (2) above.

This activity is based on the article "Bargaining Theory, or Zeno's Used Cars," by James C. Kirby, *The College Mathematics Journal*, 27(4), September, 1996.

Activity 20: Interest Rates on Credit Cards

Focus: Comparing interest rates

Comparing interest rates is essential if one is to become financially responsible. A small change in an interest rate can make a large difference in the cost of a loan. When you make a payment on a loan, do you know how much of that payment is interest and how much is applied to reducing the principal? A balance carried on a credit card is a type of loan. The money you obtain through the use of a credit card is not "free" money. There is a price (interest) to be paid for the privilege.

1. After the holidays, Madison has a balance of $3216.28, on a credit card with an annual percentage rate (APR*) of 19.7%. She decides not to make additional purchases with this card until she has paid off the balance. Many credit cards require a minimum monthly payment of 2% of the balance. What is Madison's minimum payment on a balance of $3216.28? Round the answer to the nearest dollar.

2. Find the amount of interest and the amount applied to reduce the principal in the minimum payment found in Exercise 1. $\left(\text{The interest for the first month is}\right.$

$$\left.\$3216.28 \times \frac{19.7\%}{12}.\right)$$

3. If Madison had transferred her balance to a card with an APR of 12.5% and made the minimum payment, how much of the payment would be interest and how much would be applied to reduce the principal?

4. Organize the amounts for both rates in the following chart.

Balance Before First Payment	First Month's Payment	% APR	Amount of Interest	Amount Applied to Principal	Balance After First Payment
$3216.28		19.7%			
$3216.28		12.5%			

*Actually, the interest on a credit card is computed daily with a rate called a daily percentage rate (DPR). The DPR for $3216.28 would be $19.7\% / 365 \approx 0.054\%$. When no payments or additional purchases are made during the month, the difference in total interest for the month is minimal and we will not deal with it here.

5. Using the information in the chart in Exercise 4, fill in the blanks in the following comparison.

At 19.7%, the interest is ____________ and the principal is decreased by ____________.
At 12.5%, the interest is ____________ and the principal is decreased by ____________.
Thus the principal is decreased by ____________ − ____________, or ____________ more with the 12.5% rate than with the 19.7% rate. That is, the interest at 19.7% is ____________ greater than the interest at 12.5%.

6. Even though the mathematics of the information in the chart below is beyond the scope of this course, it is interesting to compare how long it takes to pay off the balance of $3216.28 if Madison continues to pay $64 for each payment with how long it takes if she pays double that amount, $128, for each payment. Financial consultants frequently tell clients that if they want to take control of their debt, they should pay double the minimum payment.

After filling in the blanks in this chart,

Rate	Payment Per Month	Number of Payments To Pay Off Debt	Total Paid Back	Total Interest
19.7%	$64	107, or 8 yr 11 mo	107×$64 = $6848	$6848−$3216.28 = $3631.72
19.7%	$128	33, or 2 yr 9 mo		
12.5%	$64	72, or 6 yr		
12.5%	$128	29, or 2 yr 5 mo		

a) determine how much longer it takes to pay off the balance with a monthly payment of $64 than with a monthly payment of $128 for both rates, 19.7% and 12.5%.

b) determine how much is saved in interest at 19.7% with a monthly payment of $128 compared to the interest paid at 19.7% with a monthly payment of $64.

c) determine how much is saved in interest at 12.5% with a monthly payment of $128 compared to the interest paid at 12.5% with a monthly payment of $64.

Activity 21: Home Loans

Focus: Comparing interest rates and understanding amortization tables

Before beginning the process of purchasing a home. it is helpful to understand the impact of the length of the loan and a change in an interest rate.

1. The Sawyers recently purchased their first home. They borrowed $153,000 at $6\frac{5}{8}\%$ for 30 years (360 payments). Their monthly payment (excluding insurance and taxes) is $979.68.

    a) How much of the first payment is interest and how much is applied to reduce the principal?

    b) At the time of the second payment on the 30-year loan, what is the new principal?

    c) How much of the second payment is interest and how much is applied to reduce the principal?

    d) If the Sawyers pay the entire 360 payments, how much interest will be paid on the loan?

2. If we continue calculating principal and interest for each of the 360 payments in Exercise 1, we can create an amortization table, part of which is shown below. Such tables are available online.

Mortgage Amortization: Mortgage amount: $153,000; Interest rate: 6.625%;
Number of Years: 30; Monthly Payment: $979.68

PAYMENT	PRINCIPAL	INTEREST	BALANCE
1	$134.99	$844.69	$152,865.01
2	$135.73	$843.94	$152,729.28
3	$136.48	$843.19	$152,592.80
4	$137.24	$842.44	$152,455.56
5	$137.99	$841.68	$152,317.56
6	$138.76	$840.92	$152,178.81
7	$139.52	$840.15	$152,039.29
8	$140.29	$839.38	$151,898.99
9	$141.07	$838.61	$151,757.93
10	$141.85	$837.83	$151,616.08
11	$142.63	$837.05	$151,473.45
12	$143.42	$836.26	$151,330.04
.	.	**Interest for first 12 periods = $10,086.14**	
.	.	.	
.	.	.	
349	$917.04	$62.63	$10,427.84
350	$922.11	$57.57	$9,505.74
351	$947.20	$52.48	$8,578.54
352	$932.32	$47.36	$7,646.23
353	$937.46	$42.21	$6,708.76
354	$942.64	$37.04	$5,766.13
355	$947.84	$31.83	$4,818.28
356	$953.07	$26.60	$3,865.21
357	$958.34	$21.34	$2,906.87
358	$963.63	$16.05	$1,943.24
359	$968.95	$10.73	$974.30
360	$974.30	$0.00	$0.00
		Interest for last 12 periods = $405.84	

a) How much more is the amount applied to principal with the 350th payment than with the 10th payment?

b) How much less is the interest with the 350th payment than with the 10th payment?

3. Ten months after the Sawyers buy their home financed at a rate of $6\frac{5}{8}\%$, the rates drop to $5\frac{1}{2}\%$. After much consideration, they decide to refinance even though the new loan will cost them $1200 in refinance charges. They have reduced the principal a small amount in the 10 payments they have made, but they decide to again borrow $153,000 for 30 years at the new rate. Their new monthly payment is $868.72.

a) How much of the first payment is interest and how much is applied to the principal?

b) Compare the interest and the amount applied to the principal at $5\frac{1}{2}\%$ with the amounts at $6\frac{5}{8}\%$ found in Exercise 1a).

c) With the lower house payment, how long will it take the Sawyers to recoup the refinance charge of $1200?

d) If the Sawyers pay the entire 360 payments, how much interest will be paid on this loan? How much less is the total interest at $5\frac{1}{2}\%$ than at $6\frac{5}{8}\%$?

4. Now let's compare a 15-yr loan with a 30-yr loan. If the Sawyers had chosen a 15-yr loan at $6\frac{5}{8}\%$ (180 payments) instead of a 30-yr loan at $6\frac{5}{8}\%$ (360 payments), the monthly payment would have increased to $1343.33.

a) How much of the first payment is interest and how much is applied to the principal?

b) How much larger is the amount applied to the principal in the first payment with the 15-yr loan than with the 30-yr loan?

c) If the Sawyers pay the entire 180 payments, how much interest will be paid on this loan?

d) Compare the amount of interest to pay off the 15-yr loan with the amount of interest to pay off the 30-yr loan. (See Exercise 1d.)

e) There are advantages and disadvantages to each loan period. Can you think of a situation in which the 30-yr loan would be the better choice? Can you think of a situation in which the 15-yr loan would be the better choice?

Answers: Interactive Preview Worksheets

Interactive Preview 1: OPERATIONS ON THE REAL NUMBERS

Check Your Understanding

1. C
2. E
3. H
4. B
5. G
6. D
7. A
8. F

Exercises

1. 14
2. -5
3. 240
4. -2
5. $-\dfrac{9}{4}$
6. $-\dfrac{2}{25}$
7. -4.84
8. -8
9. 8
10. -34
11. -70
12. $-\dfrac{10}{9}$
13. $\dfrac{36}{11}$
14. -5.4
15. $\dfrac{85}{24}$
16. $-\dfrac{1}{16}$
17. -27
18. $\dfrac{2}{3}$
19. -2.04
20. 28

Interactive Preview 2: ADDING, SUBTRACTING, MULTIPLYING, AND DIVIDING FRACTIONS

Exercises

1. 468
2. $\dfrac{45}{16}$
3. $\dfrac{31}{60}$
4. $\dfrac{7}{6}$
5. $\dfrac{10}{3}$
6. $\dfrac{2}{3}$
7. $\dfrac{25}{48}$
8. $\dfrac{3}{40}$
9. 2
10. $\dfrac{1}{12}$
11. $\dfrac{5}{36}$
12. $\dfrac{13}{24}$
13. $\dfrac{3}{2}$
14. $\dfrac{1}{4}$
15. $\dfrac{47}{50}$
16. $\dfrac{5}{6}$
17. $\dfrac{7}{12}$
18. 1
19. $\dfrac{7}{4}$
20. $\dfrac{41}{24}$
20. $\dfrac{1}{900}$
22. $\dfrac{19}{144}$
23. $\dfrac{10}{3}$
24. $\dfrac{13}{20}$

Interactive Preview 3: ORDER OF OPERATIONS

Check Your Understanding

1. A; in Solution B, Step (1) should have been dividing 18 by 6 instead of adding $6+3$.
2. B; in Solution A, Step (4) should have been dividing 80 by 5 instead of multiplying $5(8)$.

Exercises

1. Evaluate 2^2.
   Multiply $3 \cdot 5$.
   Divide $8 \div 4$.
   Subtract $6 - 15$.
   Add $-9 + 2$.

2. Subtract $5 - 3$.
   Evaluate 2^2.
   Divide $32 \div 4$.
   Multiply $8 \cdot 4$.
   Multiply $2 \cdot 2$.
   Subtract $32 - 4$.

3. -13	**4.** 60	**5.** 4	**6.** 4	**7.** 16
8. 187	**9.** -11	**10.** 70	**11.** 22	**12.** 6
13. 27	**14.** 292	**15.** 222	**16.** 12	**17.** -105
18. 139	**19.** 36	**20.** 216		

Interactive Preview 4: INTEGERS AS EXPONENTS

Exercises

1. $\dfrac{1}{3^5}$, or $\dfrac{1}{243}$	**2.** a^{18}	**3.** x^{12}	**4.** 2^{21}
5. $\dfrac{5^3}{3^3}$, or $\dfrac{125}{27}$	**6.** 100	**7.** 1	**8.** $\dfrac{1}{x^2}$
9. y^{10}	**10.** $7^6 x^{12}$	**11.** 4^{11}	**12.** 3^4, or 81
13. $\dfrac{1}{t^{12}}$	**14.** $\dfrac{1}{z^2}$	**15.** $\dfrac{1}{2^5}$, or $\dfrac{1}{32}$	**16.** s^6
17. p	**18.** 1	**19.** $\dfrac{y^{12}}{2^{20}}$	**20.** 10^{10}
21. w^9	**22.** c^7	**23.** $9x^4$	**24.** $\dfrac{1}{y^6}$

Interactive Preview 5: SOLVING LINEAR EQUATIONS

Exercises

1. 3	**2.** 1	**3.** -32	**4.** 147	**5.** -2.5
6. -6	**7.** -12	**8.** 22	**9.** 13	**10.** $\dfrac{27}{2}$
11. -5	**12.** 5	**13.** 2	**14.** 21	**15.** $\dfrac{18}{5}$
16. $\dfrac{10}{3}$	**17.** $-\dfrac{1}{10}$, or -0.1		**18.** 6	**19.** $\dfrac{3}{10}$

21. 3120　　**21.** −4　　**22.** $-\dfrac{28}{27}$

Interactive Preview 6:　INTERVAL NOTATION

Check Your Understanding

1. D　　**2.** C　　**3.** C　　**4.** A　　**5.** B　　**6.** A

Exercises

1. $(-\infty, 6)$　　　**2.** $[-4.1, 5.6)$　　　**3.** $(-8, 21]$　　　**4.** $[3, \infty)$

5. $(0, \infty)$　　　**6.** $(-\infty, \infty)$　　　**7.** $[0, 5]$　　　**8.** $(-6, 3)$

9. $\left(-\infty, \dfrac{2}{5}\right)$　　**10.** $[17, \infty)$　　**11.** $\left[-\dfrac{1}{2}, \dfrac{3}{2}\right)$　　**12.** $(-3, -2]$

13. 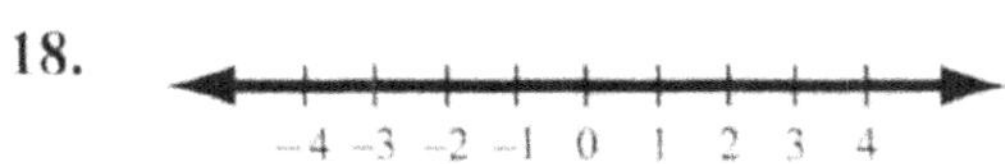　　**14.**

15. 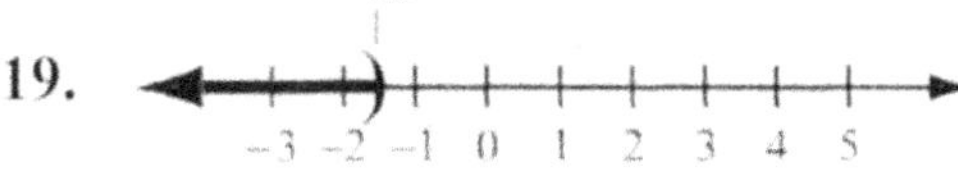　　**16.** 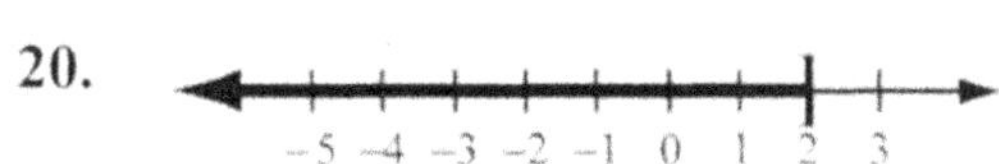

17.　　**18.**

19. 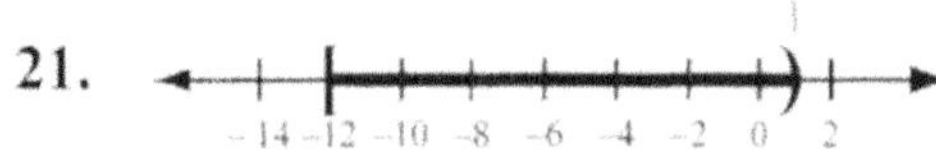　　**20.**

21. 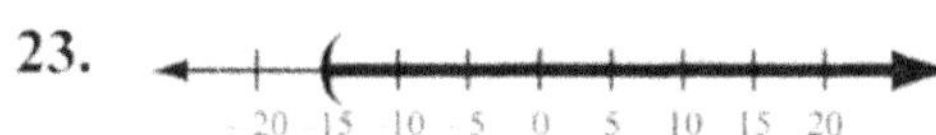　　**22.**

23.　　**24.**

Interactive Preview 7:　SOLVING LINEAR INEQUALITIES

Exercises

1. $\{y \mid y \geq -3\}$, or $[-3, \infty)$　　　**2.** $\left\{x \mid x > \dfrac{31}{20}\right\}$, or $\left(\dfrac{31}{20}, \infty\right)$

3. $\{c \mid c < -6\}$, or $(-\infty, -6)$　　　**4.** $\{q \mid q \geq 6\}$, or $[6, \infty)$

5. $\left\{x \mid x < -\dfrac{2}{3}\right\}$, or $\left(-\infty, -\dfrac{2}{3}\right)$　　　**6.** $\{z \mid z < 12\}$, or $(-\infty, 12)$

7. $\{x|x \ge 1\}$, or $[1, \infty)$

8. $\{x|x \le 5\}$, or $(-\infty, 5]$

9. $\{t|t \le 4\}$, or $(-\infty, 4]$

10. $\{x|x > -4\}$, or $(-4, \infty)$

Interactive Preview 8: SOLVING EQUATIONS AND INEQUALITIES WITH ABSOLUTE VALUE

Exercises

1. The solutions are -1 and 1.

2. $(-2, 2)$

3. $[-5, 5]$

4. $(-\infty, -4) \cup (4, \infty)$

5. $(-\infty, -3.5] \cup [3.5, \infty)$

6. $q = -8 \ or \ q = 8$

7. $-2 \le p + 5 \le 2$

8. $0.5 - x < -6.5 \ or \ 0.5 - x > 6.5$

9. $7y - 1 = -12 \ or \ 7y - 1 = 12$

10. $-21 < \dfrac{3}{4}t + 6 < 21$

11. $18q \le -9 \ or \ 18q \ge 9$

12. $-30 \le w + 10 \le 30$
$-40 \le w \le 20$
Solution set: $[-40, 20]$

13. $y - 4 < -6 \ or \ y - 4 > 6$
$y < -2 \ or \ y > 10$
Solution set: $(-\infty, -2) \cup (10, \infty)$

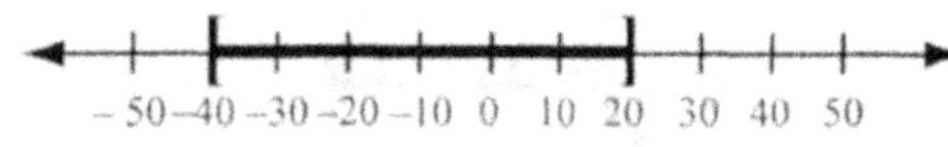
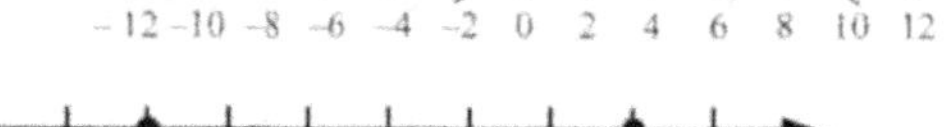

14. The solutions are -1 and 5. 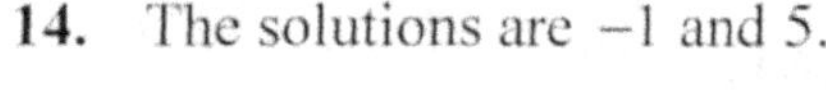

15. $(3, 5)$, 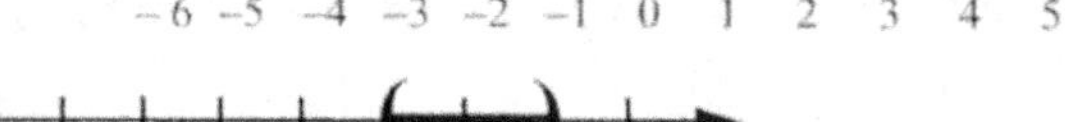

16. $(-\infty, -10] \cup [0, \infty)$, 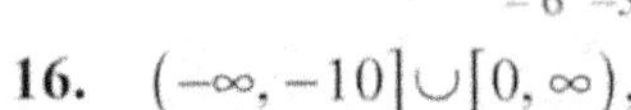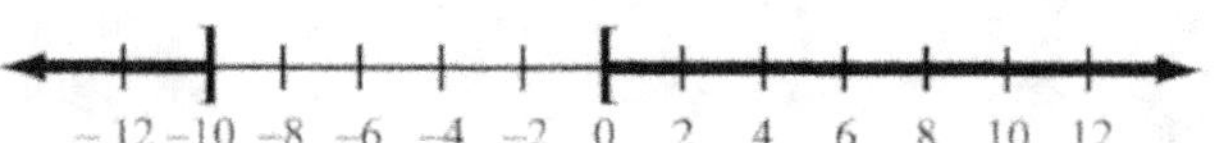

17. The solutions are -1 and $-\dfrac{3}{5}$. 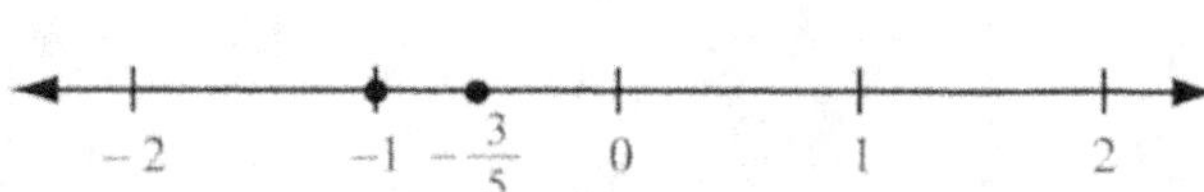

18. $(-\infty, -2) \cup (0, \infty)$, 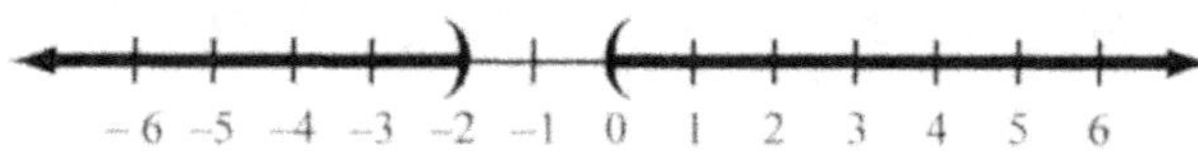

19. $[-2, 3]$.

20. $[-1, 5]$.

21. $(-\infty, -2) \cup (6, \infty)$,

22. The solutions are -10 and 20.

23. $(-\infty, -2] \cup [1, \infty)$,

24. $(-8, 4)$,

25. $[-1, 7]$,

Interactive Preview 9: GRAPHING LINEAR EQUATIONS

Check Your Understanding

1. 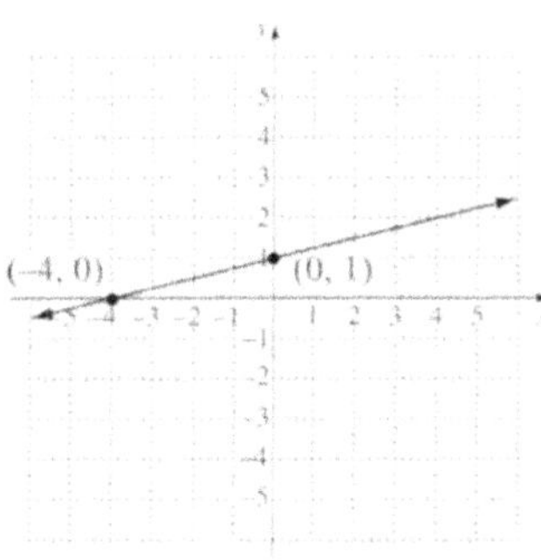

2. 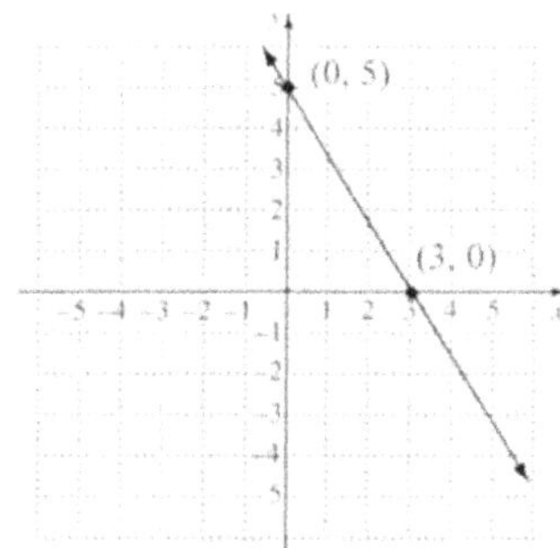

3. 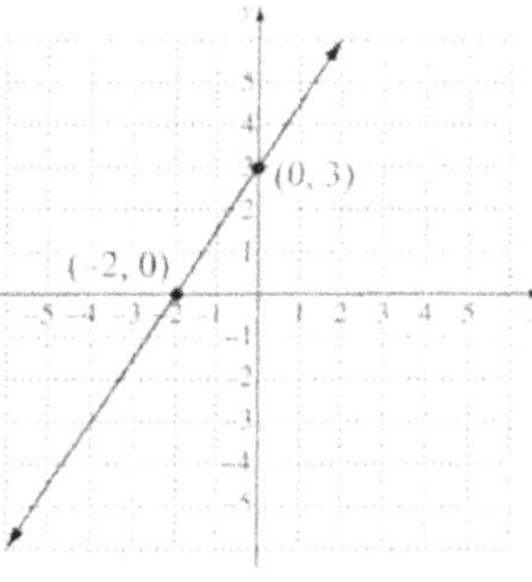

4. False   **5.** True   **6.** True   **7.** C   **8.** B   **9.** D   **10.** A

Exercises

1. 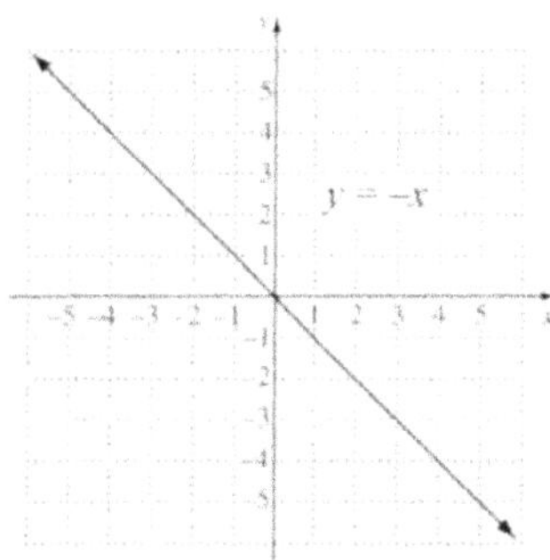

2. 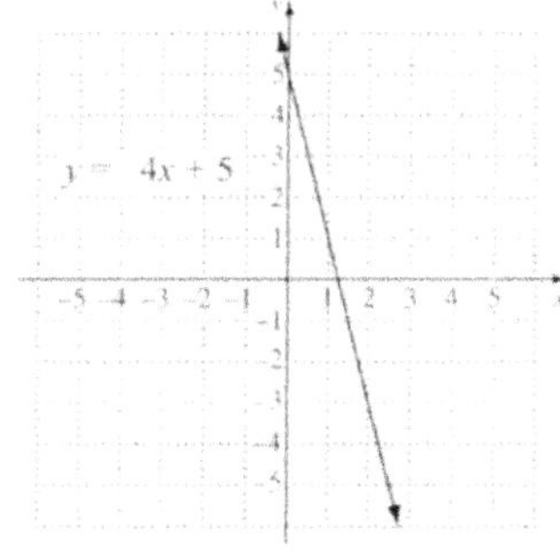

3. 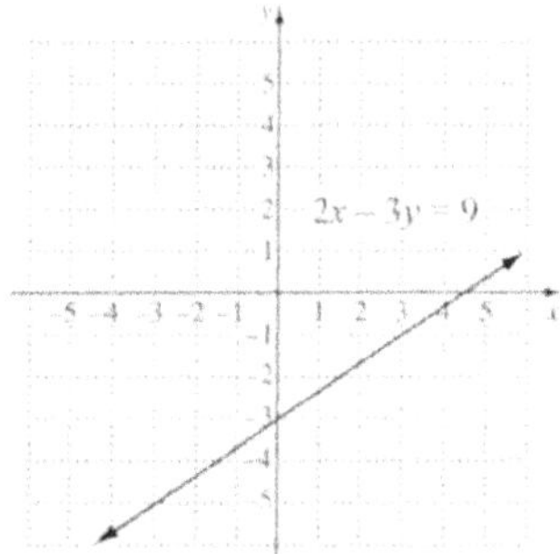

4.

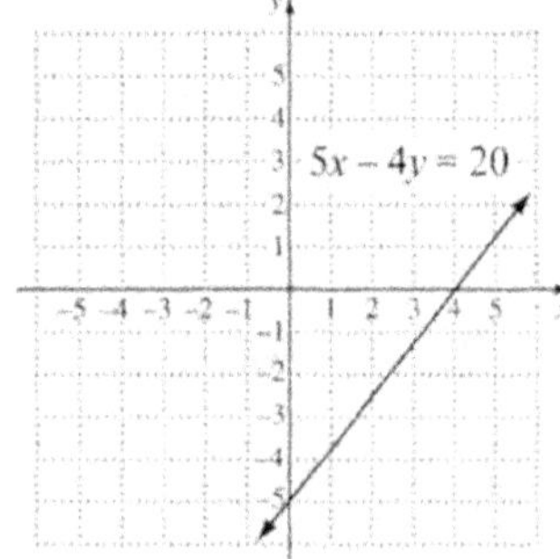

5.

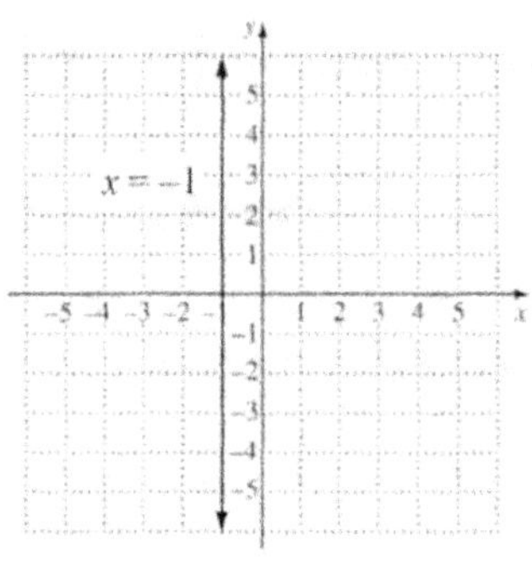

6.

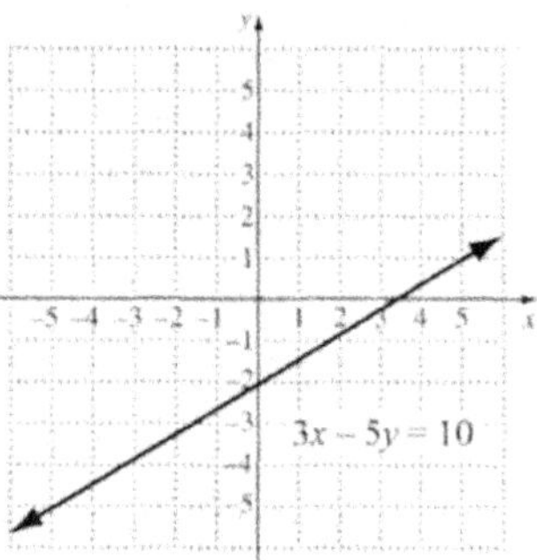

7.

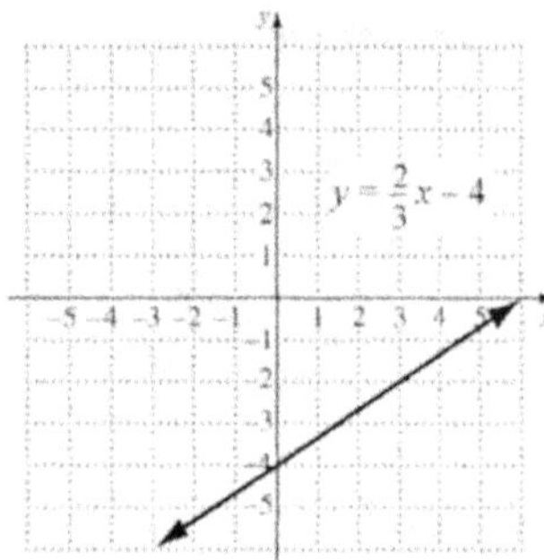

8.

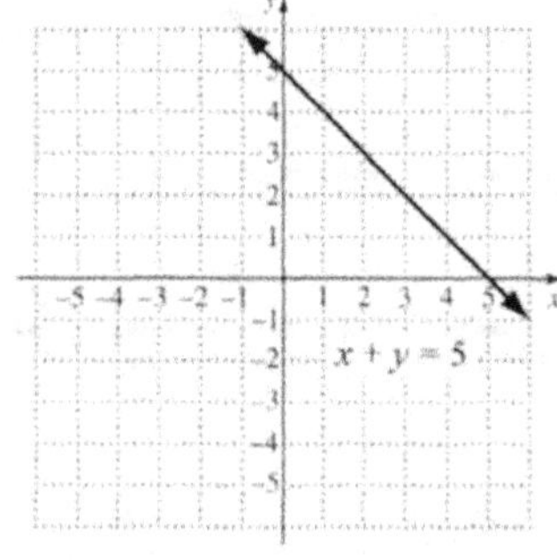

9.

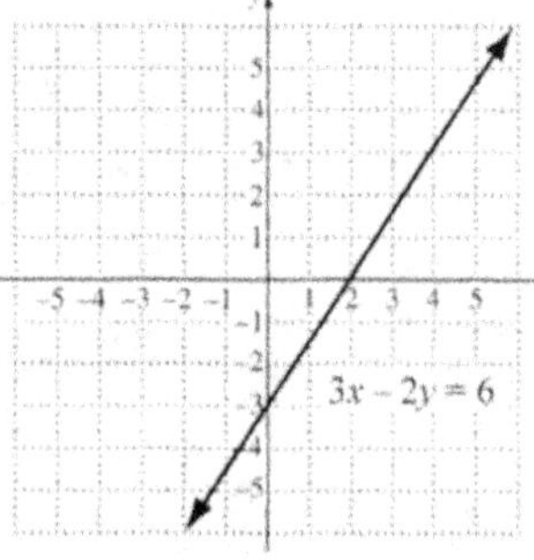

10.

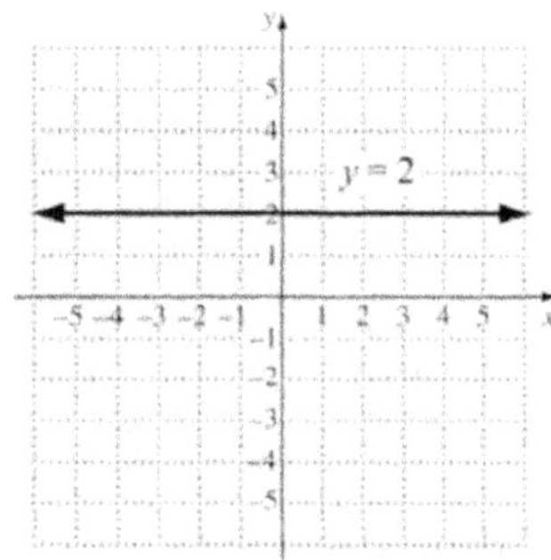

11.

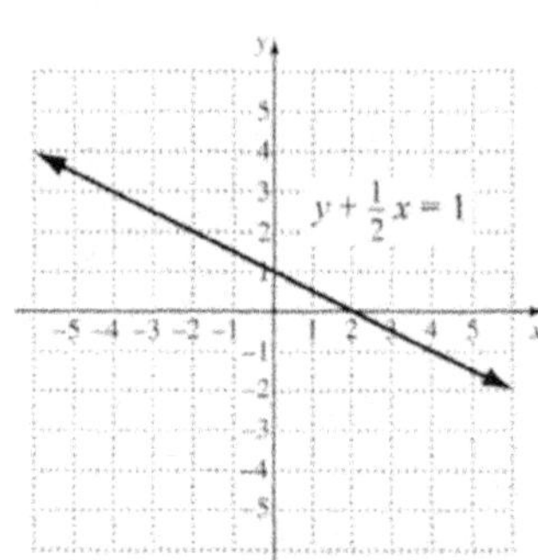

12.

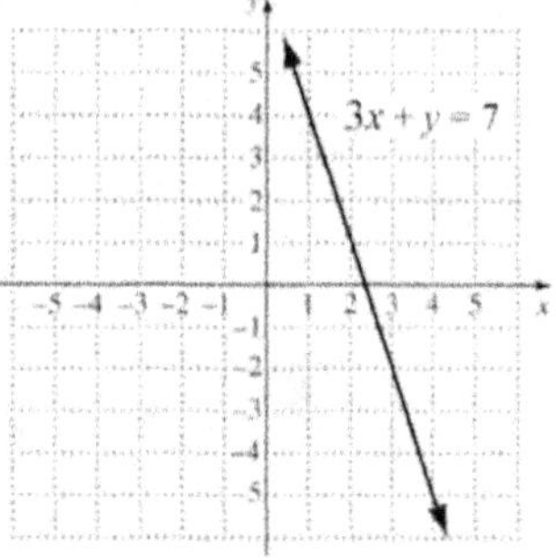

Interactive Preview 10:   FUNCTION VALUES; DOMAIN AND RANGE

Exercises

1.

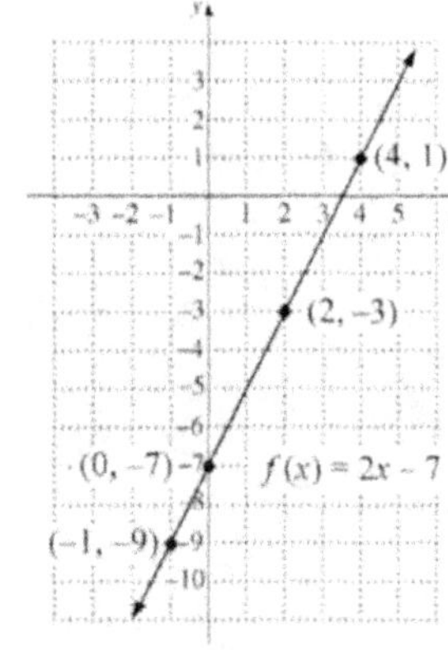

$f(2) = \underline{-3}$

$f(-1) = \underline{-9}$

$f(0) = \underline{-7}$

$f(4) = \underline{1}$

2.

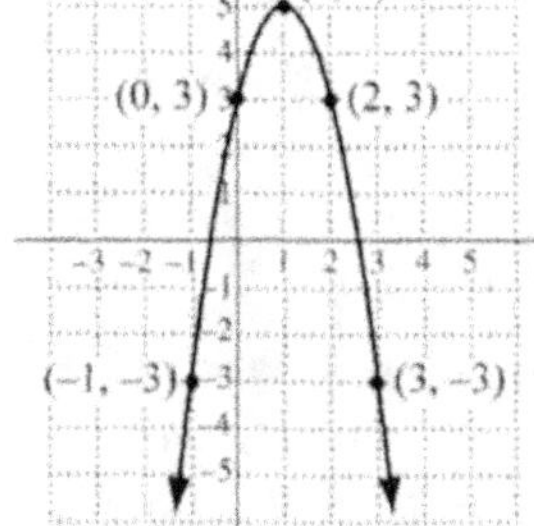

$h(0) = \underline{3}$

$h(1) = \underline{5}$

$h(-1) = \underline{-3}$

$h(3) = \underline{-3}$

$h(2) = \underline{3}$

     Copyright © 2022 Pearson Education, Inc.

3. 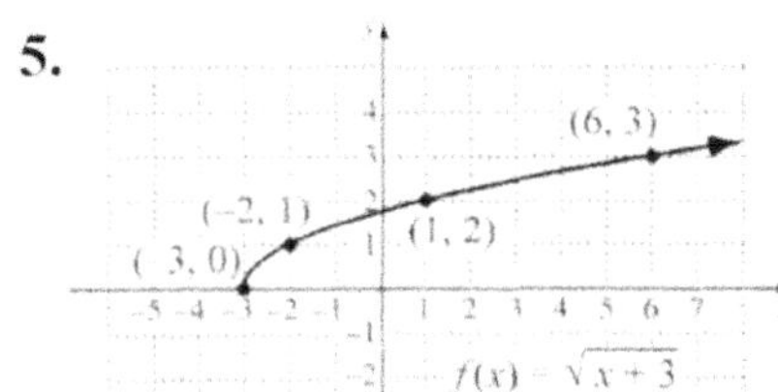

$t(-3) = \underline{\ 5\ }$

$t(4) = \underline{\ 2\ }$

$t(0) = \underline{\ 2\ }$

$t(2) = \underline{\ 0\ }$

$t(6) = \underline{\ 4\ }$

4. 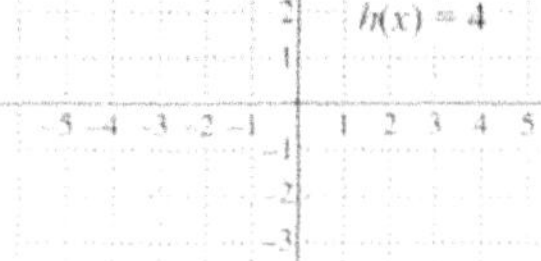

$h(3) = \underline{\ 4\ }$

$h(-2) = \underline{\ 4\ }$

$h(-5) = \underline{\ 4\ }$

$h(0) = \underline{\ 4\ }$

5.

$f(-3) = \underline{\ 0\ }$

$f(-2) = \underline{\ 1\ }$

$f(1) = \underline{\ 2\ }$

$f(6) = \underline{\ 3\ }$

6. 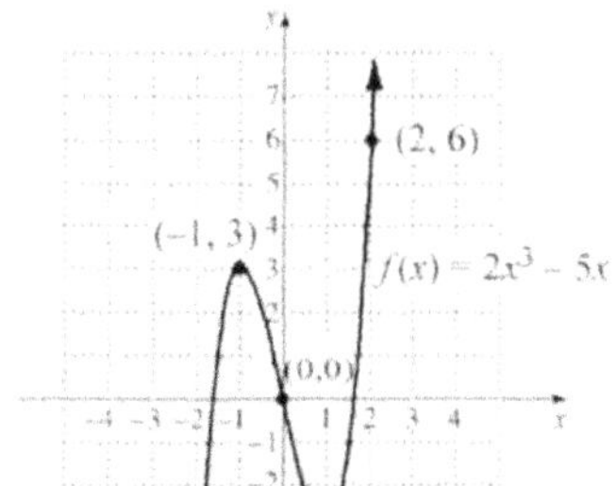

$f(-1) = \underline{\ 3\ }$

$f(2) = \underline{\ 6\ }$

$f(-2) = \underline{\ -6\ }$

$f(0) = \underline{\ 0\ }$

$f(1) = \underline{\ -3\ }$

7. Domain: $(-\infty, \infty)$
Range: $(-\infty, \infty)$

8. Domain: $(-\infty, \infty)$
Range: $\{y \mid y \le 5\}$, or $(-\infty, 5]$

9. Domain: $(-\infty, \infty)$
Range: $\{y \mid y \ge 0\}$, or $[0, \infty)$

10. Domain: $(-\infty, \infty)$
Range: $\{4\}$

11. Domain: $\{x \mid x \ge -3\}$, or $[-3, \infty)$
Range: $\{y \mid y \ge 0\}$, or $[0, \infty)$

12. Domain: $(-\infty, \infty)$
Range: $(-\infty, \infty)$

Interactive Preview 11: DETERMINE THE DOMAIN AND THE RANGE OF A FUNCTION

Exercises

1. Domain: $(-\infty, \infty)$
Range: $(-\infty, \infty)$

2. Domain: $(-\infty, \infty)$
Range: $[-4, \infty)$

3. Domain: $(-\infty, \infty)$
Range: $(-3, \infty)$

4. Domain: $[-4, \infty)$
Range: $[0, \infty)$

5. Domain: $(-1, 3]$
Range: $[-3, 5]$

6. Domain: $[-5, \infty)$
Range: $[-4, \infty)$

7. Domain: $[-4, 2]$
Range: $[-4, 1]$

8. Domain: $(-\infty, 2) \cup (2, \infty)$
Range: $(-\infty, 0) \cup (0, \infty)$

9. Domain: $(-\infty, \infty)$
Range: $[-2, 2]$

Interactive Preview 12: INTRODUCTION TO POLYNOMIALS

Exercises

1. No   **2.** Yes   **3.** Yes   **4.** Yes   **5.** No   **6.** No

7.

Term	$3y^4$	$-6y^3$	$8y^2$	$-\dfrac{2}{3}y$	5
Degree of Term	4	3	2	1	0
Degree of Polynomial	4				
Leading Term	$3y^4$				
Leading Coefficient	3				
Constant Term	5				

8.

Term	x	$-3x^4$	$4x^3$	-13	$7x^2$	$-6x^5$
Degree of Term	1	4	3	0	2	5
Degree of Polynomial	5					
Leading Term	$-6x^5$					
Leading Coefficient	-6					
Constant Term	-13					

9. Binomial   **10.** Trinomial   **11.** Monomial

12. Trinomial   **13.** Monomial   **14.** Binomial

15. $2y^3 - 8y^2 + y;\ y - 8y^2 + 2y^3$ **16.** $-a^4 + a^3 + 3;\ 3 + a^3 - a^4$

17. $-t^3 + 3t^2 - 5t + 3;\ 3 - 5t + 3t^2 - t^3$ **18.** $8x^5 - 4x^3 + 12x + 5;\ 5 + 12x - 4x^3 + 8x^5$

19. $-9b^3 - 5b^2 + 4b + 23;\ 23 + 4b - 5b^2 - 9b^3$

20. $5q^7 + \dfrac{3}{5}q - \dfrac{1}{3};\ -\dfrac{1}{3} + \dfrac{3}{5}q + 5q^7$

Interactive Preview 13: ADDING AND SUBTRACTING POLYNOMIALS

Check Your Understanding

1. f   **2.** l   **3.** i   **4.** c

Exercises

1. $-2t^2 - 6t - 3$ **2.** $4x + 5$

3. $9x^2 - 3x - 16$ **4.** $20y^2 + 4y + 6$

5. $10b^3 - b^2 - 11b - 3$ **6.** $3x^4 - 5x^3 - 11x^2 + x - 4$

7. $16x^2 + 9x - 1$ **8.** $2a^4 + 7a^3 - 9a^2 - 4a + 7$

9. $-10x^4 - 4x^3 + 2x - 8$ **10.** $w^5 - 5w^3 + 2w^2 + 4$

Interactive Preview 14:   MULTIPLYING BINOMIALS

Exercises

1. $y^2 - 5y + 6$
2. $16b^2 - 38b - 5$
3. $w^2 + 3w - 10$
4. $z^2 - 8z - 9$
5. $20x^2 - 9x - 18$
6. $3q^2 + 65q + 100$
7. $x^2 + 9x - 52$
8. $30b^2 + 19b - 28$
9. $25w^2 + 35w - 8$
10. $a^2 + 17a + 66$
11. $x^2 + 12x + 36$
12. $4x^2 - 4x + 1$
13. $9a^2 + 12a + 4$
14. $t^2 + 20t + 100$
15. $z^2 - 24z + 144$
16. $16w^2 + 40w + 25$
17. $b^2 - 6b + 9$
18. $36y^2 - 36y + 9$
19. $y^2 - 100$
20. $9c^2 - 16$
21. $t^2 - 36$
22. $w^2 - 1$
23. $4z^2 - 49$
24. $100y^2 - 25$
25. $x^2 + 7x - 30$
26. $16x^2 - 81$
27. $21x^2 - 19x + 4$
28. $y^2 + 25y + 100$
29. $s^2 - 9$
30. $a^2 - 16a + 64$
31. $25y^2 + 40y + 16$
32. $6a^2 + 64a - 22$

Interactive Preview 15:   FACTORING BY GROUPING

Checking Your Understanding

1. Yes   2. No   3. No   4. Yes

Exercises

1. $y^3 + 4y^2 + 9y + 36$
$= \left(y^3 + 4y^2\right) + (9y + 36)$
$= y^2(y + 4) + 9(y + 4)$
$= (y + 4)\left(y^2 + 9\right)$

2. $3s^3 - 15s^2 - 7s + 35$
$= 3s^3 - 15s^2 + (-7s + 35)$
$= 3s^2(s - 5) - 7(s - 5)$
$= (s - 5)\left(3s^2 - 7\right)$

3. $(c + 6)\left(c^2 - 5\right)$
4. $(x - 7)\left(x^2 + 4\right)$
5. Cannot be factored by grouping.
6. $2(y - 1)\left(y^2 + 9\right)$
7. $(x - 5)\left(x^2 + 1\right)$
8. $(2y + 3)\left(y^2 - 3\right)$
9. $(y - z)(x + w)$
10. $\left(x^2 + 3\right)\left(2x^2 - 5\right)$

Interactive Preview 16: FACTORING TRINOMIALS: ax^2+bx+c

Exercises

1. $(x+9)(x-4)$ **2.** $(y-1)(y-9)$ **3.** $(y-5)(y+2)$

4. $(s+6)(s-15)$ **5.** $(x+8)(x-1)$ **6.** $(b+6)(b+2)$

7. $(x-8)(x+7)$ **8.** $(a+6)(a-9)$ **9.** $(y-7)(y+2)$

10. $(w-11)(w+1)$ **11.** $(q+9)(q+10)$ **12.** $(x-4)(x-10)$

13. $(x+20)(x+100)$ **14.** $(z-5)(z-5)$ **15.** $(3x-2)(2x+7)$

16. $(2w+3)(2w-5)$ **17.** $(5x+2)(6x-1)$ **18.** $(3y-4)(8y+1)$

19. $(2x+5)(10x-3)$ **20.** $(8c+3)(5c+2)$ **21.** $x(x-3)(x-2)$

22. $b(b+4)(b-8)$ **23.** $2y^2(y+15)(y-2)$ **24.** $5q(q+9)(q+4)$

Interactive Preview 17: FACTORING TRINOMIAL SQUARES AND DIFFERENCES OF SQUARES

Check Your Understanding

1. Difference of squares   **2.** Neither   **3.** Trinomial square

4. Difference of squares   **5.** Neither   **6.** Trinomial square

Exercises

1. $(8x+1)(8x-1)$ **2.** $(x-6)^2$ **3.** $(9y+4)(9y-4)$

4. $(w-2)^2$ **5.** $(7x+3)^2$ **6.** $(5a+6)(5a-6)$

7. $(7+z)(7-z)$ **8.** $(x+15)(x-15)$ **9.** $(y+8)^2$

10. $(2x-5)^2$ **11.** $5(x+3)(x-3)$ **12.** $3(x+4)^2$

Interactive Preview 18: THE PRINCIPLE OF ZERO PRODUCTS

Check Your Understanding

1. False   **2.** True   **3.** False

Exercises

1. $x^2 - 7x = -10$

$x^2 - 7x + 10 = 0$

$(x-5)(x-2) = 0$

$x - 5 = 0 \quad or \quad x - 2 = 0$

$x = 5 \quad or \quad x = 2$

The solutions are 5 and 2.

2. $15z^2 = -3z$

$15z^2 + 3z = 0$

$3z(5z + 1) = 0$

$3z = 0 \quad or \quad 5z + 1 = 0$

$z = 0 \quad or \quad z = -\dfrac{1}{5}$

The solutions are 0 and $-\dfrac{1}{5}$.

3. $7, -9$

4. $0, \dfrac{1}{2}$

5. $-5, -4$

6. -10

7. $-4, 8$

8. $-2, -\dfrac{3}{4}$

9. $\dfrac{2}{3}, -\dfrac{5}{7}$

10. $0, -2, 3$

Interactive Preview 19: THE PYTHAGOREAN THEOREM

Exercises

1. $b = 8$

2. $c = 17$

3. $a = 5$

4. $c = \sqrt{32} \approx 5.657$

5. $b = \sqrt{120} \approx 10.954$

6. $a = \sqrt{175} \approx 13.229$

7. $c = \sqrt{34} \approx 5.831$

8. $b = \sqrt{132} \approx 11.489$

9. $a = \sqrt{75} \approx 8.660$

10. $c = \sqrt{98} \approx 9.899$

Interactive Preview 20: SIMPLIFYING RATIONAL EXPRESSIONS

Exercises

1. $\dfrac{4}{5}$

2. $3w$

3. $\dfrac{x+4}{9x+1}$

4. $\dfrac{x+10}{x-9}$

4. $\dfrac{1}{9y^6}$

6. $\dfrac{4z-5}{z+15}$

7. $\dfrac{y-7}{y+4}$

8. $\dfrac{2}{x}$

7. $\dfrac{a+2}{a-8}$

10. $\dfrac{8}{y}$

11. -1

12. $\dfrac{3x-1}{2x+1}$

Interactive Preview 21: FIND THE LCM OF ALGEBRAIC EXPRESSIONS

Exercises

1. 270

2. 144

3. $9x^3$

4. $4(x-6)(x+6)$

5. $a(a-8)(a+8)$

6. $(y+3)(y-10)(y+10)$

7. $(x+2)(x-2)(x+3)$ **8.** $10t^3(t-4)$ **9.** $z^7(z+1)(z-1)^2$

10. $(a-7)(a+3)(a+1)$ **11.** $210(x-1)(x+1)$ **12.** $9y^3(8y-9)$

13. $18q^6(q-6)(q+1)$ **14.** $70c^2(c+7)(c-5)$

Interactive Preview 22: SIMPLIFYING COMPLEX RATIONAL EXPRESSIONS

Exercises

1. $\dfrac{y}{14}$ **2.** $\dfrac{3}{2}$ **3.** $\dfrac{7}{20}$ **4.** $\dfrac{88}{15}$

5. $\dfrac{1-9x}{1+2x}$ **6.** $\dfrac{3z-5}{4z+3}$ **7.** $\dfrac{p^2-5}{p^2+5}$ **8.** $\dfrac{1+7a^2}{1+a^2}$

9. $\dfrac{2x-1}{x}$ **10.** $\dfrac{y}{5y+1}$ **11.** $\dfrac{ab}{b-a}$

12. $\dfrac{-1}{z(z+h)}$, or $-\dfrac{1}{z(z+h)}$ **13.** $\dfrac{-2}{x(x+h)}$, or $-\dfrac{2}{x(x+h)}$

Interactive Preview 23: SOLVING RATIONAL EQUATIONS

Exercises

1. $\dfrac{40}{9}$ **2.** 13 **3.** -11 **4.** $-\dfrac{23}{7}$

5. $-\dfrac{2}{3}, 8$ **6.** $-2, -11$ **7.** -9 **8.** 5

Interactive Preview 24: SOLVING PROPORTIONS

Exercises

1. 12 **2.** 30 **3.** 28 **4.** 56

5. 4 **6.** 0.5, or $\dfrac{1}{2}$ **7.** 10 **8.** 5

9. 2.7 **10.** $\dfrac{27}{16}$ **11.** 0.1 **12.** $\dfrac{1}{5}$

Interactive Preview 25: SIMPLIFYING RADICAL EXPRESSIONS

Exercises

1. $2\sqrt{3}$ **2.** $5\sqrt{3}$ **3.** $4\sqrt{5}$ **4.** $6\sqrt{2}$

5. $11\sqrt{3}$ **6.** $15\sqrt{2}$ **7.** $8\sqrt{5}$ **8.** $10\sqrt{6}$

9. $5\sqrt{x}$	**10.** y^{11}	**11.** $6x$	**12.** $a+1$
13. $2t^2\sqrt{21}$	**14.** $x-5$	**15.** $w^4\sqrt{w}$	**16.** $4x\sqrt{2}$
17. $x-12$	**18.** $(y+1)\sqrt{3}$	**19.** $10b^3\sqrt{14}$	**20.** $9x^3\sqrt{x}$

Interactive Preview 26: MULTIPLYING RADICAL EXPRESSIONS

Exercises

1. $3\sqrt{6}$	**2.** $10\sqrt{3}$	**3.** $15\sqrt{6}$	**4.** $6\sqrt{7x}$
5. $6\sqrt{6a}$	**6.** 13	**7.** $23\sqrt{y}$	**8.** $8w\sqrt{15}$
9. $2t$	**10.** $z+9$	**11.** $\sqrt{28x+7}$	**12.** $50\sqrt{10}$
13. $9x\sqrt{2}$	**14.** $x^6\sqrt{x}$	**15.** y^8	

Interactive Preview 27: RATIONALIZING DENOMINATORS

Exercises

1. $\dfrac{\sqrt{65}}{5}$	**2.** $\dfrac{\sqrt{5}}{5}$	**3.** $\dfrac{2\sqrt{3}}{3}$	**4.** $\dfrac{\sqrt{130}}{45}$	**5.** $\dfrac{\sqrt[3]{6}}{3}$
6. $\dfrac{\sqrt{13}}{13}$	**7.** $\dfrac{3\sqrt{22}}{88}$	**8.** $\dfrac{7\sqrt{2}}{2}$	**9.** $\dfrac{\sqrt{21}}{21}$	**10.** $\dfrac{\sqrt[4]{10}}{2}$

Interactive Preview 28: CHECKING SOLUTIONS OF RADICAL EQUATIONS

Exercises

1. Yes   **2.** No   **3.** Yes   **4.** Yes

5. Only -1 is a solution.   **6.** Only 15 is a solution.

7. Only 1 is a solution.   **8.** Only 7 is a solution.

9. Both numbers are solutions.

Interactive Preview 29: GRAPHING PIECEWISE FUNCTIONS

Exercises

1. Domain of Equation 1: $(-\infty, 2)$

Domain of Equation 2: $[2, \infty)$

Equation 1

$x < 2$	$f(x) = 3x$
1	3
0	0
-1	-3

Equation 2

$x \geq 2$	$f(x) = x-1$
2	1
4	3
5	4

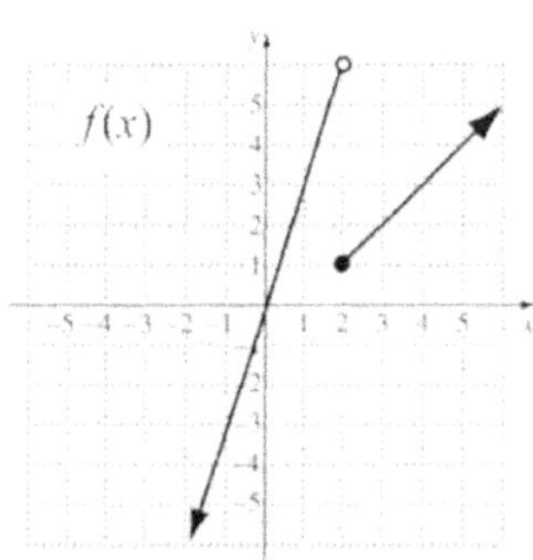

- If $f(x) = 3x$ and $x = 2$, $f(2) = 3 \cdot 2 = 6$. Since 2 is not in the domain of $f(x) = 3x$, we use an open circle at $(2, 6)$.

- If $f(x) = x - 1$ and $x = 2$, $f(2) = 2 - 1 = 1$. Since 2 is in the domain of $f(x) = x - 1$, we use a solid dot at $(2, 1)$.

2. Domain of Equation 1: $(-\infty, -1)$

   Domain of Equation 2: $[-1, 4)$

   Domain of Equation 3: $[4, \infty)$

Equation 1			Equation 2			Equation 3	
$x < -1$	$g(x) = -x + 1$		$-1 \le x < 4$	$g(x) = -3$		$x \ge 4$	$g(x) = 2x - 4$
-2	3		-1	-3		4	4
-3	4		1	-3		5	6
-4	5		3	-3		6	8

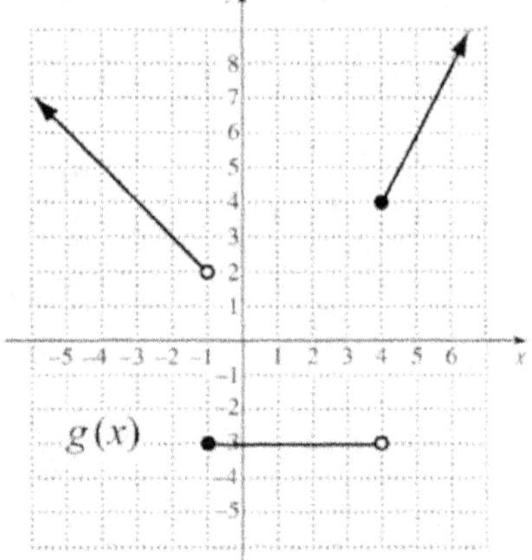

- If $g(x) = -x + 1$ and $x = -1$, $g(-1) = -(-1) + 1 = 2$. Since -1 is not in the domain of $g(x) = -x + 1$, we use an open circle at $(-1, 2)$.

- If $g(x) = -3$ and $x = -1$, $g(-1) = -3$. Since -1 is in the domain of $g(x) = -3$, we use a solid dot at $(-1, -3)$.

- If $g(x) = -3$ and $x = 4$, $g(4) = -3$. Since 4 is not in the domain of $g(x) = -3$, we use an open circle at $(4, -3)$.

- If $g(x) = 2x - 4$ and $x = 4$, $g(4) = 2 \cdot 4 - 4 = 4$. Since 4 is in the domain of $g(x) = 2x - 4$, we use a solid dot at $(4, 4)$.

3.

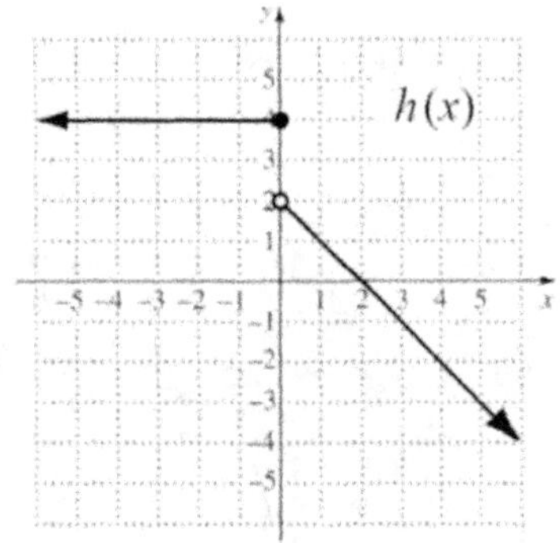

4.

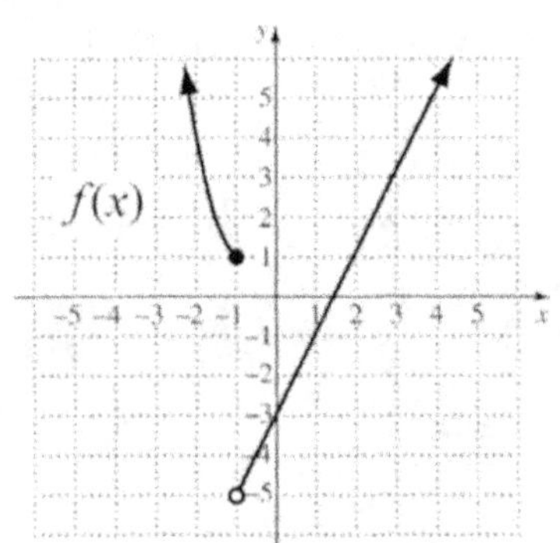

5. 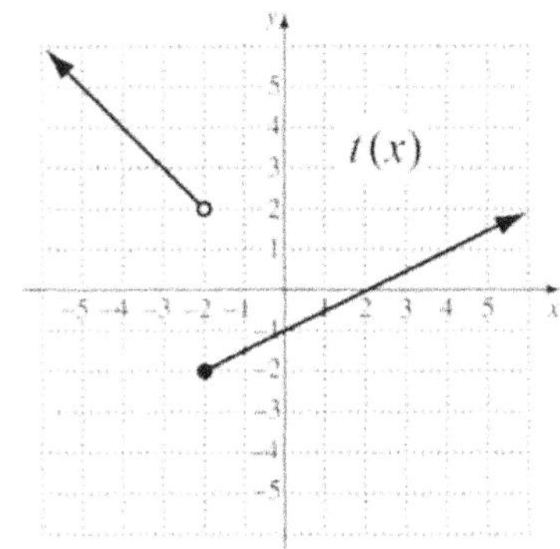

6. 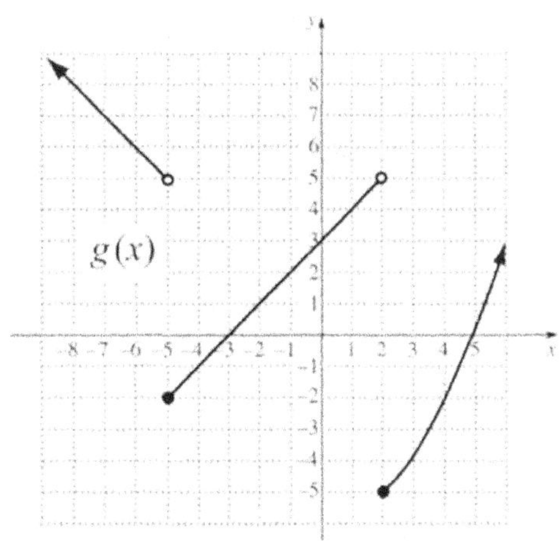

Interactive Preview 30: TRANSFORMATIONS

Exercises

1. (g)	**2.** (h)	**3.** (d)	**4.** (a)	**5.** (b)
6. (c)	**7.** (b)	**8.** (e)	**9.** (h)	**10.** (f)
11. (e)	**12.** (g)	**13.** (h)	**14.** (k)	**15.** (c)
16. (i)	**17.** (b)	**18.** (j)	**19.** (d)	**20.** (f)

Interactive Preview 31: COMPLETING THE SQUARE

Check Your Understanding

1. $x^2 + 8x + 16 = 3 + 16$

2. $y^2 - 6y + 9 = -1 + 9$

3. $t^2 - 5t + \dfrac{25}{4} = -4 + \dfrac{25}{4}$

4. $x^2 + x + \dfrac{1}{4} = 2 + \dfrac{1}{4}$

Exercises

1. $1 \pm \sqrt{2}$

2. $-7 \pm \sqrt{23}$

3. $-\dfrac{5}{2} \pm \dfrac{\sqrt{37}}{2}$

4. $\dfrac{1}{2} \pm \dfrac{\sqrt{33}}{2}$

5. $4 \pm \sqrt{13}$

6. $-3 \pm \sqrt{5}$

7. $-\dfrac{11}{2} \pm \dfrac{\sqrt{129}}{2}$

8. $-\dfrac{3}{4} \pm \dfrac{\sqrt{89}}{4}$

Interactive Preview 32:   INTRODUCTION TO QUADRATIC FUNCTIONS

Exercises

1. $h(x) = x^2 - 8x + 15$

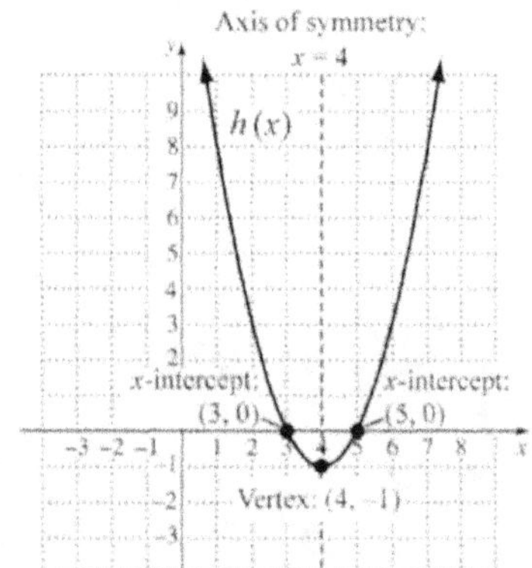

- $a = 1;\ a > 0$, thus the graph opens up.
- The zeros of $h(x)$ are 3 and 5.
- The minimum value of $h(x)$ is -1.

2. $f(x) = -x^2 - 4x - 3$

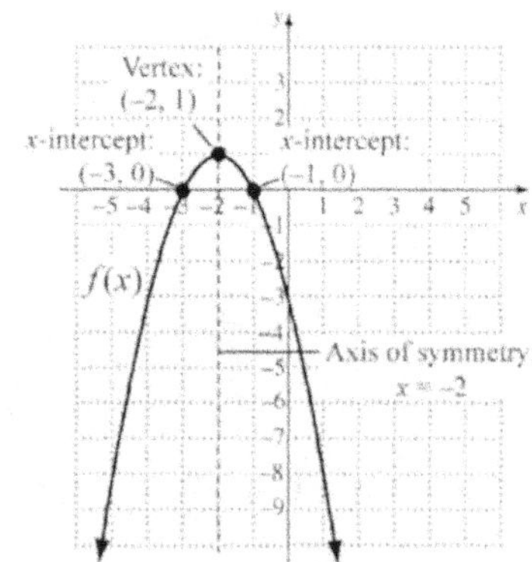

- $a = -1;\ a < 0$, thus the graph opens down.
- The zeros of $f(x)$ are -3 and -1.
- The maximum value of $f(x)$ is 1.

3. $g(x) = -x^2 - 2x + 3$

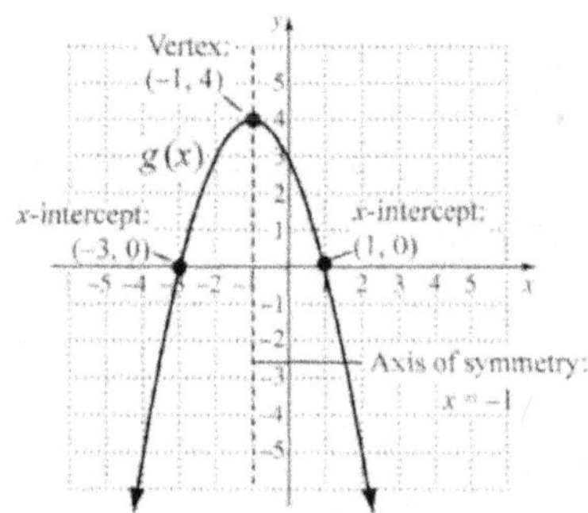

- $a = -1;\ a < 0$, thus the graph opens down.
- The zeros of $g(x)$ are -3 and 1.
- The maximum value of $g(x)$ is 4.

4. $f(x) = x^2 - 4x$

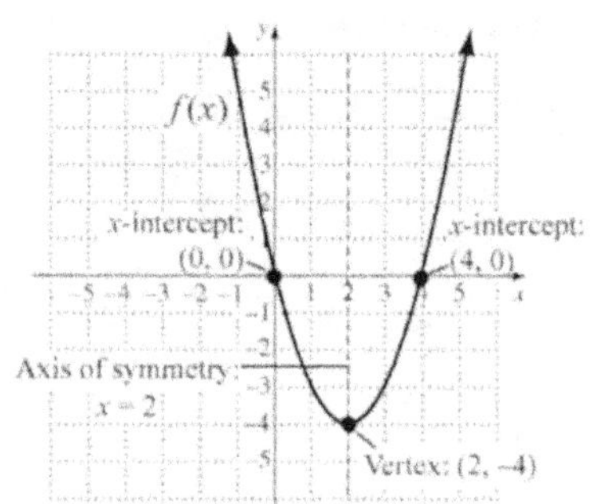

- $a = 1;\ a > 0$, thus the graph opens up.
- The zeros of $f(x)$ are 0 and 4.
- The minimum value of $f(x)$ is -4 .

Interactive Preview 33: ZEROS OF POLYNOMIAL FUNCTIONS

Exercises

1. x-intercept of graph: $(1, 0)$

Solution of $3x - 3 = 0$: 1

Zero of $f(x) = 3x - 3$: 1

2. x-intercepts of graph: $(-1, 0)$ and $(4, 0)$

Solutions of $x^2 - 3x - 4 = 0$: -1 and 4

Zeros of $f(x) = x^2 - 3x - 4$: -1 and 4

3. x-intercepts of graph:
$(-2, 0)$, $(0, 0)$, and $(3, 0)$

Solutions of $x^3 - x^2 - 6x = 0$:
-2, 0, and 3

Zeros of $f(x) = x^3 - x^2 - 6x$:
-2, 0, and 3

4.
$$3x - 3 = 0$$
$$3x = 3$$
$$x = 1$$

The zero of $f(x)$ is 1.

5. $(x+1)(x-4) = 0$

$x + 1 = 0 \quad or \quad x - 4 = 0$

$x = -1 \quad or \quad x = 4$

The zeros of $f(x)$ are -1 and 4.

6. $x(x+2)(x-3) = 0$

$x = 0 \quad or \quad x + 2 = 0 \quad or \quad x - 3 = 0$

$x = 0 \quad or \quad x = -2 \quad or \quad x = 3$

The zeros of $f(x)$ are 0, -2, and 3.

Interactive Preview 34: ASYMPTOTES OF RATIONAL FUNCTIONS

Exercises

1. $f(x) = \dfrac{5x - 1}{5x - 10}$

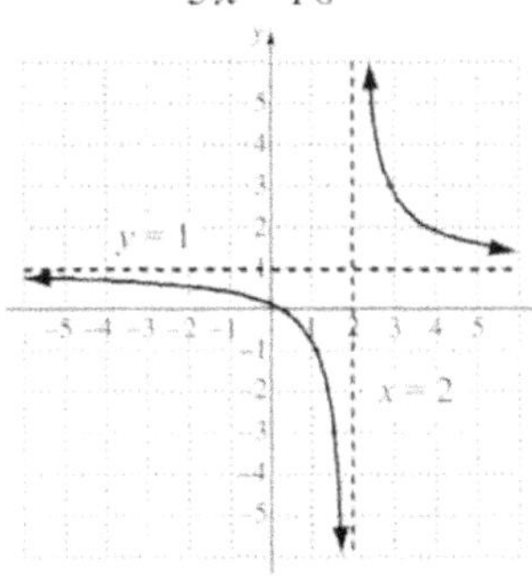

2. $f(x) = \dfrac{1}{x^2}$

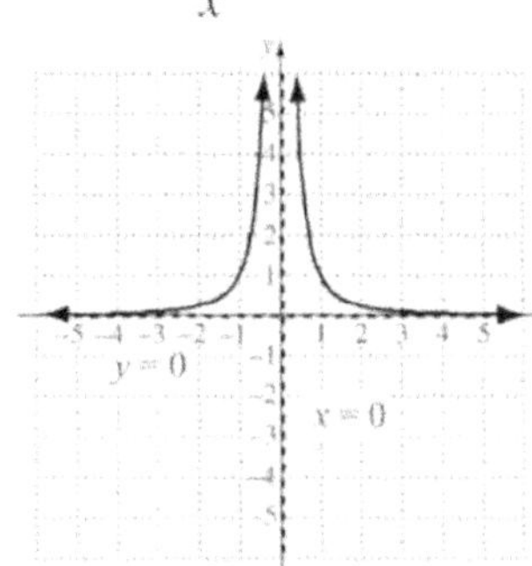

3. $f(x) = \dfrac{4}{x + 2}$

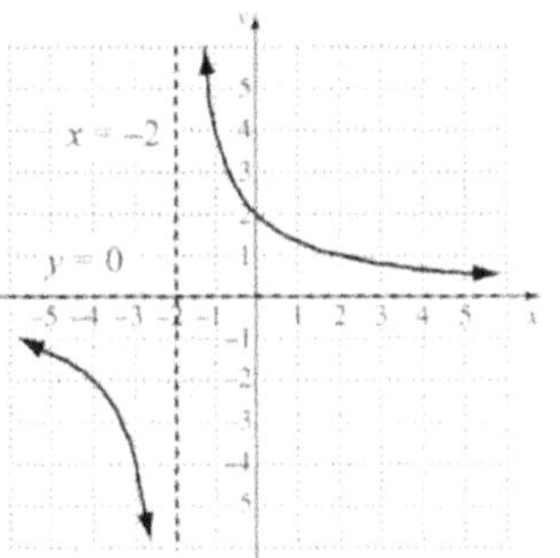

4. $f(x) = \dfrac{1 + 6x}{6 + 3x}$

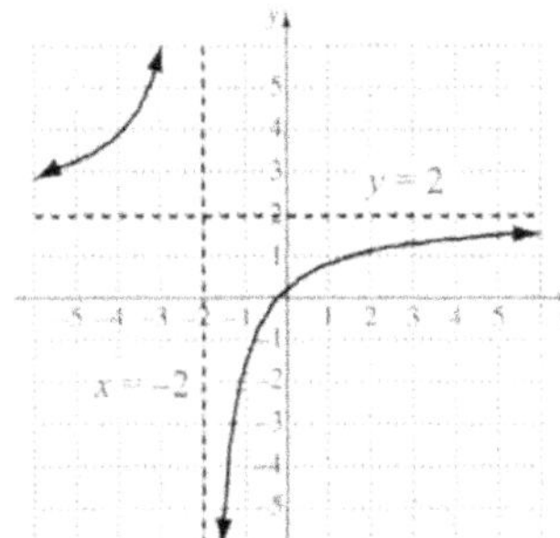

5. $f(x) = \dfrac{5}{x^2 + 3x}$

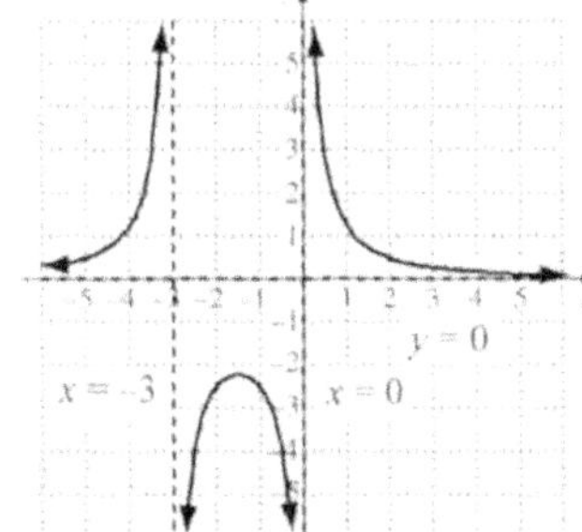

6. $f(x) = \dfrac{-2x^2 + 4x + 3}{x^2 - x - 2}$

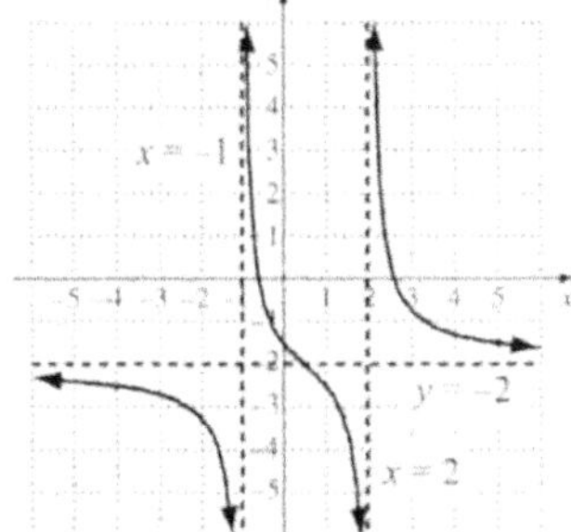

7. Zero of the denominator: 2; vertical asymptote: $x = 2$

8. Zero of the denominator: 0; vertical asymptote: $x = 0$

9. Zero of the denominator: -2; vertical asymptote: $x = -2$

10. Zero of the denominator: -2; vertical asymptote: $x = -2$

11. Zeros of the denominator: $-3, 0$; vertical asymptotes: $x = -3, x = 0$

12. Zeros of the denominator: $-1, 2$; vertical asymptotes: $x = -1, x = 2$

13. $1; 1;$ same as; $5; 5; \dfrac{5}{5},$ or $1; y = 1$

14. $0; 2;$ less than; $y = 0$

15. $y = 0$ 16. $y = 2$ 17. $y = 0$ 18. $y = -2$

Interactive Preview 35: GRAPHING INVERSE FUNCTIONS

Exercises

1.

x	$f(x) = 2x + 4$
-4	-4
-3	-2
-1	2
0	4

x	$f^{-1}(x) = \dfrac{x-4}{2}$
-4	-4
-2	-3
2	-1
4	0

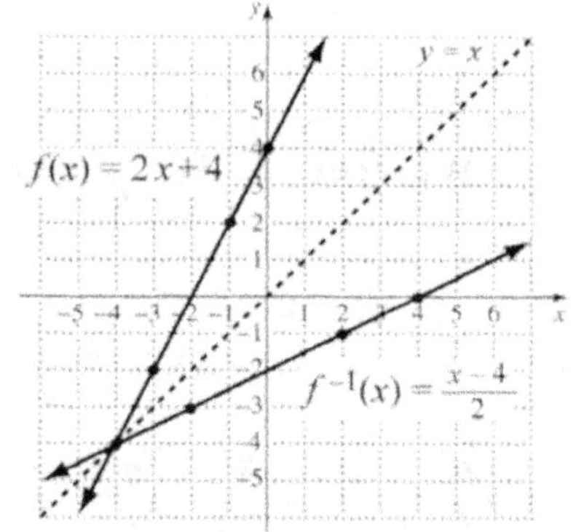

2.

x	$h(x) = 2 - 3x$
-1	5
0	2
1	-1
2	-4

x	$h^{-1}(x) = \dfrac{2-x}{3}$
5	-1
2	0
-1	1
-4	2

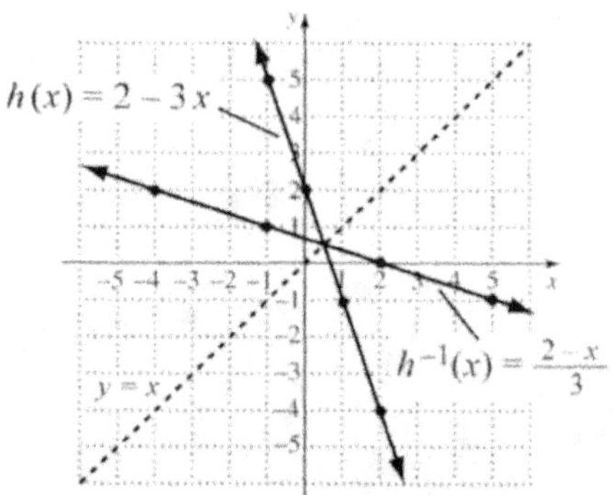

3.

x	$g(x) = x^3 - 3$
-2	-11
-1	-4
0	-3
1	-2
2	5

x	$g^{-1}(x) = \sqrt[3]{x+3}$
-11	-2
-4	-1
-3	0
-2	1
5	2

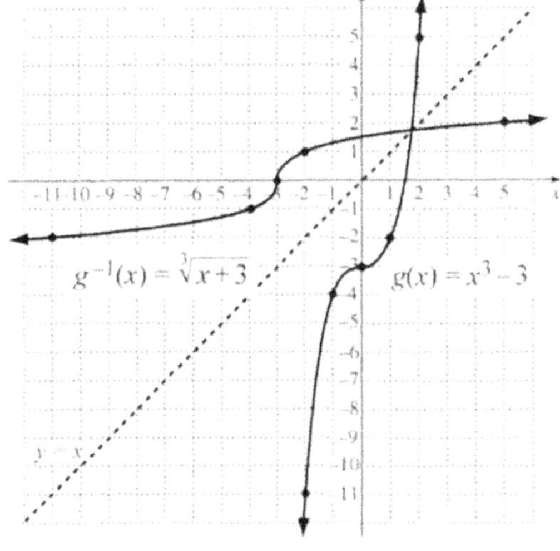

4.

x	$f(x) = x^2 + 4,\ x \geq 0$
0	4
1	5
2	8
3	13

x	$f^{-1}(x) = \sqrt{x-4}$
4	0
5	1
8	2
13	3

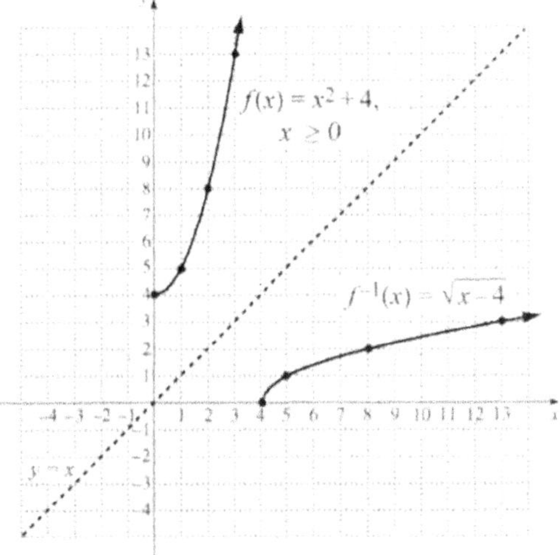

5.

x	$h(x) = x^2 - 3,\ x \geq 0$
0	-3
1	-2
2	1
3	6

x	$h^{-1}(x) = \sqrt{x+3}$
-3	0
-2	1
1	2
6	3

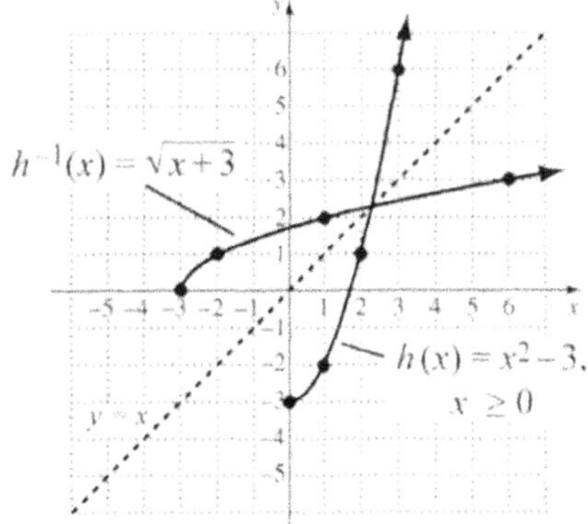

Interactive Preview 36:   INTRODUCTION TO LOGARITHMS

Exercises

1. 2   **2.** 3   **3.** 1   **4.** 0   **5.** -1

| **6.** 6 | **7.** 1 | **8.** 4 | **9.** -2 | **10.** -2 |
| **11.** -5 | **12.** 0 | | | |

Interactive Preview 37: SOLVING EXPONENTIAL EQUATIONS

Exercises

1.
$$5^x = 625$$

| Write each side as a power of 5. | $5^x = 5^4$ |
| Use the base-exponent property. | $x = 4$ |

2.
$$2^{3x+1} = 128$$

Write each side as a power of 2.	$2^{3x+1} = 2^7$
Use the base-exponent property.	$3x + 1 = 7$
Subtract 1.	$3x = 6$
Divide by 3.	$x = 2$

3.
$$8^x = 25$$

Take the common logarithm on both sides.	$\log 8^x = \log 25$
Use the power rule.	$x \log 8 = \log 25$
Divide by $\log 8$.	$x = \dfrac{\log 25}{\log 8}$
Approximate using a calculator.	$x \approx 1.5480$

4.
$$e^x = 350$$

Take the natural logarithm on both sides.	$\ln e^x = \ln 350$
Use the power rule.	$x \ln e = \ln 350$
$\ln e = 1$	$x = \ln 350$
Approximate using a calculator.	$x \approx 5.8579$

5. 6	**6.** 1	**7.** 2.4849	**8.** $\dfrac{4}{3}$	**9.** 0.3155
10. 0.9575	**11.** 8	**12.** $\dfrac{1}{5}$	**13.** $\dfrac{3}{2}$	**14.** 1.6335
15. 0.8550	**16.** 2.0437	**17.** 4	**18.** 9.2103	**19.** 16
20. 69.3147	**21.** 0.2231	**22.** -3		

Exercises

1. 243 **2.** 1 **3.** -107 **4.** $\dfrac{23}{3}$

5. 100 **6.** 24 **7.** $\dfrac{1}{1000}$ **8.** 92

9. 2 **10.** 3 **11.** $\dfrac{13}{6}$ **12.** $\dfrac{5}{22}$

13. 13 **14.** 8 **15.** 25 **16.** $\sqrt{28}$, or $2\sqrt{7}$

Interactive Preview 39: USING AN INVERSE MATRIX TO SOLVE A SYSTEM OF EQUATIONS

Exercises

1.
$$\frac{4}{9}y = -16$$
$$\frac{9}{4}\cdot\frac{4}{9}y = \frac{9}{4}(-16)$$
$$1\cdot y = -\frac{144}{4}$$
$$y = -36$$

2.
$$-\frac{5}{2}t = -\frac{7}{10}$$
$$-\frac{2}{5}\left(-\frac{5}{2}t\right) = -\frac{2}{5}\left(-\frac{7}{10}\right)$$
$$1\cdot t = \frac{14}{50}$$
$$t = \frac{7}{25}$$

3.
$$\begin{bmatrix} -3 & 4 \\ 5 & -7 \end{bmatrix}\cdot\begin{bmatrix} x \\ y \end{bmatrix} = \begin{bmatrix} -9 \\ 16 \end{bmatrix}$$
$$\begin{bmatrix} -7 & -4 \\ -5 & -3 \end{bmatrix}\begin{bmatrix} -3 & 4 \\ 5 & -7 \end{bmatrix}\cdot\begin{bmatrix} x \\ y \end{bmatrix} = \begin{bmatrix} -7 & -4 \\ -5 & -3 \end{bmatrix}\cdot\begin{bmatrix} -9 \\ 16 \end{bmatrix}$$
$$\begin{bmatrix} 1 & 0 \\ 0 & 1 \end{bmatrix}\cdot\begin{bmatrix} x \\ y \end{bmatrix} = \begin{bmatrix} -1 \\ -3 \end{bmatrix}$$
$$\begin{bmatrix} x \\ y \end{bmatrix} = \begin{bmatrix} -1 \\ -3 \end{bmatrix}$$

Check:
$$\underline{-3x + 4y = -9}$$
$$\begin{array}{c|c} -3(-1) + 4(-3) & -9 \\ 3 - 12 & \\ -9 & -9 \end{array} \quad \text{True}$$

$$\underline{5x - 7y = 16}$$
$$\begin{array}{c|c} 5(-1) - 7(-3) & 16 \\ -5 + 21 & \\ 16 & 16 \end{array} \quad \text{True}$$

The solution is $(-1, -3)$.

4.

$$\begin{bmatrix} 5 & -4 \\ 7 & -6 \end{bmatrix} \cdot \begin{bmatrix} a \\ b \end{bmatrix} = \begin{bmatrix} -6 \\ -10 \end{bmatrix}$$

$$\begin{bmatrix} 3 & -2 \\ \dfrac{7}{2} & -\dfrac{5}{2} \end{bmatrix} \cdot \begin{bmatrix} 5 & -4 \\ 7 & -6 \end{bmatrix} \cdot \begin{bmatrix} a \\ b \end{bmatrix} = \begin{bmatrix} 3 & -2 \\ \dfrac{7}{2} & -\dfrac{5}{2} \end{bmatrix} \cdot \begin{bmatrix} -6 \\ -10 \end{bmatrix}$$

$$\begin{bmatrix} 1 & 0 \\ 0 & 1 \end{bmatrix} \cdot \begin{bmatrix} a \\ b \end{bmatrix} = \begin{bmatrix} 2 \\ 4 \end{bmatrix}$$

$$\begin{bmatrix} a \\ b \end{bmatrix} = \begin{bmatrix} 2 \\ 4 \end{bmatrix}$$

The solution is $(2, 4)$.

5. $-\dfrac{2}{11}$　　　　**6.** -600　　　　**7.** $(5, -2)$　　　　**8.** $(5, -3)$

Interactive Preview 40:　THE ELLIPSE

Exercises

1.

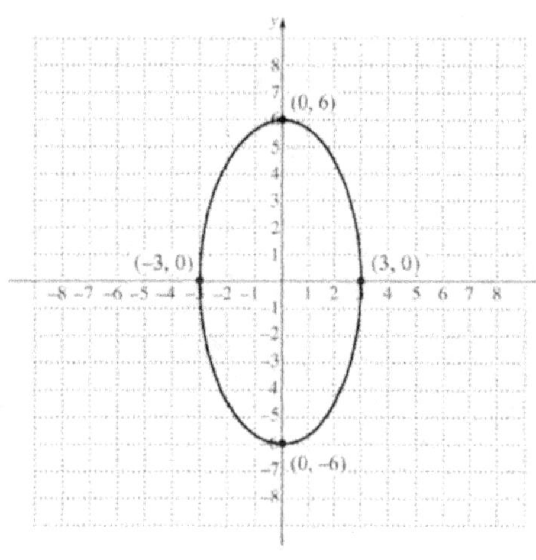

Center: $(0, 0)$

The major axis is vertical.
The minor axis is horizontal.

Vertices: $(0, -6)$ and $(0, 6)$

Endpoints of major axis: $(0, -6)$ and $(0, 6)$

Endpoints of minor axis: $(-3, 0)$ and $(3, 0)$

x-intercepts: $(-3, 0)$ and $(3, 0)$

y-intercepts: $(0, -6)$ and $(0, 6)$

Equation: $\dfrac{x^2}{3^2} + \dfrac{y^2}{6^2} = 1$, or $\dfrac{x^2}{9} + \dfrac{y^2}{36} = 1$

2.

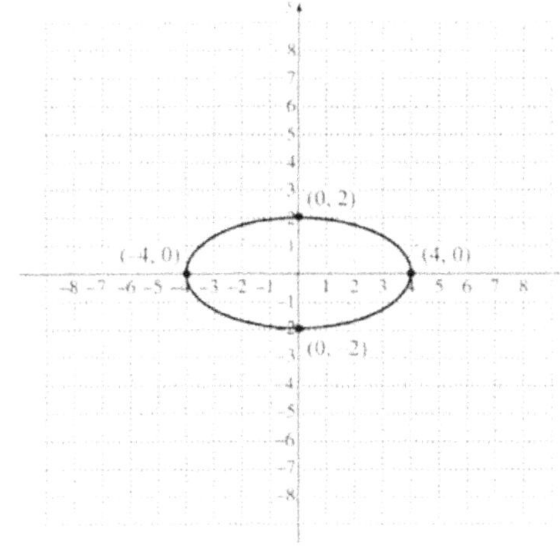

Center: $(0, 0)$

The major axis is horizontal.
The minor axis is vertical.

Vertices: $(-4, 0)$ and $(4, 0)$

Endpoints of major axis: $(-4, 0)$ and $(4, 0)$

Endpoints of minor axis: $(0, -2)$ and $(0, 2)$

x-intercepts: $(-4, 0)$ and $(4, 0)$

y-intercepts: $(0, -2)$ and $(0, 2)$

Equation: $\dfrac{x^2}{4^2} + \dfrac{y^2}{2^2} = 1$, or $\dfrac{x^2}{16} + \dfrac{y^2}{4} = 1$

Interactive Preview 41: THE HYPERBOLA

Exercises

1.
- The x^2-term is the first term, thus the transverse axis is on the x-axis.
- The endpoints of the transverse axis are $(-6, 0)$ and $(6, 0)$.
- The endpoints of the conjugate axis are $(0, -3)$ and $(0, 3)$.
- The vertices are $(-6, 0)$ and $(6, 0)$.

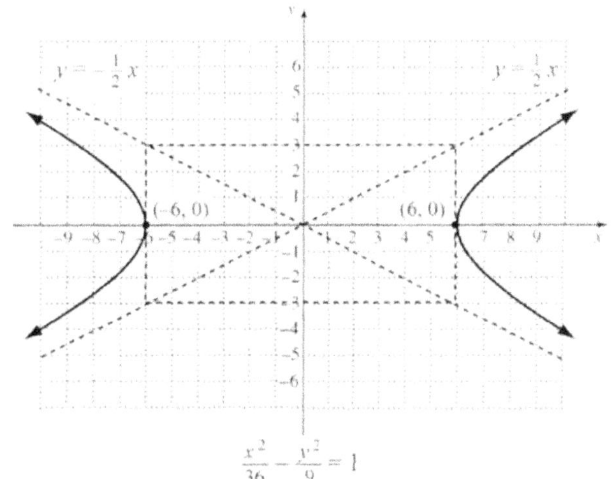

$$\frac{x^2}{36} - \frac{y^2}{9} = 1$$

2.
- The y^2-term is the first term, thus the transverse axis is on the y-axis.
- The endpoints of the transverse axis are $(0, -2)$ and $(0, 2)$.
- The endpoints of the conjugate axis are $(-5, 0)$ and $(5, 0)$.
- The vertices are $(0, -2)$ and $(0, 2)$.

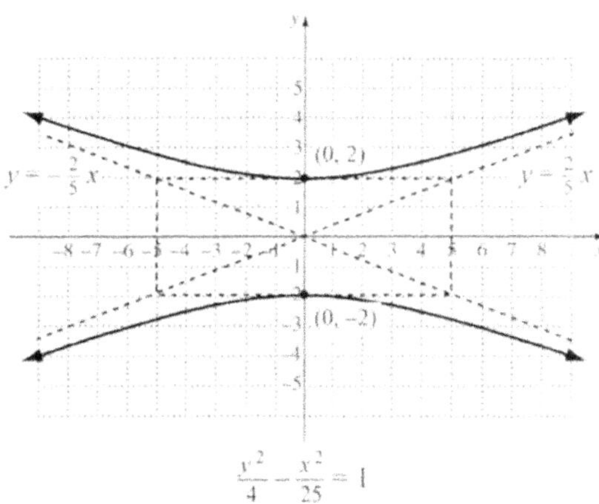

$$\frac{y^2}{4} - \frac{x^2}{25} = 1$$

| **3.** E | **4.** D | **5.** A | **6.** B | **7.** F | **8.** C |

Interactive Preview 42: CLASSIFYING EQUATIONS OF CONIC SECTIONS

Exercises

1. Parabola	**2.** Ellipse	**3.** Hyperbola	**4.** Hyperbola
5. Circle	**6.** Parabola	**7.** Parabola	**8.** Ellipse
9. Hyperbola	**10.** Circle	**11.** Parabola	**12.** Hyperbola
13. Ellipse	**14.** Circle	**15.** Ellipse	**16.** Parabola

Answers: Video Worksheets

Chapter R

Section R.1

Your Turn: Set Notation and the Set of Real Numbers: **1.** 6 **2.** $\{7, 8, 9, 10, 11, 12, 13, 14\}$
3. $\{x | x \text{ is a natural number greater than 6 and less than 11}\}$ **4. (a)** 125, 1; **(b)** 0, 125, 1;
(c) 0, −16, 125, 1; **(d)** $0.2, 0, -16, -\frac{1}{9}, 125, 1$; **(e)** $\sqrt{7}$; **(f)** $0.2, 0, -16, -\frac{1}{9}, \sqrt{7}, 125, 1$

Order for the Real Numbers: **1.** $-11 < -6$ **2.** $1 > -15$ **3.** $\frac{2}{5} < \frac{4}{7}$ **4.** $10 > y$ **5.** True

Graphing Inequalities on the Number Line: **1.** 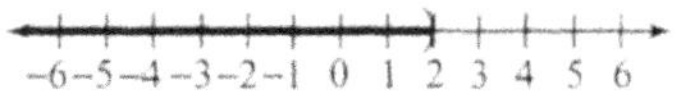

2. 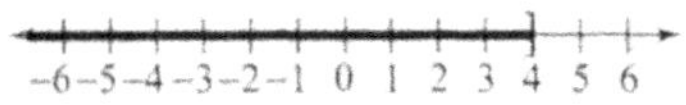 **Absolute Value:** **1.** 39 **2.** $\frac{7}{8}$

Practice Exercises: **1.** False **2.** False **3.** True **4.** True **5. (a)** 4; **(b)** 0, 4; **(c)** −1, 0, 4;
(d) $-3.1, -1, 0, \frac{75}{2}, 4$; **(e)** $\sqrt{10}$; **(f)** $-3.1, -1, 0, \sqrt{10}, \frac{75}{2}, 4$ **6. (a)** 51; **(b)** 51; **(c)** 51;
(d) $-\frac{3}{8}, \frac{1}{2}, \frac{43}{2}, 51$; **(e)** $\sqrt{12}$; **(f)** $-\frac{3}{8}, \frac{1}{2}, \sqrt{12}, \frac{43}{2}, 51$ **7.** {h, i, s, t, o, r, y}
8. $\{5, 10, 15, 20, 25, \ldots\}$ **9.** $\{x | x \text{ is a multiple of 10 between 20 and 100}\}$
10. $\{x | x \text{ is an even number between 71 and 81}\}$ **11.** > **12.** < **13.** > **14.** $-7 \le x$ **15.** True
16. True **17.** False **18.** 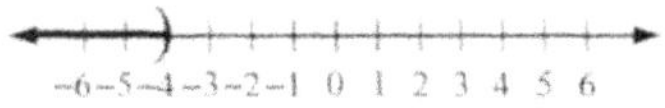 **19.** 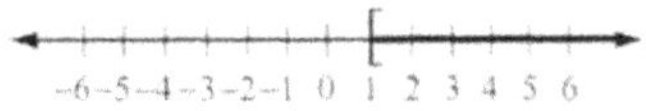 **20.** 5
21. 3.8 **22.** 0

Section R.2

Your Turn: Addition: **1.** 3 **2.** −17 **3.** 0 **4.** 8 **5.** $\frac{1}{10}$ **Opposites, or Additive Inverses:**
1. 15 **2.** −7; 7 **3.** $-\frac{2}{25}$ **Subtraction:** **1.** −5 **2.** $\frac{3}{4}$ **Multiplication:** **1.** −88 **2.** $\frac{4}{15}$

Division: **1.** 7.4 **2.** 0 **3.** $-\frac{7}{4}$ **4.** $-\frac{3}{5}$

Practice Exercises: **1.** add **2.** positive **3.** not defined **4.** multiply **5.** −23 **6.** −20.2
7. $\frac{3}{14}$ **8.** −13.6 **9.** $\frac{7}{2}$ **10.** 0 **11.** 14 **12.** 2.5 **13.** 32 **14.** −47.4 **15.** $-\frac{2}{15}$ **16.** −15
17. 48.8 **18.** $-\frac{3}{8}$ **19.** 2 **20.** −7 **21.** Not defined **22.** $-\frac{5}{7}$ **23.** $-\frac{1}{8}$ **24.** $\frac{8}{5}$ **25.** −4

Section R.3

Your Turn: Exponential Notation: 1. m^6 **2.** 49 **3.** $\dfrac{25}{64}$ **4.** 102.01 **5.** 16

Negative Integers as Exponents: 1. $\dfrac{1}{t^9}$ **2.** $\dfrac{1}{25}$ **3.** $\dfrac{125}{27}$ **4.** $(-9)^{-2}$

Order of Operations: 1. -8 **2.** -9 **3.** $\dfrac{1}{4}$

Practice Exercises: 1. division   **2.** multiplication   **3.** division   **4.** subtraction

5. multiplication   **6.** addition   **7.** t^3 **8.** $\left(\dfrac{5}{11}\right)^5$ **9.** $(-54.9)^4$ **10.** 16 **11.** -125 **12.** 1.2

13. 1 **14.** $\dfrac{8}{27}$ **15.** y^5 **16.** $-\dfrac{1}{27}$ **17.** 5^{-3} **18.** x^{-5} **19.** $(-6)^{-3}$ **20.** 1 **21.** -43 **22.** -108

23. -7 **24.** $-\dfrac{1}{2}$ **25.** 6

Section R.4

Your Turn: Translating to Algebraic Expressions: 1. $s-5$ **2.** $\dfrac{1}{3}n-4$

3. $42\%\cdot w$, or $0.42w$ **4.** $2+\dfrac{a}{b}$, or $\dfrac{a}{b}+2$ **Evaluating Algebraic Expressions: 1.** -130

2. $12.6\,\text{ft}^2$ **3.** 57

Practice Exercises: 1. (c)   **2.** (a)   **3.** (b)   **4.** (d)   **5.** $3d$ **6.** $5+2x$, or $2x+5$

7. $a-b+6$, or $6+a-b$ **8.** $20\%y-7$, or $0.20y-7$ **9.** $10+m$, or $m+10$ **10.** $\dfrac{1}{5}(3q)$

11. $25y-1$ **12.** $r-34$ **13.** $P-15\%P$, or $P-0.15P$ **14.** $\$8.65h$ **15.** 6 **16.** 54 **17.** 26

18. -33 **19.** -20 **20.** 14 **21.** $9.45\,\text{ft}^2$

Section R.5

Your Turn: Equivalent Expressions: 1. $19,-20;\,4,5;\,14.56,17;$ The expressions are not

equivalent.   **Equivalent Fraction Expressions: 1.** $\dfrac{12}{21t}$ **2.** $\dfrac{b}{5}$ **3.** $-\dfrac{7}{2}$

The Commutative Laws and the Associative Laws: 1. $-70;-70$ **2.** $15;15$

3. $(3\cdot y)\cdot x;\,(x\cdot 3)\cdot y;\,3\cdot(x\cdot y);$ answers may vary   **The Distributive Laws: 1.** $5x+10$

2. $-10m+14n-18p$ **3.** $4(2a-3b+5)$ **4.** $m(x+y+z)$

Practice Exercises: 1. (c)   **2.** (d)   **3.** (b)   **4.** (f)   **5.** (a)

6. $12,-3,12;\,27.2,12.2,27.2;\,20,5,20;\ 4(x+5)$ and $4x+20$ are equivalent expressions.

7. $\dfrac{3y}{7y}$ **8.** $\dfrac{6a}{15a}$ **9.** $-\dfrac{8}{3}$ **10.** $\dfrac{9a}{4}$ **11.** $n+6$ **12.** yx **13.** $u+(v+3)$ **14.** $5\cdot(c\cdot p)$

15. $y+(5+z),\,(z+5)+y,\,(y+5)+z;$ answers may vary

16. $(4\cdot x)\cdot y,\,y\cdot(x\cdot 4),\,4\cdot(y\cdot x);$ answers may vary   **17.** $3x+30$ **18.** $-21x+6y+3z$

19. $5x, -3y$ **20.** $-3a, -b, 18c$ **21.** $7(a+b)$ **22.** $2(2x-1)$ **23.** $3(2m+4n-5)$
24. $2 \cdot \pi \cdot r \cdot (h+r)$

Section R.6

Your Turn: Collecting Like Terms: 1. $13x$ **2.** $5x+6.9$ **3.** $\dfrac{1}{14}x - \dfrac{5}{24}y$

Multiplying by −1 and Removing Parentheses: 1. $19a+8b+12$ **2.** $10x-5y-6$
3. $-5d+10$

Practice Exercises: 1. equal **2.** opposite **3.** opposite **4.** equal **5.** $9m$ **6.** $-3n$ **7.** $13p$

8. $6a+13b$ **9.** $5.3s-7.6t$ **10.** $-\dfrac{7}{15}x+\dfrac{1}{10}y$ **11.** $5a$ **12.** $-d-14$ **13.** $-3x-2y+8$

14. $3f-4g-2h+j$ **15.** $x+\dfrac{1}{5}w+4.7y-15.4z$ **16.** $9b-7$ **17.** $8x-23$ **18.** $11y-57$
19. $-a-12$ **20.** $354+198x$ **21.** $6-34y$

Section R.7

Your Turn: Multiplication and Division: 1. $-24x^{6n}$ **2.** $-14a^{11}b$ **3.** $\dfrac{-3}{y^{4n}}$ **4.** $-9m^3n$

Raising Powers to Powers and Products and Quotients to Powers: 1. x^{-24p} **2.** $\dfrac{x^{28}y^8}{81z^{20}}$

3. $\dfrac{16}{x^4y^{12}}$ **Scientific Notation: 1.** 7.01×10^{13} **2.** 0.004078 **3.** 7.92×10^3 **4.** 8.0×10^{10}

Practice Exercises: 1. add **2.** multiply **3.** scientific **4.** negative **5.** 3^{11} **6.** $32x^4$ **7.** $-\dfrac{18}{a^2}$

8. x^7 **9.** $\dfrac{5}{x^{20}}$ **10.** $-9x^3y^4$ **11.** $36x^2y^4$ **12.** $\dfrac{x^4}{4y^6}$ **13.** $\dfrac{27x^9}{64y^3}$ **14.** $\dfrac{81}{16x^{44}y^{24}}$

15. 4.7×10^{-2} **16.** 5.0179×10^{11} **17.** 1.9×10^5 **18.** $81,300,000$ **19.** 0.00204 **20.** $24,000$
21. 6×10^9 **22.** 1.305×10^{-1} **23.** 4.0×10^{-20} **24.** 2.5×10^{15}

Chapter 1

Section 1.1
Your Turn: Equations and Solutions: 1. True **2.** False **3.** Neither **4.** No **5.** Yes
The Addition Principle: 1. -24 **2.** -3.4 **3.** $-\dfrac{17}{12}$ **The Multiplication Principle: 1.** -6

2. 13 **3.** 2200 **4.** 32 **Using the Principles Together: 1.** $\dfrac{9}{2}$ **2.** 4.1 **3.** No solution

4. All real numbers are solutions.

Practice Exercises: **1.** (f) **2.** (a) **3.** (d) **4.** (e) **5.** No **6.** Yes **7.** No **8.** 2 **9.** 13 **10.** 10.9 **11.** $-\dfrac{1}{12}$ **12.** 8 **13.** -17 **14.** 0.9 **15.** $-\dfrac{3}{2}$ **16.** 6 **17.** $-\dfrac{1}{2}$ **18.** 7.6
19. All real numbers **20.** No solution **21.** -6

Section 1.2

Your Turn: Evaluating and Solving Formulas: **1.** 227.5 miles **2.** $y = 2A - x$
3. $n = \dfrac{B - 3m}{4}$ **4.** $b = \dfrac{2a}{ac + 1}$

Practice Exercises: **1.** False **2.** True **3.** True **4.** False **5.** 10.75 meters per cycle **6.** $t = \dfrac{d}{60}$

7. $x = 26 - y$ **8.** $a = Tb$ **9.** $x = \dfrac{3y}{2}$ **10.** $x = \dfrac{y - b}{m}$ **11.** $q = 3A - p - r$ **12.** $w = \dfrac{P - 2l}{2}$

13. $r^2 = \dfrac{S}{4\pi}$ **14.** $r = \dfrac{d}{t}$ **15.** $y = \dfrac{x}{a^2 + z}$ **16.** 62 inches **17.** 12 ft

Section 1.3

Your Turn: Five Steps for Problem Solving: **1.** 20.500 thousand metric tons
2. $-13, -12, -11$ **Basic Motion Problems:** **1.** $\dfrac{2}{3}$ hr, or 40 min

Practice Exercises: **1.** Familiarize **2.** Translate **3.** Solve **4.** Check **5.** State **6.** $68.68
7. 6, 8 **8.** 215 units **9.** 42, 43 **10.** $12°, 60°, 108°$ **11.** 110 sec

Section 1.4

Your Turn: Inequalities: **1.** No **2.** No **3.** Yes **Inequalities and Interval Notation:**
1. $(-4, 3]$ **2.** $(-\infty, -7)$ **3.** $(6, 11]$ **4.** $(8, \infty)$ **Solving Inequalities:**
1. $\{x \mid x > -6\}$, or $(-6, \infty)$; **2.** $\{x \mid x > 7\}$, or $(7, \infty)$;
3. $\{x \mid x \geq 5\}$, or $[5, \infty)$; **4.** $\{x \mid x \leq 9\}$, or $(-\infty, 9]$;
5. $\{x \mid x \leq 4\}$, or $(-\infty, 4]$; **Applications and Problem Solving:**
1. More than $1046.02 in sales; $\{S \mid S > 1046.02\}$ **2.** More than 25 guests; $\{g \mid g > 25\}$

Practice Exercises: **1.** (d) **2.** (a) **3.** (e) **4.** (f) **5.** (c) **6.** (b) **7.** No **8.** Yes **9.** No
10. $[-2, \infty)$ **11.** $[-3, 5]$ **12.** $(-\infty, 6)$ **13.** $(0, 5]$ **14.** $\{x \mid x > 4\}$, or $(4, \infty)$;
15. $\{x \mid x > -6\}$, or $(-6, \infty)$; **16.** $\{x \mid x \leq 1\}$, or $(-\infty, 1]$;
17. $\{x \mid x \geq -3\}$, or $[-3, \infty)$; **18.** $\left\{x \mid x < \dfrac{5}{3}\right\}$, or $\left(-\infty, \dfrac{5}{3}\right)$;

19. Times more than 170 hours; $\{t \mid t > 170\}$

20. Scores greater than or equal to 92; $\{S \mid S \geq 92\}$

21. Times less than 40 hours; $\{t \mid t < 40\}$

Section 1.5

Your Turn: Intersections of Sets and Conjunctions of Inequalities: 1. $\{0, 1, 8\}$

2. $\{y \mid -1 \le y < 4\}$, or $[-1, 4)$; **3.** $\{x \mid x > 4\}$, or $(4, \infty)$;

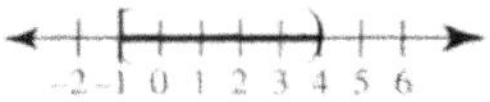

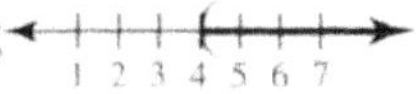

Unions of Sets and Disjunctions of Inequalities: 1. $\{0, 1, 3, 9\}$ **2.** $\{x \mid x < 2 \text{ or } x > 4\}$,

or $(-\infty, 2) \cup (4, \infty)$; 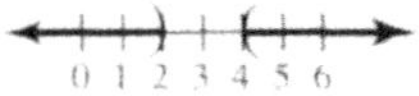 **3.** $\{x \mid x \le 8\}$, or $(-\infty, 8]$;

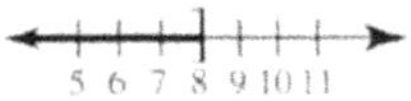

Applications and Problem Solving: 1. Temperatures from $167°F$ to $176°F$:
$\{F \mid 167° \le F \le 176°\}$

Practice Exercises: 1. False **2.** True **3.** False **4.** False **5.** True **6.** True **7.** $\varnothing$ **8.** $\{a, c\}$

9. $\{x \mid 3 \le x < 7\}$, or $[3, 7)$; **10.** $\{x \mid -3 < x < 2\}$, or $(-3, 2)$;

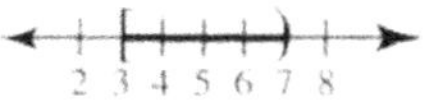

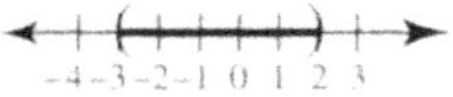

11. $\{x \mid x > -3\}$, or $(-3, \infty)$; 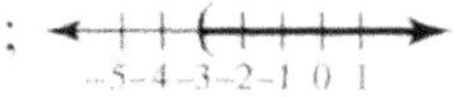 **12.** $\varnothing$ **13.** $\{1, 2, 3, a, b, c\}$ **14.** $\{1, 2, 3\}$

15. $\{x \mid x \le 0 \text{ or } x \ge 3\}$, or $(-\infty, 0] \cup [3, \infty)$; **16.** $\{x \mid x < -1 \text{ or } x \ge 4\}$, or

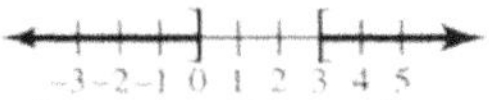

$(-\infty, -1) \cup [4, \infty)$; **17.** $\{x \mid x \ge 2\}$, or $[2, \infty)$;

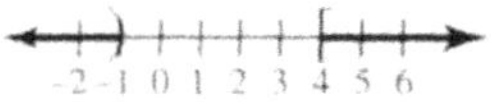

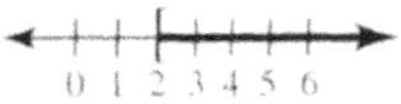

18. $\{x \mid x \text{ is a real number}\}$, or $(-\infty, \infty)$;

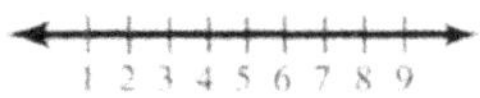

19. $-56.2° \le F \le 44.6°$ **20.** Scores greater than or equal to 86; $\{x \mid x \ge 86\}$

Section 1.6

Your Turn: Properties of Absolute Value: 1. $5x^4$ **2.** $4|y|$ **3.** $\dfrac{x^2}{3}$ **Distance on the**

Number Line: 1. 21 **2.** 11 **Equations with Absolute Value: 1.** $\varnothing$ **2.** $\{-10, 10\}$

3. $\{-2, 4\}$ **Equations with Two Absolute-Value Expressions: 1.** $\{-1, 1\}$ **2.** $\{0\}$

Inequalities with Absolute Value: 1. $\{x \mid -2 < x < 2\}$, or $(-2, 2)$;

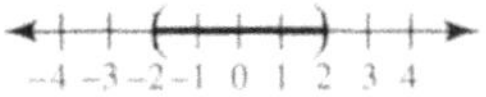

2. $\{y \mid y \le -1 \text{ or } y \ge 1\}$, or $(-\infty, -1] \cup [1, \infty)$;

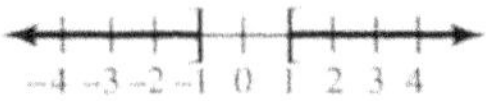

3. $\{x \mid -1 < x < 5\}$, or $(-1, 5)$;

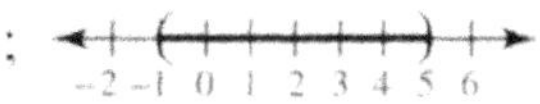

4. $\{x \mid x \le 1 \text{ or } x \ge 7\}$, or $(-\infty, 1] \cup [7, \infty)$;

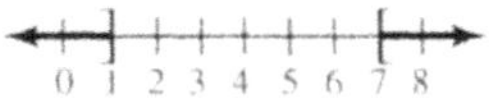

Practice Exercises: 1. True **2.** False **3.** False **4.** True **5.** True **6.** $3|x|$ **7.** $6x^2$ **8.** 27

9. 5 **10.** $\{-16, 16\}$ **11.** $\{-6, 10\}$ **12.** $\left\{-\dfrac{2}{5}, 0\right\}$ **13.** $\left\{-\dfrac{2}{7}\right\}$ **14.** $\left\{-1, -\dfrac{1}{7}\right\}$ **15.** $\{-4\}$

16. $\left\{-4, -\dfrac{4}{7}\right\}$ **17.** $\{x \mid x \text{ is a real number}\}$

18. $\{x \mid -4 \le x \le 4\}$, or $[-4, 4]$;

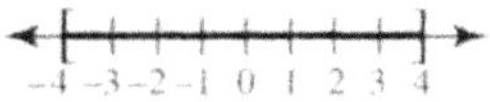

19. $\{x \mid -7 < x < -1\}$, or $(-7, -1)$;

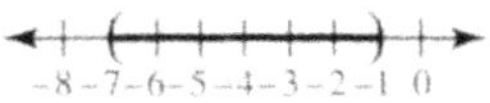

20. $\{x \mid x < -3 \text{ or } x > 3\}$, or $(-\infty, -3) \cup (3, \infty)$;

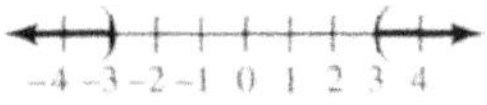

21. $\{x \mid x \le -3 \text{ or } x \ge -1\}$, or $(-\infty, -3] \cup [-1, \infty)$;

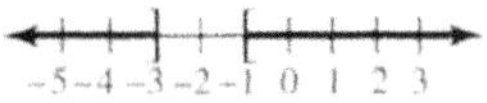

Chapter 2

Section 2.1

Your Turn: Plotting Ordered Pairs: 1.

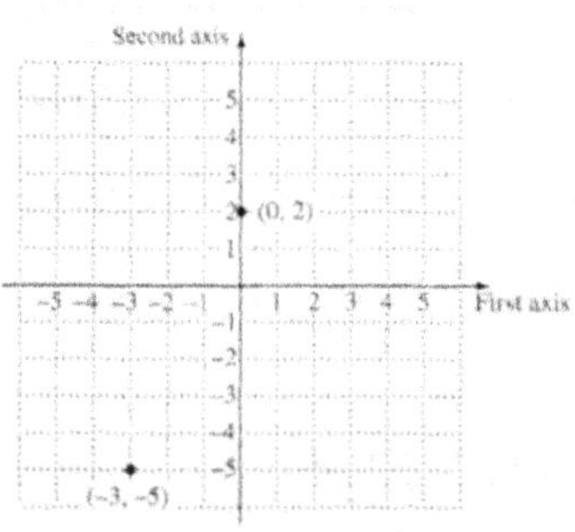

2. $(-3,-1)$: III

$(5,-2)$: IV

$(6,0)$: On an axis,
not in a quadrant

$(-1,4)$: II

Solutions of Equations:

1. $(0,-2)$ is a solution; $(5,1)$ is not a solution.

2.

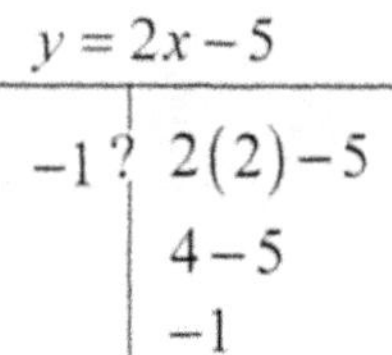

$$\frac{y = 2x - 5}{\begin{array}{c|c} -1 \;?\; 2(2)-5 \\ 4-5 \\ -1 \quad \text{TRUE} \end{array}}$$

$$\frac{y = 2x - 5}{\begin{array}{c|c} -5 \;?\; 2(0)-5 \\ 0-5 \\ -5 \quad \text{TRUE} \end{array}}$$

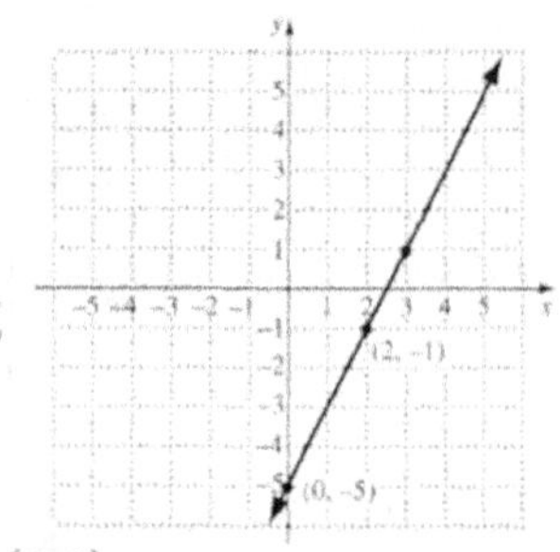

$(3,1)$: answers may vary

Graphs of Linear Equations: 1.

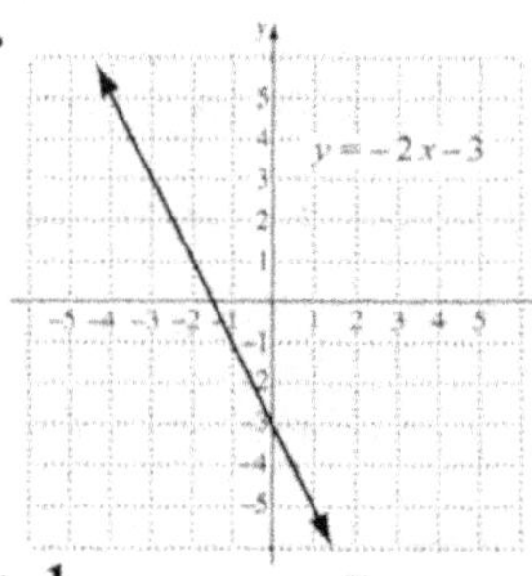

2.

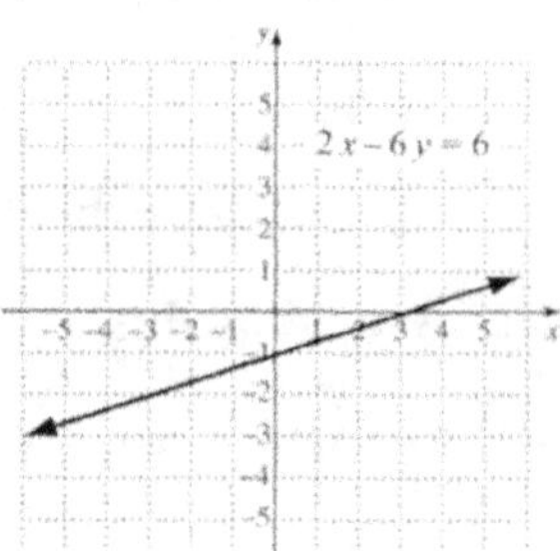

Graphs of Nonlinear Equations: 1.

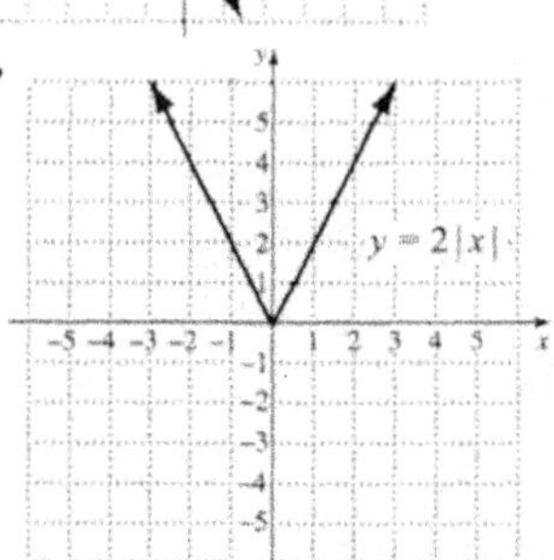

2.

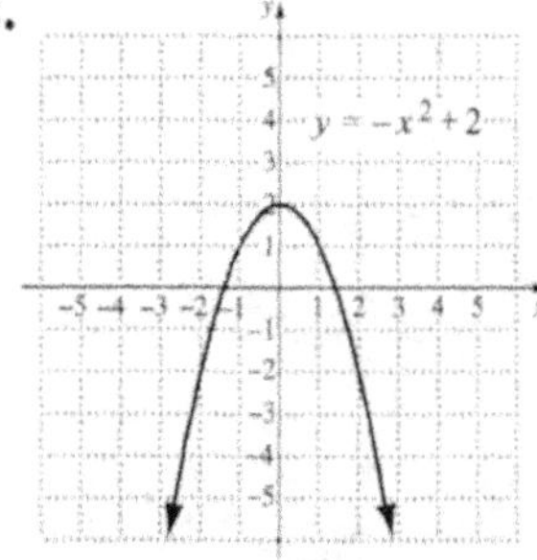

Practice Exercises: **1.** True **2.** False **3.** True **4.** True **5.** False

6.

7. Yes **8.** Yes **9.** No

10.

$$\begin{array}{c|l} \multicolumn{2}{c}{y = 1-2x} \\ \hline 5 \;?\; & 1-2(-2) \\ & 1+4 \\ & 5 \quad \text{TRUE} \end{array} \qquad \begin{array}{c|l} \multicolumn{2}{c}{y = 1-2x} \\ \hline -1 \;?\; & 1-2(1) \\ & 1-2 \\ & -1 \quad \text{TRUE} \end{array}$$

$(-2, 5)$ is a solution $\qquad$ $(1, -1)$ is a solution

$(2, -3)$; answers may vary

11.

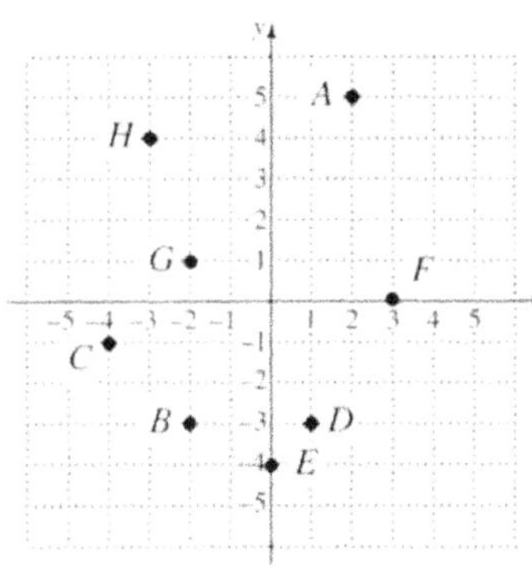

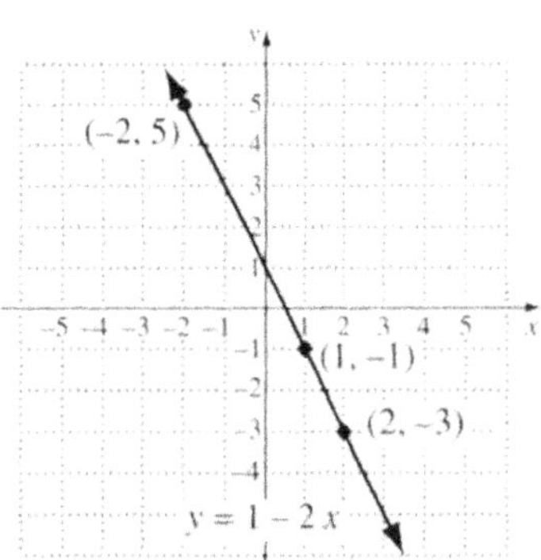

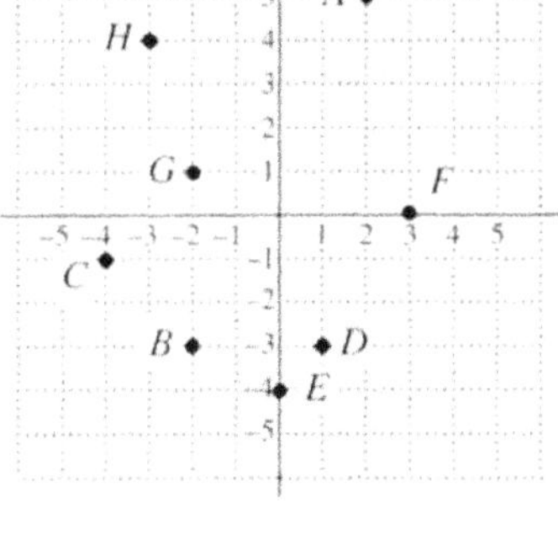

12.

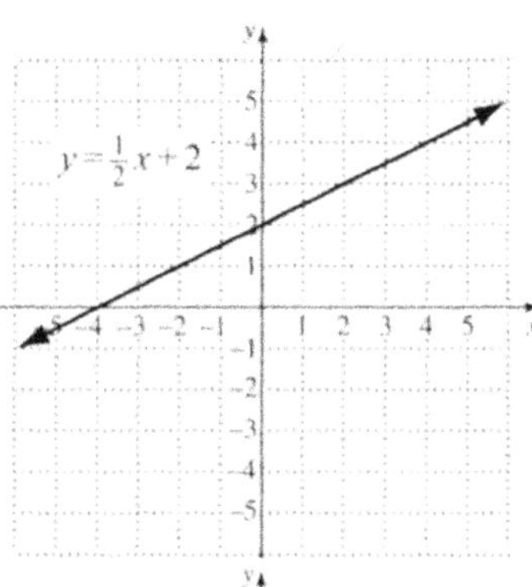

13.

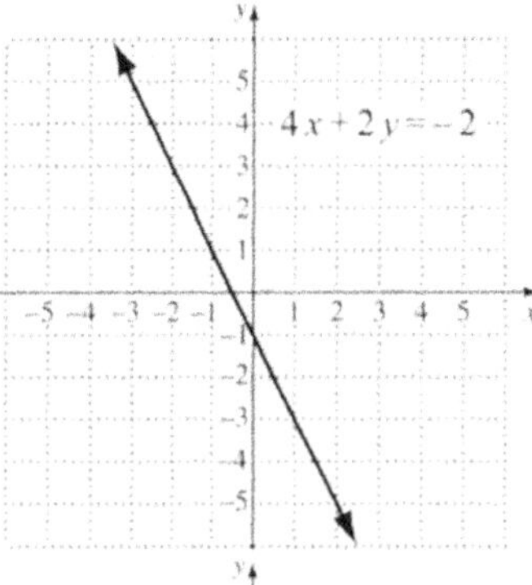

14.

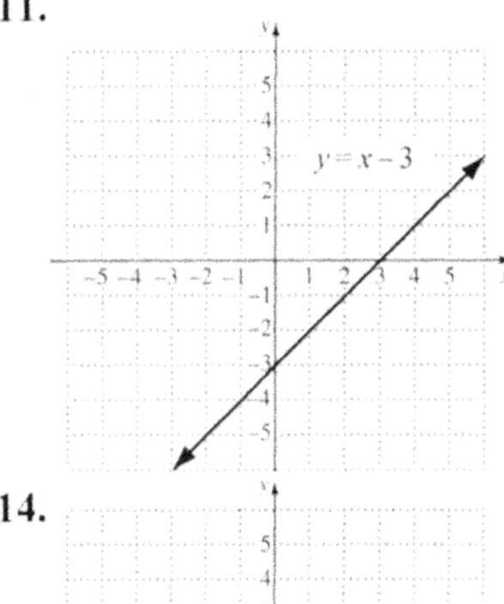

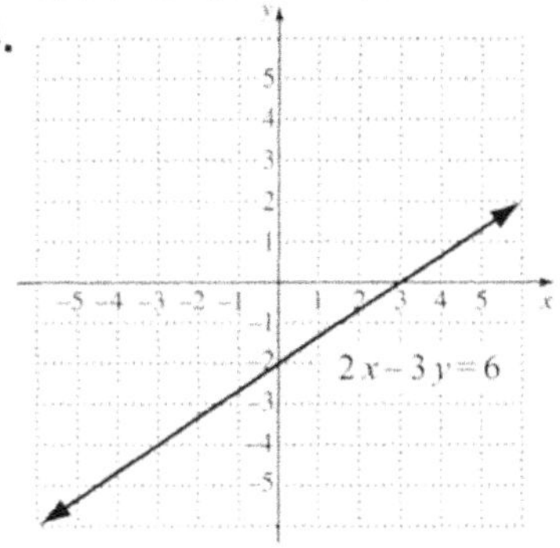

15.

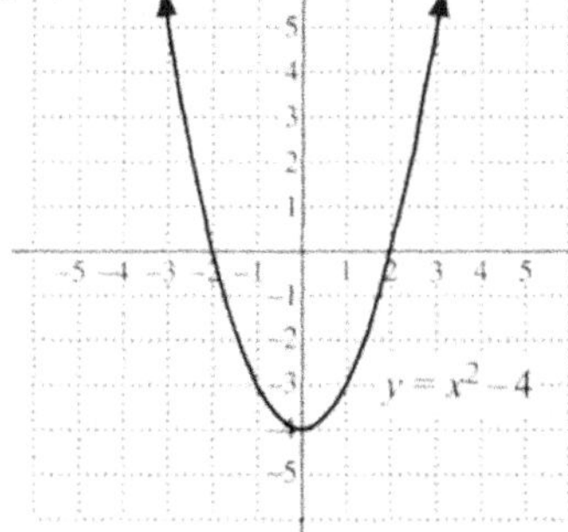

16.

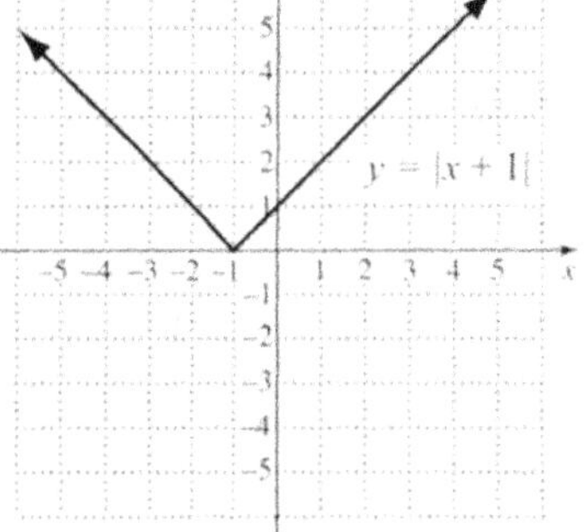

Section 2.2

Your Turn: Identifying Functions: 1. Not a function **2.** A function **3.** Not a function
4. A function **Finding Function Values: 1.** $g(0) = 3$, $g(-1) = 7$, $g(a+2) = -4a - 5$

2. 7 **3.** $h(1) = -2$, $h(-1) = 4$ **4.** 36π cm$^2 \approx 113.04$ cm^2 **Graphs of Functions:**

1. 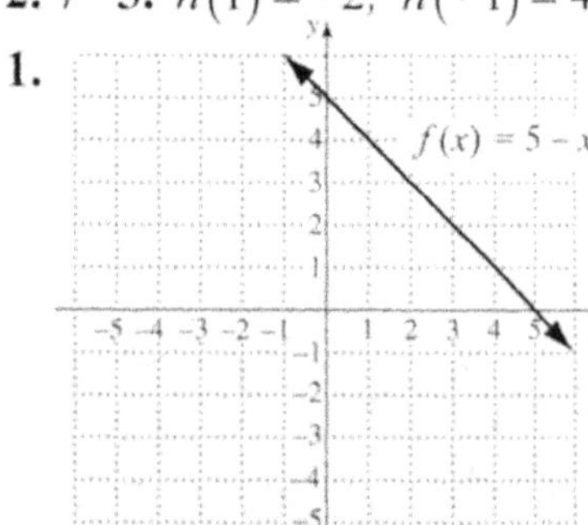

2. 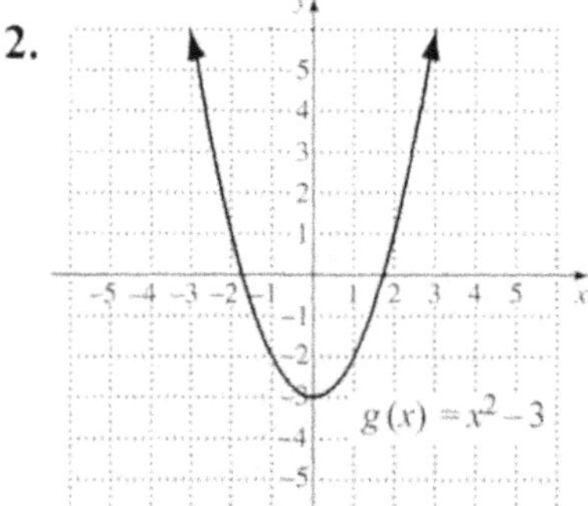

The Vertical-Line Test: 1. Yes **Applications of Functions and Their Graphs:**
1. Approximately 1000 for-profit hospitals

Practice Exercises:

1. function; domain; range **2.** vertical; function **3.** output **4.** A function
5. Not a function **6.** 14 **7.** 14 **8.** $4a - 11$ **9.** 5 **10.** 150 cm^2

11. 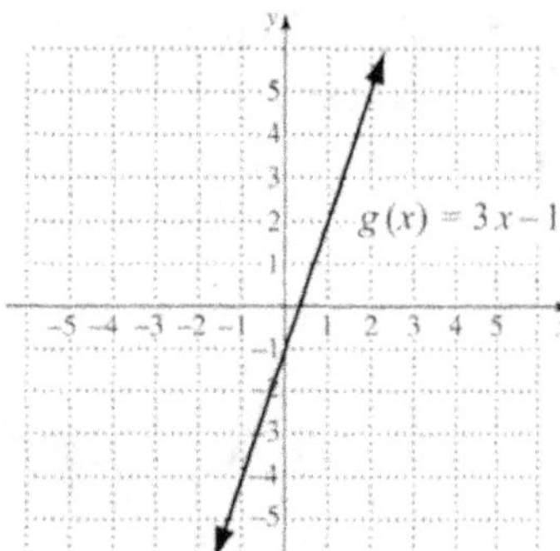

12. 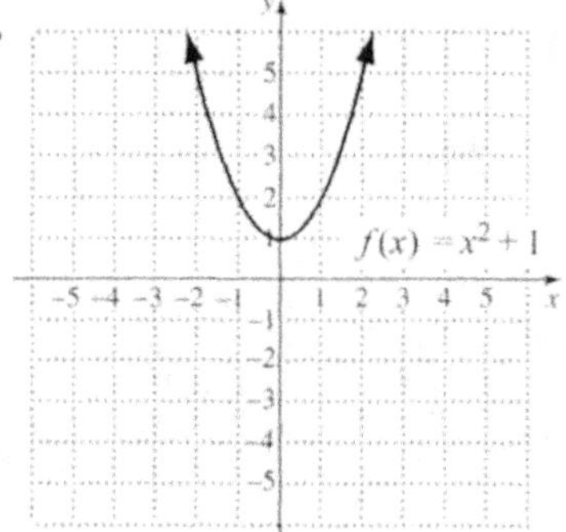

13. A function **14.** Not a function **15.** Approximately 248 hospitals

Section 2.3

Your Turn: Finding Domain and Range: 1. Domain: $\{-4, -2, 2, 4\}$; range: $\{-2, 0, 1, 3\}$

2. Domain: $\{x \mid -5 \le x \le 4\}$; range: $\{y \mid -5 \le y \le 1\}$ **3. a)** 1; **b)** $\{x \mid x \ge -3\}$; **c)** 1; **d)** $\{y \mid y \ge 0\}$

4. $\{x \mid x$ is a real number *and* $x \ne 4\}$ **5.** All real numbers

Practice Exercises: 1. function **2.** relation **3.** domain **4.** range **5. a)** 3; **b)** $\{-3, 0, 2, 4\}$;

c) -3; **d)** $\{-3, -2, 1, 3\}$ **6. a)** 1; **b)** all real numbers; **c)** 2, 4 **d)** $\{y \mid y \le 2\}$

7. a) 4; **b)** $\{x \mid -4 \le x \le 5\}$; **c)** 0; **d)** $\{y \mid -2 \le y \le 4\}$ **8.** All real numbers

9. $\{x|x$ is a real number *and* $x \neq 5\}$ **10.** $\left\{x\middle|x \text{ is a real number } and \ x \neq -\dfrac{3}{2}\right\}$

11. All real numbers   **12.** All real numbers

Section 2.4

Your Turn: The Sum, Difference, Product, or Quotient of Two Functions:

1. $3x^2 - x + 4$ **2.** -8 **3.** $2t^2 + 7t - 4$ **4.** $-\dfrac{7}{2}$ **Determining Domain:**

1. $\{x|x$ is a real number *and* $x \neq 0\}$ **2.** $\{x|x$ is a real number *and* $x \neq 2$ *and* $x \neq 7\}$

Practice Exercises: 1. (c)   **2.** (d)   **3.** (a)   **4.** (b)   **5.** 6   **6.** 2   **7.** $-x^3 + 3x^2 - x + 3$ **8.** $\dfrac{3}{5}$

9. 18   **10.** $9 - 6x + x^2$ **11.** $\dfrac{a^2 + 1}{3 - a}$, $a \neq 3$ **12.** -14 **13.** $\{x|x$ is a real number$\}$

14. $\{x|x$ is a real number$\}$ **15.** $\{x|x$ is a real number *and* $x \neq 0\}$

16. $\{x|x$ is a real number *and* $x \neq 1\}$ **17.** $\{x|x$ is a real number *and* $x \neq -2\}$

18. $\{x|x$ is a real number *and* $x \neq 3\}$ **19.** $\{x|x$ is a real number *and* $x \neq -4$ *and* $x \neq 8\}$

20. $\{x|x$ is a real number *and* $x \neq 1$ *and* $x \neq 3\}$

Section 2.5

Your Turn: The Constant b: The y-Intercept: 1. $(0, 13)$ **2.** $(0, -9.5)$ **3.** $(0, 3)$

The Constant m: Slope: 1. $\dfrac{2}{5}$ **2.** 9   **3.** $-\dfrac{1}{2}$ **Applications: 1.** 15 min/page

2. 1.5 cups/pie

Practice Exercises: 1. y-intercept   **2.** slope   **3.** down   **4.** slope-intercept form

5. $\left(0, -\dfrac{1}{2}\right)$ **6.** $(0, -1)$ **7.** 3   **8.** 1   **9.** -5 **10.** Slope: $\dfrac{3}{8}$; y-intercept: $(0, -6)$

11. Slope: $\dfrac{3}{2}$; y-intercept: $(0, 3)$ **12.** Slope: 1; y-intercept: $(0, -9)$

13. Slope: $-\dfrac{1}{3}$; y-intercept: $\left(0, \dfrac{2}{3}\right)$ **14.** $1\dfrac{1}{6}$ miles per minute   **15.** $\dfrac{1}{20}$ mile per minute

Section 2.6

Your Turn: Graphing Using Intercepts:

1.

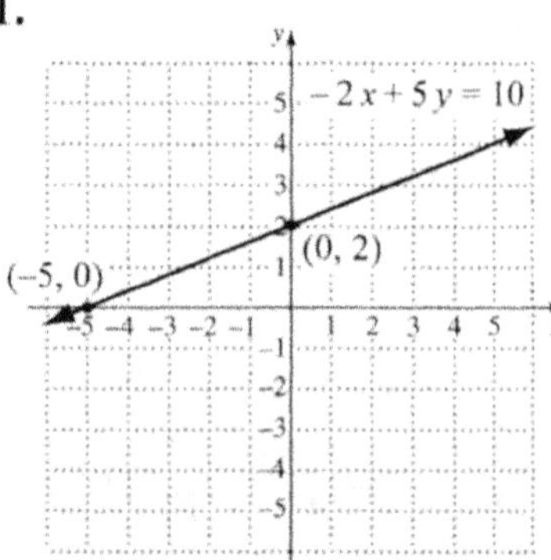

2.

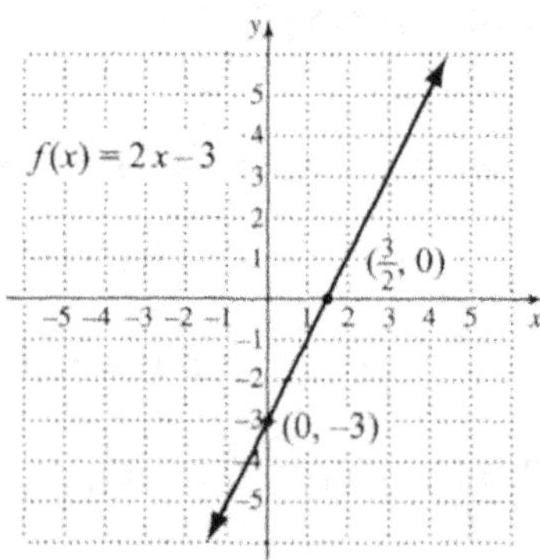

Graphing Using the Slope and the y-Intercept:

1.

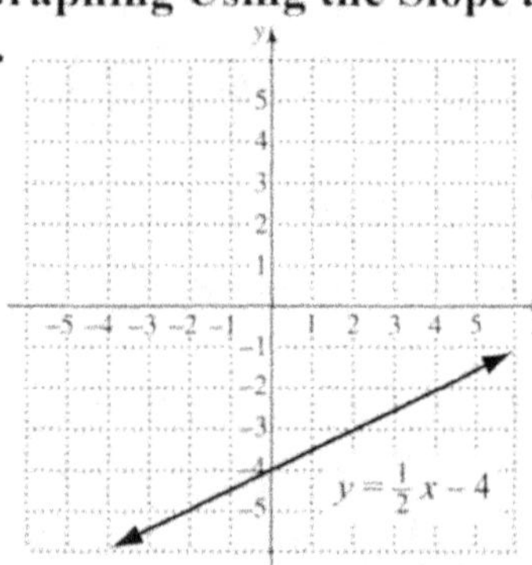

2.

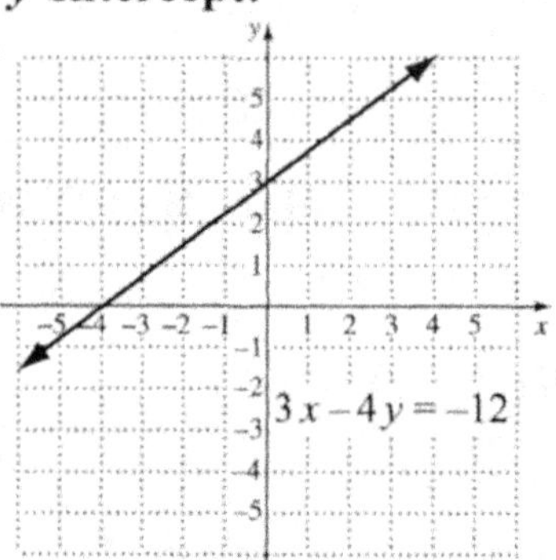

Horizontal Lines and Vertical Lines:

1.

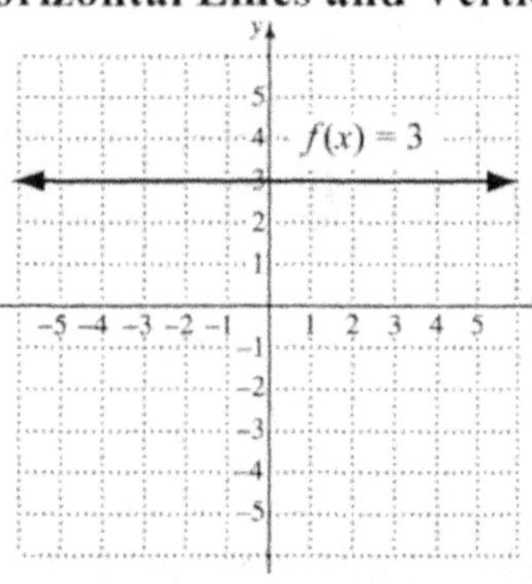

Slope is 0.

2.

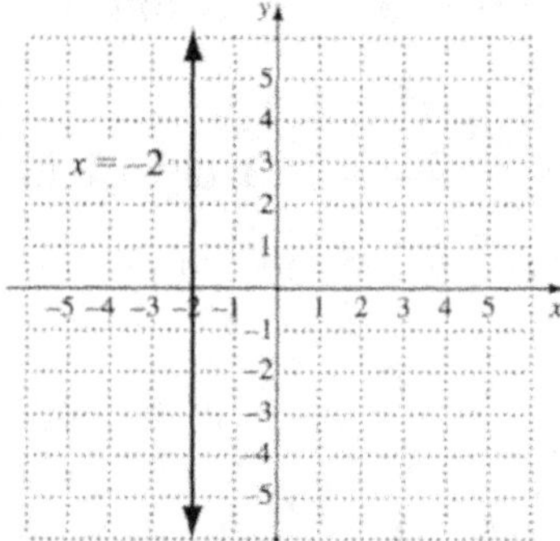

Slope is not defined.

3.

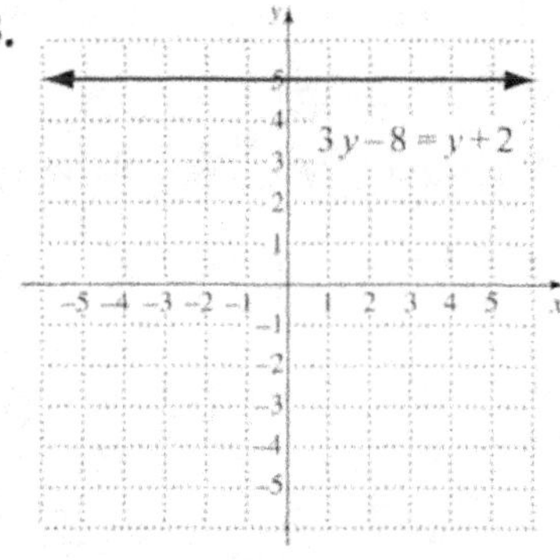

Slope is 0.

Parallel Lines and Perpendicular Lines: **1.** Not parallel **2.** Perpendicular

Practice Exercises: **1.** vertical **2.** parallel **3.** -1 **4.** rise; run

5. 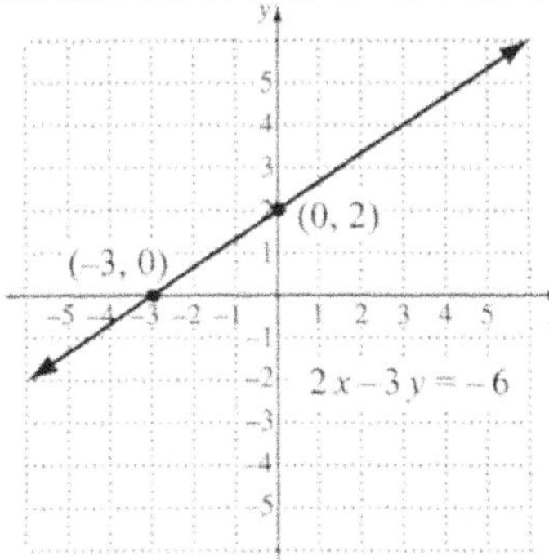

6. 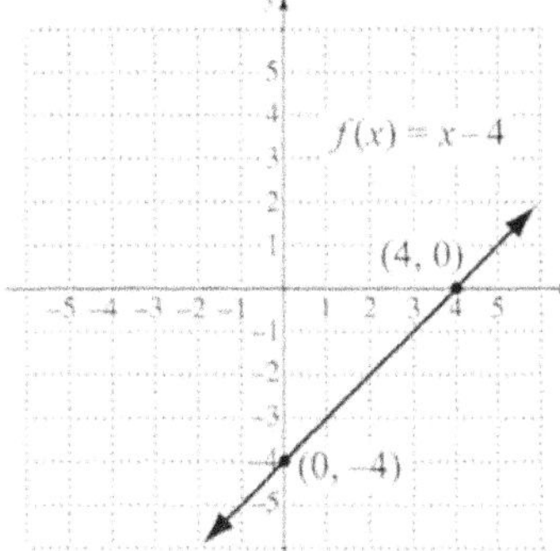

7. 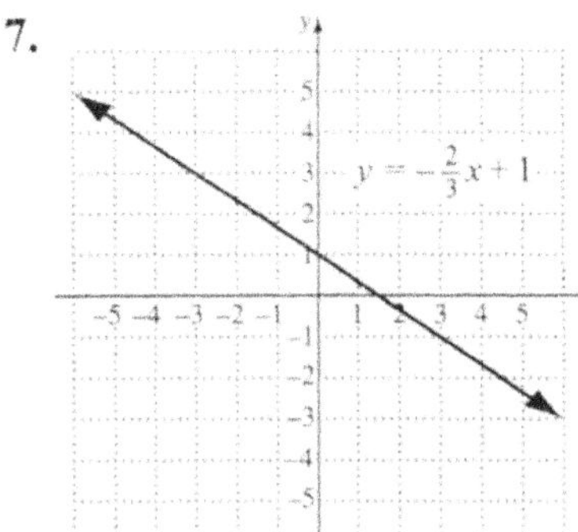

8. 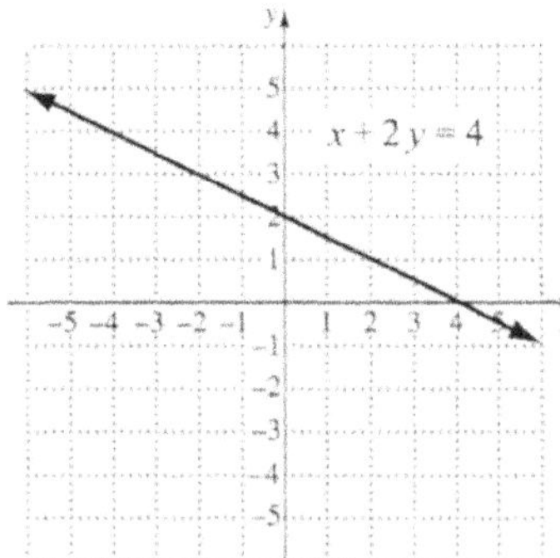

9. 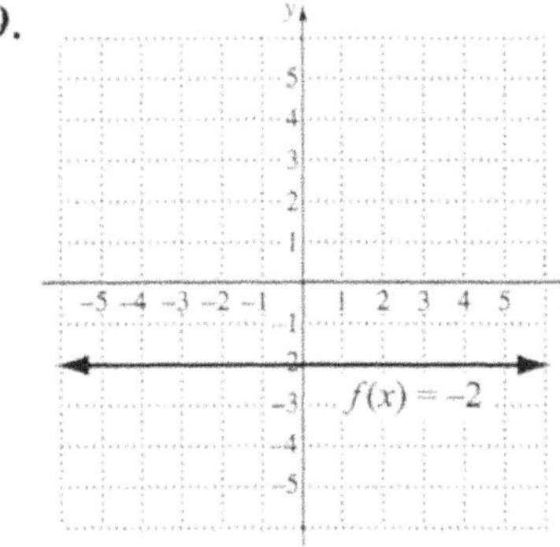

Slope is 0.

10. 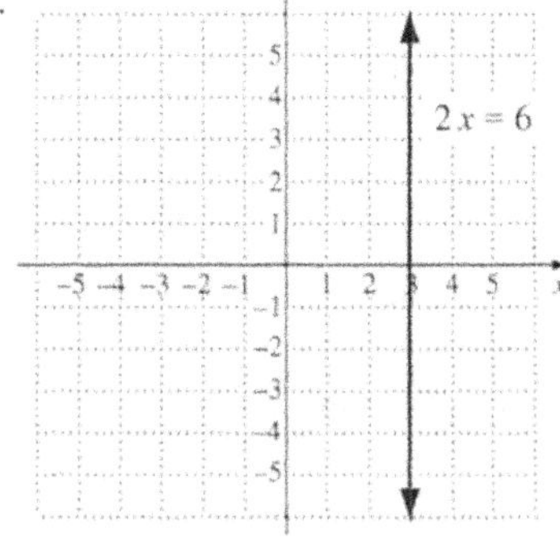

Slope is not defined.

11. Parallel **12.** Not parallel **13.** Perpendicular **14.** Perpendicular

Section 2.7

Your Turn: Finding an Equation of a Line When the Slope and the y-Intercept Are Given: 1. $y = -7x + 5$ **2.** $f(x) = 4x - 10$ **Finding an Equation of a Line When the Slope and a Point Are Given: 1.** $y = \dfrac{2}{3}x - \dfrac{35}{3}$ **2.** $y = -x + 7$ **Finding an Equation of a Line When Two Points Are Given: 1.** $y = -\dfrac{1}{2}x + \dfrac{1}{2}$ **2.** $y = -x + 5$

Finding an Equation of a Line Parallel or Perpendicular to a Given Line Through a Point Not on the Line: 1. $y = -\dfrac{2}{5}x + \dfrac{33}{5}$ **2.** $y = -2x - 9$

Copyright © 2022 Pearson Education, Inc.

A-35

Applications of Linear Functions: 1. $e(t) = -\dfrac{1}{2}t + \dfrac{49}{2}$; \$17,500

2. $h(p) = -15p + 250$; 25 headbands

Practice Exercises: 1. (d) **2.** (b) **3.** (a) **4.** (c) **5.** $y = -4x + 8$ **6.** $f(x) = \dfrac{1}{2}x - 1$

7. $y = 6x - 18$ **8.** $y = -\dfrac{1}{2}x - 7$ **9.** $y = \dfrac{1}{3}x + \dfrac{13}{3}$ **10.** $y = 2x - 5$ **11.** $y = x - 2$

12. $y = \dfrac{4}{5}x - 9$ **13. a)** $C(t) = 22.95t + 150$; **b)** \$517.20 **14. a)** $N(x) = 34x + 150$;
b) 524 students

Chapter 3

Section 3.1
Your Turn: Solving Systems of Equations Graphically:

<table>
<tr><td>1. 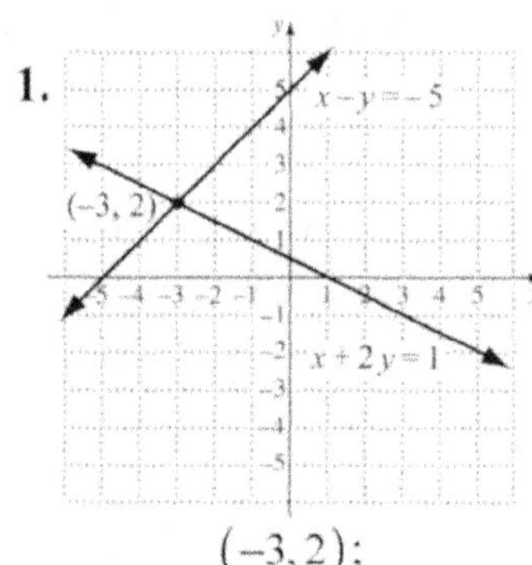</td><td>2. 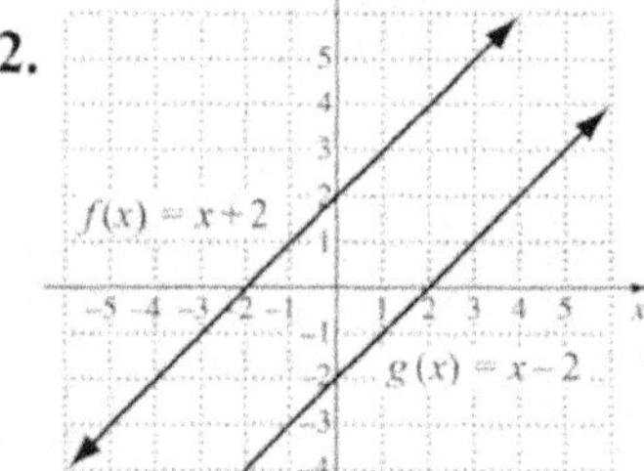</td><td>3. 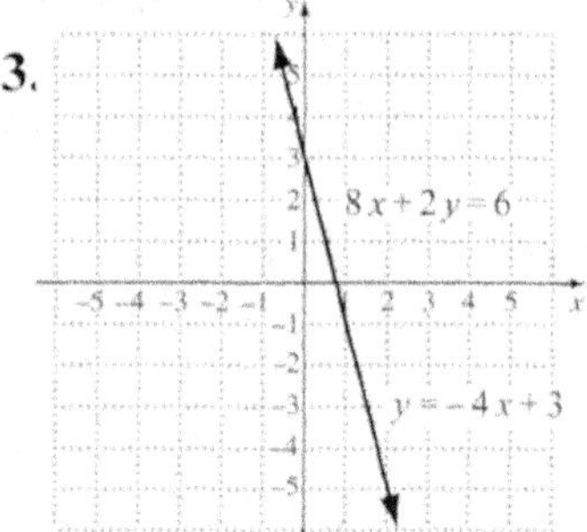</td></tr>
<tr><td align="center">$(-3, 2)$;
consistent; independent</td><td align="center">No solution;
inconsistent; independent</td><td align="center">Infinitely many solutions;
consistent; dependent</td></tr>
</table>

Practice Exercises: 1. (a), (c), (d) **2.** (b) **3.** (d) **4.** (a), (b), (c) **5.** (b) **6.** (c)

<table>
<tr><td>7. 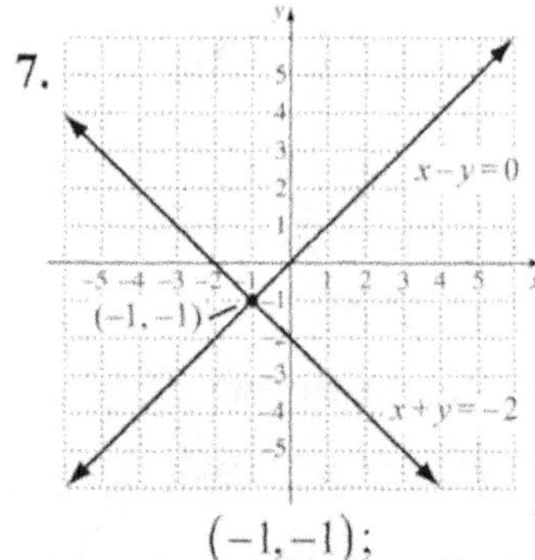</td><td>8. 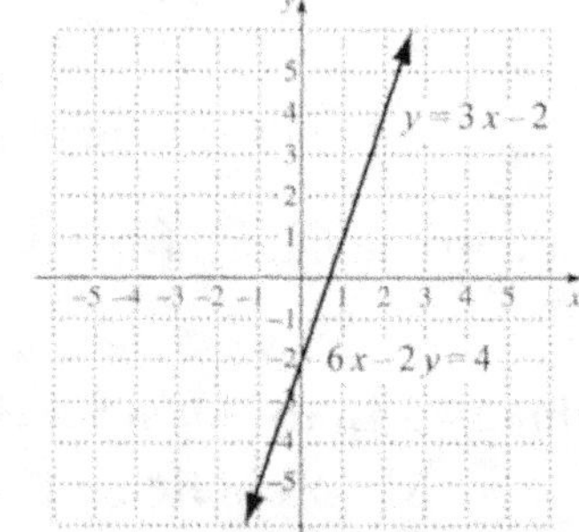</td><td>9. 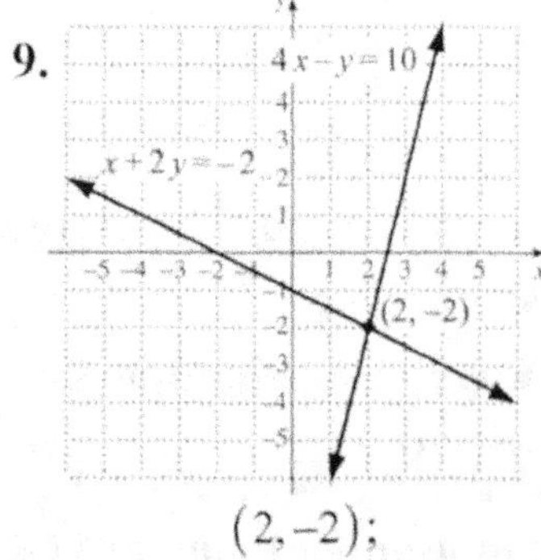</td></tr>
<tr><td align="center">$(-1, -1)$;
consistent; independent</td><td align="center">Infinitely many solutions;
consistent; dependent</td><td align="center">$(2, -2)$;
consistent; independent</td></tr>
</table>

   Copyright © 2022 Pearson Education, Inc.

10. 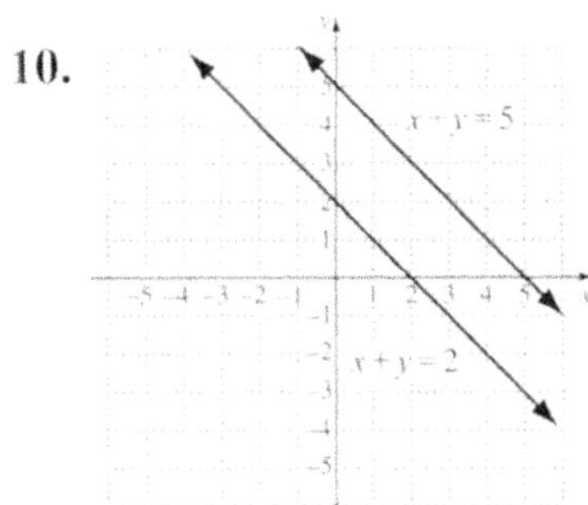

No solution;
inconsistent: independent

11. 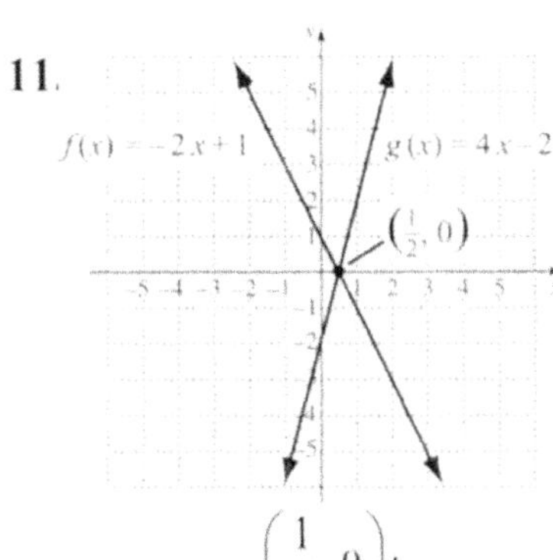

$\left(\dfrac{1}{2},0\right)$;

consistent: independent

12. 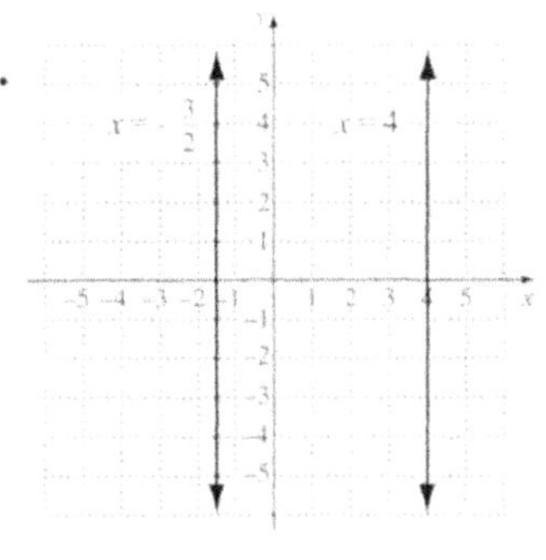

No solution;
inconsistent: independent

Section 3.2

Your Turn: The Substitution Method: 1. $(2,-2)$ **2.** No solution

3. Infinitely many solutions
Solving Applied Problems Involving Two Equations: 1. Length: 35 ft ; width: 20 ft

Practice Exercises: 1. False   **2.** False   **3.** True   **4.** False   **5.** $(3,1)$ **6.** $(2,5)$ **7.** $(-5,-3)$

8. No solution   **9.** Infinitely many solutions   **10.** $\left(\dfrac{3}{4},-\dfrac{1}{4}\right)$ **11.** $(-4,3)$

12. Infinitely many solutions   **13.** Length 9 in.; width 7 in.   **14.** $11°,79°$

Section 3.3

Your Turn: The Elimination Method: 1. $(1,-2)$ **2.** $(4,-1)$ **3.** $(3,-3)$ **4.** No solution
Solving Applied Problems Using Elimination: 1. Two-point shots: 24; three-point shots: 9

Practice Exercises: 1. True   **2.** False   **3.** True   **4.** True   **5.** $(4,2)$ **6.** $(-3,1)$

7. Infinitely many solutions   **8.** $(-2,-3)$ **9.** No solution   **10.** Infinitely many solutions

11. $\left(\dfrac{2}{5},-\dfrac{7}{5}\right)$ **12.** No solution   **13.** $15,-11$ **14.** Balcony seats: 80; floor seats: 295

Section 3.4

Your Turn: Total-Value Problems and Mixture Problems:
1. One-scoop: 8 cones; two-scoop: 17 cones   **2.** 96 L of 60%; 24 L of 20%
Motion Problems: 1. 30 mi

Practice Exercises: 1. (b)   **2.** (a)   **3.** (d)   **4.** (c)   **5.** 9 adults; 18 children
6. $3.89: 14 gal; $3.49: 11 gal   **7.** 10 L of 35%; 5 L of 20%
8. Fruit punch: 24.5 L; sparkling water: 10.5 L   **9.** 15 km/h   **10.** 650 mi

Section 3.5

Your Turn: Solving Systems in Three Variables: 1. $(-2,4,1)$

Practice Exercises: 1. three; three **2.** linear **3.** triple; all three equations **4.** $(2, -1, 3)$

5. $(-2, 0, -1)$ **6.** $\left(\dfrac{1}{2}, 3, 2\right)$ **7.** $(12, 20, -9)$ **8.** $\left(4, \dfrac{1}{2}, -\dfrac{1}{2}\right)$ **9.** $(18, -11, 3)$

Section 3.6

Your Turn: Using Systems of Three Equations:
1. Adult tickets: 296; child tickets: 102; senior tickets: 46

Practice Exercises: 1. (b) **2.** (d) **3.** (c) **4.** (a) **5.** 24, 48, 72
6. 2-point field goals: 24; 3-point field goals: 6; foul shots: 18
7. \$20 gift cards: 6; \$50 gift cards: 3; \$100 gift cards: 1 **8.** 6, 24, 10

Section 3.7

Your Turn: Graphing Inequalities in Two Variables:

1. 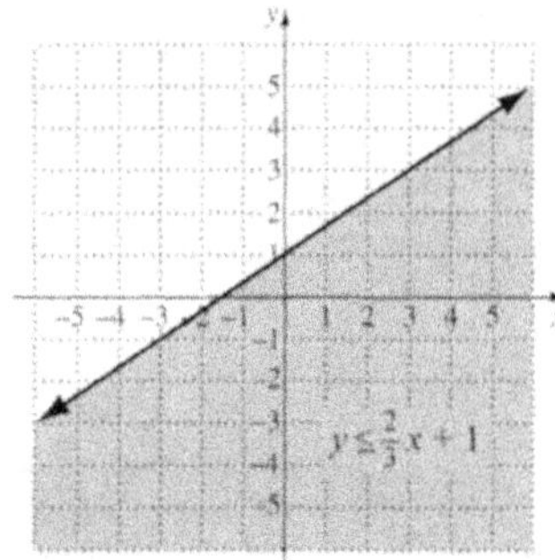

2. 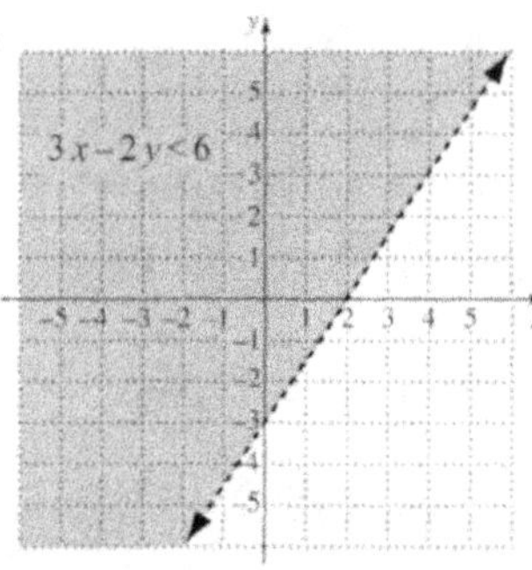

3. 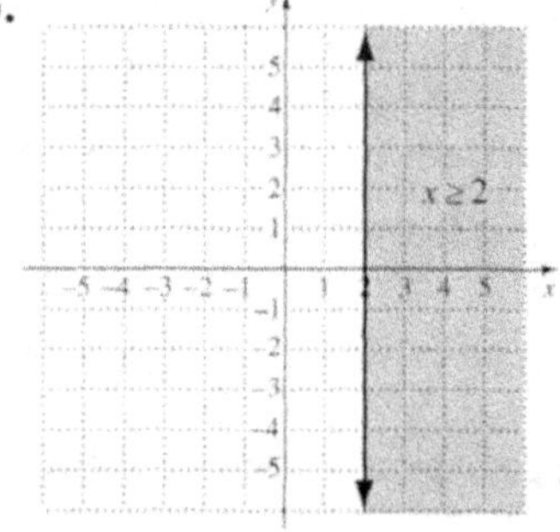

Systems of Linear Inequalities: 1. 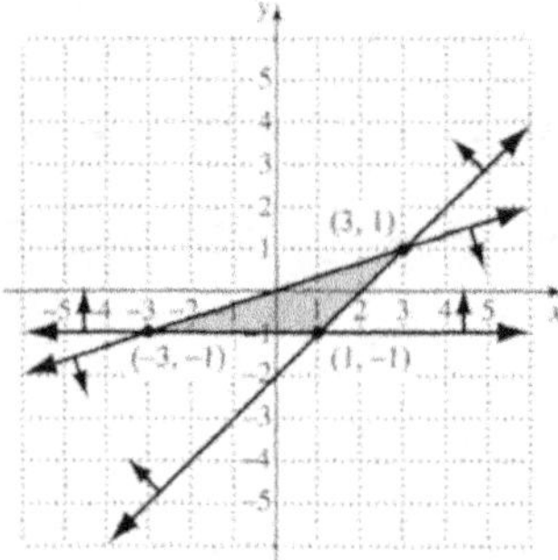

Applications: Linear Programming: 1. Maximum profit of \$11,500 occurs when 50 tuxedos and 150 dresses are made.

Practice Exercises: 1. is not **2.** dashed **3.** same **4.** a plane **5.** intersection

6. 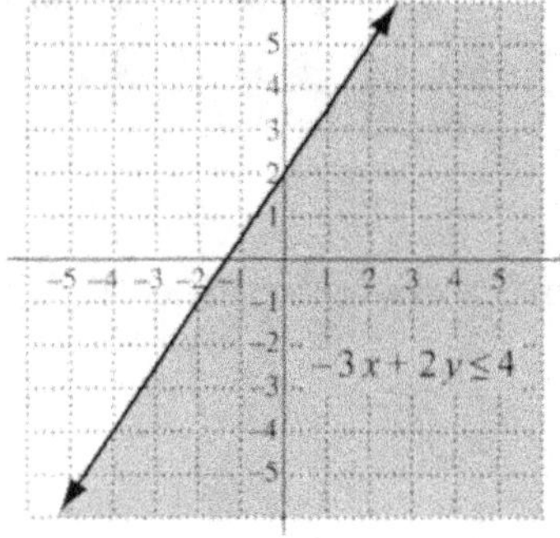

7.

8. 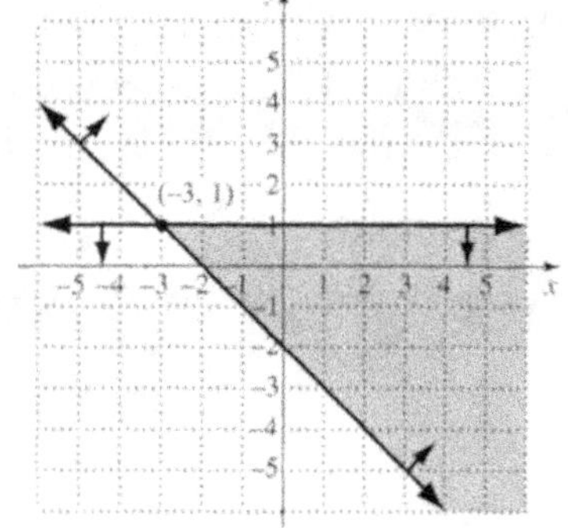

  Copyright © 2022 Pearson Education, Inc.

9. 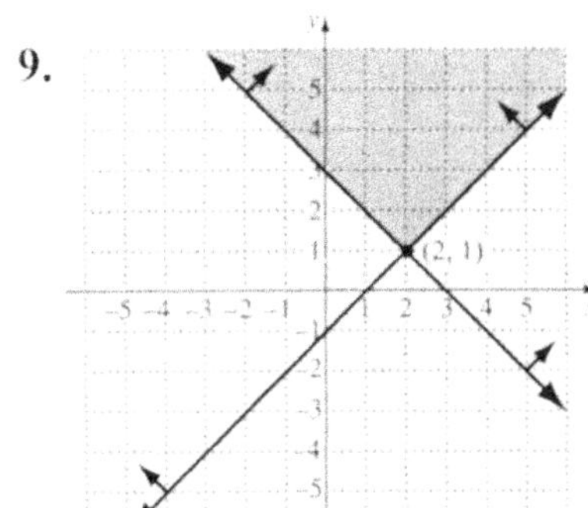

10. 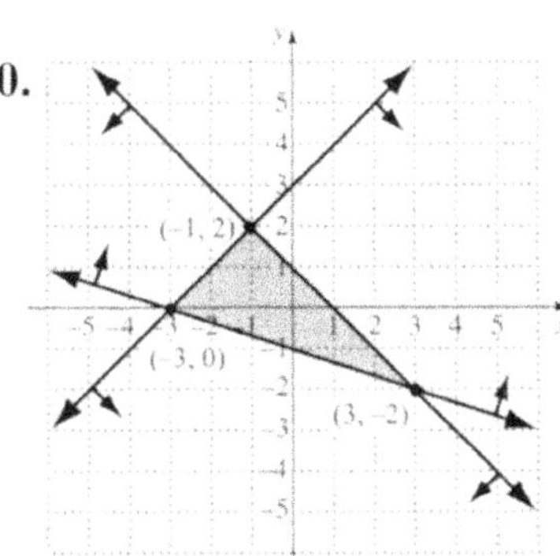

11. Maximum profit of $1300 occurs when 3 princess playhouses and 2 pirate playhouses are made and sold.

Chapter 4

Section 4.1

Your Turn: Polynomial Expressions: 1. $12y^4, -7y^7, -6y^3, -8;\ 4, 7, 3, 0;\ 7; -7y^7;\ -7; -8$
2. $-7y^7 + 12y^4 - 6y^3 - 8; -8 - 6y^3 + 12y^4 - 7y^7$ **Evaluating Polynomial Functions:**
1. -70 **2.** 30.6 mpg **Adding Polynomials: 1.** $-12x^2 - 7x + 5$ **2.** $17x^2y^3 - 7x^2y - 3x^4y$
3. $12y^4 + 2y^3 - 3y + 5$ **Subtracting Polynomials: 1.** $-\left(-6xy^2 + 5xy - 16y + 20\right);$
$6xy^2 - 5xy + 16y - 20$ **2.** $-2x^2 + 8x^3y^3 - 11$ **3.** $x^2 - x + 1$

Practice Exercises: 1. binomial; 4, 0 **2.** 4; 5; 4 **3.** ascending **4.** like terms

5. $-6x,\ x^3,\ 8,\ 3x^4;\ 1, 3, 0, 4;\ 4;\ 3x^4;\ 3;\ 8$

6. 5, $-12x^4,\ -6x^7,\ 3x^3,\ -2x;\ 0, 4, 7, 3, 1;\ 7;\ -6x^7,\ -6;\ 5$

7. $-3xy,\ -6xyz,\ 2x^2,\ -7;\ 2, 3, 2, 0;\ 3;\ -6xyz;\ -6;\ -7$ **8.** $-3x^4 + 2x^2 - x + 15$

9. $12x^6 - 8x^3 - 3x^2 + 4x - 10$ **10.** $9 + 2y^2 - 3y^3 + y^4$ **11.** $-2x^3 + 3x^2y + 6y^2 - 5xy^4$

12. 220; 4 **13.** 554; -4 **14.** $15,822.40 **15.** $2x^3 + x - 4$ **16.** $-13x^2y^2 - 10x^3$

17. $6xyz - 3x^2y + 6xy$ **18.** $2x^2 + 2x - 1$ **19.** $-\left(9x^4 - 3x^3 + 2x^2 - 5\right),\ -9x^4 + 3x^3 - 2x^2 + 5$

20. $-\left(-12xy - 7x^2y^2 + xyz - 6\right),\ 12xy + 7x^2y^2 - xyz + 6$ **21.** $2x^2 + 9x - 11$

22. $10a^3 - 3a^2 - 12$ **23.** $8x^7yz - x^2y^2z^2 - x$ **24.** $-15x^5 - 7xy + 3$

Section 4.2

Your Turn: Multiplication of Any Two Polynomials: 1. $-48a^7b^6c^2$ **2.** $36a^4 - 72a$
3. $3x^9 - 3x^5 - 6x^4 + 6$ **4.** $-10x^5 + 2x^4 + 22x^3 - 34x^2 + 2x + 6$ **Product of Two Binomials**
Using the FOIL Method: 1. $18x^2 - 42x + 12$ **2.** $m^3 + 9m^2 - 12m - 160$ **Squares of**
Binomials: 1. $9y^2 - 30y + 25$ **2.** $\dfrac{1}{4}x^2 - 5x + 25$ **Products of Sums and Differences:**

1. $25x^4y^2-1$ **2.** $0.36x^2-1.44y^2$ **Using Function Notation: 1.** (a) a^2+5a-4;

(b) a^2-3a-4 **2.** $16x^4y^{12}-24x^2y^6+9$

Practice Exercises: 1. True **2.** False **3.** False **4.** True **5.** $8x^8$ **6.** $15x^8y^9$ **7.** $-27a^3b^2c^5$

8. $24-12x$ **9.** $-12a^3-28a^2$ **10.** $-35x^3y^4+30x^5y^2$ **11.** $10x^2+13x-3$

12. $x^3+6x^2+3x-10$ **13.** $3x^3+13x^2-x-20$ **14.** $x^2-\dfrac{2}{3}x+\dfrac{1}{12}$ **15.** $12x^2-2x-2$

16. $3.12x^2+0.7x-15$ **17.** $x^2-12x+36$ **18.** $4x^2+12xy+9y^2$ **19.** $x^6y^2-4x^3y+4$

20. a^2-100 **21.** $9-36x^2$ **22.** $-16a^2+b^4$ **23.** $8x^3-26x^2+17x-5$ **24.** t^2-5t-5

25. $2a^2+4ah+2h^2+3$ **26.** $-a^2+4a+13$

Section 4.3

Your Turn: Terms with Common Factors: 1. $2xy\left(4y+2x^4-x^2\right)$ **2.** $-7x\left(x^2-2x+3\right)$

3. $2(lw+lh+wh)$ **Factoring by Grouping: 1.** $(3x-2)(2x+3)$ **2.** $(x-1)\left(2x^2+3\right)$

3. $(a+4)(2+b)$ **4.** $(x-y)(2x-3y)$

Practice Exercises: 1. $25x^5y^3z^2$ **2.** prime **3.** $x-2$ **4.** $-1(y-x)$ **5.** $6\left(2x^2-3x+4\right)$

6. $3a^3\left(2a^3+3a^2-4\right)$ **7.** $4x^4y\left(4y-2xy^2+x^2\right)$ **8.** $a^3\left(4a^5-a^2-1\right)$ **9.** $-9(x+5)$

10. $-3\left(x^2-3x+4\right)$ **11.** $-1\left(x^3+5x^2+6x-7\right)$ **12.** $-6a(-a+3b)$ **13.** $(r+2)(t+s)$

14. $\left(3a^2+2\right)\left(4a^3+1\right)$ **15.** $\left(4x^2+3\right)(x+2)$ **16.** $(x-y)(5x-6y)$ **17.** $P(x)=x(x-4)$

18. $R(x)=-0.5x(-700+x)$

Section 4.4

Your Turn: Factoring Trinomials: x^2+bx+c **: 1.** $(x-9)(x+1)$ **2.** $3(x+1)(x+4)$

Practice Exercises: 1. (a) **2.** (d) **3.** (b) **4.** (c) **5.** $(x+1)(x+6)$ **6.** $(t+1)(t-12)$

7. $(x-3)(x-5)$ **8.** $(x+6)(x-2)$ **9.** Not factorable **10.** $3(x+2)(x-8)$

11. $\left(y-\dfrac{1}{4}\right)\left(y-\dfrac{1}{4}\right)$ **12.** $x^5(x+1)(x+27)$ **13.** $-1(x+4)(x-9)$,

or $(-x-4)(x-9)$, or $(x+4)(-x+9)$ **14.** $(a+b)(a+5b)$ **15.** $\left(x^2-8\right)\left(x^2+1\right)$

16. $(t+0.3)(t+0.3)$ **17.** $-x(x-1)(x-2)$

Section 4.5

Your Turn: The FOIL Method: 1. $(2x-3)(x+4)$ **2.** $-1(5a+6)(5a-3)$, or

$(-5a-6)(5a-3)$, or $(5a+6)(-5a+3)$ **3.** $a^2(4a+3b)(2a+3b)$ **The *ac*-Method:**

1. $(2x-1)(4x+3)$ **2.** $a^3(5a-2)(2a-3)$

Practice Exercises: 1. common **2.** last **3.** positive **4.** first **5.** $(2x-1)(5x+3)$

6. $(2a-5)(3a+2)$ **7.** $a(3a+2)(5a+1)$ **8.** $(2t-1)(t+1)$ **9.** $-5(4x-3)(x+1)$

10. $2(2p-3w)(3p-5w)$ **11.** $(5x-4)(x+1)$ **12.** $-2z(3z+1)(3z-5)$

13. $(4r-7)(r-7)$ **14.** $(2t-5)(5t-2)$

Section 4.6

Your Turn: Trinomial Squares: 1. $(y+6)^2$ **2.** $(4y+7)^2$ **3.** $(x-3y)^2$ **4.** $3x^2(2x+3)^2$

Differences of Squares: 1. $(x+2)(x-2)$ **2.** $(7t+3r)(7t-3r)$ **3.** $2(x+9y)(x-9y)$

More Factoring by Grouping: 1. $(a-1)(a+7)(a-7)$ **2.** $(4x+5+y)(4x+5-y)$

Sums or Differences of Cubes: 1. $(t^5-3)(t^{10}+3t^5+9)$ **2.** $(3x+1)(9x^2-3x+1)$

3. $2(4x^4+3y^3)(16x^8-12x^4y^3+9y^6)$ **4.** $(t+1)(t^2-t+1)(t-1)(t^2+t+1)$

Practice Exercises: 1. (b)   **2.** (e)   **3.** (c)   **4.** (d)   **5.** (a)   **6.** $(x+3)^2$ **7.** $(t-9)^2$

8. $-y(y+5)^2$ **9.** $(3m+2n)^2$ **10.** $(5x-2)^2$ **11.** $3(x+6)^2$ **12.** $(x+12)(x-12)$

13. $(5a+8)(5a-8)$ **14.** $4(5m+1)(5m-1)$ **15.** $x^2(3x+4)(3x-4)$

16. $8x^2(y^2+z^2)(y+z)(y-z)$ **17.** $\left(\dfrac{1}{5}+x\right)\left(\dfrac{1}{5}-x\right)$ **18.** $(c+d+6)(c+d-6)$

19. $(y+1)(y+2)(y-2)$ **20.** $(12z+x-y)(12z-x+y)$ **21.** $(x+3)(x^2-3x+9)$

22. $(p-2y)(p^2+2py+4y^2)$ **23.** $(1-5a)(1+5a+25a^2)$ **24.** $y^3(xy-1)(x^2y^2+xy+1)$

25. $\left(a-\dfrac{1}{2}\right)\left(a^2+\dfrac{1}{2}a+\dfrac{1}{4}\right)$ **26.** $(a+1)(a^2-a+1)(a-1)(a^2+a+1)$

Section 4.7

Your Turn: A General Factoring Strategy: 1. $10(x+2)(x-6)$

2. $(m+3+5n)(m+3-5n)$

Practice Exercises: 1. True   **2.** False   **3.** False   **4.** True   **5.** $-7x(x+2)(x-4)$

6. $2y(3x+1)(3x-1)$ **7.** $-2(y+3)^2$ **8.** $(m+2)(3m+5)(3m-5)$

9. $4x(x-2)(x^2+2x+4)$ **10.** $a^2(3a-b)(9a^2+3ab+b^2)$ **11.** $(x-9+6y)(x-9-6y)$

12. $2p^3(2p+7)(2p-7)$ **13.** $-4t^2(t^3+3)$ **14.** $24(x+2y)(x^2-2xy+4y^2)$

Section 4.8

Your Turn: The Principle of Zero Products: 1. $-6, -\dfrac{5}{3}$ **2.** $-4, 2$

3. $\{x\,|\,x$ is a real number *and* $x\neq 0$ *and* $x\neq 5\}$ **Applications and Problem Solving: 1.** 4 in.

Practice Exercises: 1. False   **2.** True   **3.** False   **4.** False   **5.** $0, \dfrac{1}{3}$ **6.** $-12, 2$ **7.** $0, \dfrac{5}{2}$

8. $-4, 4$ **9.** $-\dfrac{4}{3}, \dfrac{2}{5}$ **10.** $-1, 0, 3$ **11.** $-9, -5$ **12.** $-7, -2, 2$

13. $\{x|x$ is a real number *and* $x \neq -4$ *and* $x \neq 4\}$

14. $\{x|x$ is a real number *and* $x \neq 0$ *and* $x \neq 2$ *and* $x \neq 3\}$ **15.** 5 hundred televisions

16. 15 m

Chapter 5

Section 5.1

Your Turn: Rational Expressions and Functions: 1. -3

2. $\{x|x$ is a real number *and* $x \neq -4$ *and* $x \neq 1\}$, or $(-\infty, -4) \cup (-4, 1) \cup (1, \infty)$ **Finding**

Equivalent Rational Expressions: 1. $\dfrac{3t(t-1)}{3t(2t+5)}$ **Simplifying Rational Expressions:**

1. $\dfrac{3}{y}$ **2.** $\dfrac{x+4}{5}$ **3.** $\dfrac{x+2}{x-5}$ **4.** $\dfrac{1}{y-3}$ **5.** -1 **Multiplying and Simplifying: 1.** $\dfrac{2(x+3)}{3}$

2. $\dfrac{(x-8)(x+1)}{(x+3)(x-5)}$ **Dividing and Simplifying: 1.** $\dfrac{2(x+2)}{5x}$ **2.** $\dfrac{(x+3)^2}{x+2}$ **3.** $\dfrac{x-y}{x^2 y^2}$

Practice Exercises: 1. True   **2.** False   **3.** False   **4.** True   **5.** $-\dfrac{2}{3}$

6. $\{x|x$ is a real number *and* $x \neq -1$ *and* $x \neq 11\}$, or $(-\infty, -1) \cup (-1, 11) \cup (11, \infty)$

7. $\dfrac{(a+3)(a-1)}{(a+3)(2a+5)}$ **8.** $\dfrac{3a^2}{5}$ **9.** $\dfrac{x-4}{3}$ **10.** $\dfrac{5x}{2}$ **11.** $\dfrac{t-6}{t+3}$ **12.** $\dfrac{2x+1}{5xy}$ **13.** $\dfrac{x(x-1)}{(x+4)(x+1)}$

14. $-\dfrac{3}{2}$ **15.** y^2+2y+4 **16.** $\dfrac{4a}{(a+3)(a+6)}$ **17.** $\dfrac{-5(2x-5y)(x-2y)}{x+5y}$

Section 5.2

Your Turn: Finding LCMs by Factoring: 1. 60   **2.** $45x^2 yz$ **3.** $(2x+1)(x-5)(x+5)$

4. $t(t+1)(t-1)$ **5.** $2y^2(y^2+1)(y+1)^2(y-1)$ **Adding and Subtracting Rational**

Expressions: 1. $\dfrac{9+2a}{a}$ **2.** $\dfrac{2y^2+9}{15y^3}$ **3.** $\dfrac{x^2+2xy-y^2}{(x+y)(x+2y)}$ **4.** $\dfrac{5m+1}{m-5}$ **5.** $\dfrac{x}{(x-1)(x+2)}$

Combined Additions and Subtractions: 1. $\dfrac{2(2x+1)}{(x-2)(x+2)}$

Practice Exercises: 1. True   **2.** False   **3.** False   **4.** True   **5.** 120   **6.** $30x^4 y$

7. $y(y+5)(y-5)$ **8.** $(x+1)(x-1)(x-7)$ **9.** $12a^2(a-2)^2$ **10.** $18x(x+3)^2(x-3)$

11. $\dfrac{3}{k}$ **12.** $\dfrac{y-13}{(y-4)(y+1)}$ **13.** $\dfrac{3x^2+5x-39}{(x+5)(x-6)}$ **14.** $\dfrac{x^2-8x+3}{(x-1)^2(x-5)}$ **15.** $\dfrac{-c^2-cd-d^2}{(c+d)(c-d)}$

16. $\dfrac{2y+9}{(y-3)(y+2)}$ **17.** $\dfrac{5(2m+1)}{(m-2)(m+3)}$

Section 5.3

Your Turn: Divisor a Monomial: 1. $-12x^3+4x-\dfrac{2}{x}$ **2.** $-4x^2y^3+6x^3y-10y^4$ **Divisor Not a Monomial: 1.** $x-1+\dfrac{6}{x+5}$ **2.** $y+3+\dfrac{-y-7}{y^2+1}$ **Synthetic Division:**

1. $x^2-8x+18+\dfrac{-41}{x+2}$ **2.** $5x^2+12x+24+\dfrac{47}{x-2}$

Practice Exercises: 1. False   **2.** True   **3.** False   **4.** True   **5.** $6x^3+2x-5$

6. $8mn^5-5n+6m$ **7.** $-4y^7+y-2$ **8.** $\dfrac{a^3b^2}{2}+a-\dfrac{5}{b}$ **9.** $x+5$ **10.** $z-11+\dfrac{46}{z+6}$

11. $2x^2+5x+1+\dfrac{6}{2x-1}$ **12.** $y^2-2y+5+\dfrac{-15}{y+2}$ **13.** $x^2-2x-4+\dfrac{-13}{x-3}$

14. $4x^2-13x+47+\dfrac{-191}{x+4}$ **15.** $y^2-y-1+\dfrac{9}{y+1}$ **16.** $n^4-2n^3+4n^2-8n+16+\dfrac{-64}{n+2}$

17. $9x^3+6x^2-21x+6$

Section 5.4

Your Turn: Simplifying Complex Rational Expressions: 1. $\dfrac{12a+m^5}{4am^2\left(3+m^3\right)}$

2. $\dfrac{k(1+k)(1-k)}{2(6-k)}$ **3.** $\dfrac{n^2+nm+m^2}{mn(n+m)}$ **Difference Quotients: 1.** 4   **2.** $-2x-h$

Practice Exercises: 1. complex rational expression   **2.** LCM   **3.** numerator, denominator   **4.** numerator, denominator   **5.** $-\dfrac{2}{5}$ **6.** $\dfrac{x}{3-x}$ **7.** $\dfrac{2(x-6)}{3(x+5)}$ **8.** $\dfrac{x-5}{(x-1)\left(x^2-x+3\right)}$

9. $\dfrac{cd(d-c)}{d^2-cd+c^2}$ **10.** $\dfrac{(x+8)(x-2)}{x(x+1)}$ **11.** 4   **12.** $6x+3h$ **13.** $\dfrac{-1}{6x(x+h)}$, or $-\dfrac{1}{6x(x+h)}$

Section 5.5

Your Turn: Rational Equations: 1. $-1, \dfrac{1}{3}$ **2.** No solution   **3.** 0

Practice Exercises: 1. False **2.** True **3.** False **4.** True **5.** $\dfrac{15}{2}$ **6.** -32 **7.** $\dfrac{15}{16}$ **8.** $\dfrac{5}{9}, 3$

9. No solution **10.** $-\dfrac{2}{11}$ **11.** No solution **12.** 1, 8 **13.** -2 **14.** 3 **15.** $-3, 6$

Section 5.6

Your Turn: Work Problems: 1. $1\dfrac{5}{7}$ hr **2.** Gavin: 15 hr; Jeff: 10 hr **Applications**

Involving Proportions: 1. 10 gal **2.** 30 whales **Motion Problems: 1.** 4.5 ft/sec
2. Highway: 70 mph; country roads: 40 mph

Practice Exercises: 1. rate; time, or time; rate **2.** distance; time **3.** distance; rate **4.** 1
5. $5\dfrac{5}{11}$ hr **6.** Robert: 12 days; Georgia: 15 days **7.** 18 hr **8.** 54 hits **9.** 135 deer
10. 30 ft/sec **11.** 5 km/h

Section 5.7

Your Turn: Formulas: 1. $y = \dfrac{k}{3k-2}$ **2.** $t_2 = \dfrac{mSt_1 - H}{mS}$

Practice Exercises: 1. $yx = k$ **2.** $b = ac + c - b$ **3.** $w + p = t$ **4.** $b = \dfrac{ay}{x}$ **5.** $a = \dfrac{wc}{w-c}$

6. $c = \dfrac{ab - bx}{ax}$ **7.** $x_2 = \dfrac{2A - x_1 w}{w}$ **8.** $g = \dfrac{AG}{a + A}$ **9.** $t = \dfrac{1}{y_1 + y_2}$ **10.** $P = \dfrac{A}{rt + 1}$

Section 5.8
Your Turn: Equations of Direct Variation: 1. 6; $y = 6x$ **Applications of Direct**

Variation: 1. \$13 **Equations of Inverse Variation: 1.** 7.5; $y = \dfrac{7.5}{x}$ **2.** 288; $y = \dfrac{288}{x}$

Applications of Inverse Variation: 1. 5 workers **Other Kinds of Variation: 1.** $y = 4x^2$
2. $W = \dfrac{2}{d^2}$ **3.** $y = 6xz$ **Other Applications of Variation: 1.** 12.8 W/m^2

Practice Exercises: 1. (d) **2.** (a) **3.** (b) **4.** (c) **5.** 5; $y = 5x$ **6.** 8; $y = 8x$ **7.** $\dfrac{2}{7}$; $y = \dfrac{2}{7}x$

8. $\dfrac{1}{2}$; $y = \dfrac{1}{2}x$ **9.** 48 mi **10.** 35; $y = \dfrac{35}{x}$ **11.** 9; $y = \dfrac{9}{x}$ **12.** 10; $y = \dfrac{10}{x}$ **13.** 330; $y = \dfrac{330}{x}$

14. 3 hr **15.** $y = \dfrac{0.64}{x^2}$ **16.** $y = \dfrac{8xz}{w}$ **17.** 1200 gal

Chapter 6

Section 6.1

Your Turn: Square Roots and Square-Root Functions: 1. 3 **2.** $\dfrac{5}{9}$ **3.** -11 **4.** 0.05

5. 9.110 **6.** $y+8$ **7.** $\sqrt{14}$ **8.** $\left\{x \mid x \ge -\dfrac{5}{2}\right\}$, or $\left[-\dfrac{5}{2}, \infty\right)$ **9.**

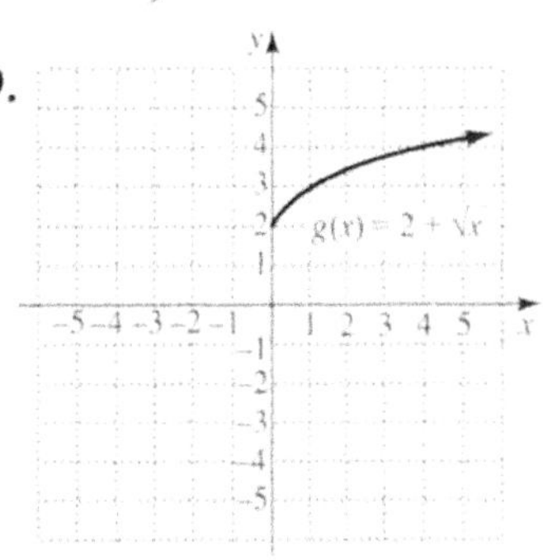

Finding $\sqrt{a^2}$: **1.** $9|y|$ **2.** $|x+2|$ **3.** $x+2$ **Cube Roots: 1.** -2 **2.** $2x$ **3.** 3
Odd and Even kth Roots: 1. 10 **2.** -2 **3.** Does not exist as a real number **4.** $2x+1$
5. $-|3x|$, or $-3|x|$

Practice Exercises: 1. True **2.** True **3.** False **4.** True **5.** $10, -10$ **6.** $16, -16$ **7.** $-\dfrac{10}{9}$

8. 0.5 **9.** 12.369 **10.** $3x-7$ **11.** 3 **12.** Does not exist as a real number **13.** 1
14. $|3x|$, or $3|x|$ **15.** $|-5x|$, or $5|x|$ **16.** $|y-6|$ **17.** 5 **18.** 1 **19.** $-6x$ **20.** -1 **21.** $|x|$
22. $-\dfrac{1}{2}$ **23.** -1 **24.** a **25.** $|y+8|$ **26.** $2x+5$

Section 6.2

Your Turn: Rational Exponents: 1. 7 **2.** $\left(-3x^2y\right)^{1/5}$ **3.** 729 **4.** $(9xy)^{5/4}$

Negative Rational Exponents: 1. $\dfrac{1}{5}$ **2.** $\dfrac{1}{8}$ **3.** $\dfrac{1}{\left(11xyz\right)^{2/3}}$ **4.** $\dfrac{7y^{4/5}}{x^{3/2}}$ **5.** $\left(\dfrac{5ab}{2xy}\right)^{3/5}$

6. $a^{2/7}$ **Laws of Exponents: 1.** $x^{9/10}$ **2.** $a^{-1/2}$, or $\dfrac{1}{a^{1/2}}$ **3.** $3.4^{1/2}$ **4.** $x^{-2/15}y^{1/10}$, or $\dfrac{y^{1/10}}{x^{2/15}}$

Simplifying Radical Expressions: 1. t^3 **2.** $\sqrt[5]{x^2}$ **3.** $\sqrt{3}$ **4.** $\sqrt[24]{x^7}$ **5.** $x^{25}y^5z^{10}$ **6.** $\sqrt[15]{x}$

Practice Exercises: 1. (d) **2.** (a) **3.** (c) **4.** (b) **5.** $\sqrt[3]{x^2}$, or $\left(\sqrt[3]{x}\right)^2$ **6.** 9 **7.** $\sqrt[3]{x^2y^2}$

8. 32 **9.** $5^{1/3}$ **10.** $m^{3/4}$ **11.** $(4xy)^{5/2}$ **12.** $\left(2x^2y\right)^{8/3}$ **13.** $\dfrac{1}{9}$ **14.** $\dfrac{x^2z^4}{y^{1/3}}$ **15.** $12ac^{4/7}$ **16.** 9

17. $\dfrac{5c^{3/4}}{a^{3/2}b^{1/5}}$ **18.** $\dfrac{1}{\left(6xyz\right)^{3/5}}$ **19.** $7^{1/2}$ **20.** $12^{3/8}$ **21.** $2.1^{4/3}$ **22.** $x^{13/15}$ **23.** $\dfrac{b^{1/28}}{a^{1/6}}$ **24.** $\dfrac{1}{5}$

25. $\sqrt[5]{x}$ **26.** a^2b^2 **27.** $\sqrt{3}$ **28.** $\sqrt[12]{ab}$ **29.** x^8y^{12} **30.** $\sqrt[4]{8}$

Section 6.3

Your Turn: Multiplying and Simplifying Radical Expressions: 1. $\sqrt{65}$ **2.** $\sqrt[3]{20x^2y}$
3. $\sqrt[6]{200}$ **4.** $5\sqrt{5}$ **5.** $4y^2\sqrt{3x}$ **6.** $2a^3b^2\sqrt[3]{4ab}$ **7.** $2\sqrt[3]{9}$ **8.** $-24\sqrt{10}$ **9.** $3x^2\sqrt[4]{x^3y^2}$

Dividing and Simplifying Radical Expressions: 1. 5 **2.** 16 **3.** $\dfrac{2}{3}$ **4.** $\dfrac{x^4\sqrt{3x}}{5y^3}$

5. $\sqrt[10]{a^9b}$, or $a\sqrt[10]{\dfrac{b}{a}}$

Practice Exercises: 1. $81x^4$ **2.** simplify **3.** perfect cube **4.** the same **5.** $5\sqrt{2}$ **6.** $2\sqrt[3]{5}$
7. $2x^2\sqrt{30x}$ **8.** $-2x^3\sqrt[3]{3}$ **9.** $5b^3\sqrt{2ab}$ **10.** $2x^3y\sqrt[4]{2y}$ **11.** $2\sqrt{15}$ **12.** $7a^4\sqrt{2}$
13. $a\sqrt[3]{18}$ **14.** $x^2y\sqrt[4]{x^2y^3}$ **15.** $2xy\sqrt[4]{180x^3y^3}$ **16.** 5 **17.** $x^2\sqrt{2}$ **18.** $5y^2\sqrt[3]{x^2}$ **19.** $\sqrt[12]{x^5y^2}$
20. $\dfrac{4}{5}$ **21.** $\dfrac{7}{x^2}$ **22.** $\dfrac{xy^2\sqrt[4]{3}}{z}$ **23.** $\dfrac{5\sqrt[3]{a}}{2b}$ **24.** $\dfrac{3x\sqrt[3]{3x^2}}{y^2}$ **25.** $\dfrac{ac\sqrt[5]{a^4b^2c}}{z^3}$

Section 6.4

Your Turn: Addition and Subtraction: 1. $10\sqrt{3}$ **2.** Cannot be simplified
3. $9\sqrt[4]{2xy}-5\sqrt[3]{2xy}$ **4.** $20\sqrt{2}$ **5.** $\left(y^3+1\right)\sqrt[3]{5x}$ **More Multiplication: 1.** $\sqrt{15}+y\sqrt{5}$
2. $5y\sqrt[3]{4}$ **3.** $16-6\sqrt{7}$ **4.** $20-14\sqrt{3}$ **5.** 23

Practice Exercises: 1. True **2.** False **3.** False **4.** True **5.** $6\sqrt{3}$ **6.** $3\sqrt[3]{5}$
7. $4\sqrt{6}+4\sqrt[4]{7}$ **8.** $-4\sqrt{5}$ **9.** $\left(7x+5\right)\sqrt{x}$ **10.** $12\sqrt[3]{2}$ **11.** $2\sqrt{x+4}$ **12.** $5x\sqrt[3]{x^2}$
13. $6\sqrt{5}-15$ **14.** $3y-3\sqrt[3]{2y^2}$ **15.** $3+2x\sqrt{3}+x^2$ **16.** $a-5\sqrt{a}-36$ **17.** $17-\sqrt{5}$
18. $2\sqrt[3]{25}-16\sqrt[3]{35}+24\sqrt[3]{49}$ **19.** 33 **20.** $x-w$

Section 6.5

Your Turn: Rationalizing Denominators: 1. $\dfrac{\sqrt{2}}{2}$ **2.** $\dfrac{\sqrt[3]{36}}{2}$ **3.** $\dfrac{\sqrt{5x}}{2x}$ **4.** $\dfrac{\sqrt[3]{4x^2y^2z^2}}{4xyz}$

Rationalizing When There are Two Terms: 1. $\dfrac{4x-4\sqrt{3}}{x^2-3}$ **2.** $\dfrac{4\sqrt{2}-4\sqrt{11}-\sqrt{22}+11}{-9}$

Practice Exercises: 1. True **2.** False **3.** False **4.** True **5.** $\dfrac{\sqrt{3}}{3}$ **6.** $\dfrac{\sqrt{42}}{7}$ **7.** $\dfrac{\sqrt{30}}{9}$
8. $\dfrac{\sqrt[3]{3x^2y}}{xy}$ **9.** $\dfrac{\sqrt{21x}}{6}$ **10.** $\dfrac{\sqrt[4]{x^2y^3}}{2yz}$ **11.** $\dfrac{5-\sqrt{3}}{11}$ **12.** $\dfrac{8\sqrt{y}+4y}{4-y}$ **13.** $\dfrac{2+\sqrt{14}+\sqrt{6}+\sqrt{21}}{-5}$
14. $\dfrac{6-13\sqrt{x}+6x}{4-9x}$

Section 6.6

Your Turn: The Principles of Powers: 1. 25 **2.** 2 **3.** -12 **Equations with Two Radical Terms: 1.** 5 **2.** 1 **3.** 2 **Applications: 1.** 45 ft

Practice Exercises: 1. isolate **2.** powers **3.** radical **4.** solve **5.** original **6.** 121 **7.** -16
8. 3 **9.** 2, 5 **10.** 1 **11.** No solution **12.** $\dfrac{1}{2}$ **13.** No solution **14.** 5 **15.** 3 **16.** -20 **17.** 10
18. 16 m

Section 6.7

Your Turn: Applications: 1. $\sqrt{2}$; 1.414 **2.** $\sqrt{409}$ ft; 20.224 ft

Practice Exercises: 1. True **2.** False **3.** True **4.** False **5.** $\sqrt{85}$; 9.220 **6.** 8
7. $\sqrt{82}$ cm; 9.055 cm **8.** $\sqrt{2}$ m; 1.414 m **9.** $\sqrt{160{,}900}$ ft, or $10\sqrt{1609}$ ft; 401.123 ft
10. $\sqrt{4525}$ m, or $5\sqrt{181}$ m; 67.268 m

Section 6.8

Your Turn: Increasing, Decreasing, and Constant Functions: 1. (a) $(-2, 2)$;
(b) $(-5, -2)$; **(c)** $(2, \infty)$ **Relative Maximum and Minimum Values: 1.** Relative
maximum: 5.500 at $x = -1.500$; relative minimum: 0.870 at $x = 0.167$; increasing:
$(-\infty, -1.500)$, $(0.167, \infty)$; decreasing: $(-1.500, 0.167)$ **Applications of Functions:**
1. (a) $A(x) = x(36 - x)$, or $36x - x^2$; **(b)** $(0, 36)$; **(c)** 18 ft by 18 ft   **Functions Defined**
Piecewise: 1. $f(-5) = -7$, $f(-2) = 1$, $f(0) = 1$, $f(3) = 9$

2. 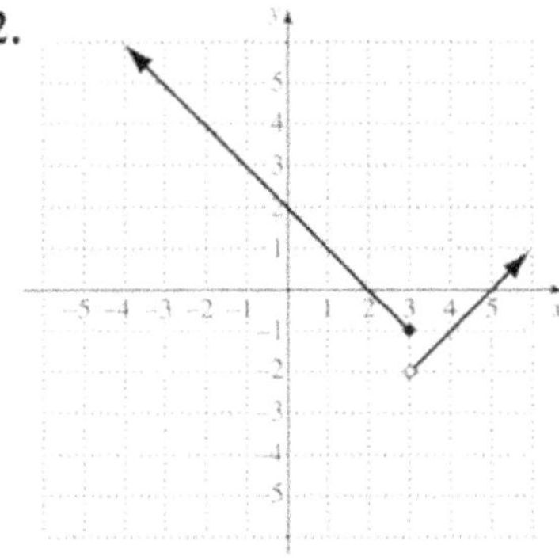
3. 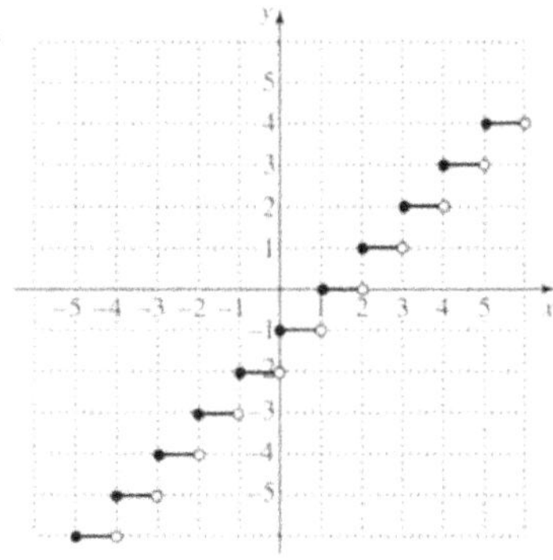

Practice Exercises: 1. False **2.** True **3.** True **4.** False **5. (a)** $(-\infty, -3)$, $(-1, 2)$;
(b) $(-3, -1)$; **(c)** $(2, \infty)$ **6.** Relative maximum: 4.902 at $x = 0.975$; relative minimum:
-4.766 at $x = -1.709$; increasing: $(-1.709, 0.975)$; decreasing: $(-\infty, -1.709)$, $(0.975, \infty)$
7. $A(b) = b^2 + 3b$ **8. (a)** $A(x) = x(360 - 3x)$, or $360x - 3x^2$; **(b)** $(0, 120)$;
(c) 180 yd by 60 yd, where the side opposite the river is 180 yd **9.** $f(-3) = -9$, $f(-1) = 1$,
$f(0) = 2$, $f(4) = 18$

10. 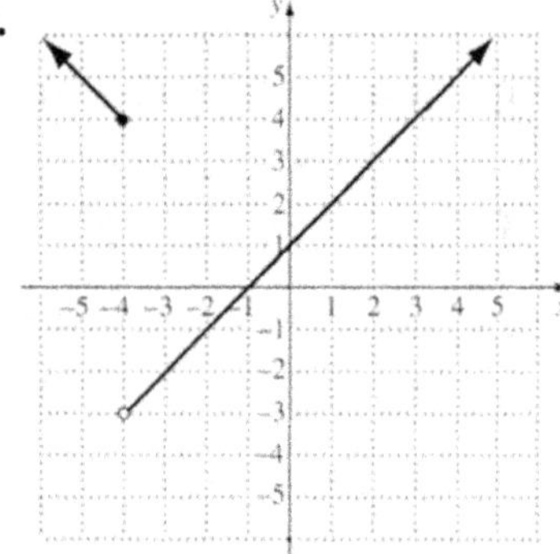

11. 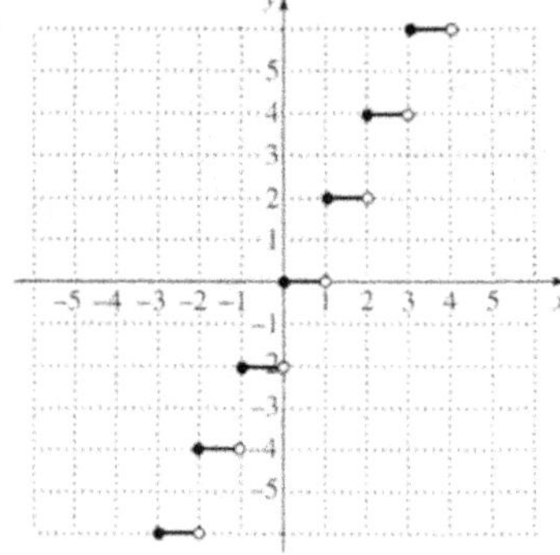

Chapter 7

Section 7.1

Your Turn: **1.** x-axis: yes; y-axis: no; origin: no   **2.** x-axis: yes; y-axis: yes; origin: yes
3. (a) Neither; (b) Even

Practice Exercises: **1.** x-axis: no; y-axis: no; origin: no
2. x-axis: yes; y-axis: yes; origin: yes   **3.** x-axis: no; y-axis: no; origin: no
4. x-axis: no; y-axis: no; origin: yes   **5.** x-axis: yes; y-axis: no; origin: no
6. x-axis: no; y-axis: no; origin: no   **7.** Even   **8.** Neither   **9.** Odd   **10.** Odd   **11.** Even
12. Even

Section 7.2

Your Turn:

1. (a) The graph of $g(x) = x^2 + 4$ is the graph of $f(x) = x^2$ translated up 4 units.

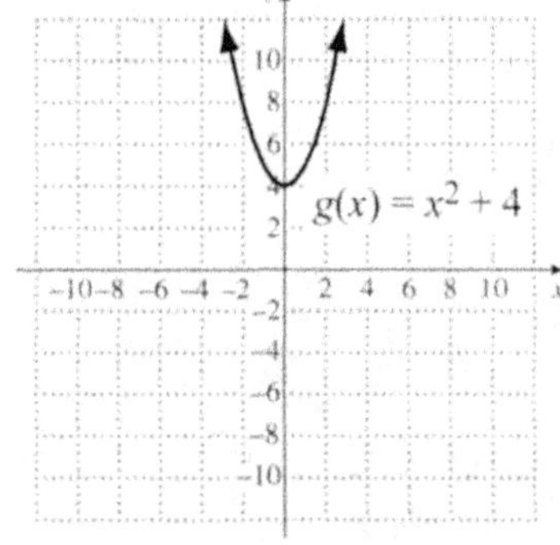

1. (b) The graph of $g(x) = |x + 3|$ is the graph of $f(x) = |x|$ shifted left 3 units.

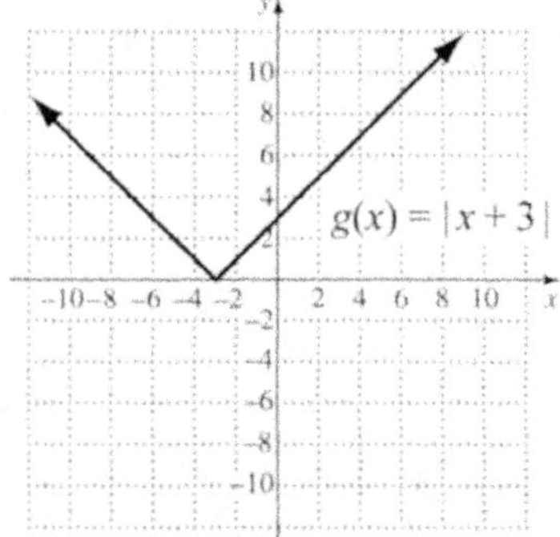

1. (c) The graph of $g(x) = \sqrt{x-4}$ is the graph of $f(x) = \sqrt{x}$ shifted right 4 units.

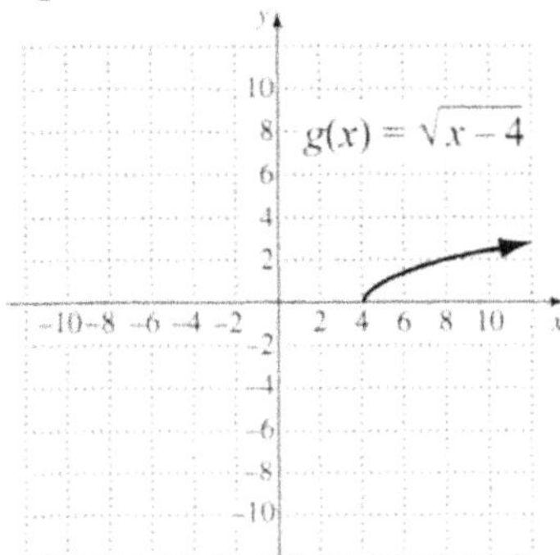

1. (d) The graph of $h(x) = \sqrt{x-2} + 5$ is the graph of $f(x) = \sqrt{x}$ shifted right 2 units and up 5 units.

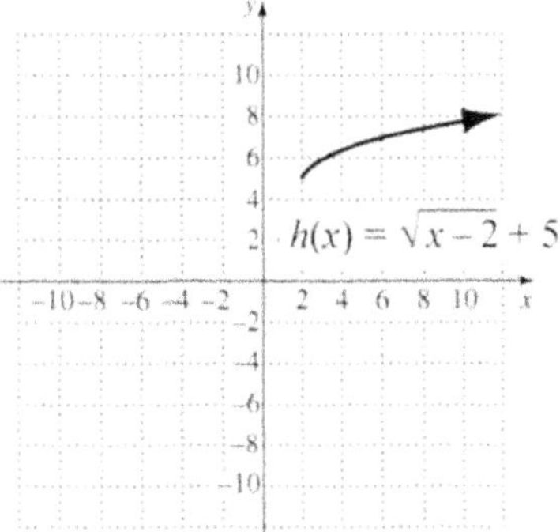

2. (a) $f(-x) = (-x)^3 - 2(-x)^2 = g(x)$ so the graph of $g(x) = (-x)^3 - 2(-x)^2$ is the reflection of the graph of $f(x) = x^3 - 2x^2$ across the y-axis.

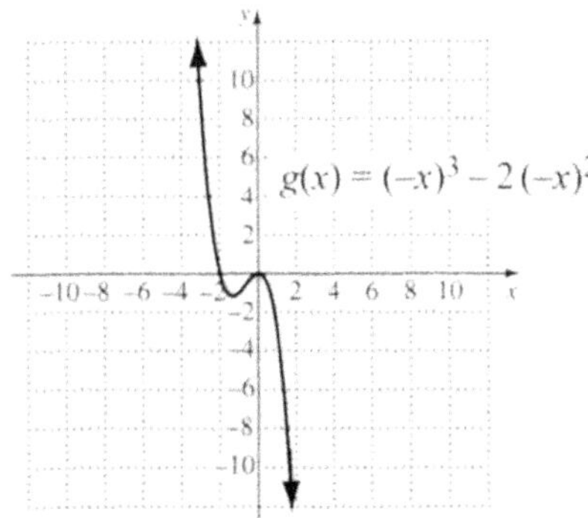

2. (b) $-f(x) = -(x^3 - 2x^2) = -x^3 + 2x^2 = 2x^2 - x^3 = h(x)$ so the graph of $h(x) = 2x^2 - x^3$ is the reflection of the graph of $f(x) = x^3 - 2x^2$ across the x-axis.

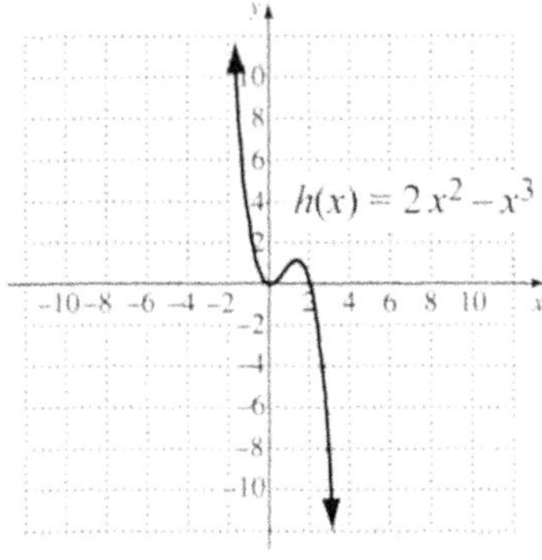

3. (a) $h(x) = 2f(x)$

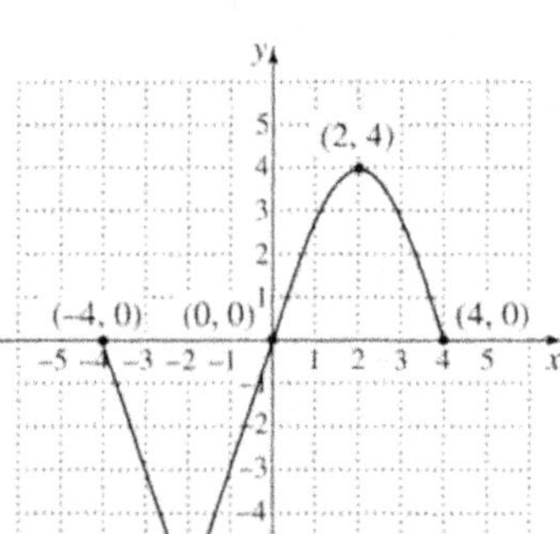

3. (b) $s(x) = \dfrac{1}{2} f(x)$

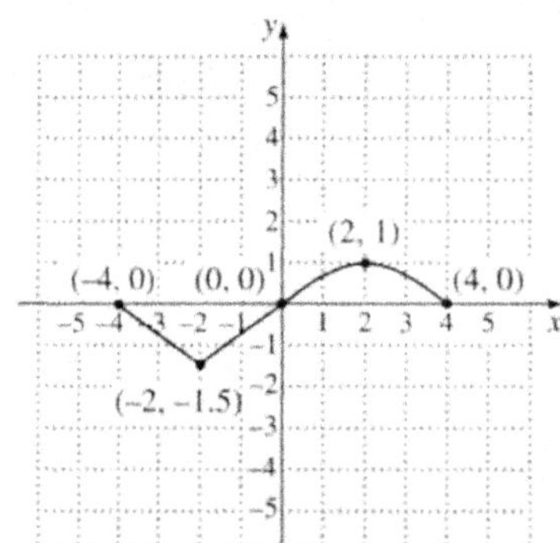

3. (c) $t(x) = f(2x)$

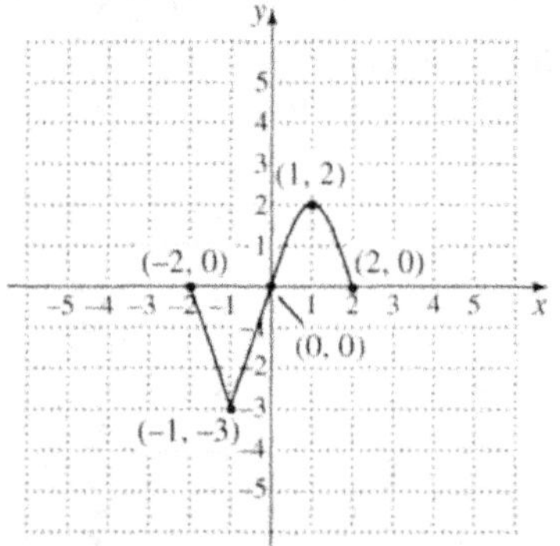

3. (d) $g(x) = f\left(\dfrac{1}{2}x\right)$

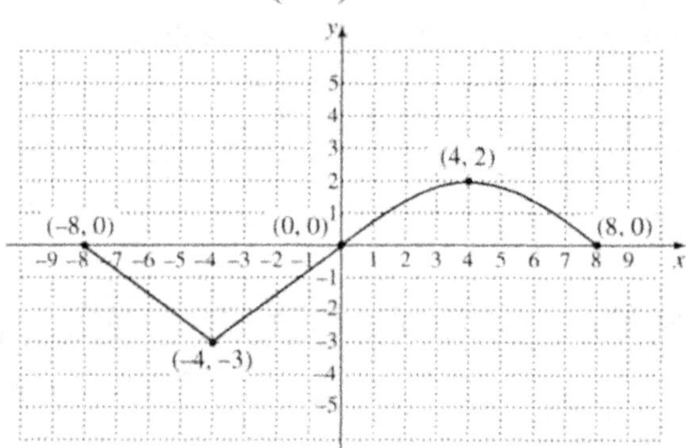

3. (e) $r(x) = f\left(-\dfrac{1}{2}x\right)$

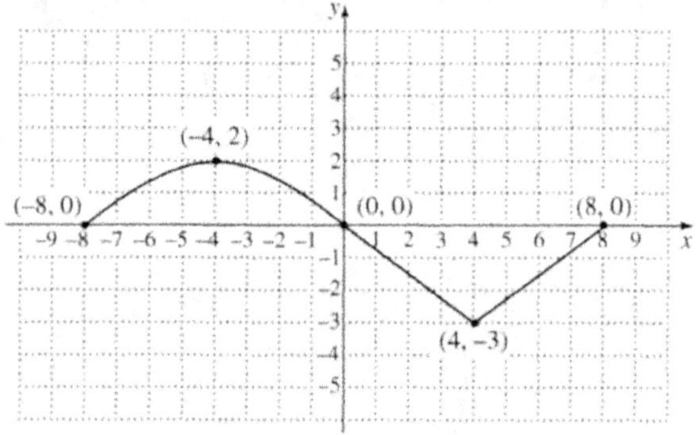

4. $y = f(x+3)$

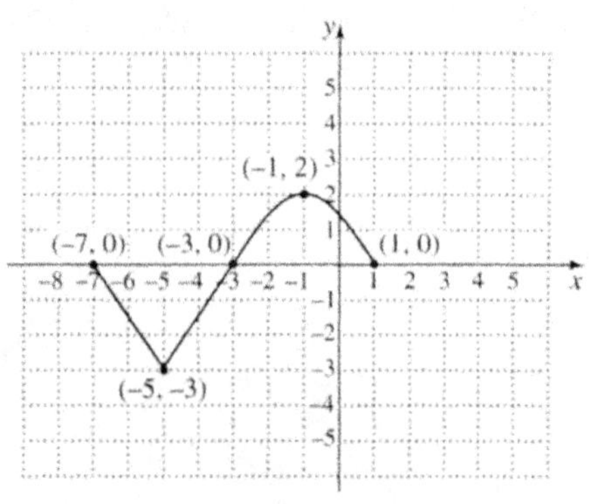

$y = \dfrac{1}{2} f(x+3)$

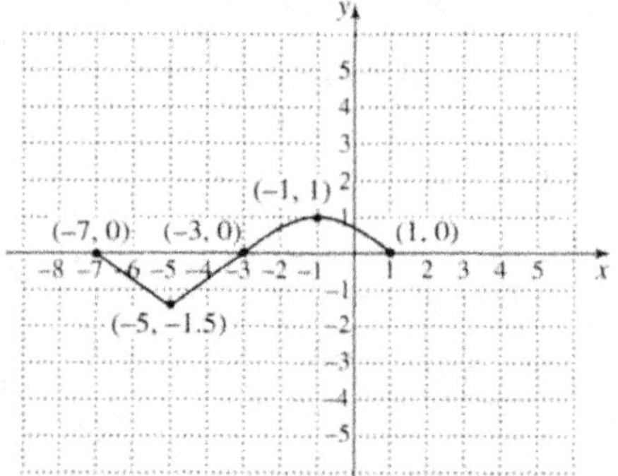

$$y = -\frac{1}{2} f(x+3)$$

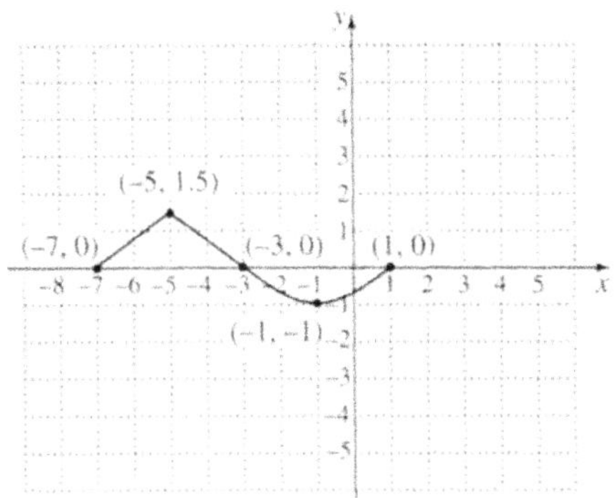

$$y = -\frac{1}{2} f(x+3) - 1$$

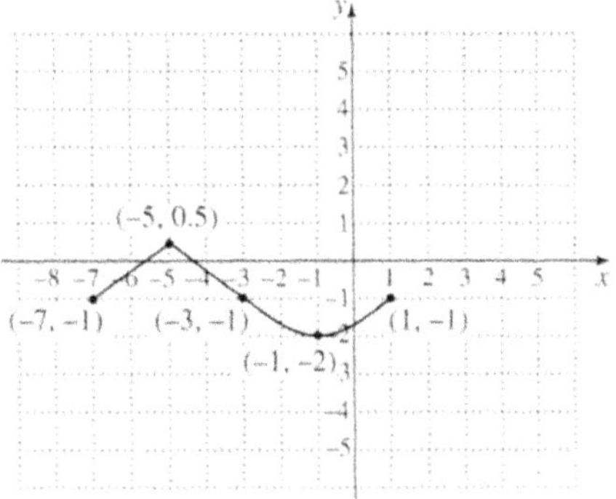

Practice Exercises:

1. The graph of $g(x) = x^2 - 2$ is the graph of $f(x) = x^2$ shifted down 2 units.

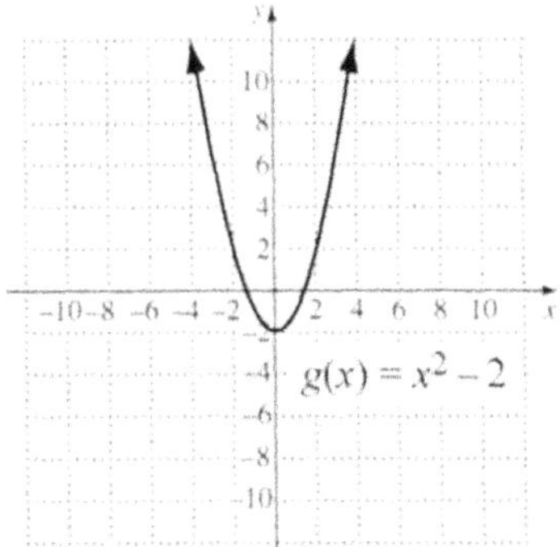

2. The graph of $g(x) = |x+4|$ is the graph of $f(x) = |x|$ shifted left 4 units.

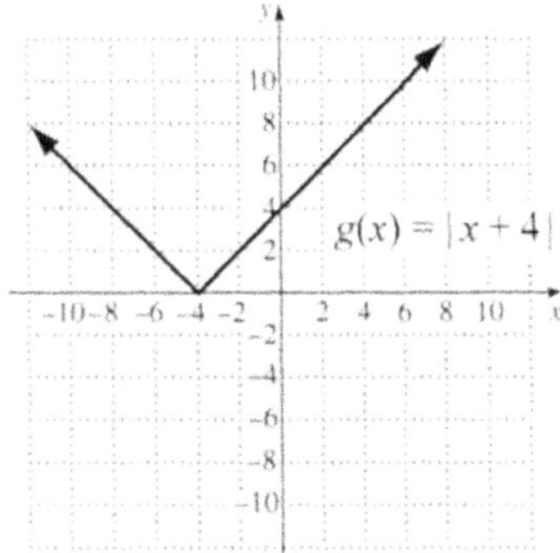

3. The graph of $g(x) = \sqrt{x+4}$ is the graph of $f(x) = \sqrt{x}$ shifted left 4 units.

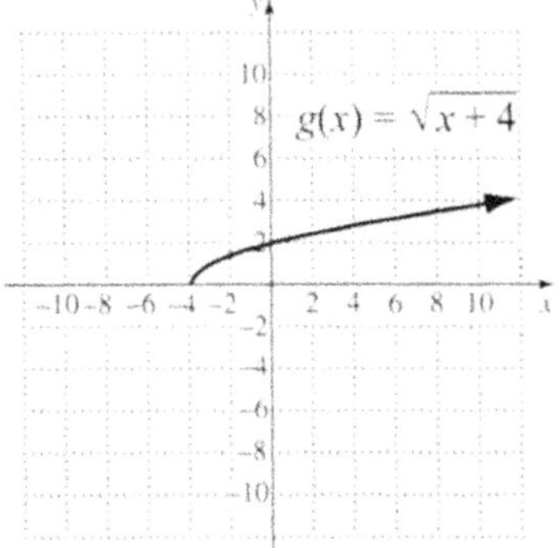

4. The graph of $g(x) = \sqrt{x-2} + 3$ is the graph of $f(x) = \sqrt{x}$ shifted right 2 units and up 3 units.

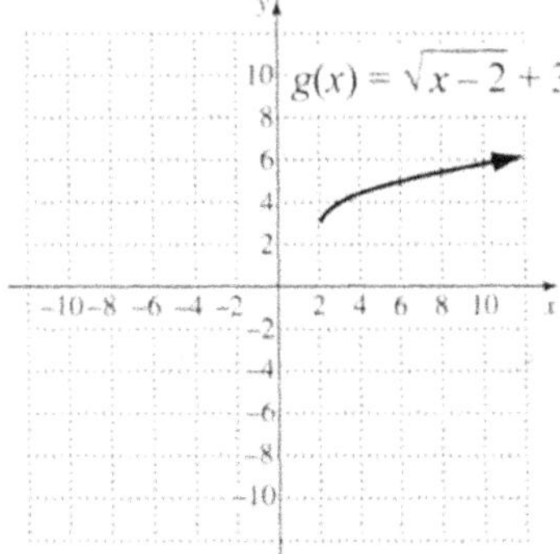

5. The graph of $g(x)$ is the reflection of $f(x)$ across the y-axis.

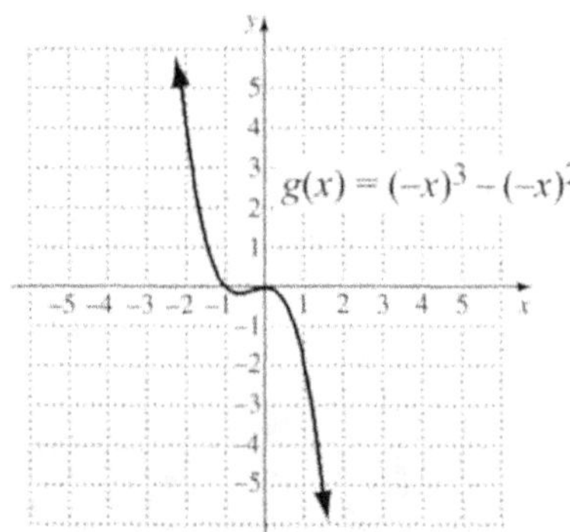

6. The graph of $h(x)$ is the reflection of $f(x)$ across the x-axis.

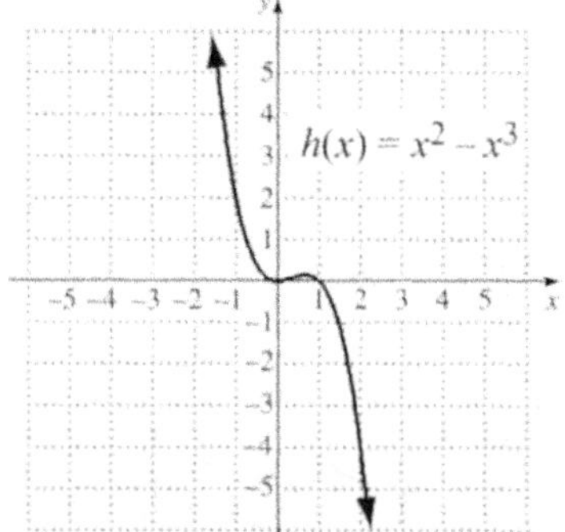

7. (a)

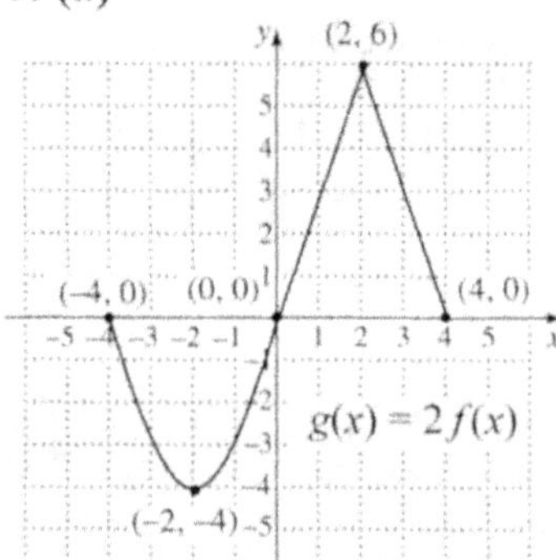

7. (b)

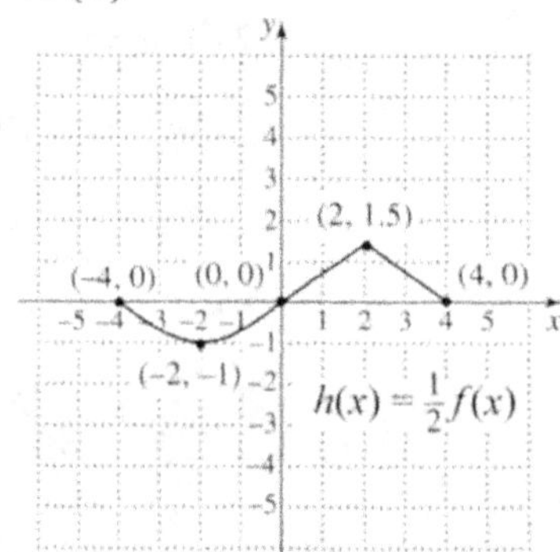

7. (c)

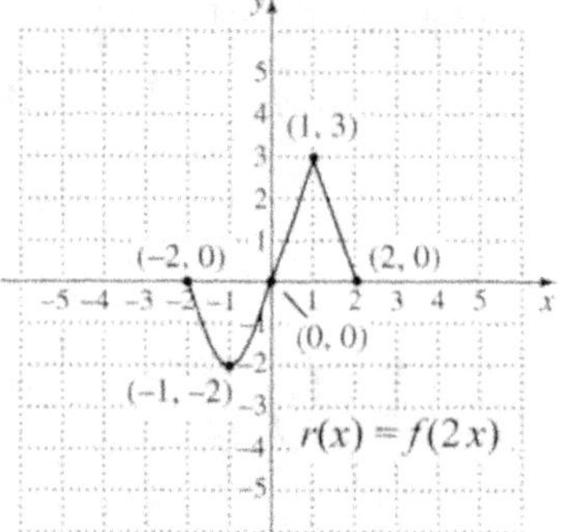

7. (d)

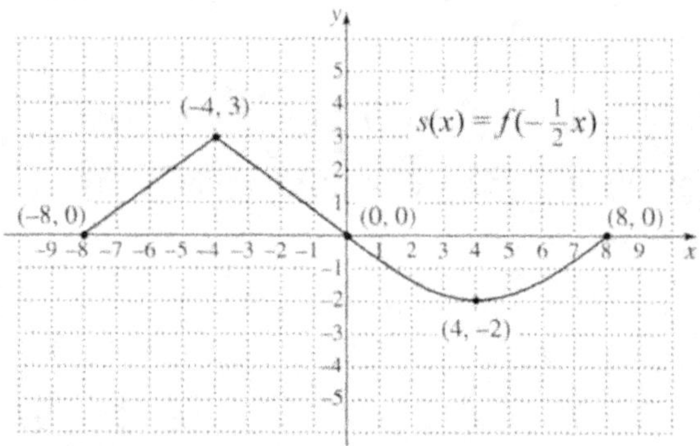

Section 7.3

Your Turn: 1. (a) $\sqrt{3}i$; **(b)** $6i$; **(c)** $-\sqrt{41}i$; **(d)** $-10i$; **(e)** $2\sqrt{21}i$

2. (a) $9+4i$; **(b)** $4+12i$ **3. (a)** -14; **(b)** $23+14i$; **(c)** $-21-20i$

4. (a) i; **(b)** 1; **(c)** $-i$; **(d)** -1 **5. (a)** 109; **(b)** 36 **6.** $\dfrac{2}{41}-\dfrac{23}{41}i$

Practice Exercises: 1. $\sqrt{29}i$ **2.** $9i$ **3.** $2\sqrt{31}i$ **4.** $-5i$ **5.** $5\sqrt{7}i$ **6.** $-\sqrt{37}i$ **7.** $12+7i$
8. $2-7i$ **9.** $-1+15i$ **10.** $9+6i$ **11.** -90 **12.** $42+9i$ **13.** $61-33i$ **14.** $-65-72i$

15. $-i$ **16.** -1 **17.** 1 **18.** i **19.** 144 **20.** 13 **21.** 101 **22.** 25 **23.** $-\dfrac{2}{5}-\dfrac{11}{5}i$

24. $-\dfrac{24}{37}-\dfrac{4}{37}i$ **25.** $-\dfrac{1}{50}+\dfrac{7}{50}i$ **26.** $\dfrac{11}{20}+\dfrac{23}{20}i$

Section 7.4

Your Turn: 1. $-\dfrac{2}{3}, 4$ **2.** $\sqrt{7}$ and $-\sqrt{7}$, or $\pm\sqrt{7}$ **3.** $4+\sqrt{29}$ and $4-\sqrt{29}$, or $4\pm\sqrt{29}$

4. $\dfrac{5}{4}+\dfrac{\sqrt{17}}{4}$ and $\dfrac{5}{4}-\dfrac{\sqrt{17}}{4}$, or $\dfrac{5\pm\sqrt{17}}{4}$ **5.** Exact: $\dfrac{-3-\sqrt{89}}{8}$ and $\dfrac{-3+\sqrt{89}}{8}$;

approximate: -1.554 and 0.804 **6.** $-\dfrac{3}{2}-\dfrac{\sqrt{11}}{2}i$ and $-\dfrac{3}{2}+\dfrac{\sqrt{11}}{2}i$ **7.** $-3,-1,1,3$ **8.** $-8, 64$

9. (a) Approximately 35 million; **(b)** 2015 **10.** 2015 **11.** Train heading south: 75 km/h; train heading west: 100 km/h

Practice Exercises: 1. $-\dfrac{5}{3}, 1$ **2.** $-2, -\dfrac{3}{4}$ **3.** $\sqrt{11}$ and $\sqrt{-11}$, or $\pm\sqrt{11}$

4. $\sqrt{17}$ and $-\sqrt{17}$, or $\pm\sqrt{17}$ **5.** $5+\sqrt{31}$ and $5-\sqrt{31}$, or $5\pm\sqrt{31}$

6. $\dfrac{-5+\sqrt{30}}{2}$ and $\dfrac{-5-\sqrt{30}}{2}$, or $\dfrac{-5\pm\sqrt{30}}{2}$ **7.** $\dfrac{9+\sqrt{21}}{10}$ and $\dfrac{9-\sqrt{21}}{10}$, or $\dfrac{9\pm\sqrt{21}}{10}$

8. $-\dfrac{3}{2}+\dfrac{\sqrt{35}}{2}i$ and $-\dfrac{3}{2}-\dfrac{\sqrt{35}}{2}i$

9. Exact: $1-\sqrt{10}$ and $1+\sqrt{10}$; approximate: -2.162 and 4.162

10. $-\dfrac{3}{2}+\dfrac{\sqrt{39}}{6}i$ and $-\dfrac{3}{2}-\dfrac{\sqrt{39}}{6}i$ **11.** $-3,-2,2,3$ **12.** $-125, 8$

13. (a) Approximately 50 million; **(b)** 2017 **14.** Train heading south: 45 mph; train heading west: 60 mph

Section 7.5

Your Turn:

1. Vertex: $(3, 1)$

Axis of symmetry: $x = 3$
Minimum: 1

2. Vertex: $(-4, -10)$

Axis of symmetry: $x = -4$
Minimum: -10

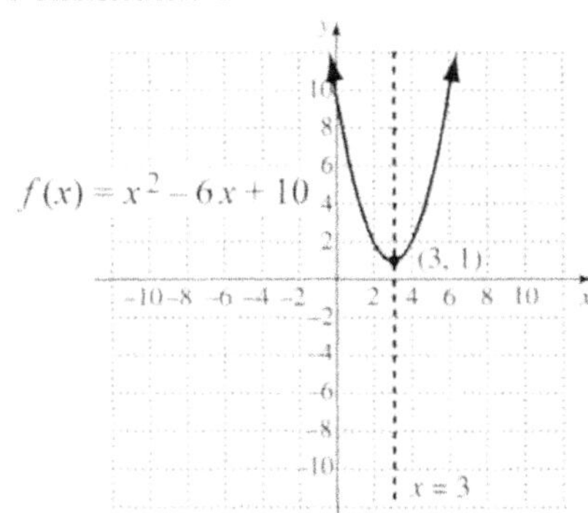

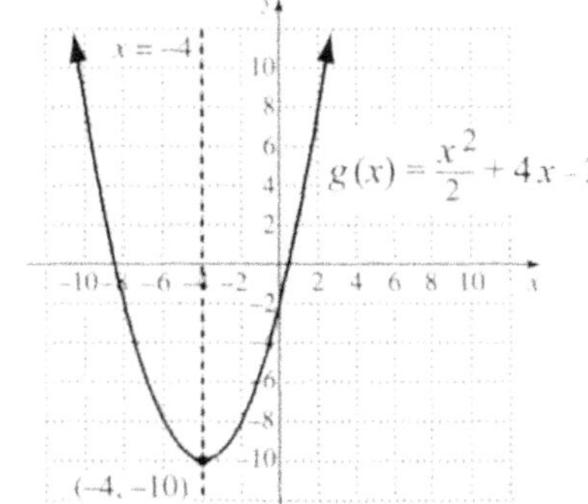

3. Vertex: $\left(\dfrac{3}{2}, 3\right)$

Axis of symmetry: $x = \dfrac{3}{2}$

Maximum: 3

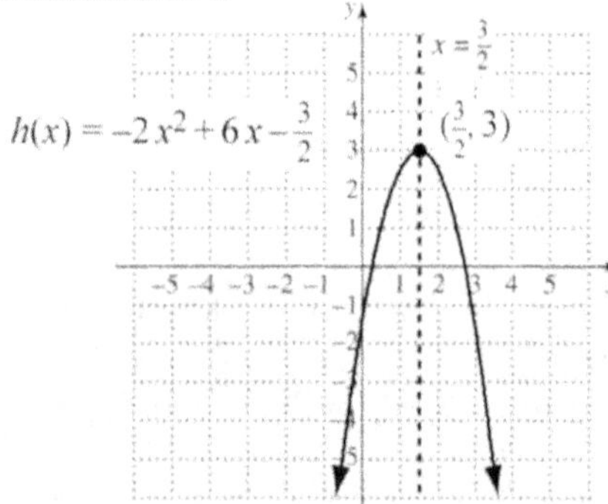

4. (a) Vertex: $(6, 1)$

    (b) Maximum value: 1;

    (c) Range: $(-\infty, 1]$;

    (d) Increasing: $(-\infty, 6)$;

        decreasing: $(6, \infty)$

5. The maximum possible area is 200 ft^2 when the pond is 10 ft wide and 20 ft long.

6. The rocket reaches a maximum height of 115 ft 2.5 sec after it has been launched.

7. 229.9 ft

Practice Exercises:

1. Vertex: $(-4, -1)$

Axis of symmetry: $x = -4$

Minimum: -1

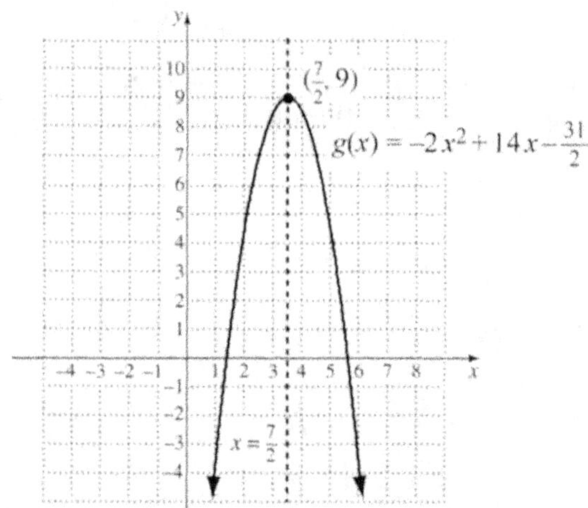

2. Vertex: $\left(\dfrac{7}{2}, 9\right)$

Axis of symmetry: $x = \dfrac{7}{2}$

Maximum: 9

3. Vertex: $(-1, 12)$

Maximum value: 12

Range: $(-\infty, 12]$

Increasing: $(-\infty, -1)$

Decreasing: $(-1, \infty)$

4. Vertex: $(5, -9)$

Minimum value: -9

Range: $[-9, \infty)$

Increasing: $(5, \infty)$

Decreasing: $(-\infty, 5)$

5. The maximum possible area is 1058 ft² when the plot is 23 ft wide and 46 ft long.
6. The rocket reaches a maximum height of 86.25 ft 1.875 sec after it has been launched.
7. The height of the bin is about 60.5 ft.

Chapter 8

Section 8.1

Your Turn: 1. (a) C; **(b)** A; **(c)** D; **(d)** B   **2.** Yes; no   **3.** 4, −2   **4.** 5, −3   **5.** 2, −2, −1

6. $\pm\sqrt{6}$, $\pm 2i$ **7.** 131 ft

Practice Exercises: 1. (a) D; **(b)** B; **(c)** A; **(d)** C   **2.** No; yes   **3.** Yes; no   **4.** 3, −4   **5.** −1, 6

6. 1, −5, −3   **7.** 2, −8   **8.** −1, 1, −4   **9.** 4, −4, −1   **10.** $\pm\sqrt{3}$, $\pm i$ **11.** $\pm\sqrt{6}$, $\pm i\sqrt{2}$

12. 2.3 mcg/mL; 6.5 mcg/mL

Section 8.2

Your Turn:

1.
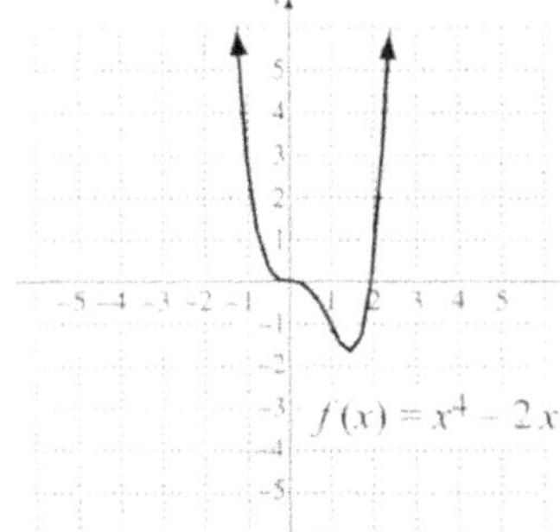

2.
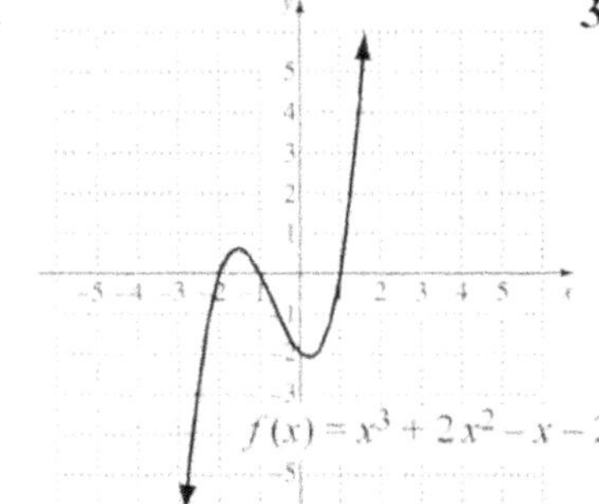

3.
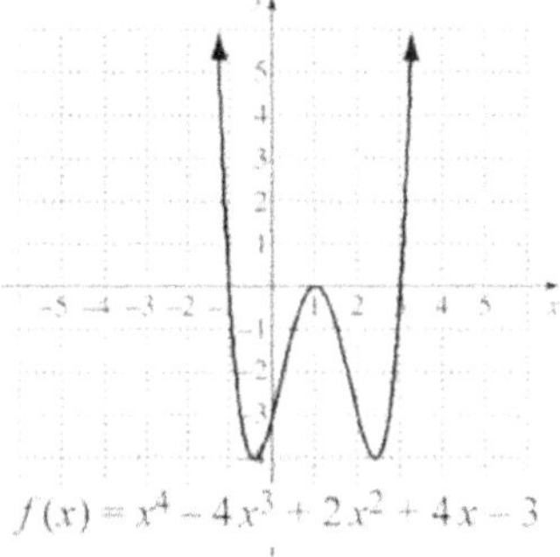

4. (a) $f(3) = 3$ and $f(4) = 16$. Since $f(3)$ and $f(4)$ have the same sign, we cannot use the intermediate value theorem to determine if there is a zero between 3 and 4. **(b)** $f(-1) = -9$ and $f(1) = 1$. By the intermediate value theorem, since $f(-1)$ and $f(1)$ have opposite signs, then $f(x)$ has a zero between −1 and 1. **(c)** $f(2) = -4$ and $f(3) = 81$. By the intermediate value theorem, since $f(2)$ and $f(3)$ have opposite signs, then $f(x)$ has a zero between 2 and 3. **(d)** $f(1) = -7$ and $f(2) = -4$. Since $f(1)$ and $f(2)$ have the same sign, we cannot use the intermediate value theorem to determine if there is a zero between 1 and 2.

Practice Exercises:

1. 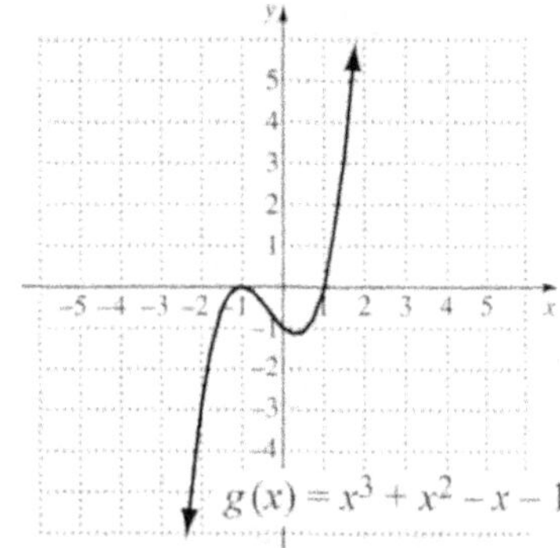

2. 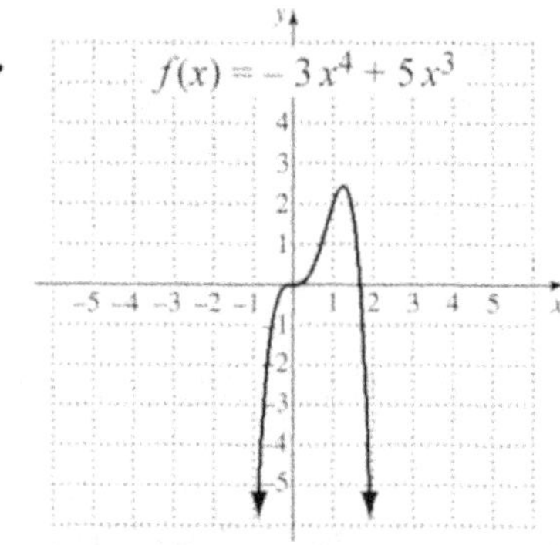

3. 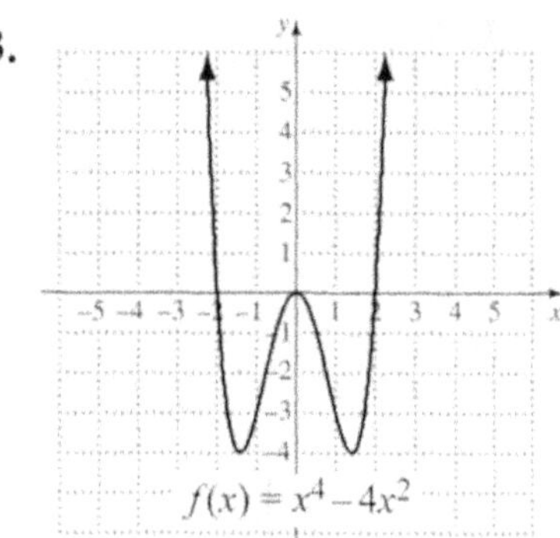

4. 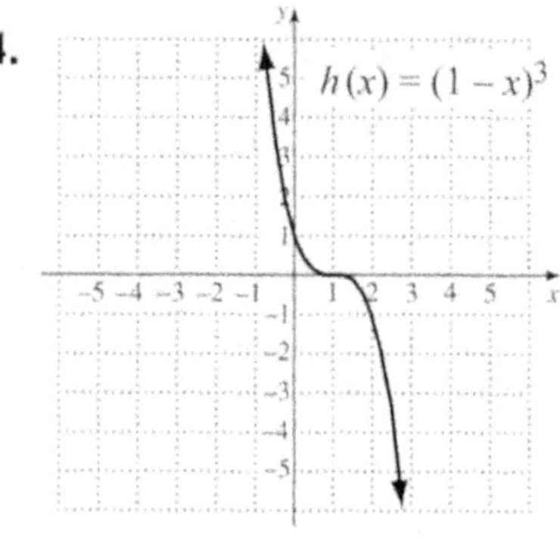

5. $f(2) = -1$ and $f(3) = 5$. By the intermediate value theorem, since $f(2)$ and $f(3)$ have opposite signs, then $f(x)$ has a zero between 2 and 3. **6.** $f(-1) = -7$ and $f(1) = -1$. Since $f(-1)$ and $f(1)$ have the same sign, we cannot use the intermediate value theorem to determine if there is a zero between -1 and 1. **7.** $f(-1) = -4$ and $f(0) = -4$. Since $f(-1)$ and $f(0)$ have the same sign, we cannot use the intermediate value theorem to determine if there is a zero between -1 and 0. **8.** $f(1) = -4$ and $f(2) = 8$. By the intermediate value theorem, since $f(1)$ and $f(2)$ have opposite signs, then $f(x)$ has a zero between 1 and 2. **9.** $f(0) = -3$ and $f(2) = 13$. By the intermediate value theorem, since $f(0)$ and $f(2)$ have opposite signs, then $f(x)$ has a zero between 0 and 2. **10.** $f(-2) = -3$ and $f(-1) = -2$. Since $f(-2)$ and $f(-1)$ have the same sign, we cannot use the intermediate value theorem to determine if there is a zero between -2 and -1.

Section 8.3

Your Turn: 1. Yes; no **2.** Quotient: $x^2 + 2x - 7$; remainder: 11
3. 149 **4.** No **5.** Yes **6.** $f(x) = (x+2)(x+3)(x-1)$; $-2,\ -3,\ 1$

Practice Exercises: 1. (a) Yes; **(b)** no; **(c)** yes **2.** Quotient: $x^3 - x^2 - 4x - 4$, remainder: -2
3. Quotient: $3x^2 + 6x + 8$; remainder: 17 **4.** 6; -18 **5.** 13; -5 **6.** Yes; no **7.** No; yes
8. No; yes **9.** $f(x) = (x+1)(x-2)(x+4)$; $-1,\ 2,\ -4$
10. $f(x) = (x+1)(x-4)(x+3)$; $-1,\ 4,\ -3$

Section 8.4

Your Turn: **1.** $f(x) = x^3 - 3x^2 + 4x - 12$ **2.** $f(x) = x^4 - 3x^2 + 2x$ **3.** $1 + 2i,\ -\sqrt{6}$

4. $f(x) = x^4 - 6x^3 + 8x^2 + 12x - 20$ **5. (a)** Rational: $1,\ -\dfrac{1}{2}$; other: $-1 \pm \sqrt{3}$;

(b) $f(x) = 2(x-1)\left(x + \dfrac{1}{2}\right)(x + 1 - \sqrt{3})(x + 1 + \sqrt{3})$ **6.** Rational: $\dfrac{1}{3},\ \pm 1$; other: $\pm 2i$

(b) $f(x) = 3\left(x - \dfrac{1}{3}\right)(x+1)(x-1)(x-2i)(x+2i)$ **7.** 3 **8.** 2 or 0; 1 **9.** 1; 3 or 1

10. 2 or 0; 1

Practice Exercises: **1.** $f(x) = x^3 + 3x^2 + x + 3$ **2.** $f(x) = x^3 - 2x^2 - 5x + 10$

3. $f(x) = x^4 - 9x^2 + 4x + 12$ **4.** $-\sqrt{5}$ **5.** $1 + 2i,\ 3 - \sqrt{2}$ **6.** $f(x) = x^3 - 7x^2 + 16x - 10$

7. (a) Rational: 2; other: $\pm i$; **(b)** $f(x) = (x-2)(x-i)(x+i)$

8. (a) Rational: $-1,\ \dfrac{3}{2}$; other: $\pm\sqrt{2}$; **(b)** $f(x) = 2(x+1)\left(x - \dfrac{3}{2}\right)(x + \sqrt{2})(x - \sqrt{2})$

9. 3 or 1; 1 **10.** 2 or 0; 3 or 1 **11.** 0; 2 or 0

Section 8.5

Your Turn: **1.** Domain: $\{x \mid x \neq -2\}$, or $(-\infty, -2) \cup (-2, \infty)$

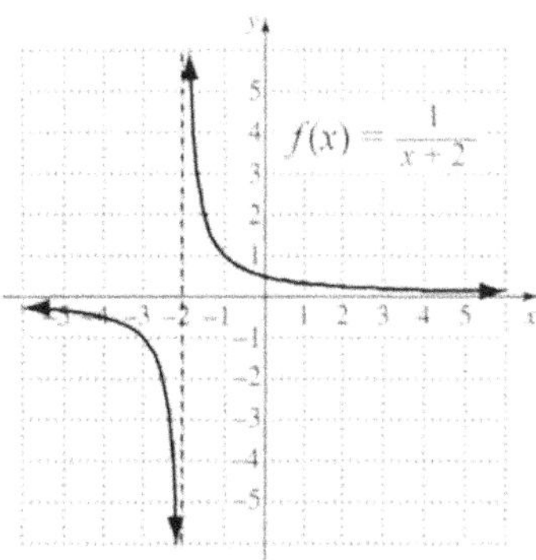

2. $\{x \mid x \neq 0\}$, or $(-\infty, 0) \cup (0, \infty)$; $\{x \mid x \neq 0\}$, or $(-\infty, 0) \cup (0, \infty)$;

$\{x \mid x \neq -2 \text{ and } x \neq 2\}$, or $(-\infty, -2) \cup (-2, 2) \cup (2, \infty)$; $\{x \mid x \neq -2\}$, or $(-\infty, -2) \cup (-2, \infty)$;

$\{x \mid x \neq -2 \text{ and } x \neq 5\}$, or $(-\infty, -2) \cup (-2, 5) \cup (5, \infty)$; $\{x \mid x \neq 4\}$, or $(-\infty, 4) \cup (4, \infty)$

3. (a) $x = -1,\ x = 5$; **(b)** $x = -2,\ x = 0,\ x = 2$; **(c)** $x = -\sqrt{3},\ x = 0,\ x = \sqrt{3}$ **4.** $y = \dfrac{3}{5}$ **5.** $y = 0$

6. 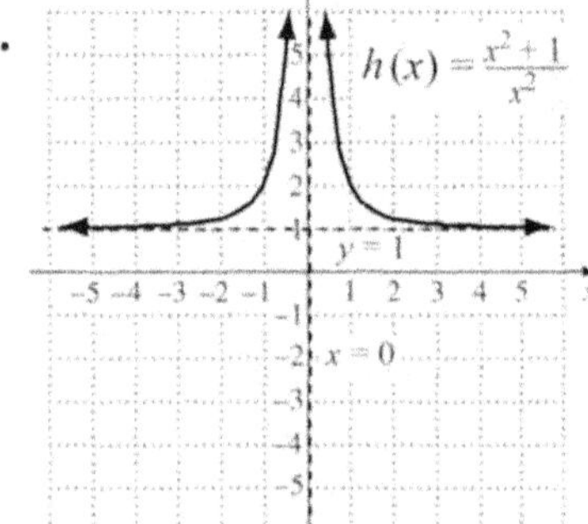

7. $x = -1,\ y = 3x - 4$

8. 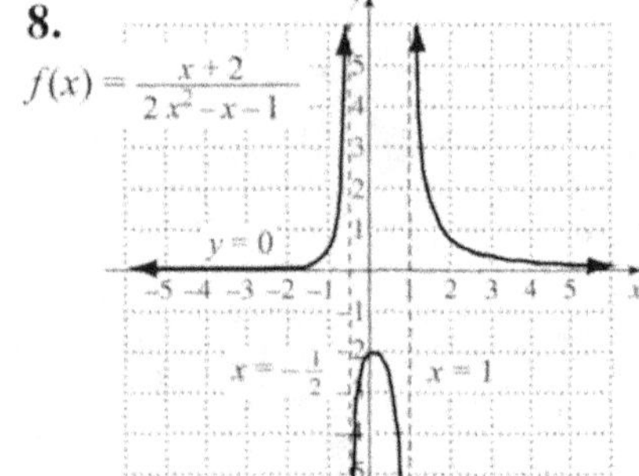

9. 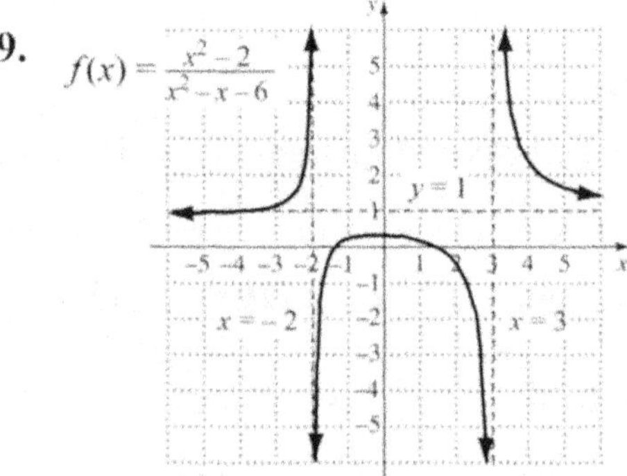

10. 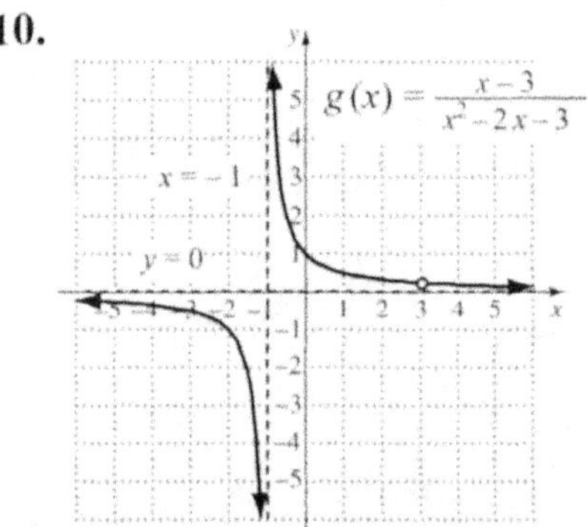

11. 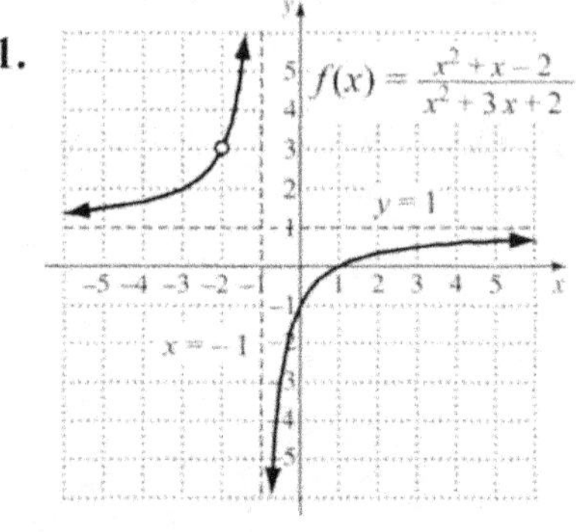

12. (a) 27.3 parts per million, 23.1 parts per million, 21.4 parts per million; **(b)** $y = 0.05$; $C(t) \to 0.05$ as $t \to \infty$; **(c)** The medication never completely leaves the body; a trace amount remains.

Practice Exercises: 1. $\{x \mid x \neq 3\}$, or $(-\infty, 3) \cup (3, \infty)$

2. $\{x \mid x \neq -6 \text{ and } x \neq 3\}$, or $(-\infty, -6) \cup (-6, 3) \cup (3, \infty)$ **3.** $x = 1,\ x = 5$ **4.** $x = -2,\ x = \dfrac{3}{2}$

5. $y = 3$ **6.** $y = 0$ **7.** $y = 2x + 1$ **8.** $y = x$

9. 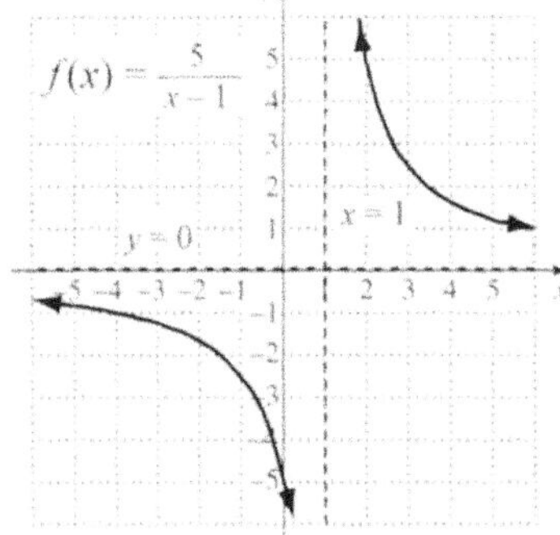

10. 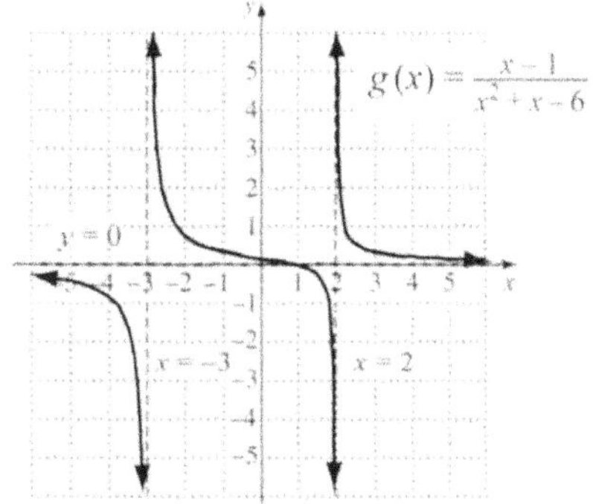

11. 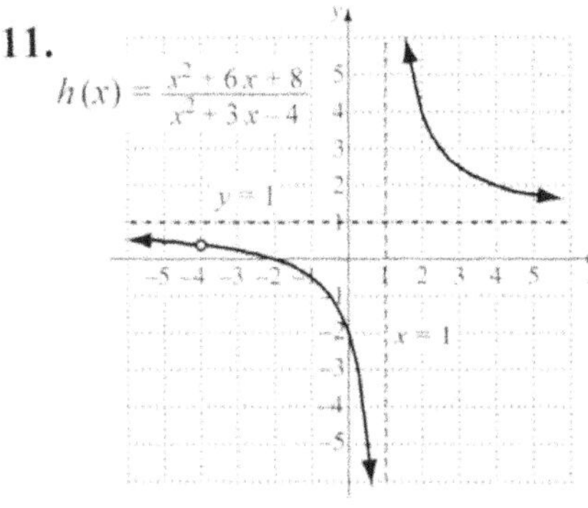

12. 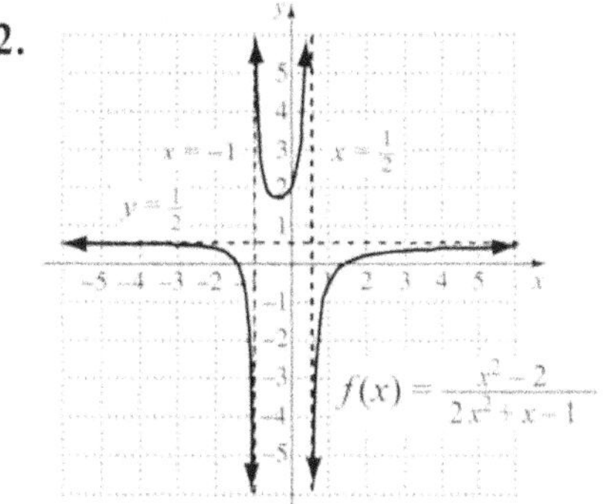

Section 8.6

Your Turn: **1.** $(-\infty, -3) \cup (2, \infty)$ **2.** $(-5, 2)$ **3.** $(-\infty, -2] \cup [0, 2]$

4. $(-\infty, -2] \cup \left[1 - \sqrt{2}, \dfrac{1}{2}\right] \cup \left[1 + \sqrt{2}, \infty\right)$ **5.** $(0, 3)$ **6.** $(2, 4]$ **7.** $\left[\dfrac{1}{2}, 2\right) \cup (4, \infty)$

Practice Exercises: **1.** $[-5, 3]$ **2.** $(-\infty, -1) \cup (1, \infty)$ **3.** $(-\infty, -4] \cup [3, \infty)$ **4.** $(1, 2)$

5. $(-\infty, -3) \cup (0, 3)$ **6.** $[-2, -1] \cup [2, \infty)$ **7.** $[0, 2)$ **8.** $(-\infty, -4) \cup [0, \infty)$ **9.** $\left(-\dfrac{7}{3}, -1\right)$

10. $(-\infty, -9] \cup (-1, \infty)$ **11.** $(-\infty, -3) \cup \left[-\dfrac{1}{3}, 0\right)$ **12.** $\left(-2, -\dfrac{5}{4}\right] \cup (4, \infty)$

Chapter 9

Section 9.1

Your Turn: **1. (a)** $(f \circ g)(x) = 3x^2 - 6x + 23,\ (g \circ f)(x) = 9x^2 + 6x + 7;$

(b) $(f \circ g)(2) = 23,\ (g \circ f)(2) = 55;$ **(c)** $(g \circ g)(1) = 31;$ **(d)** $(f \circ f)(x) = 9x + 8$

2. (a) $(f \circ g)(x) = \sqrt{x} + 4,\ (g \circ f)(x) = \sqrt{x + 4};$ **(b)** Domain of $f \circ g = [0, \infty),$

domain of $g \circ f = [-4, \infty)$ **3.** $(f \circ g)(x) = \dfrac{x}{3 - 5x};\ (g \circ f)(x) = 3(x - 5);$

domain of $f \circ g = (-\infty, 0) \cup \left(0, \dfrac{3}{5}\right) \cup \left(\dfrac{3}{5}, \infty\right);$ domain of $g \circ f = (-\infty, 5) \cup (5, \infty)$

4. $f(x) = x^7$ and $g(x) = 3x - 4$, answers may vary.

5. $f(x) = \dfrac{1}{x^5}$ and $g(x) = x - 4$, answers may vary.

Practice Exercises: 1. (a) $(f \circ g)(x) = 4x^2 - 12x + 7$, $(g \circ f)(x) = 16x^2 - 20x + 6$;
(b) $(f \circ g)(3) = 7$, $(g \circ f)(3) = 90$; **(c)** $(g \circ g)(-2) = 110$; **(d)** $(f \circ f)(x) = 16x - 5$
2. (a) $(f \circ g)(x) = \sqrt{2x - 2}$, $(g \circ f)(x) = 2\sqrt{x + 1} - 3$; **(b)** domain of $f \circ g = [1, \infty)$,

domain of $g \circ f = [-1, \infty)$ **3. (a)** $(f \circ g)(x) = 4(3x + 5)$, $(g \circ f)(x) = \dfrac{x}{12 + 5x}$;

(b) domain of $f \circ g = \left(-\infty, -\dfrac{5}{3}\right) \cup \left(-\dfrac{5}{3}, \infty\right)$,

domain of $g \circ f = \left(-\infty, -\dfrac{12}{5}\right) \cup \left(-\dfrac{12}{5}, 0\right) \cup (0, \infty)$

4. $f(x) = \dfrac{1}{x^5}$ and $g(x) = x + 7$, answers may vary.

5. $f(x) = \sqrt[3]{x}$ and $g(x) = x^2 - x + 3$, answers may vary.

Section 9.2

Your Turn: 1. Inverse of $h = \{(-4, 3), (-2, -1), (1, 5), (4, 0)\}$,

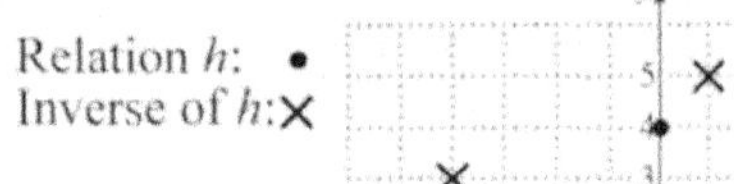
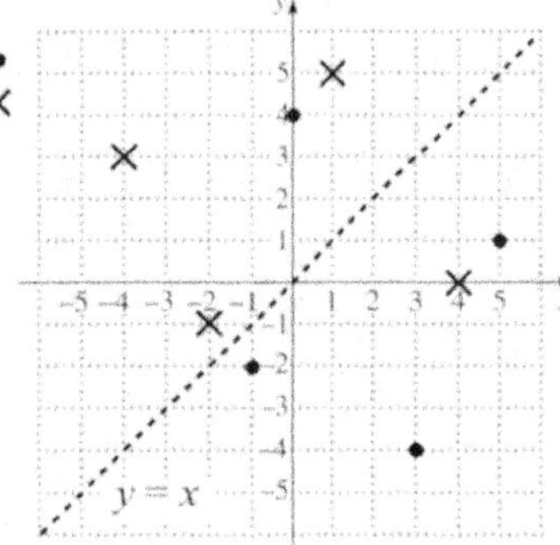

2. $x = y^2 - 2y + 3$

3. Assume $f(a) = f(b)$ for any numbers a and b in the domain of f. Since $f(a) = 7a + 10$
and $f(b) = 7b + 10$, we have

$$7a + 10 = 7b + 10$$
$$7a = 7b \qquad \text{Subtracting 10}$$
$$a = b \qquad \text{Dividing by 7}$$

Thus, if $f(a) = f(b)$, then $a = b$ and f is one-to-one. **4.** Find two numbers a and b for
which $a \neq b$ and $f(a) = f(b)$. Two such numbers are -3 and 3, because
$f(-3) = f(3) = 82$. Thus, f is not one-to-one. **5. (a)** Not one-to-one; **(b)** one-to-one;
(c) not one-to-one; **(d)** one-to-one **6.** The graph passes the horizontal-line test, so it is
one-to-one. $f^{-1}(x) = \dfrac{5 - x}{2}$

7. $f^{-1}(x)=\dfrac{1-x}{2}$

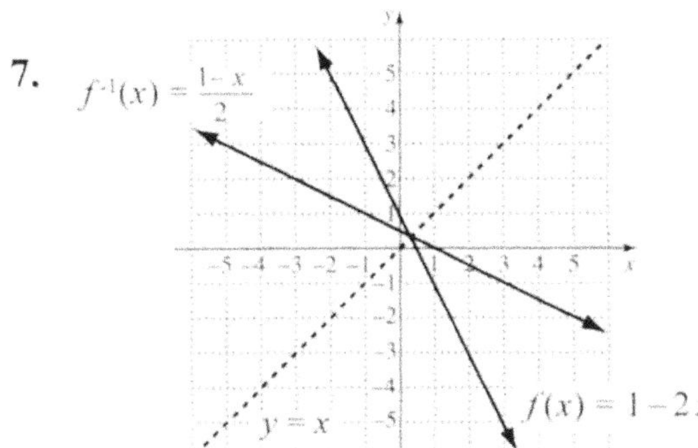

The graphs are reflections of each other across the line $y = x$.

8. (a) $h(x)$ passes the horizontal-line test, so it is one-to-one; **(b)** $h^{-1}(x)=\sqrt[3]{3-x}$;
(c)

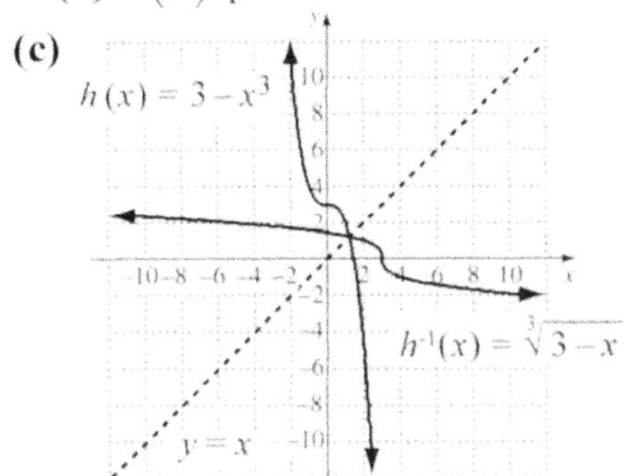

9. $\left(g^{-1}\circ g\right)(x)=g^{-1}\left(g(x)\right)=g^{-1}(3x-4)=\dfrac{(3x-4)+4}{3}=\dfrac{3x}{3}=x,$

$\left(g\circ g^{-1}\right)(x)=g\left(g^{-1}(x)\right)=g\left(\dfrac{x+4}{3}\right)=3\left(\dfrac{x+4}{3}\right)-4=x+4-4=x.$

These two compositions show that $g^{-1}(x)=\dfrac{x+4}{3}$ is the inverse of $g(x)=3x-4$.

Practice Exercises: 1. $\{(0,-6),\,(3,2),\,(-4,5)\}$ **2.** $x=2y^{2}+4y-7$ **3.** Assume $g(a)=g(b)$ for any numbers a and b in the domain of g. Since $g(a)=-2a+7$ and $g(b)=-2b+7$, we have $-2a+7=-2b+7$

$$-2a=-2b \qquad \text{Subtracting 7}$$
$$a=b \qquad \text{Dividing by } -2$$

Thus, if $g(a)=g(b)$, then $a=b$ and g is one-to-one. **4.** Find two numbers a and b for which $a\neq b$ and $f(a)=f(b)$. Two such numbers are -2 and 2, because $f(-2)=f(2)=-1$. Thus, f is not one-to-one. **5.** Not one-to-one **6.** One-to-one

7. $g^{-1}(x) = \dfrac{x+1}{3}$

8. $f^{-1}(x) = \sqrt[3]{2x-2}$

9. $h^{-1}(x) = \dfrac{-4}{x}$

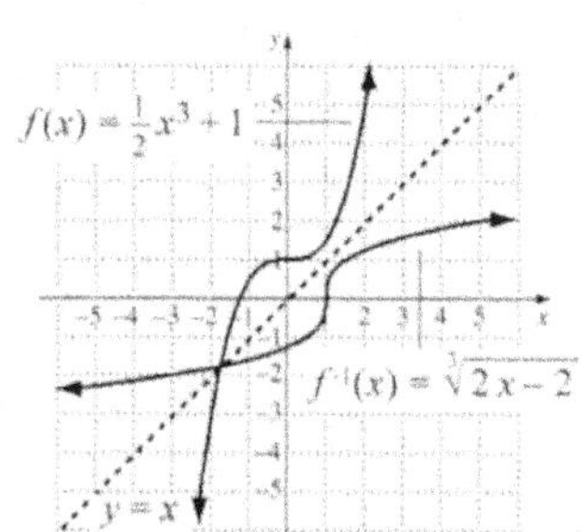

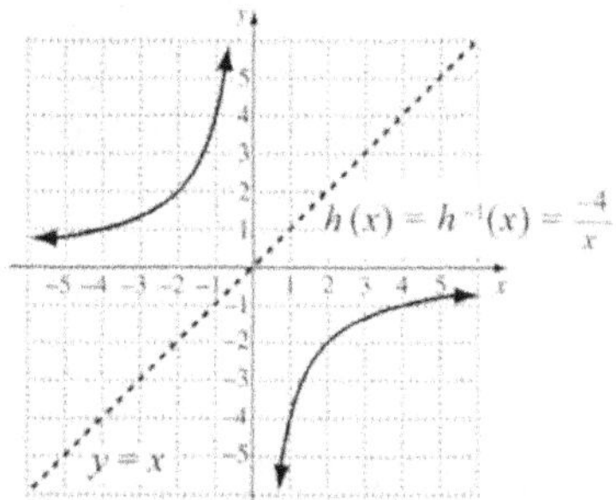

10. $\left(h^{-1} \circ h\right)(x) = h^{-1}\left(h(x)\right) = h^{-1}(5-6x) = \dfrac{5-(5-6x)}{6} = \dfrac{5-5+6x}{6} = \dfrac{6x}{6} = x,$

$$\left(h \circ h^{-1}\right)(x) = h\left(h^{-1}(x)\right) = h\left(\dfrac{5-x}{6}\right) = 5 - 6\left(\dfrac{5-x}{6}\right) = 5 - (5-x) = 5 - 5 + x = x.$$

These two compositions show that $h^{-1}(x) = \dfrac{5-x}{6}$ is the inverse of $h(x) = 5 - 6x$.

Section 9.3
Your Turn:

1.

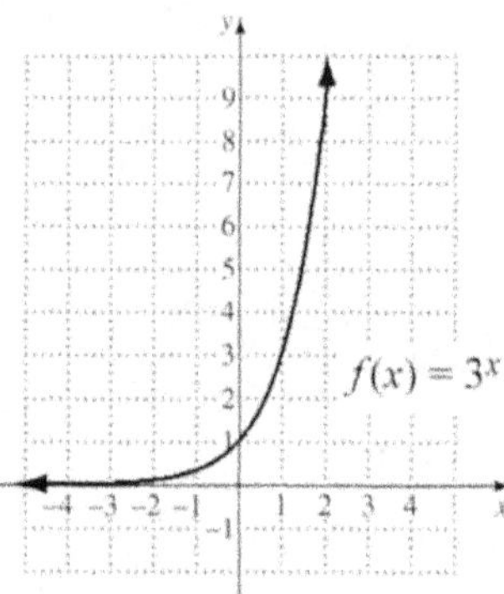

2.

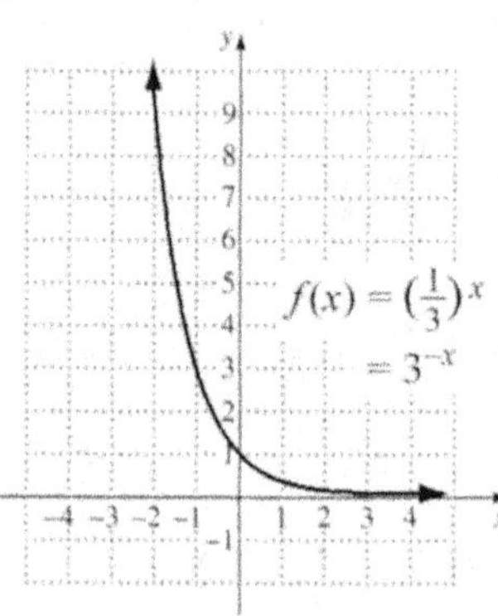

3. (a) left,

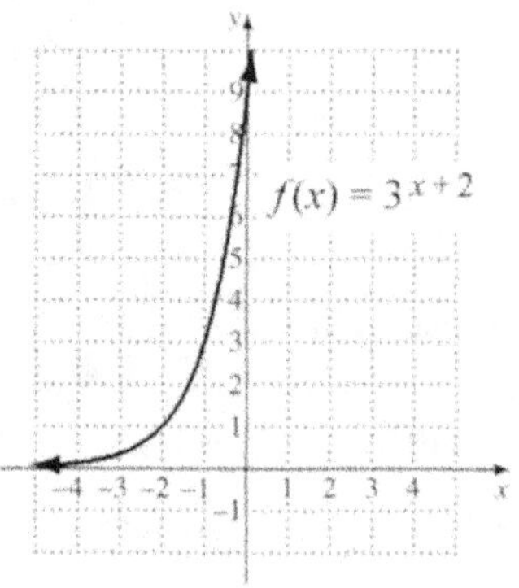

3. (b) down,

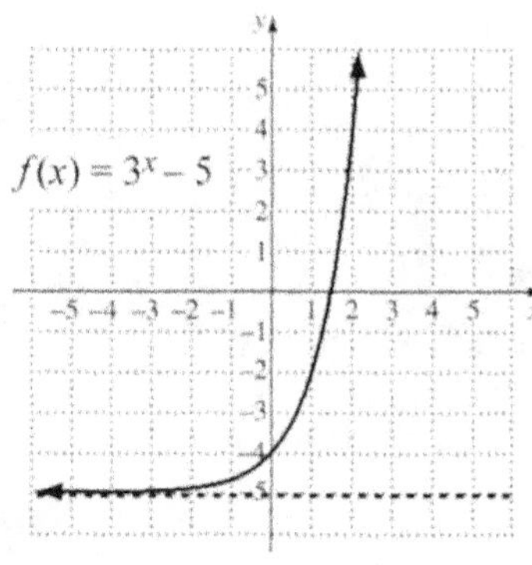

3. (c) y-axis, x-axis, up,

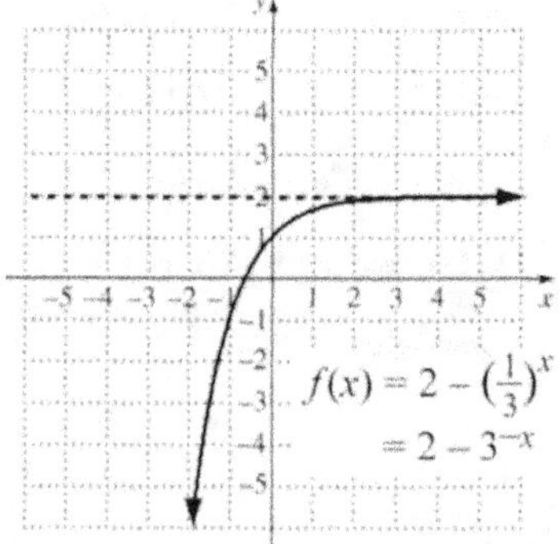

4. (a) $A(t) = \$200{,}000\left(1+\dfrac{0.05}{4}\right)^{4t} = \$200{,}000(1.0125)^{4t}$;

(b) $A(0) = \$200{,}000$

$A(3) = \$232{,}150.90$

$A(6) = \$269{,}470.21$

$A(8) = \$297{,}626.10$

$A(10) = \$328{,}723.89$;

(c) 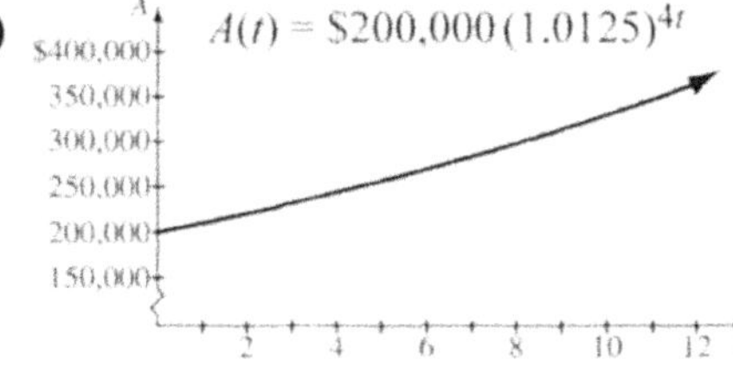

5. (a) $e^4 \approx 54.5982$; **(b)** $e^{-0.25} \approx 0.7788$; **(c)** $e^0 = 1$

6.

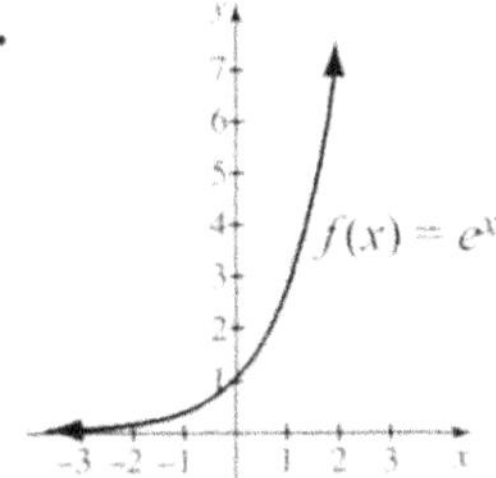

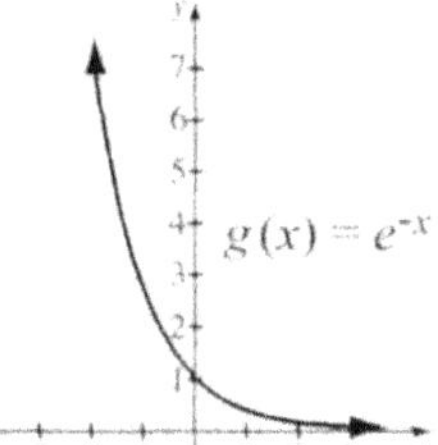

7. (a) right, **(b)** shrunk, y-axis, **(c)** vertically, x-axis, up,

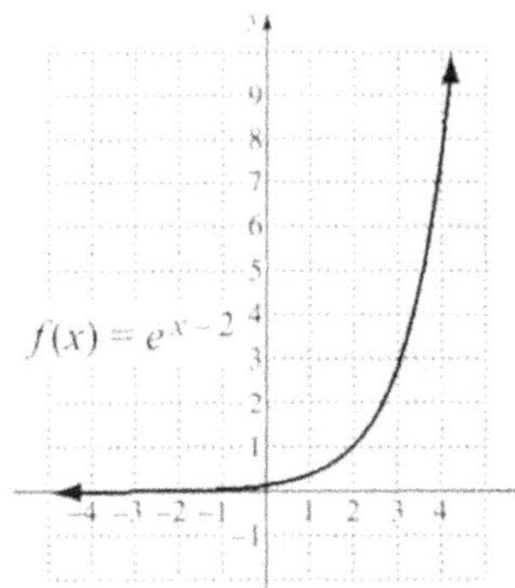

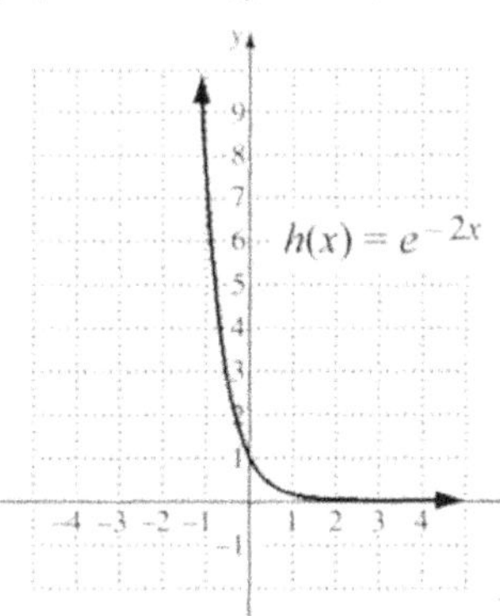

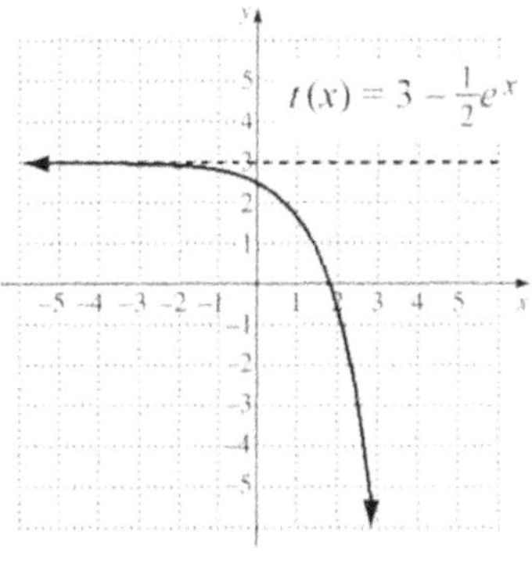

Practice Exercises: 1. $e^0 = 1$ **2.** $e^{-6} = 0.0025$ **3.** $e^3 = 20.0855$ **4.** $e^{-0.9} = 0.4066$

5. x-axis, up, **6.** stretched, stretched, **7.** left, down,

8. horizontally, y-axis,

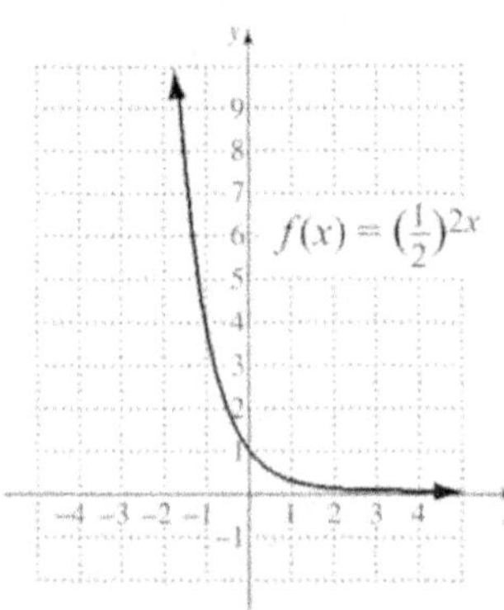

9. right, x-axis, down,

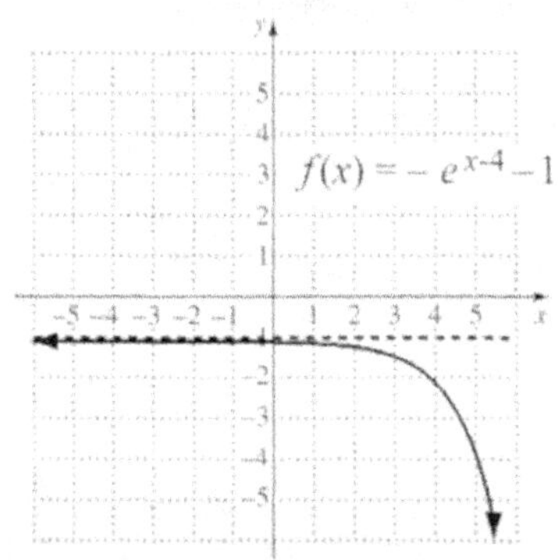

10. (a) $A(t) = \$150,000\left(1 + \dfrac{0.04}{2}\right)^{2t} = \$150,000(1.02)^{2t}$;

(b) $A(0) = \$150,000$

$A(8) = \$205,917.86$

$A(15) = \$271,704.24$

$A(25) = \$403,738.20;$

(c)

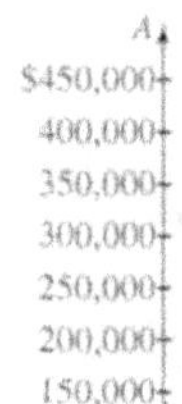

Section 9.4

Your Turn: 1.

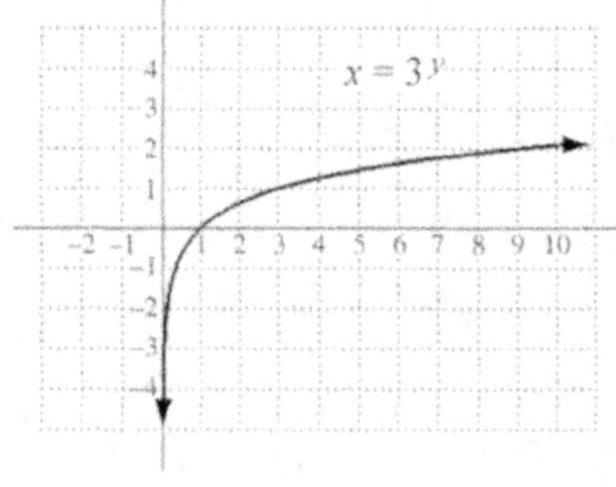

2. (a) 5; **(b)** −3; **(c)** $\dfrac{1}{2}$ **(d)** 4; **(e)** 1; **(f)** 0

3. (a) $\log_3 27 = x$; **(b)** $\log_{10} 0.000001 = -6$;

(c) $\log_e 15 = y$ **4. (a)** $4^3 = 64$; **(b)** $c^a = B$;

(c) $e^{45} = W$ **5. (a)** −3.7773;

(b) 4.3631; **(c)** $\log(-6)$ does not exist

6. (a) −8.6975; **(b)** 10.0464; **(c)** $\ln(-6)$ does not exist; **(d)** 0; **(e)** 1 **7.** 1.7712 **8.** 1.7712

9.

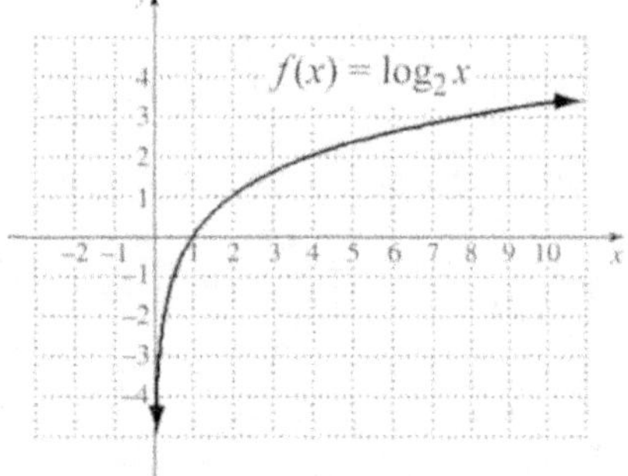

10.

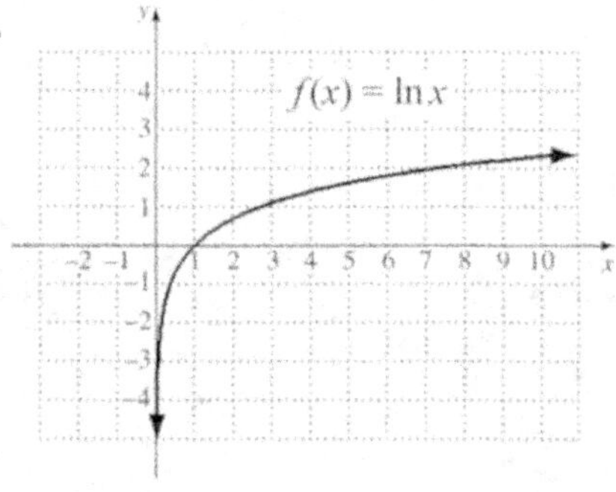

11. (a) right.

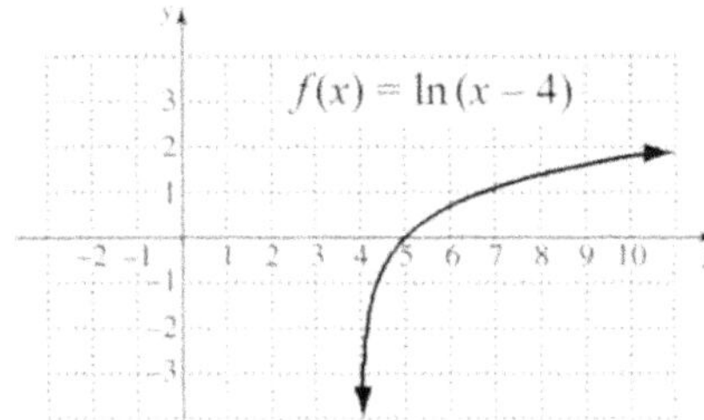

Domain: $(4, \infty)$

Asymptote: $x = 4$

(c) left, x-axis

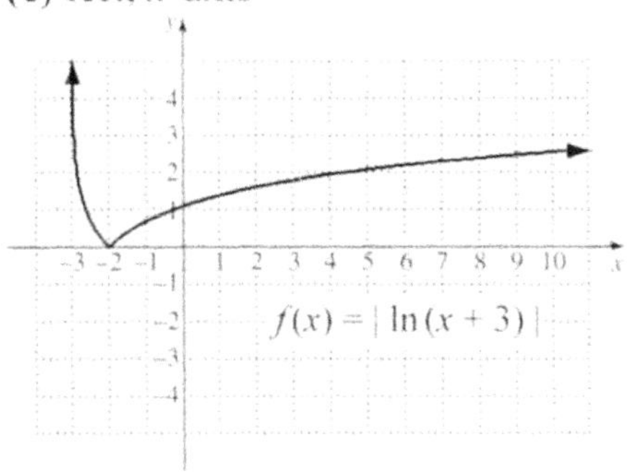

Domain: $(-3, \infty)$

Asymptote: $x = -3$

(b) vertical, x-axis, up,

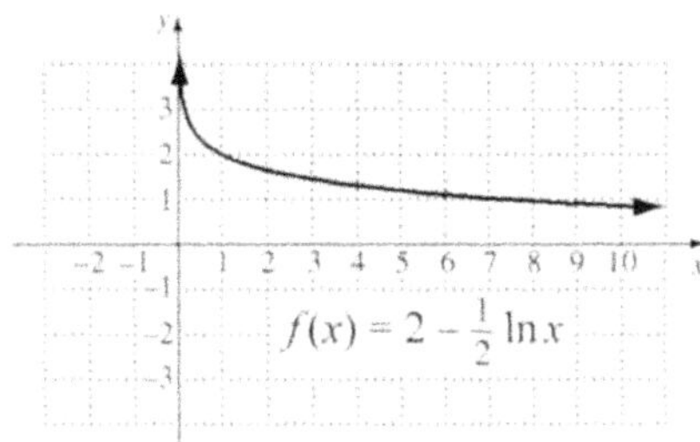

Domain: $(0, \infty)$

Asymptote: $x = 0$

12. (a) About 3.1 ft/sec; **(b)** about 1.7 ft/sec

13. 7.0

Practice Exercises: 1. 7 **2.** -4 **3.** 0 **4.** $\dfrac{1}{3}$ **5.** 6 **6.** $\dfrac{1}{2}$ **7.** $\log_y 7 = a$ **8.** $\log_4 \dfrac{1}{64} = -3$

9. $10^{-4} = 0.0001$ **10.** $c^s = Q$ **11.** 3.7038 **12.** $\log(-1000)$ does not exist. **13.** -1.4437

14. 5.3799 **15.** 1.7227 **16.** 4.1918 **17.** 1.7227 **18.** 4.1918

19. The graph of $f(x) = -\log(x+1)$ is a shift of the graph of $y = \log x$ left 1 unit and then a reflection across the x-axis.

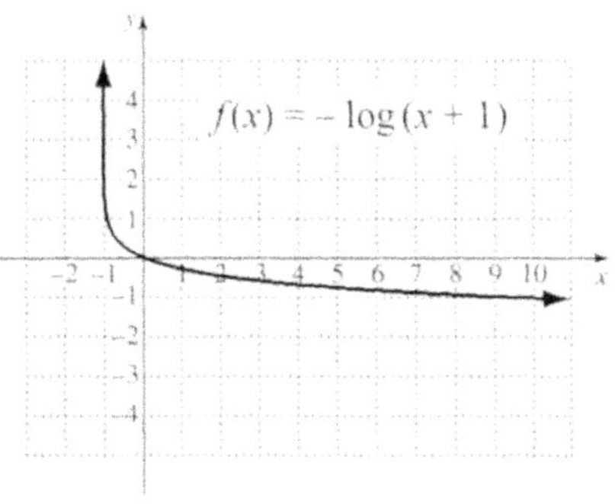

Domain: $(-1, \infty)$

Asymptote: $x = -1$

20. The graph of $f(x) = \ln 4x - 1$ is a horizontal shrinking of the graph of $y = \ln x$ followed by a shift down 1 unit.

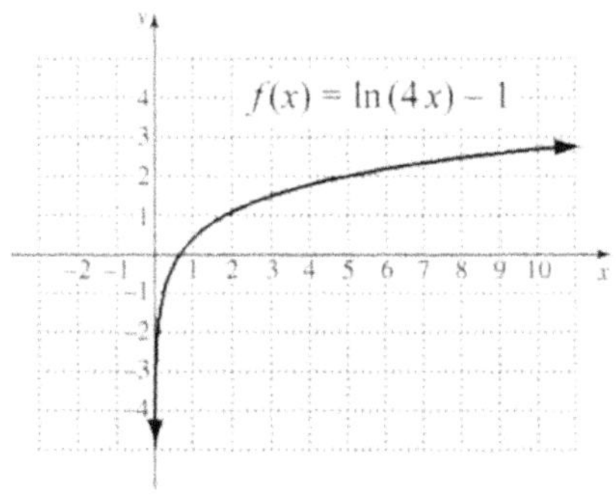

Domain: $(0, \infty)$

Asymptote: $x = 0$

21. The graph of $f(x) = |\ln(x-3)|$ is a shift of the graph of $y = \ln x$ right 3 units. Then the absolute value has the effect of reflecting negative outputs across the x-axis.

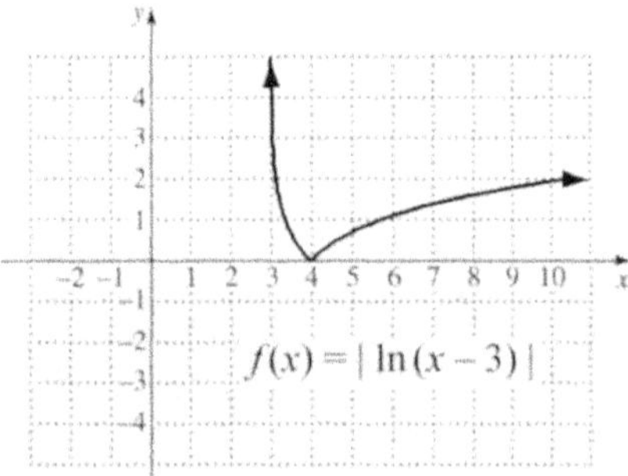

Domain: $(3, \infty)$

Asymptote: $x = 3$

22. The graph of $f(x) = \dfrac{1}{2}\log(x-2) + 3$ is a shift of the graph of $y = \log x$ right 2 units, followed by a vertical shrinking, and then a shift up 3 units.

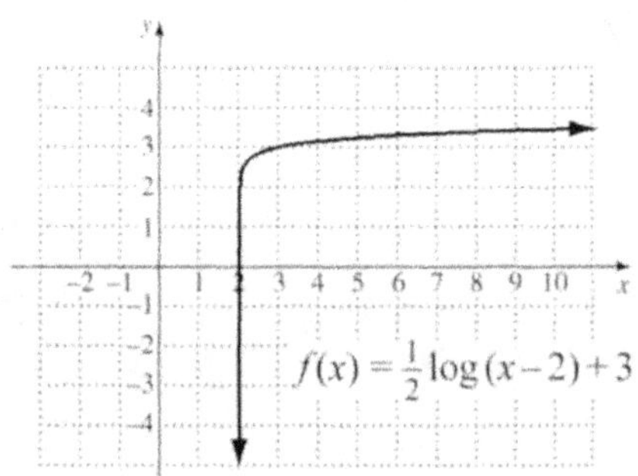

Domain: $(2, \infty)$

Asymptote: $x = 2$

23. About 2.47 ft/sec   **24.** 8.6

Section 9.5

Your Turn: **1.** $\log_4 16 + \log_4 64$ **2.** $\log_6\left(ts^2\right)$ **3. (a)** $5\log_a 9$; **(b)** $\dfrac{1}{3}\log_a 26$; **(c)** $-2\ln x$

4. $\log_c a - \log_c 9$ **5.** $\log_a \dfrac{81}{9} = \log_a 9$ **6. (a)** $4\log_a y + 3\log_a z - 6\log_a x$;

(b) $\dfrac{4}{3} + \dfrac{5}{3}\log_c a - \dfrac{2}{3}\log_c b$; **(c)** $3\log_b c + 2\log_b m - 2\log_b a - 5\log_b n$ **7.** $\log_a \dfrac{y^4 \sqrt[3]{x}}{z}$

8. $\ln\dfrac{1}{2y-7}$ **9. (a)** 1.38; **(b)** -0.426; **(c)** 1.806; **(d)** -0.954;

(e) Finding $\log_a 11$ using the given logarithms is not possible.; **(f)** 1.893   **10. (a)** $9t$; **(b)** 10;

(c) $-2k$ **11. (a)** 3; **(b)** b; **(c)** $3y$

Practice Exercises: **1.** $\log_5 25 + \log_5 125 = 2 + 3 = 5$ **2. (a)** $4\log_a 3$; **(b)** $-5\ln y$;

(c) $\dfrac{1}{5}\log_t 40$ **3.** $\log_c a - \log_c 7$ **4.** $\log_b 4 + 3\log_b x + \log_b y + 2\log_b z$

5. $\ln 7 - \ln 2 - 4\ln w - \ln z$ **6.** $\dfrac{5}{2}\log s + \dfrac{1}{2}\log z$ **7.** $\dfrac{2}{3}\ln a - \dfrac{5}{3}\ln b - 2\ln c$ **8.** $\log_a 16$

9. $\log_t \dfrac{\sqrt[4]{w}}{x^8}$ **10.** $\ln(x-5)$ **11.** 3   **12.** $\log_c \dfrac{y^6 \sqrt[3]{x}}{b}$ **13.** $\ln(x+6)$ **14.** -0.653 **15.** 3.169

16. Finding $\log_a 11$ using the given logarithms is not possible.   **17.** 0.602   **18.** 1.255

19. -1.908 **20.** k **21.** $-a$ **22.** $5c$ **23.** 13

Section 9.6

Your Turn: 1. -9 **2.** 2.3652 **3.** 46.2 **4.** 1.1679 **5.** $-2.2924,\ 2.2924$ **6.** $\dfrac{1}{64}$ **7.** 4

8. $\dfrac{28}{5}$ **9.** 4

Practice Exercises: 1. -3 **2.** 1.9534 **3.** 0.9464 **4.** 9.2103 **5.** 32 **6.** 1 **7.** $\dfrac{3}{2}$ **8.** 0.0498

9. $\dfrac{1}{125}$ **10.** $\dfrac{1}{2}$ **11.** 0.8142 **12.** 3.1623 **13.** $-0.9624,\ 0.9624$ **14.** 7

Section 9.7

Your Turn: 1. (a) $P(t) = 331.0e^{0.0059t}$, where t is the number of years after 2020 and P is in millions; **(b)** about 340.9 million; **(c)** about 117.5 years after 2020; **(d)** The population of the United States will be 420 million about 40 years after 2020.   **2. (a)** About 3.2%;
(b) $P(t) = 10,000e^{0.032t}$; **(c)** \$13,771.28; **(d)** about 21.7 years   **3.** About 0.91% per year
4. $P(0) \approx 325,\ P(4) \approx 809,\ P(12) \approx 1721,\ P(18) \approx 1871$ **5.** About 151 years

Practice Exercises: 1. (a) $P(t) = 128.9e^{0.0106t}$, where t is the number of years after 2020 and P is in millions; **(b)** about 135.9 million; **(c)** about 65.4 years after 2020   **2. (a)** About 2.8%;
(b) $P(t) = 22,000e^{0.028t}$; **(c)** \$33,483.15; **(d)** about 24.8 years   **3.** About 0.99% per year
4. $P(0) \approx 440,\ P(5) \approx 1769,\ P(10) \approx 2898,\ P(15) \approx 3158$ **5.** About 516 years

Answers: Student Activities

Activity 1: American Football

1. 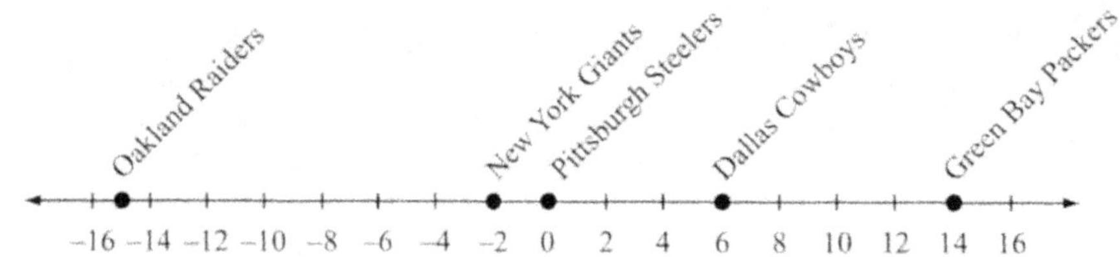

2.

Play 1:
Play 2:
Play 3:
Play 4:

3. At the 11 yard mark on the number line

4. Yes

5. $-4+6+(-3)+12$

6. $-4+6+12+(-3)$

7. The position was the same.

8. Yes; in the second set of plays, the ball would have moved more than 10 yards on the third down, so there would not have been a fourth down.

9. The direction is important, not just the distance.

10. $m = g - l$

11. Positive is good; 1 is a better turnover margin than -8

12.

Activity 2: Finding the Magic Number

1. Jaguars' magic number with respect to the Catamounts: 5; Jaguars' magic number with respect to the Wildcats: 5

2. Standings will vary.

3. Magic numbers will vary.

4. Magic numbers will vary.

5. In the formula for the magic number, $G - P$ represents the number of games that the leading team will still play in the season and L represents how many more games the second-place team has lost than the leading team has lost. A magic number of 0 means that $G - P - L = -1$, so L is greater than $G - P$. Thus, there are not enough games remaining in the season for the leading team to lose as many games as the second-place team has lost. This indicates that the second-place team has been eliminated from contention.

Activity 3: Pets in the United States

1. The number of households with one or more of the pets in 2012 and in 2017 and total pet population in 2012 and in 2017 are given; each number should be multiplied by 1,000,000. The total population of pet fish was 76,323,000 in 2017. In 2017, there were 893,000 households with pet horses.

2. The number of households increased for fish, rabbits, and poultry. The total pet population increased for fish and poultry. The lists differ because the number of pets per household changed.

3. $a = \dfrac{p}{h}$; birds, horses, fish, poultry

4. Fill-in answers are shown in **bold**.

	Households (in millions) in 2012	Households (in millions) in 2017	Change in the Number of Households (in millions)	Percent Change in the Number of Households
Birds	3.671	3.509	–0.162	–4.4%
Horses	1.780	0.893	**–0.887**	**–49.8%**
Fish	7.738	10.475	+2.737	+35.4%
Rabbits	1.408	1.534	**+0.126**	**+8.9%**
Reptiles	2.966	**3.669**	+0.703	+23.7%
Ferrets	**0.334**	0.326	–0.008	–2.4%
Livestock	**0.660**	**0.493**	–0.167	–25.3%

5. The number of households with fish increased the most. The greatest percent increase was in the number of households with pet poultry. There were many more households with fish than pet poultry in 2012, so a larger increase in the number of households with fish did not result in a percent increase as large as the one for households with pet poultry.

6. In the 5 years from 2012 to 2017, the number of households with birds as pets decreased by 0.162 million, or 162,000. If the trend remains the same, in another 5 years (2022) there will be $3,509,000 - 162,000$, or 3,347,000 households with birds as pets. In the 5 years from 2012 to 2017, the number of households with birds as pets decreased by 4.4%. If the trend remains the same, in another 5 years (2022) there will be $3,509,000 - 0.044(3,509,000)$, or 3,354,604 households with birds as pets. The percent change resulted in the smaller decrease, because this change was calculated on a smaller amount as a basis (3,509,000 rather than 3,671,000).

7. About 734,200 lizards

8. About 1.6 dogs per household

9. Answers will vary.

Activity 4: Mobile Data

1. No; the relative positions of Central and Eastern Europe, Middle East and Africa, Western Europe, and Latin America have changed.

2. Total row for 2016 to 2019: 6,766; 10,666; 16,140; 24,220
   Amount of Increase column missing data: 2,852; 2,051; 1,831; 21,697
   Percent Increase column missing data: 1341%; 1433%; 601%; 911%; 860%.
   Asia Pacific saw the greatest amount of increase, and Middle East and Africa saw the greatest percent increase.

3. (a) $W(2019) = 2,392$; this represents the average monthly mobile data usage in Western Europe in 2019

   (b) $(C+W)(2014) = 583$; this represents the average monthly mobile data usage in Central, Eastern, and Western Europe in 2014

   (c) $(N-L)(2019) = 1,766$; this represents how much greater the average monthly mobile data usage was in North American than in Latin America in 2019

   (d) $(A+N+C+M+W+L)(2019) = 24,220$; this represents the total average monthly mobile data usage in 2019 and is the same as $T(2019)$.

4. The heights of the regions represent the regional monthly mobile data functions.

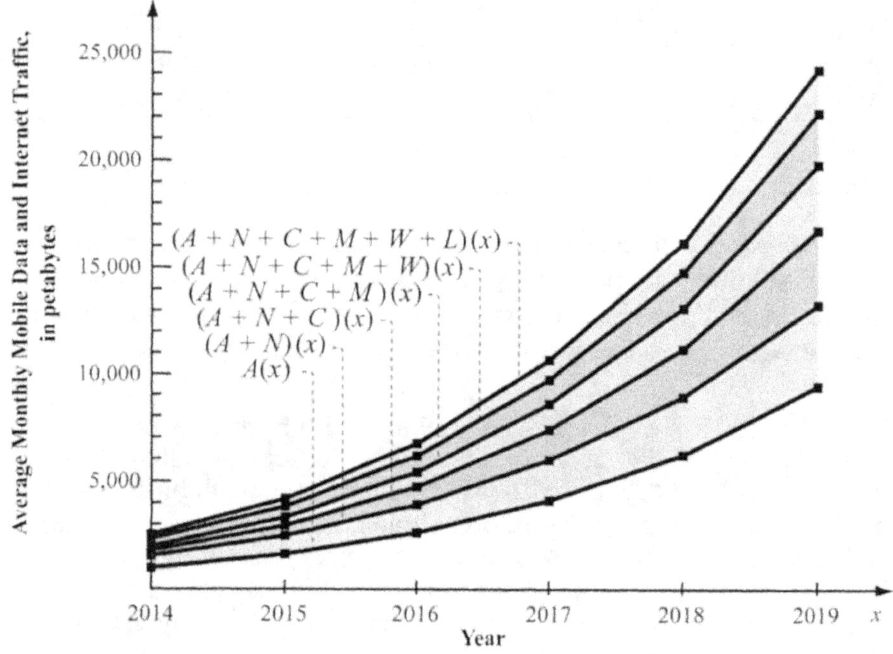

5. Answers may vary

Activity 5: Linear Regression on a Graphing Calculator

1. The coefficient a represents the slope of the regression line. This is the rate of change in TV viewing in the given age group. The constant b represents the y-intercept of the regression line.

2. The line will slant down from left to right because the slope is negative. The graph confirms this.

3. 2019: 14.82%; 2020: 12.46%

4. About 11 years after 2014, or in 2025; answers regarding whether this seems reasonable will vary.

5. $y = 0.44x + 17.44$

6. 2020: $19.64 billion; 2021: $20.08 billion

7. Answers will vary.

8. Answers will vary.

Activity 6: Going Beyond High School

1. Each number represents a percent of high school completers who enrolled in a college; the column heads reference "high school completers" because some students have a High School Equivalency instead of having graduated from high school with a diploma.

2. 1,917,160; 648,620; 1,268,540; 952,840

3. Graphs may vary; a circle graph may be the best choice.

4. Graphs may vary; a bar graph may be the best choice.

5. Graphs may vary; a line graph may be the best choice. The graphs will show that, overall, enrollment in two-year colleges and in four-year colleges is increasing.

6. 0.0781% per year; 0.4656% per year; on average, the percent of male (or female) high school completers increased by the given amount each year; answers may vary.

7. Males: $y = 0.0781x + 58.6$, females: $y = 0.4656x + 56.8$, where $x =$ the number of years after 1985 and $y =$ the percent enrolling in college; answers may vary.

8. Males: $y = 0.1697x + 59.0067$, females: $y = 0.5046x + 58.2464$, where $x =$ the number of years after 1985, $y =$ the percent enrolling in college, and coefficients are rounded to four decimal places; answers may vary.

9. The third and fourth columns of the first table are further broken down into those attending two- and four-year colleges.

10. Answers may vary.

11. Answers may vary.

Activity 7: Fruit Juice Consumption

1. Answers may vary.

2. Answers may vary.

3. Answers may vary; using linear regression and all the data points, letting x represent the number of years after 1985, g represent grape juice consumption, f represent grapefruit juice consumption, and p represent pineapple juice consumption, and rounding coefficients to 3 decimal places:
$g = 0.006x + 0.279$; $f = -0.022x + 0.797$; $p = -0.008x + 0.432$

4. Answers may vary; g is increasing, and f and p are decreasing.

5. g and f intersect at about $(18, 0.4)$; this means that grape juice consumption and grapefruit juice consumption were the same (0.4 gallons per person) 18 years after 1985, or in 2003. g and p intersect at about $(11, 0.3)$; this means that grape juice consumption and pineapple juice consumption were the same (0.3 gallons per person) 11 years after 1985, or in 1996. f and p intersect at about $(25, 0.2)$; this means that grapefruit juice consumption and pineapple juice consumption were the same (0.2 gallons per person) 25 years after 1985, or in 2010.

6. Answers may vary; using linear regression and all the data points, letting x represent the number of years after 1985, r represent orange juice consumption, and a represent apple juice and cider consumption, and rounding coefficients to 3 decimal places:
$r = -0.014x + 4.573$; $a = 0.015x + 1.573$. Predicted orange juice consumption in 2015 is 4.15 gallons per person; predicted apple juice and cider consumption in 2015 is 2.02 gallons per person. Orange juice consumption will equal apple juice and cider consumption in 2088.

7. Answers may vary.

8. Answers may vary.

9. Answers may vary.

10. Using linear regression and all the data points, letting x represent the number of years after 2000, r represent orange juice consumption, and a represent apple juice and cider consumption, and rounding coefficients to 3 decimal places: $r = -0.178x + 5.359$; $a = 0.045x + 1.799$. Predicted orange juice consumption in 2015 is 2.70 gallons per person; predicted apple juice and cider consumption in 2015 is 2.47 gallons per person. Orange juice consumption will equal apple juice and cider consumption in 2016.

11. Answers may vary.

Activity 8: Waterfalls

1. Fan: $c \le \dfrac{1}{3}b$;

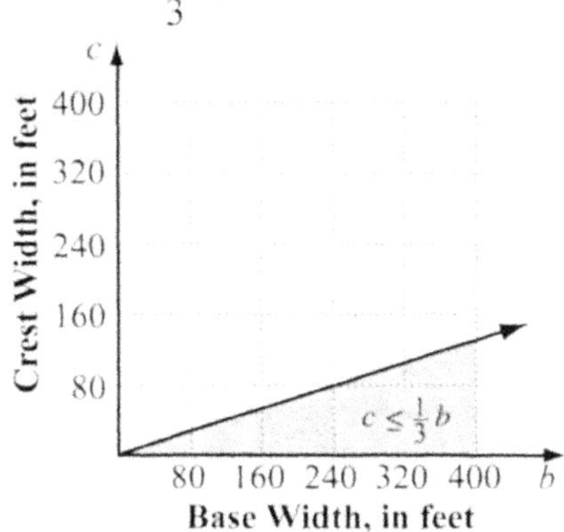

Funnel: $c > 3b$;

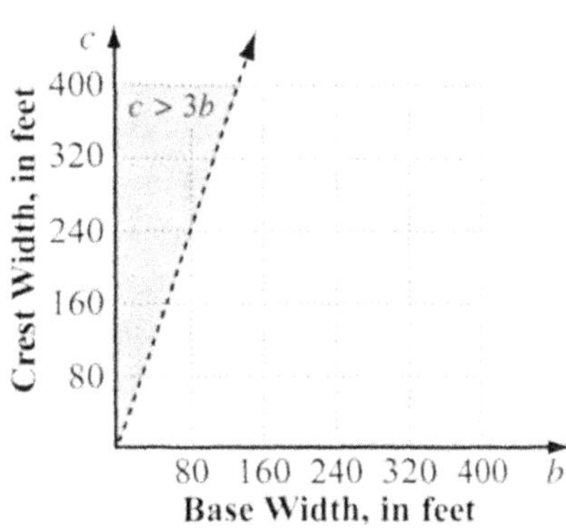

Curtain: $c > \dfrac{3}{2}h$;

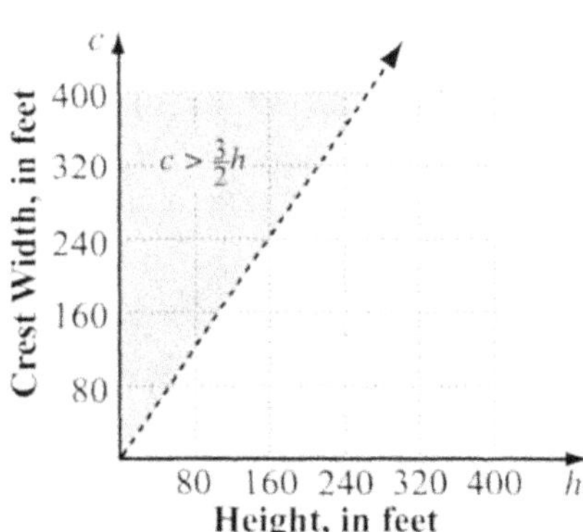

Ribbon: $c \le \dfrac{1}{2}h$;

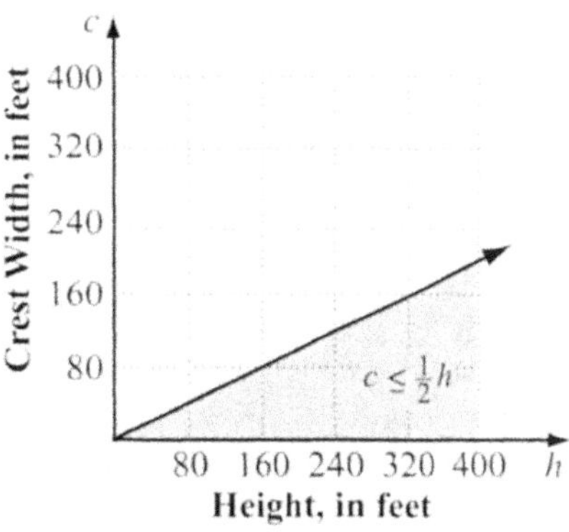

2. (a) Ribbon; (b) fan and ribbon; (c) curtain

Activity 9: Construction

1. (a) and (d)

2. $h \ge 30$

3. $d + h \ge 48$

4. Answers may vary.

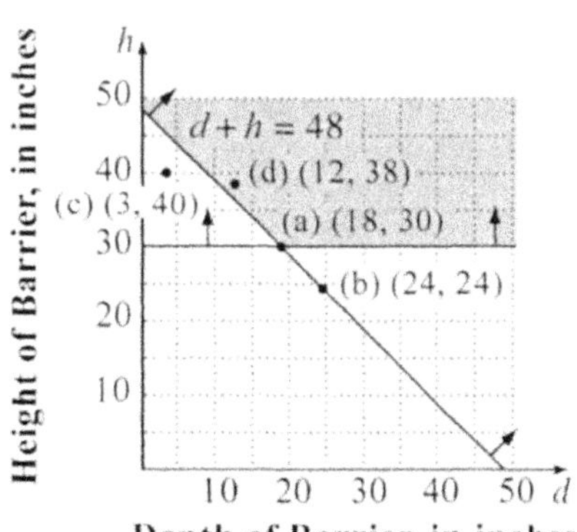

Activity 10: Visualizing Factoring

1. (a) The student should have a total of one square and 13 rectangles.

    (b) Only the rectangle that is 1 unit by 10 units will work.

    (c) $x^2 + 11x + 10 = (x+1)(x+10)$; the dimensions of the rectangle are $x+1$ and $x+10$.

2. $x^2 + 7x + 10 = (x+2)(x+5)$.

3. 2 units by 6 units

Activity 11: Matching Factorizations

$x^2 - x - 6 = (x+2)(x-3)$

$2x^2 + 5x + 3 = (2x+3)(x+1)$

$6x^2 + x - 1 = (2x+1)(3x-1)$

$6x^2 + 5x - 6 = (2x+3)(3x-2)$

$6x^2 - 5x - 6 = (2x-3)(3x+2)$

$6x^2 - 5x + 1 = (2x-1)(3x-1)$

$6x^2 + 5x - 1 = (6x-1)(x+1)$

$6x^2 - x - 1 = (2x-1)(3x+1)$

$x^2 + 5x + 6 = (x+2)(x+3)$

$x^2 - 5x + 6 = (x-2)(x-3)$

$x^2 + x - 6 = (x-2)(x+3)$

$6x^2 - 5x - 1 = (6x+1)(x-1)$

$2x^2 - 5x + 3 = (2x-3)(x-1)$

$6x^2 + 13x + 6 = (2x+3)(3x+2)$

$2x^2 + x - 3 = (2x+3)(x-1)$

$6x^2 - 7x + 1 = (6x-1)(x-1)$

Activity 12: Data and Downloading

1. As the file size increases, the download time increases and the number of downloads decreases.

2. The common ratio is 0.1; this represents 0.1 sec/MB, the time it takes to download one MB of data.

3. The common product is 5000; this represents 5000 MB, or 5 GB, the monthly plan limit.

4.

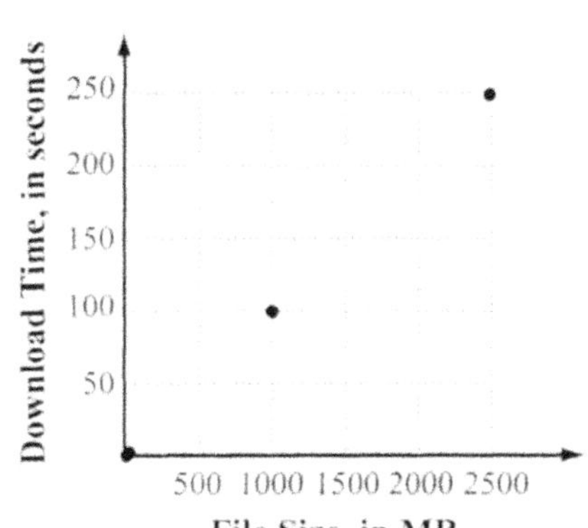 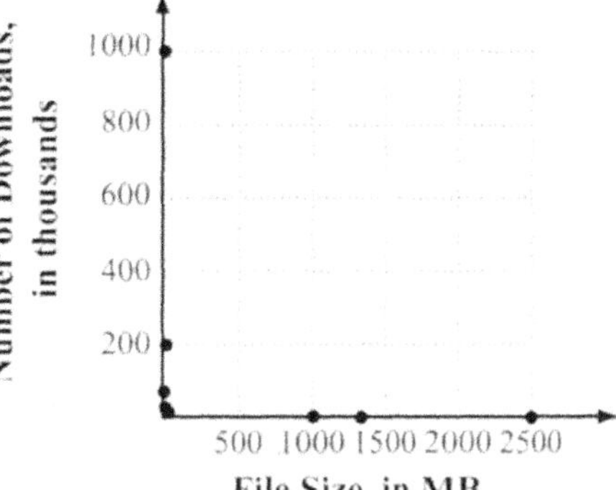

Note: In this graph, there are four points that lie very close to $(0, 0)$.

5. File size and download time vary directly; file size and number of downloads vary inversely.

6. $t = 0.1x$, where t is in seconds and x is in MB; the variation constant is the common ratio.

7. $n = \dfrac{5000}{x}$, where n is the number of downloads and x is in MB; the variation constant is the common product.

8. $t = 0.1(2) = 0.2$ seconds; $n = 5000 / 2 = 2500$ songs

9. Answers may vary.

Activity 13: Firefighting Formulas

1. Increase; answers may vary.

2., 3.

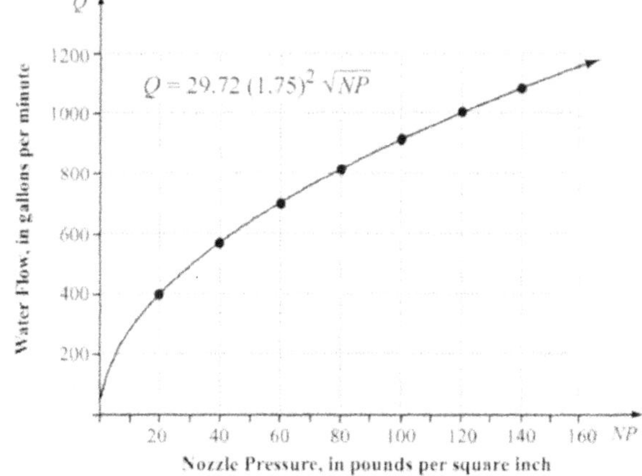

4. $NR = 0.05Q \times \sqrt{NP}$

5. $Q = 518$ gallons per minute; $NR = 212$ pounds; $V = 94$ feet per second

6. 96, 192, 288, 385, 481, 577, 673

7. Answers may vary.

Activity 14: Music Downloads

1. 428,000,000 albums; 257,000,000 albums; 1,070,000,000 single tracks; 1,108,000,000 single tracks

2. Decrease; increase

3. Overall, there was a decreasing trend in the sales of digital albums; and a linear function appears to be a good choice to model the data. Answers may vary.

4. The sales of digital single tracks increased from 2008 to 2012 and then began to decrease. A quadratic function might be a good choice to model the data. Answers may vary.

5. Answers may vary.

6. Answers may vary.

7. $A(x) = -25.1x + 407.6$, where $A(x)$ is the number of digital albums sold, in millions, x years after 2008 and coefficients are rounded to the nearest tenth; answers may vary.

8. $T(x) = -20.5x^2 + 139.8x + 1043.0$, where $T(x)$ is the number of digital single tracks sold, in millions, x years after 2008 and coefficients are rounded to the nearest tenth; answers may vary.

9. Yes; using the functions from Questions 7 and 8, sales will be the same 10.9 years after 2008, or in about 2019; answers may vary.

10. Sales of single tracks were a maximum in 2012. Answers may vary.

Activity 15: Let's Go to the Movies

1. Answers may vary.

2. Answers may vary.

3. Answers may vary; the function found using quadratic regression on a graphing utility is $E(x) = -0.0012x^2 + 0.0420x + 1.0565$, where x is the number of years after 1988 and coefficients are rounded to 4 decimal places.

4. Answers may vary; the linear function found using regression on a graphing utility is $F(x) = 0.0290x + 1.0686$, where x is the number of years after 1988 and coefficients are rounded to 4 decimal places.

5. Answers may vary; the quadratic function found using regression on a graphing utility is $S(x) = -0.0018x^2 + 0.0599x + 0.9584$, where x is the number of years after 1988 and coefficients are rounded to 4 decimal places.

6. No; S was the closest; answers may vary.

7. Answers may vary.

8. Answers may vary.

Activity 16: Teaching About Zeros: Creating a Video

1-4. Answers will vary.

Activity 17: Earthquake Magnitude

1. Strongest: September 16, 2015; weakest: March 18, 2015. Answers may vary.

2. $M_0 = 10^{\frac{3}{2}(M_0 + 10.7)}$; 3.5×10^{26}, 2.2×10^{28}, 2.0×10^{27}, 3.5×10^{26}, 4.0×10^{27}, 4.5×10^{25}, 3.5×10^{26}, 3.2×10^{25}, 4.5×10^{25}, 3.2×10^{28}

3. 1000 times stronger

4. Yes

5. Let m = the moment of magnitude of the weaker earthquake. Then $m + 1$ = the moment of magnitude of the stronger earthquake. Divide the seismic moments to determine how many times stronger the stronger earthquake is than the weaker earthquake:

$$\frac{10^{\frac{3}{2}(m+1+10.7)}}{10^{\frac{3}{2}(m+10.7)}} = 10^{\frac{3}{2}(m+1+10.7) - \frac{3}{2}(m+10.7)} = 10^{\frac{3}{2}m + \frac{3}{2} + \frac{3}{2}(10.7) - \frac{3}{2}m - \frac{3}{2}(10.7)} = 10^{\frac{3}{2}}.$$

So the stronger earthquake is $10^{\frac{3}{2}}$ or about 31.6 times as strong as the weaker earthquake.

6. 3.5×10^{17}, 1.1×10^{19}, 3.5×10^{20}, 1.1×10^{22}, 3.5×10^{23}, 1.1×10^{25}, 3.5×10^{26}, 1.1×10^{28}, 3.5×10^{29}. Answers may vary.

7. The relative magnitudes are easier to represent and discuss. Answers may vary.

Activity 18: Cosmic Path

1. 3.618414 AU; 337 million mi

2. $b^2 = a^2 - (a - 2.053218)^2$; $b \approx 3.262374$ AU; $b^2 = a^2 - (a - 191)^2$; $b \approx 304$ million mi

3. 21.631105 AU; 2012 million mi

4. 3.144056 AU/yr; 33,400 mph

5. Answers may vary.

Activity 19: Bargaining for a Used Car

1. $2833.33

2. $1000 or less

3. $2833.33; $1000 or less

4. $P = 5500 - \dfrac{2}{3}d$

Activity 20: Interest Rates on Credit Cards

1. $64

2. Interest: $52.80; amount applied to the principal: $11.20

3. Interest: $33.50; amount applied to the principal: $30.50

4.

Balance Before First Payment	First Month's Payment	% APR	Amount of Interest	Amount Applied to Principal	Balance After First Payment
$3216.28	$64	19.7%	$52.80	$11.20	$3205.08
$3216.28	$64	12.5%	$33.50	$30.50	$3185.78

5. At 19.7%, the interest is $52.80 and the principal is decreased by $11.20. At 12.5%, the interest is $33.50 and the principal is decreased by $30.50. Thus the principal is decreased by $30.50 − $11.20, or $19.30 more with the 12.5% rate than with the 19.7% rate. Thus the interest at 19.7% is $19.30 greater than the interest at 12.5%.

6. a) At 19.7%, it takes 74 months, or 6 yr 2 months longer with a $64 monthly payment than with a $128 monthly payment. At 12.5%, it takes 43 months, or 3 yr 7 months longer than with a $64 monthly payment than with a $128 monthly payment.

   b) At 19.7%, $2624 is saved with a monthly payment of $128.

   c) At 12.5%, $896 is saved with a monthly payment of $128.

Activity 21: Home Loans

1. a) Interest: $844.69; amount applied to principal: $134.99; b) $152,865.01;
   c) Interest: $843.94; amount applied to principal: $135.74; d) $199,684.80

2. a) $780.26; b) $780.26

3. a) Interest: $701.25; amount applied to principal: $167.47;

   b) The amount of interest at $5\dfrac{1}{2}\%$ in the first payment is $143.44 less than at $6\dfrac{5}{8}\%$.

   The amount applied to principal at $5\dfrac{1}{2}\%$ is $32.48 more than at $6\dfrac{5}{8}\%$.

   c) It will take approximately 11 months to recoup the refinance charge.
   d) $159,739.20; $39,945.60

4. a) Interest: $844.69; amount applied to principal: $498.64; b) $363.65;
   c) $88,799.40;
   d) The Sawyers will pay $110,885.40 less in interest with the 15-yr loan than with the
      30-yr loan.
   e) Answers will vary.